Carbohydrates

Momčilo Miljković

Carbohydrates

Synthesis, Mechanisms,
and Stereoelectronic Effects

Springer

Momčilo Miljković
Department of Biochemistry &
  Molecular Biology
Pennsylvania State University
Milton S. Hershey Medical Center
500 University Drive
Hershey PA 17033
H171
USA
mxm60@psu.edu

ISBN 978-0-387-92264-5   e-ISBN 978-0-387-92265-2
DOI 10.1007/978-0-387-92265-2
Springer New York Dordrecht Heidelberg London

Library of Congress Control Number: 2009933276

© Springer Science+Business Media, LLC 2010
All rights reserved. This work may not be translated or copied in whole or in part without the written permission of the publisher (Springer Science+Business Media, LLC, 233 Spring Street, New York, NY 10013, USA), except for brief excerpts in connection with reviews or scholarly analysis. Use in connection with any form of information storage and retrieval, electronic adaptation, computer software, or by similar or dissimilar methodology now known or hereafter developed is forbidden.
The use in this publication of trade names, trademarks, service marks, and similar terms, even if they are not identified as such, is not to be taken as an expression of opinion as to whether or not they are subject to proprietary rights.

Printed on acid-free paper

Springer is part of Springer Science+Business Media (www.springer.com)

Dedicated to the memory of Professors Milivoje S. Lozanić and Djordje Stefanović University of Belgrade, Serbia

Foreword

The development of organic chemistry over the last 40 years has been absolutely phenomenal, particularly the deepened understanding of chemical reactivity, molecular construction, and tools for analysis and purification. Without doubt, carbohydrate chemistry has played a major role in this historic advance and in the future will have crucial ramifications in most areas of biomedical research into the functioning of Nature at the molecular level.

This book covers all basic carbohydrate chemistry, including most important synthetic methods and reaction mechanisms and notably takes into account the principle of stereoelectronic effect which has played a key role in understanding the conformation and chemical reactivity of this important class of natural product. Even nomenclature has been properly covered and it is fair to say that all the key references of carbohydrates have been cited. Such a book could only have been written by an expert who has spent his entire research career in this area.

This book of Momčilo Miljković will be of interest not only to specialists in the field, but also to synthetic chemists in general. This book also contains most of the material needed for a graduate course in carbohydrate chemistry. Furthermore, it should be particularly valuable for investigators working in various aspects of bioorganic chemistry including the discovery of new medicines.

Quebec, Canada   Pierre Deslongchamps, FRS, FRSC

Preface

Carbohydrates are one of the three most important components of living cells (the other two being amino acids and lipids). In order to understand their biochemical behavior one must understand steric and electronic factors that control their reactivity and chemistry. Two properties of carbohydrates that are most important for their chemical behavior are their shape (conformation) and stereoelectronic interactions that are unique and characteristic for each carbohydrate structure.

So chapters on anomeric effect, a very important electronic effect first discovered in studies of carbohydrates and later found to be of general importance in many other organic molecules, glycosidic bond hydrolysis, isomerization of free carbohydrates in aqueous solution, relative reactivity of hydroxyl groups in a carbohydrate molecule, nucleophilic displacement with or without change of the configuration at the reacting carbon atom, addition of nucleophiles to glycopyranosiduloses, etc., are all to a various extent related to stereoelectronic effects that exist in carbohydrate structures.

Cyclic acetals and ketals and anhydrosugars are both very important intermediates in synthetic carbohydrate chemistry, first being used for protection of hydroxyl groups that are not supposed to take part in further chemical transformation of the intermediate and second being used as synthetic intermediates in carbohydrate chemistry because they can serve as starting materials for the synthesis of many different carbohydrate derivatives, for example, the amino sugars, the branched chain sugars, oligosaccharides. The amino sugars, being important components of many biomolecules such as glycosaminoglycans, heparin, chondroitin as well as many natural products, such as sugar-based antibiotics, macrolide antibiotics, are discussed in a separate chapter.

The last three chapters of the book deal with topics not usually found in carbohydrate chemistry texts like this one, although according to the author's opinion they are very important and they are unjustly neglected. These are carbohydrate-based antibiotics, synthesis of polychiral natural products from carbohydrates, and chemistry of higher-carbon carbohydrates.

Much attention has been paid to the mechanisms of various carbohydrate reactions as well as to the role of stereoelectronic effects that they play in the reactivity of carbohydrates and the stereochemical outcome of various carbohydrate reactions.

In the end, I wish to express my deep gratitude to Professor Pierre Deslongchamps for taking time to read the entire book and provide me with invaluable comments and critique. I would also like to thank my wife, Irina Miljković, for her patience and understanding throughout my scientific life.

Contents

1	**Introduction**	1
	Stereochemistry	2
	Representation of Monosaccharides	2
	Acyclic Form of Monosaccharides	2
	Cyclic Forms of Monosaccharides	4
	The Nomenclature of Carbohydrates	9
	Trivial Names	9
	Stem and Systematic Names	9
	Conventions	10
	Choice of Parent Monosaccharides	10
	Choice Between Alternative Names for Substituted Derivatives	11
	Configurational Symbols and Prefixes	11
	Ketoses	13
	Deoxy-monosaccharides	14
	Amino-monosaccharides	15
	O-Substitution	15
	Acyclic Forms	17
	Anomers and the Anomeric Configurational Symbols ("α" or "β")	18
	Glycosides	19
	Glycosyl Radicals and Glycosylamines	20
	Aldonic Acids	21
	Uronic Acids	22
	Aldaric Acids	22
	Cyclic Acetals	24
	Intramolecular Anhydrides	24
	References	25
2	**Conformational Analysis of Monosaccharides**	27
	Conformational Analysis of Acyclic Hydrocarbons	28
	Conformational Analysis of Acyclic (Aldehydo) Forms of Monosaccharides	31

Conformational Analysis of Cyclic (Lactol, Hemiacetal)
Forms of Monosaccharides 33
  Furanoses 33
  Pyranoses 36
Calculation of Conformational Energies of Pyranoses 41
References 53

3 Anomeric Effect 57
Endo-anomeric Effect 57
The Quantum-Mechanical Explanation 63
Exo-anomeric Effect 67
Generalized Anomeric Effect 69
Reverse Anomeric Effect 72
Anomeric Effect in Systems O–C–N 84
References 88

4 Isomerization of Sugars 95
Mutarotation 95
Anomerization 105
Lobry de Bruyn–Alberda van Ekenstein Transformation 108
References 109

5 Relative Reactivity of Hydroxyl Groups in Monosaccharides . . . 113
Introduction 113
Selective Acylation (Esterification) 114
  Selective p-Toluenesulfonylation (Tosylation)
  and Methanesulfonylation (Mesylation) 116
  Selective Benzoylation 120
  Selective Acetylation 127
  Other Acylating Reagents 129
  Acyl Migrations 133
Selective Alkylation and/or Arylation of Glycopyranosides 136
  Tritylation of Monosaccharides (Triphenylmethyl Ethers) 136
  Selective Benzylation of Monosaccharides 137
  Selective Alkylation of Metal Complexes of Monosaccharides . . . 138
References 139

6 Cyclic Acetals and Ketals 143
Acetalation 148
  Benzylidenation 148
  Ethylidenation 149
Ketalation 150
  Isopropylidenation (Acetonation) 150
Transacetalation and Transketalation 152
The Isomerization of Cyclic Acetals and Ketals 154
  The Migration of Acetal or Ketal Group 155
Removal of Acetal and Ketal Groups 157

	Benzylidene Group	157
	Isopropylidene Group	163
	References	163
7	**Nucleophilic Displacement and the Neighboring Group Participation**	169
	Nucleophilic Displacement	169
	Nucleophilic Displacements with Neighboring Group Participation	179
	References	188
8	**Anhydrosugars**	191
	1,6-Anhydrosugars (Glycosanes)	191
	1,4-Anhydrosugars	198
	1,2-Anhydrosugars (Brigl's Anhydrides)	200
	Anhydrosugars Not Involving the Anomeric Carbon	203
	Epoxides or Oxiranes	203
	Rearrangements of Anhydrosugars	211
	Epoxide Migration	211
	Other Isomerizations of Epoxides	212
	References	214
9	**Amino Sugars**	221
	Ammonolysis of Oxiranes	221
	Nucleophilic Displacement of Sulfonates (or Halides) with Nitrogen Nucleophiles	226
	Glycosylamines and N-Glycosides	231
	Acid-Catalyzed Hydrolysis of Purine and Pyrimidine Nucleosides	237
	References	240
10	**Oxidation of Monosaccharides**	245
	Selective Oxidations of Monosaccharides	245
	Catalytic Oxidation	245
	Bromine Oxidation	247
	Nonselective Oxidation of Secondary Hydroxyl Groups	248
	Ruthenium Tetroxide (RuO_4) Oxidation	249
	Dimethyl Sulfoxide Oxidation	254
	DMSO–DCC Method (Pfitzner–Moffatt Oxidation)	255
	DMSO–Acetic Anhydride Method	258
	DMSO–Phosphorus Pentoxide	263
	DMSO–Sulfurtrioxide Pyridine ("Parikh–Doering" Oxidation)	264
	Chromium Trioxide Oxidation	266
	Chromium Trioxide–Pyridine Oxidation	266
	Chromium Trioxide–Acetic Acid	270
	Pyridinium Chlorochromate	270
	Nicotine Dichromate	272
	Pyridinium Dichromate–Acetic Anhydride	272

Oxidation of Carbohydrates with the Cleavage
of Carbohydrate Chain 273
    Periodate Oxidation 273
    Lead Tetraacetate Oxidation 277
    Pentavalent Organobismuth Reagents 283
References 284

11 Addition of Nucleophiles to Glycopyranosiduloses 291
The Addition of a Hydride Ion (Reduction) 291
The Addition of Carbon Nucleophiles: Synthesis of Branched
Chain Sugars 297
    The Addition of Organometals 298
    Addition of Diazomethane 306
Synthesis of Branched Chain Sugars with Functionalized
Branched Chain 308
    2-Lithio-1,3-Dithiane as the Nucleophile 309
    Vinyl Carbanion as the Nucleophile 312
    Methoxyvinyl Lithium and 1,1-Dimethoxy-2-Lithio-2-Propene . . 312
    Reformatsky Reaction 314
    Opening of Oxiranes with Nucleophiles 316
References 318

12 Chemistry of the Glycosidic Bond 323
Introduction 323
Glycoside Synthesis 324
    Fischer Glycosidation 325
    Königs – Knorr Synthesis 330
    Synthesis of Acylated Glycosyl Chlorides and Bromides 335
    Glycosyl Fluorides in Glycosylation 336
    Synthesis of Glycosyl Fluorides 338
    Orthoester Method of Glycosidation 340
    Trichloroacetimidate Method of Glycosidation 349
    Glycoside Synthesis via Remote Activation 353
    n-Pentenyl Glycosides as Glycosyl Donors 354
    Glycals as Glycosyl Donors 357
    Thioglycosides as Glycoside Donors 364
    Synthesis of Thioglycosides 368
    Glycosyl Sulfoxides as Glycosyl Donors 368
    Solid-Phase Synthesis of Oligosaccharides 369
    Automated Oligosaccharide Synthesis 374
Cleavage of Glycosidic Bonds 374
    Acid-Catalyzed Hydrolysis of Glycosides 374
    The Acid-Catalyzed Hydrolysis of Glycopyranosides 377
    Acid-Catalyzed Hydrolysis of Glycofuranosides 385
    Some Recent Developments Regarding the Mechanism
    of Glycoside Hydrolysis 389

Contents

      Acetolysis of Glycosides 396
      References 406

13 Synthesis of Polychiral Natural Products from Carbohydrates . . 423
    Macrolide Antibiotics: Erythronolides A and B 423
    Thromboxane B_2 438
    Swainsonine 441
    Biotin 443
    Pseudomonic Acid C 445
    Aplasmomycin 456
    References 462

14 Carbohydrate-Based Antibiotics 469
    Aminoglycoside Antibiotics 469
      Kanamycin 470
      Amikacin 472
      Gentamicins 473
      Tobramycin (Nebramycin) 474
      Neomycin B (Actilin, Enterfram, Framecetin, Soframycin) 476
      Paromomycin 476
      Butirosins A and B 477
      Streptomycin A 477
      Orthosomycins 477
      Destomycin A 479
      Flambamycin 481
      Everninomicin 482
    References 482

15 Higher-Carbon Monosaccharides 487
    Introduction 487
    Synthesis of Higher-Carbon Sugars 490
      Wittig Olefination 490
      Aldol Condensation 498
      The Butenolide Approach 503
      Total Synthesis of Higher-Carbon Monosaccharides 503
    References 512

Author Index 517

Subject Index 539

Chapter 1
Introduction

Monosaccharides are polyhydroxy aldehydes or ketones having four to nine carbon atoms in their carbon chains – most often, five or six (Fig. 1.1).

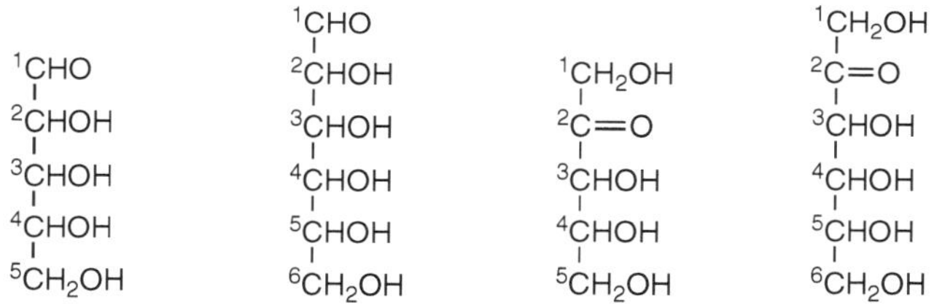

```
                    ¹CHO                              ¹CH₂OH
   ¹CHO            ²CHOH           ¹CH₂OH            ²C=O
   ²CHOH           ³CHOH           ²C=O              ³CHOH
   ³CHOH           ⁴CHOH           ³CHOH             ⁴CHOH
   ⁴CHOH           ⁵CHOH           ⁴CHOH             ⁵CHOH
   ⁵CH₂OH          ⁶CH₂OH          ⁵CH₂OH            ⁶CH₂OH

     1               2                3                 4
 aldopentose     aldohexose       ketopentose       ketohexose
```

Fig. 1.1

Depending on the number of carbon atoms in their skeleton, monosaccharides are named tetroses (four carbon atoms), pentoses (five carbon atoms), hexoses (six carbon atoms), heptoses (seven carbon atoms), etc., and depending on the nature of their carbonyl group, they are named *aldoses* (when their carbonyl group is an aldehydo group) or *ketoses* (when their carbonyl group is a keto group). Hence, there are aldo-tetroses, aldopentoses, aldo-hexoses, aldo-heptoses, etc. and keto-tetroses, keto-pentoses, keto-hexoses, keto-heptoses, etc.

The simplest monosaccharides consist of carbon, hydrogen, and oxygen and have the general molecular formula $C_nH_{2n}O_n$ or $C_n(H_2O)_n$ (hence the name *carbohydrates – hydrates* of carbon). However, they may often have fewer hydrogen and/or oxygen atoms or more oxygen atoms than required by the above general formula (e.g., unsaturated sugars, deoxy sugars, aldonic acids, uronic acids, aldaric acids), or they may also contain, in addition to carbon, hydrogen and, oxygen atoms, other elements such as nitrogen, sulfur, halogen.

The carbon atoms of a monosaccharide chain are numbered in such a way that the carbonyl carbon has always the lowest possible number, i.e., # **1** in aldoses (since the aldehydo group always occupies the terminal position) and # **2** (or higher) in ketoses (since the keto-group can occupy any position in a carbohydrate chain, except the terminal position).

Stereochemistry

Monosaccharides are *polychiral* molecules, i.e., they have two or more chiral carbon atoms in their skeleton, which are most often, but not always, hydroxymethylene carbons. The term *chiral* carbon (from the Greek word *chiros* = *hand*) has replaced the older term *asymmetric* carbon for practical reasons [1]. The configuration of the highest numbered chiral carbon of a monosaccharide (i.e., the chiral carbon) that is furthest away from the carbonyl (*anomeric*) carbon determines whether the monosaccharide belongs to a D- or to an L-series (Fig. 1.2). If the configuration of this carbon (e.g., *5* in Fig. 1.2) is identical to the configuration of D-glyceraldehyde (*6* in Fig. 1.2), the monosaccharide belongs to D-series; if it is identical to the configuration of L-glyceraldehyde (*8* in Fig. 1.2), the monosaccharide belongs to L-series (*7* in Fig. 1.2).

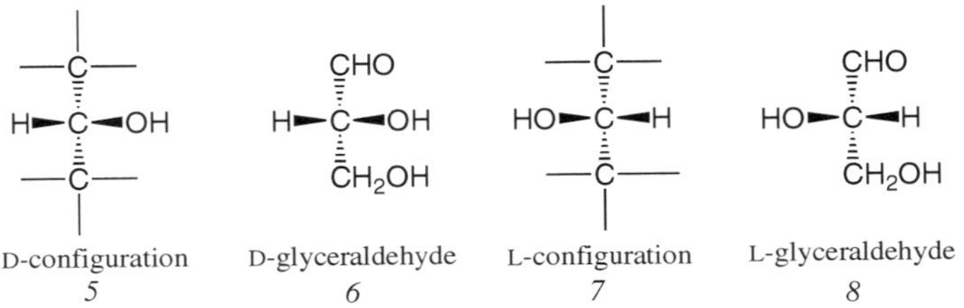

D-configuration	D-glyceraldehyde	L-configuration	L-glyceraldehyde
5	*6*	*7*	*8*

Fig. 1.2

Structures in Fig. 1.2 are drawn in a special way: the chiral carbon is placed in the plane of the paper so that the heavy tapered horizontal lines are projected above the plane of the paper (toward an onlooker) and represent the valences linking a chiral carbon to its substituents (H, OH, etc.); the vertical dotted tapered lines are projected below the plane of the drawing (away from an onlooker) and represent the valences linking that chiral carbon to the two neighboring carbon atoms of a carbohydrate carbon chain (vide infra).

Representation of Monosaccharides

Acyclic Form of Monosaccharides

In 1891 Emil Fischer [2] proposed an ingenious method for accurate two-dimensional representation of polychiral monosaccharide molecules by projecting their three-dimensional structures onto the plane of paper. In order to draw these, so-called, *Fischer's projections*, the following rules must be observed. A chiral carbon atom of a monosaccharide carbon chain must be placed in the plane of paper, with the two neighboring carbons that are *positioned below the plane of the paper*, i.e., away from an onlooker projected on a vertical line. The hydrogen atom and the hydroxyl group positioned *above the plane of paper* and toward an onlooker are

Representation of Monosaccharides

projected on a horizontal line as illustrated for D- and L-glyceraldehyde in Fig. 1.3. So the vertical line contains the chiral carbon and the two adjacent carbon atoms of a monosaccharide chain, whereas the horizontal line that is perpendicular to the vertical line contains the chiral carbon and its two substituents: the hydroxyl group and the hydrogen atom. The chiral carbon lies at the intersection of these two perpendicular lines, and it is customary not to write the symbol for this carbon. This operation has to be repeated for each chiral carbon up or down the monosaccharide chain in exactly the same manner as described above. The result of these operations is an

```
      CHO              CHO              CHO              CHO
   H—C—OH          H——OH           HO—C—H           HO——H
      CH₂OH            CH₂OH            CH₂OH            CH₂OH
        9               10                11               12
                   Fischer's projection              Fischer's projection
                   D-glyceraldehyde                  L-glyceraldehyde
```

Fig. 1.3

elongated vertical line intersected by two or more perpendicular horizontal lines, whose number depends on the length of a monosaccharide chain (tetrose, pentose, hexose, etc.).

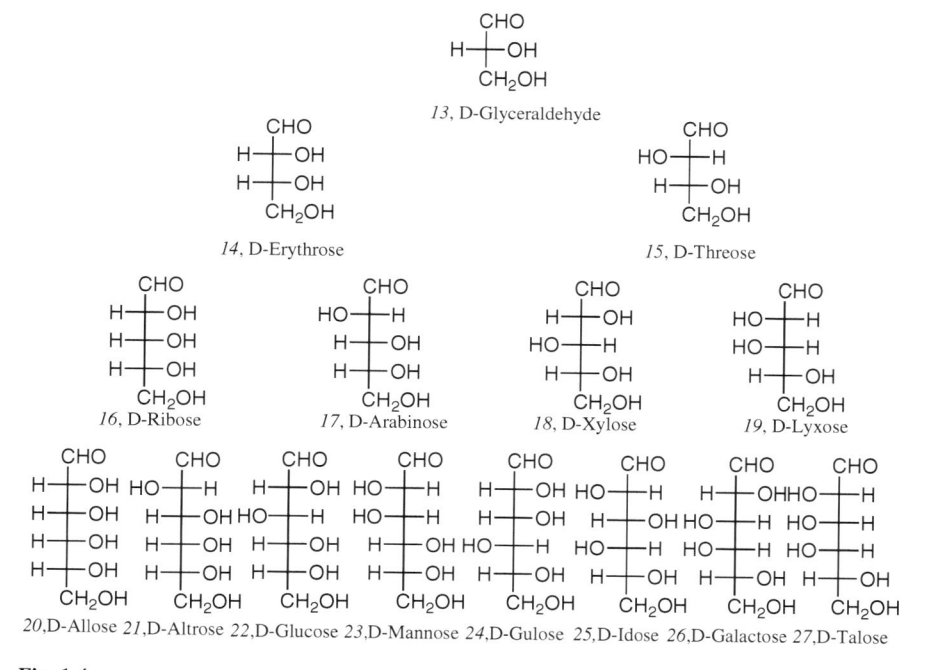

```
                             CHO
                          H——OH
                            CH₂OH
                       13, D-Glyceraldehyde
          CHO                                      CHO
        H——OH                                    HO——H
        H——OH                                    H——OH
          CH₂OH                                    CH₂OH
       14, D-Erythrose                          15, D-Threose

    CHO          CHO          CHO          CHO
  H——OH       HO——H         H——OH        HO——H
  H——OH       H——OH         HO——H        HO——H
  H——OH       H——OH         H——OH        H——OH
   CH₂OH       CH₂OH         CH₂OH        CH₂OH
 16, D-Ribose  17, D-Arabinose 18, D-Xylose 19, D-Lyxose

 CHO     CHO    CHO     CHO    CHO     CHO    CHO     CHO
H—OH  HO—H   H—OH   HO—H   H—OH  HO—H   H—OH   HO—H
H—OH  H—OH   HO—H   HO—H   H—OH  H—OH   HO—H   HO—H
H—OH  H—OH   H—OH   H—OH   HO—H  HO—H   HO—H   HO—H
H—OH  H—OH   H—OH   H—OH   H—OH  H—OH   H—OH   H—OH
 CH₂OH CH₂OH  CH₂OH  CH₂OH  CH₂OH CH₂OH  CH₂OH  CH₂OH
20,D-Allose 21,D-Altrose 22,D-Glucose 23,D-Mannose 24,D-Gulose 25,D-Idose 26,D-Galactose 27,D-Talose
```

Fig. 1.4

At the intersections of vertical and horizontal lines lie the chiral carbons and at the end of horizontal lines are placed ligands (hydrogen atoms, hydroxyl groups, amino group, etc.). The Fischer projection must be so oriented that the *anomeric* (carbonyl) carbon is always at the top (or near the top) of the vertical line, whereas the hydroxymethyl group is at the bottom. Figure 1.4 illustrates Fischer projections of aldoses in their *acyclic* (*aldehydo*) forms having three to six carbon atoms in their carbon skeleton. Only D-forms are shown; the L-forms are the mirror images of shown Fisher projections.

If transposition of a substituent must be performed in order to compare the Fischer projections of different stereoisomers, a number of rules must be obeyed, or otherwise the configurational change may take place and erroneous conclusions may be drawn.

1. The Fischer projection must never be taken out of the plane of drawing for manipulation, since this operation may result in change of the configuration of chiral carbons.
2. The rotation of a Fischer projection in the plane of drawing by 180° does not change the configuration of chiral carbons in a molecule.
3. The rotation of a Fischer projection in plane of drawing by 90° converts a given chiral carbon to its enantiomer.
4. One interchange of two chiral carbon ligands converts that chiral carbon to its enantiomer.
5. Two consecutive interchanges of two pairs of ligands do not change the configuration of that carbon

Cyclic Forms of Monosaccharides

The existence of cyclic forms of sugars was recognized very early because some of the observed chemistry of monosaccharides could not be explained by the open-chain structure. Thus, the acetylation of aldoses gave two isomeric penta-*O*-acetyl derivatives [3, 4] whereas the acyclic form should give only one penta-*O*-acetyl derivative. The reaction of methanolic hydrogen chloride with aldoses produced two isomeric mono-methoxy derivatives; so these compounds were not dimethyl acetals [5]. Further, the reactivity of the aldehydo group of aldoses was unusually low. For example, aldoses do not react with Schiff reagent (fuchsin-sulfurous acid), a reaction typical for aldehydes [6] except under special conditions and with a specially adjusted reagent [7, 8]. Finally, Tanret was able to isolate two forms of D-glucose that interconverted in aqueous solution [9].

The unusually low reactivity of the carbonyl group of aldoses is explained by Tollens [10] by postulating cyclic structure, and proposed, without any proof, the existence of tetrahydrofuran ring. Haworth later corrected this tetrahydrofuran formula [11] to a tetrahydropyran and introduced a new way for representation of cyclic structures (vide infra).

Representation of Monosaccharides

This early conclusion that monosaccharides exist in cyclic form was fully substantiated a half century later. In 1949, it was determined that the molar heat of reaction of gaseous monomeric formaldehyde (*28*) with alcohol vapors (*29*) is about 14.8 kcal/mol [12], indicating that in mixtures of alcohols and carbonyl compounds there indeed exist a strong tendency for the formation of hemiacetal structure (*30*). Contrary to expectation that substitution of a hydrogen atom of formaldehyde with a larger group would hinder the hemiacetal formation, due to steric reasons, the trichloroacetaldehyde (chloral) that contains a large trichloromethyl

Fig. 1.5

group, forms a very stable hemiacetal. Further, both, the 4-hydroxybutanal and the 5-hydroxypentanal exist predominantly as cyclic hemiacetals (*lactols*) [13] suggesting that there is indeed a strong driving force for the hemiacetal formation in mixtures of alcohols and aldehydes or ketones [14] (Figs. 1.5, 1.6).

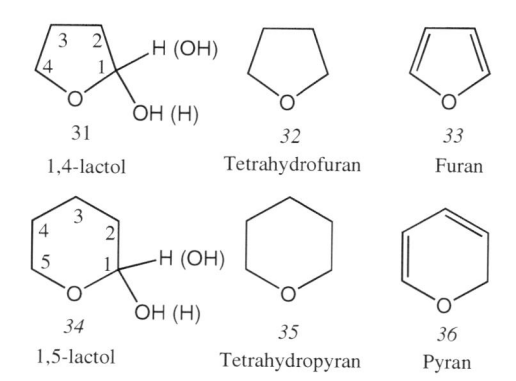

Fig. 1.6

It is therefore no surprise that intramolecular formation of hemiacetal (*lactol*) ring is so much favored in aqueous solutions of free monosaccharides since they have in the same molecule both the electrophilic *anomeric* (carbonyl) carbon and the C4 and/or the C5 nucleophilic hydroxyl groups. It has been polarographically determined that an aqueous solution of D-glucose (at pH = 6.9) contains, at equilibrium, only 0.0026% of the acyclic (aldehydo) form; the rest of the sugar is in cyclic (lactol) form [15].

The addition of C4 hydroxyl group to the anomeric carbonyl carbon of an acyclic aldose results in the formation of a five-membered ring, whereas the addition of C5 hydroxyl group to the anomeric carbonyl carbon of an acyclic aldose results in the formation of a six-membered ring. In Fig. 1.7, the cyclic forms of monosaccharides are represented using Fischer projections.

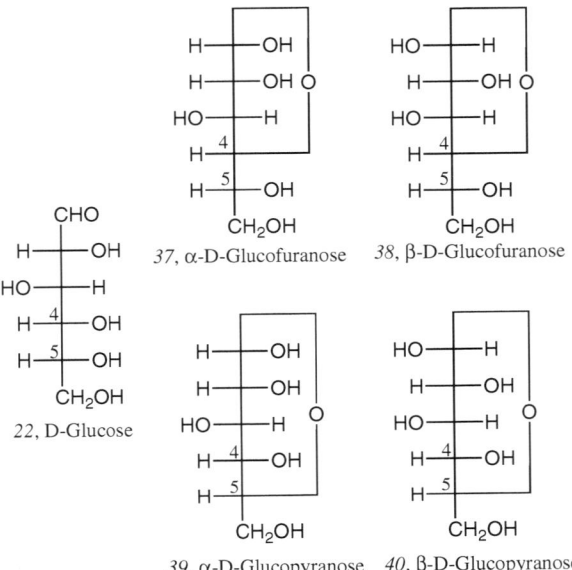

Fig. 1.7

Haworth proposed the terms *furanoses* and *pyranoses* [11] for these two monosaccharide forms due to their apparent similarity to furan and pyran molecules (Fig. 1.6) and at the same time introduced also a different way of representing the cyclic structures of monosaccharides.

In Fig. 1.8, the two cyclic forms of D-glucose are represented with the so-called *Haworth projection* formulae (β-D-glucopyranose *41* and α-D-glucofuranose *42*), which are still in use today despite their serious and obvious shortcomings (in these representations the carbon valence angle of 109° is totally ignored since the C–C bond angles in carbohydrate ring are 120°, and the C–H and the C–OH bond angles

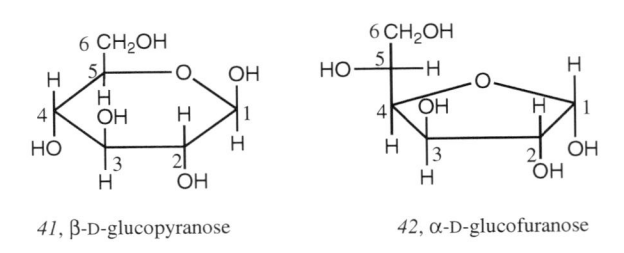

Fig. 1.8

are 90°). However, Haworth, like Fischer before him, never intended to represent the actual shape of a molecule, but rather to accurately describe the configurations of individual chiral carbons and their configurational relationships. In Fig. 1.9, D-glucose is represented in the acyclic form *22*, in the cyclic pyranose form using Fischer projection *39*, and in the cyclic pyranose form using Haworth projection *43*. It is obvious that the Haworth projection of cyclic pyranoses has fewer shortcomings than Fischer projection, but it is nevertheless still very far from realistic representation.

22, aldehydo D-Glucose   *39*, α-D-glucopyranose   *43*, α-D-glucopyranose
                                Fischer Projection   Haworth Projection

Fig. 1.9

The Haworth projection can be obtained from a Fischer projection in the following way. The C5 hydrogen exchange with the C5 hydroxymethyl group of α-D-glucopyranose written as Fischer projection (*39* in Fig. 1.10) will produce *44* (Fig. 1.10), whereas the C5 hydrogen exchange with the C5 ring oxygen will produce *45* (Fig. 1.10). Since described operation consisted of two consecutive sub-

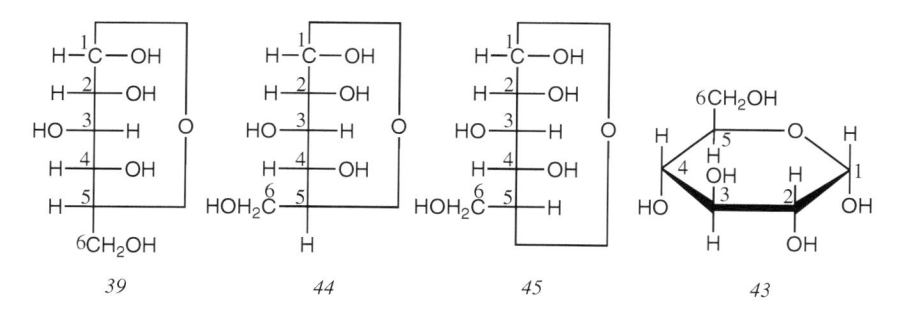

39   *44*   *45*   *43*

Fig. 1.10

stituent exchanges at the C5 carbon atom, the configuration of the C5 carbon atom did not change (vide supra, Rule 4 in handling the Fischer projections). In this way, the Fischer projection *39* is easily converted into Haworth projection *43*.

In Haworth projection the six-membered pyranose ring is written in such a way that the plane of the *planar* six-membered ring is positioned perpendicularly to the plane of paper with carbon atoms 1 and 4 lying in the plain of paper. The ring oxygen and the C5 carbon are placed behind the plane of paper so that the ring oxygen is placed on the right and the C5 carbon on the left side; the C2 and C3 carbons are projected above the plain of the paper. The C1–C2–C3–C4 bonds are represented with heavy lines (indicating that they are closer to an onlooker) and the C1–O5–C5–C4 bonds are represented with lighter lines (indicating that they are away from an onlooker). Ligands that are in the Fischer projection on the right-hand side from the vertical line representing the carbon skeleton are written in Haworth projection below the plane of the six-membered pyranose ring, and the ligands that are, in Fischer projection, on the left-hand side from the vertical line representing the carbon skeleton of a monosaccharide are written above the six-membered pyranose ring of Haworth projection (*45* and *43* in Fig. 1.10).

The Haworth projection of furanose form of a monosaccharide is obtained from Fischer projections similarly to pyranoses, namely the planar five-membered ring is again positioned perpendicularly to the plane of paper with the C1 and the C4 carbon atoms lying in the plane of a paper. The carbon atoms 2 and 3 are placed above the plane of the paper whereas the ring oxygen is placed behind the plane of the paper. The atoms positioned above the plane of the paper are connected with heavy lines (C1–C2–C3–C4 bonds) indicating that they are oriented toward an onlooker; C1–O5–C4 bonds are represented with lighter lines indicating that they are positioned behind the plane of the paper, and away from an onlooker (see Fig. 1.11). The ligands that are in the Fischer projection positioned on the right-hand side from the vertical line representing the carbon skeleton of a monosaccharide are again written in Haworth projection below the furanose five-membered ring, and the ligands that are in the Fischer projection positioned on the left-hand side from the vertical line representing the carbon skeleton of a monosaccharide are written in Haworth projection above the furanose five-membered ring (Fig. 1.11)

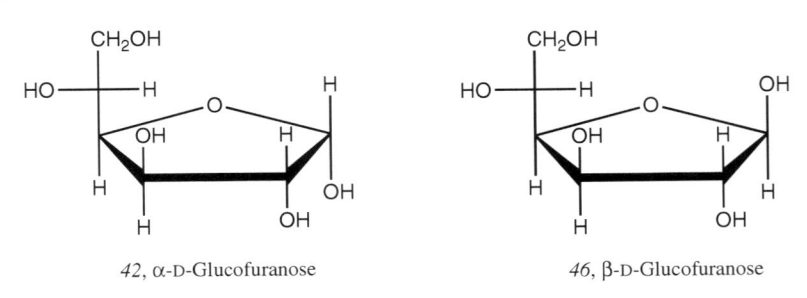

42, α-D-Glucofuranose   *46*, β-D-Glucofuranose

Fig. 1.11

The conversion of the Fischer projection of a furanose to Haworth projection is accomplished similarly as described for pyranoses, just the two consecutive substituent exchanges are taking place at the C4 carbon atom of the monosaccharide chain.

The cyclization of an acyclic sugar, in addition of converting the acyclic chain of a polyhydroxy aldehyde or ketone into a five-membered tetrahydrofuran or six-membered tetrahydropyran ring, introduces yet another very important structural change in a monosaccharide molecule and that is that the achiral carbonyl carbon becomes a chiral *hemiacetal (lactol)* carbon, adding thus two more stereoisomers to the parent monosaccharide. These two stereoisomers are called *anomers* and are referred to as α- and β-isomers.

The configuration of this new chiral carbon, the anomeric carbon, is determined by the orientation of its hydroxyl group with regard to the orientation of hydroxyl group at the highest numbered chiral carbon in Fischer's projection (Fig. 1.7). If the two oxygen atoms are positioned on the same side relative to the vertical line representing the carbon skeleton (*cis*), the anomer is α (*37*); if they are positioned on the opposite side (*trans*), it is β (*38*). It should be noted that the ring oxygen, although written in Fischer projection in the middle of a long line shaped like a rectangle and connecting the C5 and the C1 carbon atoms is nevertheless the oxygen linked to the highest numbered chiral carbon of D-glycopyranose (C5), and it is projected to the right of the vertical line (carbon chain).

The Nomenclature of Carbohydrates

For a comprehensive description of the nomenclature of carbohydrates, see Reference [16].

The nomenclature described here is based on these rules, but it is significantly abbreviated and limited only to monosaccharides and some of their derivatives.

Trivial Names

The trivial names of the acyclic aldoses with three, four, five, or six carbon atoms are preferred to their systematic (IUPAC) names. The trivial names of most common aldoses are

Triose: glyceraldehyde
Tetroses: erythrose, threose
Pentoses: ribose, arabinose, xylose, lyxose
Hexoses: allose, altrose, glucose, mannose, gulose, idose, galactose, talose

Stem and Systematic Names

The *stem name* indicates the *length of the carbon chain of a monosaccharide*. Thus the acyclic aldoses having three, four, five, six, seven, eight, nine, ten, etc., carbon atoms in the chain are *triose, tetrose, pentose, hexose, heptose, octose, nonose, decose, etc.*

The *stem names* of the acyclic ketoses having four, five, six, seven, eight, nine, ten, etc. carbon atoms in the chain are *tetrulose, pentulose, hexulose, heptulose, octulose, nonulose, deculose, etc. Systematic names* are formed from a stem name and a configurational prefix, which are always written in italicized lower case letters.

Each prefix is preceded by configurational symbol D or L. Configurational prefixes are obtained from trivial names of sugars by omitting the ending -se. Examples: D-*ribo*-pentose for D-ribose; D-galacto-hexose for D-galactose.

Conventions

The following abbreviations are commonly used for substituents in carbohydrate structures: Ac (acetyl), Bn (benzyl), Bz (benzoyl), Et (ethyl), Me (methyl), Me$_3$Si (trimethylsilyl), Ms (mesyl = methanesulfonyl), Ph (phenyl). Tf (triflyl = trifluoromethanesulfonyl), Tr (trityl = triphenylmethyl), Ts (tosyl = *p*-toluenesulfonyl).

Choice of Parent Monosaccharides

In cases where more than one monosaccharide structure is embedded in a larger molecule, a *parent structure* is chosen on the basis of the following *order of preference*.

(a) The *parent* with the greatest number of carbon atoms in the chain has the preference, e.g., heptose rather than hexose.
(b) The *parent*, of which: the first letter of the trivial name or of the configurational prefix occurs earliest in the alphabet (e.g., *47* in Fig. 1.12). If two possible parents have the same initial letter, then the choice will be made according to the letter at the first point of difference in the trivial name. Examples: allose before altrose, galactose before glucose, glucose before gulose and mannose, *allo*- before *altro*-, *gluco*- before *gulo*- (*47* in Fig. 1.12).
(c) The configurational symbol D- before L- (*49* in Fig. 1.12).
(d) The anomeric symbol α- before β-

Examples:

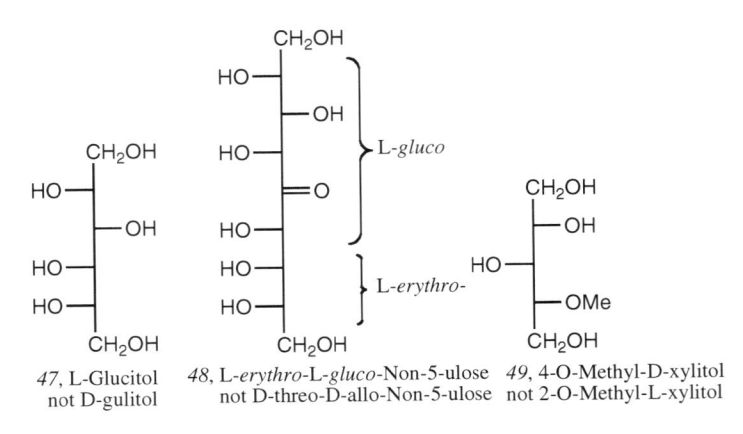

47, L-Glucitol
not D-gulitol

48, L-*erythro*-L-*gluco*-Non-5-ulose
not D-threo-D-allo-Non-5-ulose

49, 4-O-Methyl-D-xylitol
not 2-O-Methyl-L-xylitol

Fig. 1.12

The Nomenclature of Carbohydrates

Choice Between Alternative Names for Substituted Derivatives

When the parent structure is symmetrical, preference between alternative names for derivatives should be given according to the following criteria, taken in order:

(a) The name that includes the configurational symbol D rather than L as is for example shown with *49* in Fig. 1.12.
(b) The name that gives the lowest set of locants to the substituents present (see *50* in Fig. 1.13)

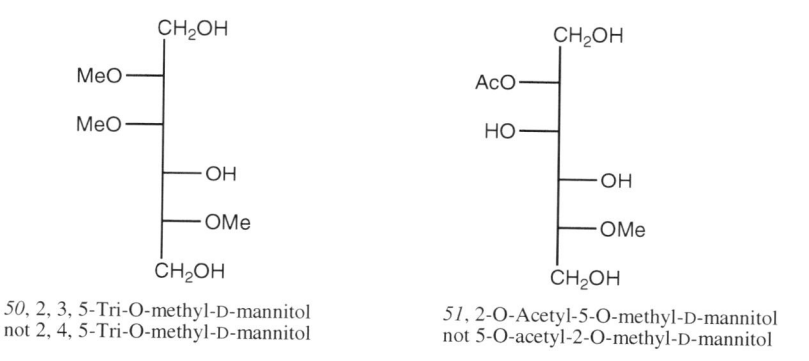

```
        CH₂OH                          CH₂OH
MeO  ─┤                        AcO  ─┤
MeO  ─┤                        HO   ─┤
      ├─ OH                          ├─ OH
      ├─ OMe                         ├─ OMe
        CH₂OH                          CH₂OH
```

50, 2,3,5-Tri-O-methyl-D-mannitol   *51*, 2-O-Acetyl-5-O-methyl-D-mannitol
not 2,4,5-Tri-O-methyl-D-mannitol   not 5-O-acetyl-2-O-methyl-D-mannitol

Fig. 1.13

Configurational Symbols and Prefixes

Configurational prefixes are given in Figs. 1.14 and 1.15.

Each prefix is D or L according to whether the configuration at the reference carbon atom in the Fischer projection is the same as, or opposite of, that in D(+)-glyceraldehyde. Only Fischer projections of D-prefixes are given above; X is the group with the lowest numbered carbon atom.

The configuration of >CHOH group or a set of two, three, or four contiguous >CHOH groups (fully or partially derivatized, such as >CHOMe, >CHOAc, etc., or fully substituted hydroxyl groups, such as >CHNH₂, >CHCl) are designated by one of the following configurational prefixes which are derived from the trivial names of the aldoses (Fig. 1.14 and 1.15).

A monosaccharide is assigned to the D or to the L series according to configuration of the highest numbered chiral carbon atom (configurational carbon atom). This was discussed earlier (vide supra).

Configurational classification is denoted by the symbols D and L which in print will be small capital roman letters and which are unrelated to terms *dextro* and *levo* that denote the sign of optical rotation of a sugar. Racemic forms are indicated by DL. Such symbols are affixed by a hyphen immediately before the monosaccharide trivial name. If the sign of optical rotation is to be indicated, this is done by adding

One >CHOH group

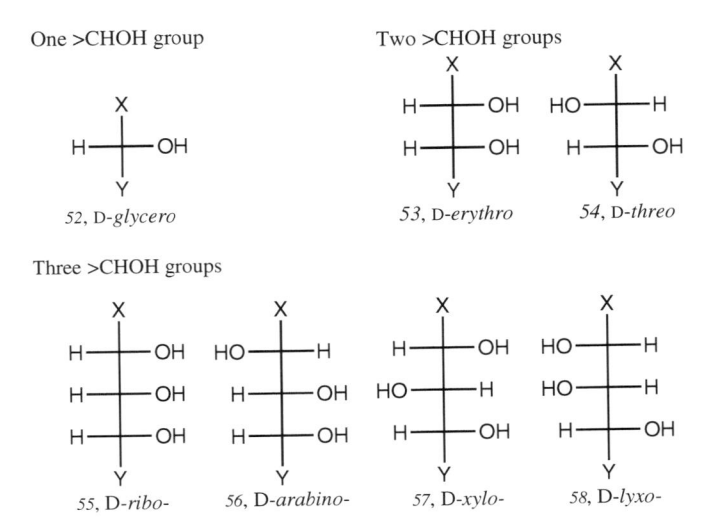

$$\begin{array}{c} X \\ H-\!\!\!-\!\!\!-\!\!\!-OH \\ Y \end{array}$$

52, D-*glycero*

Two >CHOH groups

$$\begin{array}{c} X \\ H-\!\!\!-\!\!\!-\!\!\!-OH \\ H-\!\!\!-\!\!\!-\!\!\!-OH \\ Y \end{array} \quad \begin{array}{c} X \\ HO-\!\!\!-\!\!\!-\!\!\!-H \\ H-\!\!\!-\!\!\!-\!\!\!-OH \\ Y \end{array}$$

53, D-*erythro*   54, D-*threo*

Three >CHOH groups

55, D-*ribo*-   56, D-*arabino*-   57, D-*xylo*-   58, D-*lyxo*-

Fig. 1.14

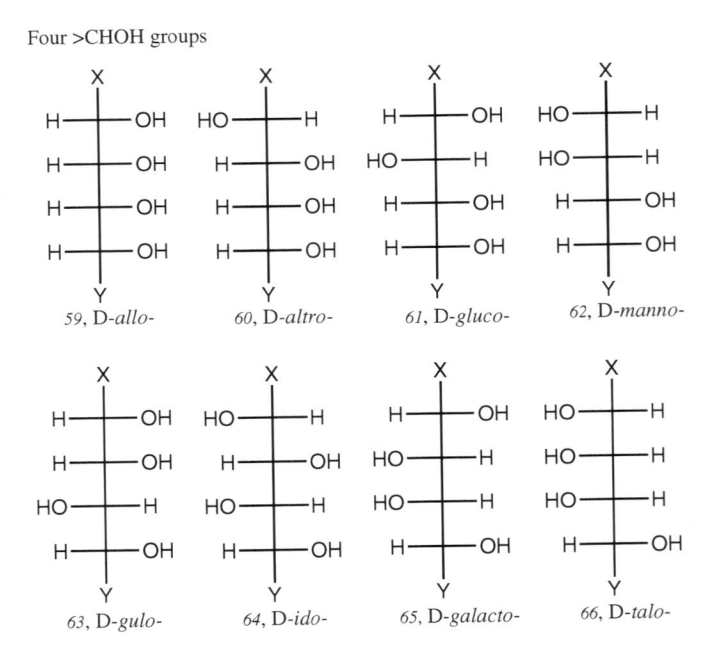

Four >CHOH groups

59, D-*allo*-   60, D-*altro*-   61, D-*gluco*-   62, D-*manno*-

63, D-*gulo*-   64, D-*ido*-   65, D-*galacto*-   66, D-*talo*-

Fig. 1.15

(+) or (−) immediately after the configurational symbols D and L. Racemic forms may be indicated by (±). Examples: D-glucose or D (+)-glucose, D-fructose or D (−)-fructose, DL-glucose or (±)-glucose.

When monosaccharides have a plane of symmetry as a consequence of which they are optically inactive, the prefix *meso* is used where appropriate.

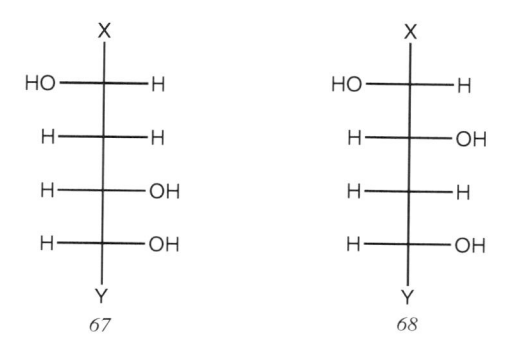

both D-*arabino*

Fig. 1.16

The monosaccharides that have a sequence of consecutive but not contiguous chiral carbons are named so that the interrupting achiral carbons are ignored as if they are not there (for examples see Fig. 1.16). The achiral carbon can be methylene carbon (as in case of deoxy-sugar), but it can also be a keto group of a ketose containing not more than four chiral carbons (see Fig. 1.17).

Ketoses

Ketoses are classified as 2-ketoses, 3-ketoses, etc., according to the position of the carbonyl or potential carbonyl group. When the carbonyl group is at the middle

69
D-*glycero*-2-tetrulose

70
D-*erythro*-2-pentulose

71
4-O-acetyl-D-*erythro*-3-pentulose

72
2-O-acetyl-L-*erythro*-3-pentulose

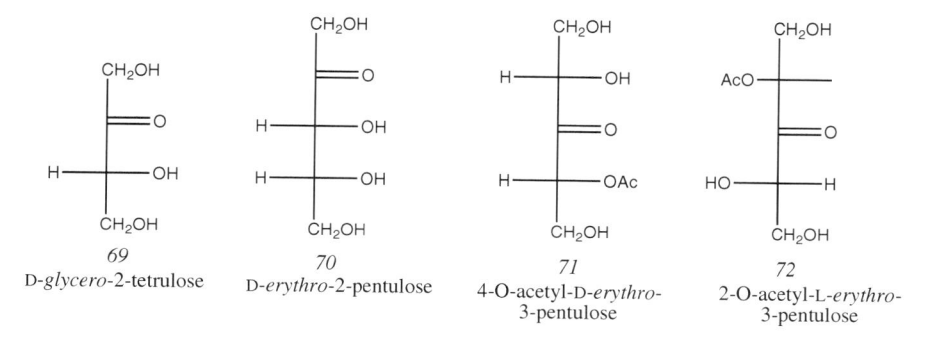

Fig. 1.17

carbon atom of a ketose containing an uneven number of carbon atoms in the chain, two names are possible. The name that will be selected must be in accordance with the order of precedence given in rule on *parent names*. For example, the rotation by 180° of the 4-*O*-acetyl-D-*erythro*-3-pentulose *(71*, Fig. 1.17) in the plane of the paper will give the 2-*O*-acetyl-L-*erythro*-3-pentulose *(72*, Fig. 1.17). However, the 4-*O*-acetyl-D-*erythro*-3-pentulose will be the correct name for this ketose, since the prefix D has the preference over the prefix L. Some other examples for naming ketoses are given in Fig. 1.17.

Deoxy-monosaccharides

The replacement of an alcoholic hydroxyl group of a monosaccharide, or a monosaccharide derivative, by a hydrogen atom is expressed by using the prefix *deoxy*, preceded by the locant and followed by a stem or trivial name separated by a hyphen. The stem name must include configurational prefixes necessary to describe the configurations of the chiral carbons present in the deoxy-compound. The order of citation of the configurational prefixes starts at the end farthest from the carbon atom number one (See Fig. 1.15). "Deoxy" is regarded as a detachable prefix, i.e., it is placed in alphabetical order with other substituent prefixes – for example, acetyl before anhydro, benzyl before benzoyl, deoxy before fluoro. Several examples for naming deoxy sugars are given in Fig. 1.18.

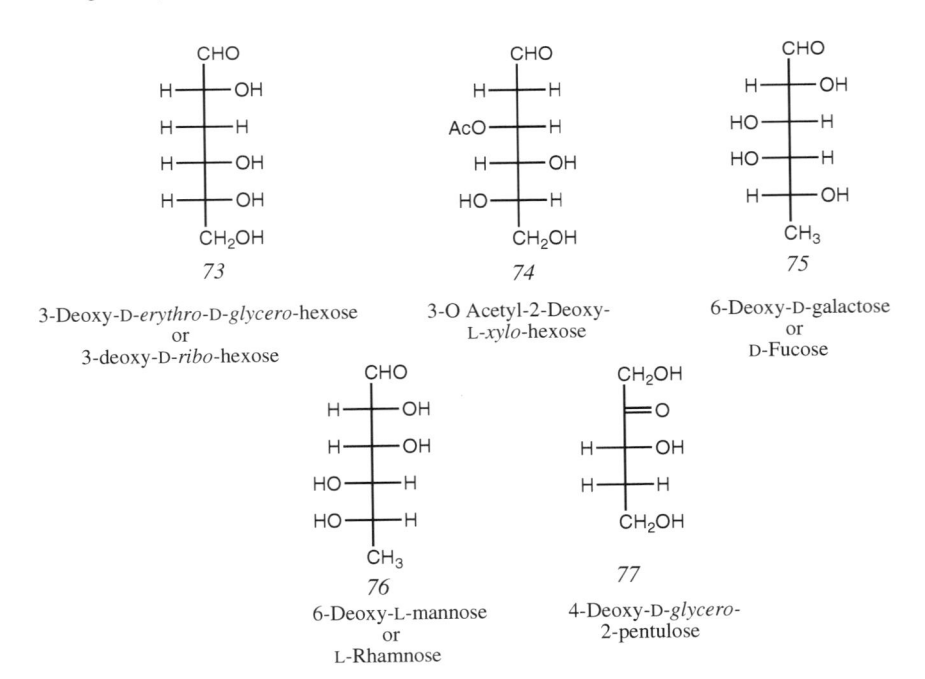

```
        CHO                  CHO                  CHO
   H ─┬─ OH              H ─┬─ H              H ─┬─ OH
   H ─┼─ H             AcO ─┼─ H             HO ─┼─ H
   H ─┼─ OH              H ─┼─ OH             HO ─┼─ H
   H ─┼─ OH             HO ─┼─ H              H ─┼─ OH
      CH₂OH                 CH₂OH                 CH₃
        73                   74                   75
```

3-Deoxy-D-*erythro*-D-*glycero*-hexose 3-O Acetyl-2-Deoxy- 6-Deoxy-D-galactose
or L-*xylo*-hexose or
3-deoxy-D-*ribo*-hexose   D-Fucose

```
             CHO                   CH₂OH
        H ─┬─ OH                     ═O
        H ─┼─ OH              H ─┬─ OH
       HO ─┼─ H               H ─┼─ H
       HO ─┼─ H                   CH₂OH
            CH₃                     77
             76
    6-Deoxy-L-mannose         4-Deoxy-D-*glycero*-
            or                     2-pentulose
        L-Rhamnose
```

Fig. 1.18

Amino-monosaccharides

The replacement of an alcoholic hydroxyl group of a monosaccharide, or a monosaccharide derivative, by an amino group is treated as substitution of the appropriate hydrogen atom of the corresponding deoxy-monosaccharide by the amino group (Fig. 1.19).

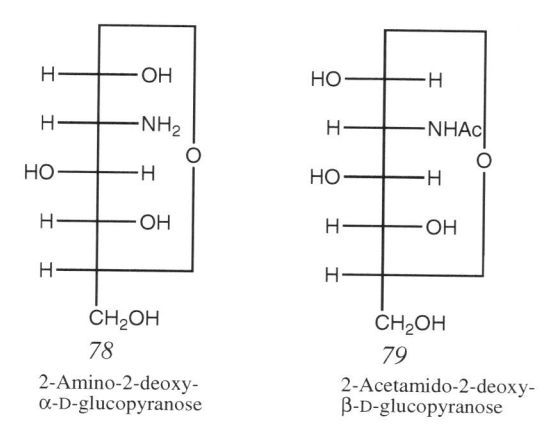

```
H ──── OH            HO ──── H
H ──── NH₂           H ──── NHAc
HO ──── H   O        HO ──── H    O
H ──── OH            H ──── OH
H ────               H ────
     CH₂OH                CH₂OH
      78                   79
2-Amino-2-deoxy-      2-Acetamido-2-deoxy-
α-D-glucopyranose     β-D-glucopyranose
```

Fig. 1.19

Substitution in the amino group is indicated by use of the prefix N unless the substituted amino group has a trivial name (for example, CH_3CONH-, acetamido). The trivial names accepted for biochemical usage are D-galactosamine for 2-amino-2-deoxy-D galactopyranose, D-glucosamine for 2-amino-2-deoxy-D-glucopyranose, D-mannosamine for 2-amino-2-deoxy-D-mannopyranose, etc.

When the complete name of a derivative includes other prefixes, *deoxy* takes its place in the alphabetical order of detachable prefixes; in citation the alphabetical order is preferred to numerical order – for example, 4-amino-4-deoxy-3-*O*-methyl-D-*erythro*-2-pentulose *80* and 4-deoxy-4-(ethylamino)-D-*erythro*-2-pentulose *81* (see Fig. 1.20).

O-Substitution

Replacement of a hydrogen atom of an alcoholic hydroxyl group of a monosaccharide or its derivative by another atom or group is denoted by placing the name of this atom or group before the name of the parent monosaccharide. The name of the atom or group is preceded by an italic capital letter *O* (for oxygen), followed by a hyphen in order to make clear that substitution is on oxygen. The *O* prefix need not be repeated for multiple replacements by the same atom or group. Instead a prefix, di- tri-, etc., is used.

Replacement of hydrogen attached to nitrogen or sulfur by another atom or group is indicated in a similar way with the use of italic capital letters *N* or *S*.

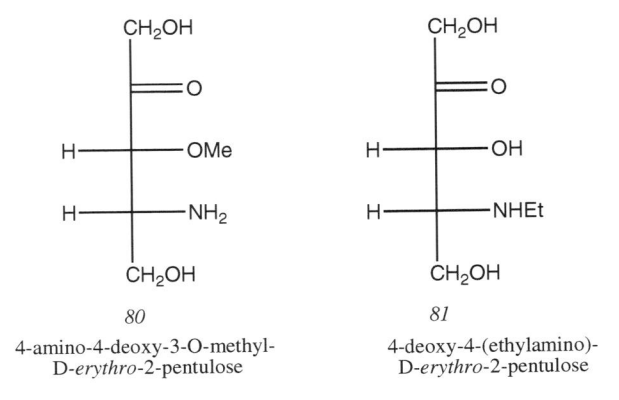

```
        CH₂OH                    CH₂OH
         ‖                        ‖
         O                        O
  H ──┼── OMe              H ──┼── OH
  H ──┼── NH₂              H ──┼── NHEt
        CH₂OH                    CH₂OH
         80                       81
 4-amino-4-deoxy-3-O-methyl-  4-deoxy-4-(ethylamino)-
    D-erythro-2-pentulose      D-erythro-2-pentulose
```

Fig. 1.20

The italic capital letter *C* may be used to indicate replacement of hydrogen attached to carbon (for example in branched chain monosaccharides), to avoid possible ambiguity. Examples are given in Fig. 1.21.

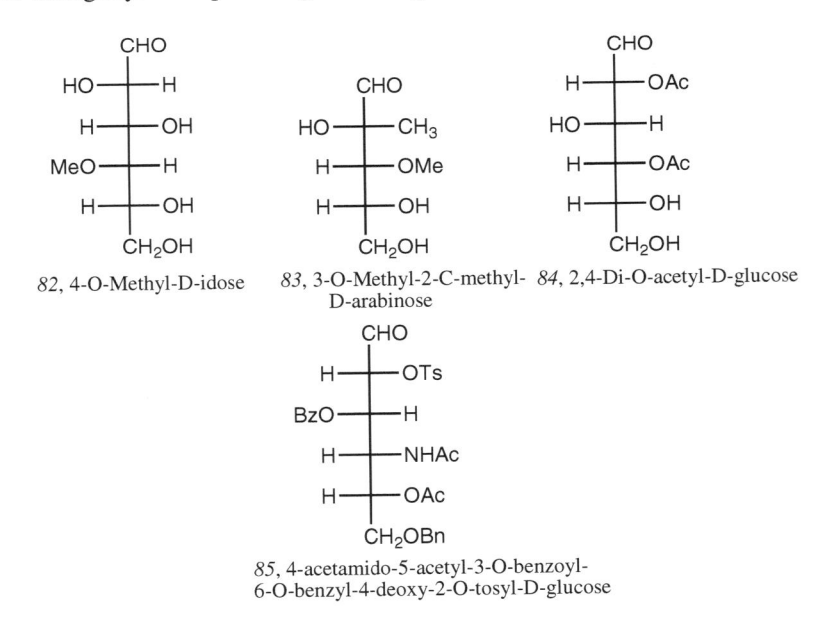

```
      CHO                    CHO                    CHO
 HO──┼──H                                     H ──┼── OAc
 H ──┼── OH            HO──┼──CH₃            HO──┼──H
 MeO──┼──H             H──┼──OMe             H ──┼── OAc
 H ──┼── OH            H ──┼── OH            H ──┼── OH
     CH₂OH                  CH₂OH                 CH₂OH
 82, 4-O-Methyl-D-idose  83, 3-O-Methyl-2-C-methyl-  84, 2,4-Di-O-acetyl-D-glucose
                              D-arabinose

                            CHO
                       H ──┼── OTs
                      BzO──┼──H
                       H ──┼── NHAc
                       H ──┼── OAc
                           CH₂OBn
              85, 4-acetamido-5-acetyl-3-O-benzoyl-
               6-O-benzyl-4-deoxy-2-O-tosyl-D-glucose
```

Fig. 1.21

O-Substitution products of monosaccharides or their derivatives may be named as esters, ethers, etc., following the procedures presented for that purpose in IUPAC Nomenclature for Organic Chemistry, Section C, 1963. Examples are given in Fig. 1.22.

The Nomenclature of Carbohydrates

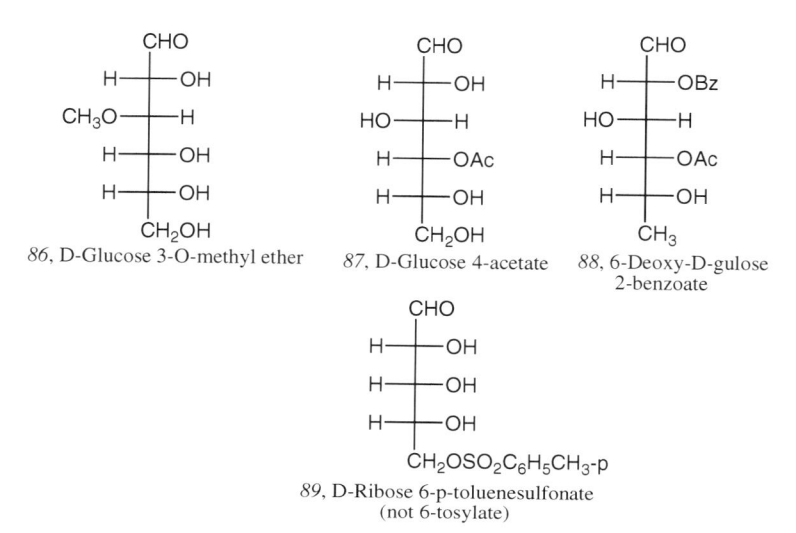

```
        CHO                    CHO                    CHO
   H ──┼── OH            H ──┼── OH            H ──┼── OBz
 CH₃O ──┼── H          HO ──┼── H            HO ──┼── H
   H ──┼── OH            H ──┼── OAc           H ──┼── OAc
   H ──┼── OH            H ──┼── OH            H ──┼── OH
       CH₂OH                  CH₂OH                  CH₃
86, D-Glucose 3-O-methyl ether   87, D-Glucose 4-acetate   88, 6-Deoxy-D-gulose
                                                              2-benzoate
```

```
              CHO
        H ──┼── OH
        H ──┼── OH
        H ──┼── OH
           CH₂OSO₂C₆H₅CH₃-p
    89, D-Ribose 6-p-toluenesulfonate
         (not 6-tosylate)
```

Fig. 1.22

Acyclic Forms

The acyclic nature of a monosaccharide or monosaccharide derivative containing an uncyclized carbonyl group may be stressed by inserting the italicized prefix *aldehydo* or *keto* respectively, immediately before the configurational prefix(es) or before trivial name. These prefixes may be abbreviated to *aldehydo* and *keto*. Examples are given in Fig. 1.23.

```
        CHO                          CH₂OAc
   H ──┼── OAc                         ║O
  AcO ──┼── H                  AcO ──┼── H
   H ──┼── OAc                   H ──┼── OAc
   H ──┼── OAc                   H ──┼── OAc
       CH₂OAc                        CH₂OAc
```

90, 2,3,4,5,6-Penta-O-acetyl-aldehydo-D-glucose, or *aldehydo*-D-Glucose 2,3,4,5,6-pentaacetate

91, 1,3,4,5,6-Penta-O-acetyl-*keto*-D-fructose
or
keto-D-Fructose 1,3,4,5,6-pentaacetate

Fig. 1.23

Anomers and the Anomeric Configurational Symbols ("α" or "β").

The free hemiacetal hydroxyl group of a cyclic form of monosaccharide or monosaccharide derivative is termed the *anomeric* or *glycosidic* hydroxyl group (the word *anomer* comes from the Greek words ανω = *ano* and μερωσ = *meros*. The first word means *up, top* and the second word means *part*, or in this case *position*).

The two stereoisomers of an aldose or ketose, or their derivatives that differ in the configuration of anomeric hydroxyl group (termed *anomers*), are distinguished with the aid of the two anomeric prefixes α and β, relating the configuration of the anomeric carbon to that of the reference chiral carbon atom of the monosaccharide;

92, β-D-Galactopyranose
or
β-D-Galactose-(1,5)

93, α-D-Glucoseptanose
or
α-D-Glucose-(1,6)

94, β-D-Fructopyranose
or
β-D-Fructose-(2,6)

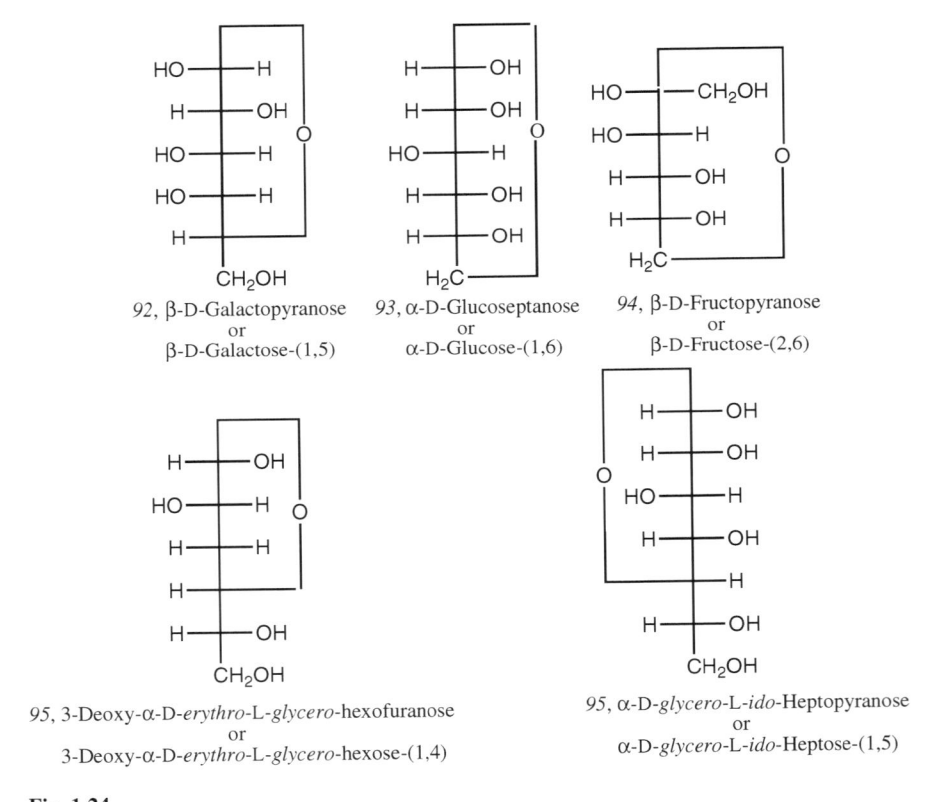

95, 3-Deoxy-α-D-*erythro*-L-*glycero*-hexofuranose
or
3-Deoxy-α-D-*erythro*-L-*glycero*-hexose-(1,4)

95, α-D-*glycero*-L-*ido*-Heptopyranose
or
α-D-*glycero*-L-*ido*-Heptose-(1,5)

Fig. 1.24

the anomer having the same configuration, in the Fischer projection, at the anomeric and the reference carbon atom is designated α. In the α-anomer, the exocyclic oxygen atom at the anomeric center is formally *cis*, in the Fischer projection, to the oxygen atom attached to the anomeric reference atom; in the β anomer, these oxygen atoms are formally *trans*.

The Nomenclature of Carbohydrates

The anomeric prefix, α or β, followed by a hyphen, is placed immediately in front of the configurational symbol, D or L, of the trivial name, or of the configurational prefix denoting the group of achiral carbon atoms that include the reference carbon atom.

Examples are given in Fig. 1.24.

Glycosides

Mixed acetals, resulting from replacement of the hydrogen atom of the anomeric or glycosidic hydroxyl group by a group R, derived from an alcohol or phenol (ROH), are named *glycosides*. The term glycoside is used in a generic sense only and may not be applied to specific compounds. Glycosides are named by replacing the terminal "e" of the name of the corresponding cyclic form of the saccharide derivative by "ide" and placing before the word thus obtained, as a separate word, the name of the group R. Examples are given in Figs. 1.25, 1.26, and 1.27.

97, Methyl α–L-arabinopyranoside   *98*, Methyl β-L-threofuranoside

Fig. 1.25

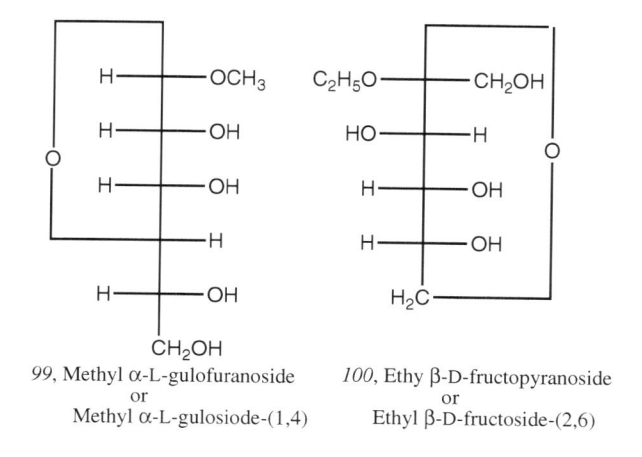

99, Methyl α-L-gulofuranoside
or
Methyl α-L-gulosiode-(1,4)

100, Ethy β-D-fructopyranoside
or
Ethyl β-D-fructoside-(2,6)

Fig. 1.26

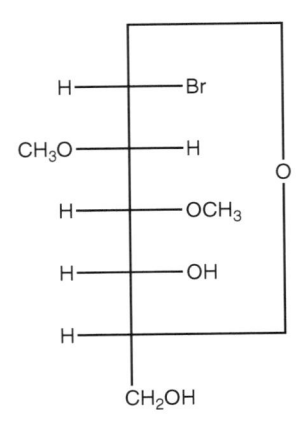

101, 2,3-Di-O-methyl-α-D-altropyranosyl bromide
or
2,3-Di-O-methyl-α-D-altrosyl-(1,5) bromide

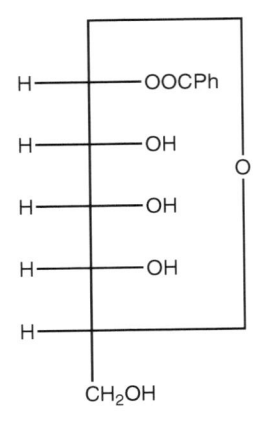

102, α-D-Allopyranosyl benzoate,
1-O-Benzoyl-α-D-allopyranose
or
α-D-allosyl-(1,5) benzoate.
1-O-Benzoyl-α-D-allose-(1,5)

Fig. 1.27

Glycosyl Radicals and Glycosylamines

(a) The radical formed by detaching the anomeric or glycosidic hydroxyl group from the cyclic form of a monosaccharide or monosaccharide derivative is named by replacing the terminal "e" of the name of the monosaccharide or monosaccharide derivative by "yl." The general name of these radicals is *glycosyl* (glucofuranosyl, glucopyranosyl, mannopyranosyl, galactopyranosyl, fructofuranosyl, etc.) radical (Fig. 1.28).

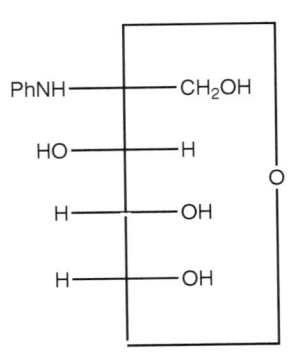

103, N-Phenyl-β-D-fructopyranosylamine
or
N-Phenyl-β-D-fructosyl-(2,6) amine

Fig. 1.28

The Nomenclature of Carbohydrates

(b) The replacement of the glycosidic hydroxyl group of a cyclic form of a monosaccharide derivative by an *amino* group is indicated by adding the suffix "amine" to the name of the glycosyl radical (Fig. 1.28).

Aldonic Acids

Monocarboxylic acids formally derived from aldoses having three or more carbon atoms in the chain, by oxidation of the aldehydic group, are named *aldonic acids* and are divided into aldotrionic acids, aldotetronic acids, aldopentonic acids, aldohexonic acids, etc., according to the number of carbon atoms in the chain. Names of the individual compounds of this type are formed by replacing the ending "ose" of the systematic or trivial name of the aldose by "onic acid" (Fig. 1.29).

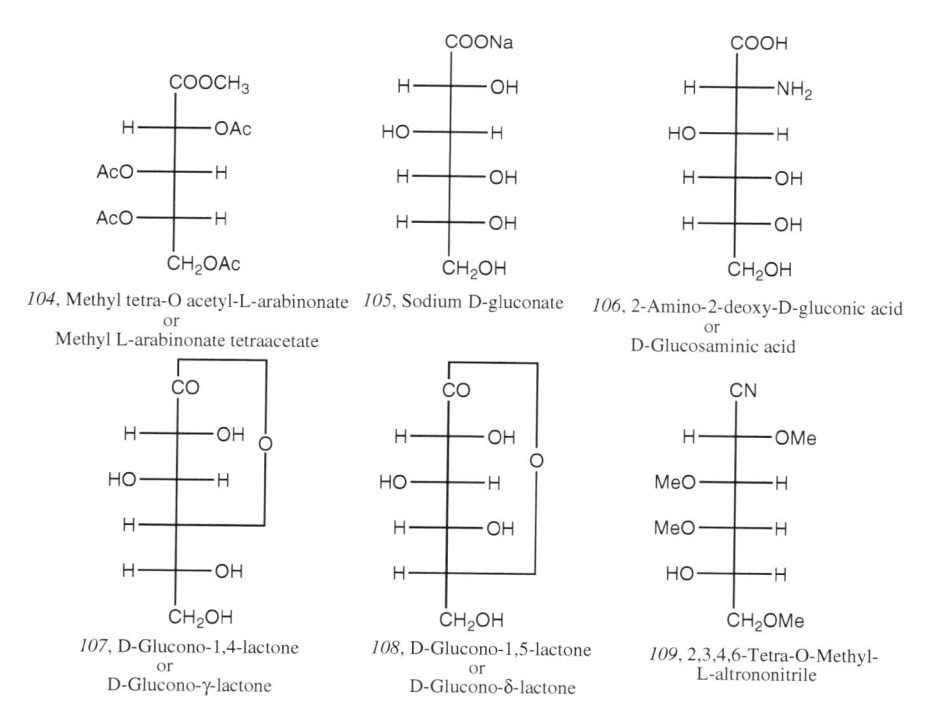

104, Methyl tetra-O acetyl-L-arabinonate
or
Methyl L-arabinonate tetraacetate

105, Sodium D-gluconate

106, 2-Amino-2-deoxy-D-gluconic acid
or
D-Glucosaminic acid

107, D-Glucono-1,4-lactone
or
D-Glucono-γ-lactone

108, D-Glucono-1,5-lactone
or
D-Glucono-δ-lactone

109, 2,3,4,6-Tetra-O-Methyl-L-altrononitrile

Fig. 1.29

Derivatives of these acids formed by change in the carboxyl group (salts, esters, lactones, acyl halides, amides, nitriles, etc.) are named according to the IUPAC Nomenclature of Organic Chemistry, Section C, 1965, Rules C-4. Examples are given in Fig. 1.29.

Uronic Acids

The monocarboxylic acids formally derived by oxidation of the terminal CH_2OH group of aldoses having four or more carbon atoms in the chain, or of glycosides derived from these aldoses, to a carboxyl group are named "uronic acids." The names of the individual compounds of this type are formed by replacing the (a) ending "ose" of the systematic or trivial name of the aldose by "uronic acid" or (b) ending "oside" of the name of the glycoside by "osiduronic acid."

The carbon atom of the (potential) aldehydic carbonyl group (not that of the carboxyl group) is numbered 1.

Derivatives of these acids formed by change in the carboxyl group (salts, esters, lactones, acyl halides, amides, nitriles, etc.) are named according to the IUPAC Nomenclature of Organic Compounds, Section C, 1965, Rules C-4.

Examples are shown in Fig. 1.30.

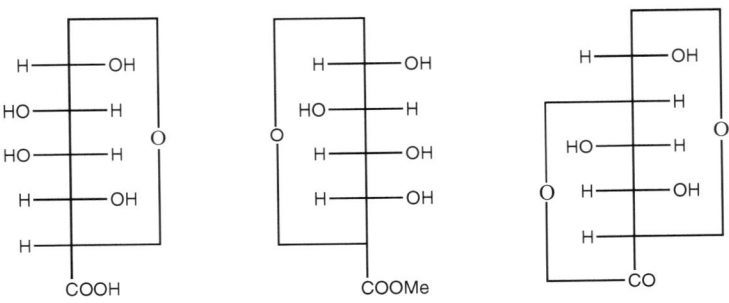

110, α-D-Mannopyranuronic acid   *111*, Methyl β-L- galactopyranuronate   *112*, α-D-Mannopyranurono-6,2-lactone

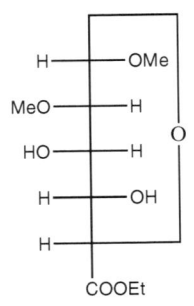

113, Ethyl (methyl 2-O-methyl-α-D-mannopyranosid)uronate

Fig. 1.30

Aldaric Acids

The dicarboxylic acids formed by oxidation of both terminal groups of an aldose to carboxyl groups are called "aldaric acids." Names of individual components of this

The Nomenclature of Carbohydrates

type are formed by replacing the ending "ose" of the systematic or trivial name of the corresponding aldose by "aric acid." Choice between the several possible names is based on the order of precedence given in Rule 3. Examples:

a) Names requiring D or L:

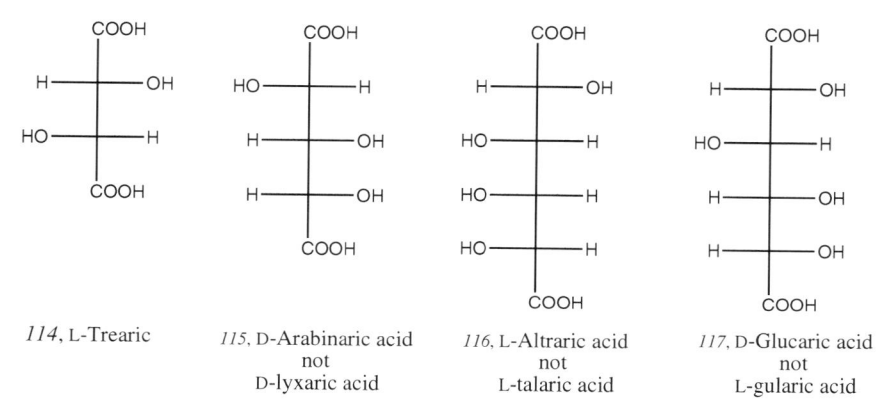

```
      COOH              COOH              COOH              COOH
   H──┼──OH          HO──┼──H           H──┼──OH           H──┼──OH
  HO──┼──H            H──┼──OH         HO──┼──H           HO──┼──H
      COOH            H──┼──OH         HO──┼──H            H──┼──OH
                         COOH          HO──┼──H            H──┼──OH
                                           COOH                COOH

  114, L-Trearic    115, D-Arabinaric acid   116, L-Altraric acid   117, D-Glucaric acid
                          not                      not                    not
                      D-lyxaric acid           L-talaric acid         L-gularic acid
```

Fig. 1.31

b) To the names of aldaric acids optically compensated intramolecularly, which therefore have no D or L prefix, the prefix *meso-* may be added for the sake of clarity. Examples: *meso*-erythraric acid, *meso*-ribaric acid, *meso*-xylaric acid, *meso*-allaric acid, *meso*-galactaric acid.

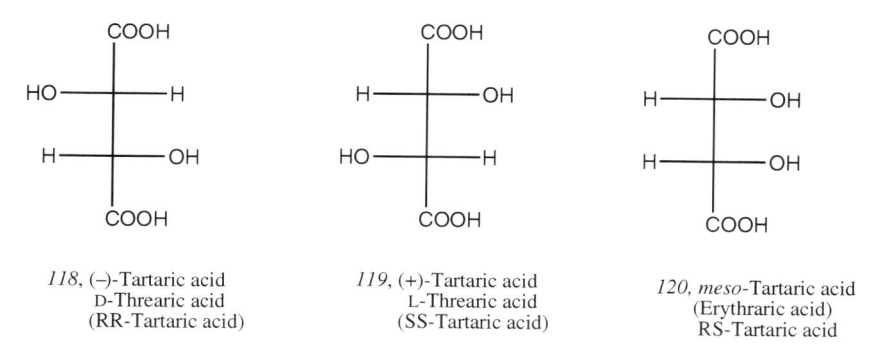

```
      COOH              COOH              COOH
  HO──┼──H           H──┼──OH           H──┼──OH
   H──┼──OH         HO──┼──H            H──┼──OH
      COOH              COOH              COOH

 118, (−)-Tartaric acid   119, (+)-Tartaric acid   120, meso-Tartaric acid
    D-Threaric acid         L-Threaric acid           (Erythraric acid)
   (RR-Tartaric acid)      (SS-Tartaric acid)        RS-Tartaric acid
```

Fig. 1.32

The D and L prefix must however be used when *meso*-aldaric acid has become asymmetric as a result of substitution (Fig. 1.31).

c) The trivial names that are preferred to the systematic names are given in Fig. 1.32.

Cyclic Acetals

Cyclic acetals formed by reaction of monosaccharides or their derivatives with aldehydes or ketones are named in accordance with Rule 7 for O-substitution, bivalent radical names being used as prefixes, the names of such radicals following the rules of general organic chemical nomenclature. In indicating more than one cyclic acetal grouping of the same kind, the appropriate pairs of locants are separated typographically when the exact placement of the acetal groups is known (Fig. 1.33).

Examples:

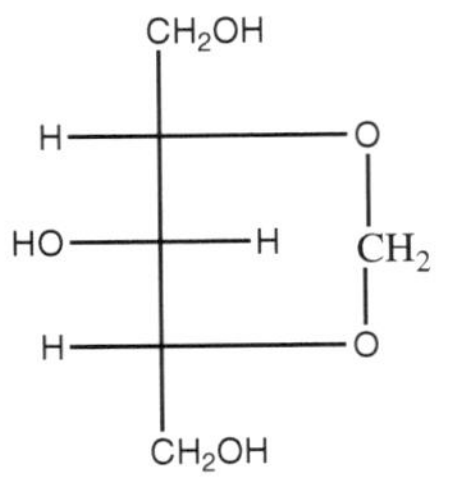

121, 2,4-O-Methylenexylitol 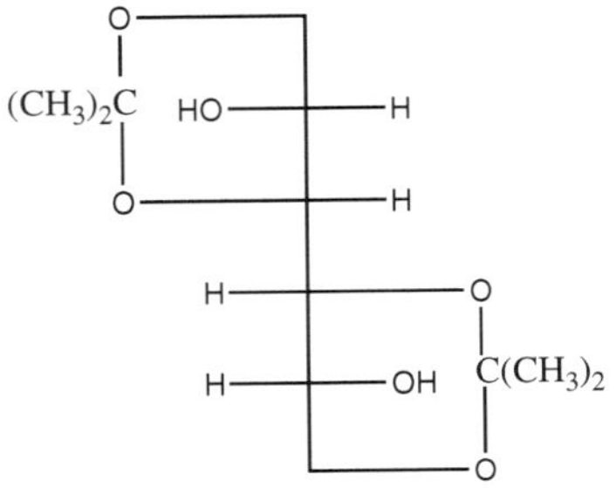   *122*, 1,3:4,6-Di-O-isopropylidene-D-mannitol 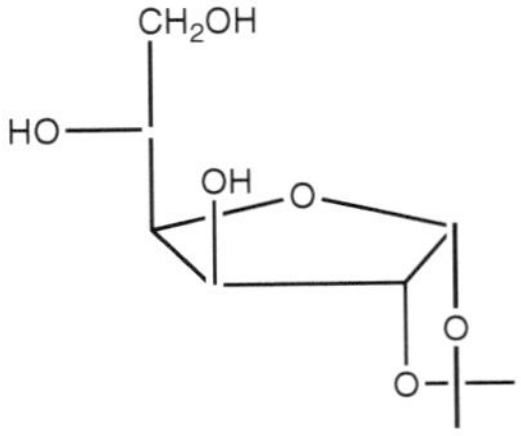   *123*, 1,2-O-Isopropylidene-α-D-glucofuranose

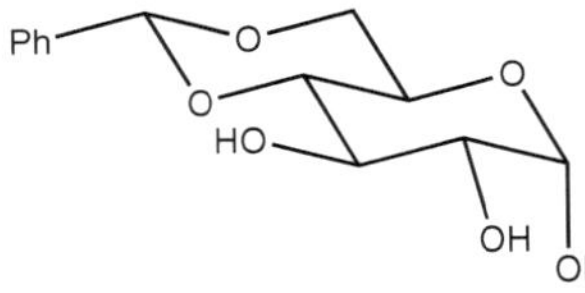

124, Methyl 4,6-O-benzylidene-α-D-glucopyranoside

Fig. 1.33

Intramolecular Anhydrides

An intramolecular ether (commonly called an intramolecular anhydride), formed by elimination of water from two alcoholic groups of a single molecule of a monosaccharide (aldose or ketose) or monosaccharide derivative, is named by attaching the (detachable) prefix "anhydro" by a hyphen before the monosaccharide name; this prefix, in turn, is preceded by a pair of locants identifying the two hydroxyl groups involved. Examples are given in Fig. 1.34.

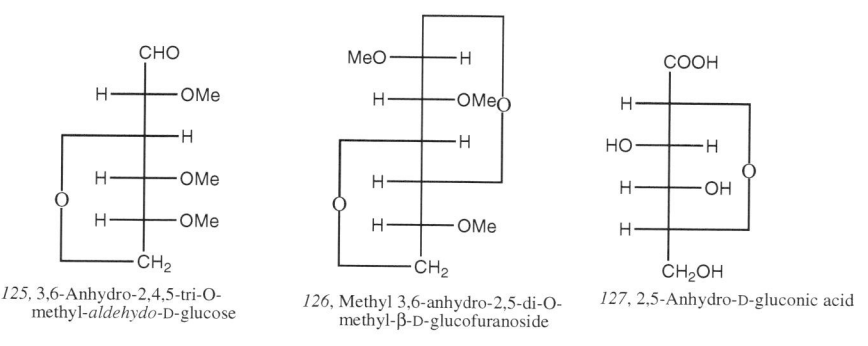

125, 3,6-Anhydro-2,4,5-tri-O-methyl-*aldehydo*-D-glucose

126, Methyl 3,6-anhydro-2,5-di-O-methyl-β-D-glucofuranoside

127, 2,5-Anhydro-D-gluconic acid

Fig. 1.34

References

1. Eliel, E. L., "*Stereochemistry of Carbon Compounds*", McGraw-Hill, New York, 1962
2. Fischer, E., "*Über die configuration des Traubenzuckers und seiner Isomeren. II*", Berichte (1891) **24**, 2683–2687
3. Erwig, E.; Koenigs, W., "*Notiz über Pentacetyldextrose*", Berichte (1889) **22**, 1464–1467
4. Erwig, E.; Koenigs, W., "*Über fünffach acetylierte Galactose und Dextrose*", Berichte (1889) **22**, 2207–2213
5. Fischer, E., "*ber die Verbindungen der Zucker mit Alkoholen und Ketonen*", Berichte (1895) **28**, 1145–1167
6. Villiers, A.; Favolle, M., Bull. Soc. Chim. France (1894) **11**, 692
7. Tobie, W. C., "*Supersensitive Schiff's aldehyde reagent. Demonstration a free aldehyde group in certain aldoses*", Ind. Eng. Chem. Anal. Edition (1942) **14**, 405–406
8. Alexander, J.; McCarty, K. S.; Alexander-Jackson, E., "*Rapid method of preparing Schiff's reagent for the Feulgen test*", Science (1950) **111**, 13
9. Tanret, C., "*Molecular modifications of glucose*", Compt. Rend. (1895) **120**, 1060–1062
10. Tollens, B., "*Über das Verhalten der Dexrose zu ammoniakalischer Silberlösung*", Berichte (1883) **16**, 921–924
11. Haworth, W. N., *The Constitution of the Sugars*, Arnold, London, 1929
12. Hall, M. W.; Piret, E. L., "*Distillation principles of formaldehyde solutions. State of formaldehyde in the vapor phase*", Ind. Eng. Chem. (1949) **41**, 1277–1286
13. Hurd, C. D.; Saunders, W. H., "*Ring-Chain Tautomerism of Hydroxy Aldehydes*", J. Am. Chem. Soc. (1952) **74**, 5324–5329
14. Adkins, H; Broderick, A. E., "*Hemiacetal formation and the refractive indices and densities of mixtures of certain Alcohols and Aldehydes*", J. Am. Chem. Soc. (1928) **50**, 499–503
15. Los, J. M.; Simpson, L. B.; Wiesner, K., "*The kinetics of mutarotation of D-Glucose with consideration of an intermediate free-aldehyde form*", J. Am. Chem. Soc. (1956) **78**, 1564–1568
16. International Union of Pure and Applied Chemistry and International Union of Bio- chemistry and Molecular Biology, Joint Commission on Biochemical Nomenclature, Nomenclature of Carbohydrates (Recommendations 1996), Pure. Appl. Chem. (1996) 68, 1919; Adv. Carbohydr. Chem., Horton, D., Ed., (1997) 52, 47–177

Chapter 2
Conformational Analysis of Monosaccharides

Molecules are dynamic assemblies of atoms chemically linked by single or multiple bonds. Hence, all atoms as well as groups of atoms are in perpetual motion, vibrating and rotating about chemical bonds. The deformation of chemical bonds due to vibration (stretching, wagging, etc.) of atoms is quantized and it is not the subject of conformational analysis. However, the rotation of atoms about the single bonds is the subject of conformational analysis.

Conformation of a molecule can be defined as a spatial arrangement of its atoms (or ligands) in a molecule that is obtained by free rotations about single bonds. Hence, the conformations are interconvertible by rotation about the single bonds.

When an atom that carries no substituent rotates about a single bond that links it to another atom, which does or does not carry a substituent, the rotation will not change the spatial arrangement of atoms in that molecule and therefore there will be no change in conformation due to rotation, as shown in Fig. 2.1 for molecules of hydrogen (*1*), chlorine (*2*), and water (*3*).

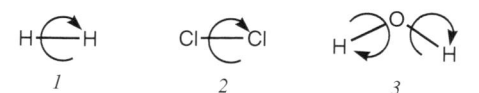

Fig. 2.1

Hence, these rotations of atoms in a molecule are not the subject of conformational analysis.

The rotation of two atoms about a single bond that connects them will, however, result, in the course of time, in formation of many nonidentical spatial arrangements of atoms in a molecule, if both atoms carry at least one substituent. Each individual spatial arrangement of atoms or group of atoms in a molecule obtained *via* the free rotation is called the *conformation* of that molecule.

Thus, for example, in hydrogen peroxide (Fig. 2.2) the rotation of oxygen atoms about the oxygen–oxygen single bond could theoretically produce an "infinite" number of conformations. However, even though the hydrogen atom is very small, some conformations will be favored over the others because the rotational barrier is not small (1.1 kcal/mol [1, 2]). Various conformations of hydrogen peroxide in

Fig. 2.2 are represented with the so-called "sawhorse" (perspective) formulas (upper raw) and with the so-called Newman's projections (lower raw) (the relative position of the two hydrogen atoms is looked upon along the oxygen–oxygen bond). It was found by millimeter-wave spectroscopy [3] and by ab initio calculation [4] that the gauche coformation 6 is the favored conformation (dihedral angle between the hydrogens is ca. 120°).

Fig. 2.2

Conformational Analysis of Acyclic Hydrocarbons

The rotation of two methyl groups of ethane about the C–C single bond (Fig. 2.3) should theoretically produce also an "infinite" number of conformations if the rotation about the C–C bond was completely free. However, due to a nonbonded interaction between the hydrogen atoms on two adjacent methyl groups (known as

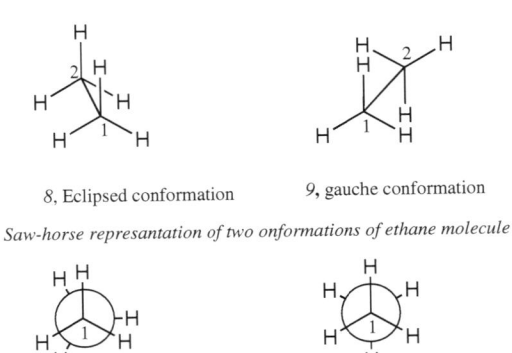

Fig. 2.3

torsional or *Pitzer strain* [1]) the rotation about the C–C bond in ethane molecule is not completely free [3] because in the *eclipsed* conformation the interaction between the hydrogen atoms is larger (ca. 2.89–2.93 kcal/mol [1–8]) than in the *gauche* conformation, since the inter-atomic distance between the vicinal hydrogen atoms of the two neighboring methyl groups in eclipsed conformation (torsional angle 0°) is significantly shorter (2.26 Å) than in the gauche conformation (torsional angle 60°) (2.50 Å). Since the conformational energy of *8, 10* is higher than that of *9, 11* the *gauche* conformation will be preferred in the conformational equilibrium mixture of ethane. In Fig. 2.3 only two conformations of ethane molecule are shown – the least stable *eclipsed* conformation (*8* and *10*) (it has the highest internal energy of all possible ethane conformations because the hydrogen atoms are closest to each other) and the most stable *gauche* (*9* or *11*) conformation because it has the lowest internal energy of all possible ethane conformations (the hydrogen atoms are at the greatest distance from each other).

It should be, however, noted that the steric (van der Waals) interactions actually account for less than 10% of the rotational barrier in ethane, since the hydrogen atoms of two methyl groups are barely within the van der Waals distance. Electrostatic interactions of the weakly polarized C–H bonds are negligent, too. The principal interaction responsible for the rotational barrier in ethane is, according to Pitzer [9], the overlap resulting in repulsion [10] of bond orbitals in the eclipsed conformation; changes in electronic structure other than those required by the changes in C–H bond overlaps are of minor importance for the existence of the barrier [9]. The existence of rotational barrier in ethane, according to Bader et al. [11] is explained by the increase of the C–C bond length during the transition from the gauche (staggered) conformation to the eclipsed one; this increase is more than 10 times larger than the accompanying decrease in the C–H separation.

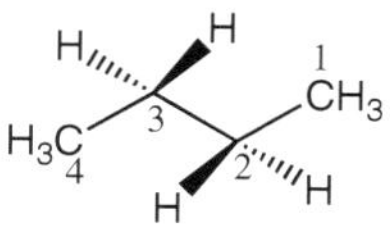

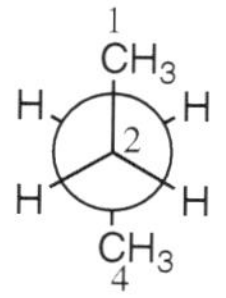

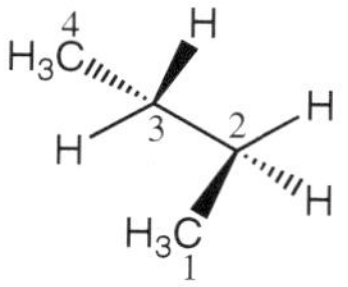

Fig. 2.4

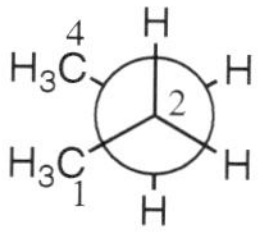

When two carbon atoms linked by a single bond carry larger atoms or group of atoms as substituents, then the conformational energy difference of *gauche* and *eclipsed* conformations becomes much larger. This difference becomes even greater if substituents are polar or if they are even electrically charged, because in addition to the nonbonded (van der Waals) interactions there now will also be present dipolar and/or electrostatic interactions between the neighboring substituents.

Since, it is well established that molecules tend to assume their most stable conformations in which the magnitude and the number of unfavorable stereoelectronic interactions are at the minimum, then all possible *eclipsed* conformations may be safely eliminated from discussions because they will not be present in conformational mixtures in significant amounts to influence the chemical behavior of a given molecule.

It should be pointed out that of all possible *gauche* conformations there is only one in which the dihedral angle between the two largest substituents is 180°. This conformation is called *anti* or antiperiplanar [12] (*ap*) conformation and is always the preferred conformation.

This is now a good place to briefly describe Klyne–Prelog proposal [12] for describing the stereochemistry across a single bond in terms of torsional angle, τ, between the ligands such as, for example, between the two methyl groups in *n*-butane. When $\tau = 60°$ the conformer is *gauche* (*staggered* or *skewed*) but according to Klyne–Prelog proposal it is either $\pm syn$-clinal ($\pm sc$) *(16, 17)* or $\pm anti$-clinal ($\pm ac$)*(18, 19)* (Fig. 2.5). When torsional (dihedral) angle between the two methyl groups is between $+30°$ and $+90°$ the conformer is $+syn$-clinal (+sc), and if it is between $-30°$ and $-90°$, the conformer is $-syn$-clinal (−sc). When the torsional (dihedral) angle between the two methyl groups is between $+90°$ and $+150°$ the conformer is $+anti$-clinal (+ac) and if it is between $-90°$ and $-150°$ the conformer is $-anti$-clinal (−ac). When torsional angle between two methyl groups is between

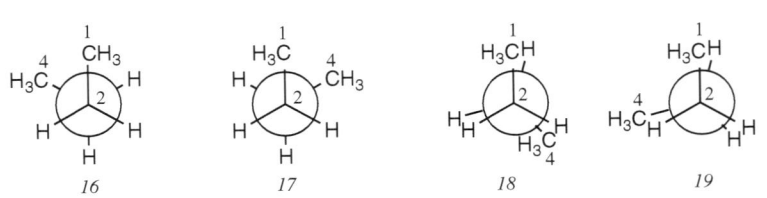

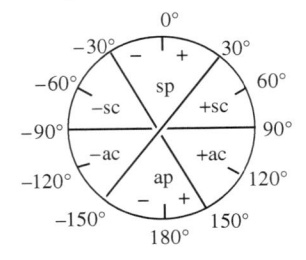

Fig. 2.5

−30° and +30° the conformer is ±*syn*-periplanar (±sp) and if it is between −150° and +150° the conformer is ±*anti*-periplanar (±ap). When torsional (dihedral) angle between two methyl groups is 0° the conformer is eclipsed and when the torsional (dihedral) angle between two methyl groups is 180° the conformer is *trans* or *anti*.

In aliphatic hydrocarbons, such as *n*-butane, for example, where only steric (van der Waals) interactions are involved, the *anti*-conformer is significantly preferred, over the *gauche* conformers, since the two methyl groups (the largest substituents on the C2 and C3 carbon atoms) are furthest apart (Fig. 2.4). The *anti/gauche* conformer ratio in *n*-butane is, at room temperature, almost 2:1 as determined by NMR spectroscopy.

Consequently, the polymethylene hydrocarbons tend to adopt planar *zigzag* conformation in which all carbon atoms lie in one plane (Fig. 2.6). In this conformation, all carbon atoms (the largest "substituents" in *n*-butane subunits) are in the *anti*-orientation (C1–C4, C2–C5, C3–C6, etc.). The small 1,3-nonbonding interactions between the hydrogen atoms (C_2H-C_4H, C_3H-C_5H, etc.) are the only destabilizing interactions. Unlike the acyclic hydrocarbons, each carbon atom of an acyclic monosaccharide has in addition to a hydrogen atom one large substituent which is most often hydroxyl group, but it can also be an amino, thio, or other group or an atom. In this case, if the acyclic form of a carbohydrate adopts the *zigzag* conformation then depending on a monosaccharide one or more 1,3-nonbonded interactions are possible between two larger ligands (oxygen atoms, for example).

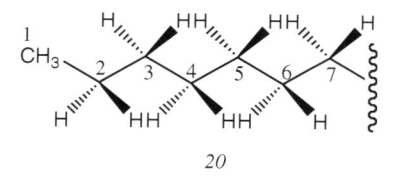

20

Fig. 2.6

Conformational Analysis of Acyclic (Aldehydo) Forms of Monosaccharides

The conformation of acyclic aldehydo sugars cannot be studied in solutions due to their spontaneous cyclization and formation of hemiacetals. However, derivatization of the C4 and the C5 hydroxyl groups (by alkylation or acylation) or derivatization of the aldehydo group of a monosaccharide (by formation of dialkylacetals or dialkyldithioacetals) prevents the hemiacetal formation permitting thus the conformational studies to be conducted in solution. To avoid derivatization the carbonyl group of a monosaccharide can simply be reduced with complex metal hydrides and the obtained sugar alcohol – *alditol* can then be used for conformational studies.

If D-glucitol (*21*), D-galactitol (*22*), and D-mannitol (*23*) (the alditols obtained by reduction of D-glucose, D-galactose, and D-mannose) are represented in their zigzag conformations (Fig. 2.7) it can be seen that only in D-glucitol there is one large 1,3-nonbonded interaction (both steric and dipolar) between the C2 and the C4 hydroxyl groups. This interaction destabilizes this conformation by 1.5 kcal/mol. In all three *zigzag* conformations there is present a number of much weaker 1,3-*syn*–axial interactions (we will explain the term "1,3-*syn*–axial interac-

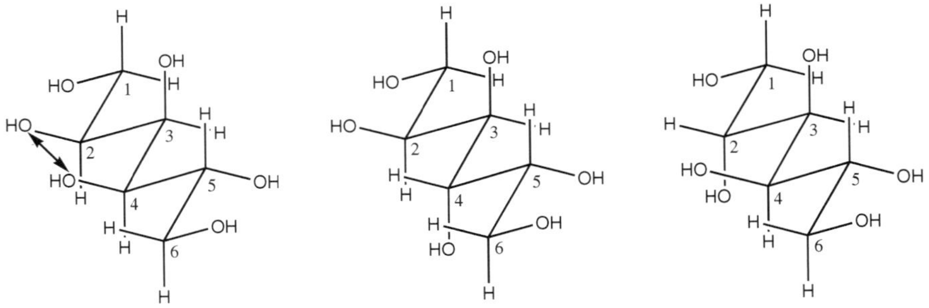

21, D-Glucitol   *22*, D-Galactitol   *23*, D-Mannitol

Fig. 2.7

tion" when we discuss the conformational analysis of pyranoid forms of monosaccharides). In D-glucitol there are four 1,3-*syn*–axial interactions (two between the C3 oxygen and the C1 and C5 hydrogens, one between the C5 oxygen and the C3 hydrogen, and one between the C4 oxygen and the C6 hydrogen) that destabilize this conformation by additional 1.8 kcal/mol (one *syn*–axial interaction between an oxygen and hydrogen atom destabilizes the given conformation by 0.45 kcal/mol). In D-galactitol (*22*) there are six 1,3-*syn*–axial interactions between the oxygen and hydrogen atoms (two between the C3 oxygen and the C1 and C5 hydrogens, two between the C4 oxygen and the C2 and C6 hydrogens, one between the C5 oxygen and the C3 hydrogen, and one between the C2 oxygen and C4 hydrogen destabilizing this conformation by a total of 2.70 kcal/mol). In D-mannitol (*23*) there are also six such interactions present (one between the C2 oxygen and the C4 hydrogen, two between the C3 oxygen and the C1 and C5 hydrogens, two between the C4 oxygen and the C2 and C6 hydrogens, and one between the C5 oxygen and C3 hydrogen) destabilizing this conformation by 2.70 kcal/mol.

Unlike D-galactitol and D-mannitol, which in solution exist predominantly in planar *zigzag* conformation, as shown [13] by ^{1}H NMR spectroscopy and in crystalline state by X-ray crystallography [14–18] D-glucitol adopts the so-called *sickle* (bent) conformation [14], in which the C2 and the C5 carbon atoms are in the *gauche* rather than in the *anti*-orientation (Fig. 2.8), to avoid the destabilizing *syn*–axial interaction between the C2 and the C4 hydroxyl groups (*21* in Fig. 2.7).

Fig. 2.8

Conformational Analysis of Cyclic (Lactol, Hemiacetal) Forms of Monosaccharides

Furanoses

The conformational analysis of furanoid forms of sugars is closely related to the conformational analysis of substituted cyclopentanes because they are both five-membered ring systems. For that reason we will start our discussion on conformational analysis of furanoid form of monosaccharides by briefly discussing the conformational analysis of cyclopentane first.

The cyclopentane can assume three distinct conformations: planar *25*, envelope *26*, and twist *27* conformation (Fig. 2.9), together with an infinite number of conformations that lie in the conformational interconversion path of cyclopentane.

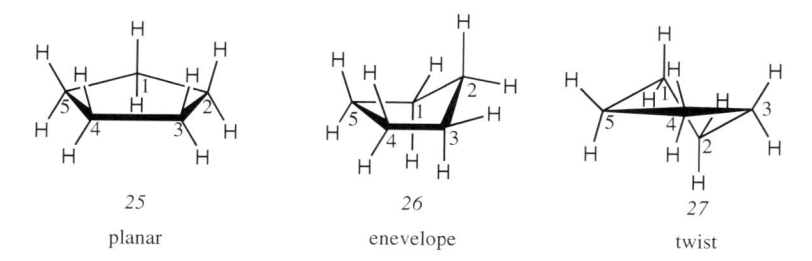

Fig. 2.9

The deviation of the endocyclic C–C bond angle of 108° in planar conformation of cyclopentane (*25* in Fig. 2.9) from tetrahedral valence angle of 109.5° of a carbon atom is relatively small (1.5°) and cannot have noticeable effect on the energy content of cyclopentane due to the Baeyer ring strain. Yet if one compares the heat of combustion per CH_2 group of acyclic hydrocarbons (157.5 kcal/mol) [19] and cyclohexane (157.4 kcal/mol) with that of cyclopentane (158.7 kcal/mol) it is estimated that the energy content of cyclopentane is higher by 6–7 kcal/mol than that of both *n*-pentane or the five-carbon fragment of cyclohexane. Since this higher internal energy of cyclopentane cannot be related to Baeyer ring strain, it was suggested by Pitzer [20–22] that the energy content of cyclopentane must be due to sterically unfavorable arrangement of adjacent methylene groups. As can be seen from Fig. 2.9 in the planar conformation *25* all 10 hydrogen atoms are completely eclipsed as well as all five-ring carbons. If one accepts the value of ca. 1 kcal/mol as the magnitude of unfavorable interaction energy between two eclipsed hydrogen atoms in ethane molecule, the internal energy of the planar conformation of cyclopentane should be greater than 10 kcal/mol. However it is 6–7 kcal/mol suggesting that in order to minimize these destabilizing interactions the cyclopentane adopts a *puckered* (a nonplanar) conformation which can be either the *envelope 26* conformation (also called the C_2 conformation – named after the symmetry point group) or the *twist* conformation *27* (also called the *half-chair* or the C_s conformation – named after symmetry point group) (Fig. 2.9). By adopting the envelope conformation the total torsional strain is only 60% of total torsional strain of planar conformation because of the presence of only six pairs of eclipsed hydrogens. According to Pitzer the cyclopentane ring is puckered to such an extent that one carbon sticks out about 0.2 Å from the plane containing the other four carbon atoms [20, 21]. The more recent calculations [22, 23] have shown that the energy minimum of cyclopentane molecule is attained when one carbon atom takes up the position 0.5 Å out of plane containing the other four carbon atoms. The carbon atom protruding from the plane is not fixed in that position [20, 21] but it "rotates around the ring" by an up and down motion of all five methylene carbons; this motion has been termed "*pseudorotation*" the result of which is the adoption of an infinite number of intermediate conformations.

In twist conformation the three neighboring carbons lie in one plane, while the other two are twisted, one lies below and the other above that plane, and they are equidistant from the plane containing the other three carbons. Since the twist conformation has only four pairs of eclipsed hydrogen atoms the total torsional strain of this conformation is 40% of the torsional strain of planar conformation. Difference in energy between the envelope and the twist conformations is very small, the envelope form being more stable by 0.5 kcal/mol.

Puckering of cyclopentane ring has been experimentally confirmed (a) thermodynamically from entropy measurements [24–27] and (b) by electron scattering [28]. Since the puckering of cyclopentane ring is not fixed and the energy barrier for conformational inter-conversions via pseudorotation is small (less than 0.6 kcal/mol at room temperature) no definite conformational energy minima and maxima can be observed for cyclopentane.

Most of what is said for cyclopentane is applicable to the conformational analysis of tetrahydrofuran. Calculation on tetrahydrofurans suggests [22] that the molecule exists in *twist* (half-chair) form with the maximum puckering occurring at carbon atoms 3 and 4, away from the heteroatom (*28* in Fig. 2.10). The replacement of one cyclopentane carbon with an oxygen atom decreases the Pitzer strain but introduces additional Baeyer strain into the molecule because the valence angle of oxygen atom is 104.45°, whereas the endocyclic valence angle of planar cyclopentane ring is 108°.

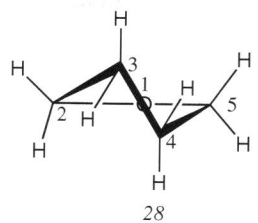

28
Tetrahydrofuran in twist conformation

Fig. 2.10

The introduction of substituents into either cyclopentane or tetrahydrofuran dramatically limits the number of possible conformations that they may adopt due to the increase of energy barriers for conformational interconversions.

The proof that furanoses do exist in nonplanar conformations has been obtained from X-ray crystallographic studies as well as from NMR spectroscopy studies in solutions.

Depending on the nature and location of its substituents, the furanose ring was shown to adopt both the C_s and the C_2 conformation [29–33]. Bulky substituents are taking up the most staggered ("equatorial") position, at the most staggered carbon atom. The oxygen atom most likely takes up the least staggered position in the furanose ring. Lemieux [34] has discussed, in considerable depth, the conformational behavior of furanoses based on the assumption that they adopt the C_s conformations.

Nonbonded interactions between the *cis*-1,2-substituents on furanoid ring are relieved by puckering of a planar conformation to the C_2 conformation by displacing the C_2 or C_3 carbons from the plane containing the other four atoms [35–40] (Fig. 2.11) or to the C_s conformation (Fig. 2.10). Hence it may be concluded that E conformations with the C_2 or C_3 carbons displaced from the plane of the ring (i.e., ^2E, E_2, ^3E, E_3) are the most stable conformations (for instance, *31* corresponds to ^2E conformation). In addition, electronegative substituents at carbon atoms other than C_1 will prefer to take up quasi-equatorial or isoclinal orientations. The C_1 electronegative substituents will tend to assume quasi-axial orientation.

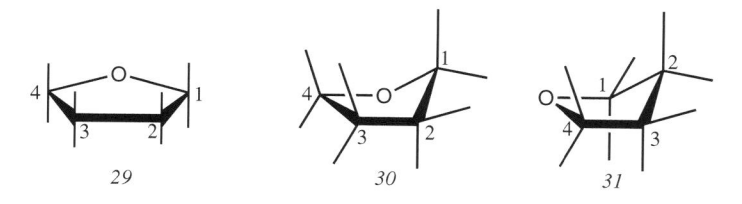

29   *30*   *31*

Fig. 2.11

The conformational analysis of furanoses is plagued by two insurmountable difficulties: (a) the flexible nature of the five-membered ring which is not allowing the accurate evaluation (or calculation) of interaction energies between the ring substituents and (b) difficulties in accurately determining the concentration of furanose forms in solutions at equilibrium.

Pyranoses

Although carbohydrate pyranoid rings are slightly flattened compared to cyclohexane rings and have no elements of symmetry, the conformational analysis of pyranoses is intimately related to the conformational analysis of cyclohexane. Therefore, we will begin our discussion of conformational analysis of pyranoses by briefly discussing the conformational analysis of cyclohexane first.

32
chair conformation

33
boat conformation

34
Skew-boat conformation

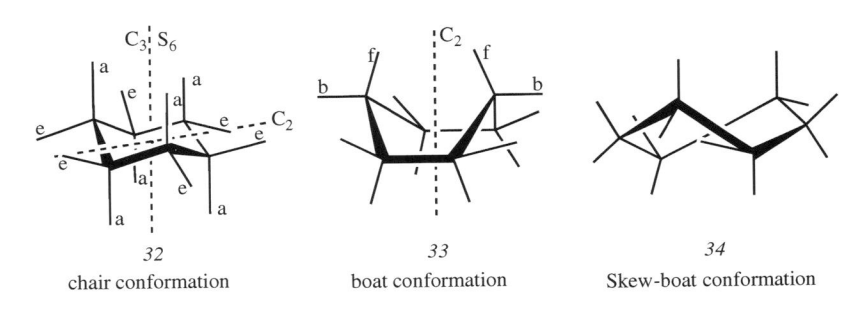

Fig. 2.12

Cyclohexane may exist in three distinct conformations (Fig. 2.12): *chair 32*, *boat 33*, and *twist-boat* (or *skew-boat*) *34* conformations. The *chair* conformation is relatively stiff and has the following elements of symmetry: C_3 axis of symmetry that coincides with the S_6 rotation–reflection axis of symmetry, three C_2 axis of symmetry, three planes of symmetry, and a center of symmetry. The *boat* conformation is flexible and has one C_2 axis of symmetry and two planes of symmetry, whereas the *skew-boat* (or *twist-boat*) *conformation* is flexible and has no symmetry elements.

Cyclohexane in chair conformation *32* has two geometrically different sets of hydrogens: six hydrogens are oriented parallel to the S_6 (that is also the C_3) axis of the molecule and are called *axial* (*a*) hydrogens and the six hydrogen atoms that alternate about the so-called "equatorial" plane that is a plane perpendicular to the S_6 (C_3) axis and are called *equatorial* (*e*) hydrogens. In the boat conformation (*33* in Fig. 2.12) there are two pairs of hydrogens that are very different from the other eight. These are the four hydrogens that are attached to two carbons that lie out of the plane of other four carbons of the boat conformation. The two hydrogens that point away from the ring and are almost parallel to the plane containing the four carbon atoms are said to be in *bowsprit* (*b*) orientation; the two hydrogens that are pointed toward the ring are said to be in *flagpole* (*f*) orientation. From other eight hydrogens four are equatorial and the other four pseudo-axial. Although the boat conformation is free from Baeyer (ring) strain, there is eclipsing of eight hydrogen atoms that are part of the planar part of the six-membered ring causing elevated torsional strain; in addition to that there is also a severe van der Waals interaction between two hydrogen atoms in the *flagpole* orientation. As a result, the boat conformation is calculated to be 6.9 kcal/mol less stable than the chair conformation [41]. To avoid these unfavorable interactions the boat adopts the *skew-boat* conformation 34 by pseudo-rotation around its ring C–C bonds (Fig. 2.12). Its energy has been calculated to be 5.3 kcal/mol, i.e., 1.6 kcal/mol less than the boat conformation, but still 5.3 kcal/mol higher than the energy of the chair conformation.

Although relatively rigid, the cyclohexane *chair* conformation 35 is able to *flip* over to the alternate chair conformation 36, as illustrated in Fig. 2.13. After flipping over, the hydrogen atoms of a cyclohexane that were axial (red in the left conformation) became equatorial and vice versa, the hydrogens that were equatorially oriented in the left conformation (blue hydrogens) became axial in the right conformation.

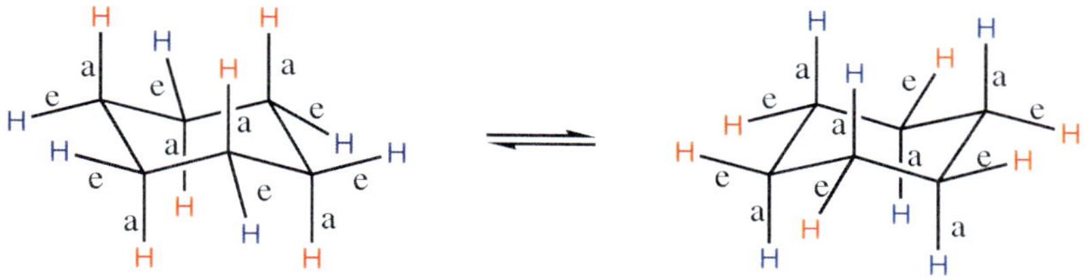

35   36
Interconversion of two alternate chair conformations

Fig. 2.13

The interconversion of the two-cyclohexane chair conformations ($37 \rightleftharpoons 42$) (Fig. 2.14) involves the formation of a *half-chair conformation 38* in the transition state (first step), which (in the second step) collapses either into flexible twist-boat conformation (39, 40) or back to the initial conformation 37. The *twist-boat* conformation 39 is in dynamic equilibrium with the boat conformation 40; in this conformational mixture the *twist-boat* conformation is at the energy minimum and the boat conformation is at the energy maximum (energy difference between these two forms is estimated to be 1.0–1.5 kcal/mol). The conversion of a *twist-boat* conformation into an alternate chair conformation 42 requires the formation of the corresponding alternate *half-chair* conformation 41 in the second transition state. Figure 2.14 illustrates the chair–alternate chair ($37 \rightleftharpoons 42$) interconversion via the half-chair 38 and alternate half-chair 41 transition state intermediates.

The activation energy of chair-to-chair interconversion (the barrier of the ring inversion in cyclohexane) has been experimentally determined by variable temperature NMR studies [42, 43] and calculated by force field method [42, 44, 45]. The obtained $\Delta G^{\#}$ values for the barrier was found to vary from 10.1 to 10.25 kcal/mol at temperatures between −50 and −70°C. Calculated and experimentally determined values are in good agreement.

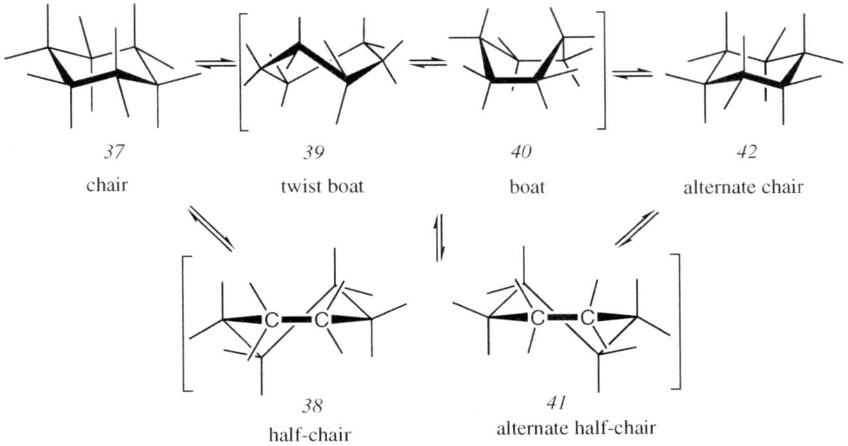

Fig. 2.14

37 chair
39 twist boat
40 boat
42 alternate chair
38 half-chair
41 alternate half-chair

Substitution of one hydrogen atom in cyclohexane with a "bulky" substituent dramatically changes the described situation. The most stable chair conformation of a mono-substituted cyclohexane is the one in which the substituent takes up the equatorial orientation.

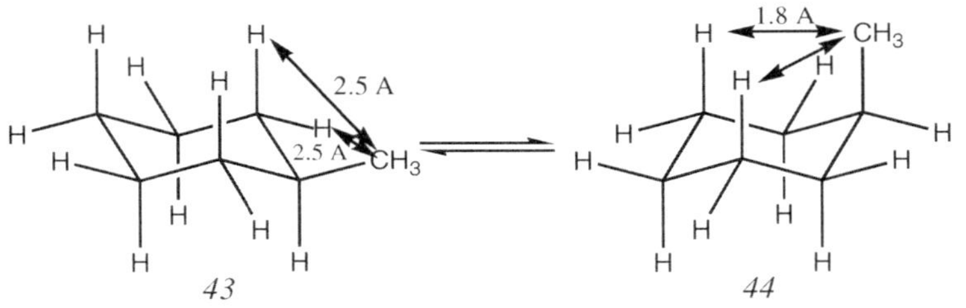

Fig. 2.15

The reason for this is the unfavorable nonbonding interaction between the axial substituent and the two 1,3-*syn*–axial hydrogen atoms. For example 1,3-*syn*–axial interaction between an axial methyl group and the two axially oriented hydrogen atoms in methylcyclohexane (*44* in Fig. 2.15) is 1.6–1.8 kcal/mol suggesting that at conformational equilibrium (Fig. 2.15) the conformation having methyl group equatorially oriented (*43* in Fig. 2.15) will strongly predominate (*43*:*44* = 19:1; i.e., 95:5%). Disubstituted and polysubstituted cyclohexanes are even more restrictive in number of possible conformations.

Replacing one carbon atom of a cyclohexane ring with an oxygen atom, converting thus the cyclohexane to tetrahydropyran, introduces several very important changes. First, the six-membered ring of a pyranoid sugar is somewhat flattened compared to the cyclohexane due to the shorter C–O bond as compared to C–C bond (1.43 Å vs. 1.53 Å, respectively) and due to the larger endocyclic C–O–C valence angle (112–114°) as compared to C–C–C valence angle (109°) of cyclohexane. Consequently the nonbonding interactions in pyranoid forms of sugars are slightly different than in cyclohexane. Second, the chair conformers of pyranoid rings are asymmetric. Third, for all pyranoid forms of sugars two distinct chair conformations are possible whereby very often one is much more stable than the other.

When describing various pyranoid conformations the following conventions are commonly used:

1. Conformations are designated: C, for the chair, B for the boat, S for the twist-boat, and H for the half-chair conformation.
2. A reference plane must always be chosen so that it contains four of the ring six atoms. If an unequivocal choice is impossible, as with the chair and twist-boat conformers, the reference plane is chosen so that the lowest numbered carbon in the ring is displaced from this plane.
3. Ring atom(s) that lie(s) above the reference plane (numbering clockwise from above) is/are written as superscript(s) and precede(s) the letter designating the conformation, while the ring atom(s) that lie(s) below the reference plane is/are written as subscript(s) and follow the letter designating the conformation.

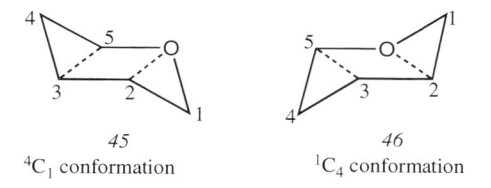

45   46

$^4C_1$ conformation $^1C_4$ conformation

Fig. 2.16

Thus as shown in Fig. 2.16, the reference plane for the two possible chair conformations of a pyranoid ring is chosen so that they contain O, C2, C3, and C5 atoms. When the C1 atom is below and the C4 atom above the reference plane the chair conformation is designated as $^4C_1$; when the C1 atom is above and the C4 atom below the reference plane, it is designated as $^1C_4$. Since enantiomeric conformers have different descriptors the $^4C_1$ (D) conformer is the enantiomer of $^1C_4$ (D) conformer. For this reason conformational descriptors should always be used in reference to either the D- or the L-series. This can be illustrated with the $^4C_1$ and $^1C_4$ conformers of methyl α-D- and α-L-ribopyranosides (*47* and *48*, respectively) (Fig. 2.17).

An alternative method to ascribe the descriptors $^4C_1$ and $^1C_4$ to two pyranoid conformers is as follows: when the atom numbering is clockwise (looking from the above) and the plane containing the C2, C4, and O5 (ring oxygen) is above the plane containing C1, C3, and C5 carbons, the ring conformer is $^4C_1$ (*49*); similarly, when the atom numbering is anticlockwise (looking from the above) and the plane

47, Methyl α-D-ribopyranoside

48, Methyl α-L-ribopyranoside

Fig. 2.17

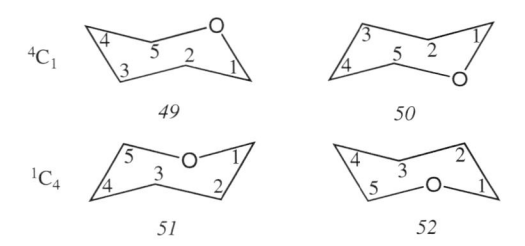

Fig. 2.18

containing the C2, C4, and O5 (ring oxygen) is below the plane containing the C1, C3, and C5 carbon atoms, the conformer is again $^4C_1$ (*50*). However, when the atom numbering is clockwise (looking from above) and the plane containing the C1, C3, and C5 carbon atoms is above the plane containing the C2, C4, and O5 (ring oxygen) the conformer is $^1C_4$ (*51*). The conformer is again $^1C_4$ (*52*), when numbering is anticlockwise and the plane containing the C1, C3, and C5 carbon atoms is below the plane containing C2, C4, and O5 (ring oxygen) (Fig. 2.18).

The hydroxymethyl group (the C6 carbon) of a hexopyranose will always assume the equatorial orientation because it is, as the largest substituent on a pyranoid ring, subjected to largest nonbonded steric interactions with other ring substituents. Consequently, whether a glycopyranose will adopt the $^4C_1$ or the $^1C_4$ conformation will depend on in which conformation the hydroxymethyl group is equatorially oriented. Thus, in D-series the hydroxymethyl group is oriented equatorially in $^4C_1$ conformation, whereas in L-series the hydroxymethyl group is oriented equatorially in $^1C_4$ conformation.

For example, β-D-glucopyranose normally exists in $^4C_1$ conformation, with the largest substituent (hydroxymethyl group) equatorially oriented (*53* in Fig. 2.19). In

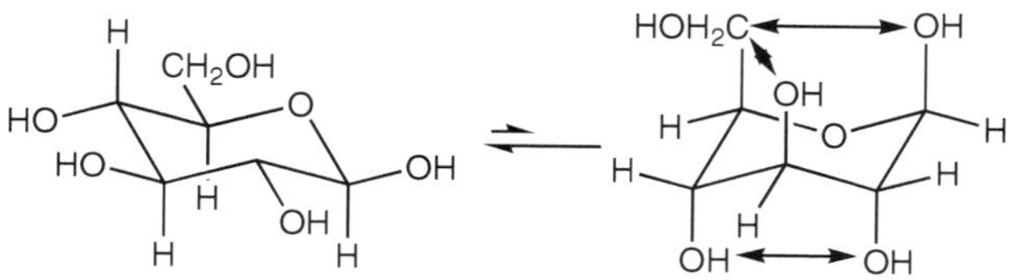

Calculation of Conformational Energies of Pyranoses   41

Fig. 2.19

$^4C_1$ Conformation of β-D-glucopyranose $^1C_4$ Conformation of β-D-glucopyranose

the $^1C_4$ conformation the hydroxymethyl group of β-D-glucopyranose will be axially oriented and subjected to two strong 1,3-*syn*–axial interactions with the C1 and the C3 hydroxyl groups (*54*); in addition, the axially oriented C2 hydroxyl group will be involved in 1,3-*syn*–axial interaction with the C4 hydroxyl group additionally destabilizing this conformation (Fig. 2.19).

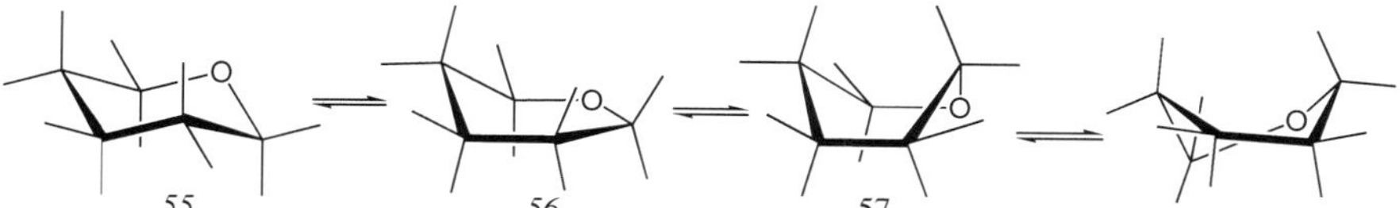

Fig. 2.20

By analogy to conformational interconversions of cyclohexane rings, the $^4C_1$ and the $^1C_4$ conformers of a pyranoid sugar can also undergo interconversion. This is shown in Fig. 2.20. The rigid $^4C_1$ chair pyranoid conformer converts, via a transition state that resembles a half-chair conformation, to flexible boat and twist-boat conformations which can then either convert back, via the same transition state, to initial $^4C_1$ conformation or to a rigid $^1C_4$ conformation. Six different twist-boat conformers and six different boat conformers may be identified in the course of boat/twist-boat pseudorotation pathway of a pyranoid ring (Fig. 2.21). However, similar to the cyclohexane ring, twist-boat conformers are found to be much less stable than $^4C_1$ and $^1C_4$ chair conformations and their presence in conformational equilibria can be considered in most instances negligible.

Calculation of Conformational Energies of Pyranoses

The prediction of conformational properties of pyranoid form of a monosaccharide can be done only if the relative free energies of two chair forms are known. Angyal [35, 46, 47] has developed a semiempirical method for obtaining the values of these relative free energies by taking both steric and electronic interactions into consideration [48].

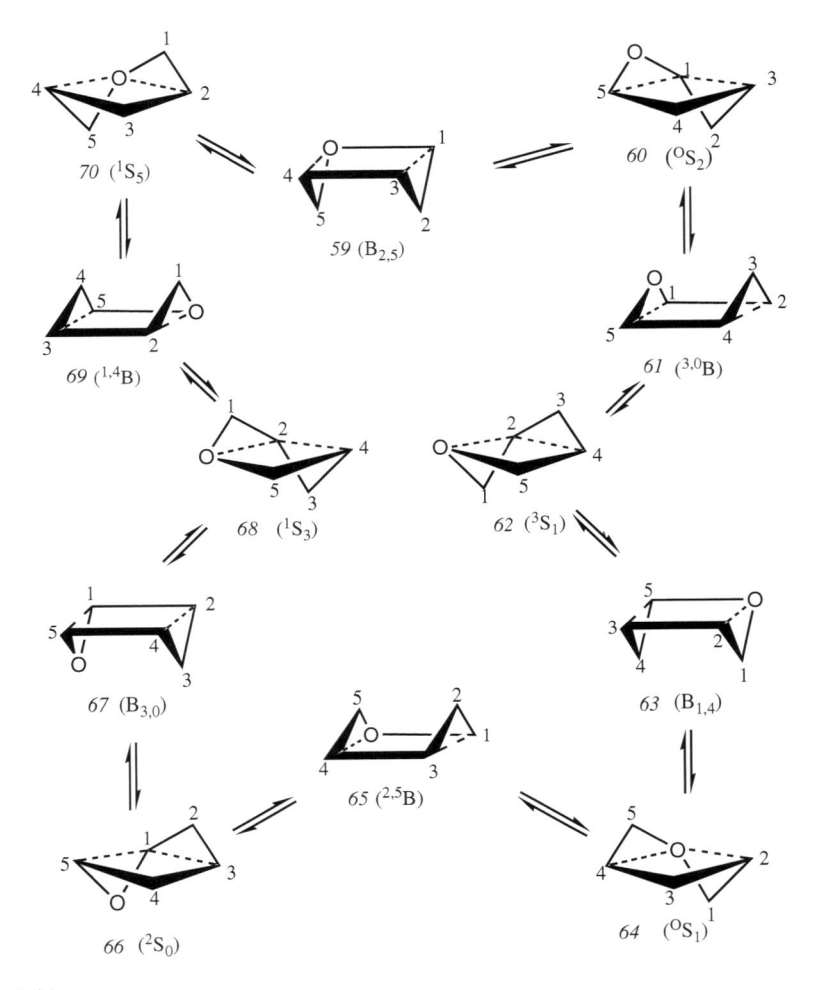

Fig. 2.21

The numerical values for nonbonding interactions in pyranoid structures of monosaccharides were obtained from conformational studies of cyclitols and some model pyranose sugars.

However, this semi-quantitative calculation of relative free energy of cyclitols and pyranoses in aqueous solutions is based on a number of assumptions, of which the following three are the most important.

1. *The pyranose and the cyclohexane rings have the same geometry.* This is only partially true since the replacement of one ring carbon with an oxygen results in considerable flattening of the six-membered ring due to (1) shorter C–O bonds (they are ca. 10% shorter than C–C bonds, i.e., 1.42 Å vs. 1.54 Å, respectively)

and (2) due to larger endocyclic O–C–O bond angle (112–114°) with regard to the endocyclic C–C–C bond angle (109°).
2. *In aqueous solutions the intramolecular hydrogen bonds do not play any role in conformational equilibria.*
3. *The relative free energy of conformers can be obtained by simple addition of (1) energies of various nonbonded interactions between ligands, (2) energies of electronic interactions, and (3) entropy differences.* The assumption is made that *the contributions of these individual thermodynamic quantities are independent of each other.*

Although none of the above assumptions is strictly true, the success of this approach in practice suggests that the errors introduced by making these assumptions are quite small.

If the energy contributions due to bond length deformation strain and bond angle bending strain are considered to be negligible, then only two types of nonbonded interactions are important for the conformational analysis of hexopyranoses:

(1) 1,3-*Syn*–axial nonbonded interactions between *syn*–axial ligands (excluding the *syn*–axial interaction between two hydrogen atoms). This interaction is designated as $X_a:Y_a$.
(2) 1,2-Nonbonded interactions between ligands on two vicinal (adjacent) carbon atoms that are in *gauche* conformation (again excluding the *gauche* interaction between two hydrogen atoms). This interaction is designated as $X_1:Y_2$.

The standard conformational free energy difference between two conformers can be calculated from the conformational equilibrium constant according to the equation:

$$\Delta G_X^0 = -RT \ln K \qquad (1)$$

where ΔG_X^0 is the standard conformational free energy difference, R is the universal gas constant (1.98726 cal/deg mol), T is the absolute temperature in Kelvin, and K is the equilibrium constant.

The most of interaction energies in pyranoid forms of sugars were determined by studying the equilibrium composition of tridentate borate complexes of cyclohexitols having three hydroxyl groups in *cis*-1,3,5 relationship (this arrangement is not possible in sugar pyranoid forms (Fig. 2.22)).

The formation of tridentate complex involves the initial inversion of more stable chair form *71*, in which the three hydroxyl groups are equatorially oriented, to the alternative chair conformation *72*, in which they are all axial (Fig. 2.22). This conformation is then capable to form the borate complex. Tridentate complexes are formed in the 1:1 ratio from their components (cyclitols and borate) and are strong acids, unlike borate complexes with vicinal diols that are weak acids (the acid strength of the latter complexes is comparable to boric acid). Consequently, the formation of tridentate complex lowers the pH of borate solutions. The boric

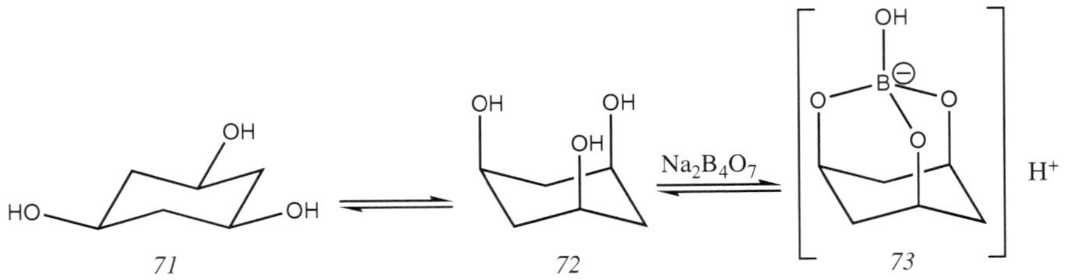

Fig. 2.22

acid itself is a weak Lewis acid with little tendency to form tetrahedral borate anion $B(OH)_4^-$, because its planar form is considerably stabilized by resonance involving the limiting structures $H^+O=B^-(OH)_2$. From the change in pH of the borate solution caused by successive addition of a cyclitol, the equilibrium constant K of complex formation can be calculated (Equation (1)) [49]

$$K = [\text{Complex}^-]/[\text{Borate}^-][\text{Cyclitol}] \qquad (2)$$

The equilibrium constants for the formation of borate complexes with *scyllo*-quercitol *74*, *myo*-inositol *75*, *epi*-quercitol *78*, *epi*-inositol *79*, *cis*-quercitol *82*, and *cis*-inositol *83* are shown in Figs. 2.23 and 2.24.

Since the equilibrium constants of tridentate borate formation depend on nonbonded interactions in the complex, the values of interaction energies can be calculated from *K*. In order to do so, two assumptions were made:

(1) Free energy of formation of tridentate anion from axial *cis*-1,3,5-hydroxyl groups and borate ion is independent of the nature of other substituents on cyclohexane ring
(2) Free energies of conformational isomers are additive functions of energy terms associated with the presence of nonbonded interactions, that is, the occurrence of one interaction in a molecule does not affect the magnitude of another one

The stability of the complex depends on steric orientation of free hydroxyl groups in the complex: the more these are in the axial position, the less stable the complex is. Since the equilibrium constants of tridentate borate formation depend on nonbonded interactions of the free hydroxyl groups, the values of these interaction energies were calculated from equilibrium constants (Figs. 2.23, 2.24, and 2.25).

The interaction energies are calculated in the following way. The energies of each nonbonded interaction in individual cyclitols, and in the tridentate borate complexes, are listed and separately added up. For borate complexes a term is added to account for the free energy change on formation of the tridentate borate complex from the three hydroxyl groups; and the difference between the totals for each cyclitols and for its borate complex is equated with the experimentally determined free energy

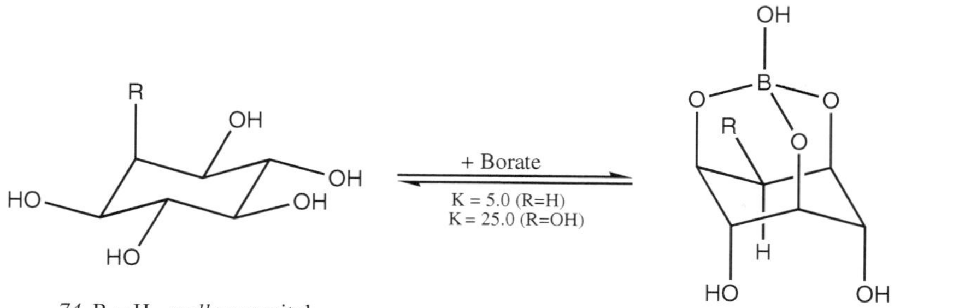

Calculation of Conformational Energies of Pyranoses 45

74, R = H, *scyllo*-quercitol
75, R = OH, *myo-i*nositol

+ Borate
K = 5.0 (R=H)
K = 25.0 (R=OH)

76, R = H, *scyllo*-quercitol
77, R = OH, *myo-i*nositol

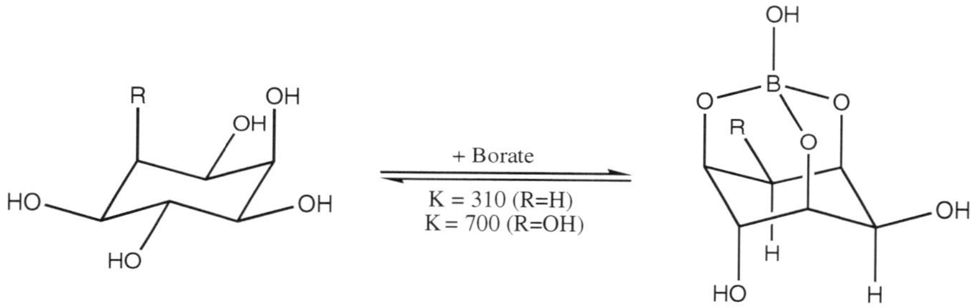

78, R = H, *epi*-iquercitol
79, R = OH, *epi*-inositol

+ Borate
K = 310 (R=H)
K = 700 (R=OH)

80, R = H, *epi*-iquercitol
81, R = OH, *epi*-inositol

Fig. 2.23

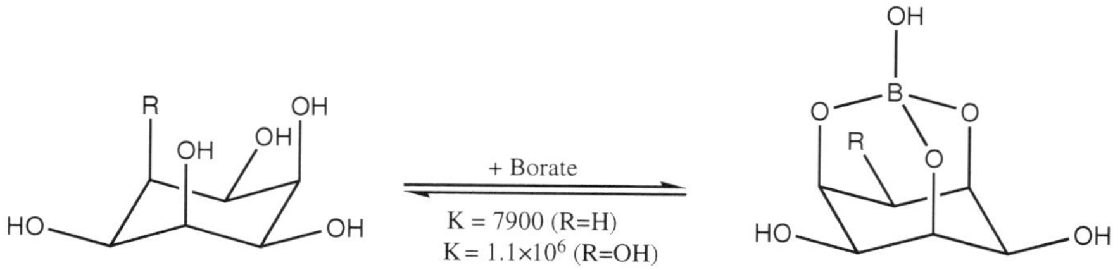

82, R = H, *cis*-quercitol
83, R = OH, *cis*-inositol

+ Borate
K = 7900 (R=H)
K = 1.1×10⁶ (R=OH)

84, R = H, *cis*-quercitol
85, R = OH, *cis*-inositol

Fig. 2.24

difference (calculated from $\Delta G^0 = -RT \ln K$). A series of equations resulted from which all unknown quantities were calculated [50, 51] (Table 2.1).

Thus, for example, from Fig. 2.23 we see that for *scyllo*-quercitol 74 the equilibrium constant K for the complex formation with borate is 5. This corresponds to ΔG^0 value of –0.95 kcal/mol. Since ΔG^0 is equal to the difference between the sum of nonbonded interactions in the complex 76 which is $2(O_a:H_a) + (O_a:O_a)$ and the

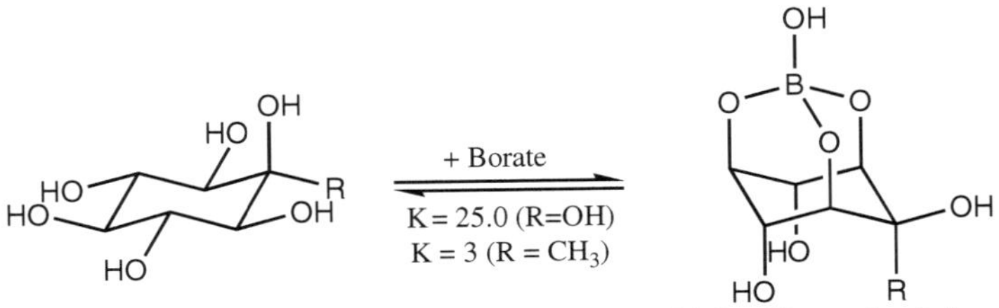

Fig. 2.25

86, R = H, *myo*-Inositol
87, R = CH$_3$, Isomytilitol

K = 25.0 (R=OH)
K = 3 (R = CH$_3$)

88, R = H, *myo*-Inositol
89, R = CH$_3$, Isomytilitol

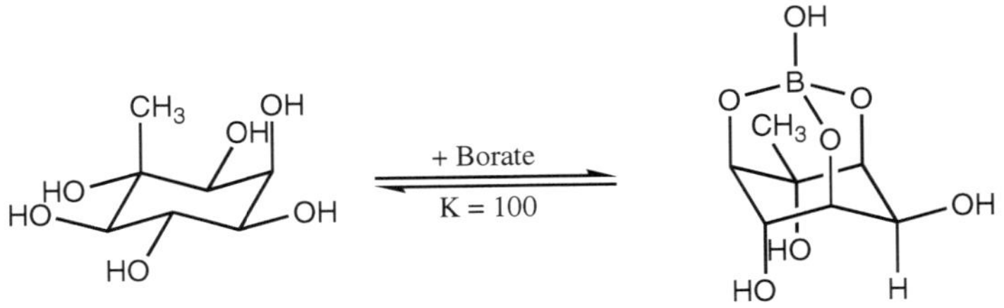

90, Laminitol

K = 100

91, Laminitol

Table 2.1 The values for 1,2-*gauche* and 1,3-*syn*–axial interactions between the ligands on a pyranoid ring (excluding the interactions between hydrogen atoms)

Interaction	ΔG^0 (kcal/mol at 25°C)
$O_1:O_2$	0.35
$C_1:O_2$	0.45
$O_a:H_a$	0.45
$C_a:H_a$	0.9
$O_a:O_a$	1.9
$C_a:O_a$	2.5*
Anomeric effect	0.55–3.2**
$\Delta 2$ effect	1.36

* Determined at 40°C
** Value of 0.55 kcal/mol is for free sugars; for OMe it is 0.9 kcal/mol, and for 1-halogens higher.

sum of nonbonded interactions in *scyllo*-quercitol 74 which is 4($O_1:O_2$), one can write

$$\Delta G^0 = 2(O_a:H_a) + (O_a:O_a) + \Delta G^0_F - 4(O_1:O_2) = -0.95 \text{ kcal/mol}$$

In analogous fashion, another five equations can be written for the other five cyclitols from Figs. 2.23 and 2.24, and the set of six simultaneous equations can be solved for ($O_1:O_2$), ($O_a:H_a$), ($O_a:O_a$), and ΔG^0_F. The value found for ΔG^0_F is –2.5 kcal/mol.

Calculation of Conformational Energies of Pyranoses   47

In order to determine the interaction energies of hydroxymethyl group that is needed for the calculation of conformational energies in pyranoid rings, it was assumed that conformational energy of hydroxymethyl group is not very different from conformational energy of methyl group in aqueous solvents. Preliminary experiments [52] have indicated that interactions of the methoxymethyl group are approximately of the same value as those of methyl group. Based on this one can safely assume that this would also apply to the hydroxymethyl group.

The values for carbon–oxygen [$(C_a:O_a)$ and $(C_1:O_2)$] and carbon–hydrogen $(C_a:H_a)$ interactions were determined similarly as was done previously, only the methyl cyclitols were used for the formation of tridentate borate complexes such as *myo*-inositol *86*, isomytilitol *87*, and laminitol *90* (Fig. 2.25). From equilibrium constants and the sums of nonbonded interactions in cyclitols and their corresponding tridentate borate complexes conformational free energies were again calculated. The obtained values are listed in Table 2.1.

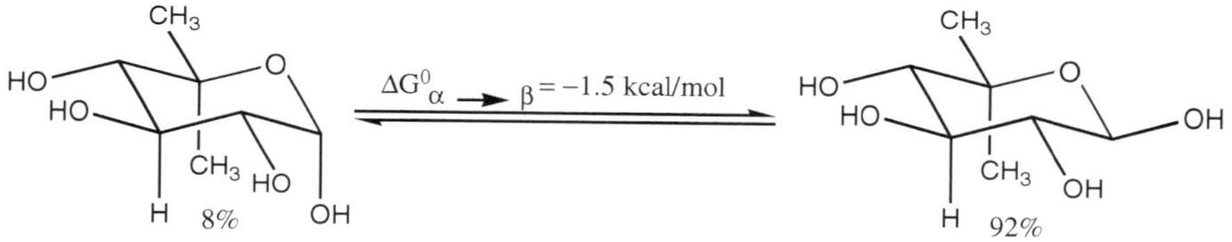

$$\Delta G^0 = (C_a:H_a) + (O:OH) - (C_a:O_a) - (O_a:H_a)$$
$$-1.5 = 0.9 + 0.55 - C_a:O_a - 0.45 \quad\quad C_a:O_a = 2.5 \text{ kcal/mol}$$

Fig. 2.26

It should be noted that the 1,2-interactions between the two vicinal oxygen atoms (equatorial–equatorial or axial–equatorial) are considered to be identical.

Alternatively, from a study of the equilibrium in aqueous solution between the α- and β-anomers of 6-deoxy-5-C-methyl-D-*xylo*-hexopyranose value of 2.5 kcal/mol was obtained for the $(C_a:O_a)$ (Fig. 2.26).

The value for $(C_a:H_a)$ is assumed to be one-half the conformational energy of a methyl group, i.e., 0.9 kcal/mol ($-\Delta G^0_{methyl} = 1.8$ kcal/mol [52]).

In addition to nonbonding interaction energies that have been so far determined there is another stereoelectronic interaction that has to be taken into account when calculating the free energies of pyranoses and that is the so-called "anomeric effect." The existence of this stereoelectronic effect was first recognized by Edwards [53] but it was given its name and exhaustively studied by Lemieux [54] and has become the subject of study of many investigators (this topic will be discussed later to a much greater detail). It is related to the composition of equilibrium mixtures of free sugars in aqueous solutions which seemed to be in violation of the classical postulates of conformational analysis of cyclohexane. The aqueous solution of D-glucose, for example, contains at equilibrium 36% of α-D-glucose and 64% of the β-D-glucose, corresponding to a free energy difference of only 0.35 kcal/mol, in spite of the fact that there are in α-D-glucose two O_aH_a interactions each one raising the free energy of α-D-glucose by 0.45 kcal/mol for a total of 0.9 kcal/mol. Furthermore, the conformational mixture of cyclohexanol at equilibrium contains 77% of conformer with equatorial hydroxyl group *94* and 23% of conformer with axially oriented hydroxyl group *95* corresponding to a free energy difference of 0.8 kcal/mol [55] (Fig. 2.27).

94 (77%)   *95* (23%)

Fig. 2.27

It should be noted that the reported *A*-values in the literature for the hydroxyl group vary significantly and depend upon the method used for its determination. Thus for example, from esterification studies Eliel [56] determined that the *A*-value for OH is 0.5 kcal/mol, whereas Subotin et al. [57] found from ^{13}C NMR studies of cyclohexanol equilibrium at low temperatures (–80°C) the *A*-value for hydroxyl group to be 1.02 kcal/mol.

Figure 2.28 illustrates how the value for the anomeric effect (O:OH) was determined. As can be seen the anomeric effect is lowest for sugars having the C2 hydroxyl group equatorially oriented and highest for sugars having the C2 hydroxyl group axially oriented. The 2-deoxy sugars have the anomeric effect between these two extremes.

96   *97*   *98*

O:OH = 0.55 kcal/mole   O:OH = 0.85 kcal/mole   O:OH = 1.0 kcal/mole

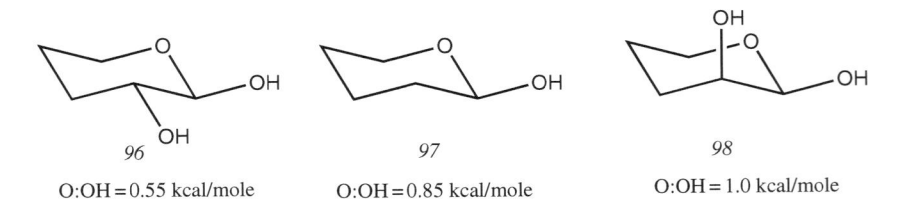

Fig. 2.28

Since the C2 carbon of 2-deoxy sugars bears no electronegative substituent (hydroxyl oxygen) which can interfere with the electronic properties of the C1 carbon and thus with the anomeric effect, then 0.85 kcal/mol can be taken as the value for the anomeric effect. The introduction of an equatorial electronegative substituent at the C2 carbon (hydroxyl group) decreases the value of anomeric effect (O:OH = 0.55 kcal/mol), whereas the introduction of an axial electronegative substituent (hydroxyl group) at the C2 carbon of a hexopyranoside increases the value of the anomeric effect (O:OH = 1.0 kcal/mol) (Fig. 2.30).

Calculation of Conformational Energies of Pyranoses

D-Glucopyranose

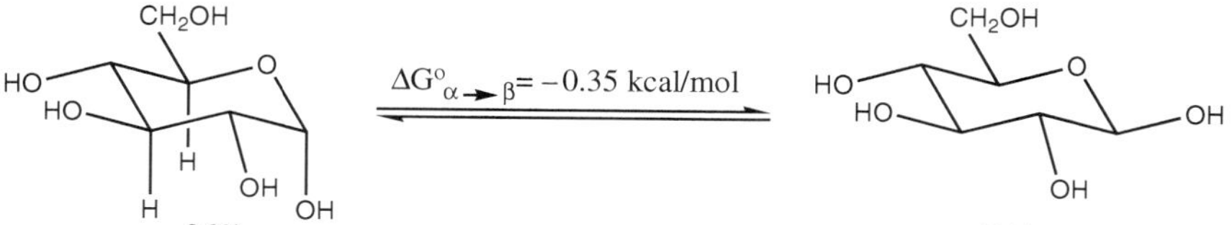

[α-D-Glucopyranose structure] ⇌ $\Delta G°_{\alpha \to \beta} = -0.35$ kcal/mol [β-D-Glucopyranose structure]

36%   64%

Interactions: 2 ($O_a:H_a$)   Interactions: (O:OH)

$\Delta G° = $ (O:OH) $- 2$ ($O_a:H_a$) $-0.35 = $ (O:OH) $- 0.9$ (O:OH) $= 0.55$ kcal/mol

D-Mannopyranose

[α-D-Mannopyranose structure] ⇌ $\Delta G°_{\alpha \to \beta} = 0.45$ kcal/mol [β-D-Mannopyranose structure]

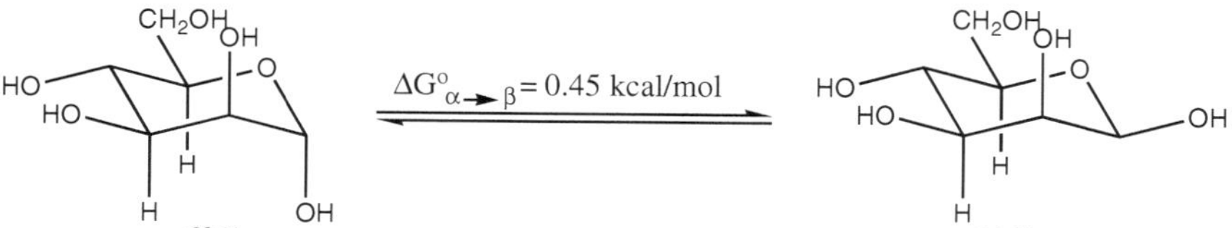

69%   31%

Interactions: 2 ($O_a:H_a$)   Interactions: (O:OH) + ($O_1:O_2$)

$\Delta G° = $ (O:OH) + ($O_1:O_2$) $- 2$ ($O_a:H_a$) $0.45 = $ (O:OH) $+ 0.35 - 0.9$ (O:OH) $= 1.0$ kcal/mol

2-Deoxy-D-glucopyranose

[α-2-Deoxy-D-glucopyranose structure] ⇌ $\Delta G°_{\alpha \to \beta} = -0.05$ kcal/mol [β-2-Deoxy-D-glucopyranose structure]

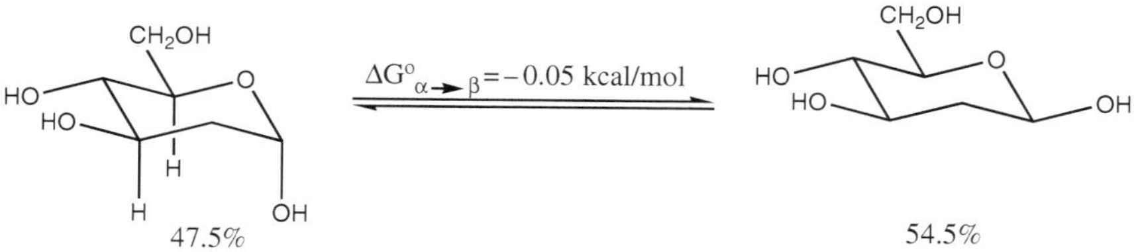

47.5%   54.5%

Interactions: 2 ($O_a:H_a$)   Interactions: (O:OH)

$\Delta G° = $ (O:OH) $- 2$ ($O_a:H_a$) $-0.05 = $ (O:OH) $- 0.9$ O:OH) $= 0.85$ kcal/mol

Fig. 2.29

The increased values for the anomeric effect of sugars having the C2 hydroxyl group axially oriented, as, for example, in D-mannose, was originally considered as a separate electronic interaction and was named by Reeves [58, 59] Δ2 effect. The originally suggested explanation for the high value of the anomeric effect of D-mannose was that the C2 axial hydroxyl group in β-D-mannopyranose lies parallel and coplanar with the resultant dipole of the C1–O5 and the C1–O1 oxygens (*100* in Fig. 2.30) introducing thus an additional destabilization.

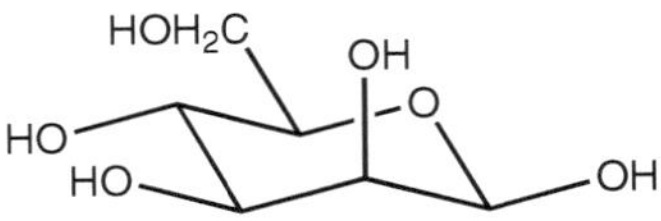

Fig. 2.30

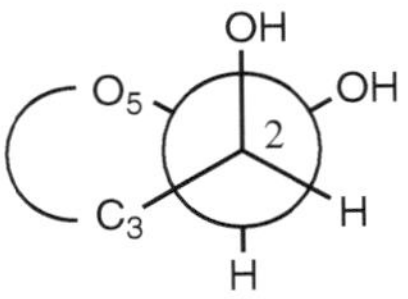

If both the C2 and C3 carbons bear an axial hydroxyl group they are considered [47] to cancel out each other and the value of 0.85 kcal/mol is used for the anomeric effect.

In Table 2.2 are given the compositions of the equilibrium mixtures of aldoses after dissolution in water as determined by NMR [61].

The equilibrium compositions of the sugars in solutions are affected by temperature, by the nature of solvent, and by the presence of substituents. In less polar solvents than water, the anomeric effect increases favoring the α-pyranose over the β-pyranose form if the sugar belongs to D-series and is in $^4C_1$ conformation. Aldoses with an axial C2 hydroxyl group (mannose, lyxose) contain much higher proportion

Table 2.2 Predominant conformations of D-aldohexo- and D-aldopentopyranoses in aqueous solution

Aldose	Conformation found by		Calculated interaction energies (kcal/mol)	
	NMR [63–65]	Calculation [66]	C1	1C
α-D-Allose	C1	C1	3.9	5.35
β-D-Allose	C1	C1	2.95	6.05
α-D-Altrose	C1, 1C	C1, 1C	3.65	3.85
β-D-Altrose	C1	C1	3.35	5.35
α-D-Glucose	C1	C1	2.4	6.55
β-D-Glucose	C1	C1	2.05	8.0
α-D-Mannose	C1	C1	2.5	5.55
β-D-Mannose	C1	C1	2.95	7.65
α-D-Gulose		C1	4.0	4.75
β-D-Gulose	C1	C1	3.05	5.45
α-D-Idose	C1, 1C	C1, 1C	4.35	3.85
β-D-Idose		C1	4.05	5.35
α-D-Galactose	C1	C1	2.85	6.3
β-D-Galactose	C1	C1	2.5	7.75
α-D-Talose	C1	C1	3.55	5.9
β-D-Talose		C1	4.0	8.0
α-D-Ribose	C1, 1C	C1, 1C	3.45	3.55
β-D-Ribose	C1, 1C	C1, 1C	2.5	3.1
α-D-Arabinose	1C	1C	3.2	2.05
β-D-Arabinose		C1, 1C	2.9	2.4
α-D-Xylose	C1	C1	1.95	3.6
β-D-Xylose	C1	C1	1.6	3.9
α-D-Lyxose	C1, 1C	C1, 1C	2.05	2.6
β-D-Lyxose	C1	C1	2.5	3.55

Calculation of Conformational Energies of Pyranoses

of the α-pyranose form in dimethyl sulfoxide than in water, whereas sugars with an equatorial C2 hydroxyl group (glucose, xylose) have approximately the same equilibrium composition in both solvents [62, 63].

The conformations of aldohexopyranoses, in aqueous solutions, are, as seen from the data in Table 2.2, controlled mainly by the orientation of hydroxymethyl group, which is the bulkiest substituent and as such tends to assume the equatorial position. Hence, all of the β-D-anomers are predominantly in the C1 ($^4C_1$) conformation, because in the 1C ($^1C_4$) form there is a large *syn*–axial interaction between the axial hydroxymethyl and the axial anomeric hydroxyl groups, as shown in Fig. 2.18. This interaction is absent from 1C conformation of α-anomers but most of them also prefer the C1 conformation; only α-D-idopyranose exists predominantly in the 1C form; the aqueous solutions of α-D-altropyranose and α-D-gulopyranose have substantial proportions of both chair forms at equilibrium.

In the absence of a hydroxymethyl group at C5, the conformations of the aldopentopyranoses are controlled by the orientation of the hydroxyl groups. Thus, the D-arabinopyranoses favor the 1C form, α-D-lyxopyranose and α-D-ribopyranose are conformational mixtures, and the other pentoses are predominantly in the C1 form.

A relationship between the percentage of the more stable conformer at equilibrium, equilibrium constant K, and standard free energy difference at 25°C for an equilibrium of isomers A ⇌ B is given in Table 2.3.

Table 2.3 Relationship between the % of the morer stable isomer in equilibrium, the equilibrium constant and the conformational energy difference

% of more stable isomer	K	ΔG^0 (kcal/mol)
50	1.00	0
60	1.50	0.119
70	2.33	0.502
80	4.00	0.973
90	9.00	1.302
95	19.00	1.744
98	49.00	2.306
99.9	999.0	4.092
99.99	9999	5.456

As it can be seen the difference in free energies between two conformers of 0.973 kcal/mol results in fourfold excess of the more stable isomer in the equilibrium mixture.

Computational studies of carbohydrate structures were reported by many groups [67–78]. The early studies were based partly on experimental data, and partly on ab initio calculations, usually at the Hartree–Fock level. The experimental data, though reliable, were normally determined in solutions, very often in water, or some other hydroxylic solvents, with sugar concentrations most often unknown. Hence, one must be concerned about solvation effects in the interpretation of these data. An additional difficulty was that ab initio calculations of carbohydrates required the use of a fairly large basis set in order to obtain what is hoped to be the chemical accuracy.

Hexopyranoses

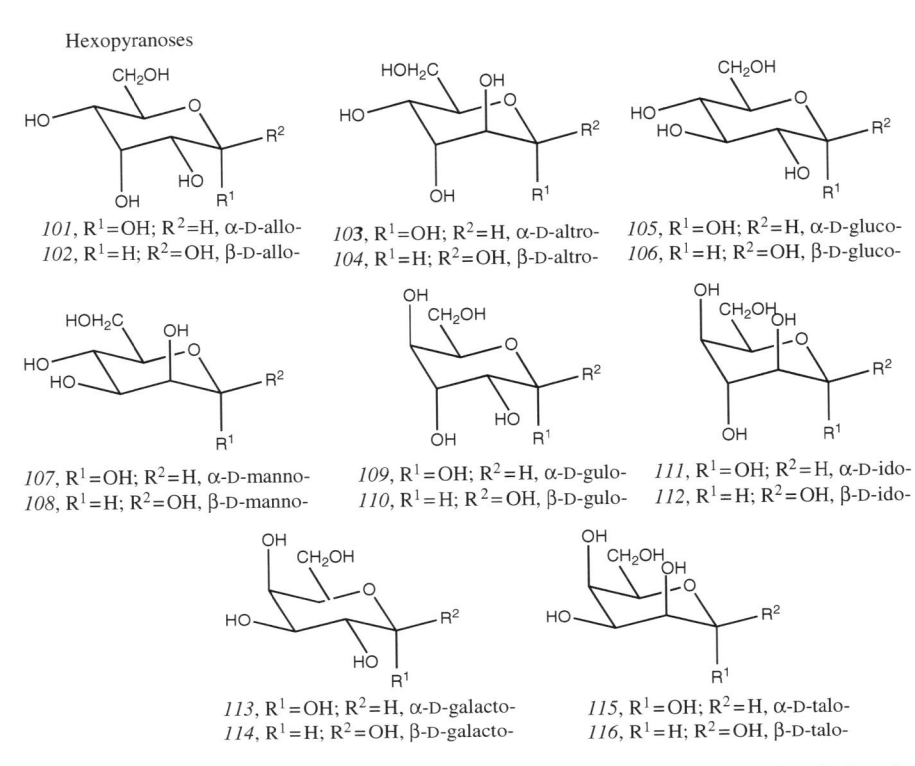

101, R¹=OH; R²=H, α-D-allo-
102, R¹=H; R²=OH, β-D-allo-

103, R¹=OH; R²=H, α-D-altro-
104, R¹=H; R²=OH, β-D-altro-

105, R¹=OH; R²=H, α-D-gluco-
106, R¹=H; R²=OH, β-D-gluco-

107, R¹=OH; R²=H, α-D-manno-
108, R¹=H; R²=OH, β-D-manno-

109, R¹=OH; R²=H, α-D-gulo-
110, R¹=H; R²=OH, β-D-gulo-

111, R¹=OH; R²=H, α-D-ido-
112, R¹=H; R²=OH, β-D-ido-

113, R¹=OH; R²=H, α-D-galacto-
114, R¹=H; R²=OH, β-D-galacto-

115, R¹=OH; R²=H, α-D-talo-
116, R¹=H; R²=OH, β-D-talo-

Fig. 2.31 Structures of hexopyranoses depicting the orientations of their secondary hydroxyl groups

Figure 2.31 shows the preferred conformations for aldohexopyranoses in aqueous solution as determined by NMR spectroscopy [60]. Allinger et al. [79] have reported the ab initio calculations of 84 conformations of 12 different sugars (hexoses), in both pyranose and furanose forms. They used the same MM4 force field calculation that they used for the calculation of simple alcohols and ethers, but for carbohydrates they had to make some important additions to take into the account the anomeric affect, gauche effect, and the Δ2-effect which are all present in sugars but not in simple alcohols and ethers (Fig. 2.32).

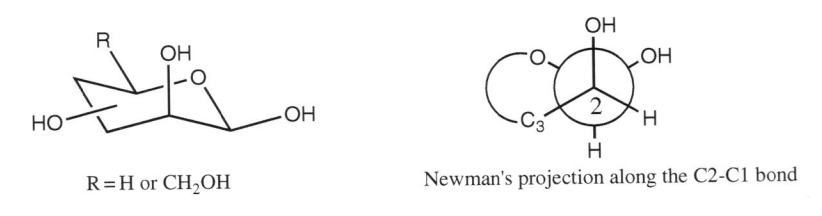

R = H or CH₂OH

Newman's projection along the C2-C1 bond

Fig. 2.32

The Δ2-effect is energetically important only in a few sugars, such as β-D-altropyranose *104*, β-D-mannopyranose *108*, β-D-idopyranose *112*, and β-D-talopyranose *116* (Fig. 2.31) and tends to stabilize the α anomer relative to β, by about 1.36 kcal/mol. It should be noted that a recent study found no evidence that Δ2-effect plays an important role in determining the conformational properties of sugars [80].

References

1. Lowe, J. P., *"Barriers to internal rotation about single bonds"*, Progr. Phys. Org. Chem. (1968) **6**, 1–80
2. Lowe, J. P., *"Barrier to internal rotation in ethane"*, Science (1973) **179**, 527–532
3. Oelfke, W. C.; Gordy, W., *"Millimeter-wave spectrum of hydrogen peroxide"*, J. Chem. Phys. (1969) **51**, 5336–5343
4. Bair, R. A.; Goddard, III, W. A., *"Ab initio studies of the structure of peroxides and peroxy radicals"*, J. Am. C hem. Soc. (1982) **104**, 2719–2724
5. Pitzer, K. S., *"Potential energies for rotations about single bonds"*, Discuss. Fara- day Soc. (1951) **10**, 66–73
6. Weiss, S.; Leroi, G. E., *"Direct observation of the infrared torsional spectrum of C_2H_6, CH_3CD_3, and C_2D_6"*, J. Chem. Phys. (1968) **48**, 962–967
7. Hirota, E.; Saito, S.; Endo, Y., *"Barrier to internal rotation in ethane from the microvawe spectrum of CH_3CHD_2"*, J. Chem. Phys. (1979) **71**, 1183–1187
8. Fantoni, R.; van Helroort, K.; Knippers, W.; Reuss, J., *"Direct observation of torsional levels in Raman spectra of C_2H_6"*, Chem. Phys. (1986) **110**, 1–16
9. Pitzer, R. M., *"The barrier to internal rotation in ethane"*, Acc. Chem. Res. (1983) **16**, 207–210
10. Sovers, O. J.; Kern, C. W.; Pitzer, R. M.; Karplus, M., *"Bond-function analysis of rotational barriers: Ethane"*, J. Chem. Phys. (1968) **49**, 2592–2599
11. Bader, R. F. W.; Cheeseman, J. R.; Laidig, K. E.; Wiberg, K. B.; Brenemann, C., *"Origin of rotation and inversion barriers"*, J. Am. Chem. Soc. (1990) 112, 6530–6536
12. Klyne, W.; Prelog, V., *"Description of steric relationship across single bond"*, Experientia (1960) **16**, 521–523
13. Horton, D.; Wander, J. D., *"Conformations of acyclic derivatives of sugars"*, Carbohydr. Res. (1969) 10, 279–288
14. Azarnia, N.; Jeffrey, G. A.; Kim, H. S.; Park, Y. J., Joint ACS/CIC Conference Abstracts of Papers, CARB 26 (1970)
15. Berman, H. M.; Rosenstein, R. D., *"The crystal structure of galactitol"*, Acta Cryst. (1968) B24, 435–441
16. Berman, H. M.; Jeffrey, G. A.; Rosenstein, R. D., *"The crystal structures of the [alpha] and [beta] forms of D-mannitol"*, Acta Cryst. (1968) B24, 442–449
17. Kim, S. H.; Jeffrey, G. A.; Rosenstein, R. D., *"The crystal structure of the K form of D-mannitol"*, Acta Cryst. (1968) B24, 1449–1455
18. Hunter, F. D.; Rosenstein, R. D., *"The crystal structure of D,L-arabinitol"*, Acta Cryst. (1968) B24, 1652–1660
19. Staab, H. A., *"Einführug in die theoretische organische Chemie"*, Verlag Chemie, 1960, p.547
20. Pitzer, K. S., *"Strain energies of cyclic hydrocarbons"*, Science, (1945) **101**, 672
21. Kilpatrick, J. E.; Pitzer, K. S.; Spitzer, R., *"The thermodynamics and molecular structure of cyclopentane"*, J. Am. Chem. Soc. (1947) **69**, 2483–2488
22. Pitzer, K. S.; Donath, W. E., *"Conformations and strain energy of cyclopentane and its derivatives"*, J. Am. Chem. Soc. (1959) **81**, 3213–3218

23. Brutcher, F. V., Jr.; Roberts, T.; Barr, S. J.; Pearson, N., "*The conformations of substituted cyclopentanes. I. The infrared analysis and structure of the α-Halocam phors, the α-Halo-2-indanones and the α-Halocyclopentanones*", J. Am. Chem. Soc. (1959) **81**, 4915–4920
24. Aston, J. G.; Schumann, S. C.; Fink, H. L.; Doty, P. M., "*The structure of Ali- cyclic compounds*", J. Am. Chem. Soc. (1941) **63**, 2029–2030
25. Aston, J. G.; Fink, H. L.; Schumann, S. C., "*The heat Capacity and entropy, heats of transition, fusion and vaporization and the vapor pressures of cyclopentane. Evidence for a non-planar structure*", J. Am. Chem. Soc., (1943) **65**, 341–346
26. McCullough, J. P., "*Pseudorotation in cyclopentane and related molecules*", J. Chem. Phys. (1958) **29**, 966–967
27. McCullough, J. P.; Pennington, R. E.; Smith, J. C.; Hossenlopp, I. A.; Waddington, G., "*Thermodynamics of cyclopentane, methylcyclopentane and 1, cis-3-Dimethyl cyclopentane: Verification of the concept of Pseudorotation*", J. Am. Chem. Soc. (1959) **81**, 5880–5883
28. Almenningen, A.; Bastiansen, O.; Skancke, P. N., "*Preliminary results of an electron diffraction reinvestigation of cyclobutane and cyclopentane*", Acta Chem. Scand. (1961) **15**, 711–712
29. Lemieux, R. U., "*The configuration and conformation of Thymidine*", Can. J. Chem. (1961) **39**, 116–120
30. Jardetzky, C. D., "*Proton magnetic resonance studies on Purines, Pyrimidines, Ribose Nucleosides and Nucleotides. III. Ribose conformation*", J. Am. Chem. Soc. (1960) **82**, 229–233
31. Jardetzky, C. D., "*N.m.r. of nucleic acid derivatives. V. Deoxyribose conformation*", J. Am. Chem. Soc. (1961) **83**, 2919–2920
32. Jardetzky, C. D., "*Proton magnetic resonance of nucleotides. IV. Ribose conformation*", J. Am. Chem. Soc. (1962) **84**, 62–66
33. Abraham, R. J.; Hall, L. D.; McLauchlan, K. A., "*A proton resonance study of the conformation of carbohydrates in solution. Part I. Derivatives of 1,2-O- Isopropylidene-α-D-xylohexofuranose*", J. Chem. Soc. (1962) 3699–3705
34. Lemieux, R. U., "*Rearrangements and Isomerizations in Carbohydrate Chemistry*", In P. de Mayo, ed., "*Molecular Rearrangements*", Interscience Division, John Wiley and Sons, New York, 1963, p. 709
35. Angyal, S. J., "*The composition and conformation of sugars in solution*", Angew. Chem. Intern. Ed. (1969) **8**, 157–166
36. Lemieux, R. U., In "*Molecular Rearrangements*", P. De Mayo, ed., Vol 2, Wiley- Interscience, New York, 1963, pp. 710–713
37. Bishop, C. T.; Cooper, F. P., "*Glycosidation of sugars. II. Methanolysis of D-xylose, D-arabinose, D-lyxose, and D-ribose*", Can. J. Chem. (1963) **41**. 2743–2758
38. Smirnyagin, V.; Bishop, C. T., "*Glycosidation of sugars. IV. Methanolysis of D- glucose, D-galactose, and D-mannose*", Can. J. Chem. (1968) **46**, 3085–3090
39. Spencer, M., Acta Cryst. (1959) **12**, 59
40. Lemieux, R. U.; Nagarajan, R.,"*The configuration of Di-D-Fructose Anhydride I*", Can. J. Chem. (1964) 42, 1270–1278
41. Hendrickson, J. B., "*Molecular geometry. I. Machine Computation of the common rings*", J. Am. Chem. Soc. (1961) **83**, 4537–4547
42. Anet, F. A. L.; Anet, R., "*Conformational Processes in Rings*", in L. M. Jackman and F. A. Cotton, eds. "*Dynamic Nuclear Magnetic Resonance Spectroscopy*", Academic Press, New York, 1975, p. 543ff
43. Sandstrom, J., "*Dynamic NMR Spectroscopy*", Academic Press, New York, 1982
44. Van de Graaf, B.; Baas, J. M. A.; Wepster, B. M., "*Studies on cyclohexane derivatives. Part XV. Force field calculations on some tert-butyl substituted cyclohexane compounds*", Rec.Trav. Chim.(Pays-Bas) (1978) **97**, 268–273
45. Van de Graaf, B.; Baas, J. M. A.; Van Veen, A., "*Empirical force field calculations. VI. Exploration of reaction paths for the interconversion of conformers. Application to the interconversion of the cyclohexane conformers*", Rec. Trav. Chim. (Pays-Bas) (1980) **99**, 175–178

References

46. Eliel, E. L.; Allinger. N. L,.Angyal, S. J.; Morrison, G. A., "*Conformational Analysis*", Wiley-Interscience, New York, 1965, pp. 351–432
47. Angyal, S. J., "*Conformational analysis in carbohydrate chemistry. I. Conformational free energies. The conformations and α: β ratios of aldopyranoses in aqueous solution*", Aust. J. Chem. (1968) **21**, 2737–2746
48. Eliel, E. L.; Knoeber, M. C., "*Conformational analysis. XVI. 1,3-Dioxanes*", J. Am. Chem. Soc. (1968) **90**, 3444–3458
49. Angyal, S. J.; McHugh, D. J., "*Cyclitols. Part V. Paper ionophoresis, complex formation with borate, and the rate of periodic acid oxidations*", J. Chem. Soc. (1957) 1423–1431
50. Angyal, S. J.; McHugh, D. J., "*Interaction energies of axial hydroxyl groups*", Chem. Ind. (London) (1956) 1147–1148
51. Angyal, S. J.; Klavins, J. E. unpublished data
52. Eliel, E. L.; Allinger, N. L.; Angyal, S. J.; Morrison, G. A., "*Conformational Analysis*", Wiley-Interscience, New York, 1965, pp. 439–440
53. Edward, J. T., "*Stability of glycosides to acid hydrolysis*", Chem. Ind. (London) (1955) 1102–1104
54. Lemieux, R. U.; Chü, N. J., Abstr. Papers Amer. Chem. Soc. Meeting (1958) **133**, 31 N
55. Winstein, S.; Holness, N. J., "*Neighboring carbon and hydrogen. XIX. t-Butylcyclohexyl Derivatives. Quantitative conformational analysis*", J. Am. Chem. Soc. (1955) **77**, 5562–5578
56. Eliel, E. L.; Lukacs, C. A., "*Conformational analysis. II. Esterification rates of Cy- clohexanol*s", J. Am. Chem. Soc. (1957) **79**, 5986–5992
57. Subbotin, O. A.; Sergeyev, N. M., "*Conformational Equilibria in Cyclohexanol, Nitro cyclohexane, and Methylcyclohexane from the Low Temperature ^{13}C Nuclear Magnetic Resonance Spectra*", J. Chem. Soc. Chem. Commun. (1976) 141–142
58. Reeves, R. E., "*The shape of pyranose rings*", J. Am. Chem. Soc. (1950) 72, 1499–1506
59. Reeves, R. E., "*Cuprammonium-glycoside complexes*", Adv. Carbohydr. Chem. (1951) **6**, 107–134
60. Angyal, S. J., "*The composition and conformation of sugars in solution*", Angew. Chem. Internat. Ed. (1969) **8**, 157–166
61. Perlin, A. S., "*Hydroxyl proton magnetic resonance in relation to ring size, Substituent groups, and mutarotation of carbohydrate*s", Canad. J. Chem. (1966) **44**, 539–550
62. Mackie, W.; Perlin, A. S., "*Pyranose-Furanose and Anomeric Equilibria: Influence of solvent and of partial methylation*", Canad. J. Chem. (1966) **44**, 2039–2049
63. Lemieux, R. U.; Stevens, J. D., "*The proton magnetic resonance spectra and tautomeric equilibria of aldoses in deuterium oxide*", Can. J. Chem. (1966) **44**, 249–262
64. Rudrum, M.; Shaw, D. F., "*The structure and conformation of some monosaccharides in solution*", J. Chem. Soc. (1965) 52–57
65. Angyal, S. J.; Pickles, V. A.; Ahluwahlia, R., "*Interaction energy between an axial methyl and an axial hydroxyl group in pyranoses*", Carbohydr. Res. (1966) **1**, 365–370
66. Angyal, S. J., "*Conformational analysis in carbohydrate chemistry. I. Conformational free energies. The conformations and α: β ratios of aldopyranoses in aqueous solution*", Aust. J. Chem. (1968) **21**, 2737–2746
67. Brady, J. W., "*Molecular dynamics simulations of alpha-D-glucose in aqueous solution*", J. Am. Chem. Soc. (1989) **111**, 5155–5165
68. Ha, S.; Gao, J.; Tidor, B.; Brady, J. W.; Karplus, M., "*Solvent effect on the anomeric equilibrium in D-glucose: a free energy simulation analysis*", J. Am. Chem. Soc. (1991) **113**, 1553–1557
69. Polavarapu, P. L.; Ewig, C. S., "*Ab Initio computed molecular structures and energies of the conformers of glucose*", J. Comput. Chem. (1992) **13**, 1255–1261
70. Zheng, Y.-J.; Le Grand, S. M.; Merz, K. M., "*Conformational preferences for hydroxyl groups in substituted tetrahydropyran*s", J. Comput. Chem. (1992) **13**, 772–791

71. Glennon, T. M.; Zheng, Y.-J.; LeGrand, S. M.; Shutzberg, B. A.; Merz, K. M., "*A force field for monosaccharides and (1 4) linked polysaccharides*", J. Comput. Chem. (1994) **9**, 1019–1040
72. Barrows, S. E.; Dulles, F. J.; Cramer, C. J.; French, A. D.; Truhlar, D. G., "*Relative stability of alternative chair forms and hydroxymethyl conformations of β-Image- glucopyranose*", Carbohydr. Res.(1995) **276**, 219–251
73. Woods, R. J.; Raymond A. D.; Edge, C. J.; Fraser-Reid, B., "*Molecular mechanical and molecular dynamic simulations of glycoproteins and oligosaccharides. 1. GLYCAM_93 parameter development*", J. Phys. Chem. (1995) **99**, 3832–3846
74. Woods, R. J.; Smith, V. H.; Szarek, W. A.; Farazdel, A., "*Ab initio LCAO-MO calculations on α-D-glucopyranose, β-D-fructopyranose, and their thiopyranoid-ring analogs. Application to a theory of sweetness*", J. Chem. Soc., Chem. Commun. (1987) **12**, 937–939
75. Senderowitz, H.; Parish, C.; Still, W. C, "*Carbohydrates: United Atom AMBER∗ Pa- rameterization of Pyranoses and Simulations Yielding Anomeric Free Energies*", J. Am. Chem. Soc. (1996) **118**, 2078–2086
76. Reiding, S.; Schlenkrich, M.; Brickman, J., "*Force field parameters for carbohydrates*", J. Comput. Chem. (1996) **17**, 450–468
77. Ott, K.-H.; Meyer, B., "*Parametrization of GROMOS force field for oligosaccharides*", J. Comput. Chem. (1996) **17**, 1068–1084
78. Damm, W.; Frontera A.; Tirado-Rives, J.; Jorgensen, W. L., "*OPLS all-atom force field for carbohydrates*", J. Comput. Chem. (1997) **18**, 1955–1970
79. Lii, J.-H.; Chen, K.-H.; Allinger, N. L., "*Alcohols, ethers, carbohydrates, and related compounds. IV. Carbohydrates*", J. Comput. Chem. (2003) **24**, 1504–1513
80. Guler, L. P.; Yu, Y.-Q.; Kenttämaa, H. I.,"*An experimental and computational study of the gas-phase structures of five-carbon Monosaccharides*", J. Phys. Chem. A (2002) **106**, 6754–6764

Chapter 3
Anomeric Effect

In substituted cyclohexanes, such as cyclohexanol or its methyl ether, the substituent will preferably assume the equatorial position as opposed to the axial one, due to fewer nonbonded interactions with other ligands (in this case the C3 and the C5 hydrogen atoms) on the cyclohexane ring [1–5]. Thus, the conformational mixture of cyclohexanol or its methyl ether contains, at equilibrium, 89% of the conformer with equatorially oriented hydroxyl or methoxy group and 11% of the conformer with axially oriented hydroxyl or methoxy group, indicating clear preference for the equatorial conformer (Fig. 3.1).

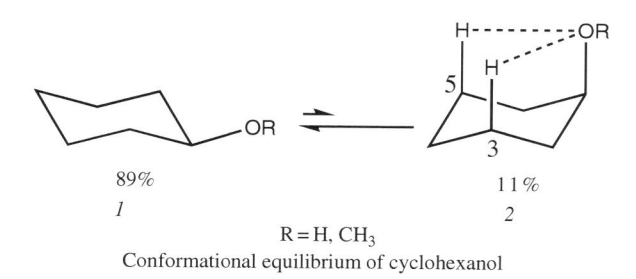

Fig. 3.1 Conformational equilibrium of cyclohexanol

Endo-anomeric Effect

The equilibrium mixture of two D-glucopyranose anomers contains 63% of equatorial (β) and 36% of axial (α) anomer (Fig. 3.2). Thus, in spite of the presence of two $O_a{:}H_a$ 1,3-*syn*–axial interactions between the anomeric oxygen and the C3 and C5 hydrogens present in α-anomer that destabilize this isomer by 0.9 kcal/mol the amount of the α-anomer in equilibrium mixture is significantly higher than in cyclohexanol. It should be noted that this amount of destabilization energy would require that the equilibrium mixture of two D-glucopyranose anomers does not contain more than ca. 20% of α-anomer.

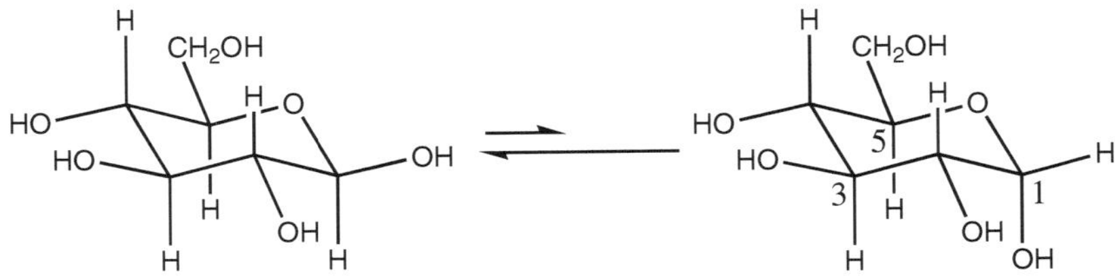

Anomeric equilibrium of D-glucopyranose

Fig. 3.2

The studies of conformational equilibria of anomers of other glycopyranoses have shown that monosaccharides with the axial anomeric oxygen (conformationally less favored than equatorial ones) are also present in higher percentages than expected (Tables 3.1 and 3.2).

Table 3.1 Composition of aqueous solution of some monosaccharides at equilibrium [6]

Sugar	Estimated from oxidation[a] (%)		Calculated from optical rotation[b] (%)	
	α	β	α	β
D-Glucose	37.4	62.6	36.2	63.8
D-Mannose	68.9	31.1	68.8	31.2
D-Galactose	31.4	68.6	29.6	70.4

[a]Oxidation of sugar solutions at 0°C with bromine water in the presence of $BaCO_3$
[b]Calculated from optical rotation, assuming that only two constituents are present in the solution.

Table 3.2 Relative free energies (kcal/mol) and the percentage of α-anomer for selected D-aldohexo- and D-aldopentopyranoses in aqueous solution at equilibrium[a]

Pyranose	G^0_α	G^0_β	$G^0_{pyranose}$	α-Anomer (%)	
				Calculated	Experimental
Glucose	2.4	2.05	1.8	36	36
Galactose	2.85	2.5	2.25	36	27
Mannose	2.5	2.95	2.25	68	67
Idose	3.65	4.0	3.4	64	46
Ribose	3.1	2.3	2.15	20.5	26
Arabinose	1.95	2.2	1.65	60	63
Xylose	1.9	1.6	1.35	37	33
Lyxose	1.85	2.4	1.65	72	71

[a]By ^1H nuclear magnetic resonance [7, 8].

3   Anomeric Effect

In the case of D-glucose and D-galactose the anomer with the equatorial C1 hydroxyl group (β) is, as expected, more stable; D-mannose represents a special case and will be discussed later.

The preference for axial orientation of the C1 substituent in D-glucopyranose was found to increase with the increase of electronegativity of the C1 substituent (Table 3.3).

Table 3.3 Anomeric equilibria of 1-*O*-substituted D-glucopyranoses

Compound	C1 substituent	Axial isomer (%)
D-Glucopyranose[a]	OH	36
Methyl D-glycopyranoside[b]	OMe	67
Penta-*O*-acetyl-D-glycopyranose[c]	OAc	86
Tetra-*O*-acetyl-D-glucopyranosyl chloride[d]	Cl	94

[a]In water at 25°C.
[b]In methanol at 25°C.
[c]In acetic acid–acetic anhydride at 25°C using perchloric acid as catalyst [9, 10].
[d]In acetonitrile [11] at 30°C.

In 1928 Pacsu discovered [12] that the β-anomers of acetylated alkyl glycopyranosides anomerize in the presence of stannic chloride or titanium tetrachloride in chloroform to α-anomers suggesting the greater thermodynamic stability of anomeric alkoxy group in the axial orientation. A possibility that the driving force for anomerization is somehow related to complexing reaction involving the Lewis acid catalysts (including boron trifluoride [13]) was excluded by observing the extensive β- to α-anomerization of several acetylated methyl glycopyranosides also during sulfuric acid catalyzed acetolysis [14, 15].

Detailed study of anomerization of peracetylated pento- and hexopyranoses [16] suggested that the preference for the axial orientation of aglycon could only be explained if one assumes that there is a special driving force which forces the 1-acetoxy group to take a *syn*-clinal orientation relative to the C5 of the pyranose ring. This driving force, named "anomeric effect" [17], was estimated to be about 1.5 kcal/mol and depended on the sugar (Fig. 3.3).

The first rationalization of a tendency for aglycons of alkyl glycopyranosides to assume axial orientation was proposed by Edward [18] and was inspired by a Corey study [19] on stereochemistry of some α-halocyclohexanones in which it was

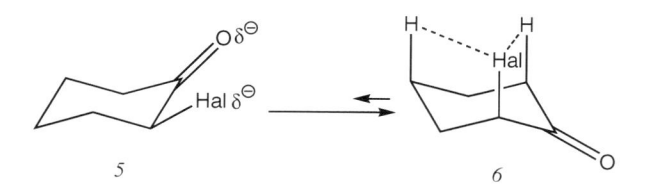

Fig. 3.3

7, R = H, β-D-xylopyranose
8, R = CH$_2$OH, β-D-glucopyranose

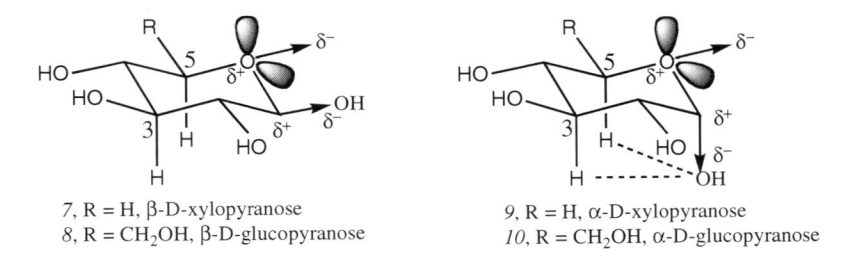

9, R = H, α-D-xylopyranose
10, R = CH$_2$OH, α-D-glucopyranose

Fig. 3.4

determined by IR spectroscopy that the most stable conformation of α-chloro- and α-bromocyclohexanone is that chair form (6) in which the halogen substituent is axial. This was explained to be a consequence of dipole–dipole interaction between the C–halogen and the C=O group when halogen atom is equatorially oriented which is considered to be more important than the steric interactions in the axial conformer. By analogy, Edward proposed that the equatorially oriented electronegative aglycon group such as O-alkyl, halogen, etc., is subjected to energetically unfavorable dipolar interaction with the resultant dipole of two unshared electron pairs of the ring oxygen atom since these two dipoles are coplanar and equally oriented (7 and 8, in Fig. 3.4). This destabilization was considered to be larger than the two *syn*–axial interaction between the axially oriented aglycone (in the α-D-anomer) and the C3 and C5 hydrogen atoms and as a result the electronegative substituents such as Cl, OH, and OR at C1 of pyranosides preferred an axial orientation (9 and 10 in Fig. 3.4). Thus the *anomeric effect was originally considered to be an electrostatic effect that destabilizes the equatorially oriented C1 electronegative substituent through dipolar interaction with the two pairs of nonbonding electrons on the ring oxygen* (Fig. 3.4)

Although recognizing that the anomeric effect is the preference of glycosides to assume that conformation in which the number of interactions between opposing unshared pairs of electrons should be at minimum (which he called e//e component of anomeric effect) [20] Lemieux [16, 17, 20] offered no explanation for the anomeric effect but did point out that the simple consideration of the effect on charge distribution arising from a change of the polar bonds from the antiperiplanar to the *syn*-clinal orientation provides a change in energy which is close to the experimentally observed value for the anomeric effect.

Thus the anomeric effect was originally considered to be an electrostatic effect that destabilizes equatorially oriented C1 electronegative substituent through dipolar interaction with the pair of nonbonding electrons on the ring oxygen (Fig. 3.4).

Comparison of two anomeric structures of D-glucopyranose (Fig. 3.4) shows that, except for the two 1,3-*syn*–axial interactions between the axial O1 and the axial H3 and H5 atoms, which are present only in α-anomer, both anomers have all other nonbonded interactions identical, including the gauche interaction between the C1 and the C2 hydroxyl groups. Consequently, the study of interconversion of two anomers

Fig. 3.5

of D-glucopyranose (*anomerization*, the isomerization of anomeric carbon) can be simplified by substituting D-glucopyranose with the 2-hydroxy-tetrahydropyran as model compound for the anomerization studies (Fig. 3.5). The free-energy difference between *11* and *12* defines the conformational free energy of hydroxyl group (the so-called A-value [21, 22] = $-\Delta G^0$) in 2-hydroxy-tetrahydropyran (Fig. 3.5). The A-value is significantly greater in protic (aqueous) solution than in aprotic solutions, possibly because of hydroxyl group solvation via hydrogen bonding that increases its effective size.

The quantitative estimate of the magnitude of anomeric effect must take into account the steric preference for equatorial orientation of an electronegative group larger than hydrogen in the corresponding cyclohexane compound. In order to do this it must be assumed that the conformational energies of hydroxyl group in 2-hydroxy-tetrahydropyran and in cyclohexanol are of the same magnitude. Although this assumption ignores the difference in geometry between cyclohexane and tetrahydropyran ring, in most cases this does not lead to significant discrepancy.

The magnitude of the anomeric effect is thus defined as the difference between the conformational free energy $(-\Delta G^0{}_X)_O$ for the equilibrium of two 2-substituted tetrahydropyran conformers (Fig. 3.6) and the conformational free energy $(\Delta G^0{}_X)$ for the equilibrium of the two analogously substituted cyclohexane (Fig. 3.7) [23].

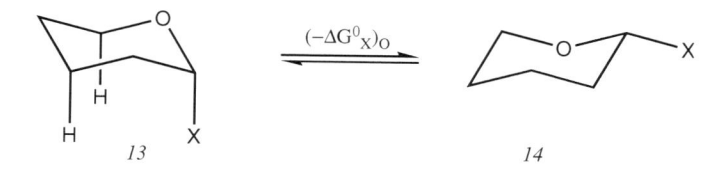

Fig. 3.6

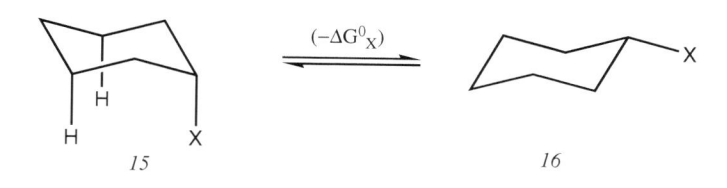

Fig. 3.7

Thus, the anomeric effect (AE or O:X, where X = OH or any other electronegative substituent such as OMe, OAc, Cl, Br, etc.) can be expressed as

$$(O:X) = (-\Delta G_X^0) - (-\Delta G_X^0)_O \qquad (1)$$

Rewriting the above equation gives the following expression for the magnitude of anomeric effect:

$$(O:X) = (\Delta G_X^0)_O - (-\Delta G_X^0) \qquad (2)$$

The value of anomeric effect in D-glucopyranose was determined in the following way. The internal energies of D-glucopyranosyl residues in both α- and β-anomer are identical since they have identical number and types of nonbonded interactions. The introduction of an electronegative anomeric substituent (hydroxyl group, halogen, etc.) into these two residues introduces the difference between their respective internal energies, depending on whether the electronegative substituent is equatorially or axially oriented. Therefore, the internal energy of α-anomer will be the sum of internal energy of D-glucopyranosyl residue, E^0, and the two *syn*–axial interactions between the axial anomeric hydroxyl group and the C3 and C5 axial hydrogen atoms of the pyranoid ring (2 × 0.45 kcal/mol = 0.9 kcal/mol). The internal energy of β-anomer will be the sum of internal energy E^0 of D-glucopyranosyl residue and the anomeric effect (AE). The number of gauche 1,2-interactions is identical in both α- and β-D-glucopyranose. From the composition of equilibrium mixture of α- and β-D-glucopyranose (36% vs. 64%, respectively) one can calculate that the β-anomer has lower free energy than the corresponding α-anomer by 0.35 kcal/mol [24].

Therefore,

$$E_\alpha - E_\beta = 0.35 \text{ kcal/mol}$$

If E_α is now substituted with (E^0 + 0.9) and E_β with (E^0 + AE), the following expression is obtained:

$$(E^0 + 0.9 \text{ kcal/mol}) - (E^0 + \text{AE}) = 0.35 \text{ kcal/mol}$$

Solving this equation for AE (O:OH) gives

$$0.9 \text{ kcal/mol} - (O + OH) = 0.35 \text{ kcal/mol}$$

$$(O:OH) = 0.9 - 0.35 \text{ kcal/mol} = 0.55 \text{ kcal/mol}$$

Hence, the difference of 0.55 kcal/mol between 0.9 kcal/mol and 0.35 kcal/mol corresponds to the anomeric effect (O:OH) and represents the electronic stabilization of the axially oriented hydroxyl group in the α-anomer. Thus, in other words, this electronic interaction was thought to be responsible for the higher percentage of axial anomer in an equilibrium mixture in spite of the unfavorable steric 1,3-*syn*–axial interactions between the C1 substituent and the C3 and the C5 hydrogens present in such anomers. This is in contrast to the Edward explanation for the anomeric effect as the electronic destabilization of the equatorially oriented anomer due to dipolar interaction between the equatorially oriented electronegative substituent and the resultant dipole of the two pairs of nonbonding electrons on the ring oxygen.

Similar calculations for the D-mannopyranose which at equilibrium contain 69% of α-anomer and 31% of β-anomer [25] gave the value for (O:OH) of 1.0 kcal/mol and for the 2-deoxy-D-arabino-hexopyranoses which at equilibrium contain 47.5% of α-anomer and 52.5% of β-anomer [25, 26] gave the value for (O:OH) of 0.85 kcal/mol.

Thus, the magnitude of *anomeric effect* determined in this way depends, among other factors, on the nature and on the configuration of substituent at the C2 carbon atom. In the case of β-D-mannopyranose, the C2–oxygen bond bisects the torsional angle between the C1–O1 and C1–O5 bonds (Fig. 3.8) and this dipolar interaction seems to introduce an additional electronic destabilization which is evident from the increased value for anomeric effect(1.0 kcal/mol). This interaction was considered as a separate electronic interaction and was named by Reeves [27–29] the Δ2 effect. It is now regarded as simpler to take as the base value for anomeric effect the value of 0.85 kcal/mol which is the value for 2-deoxy-arabino-hexopyranose and then when an elecronegative substituent at the C2 carbon is axial to increase this value by 0.15 kcal/mol and when the C2 substituent is equatorial to decrease it by 0.3 kcal/mol.

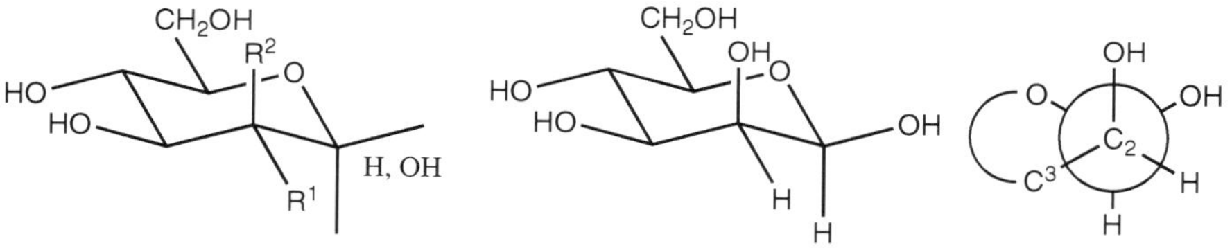

17, $R^1 = H$; $R^2 = OH$
18, $R^1 = R^2 = H$

19

20

Fig. 3.8

The Quantum-Mechanical Explanation

In halogeno-1,4-dioxanes ($X_1 = X_4 =$ oxygen), halogeno-1,4-thioxanes ($X_1 =$ oxygen, $X_4 =$ sulfur), and halogeno-1,4-dithianes ($X_1 = X_4 =$ sulfur; $Y =$ Cl, Br) *21* (Fig. 3.9), the halogen atoms were found to occupy preferentially the axial orientation, which was in contradiction to the well-known situation in monohalogenocyclohexanes [30, 31]. Similarly, in 2- or 6-monochloro or monobromo tetrahydropyran ($X =$ Cl, Br) (22 in Fig. 3.10) the halogen atom takes up the axial

Fig. 3.9

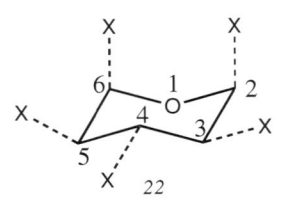

21

Fig. 3.10

22

orientation, whereas the halogen bonded to the C3, C4, or C5 carbon has a great preference for the equatorial orientation [32–34].

Study of a simple acyclic compound such as monochloromethoxymethane (*23, 24*) by electron diffraction [35, 36] (Fig. 3.11) has shown that the molecule does not exist in conformationally more stable *anti* conformation 24 (Fig. 3.11) but in *gauche* conformation 23 which is equivalent to the axial orientation in a six-membered ring. This suggests that the anomeric effect or the preference of the C–O–C–Hal system for the gauche conformations is a general phenomenon. The most intriguing finding was that the anomeric effect for Cl or Br as substituents amounts to several kcal/mol.

23, gauche conformation   *24, anti* conformation

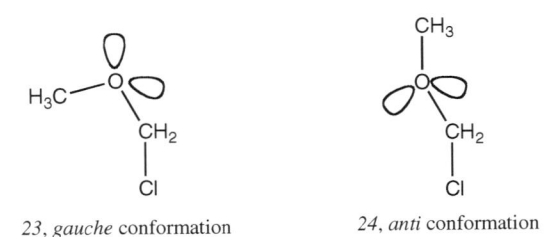

Fig. 3.11

The anomeric effect has been defined as the sum of free-energy difference between the axial (favored) and the equatorial anomer plus the conformational preference (the *A*- value) for the same substituent in cyclohexane [37]. Thus the anomeric effect measures the stability of axial over an equatorial substituent in 2-substituted tetrahydropyran relative to the expected value in cyclohexane (where the equatorial substituent is favored). The anomeric effect for chlorine, bromine, and

Fig. 3.12

iodine in 2-halo-4-methyl-tetrahydropyrans (Fig. 3.12) was found (by NMR) to be 2.65, > 3.2, and > 3.1 kcal/mol, respectively [34]. In polar solvents, such as acetonitrile, the value for chlorine seems to be smaller (2.0 kcal/mol) than in neat liquid (2.65 kcal/mol) [34]. However, all these values are much higher than those for the anomeric effect of hydroxy, alkoxy, or acyloxy groups in the 2-substituted tetrahydropyrans (0.9–1.4 kcal/mol); the values for the anomeric effect of these substituents were also found to be significantly solvent dependent [23, 38–41].

The initially proposed explanation for the anomeric effect [18] as a simple dipole–dipole interaction accounts therefore only for a part of the effect, but it does not represent the whole story. If one calculates the electrostatic interaction energy in *trans*-2,5-dichloro-1,4-dioxane (Fig. 3.13) (the molecular geometry of which is known from the X-ray analysis) using the values of $\mu = 2.2$ and 1.4 D for the dipole moments of C–Cl and C–O bonds and $\epsilon = 2.3$ for the dielectric constant, one arrives at the energy difference of about 1 kcal/mol in favor of the diaxial form [42]. This difference is clearly too small to account for a strong preference for diaxial conformation [43]. Consequently, it was proposed [44] that the *anomeric effect consists of two contributing components*. One substantial component is that in conformer with two axially oriented chlorine atoms (*29*) there are two gauche halogen–oxygen lone pair electron interactions (Figs. 3.13 and 3.14) (one at the C2 and one at the C5 carbon), whereas in the conformer having two chlorine atoms equatorially oriented

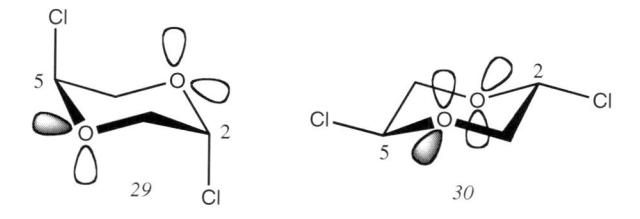

Fig. 3.13

66

3 Anomeric Effect

Fig. 3.14

(*30*) there are four gauche halogen–oxygen lone pair electron interactions (two at the C2 and two at the C5 carbon) (Figs. 3.13 and 3.14).

The other contributing component emerged from a study of geometry of halogenodioxanes *32*, halogenothioxanes *33*, and halogenodithianes *34* (Fig. 3.15) by X-ray crystallography and of chloromethoxymethane (*23*, *24* in Fig. 3.11) by electron diffraction. The result of studies of *32* and *33* was that in all cases where the accuracy of measurements was good, the C_2–O distance is significantly shorter than the C_6–O distance (*32* in Fig. 3.15). When compared to the lengths of C–O bonds in aliphatic ethers, the C_6–O_1 bond appears to be normal, whereas the O_1–C_2 bond appears to be shorter (*32* in Fig. 3.15). A second observation was that the axial C_2–Cl bond is somewhat longer than the corresponding equatorial C_3–Cl bond [in *cis*-2,3-dichloro-1,4-dioxane (Fig. 3.15)]. The axial C_2–Cl bond was measured to be 1.819 Å, and the equatorial C_3–Cl bond 1.781 Å; the accepted values for the aliphatic C–Cl bond is 1.79 Å. These bond length abnormalities in the system C–X–C–Y suggested [45–48] that the one nonbonding electron pair of the ring oxygen is delocalized by orbital mixing with a suitably oriented σ^* antibonding orbital of the C–Hal bond. As a result of this delocalization (Fig. 3.16) the C–O bond between the carbon bearing the halogen and oxygen will be strengthened (shortened) and the C–Hal bond weakened (elongated). In Fig. 3.16 two resonance forms of this structure are shown using the concept "double bond–no bond resonance." Table 3.4 lists bond distances in the C_6–X–C_2–Y system.

In Fig. 3.17 are compared the electronic distributions in chloromethoxymethane *38* and in the partial structure of *cis*-2,3-dichloro-1,4-dioxane *32* (Fig. 3.15).

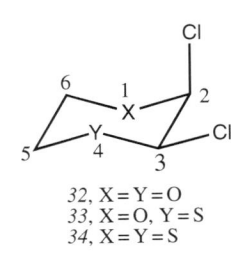

32, X = Y = O
33, X = O, Y = S
34, X = Y = S

Fig. 3.15

3 Anomeric Effect

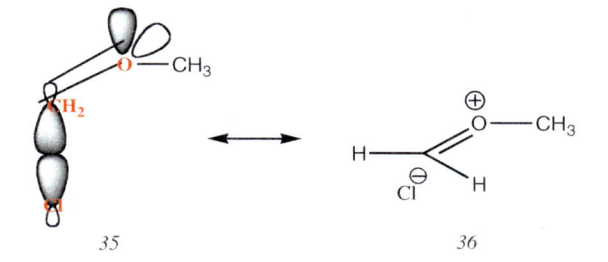

Fig. 3.16

Table 3.4 Bond distances in the group C_6–X–C_2–Y (in Angstroms) [44]

Compound	X	Y	C_6–X	C_2–X	C_2–Y
trans-2,3-Dichlorodioxane	O	Cl	1.43	1.38	1.84
cis-2,3-Dichlorodioxane	O	Cl	1.466	1.394	1.819
trans-2,5-Dichlorodioxane	O	Cl	1.428	1.388	1.845
Chloromethoxymethane	O	Cl	1.414	1.368	1.813

37
C_6-O-C_2-Cl fragment of
cis-2, 3-dichloro-1, 4-dioxane

38
chloromethoxymethane

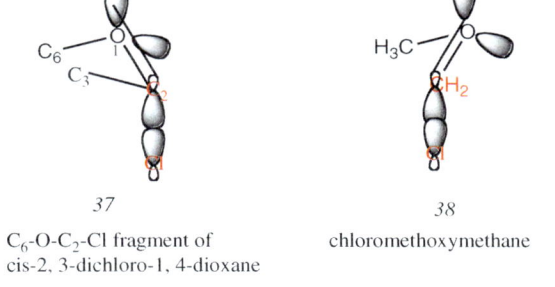

Fig. 3.17

Exo-anomeric Effect

The *exo*-anomeric effect [49] relates to the preference of the aglycon (e.g., methyl group of a methyl glycopyranoside) to be in near *syn*-clinal orientation to both the ring oxygen and the anomeric hydrogen, whereas the anomeric effect which should be more correctly called *endo-anomeric effect* relates to the preference for the axial orientation of the glycosidic oxygen of glycopyranosides. In Fig. 3.18 this is illustrated by using the C2–methoxy oxygen bond rotamers of 2-methoxy-tetrahydropyran with the methoxy group equatorially or axially oriented (*39*, *41*, *43* and *40*, *42*, *44*, respectively). The eclipsing of unshared electron pairs on glycosidic oxygen with the nonbonding electrons on the ring oxygen giving rise to destabilizing

E1 39 (1 *exo*-AE; 1 e://e:)

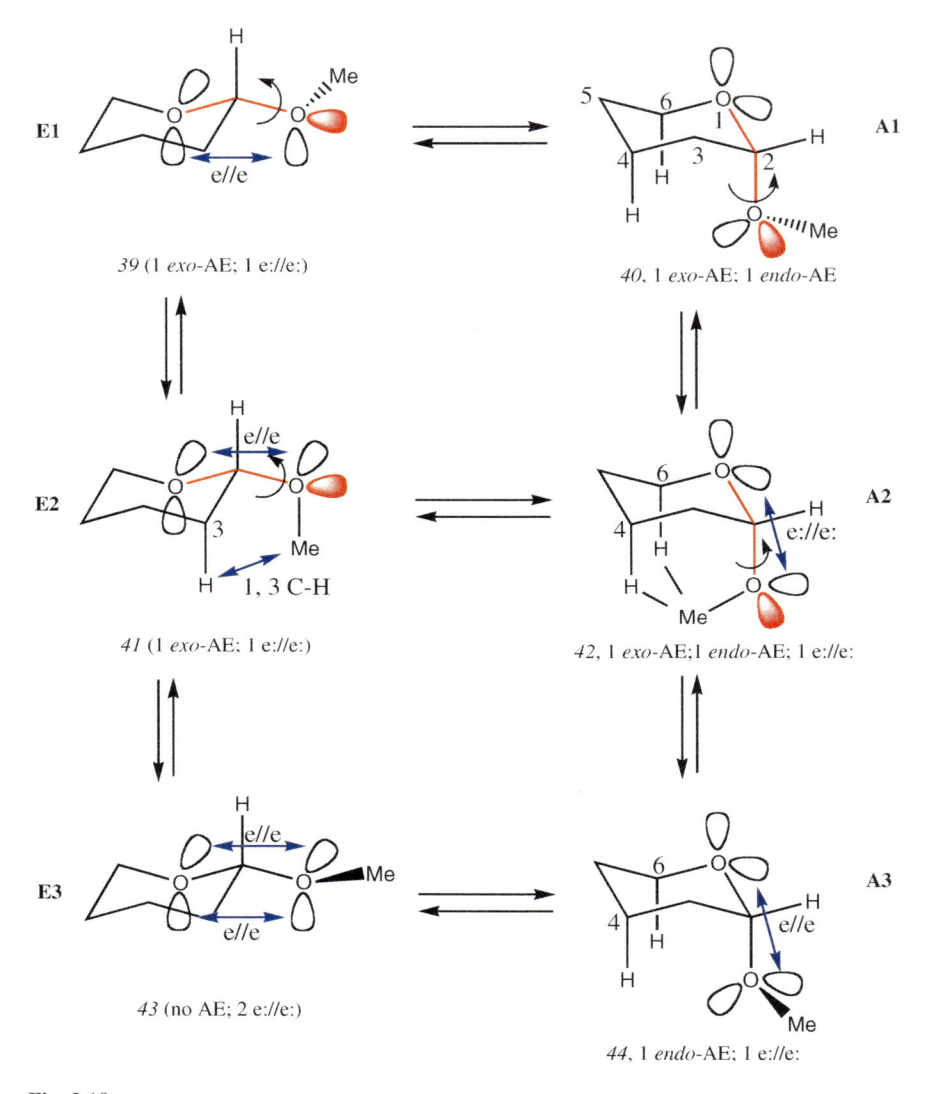

A1 40, 1 *exo*-AE; 1 *endo*-AE

E2 41 (1 *exo*-AE; 1 e://e:)

A2 42, 1 *exo*-AE; 1 *endo*-AE; 1 e://e:

E3 43 (no AE; 2 e://e:)

A3 44, 1 *endo*-AE; 1 e://e:

Fig. 3.18

syn–axial lone electron pair interactions is shown with a blue double-headed arrow and denoted e://e:. The *exo*-anomeric effect is shown by red bonds.

Three staggered conformations are possible for the rotation about the C1–O1 bond in both equatorial and axial conformers of a 2-methoxy-tetrahydropyran (Fig. 3.18). These are referred to as E1–E3 (*39, 41, 43*) and A1–A3 (*40, 42, 44*). In the E1 conformer (*39*) there are no *syn*–axial steric interaction but there is one *exo*-anomeric effect (stabilizing interaction) and one destabilizing *syn*–axial lone pair (electronic) interaction. In conformer E2 (*41*) there are several steric and electronic interactions: one 1,3-*syn*–axial interaction between the methyl group and the axial C3 hydrogen atom (steric interaction), one *endo*-anomeric effect (stabilizing electronic interaction), and one destabilizing *syn*–axial interaction between two lone pair electrons (one on the glycosidic oxygen and the other on the ring oxygen). In the conformer E3 (*43*) there are only two destabilizing *syn*–axial interaction between the four lone pair electrons (two on the glycosidic oxygen and two on the ring oxygen). In axial conformer *40* there are two stabilizing electronic interactions (one *endo*- and one *exo*-anomeric effect). In conformer *42* there is one severe steric interaction between the methyl group and the two axially oriented hydrogen atoms, one at the C3 and the other at the C5 carbon. In addition to that there is one destabilizing electronic *syn*–axial interaction between the two lone pair electrons (one on the glycosidic oxygen and the other on the ring oxygen). Finally there is one stabilizing electronic interaction: the *endo*-anomeric effect. In conformer *44* there is only one destabilizing *syn*–axial electronic interaction between two lone pair electrons (one on the glycosidic oxygen and the other on the ring oxygen).

Based on the above discussion from the three equatorial conformers E1 conformer should be favored, and from the three axial conformers the conformer A1 should be favored. Thus the *exo*-anomeric effect controls the conformation of the aglycon group.

The experimental evidence, for the *exo*-anomeric effect although initially difficult to obtain, has gradually accumulated over the years and today this phenomenon is fully accepted.

For molecules in crystalline state the evidence is unequivocal. It was determined that alkyl pyranosides adopt either the A1 or the E1 conformation [41] and analysis of over 50 carbohydrate structures reveals the following regularities: for axial methyl pyranosides the torsional angle O_5–C_1–O-CH_3 (which should be 60° in A1 conformer) lies between 61 and 74° and for equatorial anomers the range is 68–87°.

There is conflicting evidence for whether the *exo*-anomeric effect is larger for the axially or equatorially oriented groups. Even the analysis of crystal structures quoted above does not give a clear answer for glycopyranosides in the solid state, and the results in solutions are equally ambiguous, particularly for oligosaccharides. One thing is, however, clear: it is dominant short-range interaction that controls the conformation about the glycosidic linkage in both α- and β-linked oligosaccharides and therefore it is important for conformational analysis of these molecules.

Generalized Anomeric Effect

In 1968 Hutchins et al. [50] reported that there is a widespread phenomenon in structural chemistry that the conformations are strongly disfavored if the unshared electron pairs on nonadjacent atoms are parallel or *syn*-axial, as is the case for example in Fig. 3.19. This effect is thought to be the result of the repulsion of electric dipoles engendered by the unshared electron pairs. For obvious reasons Eliel

45

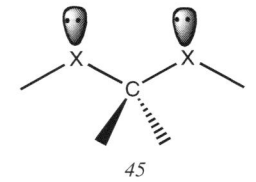

Fig. 3.19

proposed to call this phenomenon the "rabbit-ear effect." Although the existence of this effect has been pointed out earlier when we discussed the anomeric effect [it is the destabilizing component of the anomeric effect consisting of electrostatic repulsion of 1,3-*syn*–diaxial or 1,3-parallel unshared pairs of electrons (e://e: interaction)]. Support for this came from the finding that dimethoxymethane would tend to exist in the *gauche–gauche* conformation *48* (Fig. 3.20) rather than in the extended *trans–trans* conformation *46* with all large groups in *anti*-orientation or in *gauche–trans* conformation *47*. There are two reasons for this: the first destabilizing one is that there are two *syn*-parallel interactions between the four pairs of unshared electrons on two oxygen atoms (rabbit-ear effect) and in the *gauche–gauche* conformation there are two stabilizing *endo*-anomeric effects.

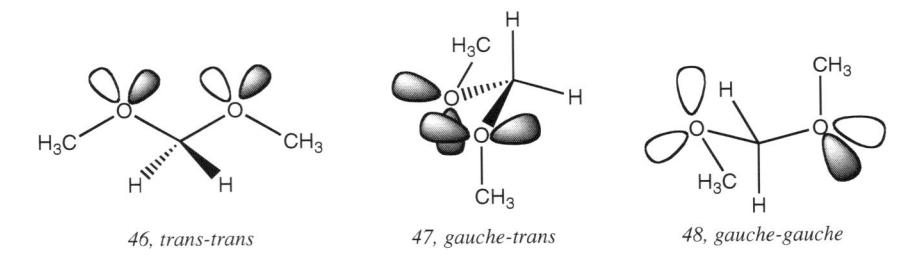

46, trans-trans   *47, gauche-trans*   *48, gauche-gauche*

Fig. 3.20

Kubo [51] obtained evidence through dipole-moment measurements that dimethoxymethane exists in *gauche–gauche* (+sc, +sc) conformation *48*. This conclusion was later substantiated by electron diffraction studies [52, 53].

By using NMR Hutchins et al. [50] studied the conformations of variously substituted 1,3-diazanes and found a striking support for the rabbit-ear effect (Fig. 3.21).

The introduction of one (equatorial) methyl group at the C5 carbon of *N,N*,2-trimethyl-1,3-diazane *49* giving the *N,N*,2,5-tetramethyl-1,3-diazane *50* slightly affects the position of the H2 chemical shift. However, the introduction of the second (axial) methyl group at the C5 carbon of *49* dramatically affects the position of the H2 chemical shift. This large upfield shift of H2 in *N,N*,2-trimethyl-1,3-diazane upon introduction of geminal methyl groups at the C5 carbon (*N,N*,2,5,5-pentamethyl-1,3-diazane *51*) was explained by assuming that in *N,N*,2-trimethyl-1,3-diazane *49* one *N*-methyl group is oriented axially and the other equatorially as

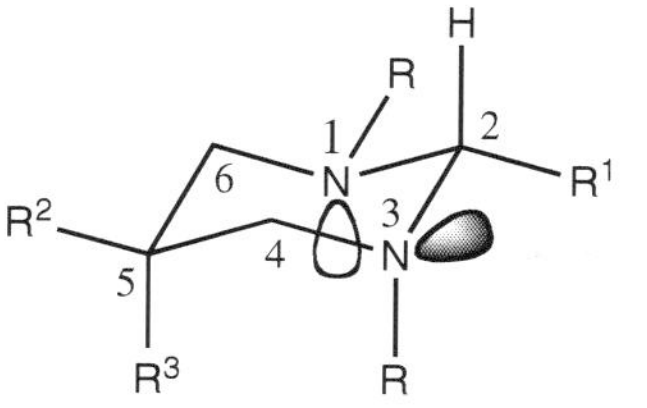

3 Anomeric Effect

49, R = R^1 = CH$_3$; R^2 = R^3 = H
50, R = R^1 = R^2 = CH$_3$; R^3 = H

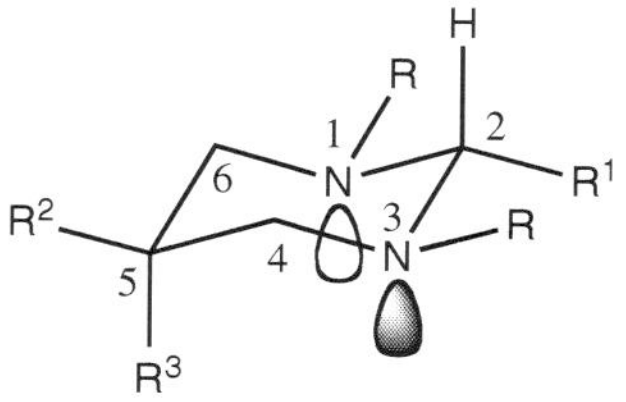

51, R = R^1 = R^2 = R^3 = CH$_3$

Fig. 3.21

shown in Fig. 3.21. Introduction of an equatorial C5 methyl group in *N,N*-trimethyl-1,3-diazane *49* does not significantly increase the conformational energy, whereas the introduction of the second axial methyl group in the *N,N*,2,5-tetramethyl-1,3-diazane encounters a very severe nonbonding steric interaction with one of the two methyl groups on the nitrogen atom suggesting that in *N,N*,2-trimethyl-1,3-diazane one methyl group must be oriented axially and the other equatorially, in spite of the 1,3-*syn*–diaxial interaction of the axially oriented *N*-methyl group and the axial C5 hydrogen. This suggests that the 1,3-*syn*–diaxial interaction of two nitrogen unshared electron pairs must exist and that is larger than the 1,3-*syn*–diaxial interaction of the axially oriented *N*-methyl group and the axial C5 hydrogen. It should be noted that the conformer *49* has also one *endo* N-anomeric effect that additionally stabilizes the axial orientation of the C3 methyl group.

Booth and Lemieux [54] have studied the conformations of six-membered perhydro-1,3-oxazoline and 1,3-diazine compounds (Fig. 3.22) with NMR and found that the conformer which avoids placing the unshared electron pair orbitals of both heteroatoms in axial orientation is more stable. This conclusion was drawn in view of the magnitude of the coupling constant between the *N*-hydrogens and the vicinal hydrogen in the axial orientation.

For historical reasons Lemieux proposed the term *generalized anomeric effect* for the general preference for the *gauche* conformation about the carbon-hetero atom bond in systems R-X-C-Y which are the results of the same kind of interactions as were proposed for explaining the anomeric effect but present in noncarbohydrate structures. This proposal has now been universally adopted.

Many cases are known where substituents on six-membered rings prefer an axial orientation [55], and not all of these are the consequence of the anomeric effect: for example, the 2-halocyclohexanone system [56] where the axial preference decreases in the order Br > Cl > F and can be explained as a combination of steric effect and dipole–dipole interactions (Fig. 3.23) and in 2-alkoxycyclohexanones [57] where the effect is comparable in magnitude to that caused by anomeric effect in 2-alkoxytetrahydropyrans (Table 3.5).

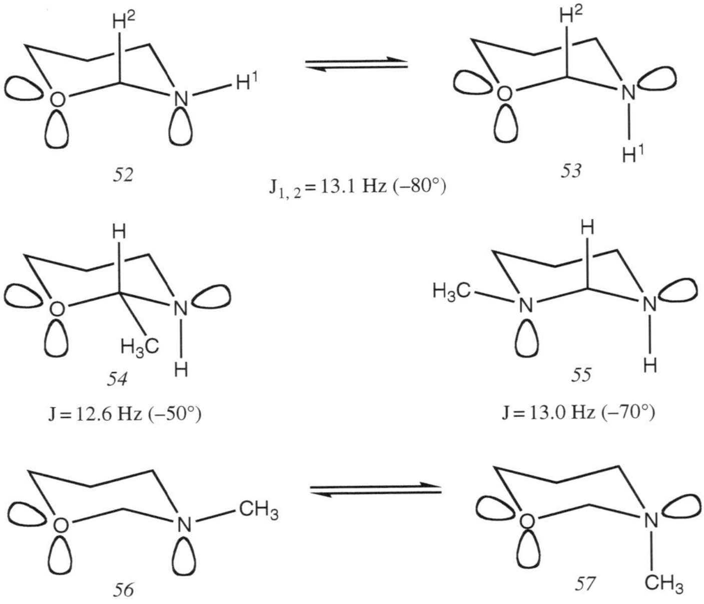

Fig. 3.22

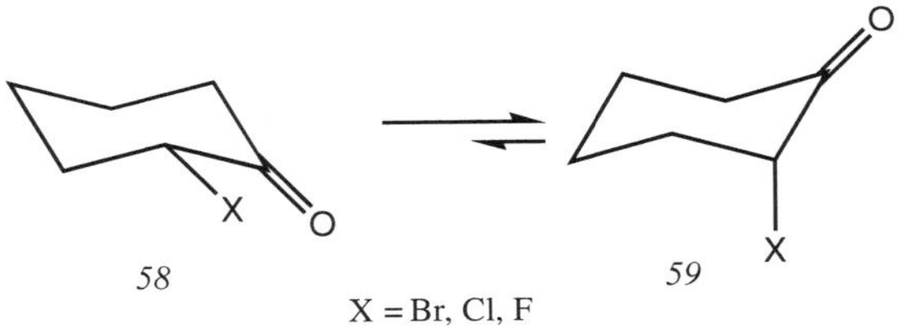

52   $J_{1,2} = 13.1$ Hz (−80°)   53

54   $J = 12.6$ Hz (−50°)

55   $J = 13.0$ Hz (−70°)

56   57

Fig. 3.23

58   X = Br, Cl, F   59

Reverse Anomeric Effect

In 1965, Lemieux and Morgan [58] studied the conformation of *N*-(tetra-*O*-acetyl-α-D-glucopyranosyl)-4-methyl-pyridinium bromide *68* by NMR spectroscopy and found that the 4-methyl-pyridinium group is equatorially oriented and have suggested that *68* exists in $^1C_4$ conformation *68e* (Fig. 3.24), forcing thus all other substituents to assume the axial orientation despite the presence of one O//O 1,3-*syn*–axial interaction between the C2 and the C4 acetyl groups and one O//C 1,3-*syn*–axial interaction between the C3 acetyl and the C5 acetoxymethyl group amounting to 1.5 + 2.5 = 4.0 kcal/mol.

3   Anomeric Effect

Table 3.5 Axial preference for methoxyl groups adjacent to sp^2-hybridized ring carbon

Compound	Equilibrium	% of axial isomer in CCl$_4$
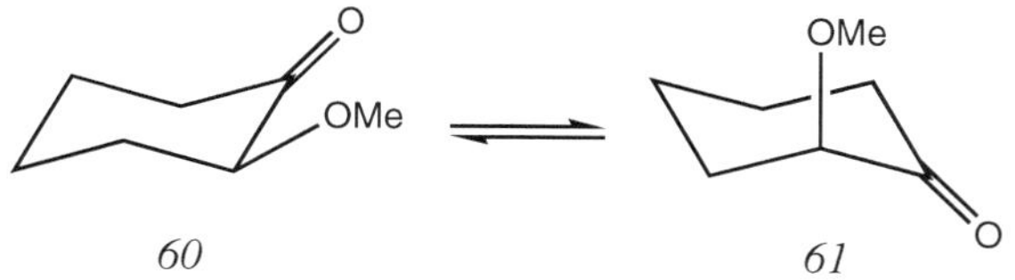 60	⇌ 61	63
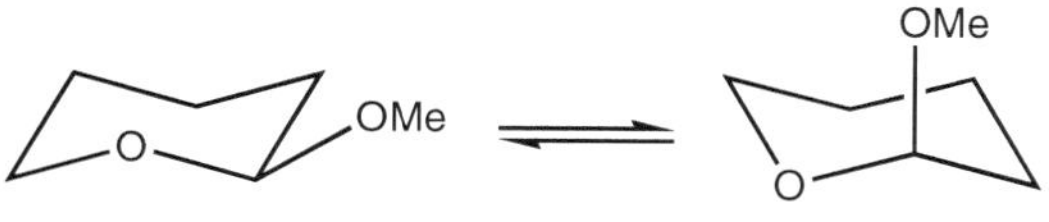 62	⇌ 63	69
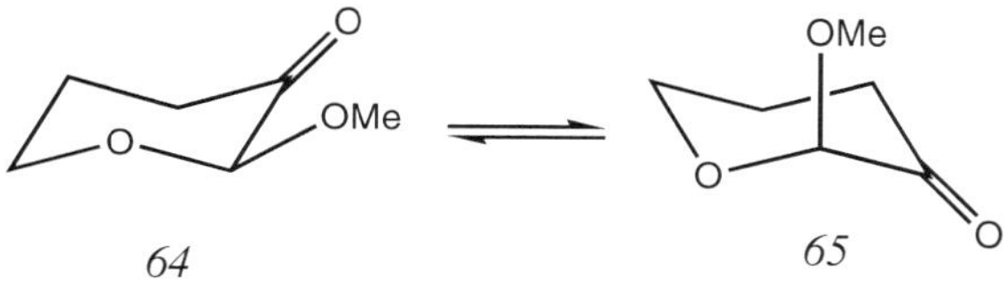 64	⇌ 65	100
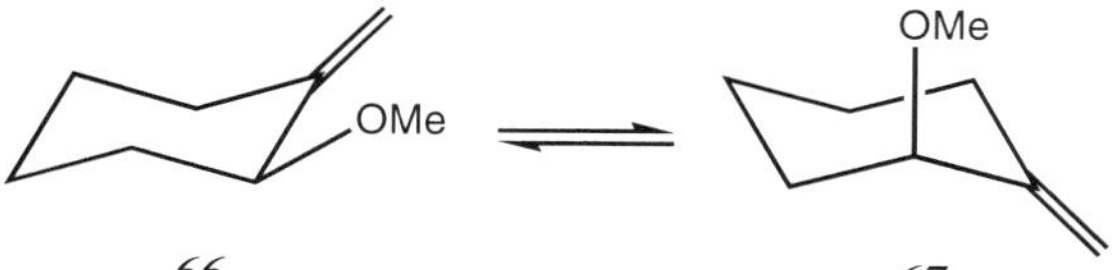 66	⇌ 67	78

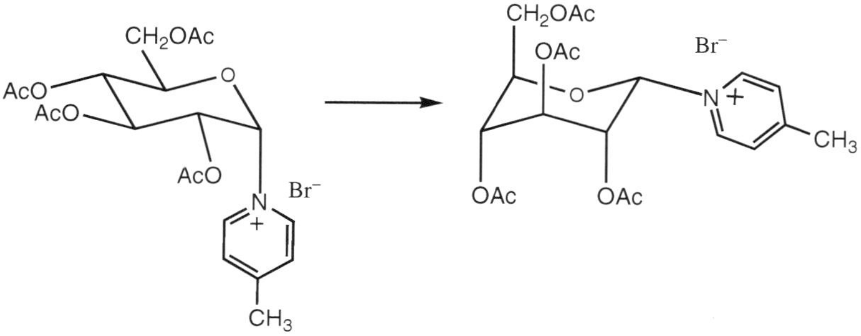

68a, aglycon axial (in the $^4C_1$ conformation)   68e, aglycon equatorial (in the $^1C_4$ orientation)

Fig. 3.24

James has, however, found [59] that the compound 68 in crystalline state does not exist in the $^1C_4$ conformation (68e in Fig. 3.24) but in the $B_{2,5}$ conformation 69 as shown in Fig. 3.25 with the methyl-pyridinium group quasi-equatorially oriented.

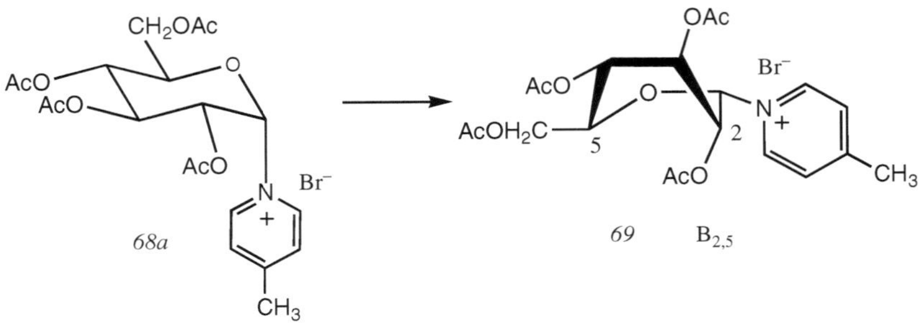

Fig. 3.25

Since both NMR and crystallographic studies showed that the conformations of aminoglycosides with anomeric nitrogen in axial orientation are strongly disfavored, particularly in cases where the nitrogen carries a positive charge, Lemieux concluded [20, 60, 61] that there must exist a powerful driving force for the pyridinium group to adopt an equatorial orientation.

Lemieux named as *reverse anomeric effect* (RAE) this driving force for the electropositive aglycon in hexopyranosides to assume the equatorial orientation. Since the reverse anomeric effect could be either the result of steric interactions due to the bulkiness of pyridinium group or due to electronic interactions stemming from the presence of positively charged nitrogen, or both, Lemieux and Saluja [20, 62] suggested that the existence of (polar) *reverse anomeric effect* can be established only if a clear distinction between the steric and polar effects can be made.

Soon thereafter two groups (Lemieux et al. [20, 62] and Paulsen et al. [63]) independently concluded that the glycosyl imidazoles (Fig. 3.26) would be more suitable substrates for these studies than pyridinium glycosides since the protonation of an imidazole ring is not expected to significantly change its size and therefore any conformational change due to protonation could be attributed to polar effect (i.e., to the reverse anomeric effect). While this argument seems likely, it is still uncertain to what extent the association of a counterion with the positively charged imidazolium

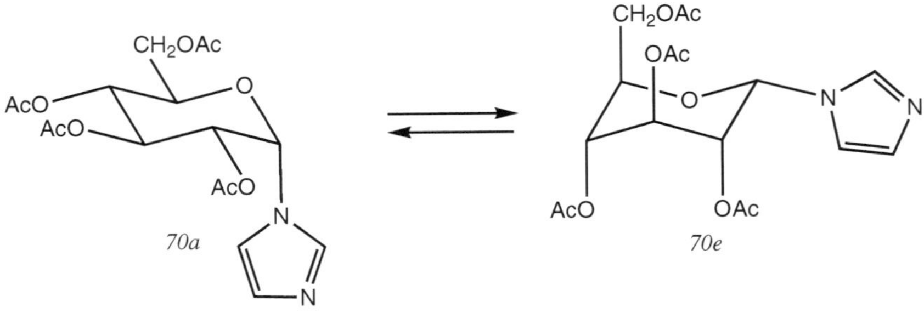

Fig. 3.26

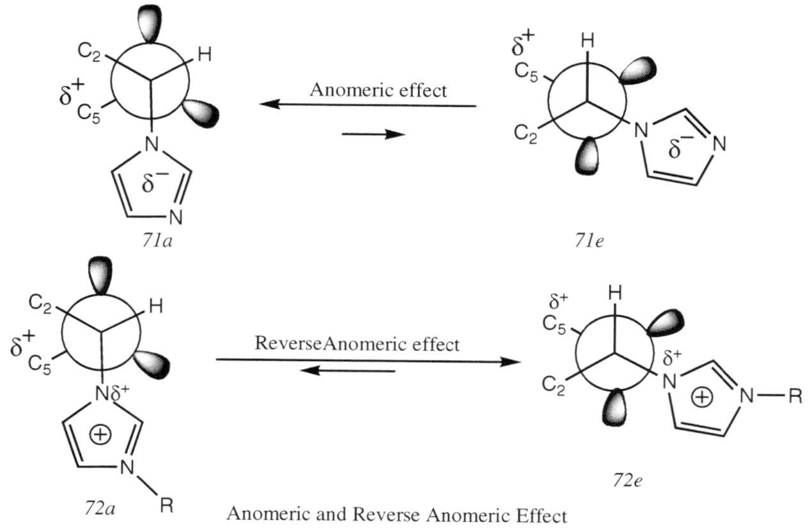

Fig. 3.27

ion affects the **A**-value of the imidazolium group, as well as what effect has the solvation of the imidazolium salt on the A-value of the imidazolium group.

Lemieux and Saluja [20, 62] studied the protonation of imidazole ring of N-(2,3,4,6-tetra-O-acetyl-α-D-glucopyranosyl)imidazole 70 (Fig. 3.26) in deuterochloroform and found that the addition of equimolar amount of a weak acid (acetic acid) produced a much smaller effect on the NMR spectrum than the addition of equimolar amount of a strong acid such as trifluoroacetic acid. The addition of a strong acid had an effect upon decreasing the magnitudes of $J_{2,3}$, $J_{3,4}$, and $J_{4,5}$ coupling constants that is nearly equivalent to the methylation of the imidazole group.

The distribution of electrical charge is more favorable with the imidazole group in the axial orientation when the nitrogen attached to the anomeric carbon carries a partial negative charge and this is the anomeric effect (Fig. 3.27). However, the distribution of electrical charge is more effective in the anomer with the imidazole group in equatorial orientation when the imidazole ring has a positive charge that did acquire either through protonation or alkylation, and this is reverse anomeric effect.

Deslongchamps and Grein [64, 65] suggested that the equatorial orientation of aglycon is favored because of electronic stabilization via dipolar interaction of the positively charged aglycon (N^+) with the two unshared pairs of electrons on the hexopyranose ring oxygen as shown in Fig. 3.28. Apparently, the lp-N^+ electrostatic attraction exceeds the desire of lp delocalization, corresponding to the *endo-*anomeric effect (Fig. 3.28).

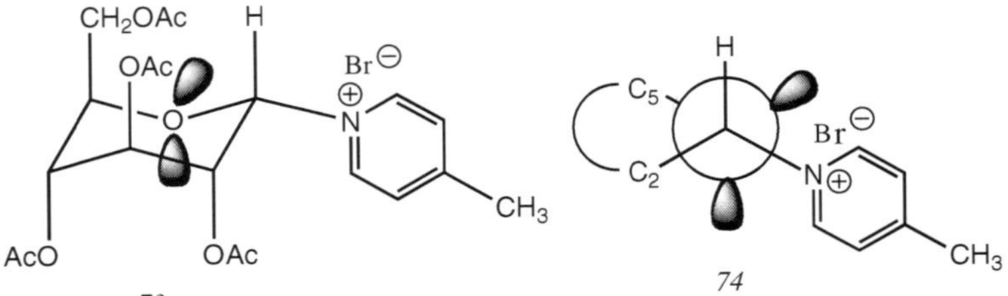

Fig. 3.28

Figure 3.28 illustrates the Deslongchamps and Grein [64, 65] explanation of the reversed anomeric effect. The imidazole ring is an electron-rich group due to the presence of two nonbonding p-electron pairs on nitrogen and therefore tends to adopt, due to anomeric effect, the axial orientation. However, the imidazole ring on protonation becomes positivelycharged and consequently adopts the equatorial orientation because in this conformation the positively charged imidazole ring is in gauche orientation relative to two nonbonding p-electrons on the ring oxygen that stabilizes the positive charge of imidazole.

The strongest support for the existence of reverse anomeric effect (RAE) comes from ^{1}H NMR study of conformational equilibrium of N-(2,3,4-tri-O-acetyl-α-D-xylopyranosyl)imidazole in CDCl$_3$ solution in the absence and presence of trifluoroacetic acid (TFA) conducted by Paulsen et al. [63]. It was found that in the absence of acid the equilibrium mixture contained 65% of the $^1C_4$ conformer 76 with imidazole aglycon equatorially oriented and 35% of the $^4C_1$ conformer 75 with imidazole aglycon axially oriented (Fig. 3.29). In the presence of acid the proportion of the $^1C_4$ conformer 77 with the imidazole aglycon equatorially oriented increased to more than 95%. This difference corresponds to free-energy change >1.4 kcal/mol. The authors [63] attributed the shift of conformational equilibrium to the presence of positive charge on imidazolium ring due to protonation assuming that N-protonation did not significantly change the size of the imidazolyl group. Thus they completely excluded steric effects as the possible cause for the observed conformational change.

Finch and Nagpurkar [66] studied the population of equatorial conformer in an equilibrium mixture of N-(α-D-glycopyranosyl)imidazole of D-glucose,

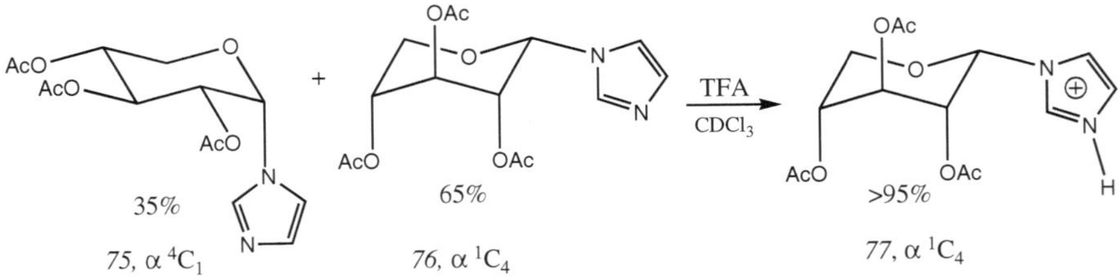

Fig. 3.29

D-mannose, and D-galactose in D_2O, of N-(α-D-glycopyranosyl)imidazole of D-glucose, D-mannose, and D-galactose tetraacetate in $CDCl_3$, and of N-(α-D-xylopyranosyl)imidazole triacetate in $CDCl_3$ in the absence and presence of acid and found that both steric factors and polar factors (reverse anomeric effect) are likely to be involved in determining the relative percentages of the two conformations at conformational equilibrium. The obtained results could in large part be accounted for by steric factors, but the operation of additional polar factors was also likely (Table 3.6).

The concept of reverse anomeric effect (RAE) has been subject to much controversy and skepticism, because the positively charged anomeric nitrogen ought to lower the energy of the σ_{C-X}^* orbital and enhance the stabilization of the axial conformer, not to destabilize it.

One of the first reports [67] that challenged the existence of RAE was the stereospecific formation of α-D-glucopyranosylacetonitrilium ion 79 when α- and β-anomers of pent-4-enyl D-glucopyranoside 78 reacted with N-bromosuccinimide in dry acetonitrile (68% from β-anomer and 64% from α-anomer). The anomeric configuration of 79 was determined by trapping the acetonitrilium ion 79, in situ, with 2-chlorobenzoic acid and by subsequent conversion of the obtained α-imide 81 with sodium methoxide to α-2-chlorobenzamide 82 (Fig. 3.30). The exclusive formation of axially oriented α-acetonitrilium ion 79 is clearly in contrast to what would be predicted by the reverse anomeric effect.

Since both RAE and steric repulsions favor the equatorial conformer, it was essential to quantitatively assess steric factors and to determine whether the preference of pyridinium and imidazolium groups to adopt the equatorial orientation is predominantly due to steric interactions of these two groups. Since pyridinium and imidazolium groups were too bulky for this assessment to be made reliably,

Table 3.6 Population of equatorial conformer in equilibrium mixture of N-(α-D-glycopyranosyl) imidazoles of D-glucose, D-mannose, D-galactose, and D-xylose and their tetra and triacetyl derivatives, respectively

Glycon residue	Conformation	X	Solvent	Average % of $^1C_4$ conformer
Gluco	$^4C_1$	N	D_2O	0
Gluco	$^4C_1$	NH^+	D_2O + TFA	0
Manno	$^4C_1 + {}^1C_4$	N	D_2O	30.4
Manno	$^4C_1 + {}^1C_4$	NH^+	D_2O +TFA	31.3
Galacto	$^4C_1$	N	D_2O	0
Galacto	$^4C_1$	NH^+	D_2O + TFA	0
Ac_4Gluco	$^4C_1$	N	$CDCl_3$	0
Ac_4Gluco	$^4C_1 + {}^1C_4$	NH^+	$CDCl_3$ + TFA	27.4
Ac_4Manno	$^4C_1 + {}^1C_4$	N	CDC_3	Not given
Ac_4Manno	$^4C_1 + {}^1C_4$	NH^+	$CDCl_3$ + TFA	67
Ac_4Manno	$^4C_1 + {}^1C_4$	N	$(CD_3)_2CO$	51.1
Ac_4Manno	$^4C_1 + {}^1C_4$	NH^+	$(CD_3)_2CO$+TFA	71.8

Fig. 3.30

a protonable cyclohexyl substituent whose steric size is known in both protonated and unprotonated forms seemed to be more suitable for probing RAE. One such substituent is NH$_2$ (Fig. 3.31). The conformational energy ($-\Delta G^0$ or A-value) [21] for NH$_3^+$ was found to be larger (2.15 kcal/mol in D$_2$O) than the A-value for NH$_2$ (1.7 kcal/mol in D$_2$O and 1.65 in aprotic solvents, e.g., cyclohexane) [68–70]. The increase of A-value on protonation is a measure of the increase in size of the protonated substituent, relative to the unprotonated substituent. This extra bulk is due to the additional proton itself and also to the additional solvent molecules attached to the positive charge needed to stabilize it. (The increase in protic solvent is due to hydrogen bonding, which clusters solvent molecules around the polar group [71, 72].) Since the C–O bond is shorter than C–C bond, steric repulsions in tetrahydropyran system with axially oriented 2-amino group are greater than in cyclohexane system with axially oriented amino group and should be corrected to 2–2.5 kcal/mol for aprotic solvents and 2.4–2.9 kcal/mol for protic solvents.

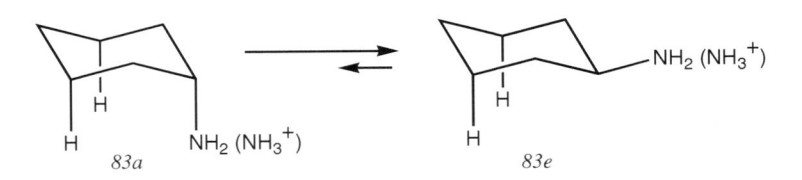

Fig. 3.31

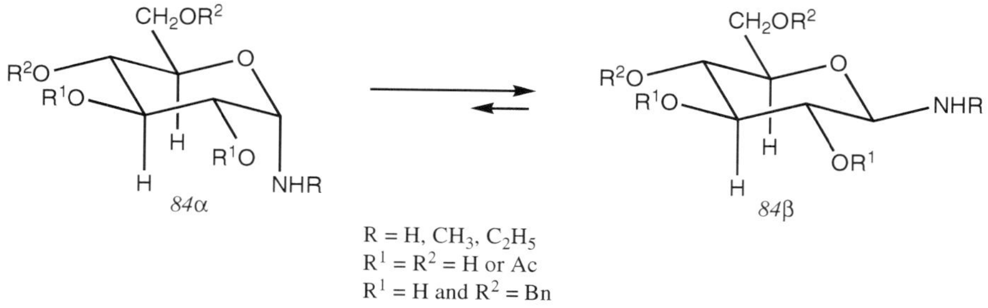

Fig. 3.32

R = H, CH$_3$, C$_2$H$_5$
R^1 = R^2 = H or Ac
R^1 = H and R^2 = Bn

Unlike previous experimental investigations of the reverse anomeric effect that involved the study of conformational equilibrium between the unprotonated and protonated aminoglycosides, Perrin and Armstrong [73] carried out a ^{1}H NMR study of composition of equilibrium mixture obtained by acid-catalyzed anomerization of glycosylamines of a wide variety of glycopyranosylamine derivatives along with their conjugate acids (Fig. 3.32).

This interconversion is known to proceed in four steps [74]: (1) the reversible protonation of the ring oxygen 84α, (2) the pyranoid ring opening to the imminium ion intermediate 85, (3) the rotation about the C1–C2 bond 85→86, and (4) reclosure of the pyranoid ring 86→84β (Fig.3.33). The greatest experimental difficulties encountered in this work were the sensitivity of glycosylamines to hydrolysis and

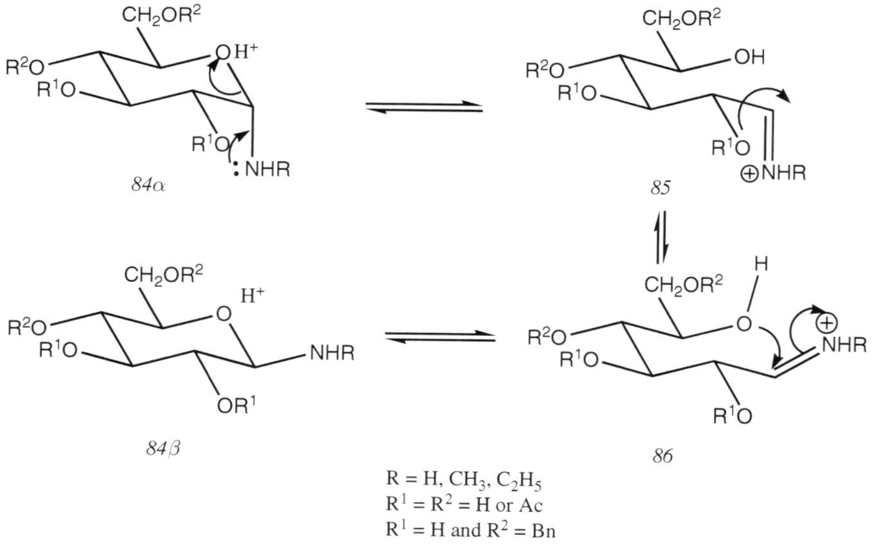

R = H, CH$_3$, C$_2$H$_5$
R^1 = R^2 = H or Ac
R^1 = H and R^2 = Bn

Fig. 3.33

the problem of assignment of ¹H NMR signals to the axial stereoisomer which was present only in low concentration.

The *84α/84β* ratio was measured by integration of corresponding ¹H NMR signals for the anomeric protons across a range of solvents, for both the unprotonated and protonated glycosylamines, and the obtained results are given in Table 3.7.

Table 3.7 Average percentage of α-anomer, free-energy change $\Delta G^0 (\beta \to \alpha)$ (kcal/mol), $A(NH_2)$ or $A(NH_3^+)$ (kcal/mol) in glycopyranosylamines and glycopyranosylammonium ions

Amine	α-Anomer (%)	$\Delta G^0\ \beta \to \alpha$	A Cyclohexane	A THP	α–Anomer (%) estimated
–NH₂	10	1.6 ± 0.4	1.6 or 1.3	2.5 or 2.0	–
–NHR	13	1.5 ± 0.3	1.6 or 1.3	2.5 or 2.0	–
:H⁺, aq.	3.5	2.0 ± 0.1	1.9	2.9	0.8
:H⁺, nonaq.	7.5	1.5 ± 0.1	Ca. 1.6	Ca. 2.4	1.7

The most important result of these studies is that the axial anomer *82α* is present in appreciable amounts even in acid solution; it is present in smaller percentage in aqueous solution perhaps because in water, the $-NH_3^+$ or $-NH_2R^+$ group is bulkier due to solvation. The fact that the $\Delta G^0(\beta \to \alpha)$ values (the free-energy change for the conversion of equatorial *84β* to the axial *84α* isomer) are considerably lower than A-values for $-NH_3^+$ or $-NH_2R^+$ in THP (Table 3.7) indicates that the preference of the $-NH_3^+$ for equatorial orientation can be accounted for chiefly by the steric effect. These results also suggest the existence of a weak anomeric effect, but not of a reverse anomeric effect.

The reverse anomeric effect can also be determined from a difference in anomerization free-energy changes between the protonated and unprotonated glycosylamines, as shown in Equation (3):

$$\Delta \Delta G^0(N \to N^+) = \Delta G^0(\beta \to \alpha)(NH^+) - \Delta G^0(\beta \to \alpha)(N) \qquad (3)$$

$\Delta \Delta G^0$ is the extent to which the N-protonation increases the preference of amino substituent for the equatorial orientation. Across all the glycosylamines examined the average $\Delta \Delta G^0(N \to N^+)$ is found to be 0.1 ± 0.1 kcal/mol [73] which is not significantly different from zero. Furthermore this value is definitely smaller than $A(NH_3^+) - A(NH_2)$ which is what would be expected from the increase in steric bulk. Even though NH_3^+ is certainly bulkier than NH_2, the proportion of axial isomer *83α* does not decrease on N-protonation. Therefore, Perrin and Armstrong [73] concluded that there is probably no reverse anomeric effect present with any cationic nitrogen substituent.

Several computational studies on reverse anomeric effect have been published [64, 65, 75–77]. Thus, conformational equilibrium of 2,3,4-tri-*O*-acetyl-D-xylopyranosylimidazol *75⇌76* (Fig. 3.34) was subjected to ab initio calculation. In order to simplify the calculations, the axial and equatorial conformers of unprotonated and protonated 2,3,4-tri-*O*-acetyl-D-xylopyranosylimidazoles (*75*, *76*, *77*, and *87*, respectively) were substituted with unprotonated and protonated model

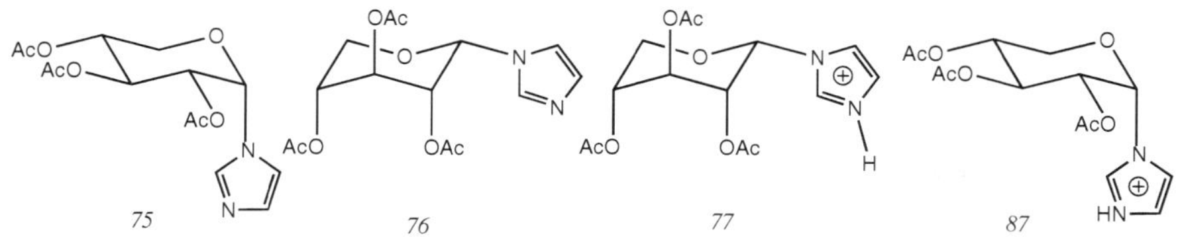

Fig. 3.34

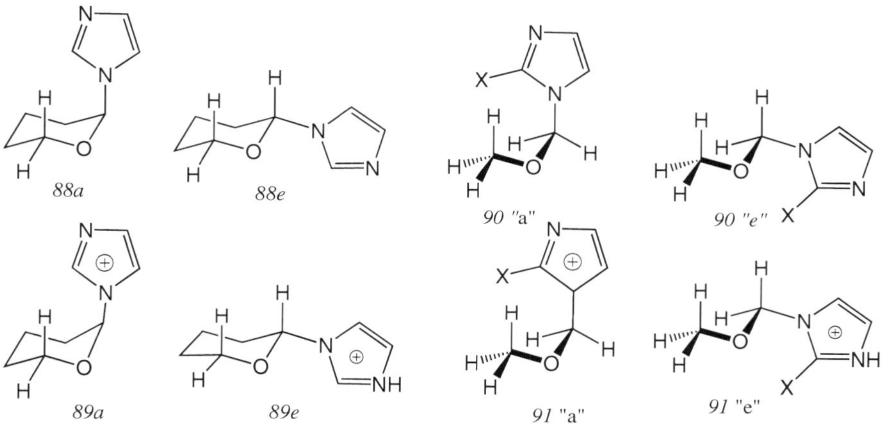

Fig. 3.35

substrates (*88a*, *88e*, and *89a* and *89e*) (Fig. 3.35) whose conformational energies were then calculated. Also the conformational energies of truncated acyclic models of *88a*, *88e*, *89a*, and *89e* (the structures *90*"a", *90*"e", *91*"a", and *91*"e" with X = H, F, and CH_3) were calculated [78].

From the results of calculations it was concluded that the dominant contributions to the conformational equilibrium of *N*-pyranosylimidazoles were stabilizing anomeric hyperconjugation and destabilizing steric 1,3-interactions in the axial unprotonated conformer (*88a*). Both effects increase on N-protonation (*89a*). It was also concluded that the fine balance between these opposing contributions would allow small intramolecular electrostatic interactions to control the position of equilibrium. Stabilizing electrostatic interactions in N-protonated equatorial conformers were identified and found to be associated with ImH2–O (ring) hydrogen bonding (Type 1) and dipole–dipole electrostatic stabilization between the nonbonding electrons on ring oxygen and the cationic imidazolium dipole (Type 2) (see Deslongchamps and Grein's proposal [64, 65]). An equatorial shift on N-protonation of 0.4–2.4 kcal/mol was predicted using models *89a* and *89e*. Since Perrin [73] suggested that only 0.024–0.089 kcal/mol should result from the steric effect related to imidazole N-protonation, it was concluded that reverse anomeric effect for the *N*-(xylopyranosyl)imidazoles is approximately 0.8–1.4 kcal/mol.

Since the anomeric effect is known to be sensitive to solvent polarity Vaino et al. [79] have reexamined the conformational equilibrium of 2,3,4-tri-*O*-acetyl-α-D-xylopyranosylimidazole 75 and 2,3,4-tri-*O*-acetyl-α-D-xylopyranosyl-2-methylimidazole 92 in the presence of trifluoroacetic acid (TFA) because protonation of imidazole by TFA increases the ionic strength of solution. The ^{1}N NMR titration of the glycosides 75 and 92 with varying amounts of TFA and/or tetra-*N*-butyl-ammonium bromide (TBAB) were undertaken in order to account for the effects of solvent ionic strength change upon equilibrium (Fig. 3.36).

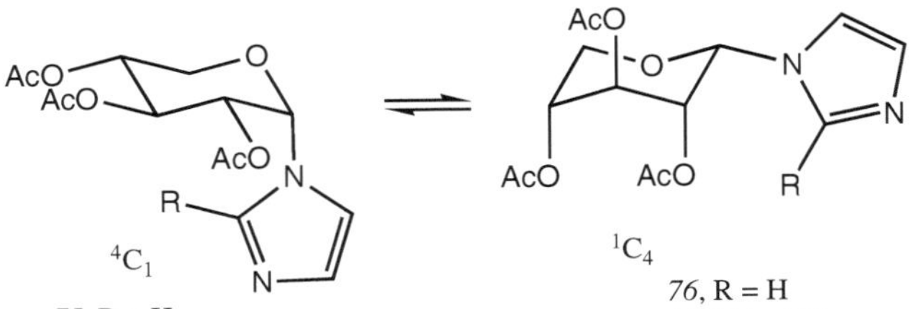

$^4C_1$
75, R = H
92, R = CH$_3$

$^1C_4$
76, R = H
93, RT = CH$_3$

Fig. 3.36

From the results obtained it was concluded that the large equatorial shifts observed for 75 and 92 on N-protonation are not the results of solvent and ionic strength effects. Interestingly the effect of increasing solution ionic strength with TBAB produces a small axial shift for 75. The authors suggest that RAE does exist and that it is the result of stabilizing intramolecular electrostatic contributions to the $^1C_4$ conformer on N-protonation [64, 78]. The size of the effect in the two xylopyranosyl systems studied is quantified as 0.8–1.4 kcal/mol. Since contributions from this electrostatic RAE may be small, there may be other contributions to conformational energy and will be diminished on transferring to solvents more polar than chloroform.

Fabian et al. [80] used an ^{1}H NMR titration method to measure with high precision the shift of anomeric equilibrium on protonation of *N*-(D-glucopyranosyl)imidazole 94 and its tetraacetyl derivative 95 (Fig. 3.37) and found that $\Delta\Delta G^0_{\beta\to\alpha} = \Delta G_{\text{N-ImidazolylH+}} - \Delta G_{\text{N-Imidazolyl}} = -0.018$ to -0.368 kcal/mol.

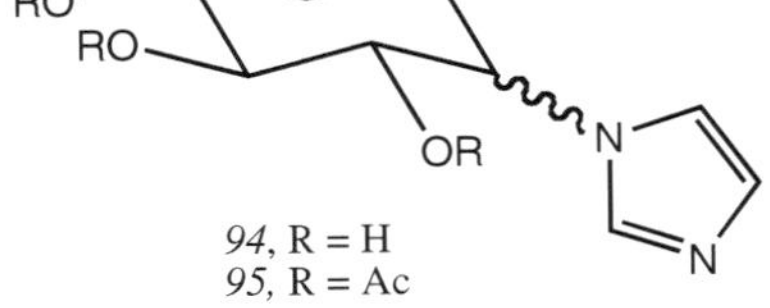

94, R = H
95, R = Ac

Fig. 3.37

This result means that the protonated imidazolyl group has a small but significantly greater preference for the axial position than does the unprotonated group. This is exactly opposite to what is expected from the existence of the reverse anomeric effect. This led authors to conclude that RAE does not exist [2, 5, 81].

Additional experimental evidence is sparse [82–86]. The geometric changes are consistent with an enhanced anomeric effect, not a reverse one [87, 88]. Molecular orbital calculations are not conclusive because it is difficult to separate the RAE from steric effects and hydrogen bonding, which also favor the equatorial conformer [76, 78, 89–91].

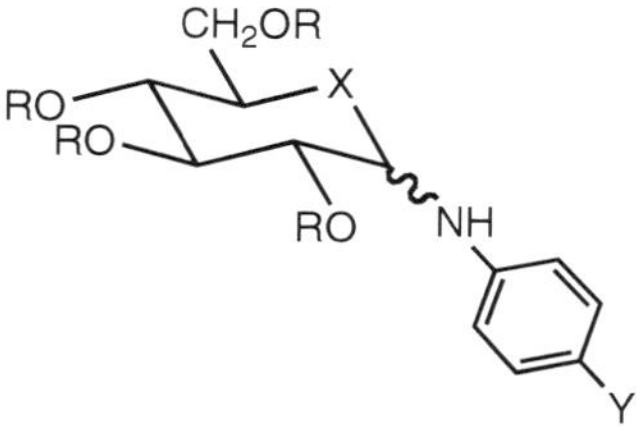

96, R = Ac; X = S; Y = OCH$_3$
97, R = Ac; X = S; Y = H
98, R = Ac; X = S; Y = CF$_3$
99, R = Ac; X = S; Y = NO$_2$
100, R = H; X = S; Y = OCH$_3$
101, R = H; X = S; Y = H
102, R = H; X = S; Y = CF$_3$
103, R = H; X = S; Y = NO$_2$

104, R = Ac; X = O; Y = OCH$_3$
105, R = Ac; X = O; Y = H
106, R = Ac; X = O; Y = NO$_2$

Fig. 3.38

The question of the existence of a generalized RAE was addressed by systematic examination of substituent and solvent effects on the configurational equilibria of N-aryl-5-thioglucopyranosylamines *96–103* and N-arylglucopyranosylamines *104–106* (Fig. 3.38) and the corresponding protonated species [92] (Fig. 3.38).

The equilibrium populations of the 5-thio compounds *96–99* and their protonated derivatives were determined by ^{1}H NMR spectroscopy at 294 K. Equilibration of neutral species *96–99* (Fig. 3.39) was achieved by the HgCl$_2$ catalysis of the

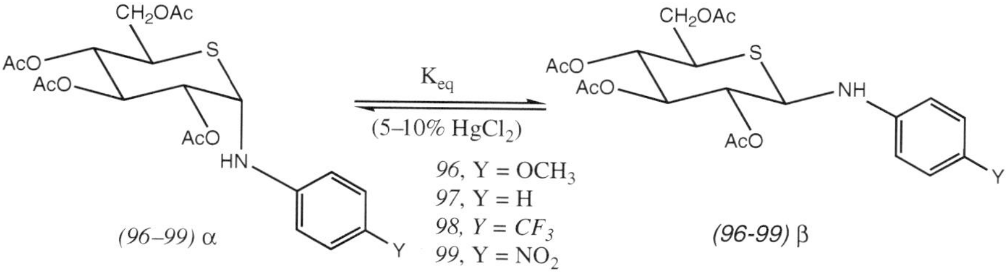

96, Y = OCH$_3$
97, Y = H
98, Y = CF$_3$
99, Y = NO$_2$

Fig. 3.39

individual isomers in polar solvents CD_3OD, CD_3NO_2, and $(CD_3)_2CO$ only, because of the limited solubility of β-anomers, to ensure that the true equilibrium had been reached.

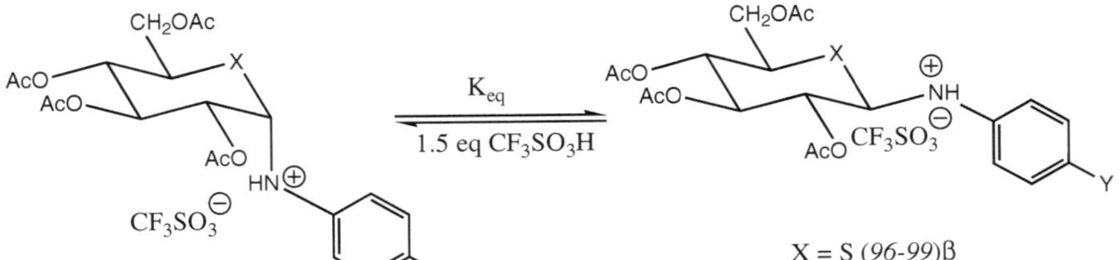

Fig. 3.40

The corresponding equilibrations of the protonated species *104–106* (Fig. 3.40) were studied in the presence of 1.5 equivalents of triflic acid, in polar and nonpolar solvents. The addition of 1.5 equivalents of triflic acid would ensure the complete protonation of amines since the pK_a of triflic acid is –5.9 [93] while the pKas of the isolated aglycons are 5.31 for *p*-anisidine (Y = CH_3O), 4.60 for aniline (Y = H), 2.45 for *p*-trifluoromethyl-aniline (Y = CF_3), and 1.00 for the weakest base, *p*-nitroaniline (Y = NO_2).

The equilibration of the oxygen analogs *104–106* (Fig. 3.40) was achieved by addition of catalytic amounts of triflic acid to the solution of individual anomers and was performed at 230 K in CD_2Cl_2 and CD_3OD. In this series the choice of solvents was restricted because of the instability of compounds or line broadening effects in the spectra that did not permit unambiguous assignment of signals or their accurate integration.

The conclusion based on the obtained results is that there is no evidence to support the existence of generalized reverse anomeric effect in neutral or protonated *N*-aryl-5-thioglucopyranosylamines and *N*-arylglucopyranosylamines.

For the neutral compounds, the anomeric effect ranges from 0.85 kcal/mol in *95* to 1.54 kcal/mol in *106*. The compounds *96–99* and *104–106* show an enhanced anomeric effect upon protonation. The anomeric effect in the protonated derivatives ranges from 1.73 kcal/mol in *98* to 2.57 kcal/mol in *105*. The values of K_{eq} in protonated *96–99* increase in the order OMe < H < CF_3 < NO_2, in agreement with the dominance of steric effects (due to counterion) over the *endo*-anomeric effect. The values of K_{eq} in protonated *104–106* show the trend OMe < H < NO_2 that is explained by the balance of the anomeric effect and steric effects in the individual compounds.

Anomeric Effect in Systems O–C–N

The NMR studies at low temperature of unsubstituted and C2 substituted 1,3-diazines such as *107* (R = H) and *109* (R_1 = CH_3) [94] strongly suggest that

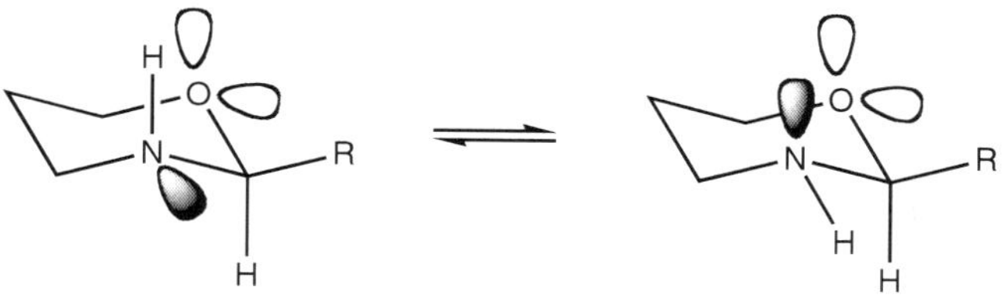

```
         H
         |-O
   ⟨N⟩   C—R
         |
         H
   107, R = H
   109, R = CH₃
```
```
         O
         |-O
   ⟨N⟩   C—R
    |    |
    H    H
   108, R = H
   110, R = CH₃
```

Fig. 3.41

the conformers *107* and *109* are the major components of the equilibrium mixtures *107* ⇌ *108* and *109* ⇌ *110* (Fig. 3.41).

These results were explained in the following way. Two anomeric effects are possible for conformers *107* and *109* (one anomeric effect where the equatorial nitrogen lone pair of electrons mixes with the antibonding C–O orbital and the other where the equatorial oxygen lone pair of electrons mixes with the antibonding C–N orbital). In conformers *108* and *110* there is only one anomeric effect possible and that is where the equatorial oxygen lone pair of electrons mixes with the C–N antibonding orbital. It should be, however, noted that because the nitrogen is less electronegative than oxygen, it is a better electron donor than oxygen and σ*C–N a weaker acceptor of electrons than σ*C–O bond. Consequently the O- and N-anomeric effects are not of equal energy. In addition, in the conformers *108* and *110* ring nitrogen and oxygen atoms would have their axially oriented lone pairs of electrons in 1,3-syn-axial orientation, which will further destabilize this conformation due to the generalized anomeric effect.

The NMR study of *N*-methyl tetrahydro-1,3-oxazine such as *111* suggested that the conformer *111* with the *N*-methyl group in the axial orientation is more stable than conformer *112* wherein the methyl group is equatorially oriented for the same reasons given above [95] (Fig. 3.42).

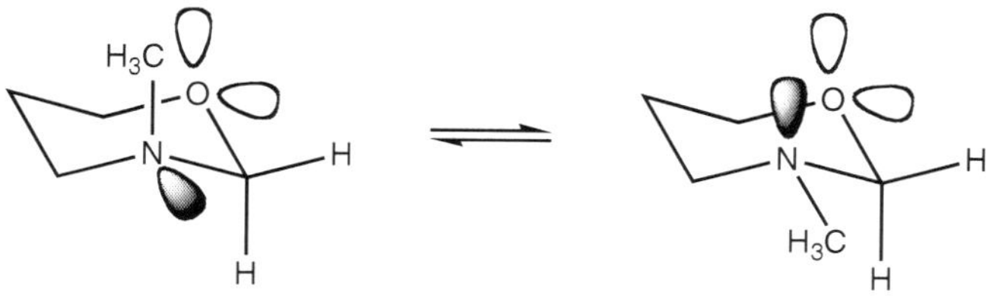

111 ⇌ *112*

Fig. 3.42

Similarly it was found that the conformer *113* is present in appreciable concentration in conformational equilibrium *113* ⇌ *114* (Fig. 3.43) again for the same reasons given above [95, 96].

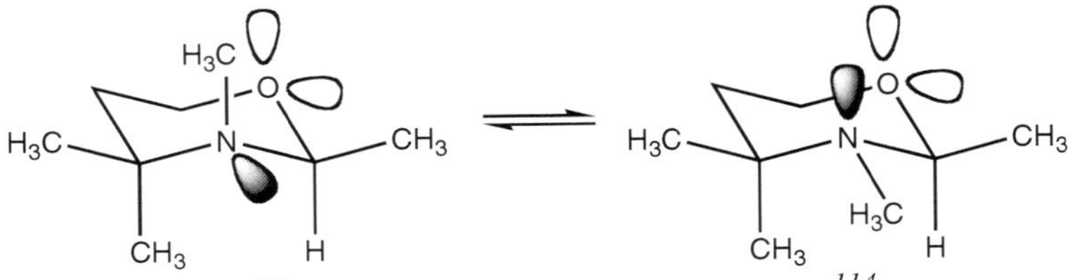

Fig. 3.43

Kirby and Wothers [98] studied conformational equilibrium of amide acetal *115e* ⇌ *115a* to determine the magnitude of the anomeric effect of dimethylamino group, by comparing ΔG^0 for ring inversion with that of its cyclohexyl analog *116e* ⇌ *116a* (Fig. 3.44).

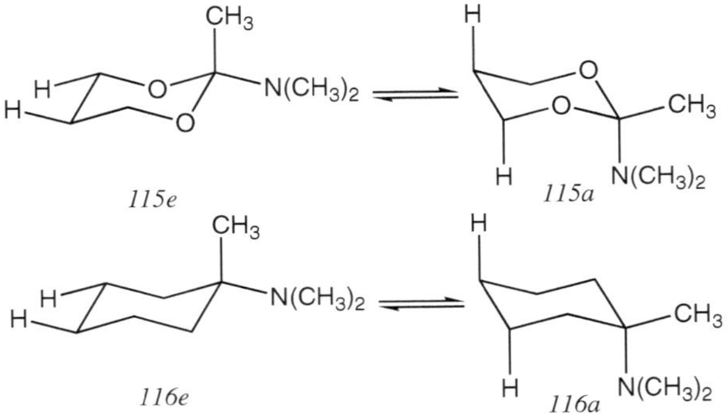

Fig. 3.44

On cooling a sample of amide acetal *115* to 140 K (in 70:30 CBr_2F_2–CD_2Cl_2), not a trace of conformer with the NMe_2 group equatorial (*115e*) was detected. It was concluded that the amide acetal exists as a single conformer *115a* and this conclusion was confirmed by NOE experiments.

If steric repulsion experienced by the methyl and dimethylamino groups increases by the same factor on going from the cyclohexane to dioxane, then the equilibrium constants for the ring inversion of the two compounds should be the same. At 185 K, the cyclohexane conformation in which the NMe_2 is axially oriented is more favorable by $\Delta G^0 = 1.59$ kJ/mol, whereas in dioxane analog the axial orientation of NMe_2 group is favored by $\Delta G^0 \geq 4.5$ kJ/mol. Thus, the conformation *115a* is at least 3 kJ/mol more stable at 185 K than what would be expected from steric factors alone.

This result can be used to shed some light on the *exo/endo*-anomeric effect in the 2-aminotetrahydropyrans studied extensively by Booth et al. [99, 100]. Because nitrogen is less electronegative than oxygen it is a better donor, and the antibonding σ^*C–N orbital a weaker acceptor of electrons than the antibonding σ^*C–O orbital.

3 Anomeric Effect

Consequently it could be expected that 2-methylaminotetrahydropyran exhibits a stronger *exo*-anomeric effect (n_N-σ^*_{C-O}) than *endo*-anomeric effect (n_O-σ^*_{C-N}). Thus, as shown by Booth et al. (loc. cit.) 2-methylaminotetrahydropyran prefers the equatorial conformation (Fig. 3.45).

117a (1)   117a (2)   117e (1)   117e (2)

Fig. 3.45

With the NHMe group equatorial rotamer *117e(1)* is stabilized by *exo*-anomeric effect and is preferred. With NHMe axial the preference is for rotamer *117a(2)*.

In contrast to the *N*-methyltetrahydropyrans, no *exo*-anomeric effect is expected when dimethylamino group is axial in the 2-position on dioxane ring, because in order for the nitrogen lone pair to be *anti* to the C–O bond one methyl group would be subjected to severe steric interaction with the axially oriented C4 and C6 hydrogen atoms. This is possible for the hydrogen but the steric demands of a methyl group are prohibitive.

Calculations of the energies of different conformers of *118* and *119* (Fig. 3.46) by using MM2 and 6–31G* basis set have shown that the conformer with the dimethylamino group axially oriented as shown in Fig. 3.46 is preferred.

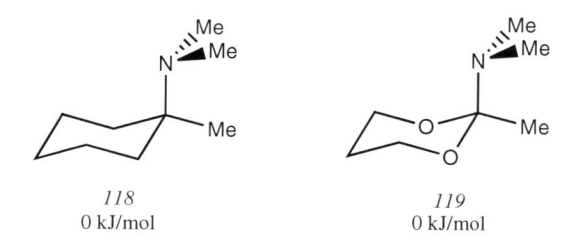

118
0 kJ/mol

119
0 kJ/mol

Fig. 3.46

The pseudorotation concept was introduced to describe the continuous interconversion of puckered forms of the cyclopentane ring [101]. The same concept is applied to furanose geometry where the C1, C4, and O4 atoms lie in one plane and the C2 and C3 atoms lie above and below that plane. A statistical analysis of X-ray crystal structures of nucleosides and nucleotides has shown that North (N) (*120*) and South (S) (*121*) conformations are the most dominant forms, which has been the basis of the assumption of the two-state N ⇌ S pseudorotational equilibrium in solution (Fig. 3.47).

The two-state N ⇌ S pseudorotational equilibrium of the sugar moiety of β-D-ribofuranosyl-*N*-nucleosides in solution is energetically controlled by various

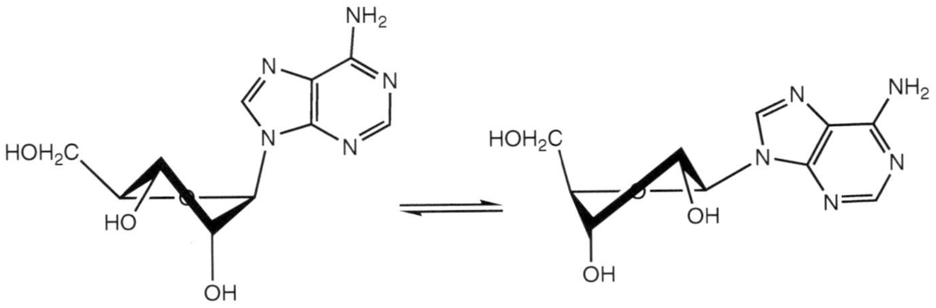

Fig. 3.47

North (N) conformation
($C_{3'}$-endo-$C_{2'}$-exo)
120

South (S) conformation
($C_{2'}$-endo-$C_{3'}$exo)
121

stereoelectronic gauche and anomeric effects [102–107]. The gauche effects [107, 108] of O4′–C4′–C3′–O3′ and O2′–C2′–C1′–N fragments drive the sugar pseudorotational equilibrium toward S^{102} whereas it is driven to N by the gauche effect of O4′–C1′–C2′–O2′ (Fig. 3.43).

The X-ray crystal structure of N-nucleosides shows the shortening of the O4′–C1′ bond relative to C4′-O4′ by about 0.03 Å which has been considered as a manifestation of the anomeric effect.

The preference of 5′-CH$_2$OH group to occupy the pseudoequatorial orientation is manifested in the positive $\Delta H^{\#}$ value for the pseudorotational equilibrium [108] (Fig. 3.47). From the determination of energetics of the two-state pseudorotational equilibrium in 36 nucleosides it was found that the combined stereoelectronic and steric contributions in the anomeric effect of the nucleobases increases in the following order: adenine ≈ guanine < thymine < uracil < cytosine. One reason for the stronger anomeric effect in pyrimidine than in purine nucleosides could be that the n(O4′)→σ*$_{C1'-N}$ delocalization is more effective in the π-deficient pyrimidine moiety compared to relatively more electron-rich purine [109–111].

The strength of the anomeric effect was enhanced upon protonation (evident by the increase in N-type sugar population) relative to the neutral state. These results are consistent with the favorable n(O4′)→σ*$_{C1'-N}$ delocalization in the electron-deficient protonated aglycone at the acidic pH and unfavorable n(O4′)→σ*$_{C1'-N}$ delocalization in the electron-rich anionic aglycone at the basic pH, compared to the neutral state, as the origin of the anomeric effect.

References

1. Szarek, W. A.; Horton, D., Eds. *"Anomeric Effect, Origin and Consequences"*, ACS Symposium Series 87, American Chemical Society, Washington, DC, 1979
2. Kirby, A. J., *"The Anomeric Effect and Related Stereoelectronic Effects at Oxygen"*, Springer-Verlag, Berlin, 1983

References

3. Deslongchamps, P., "Stereoelectronic Effects in Organic Chemistry", Pergamon Press, Oxford, 1983
4. Thatcher, G. R. J., Eds., "*The Anomeric Effect and Associated Stereoelectronic Effects*", ACS Symposium Series 539, American Chemical Society, Washington, DC, 1993
5. Juaristi, E.; Cuevas, G., "*The Anomeric Effect*", CRC Press, Boca Raton, 1995
6. Bates, F. J. and Associates, "*Polarimetry, Saccharimetry, and the Sugars*", United States Government Printing Office, Washington, DC, 1942, p.455
7. Angyal, S. J., "*Conformational analysis in carbohydrate chemistry. I. Conformational free energies. The conformations and α:β ratios of aldopyranoses in aqueous solution*", Aust. J. Chem. (1968) **21**, 2737–2746
8. Angyal, S. J., "*The composition and conformation of sugars in solution*", Angew. Chem. Intern. Ed. (1969) **8**, 157–166
9. Bonner, W. A., "*The acid-catalyzed anomerization of Acetylated Aldopyranoses*", J. Am. Chem. Soc. (1959) **81**, 1448–1452
10. Lemieux, R. U., "*Molecular Rearrangements*", Vol. 2, p 709, de Mayo, P., Ed., Interscience, New York, 1963
11. Lemieux, R. U.; Hayami, J-Y., "*The mechanism of the Anomerization of the Tetra-O-acetyl-D-glucopyranosyl chlorides*" Can. J. Chem. (1965) **43**, 2162173
12. Pacsu, E., "*Über die Einwirkung von Titan (IV)-chlorid auf Zucker-Derivate, I.: Neue Methode zur Darstellung der αβAceto-chlor-zucker und Umlagerung des β-Methyl-glucosids in seine α-Form*", Chem. Ber. (1928) **61**, 1508–1513
13. Lindberg, B., "*The Zempl.acte.en glucoside synthesis*", Arkiv Kemi, Mineral., Geol., Ser. B (1944), **18**, No. 9, 1–7
14. Lindberg, B., "*Action of strong acids on acetylated glycosides. I. Transformation of some aliphatic tetraacetyl-β-glucosides to the α-form*", Acta Chem. Scand. (1948) **2**, 426–429
15. Lindberg, B., "*Action of strong acids on acetylated glucosides. III. Strong acids and aliphatic glucoside tetraacetates in acetic anhydride-acetic acid solutions*", Acta Chem. Scand. (1949) **3**, 1153–1169
16. Chü, N. J., Ph. D. thesis, Department of Chemistry, University of Ottawa, 1959
17. Lemieux, R. U.; Chü, N. J., "*Conformation and relative stabilities of acetylated sugars as determined by nuclear magnetic resonance spectroscopy and anomerization equilibria*", Abstracts of Papers, Am. Chem. Soc. **133**, 31 N (1958)
18. Edward, J. T., "*Stability of glycosides to acid hydrolysis*", Chemistry & Industry (London) (1955) 1102–1104
19. Corey, E. J., "*The Stereochemistry of α-Haloketones. I. The molecular configurations of some Monocyclic α-Haloketones*", J. Am. Chem. Soc. (1953) **75**, 2301–2304
20. Lemieux, R. U., "*Effects of unshared pairs of electrons and their Solvation Conformational Equilibria*", Pure Appl. Chem. (1971) **27**, 527–548
21. Winstein, S.; Holness, N. J., "*Neighboring Carbon and Hydrogen. XIX. t-Butylcyclo hexyl Derivatives. Quantitative Conformational Analysis*", J. Am. Chem. Soc. (1955) **77**, 5562–5578
22. Eliel, E.L., *Stereochemistry of Carbon Compounds*, McGraw Hill, New York, 1962, p.236
23. Andersen, C.B.; Sepp, D. T., "*Conformation and the anomeric effect in 2-oxy-substituted tetrahydropyrans*", Tetrahedron (1968) **24**, 1707
24. Isbell, H. S.; Pigman, W. W., "*Bromine oxidation and mutarotation measurements of the α- and β-aldoses*", J. Res. Natl. Bur. Std. (1937) **18**, 141–194
25. Angyal, S. J., "*Conformational analysis in carbohydrate chemistry. I. Conformational free energies. The conformations and α: Ⓡ ratios of aldopyranoses in aqueous solution*", Aust. J. Chem. (1968) **21**, 2737–2746
26. Angyal, S. J., "*The composition and conformation of sugars in solution*" Angew. Chem. Intern. Ed. (1969) **8**, 157–166
27. Reeves, R. E., "*Cuprammonium–Glycoside Complexes. II. The Angle Between Hydroxyl Groups on Adjacent Carbon Atoms*", J. Am. Chem. Soc. (1949) **71**, 212–214

28. Reeves, R. E., "*The shape of pyranoside rings*", J. Am. Chem. Soc. (1950) **72**, 1499–1506
29. Reeves, R. E., "*Cuprammonium-Glycoside complexes*", Adv. Carbohydr. Chem. (1951) **6**, 107–134
30. Hageman, H. J., PhD Thesis, Leiden (1965)
31. Eliel, E. L.; Allinger, N.L.; Angyal, S. J.; Morrison, G. A., *Conformational Analysis*, Interscience, New York, 1965, p. 44
32. Planje, M. C., PhD Thesis, Leiden (1964)
33. Booth, G. E.; Ouellette, R. J., "*Conformational Analysis. V.1, 2 2-Chloro- and 2-Bromotetrahydropyran*", J. Org. Chem. (1966) **31**, 544–546
34. Anderson, C. B.; Sepp, D. T., "*Conformation and the anomeric effect in 2-halotetrahydropyrans*", J. Org. Chem. (1967) **32**, 607–611
35. Akishin, P. A.; Vilkov, L. V.; Sokolova, N. P., "*Electronographic study of the structure of the molecules of monochloro- and monobromodimethyl ethers*", Izvest. Sibir. Otdel. Akad. Nauk S.S.S.R. (1960) **5**, 59–60
36. Planje, M. C.; Toneman, L. H.; Dallinga, G., Rec. Trav. Chim. (1965) **84**, 232
37. Bishop, C. T.; Cooper, F. P., "*Glycosidation of sugars. II. Methanolysis of D-Xylose. D-Arabinose, D-Lyxose, and D-Ribose*", Can. J. Chem. (1963) **41**, 2743–2758
38. Eliel, E. L.; Giza, C. A., "*Conformational analysis. XVII. 2-Alkoxy- and 2-alkylthio-tetrahydropyrans and 2-alkoxy-1,3-dioxanes. Anomeric effect*", J. Org. Chem. (1968) **33**, 3754–3758
39. Pierson, G. O.; Runquist, O. A., "*Conformational analysis of some 2-alkoxytetrahydropyrans*", J. Org. Chem. (1968) **33**, 2572–2574
40. Sweet, F.; Brown, R. K., "*Cis- and trans-2,4-dimethoxytetrahydropyran. Models for the study of the anomeric effect*", Canad. J. Chem. (1968) **46**, 1543–1548
41. de Hoog, A. J.; Buys, H. R.; Altona, C.; Havinga, E., "*Conformation of non- aromatic ring compounds—LII: NMR spectra and dipole moments of 2-alkoxytetrahydropyrans*", Tetrahedron (1969) **25**, 3365–3375
42. Altona, C.; Havinga, E. cited as unpublished in Romers, C.; Altona, C.; Buys, H. R.; Havinga, E. *Topics in Stereochemistry*, Eliel, E. L.; Allinger, N. L., Eds., Vol. 4, Wiley-Interscience, New York 1969, pp. 39–97
43. Altona, C.; Romers, C.; Havinga, E., "*Molecular structure and conformation of some dihalogenodioxanes*", Tetrahedron Lett. (1959) **1**, 16–20
44. Romers, C.; Altona, C.; Buys, H. R., Havinga, E., "*The Anomeric Effect*", in *Topics in Stereochemistry*, Eliel, E. L.; Allinger, N. L., Eds., Vol. 4, Wiley- Interscience, New York1969, pp. 39–97
45. Altona, C., Ph. D. Thesis, Leiden, (1964)
46. Altona, C.; Romers, C, "*The conformation of non-aromatic ring compounds. VIII. The crystal structure of cis-2,3-dichloro-1,4-dioxane at -140°C*", Acta Cryst. (1963) **16**, 12251232
47. Altona, C.; Knobler, C.; Romers, C., "*The conformation of non-aromatic ring compounds. VII. Crystal structure of trans-2,5-dichloro-1,4-dioxane at 125°C*", Acta Cryst. (1963) **16**,1217–1225
48. de Wolf, N.; Romers, C.; Altona, C., "*The conformation of non-aromatic ring compounds. XXXIV. The crystal structure of trans-2,3-dichloro-1,4-thioxane at -185°C*". Acta. Cryst. (1967) **22**, 715–719
49. Lemieux, R. U.; Koto, S.; Voisin, D., "*The Exo-Anomeric Effect*", in Szarek, W. A.; Horton, D., Eds., "*The Anomeric Effect: Origin and Consequences*", Am. Chem. Soc. Symposium Series, Vol. 87, Washington 1979, pp 17–29
50. Hutchins, R. O.; Kopp, L. D.; Eliel, E. L., "*Repulsion of syn-axial electron pairs. The rabbit-ear effect*", J. Am. Chem. Soc. (1968) **90**, 7174–7175
51. Kubo, M., Sci. Papers Inst. Phys. Chem. Res. (Tokyo) (1936) **29**, 179
52. Aoki, K., J. Chem. Soc. (Japan), Pure Chem. Sect. (1953) **74**, 110; Chem. Abstr. (1953) **47**, 5191
53. Astrup, E. E., "*Molecular structure of dimethoxy-methane, MeOCH$_2$OMe*", Acta Chemica Scand. (1971) **25**, 1494–1495

References

54. Booth, H.; Lemieux, R. U., "*Anomeric effect: the conformational equilibriums of tetrahydro-1,3-oxazines and 1-methyl-1,3-diazane*", Can. J. Chem. (1971) **49**, 777–788
55. Zefirov, N. S., "*The problem of conformational effects*", Tetrahedron (1977) **33**, 3192
56. Eliel, E, L.; Allinger, N. L.; Angyal, S. J.; Morrison, G. A., Conformational Anal., p. 460, Wiley-Interscience, 1965
57. Horton, D.; Turner, W. N., "*Conformational and configurational studies on some Acetylated Aldopyranosyl Halides*", J. Org. Chem. (1965) **30**, 3387–3394
58. Lemieux, R. U.; Morgan, A. R., "*The abnormal conformations of Pyridinium α-glycopyranosides*", Can. J. Chem. (1965) **43**, 2205–2213
59. James, M. N. G., Proc. Can. Fed. Biol. Soc. (1969) **13**, 71
60. Lemieux, R. U.; Koto, S., "*The conformational properties of glycosidic linkages*", Tetrahedron (1974) **30**, 1933–1944
61. Lemieux, R. U.; Hendriks, K. B.; Stick, R. V.; James, K. "*Halide ion catalyzed glycosidation reactions. Syntheses of α-linked disaccharides*", J. Am. Chem. Soc. (1975) **97**, 4056–4062
62. Saluja, S. S., Ph. D. Thesis, University of Alberta, 1971
63. Paulsen, H.; Györgdeák, Z.; Friedmann, M., "*Konformationsanalyse, V. Einfluß des anomeren und inversen anomeren Effektes auf Konformationsgleichgewichte von N-substituierten N-Pentopyranosiden*", Chem. Ber. (1974) **107**, 1590–1613
64. Grein, F.; Deslongchamps, P., "*The anomeric and reverse anomeric effect. A simple energy decomposition model for acetals and protonated acetals*, Can. J. Chem. (1992) **70**, 1562–1572
65. Grein, F., "*Anomeric and Reverse Anomeric Effect in Acetals and Related Functions*" in "*The Anomeric Effect and Associated Stereoelectronic Effects*", Thatcher, G. R. J., Ed. ACS Symposium Series No. 539, 205–226, ACS, Washington, DC, 1993
66. Finch, P.; Nagpurkar, A. G., "*The reverse anomeric effect: further observations on N-glycosylimidazoles*", Carbohydr. Res. (1976) **49**, 275–287
67. Ratcliffe, A. J.; Fraser-Reid, B., "*Generation of -D-glucopyranosylacetonitrilium ions. Concerning the reverse anomeric effect*", J. Chem. Soc. Perkin 1(1990) 747–750
68. Batchelor, J. G., "*Conformational analysis of cyclic amines using carbon-13 chemical shift measurements: dependence of conformation upon ionisation state and solvent*", J. Soc., Perkin Trans. 2 (1976) 1585–1590
69. Booth, H.; Jozefowicz, M. L., "*The application of low temperature 13C nuclear magnetic resonance spectroscopy to the determination of the A values of amino-, methylamino-, and dimethylamino-substituents in cyclohexane*" J. Chem. Soc. Perkin Trans. 2 (1976) 895–901
70. Sicher, J.; Jonás, J.; Tichý, M., "*The a-values of the amino acid and dimethylamino groups*", Tetrahedron Lett. (1963) **4**, 825–830
71. Eliel, E. L.; Della, E. W.; Williams, T. H., "*The conformational equilibrium of the amino group*", Tetrahedron Lett. (1963) **4**, 831–835
72. Ford, R. A.; Allinger, N. L., "*Conformational analysis. LXVII. Effect of solvent on the conformational energy of the carbethoxy group*", J. Org. Chem. (1970) **35**, 3178–3181
73. Perrin, C. L.; Armstrong, K. B., "*Conformational analysis of glucopyranosylammonium ions: does the reverse anomeric effect exist?*", J. Am. Chem. Soc. (1993) **115**, 6825–6834
74. Isbell, H. S.; Frush, H. L., "*Mutarotation, Hydrolysis, and Rearrangement Reactions of Glycosylamines*", J. Org. Chem. (1958) **23**, 1309–1319
75. Pinto, B. M.; Leung, Y. N., "*The Anomeric Effect and Associated Stereoelectronic Effects*", Thatcher, G. R. J., Ed. ACS Symposium Series No. 539, 126–155, ACS, Washington, DC, 1993
76. Cramer, C. J., "*Anomeric and reverse anomeric effects in the gas phase and aqueous solution*", J. Org. Chem. (1992) **57**, 7034–7043
77. Salzner, U; Schleyer, P. v. R., "*Ab initio examination of anomeric effects in Tetrahydropyrans, 1,3-Dioxanes, and Glucose*", J. Org. Chem. (1994) **59**, 2138–2155
78. Chan, S. S. C.; Szarek, W. A.; Thatcher, G. R. J., "*The reverse anomeric effect in N-pyranosylimidazolides: a molecular orbital study*", J. Chem. Soc. Perkin Trans. 2, (1995) 45–60

79. Vaino, A. R.; S. S. C. Chan; Szarek, W. A.; Thatcher, G. R. J., "*An experimental re-examination of the reverse anomeric effect in N-Glycosylimidazoles*", J. Org. Chem. (1996) **61**, 4514–4515
80. Fabian, M. A.; Perrin, C. L.; Sinnott, M. L., "*Absence of reverse anomeric effect: Conformational analysis of Glucosylimidazolium and Glucosylimidazole*" J. Am. Chem. Soc. (1994) **116**, 8398–8399
81. Juaristi, E.; Cuevas, G, "*Recent studies of the anomeric effect*", Tetrahedron, (1992) **48**, 5019–5087
82. Mikolajczyk, M.; Graczyk, P.; Wieczorek, M. W.; Bujacz, G "*Conformational preference of 2-Triphenylphosphonio-1,3-Dithianes: Competition between steric and anomeric effects*", Angew. Chem. Int. Ed. Engl.(1991) **30**, 578–580
83. Graczyk, P. P.; Mikolajczyk, M.; Phosphorus, Sulfur Silicon Relat. Elem. (1993) **78**, 313
84. Juaristi, E.; Cuevas, G., "*Conformational analysis of 1,3-dithian-2-yltrimethylphosphonium chloride. Origin of the S-C-P anomeric effect*", J. Am. Chem. Soc. (1993) **115**, 1313–1316
85. Thibaudeau, C.; Plavec, J.; Watanabe, K. A.; Chattopadhyaya, J., "*How do the aglycons drive the pseudorotational equilibrium of the pentofuranose moiety in C-nucleosides?*", J. Chem Soc., Chem. Commun. (1994) 537–540
86. Jones, P. G.; Komarov, I. V.; Wothers, P. D.,"*A test for the reverse anomeric effect*", Chem. Commun., 1998, 1695–1696
87. Kennedy, J.; Wu, J.; Drew, K.; Carmichael, I.; Serianni, A. S., "*Carbohydrate reaction intermediates: Effect of ring-oxygen protonation on the structure and conformation of Aldofuranosyl rings*", J. Am. Chem. Soc. (1997) **119**, 8933–8945
88. Alder, R. W.; Carniero, T. M. G.; Mowlam, R. W.; Orpen, A. G.; Petillo, P. A.; Vachon, D. J.; Weisman, G. R.; White, J. M., "*Evidence for hydrogen-bond enhanced structural anomeric effects from the protonation of two aminals, 5-methyl-1,5,9-triazabicyclo[7.3.1]tridecane and 1,4,8,11-tetraazatricyclo[9.3.1.14,8]hexadecane*", J. Chem. Soc., Perkin Trans. 2 (1999) 589–599
89. Cramer, C. J., "*Hyperconjugation as it affects conformational analysis*", J. Mol. Struct. (THEOCHEM) (1996) **370**, 135–146
90. Ganguly, B.; Fuchs, B., "*Stereoelectronic Effects in Negatively and Positively (Protonated) Charged Species. Ab Initio Studies of the Anomeric Effect in 1,3-Dioxa Systems*", J. Org. Chem.(1997) **62**, 8892–8901
91. Cloran, F.; Zhu, Y.; Osborn, J.; Carmichael, I.; Serianni, A. S., "*2-Deoxy-D- ribofuranosylamine: Quantum mechanical calculations of molecular structure and NMR spin–spin coupling constants in nitrogen-containing Saccharides*", J. Am. Chem. Soc. (2000) **122**, 6435–6448
92. Randell, K. D.; Johnston, B. D.; Green, D. F.; Pinto, B. M., "*Is there a generalized reverse anomeric effect? Substituent and solvent effects on the configurational equilibria of neutral and protonated N-Arylglucopyranosylamines and N-Aryl-5- thioglucopyranosylamines*", J. Org. Chem. (2000) **65**, 220–226
93. Guthrie, J. P., "*Hydrolysis of esters of oxy acids: pKa values for strong acids; Brønsted relationship for attack of water at methyl; free energies of hydrolysis of esters of oxy acids; and a linear relationship between free energy of hydrolysis and pKa holding over a range of 20 pK units*", Can. J. Chem. (1978) **56**, 2342–2354
94. Booth, H.; Lemieux, R. U., "*Anomeric effect: the conformational equilibriums of tetrahydro-1,3-oxazines and 1-methyl-1,3-diazane*" Can. J. Chem. (1971) **49**, 777–788
95. Allingham, Y.; Cookson, R. C.; Crabb, T. A.; Vary, S., "*The NMR spectra and conformations of some tetrahydro-1, 3-oxazines*", Tetrahedron (1968) **24**, 4625–4630
96. Allingham, Y.; Cookson, R. C.; Crabb, T. A.; Vary, S., "*The NMR spectra and conformations of some tetrahydro-1, 3-oxazines*", Tetrahedron (1968) **24**, 4625–4630
97. Booth, H.; Khedhair, K. A., "*Endo-anomeric and exo-anomeric effects in 2-substituted tetrahydropyrans*", J. Chem. Soc., Chem. Commun. (1985) 467–468

98. Kirby, A. J.; Wothers, P. D., "*Conformational equilibria involving 2-amino-1, 3-dioxanes: steric control of the anomeric effect*", ARKIVOC (2001) XII, 58–71
99. Booth, H.; Readshaw, S. A., "*Experimental studies of the anomeric effect. Part IV. Conformational equilibria due to ring inversion in tetrahydropyrans substituted at position 2 by the groups ethoxy, 2'-fluoroethoxy, 2,'2'-difluoroethoxy, and 2',2',2'-trifluoroethoxy*", Tetrahedron (1990) **46**, 2097–2110
100. Booth, H.; Khedhair, K. A.; Readshaw, S. A., "*Experimental studies of the anomeric effect. I. 2-Substituted tetrahydropyrans*", Tetrahedron (1987), 43(20), 4699–4723
101. Kilpatrick, J. E.; Pitzer, K. S.; Spitzer, R., "*The thermodynamics and molecular structure of cyclopentane*", J. Am. C hem. Soc. (1947) **69**, 2483–2488
102. Plavec, J.; Tong, W.; Chattopadhyaya, J., "*How do the gauche and anomeric effects drive the pseudorotational equilibrium of the pentofuranose moiety of nucleosides?*", J. Am. Chem. Soc. (1993) **115**, 9734–9746
103. Plavec, J.; Garg, N.; Chattopadhyaya, J., "*How does the steric effect drive the sugar conformation in the 3-C-branched nucleosides?*", J. Chem. Soc. Chem. Commun. (1993), 1011–1014
104. Plavec, J.; Koole, L. H.; Chattopadhyaya, J., "Structural analysis of 2',3'-dideoxyinosine, 2',3'-dideoxyadenosine, 2',3'-dideoxyguanosine, and 2',3'-dideoxycytidine by 500-MHz proton-NMR spectroscopy and ab-initio molecular orbital calculations", J. Biochem. Biophys. Methods (1992) **25**, 253–272
105. Altona, C.; Sundaralingam, M., "*Conformational analysis of the sugar ring in nucleosides and nucleotides. New description using the concept of pseudorotation*", J. Am. Chem. Soc. (1972) **94**, 8205–8212
106. Altona, C.; Sundaralingam, M., "*Conformational analysis of the sugar ring in nucleosides and nucleotides. Improved method for the interpretation of proton magnetic resonance coupling constants*", J. Am. Chem. Soc. (1973) **95**, 2333–2344
107. Saenger, W., "*Principles of Nucleic Acid Structure*", Springer-Verlag, Berlin, 1988
108. Olson, W. K.; Sussman, J. L., "*How flexible is the furanose ring? 1. A comparison of experimental and theoretical studies*", J. Am. Chem. Soc. (1982) **104**, 270–278
109. Olson, W. K., "*How flexible is the furanose ring? 2. An updated potential energy estimate*", J. Am. Chem. Soc. (1982) **104**, 278–286
110. Plavec, J.; Tong, W.; Chattopadhyaya, J., "*How do the gauche and anomeric effects drive the pseudorotational equilibrium of the pentofuranose moiety of nucleosides?*", J. Am. Chem. Soc. (1993) **115**, 9734–9746
111. Plavec, J.; Thibaudeau, C.; Chattopadhyaya, J., "*How do the energetics of the stereoelectronic gauche and anomeric effects modulate the conformation of nucleos(t)ides*", Pure. Appl. Chem. (1996) **68**, 2137–2144

Chapter 4
Isomerization of Sugars

Mutarotation

In 1846 Dubrunfaut [1] observed that the optical rotation of a freshly dissolved α-D-glucose in water was changing with time and that after several hours it became constant.

$$\alpha\text{-D-glucose} \rightleftharpoons \gamma\text{-glucose} \rightleftharpoons \beta\text{-D-glucose}$$
$$+112° \rightarrow +52.7° \leftarrow +18.7°$$

The same equilibrium rotation of 52.7° was attained regardless of whether the starting sugar was α- or β-D-glucose. Initially, this "new form" of glucose, having the optical rotation of +52.7°, was named "γ-glucose," but very soon it was realized that the "γ-glucose" was not a new form of glucose but simply an equilibrium mixture of α- or β-D-glucose. Similar behavior was later observed with all other sugars.

This interconversion of α- and β-D-glucose (Equation 1) was named *mutarotation*, and it was kinetically described [2] with Equation 2:

$$\alpha \underset{k_2}{\overset{k_1}{\rightleftharpoons}} \beta \qquad (1)$$

$$\frac{dx}{dt} = k_1[a-x] - k_2[x] \qquad (2)$$

where k_1 is the rate constant for α → β conversion, k_2 is the rate constant for β → α conversion, "a" is the initial concentration of α form, "x" is the concentration of β form at time t, and $[a-x]$ is the concentration of α form at time t. The integration of (2) gives after rearrangement

$$k_1 + k_2 = \frac{1}{t} \ln \frac{K_a}{K_a - (1+K)x} \qquad (3)$$

where $(k_1 + k_2)$ is *mutarotation constant*, and $K = k_1/k_2$. Equation (3) can also be expressed in terms of optical rotation [2] if the sugar concentrations are substituted with concentration-dependent optical rotations:

$$k_1 + k_2 = (1/t) \ln [(r_0 - r_{eq})/(r_t - r_{eq})] \qquad (4)$$

where r_0 is the initial rotation, r_{eq} is the rotation at equilibrium, and r_t is the rotation at time t.

This type of mutarotation is called *simple mutarotation* since the result of this isomerization is the formation of another anomer (*anomerization*) of the parent sugar. The simple mutarotation is observed in solutions of gluco-, manno-, gulo-, and allopyranoses, hence in the solution of sugars that contain practically only α- and β-pyranose forms, with very small proportions of other forms of sugars (e.g., furanoses).

$$\alpha\text{-pyranose} \rightleftharpoons \text{acyclic form} \rightleftharpoons \beta\text{-pyranose}$$

Scheme 4.1

It has been suggested that simple mutarotation takes place via an acyclic intermediate (Scheme 4.1) as shown in Fig. 4.1. In the first step, the lactol ring of β-D-glucopyranose *1* opens with the formation of the acyclic D-glucose conformer *2* in which the carbonyl oxygen is pointed outward. The acyclic conformer *2* can then either cyclize back to the starting β-D-glucopyranose *1* or it can, prior to cyclization, be converted by a free rotation about the C1-C2 bond into the conformer *3* of the acyclic D-glucose intermediate in which the carbonyl oxygen is pointed down. The cyclization of this acyclic conformer will then give the α-anomer *4*. The participation of water molecules as general acid–general base catalysts is essential for this reaction.

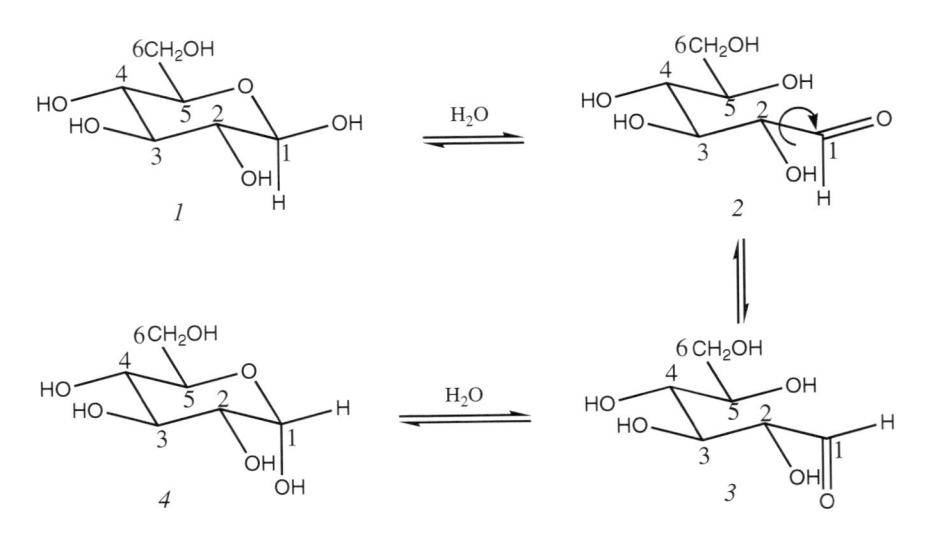

Fig. 4.1

The equilibrium solutions of galacto-, talo-, altro-, and idopyranoses have much more complex composition because they contain, in addition to α- and β-pyranose forms, substantial amounts of furanose and other forms of sugars. Apparently these sugars undergo besides anomerization also the ring isomerization. This ring isomerization, observed also in solutions of gluco-, manno-, gulo-, and allopyranoses, although to a much smaller extent, was explained in the following way.

If, after opening of the pyranose ring and after the formation of the corresponding acyclic conformer of D-glucose (2 in Figs. 4.1 and 4.2), the rotation about the C3-C4 single bond takes place prior to cyclization, the conformation 5 of acyclic glucose will be formed (2→5) (Fig. 4.2), which upon cyclization will give the α-D-glucofuranose.

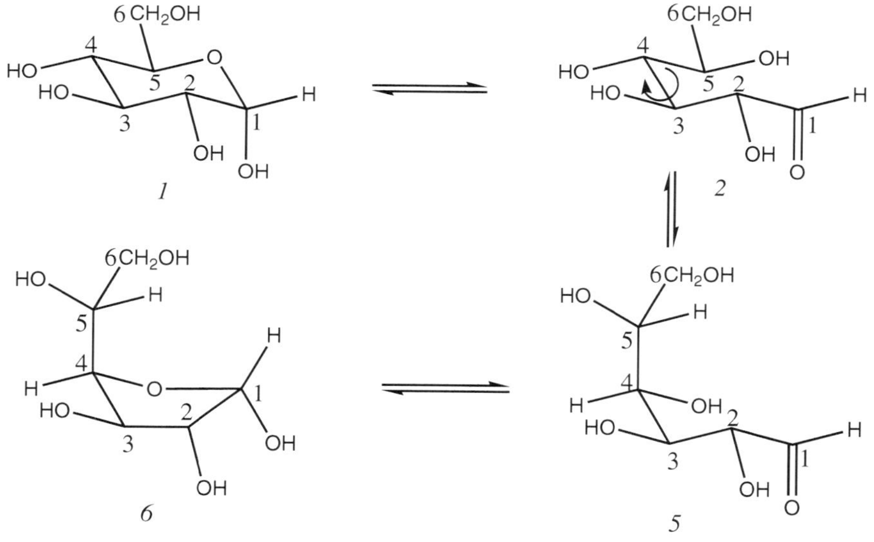

Fig. 4.2

If after rotation about the C3-C4 bond rotation about the C1-C2 bond also takes place prior to cyclization, the acyclic conformer 7 (Fig. 4.3) will be formed which on cyclization will give the β-anomer of D-glucofuranose 8 (Fig. 4.3). The ring isomerization, i.e., the interconversion of pyranose and furanose ring forms is called *complex mutarotation*.

The mutarotation of reducing sugars is catalyzed by both acids and bases. The increase of rate of mutarotation by acids was first reported by Erdmann [3] and catalysis by both acids and bases was described by Urech [4]. Hudson [5] has shown that mutarotation is catalyzed by water molecules. Lowry and coworkers [6–8] studied mutarotation of tetra-*O*-methyl-α-D-glucopyranose and found that the rate of mutarotation is low in dry pyridine or in dry cresol, but that it is high in the

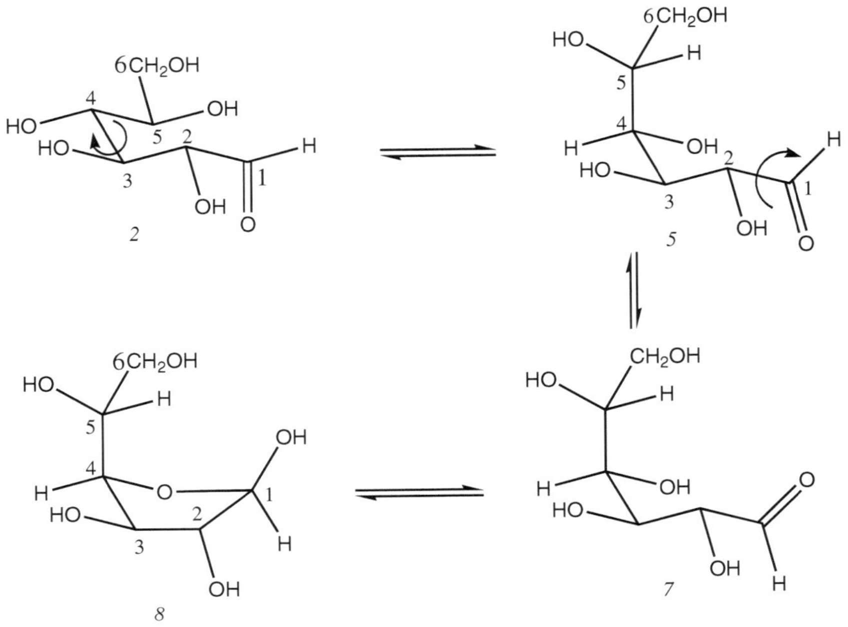

Fig. 4.3

mixture of these two solvents, or in each one of them when moist. Lowry and Smith [9] concluded that *mutarotation requires both an acid and a* base catalyst and that amphoteric solvents are perfect catalysts for mutarotation, whereas the aprotic solvents are not.

Since mutarotation is catalyzed by both acids and basis, Swain and Brown [10] designed the first bifunctional catalyst, 2-pyridinol *10*, which on its two adjacent ring atoms has an acidic and a basic functional group. It was found that the rate of mutarotation catalyzed by 2-pyridinol in benzene, as solvent, is 7000 times faster than mutarotation calculated for analogous concentration of pyridine or phenol alone. The action of bifunctional catalyst is explained by postulating a concerted mechanism that involves the formation of an eight-membered cyclic transition state, as illustrated in Fig. 4.4.

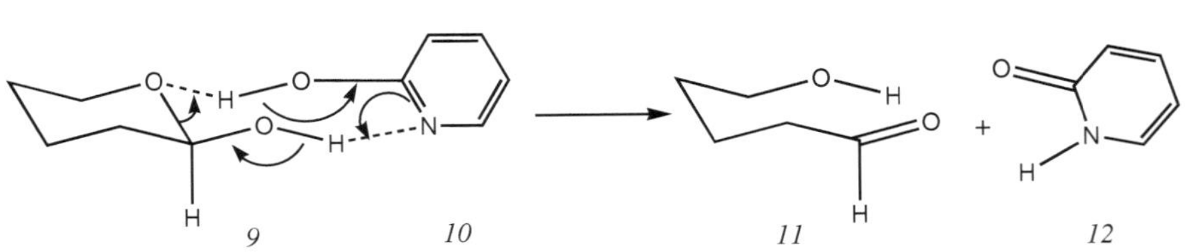

Fig. 4.4

Mutarotation

Carboxylic acids also act as bifunctional catalysts for mutarotation because the carbonyl oxygen is a hydrogen acceptor (base) whereas the carboxylic hydrogen is an acid, and they, too, are able to form an eight-membered cyclic transition state similar to the one formed by 2-pyridinol (Fig. 4.5).

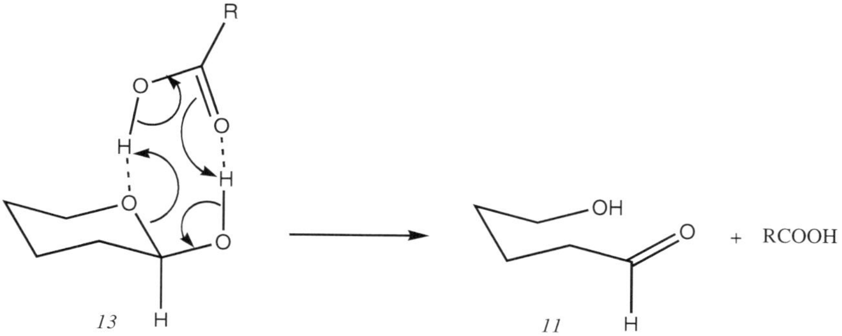

Fig. 4.5

Studies of mutarotation in water and in deuterium oxide have shown that the reaction is faster in water than in deuterium oxide, regardless of the catalyst. It was found [11] that the k_H/k_D value is 1.37 for catalysis by specific acids (H_3O^+) and 3.80 for catalysis by water molecules. From a study of large number of sugars, Nicolle and Weisbuch [12] determined that the k_H/k_D ratio lies in the range of 3.0–3.8 and is independent of pH.

Several interpretations have been put forward to explain this observation. All of them agree in the assumption that the opening of the pyranose ring with formation of the corresponding acyclic intermediate is the rate-determining step. The contested part of the mechanism is whether the formation of a pre-equilibrium state precedes the ring rupture or not. According to Bonhoeffer et al. [13, 14], Bell [15], and Purlee [16], an intermediate *15* (Fig. 4.6) is formed prior to proton transfer from an acid to the ring oxygen (protonation of the pyranose ring oxygen) and the ring cleavage (Fig. 4.7). This intermediate then undergoes slow pyranose ring opening with simultaneous proton transfer from the acid H–A and simultaneous cleavage of the H–A bond (protonation step), in aqueous solution, and D–A bond in deuterium

Fig. 4.6

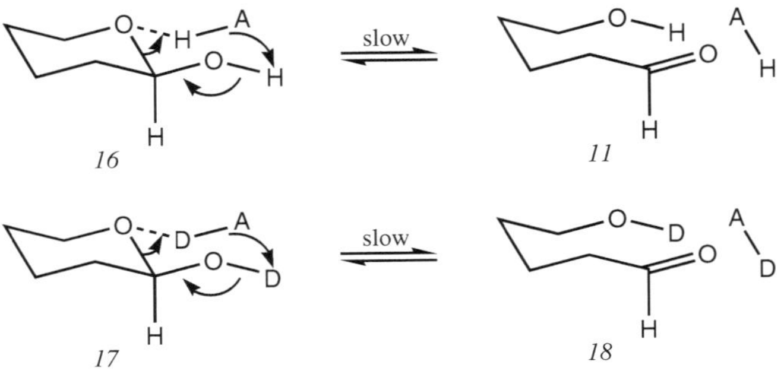

Fig. 4.7

oxide. Since the D–A bond is stronger than the H–A bond, the reaction in deuterium oxide will be slower than in H$_2$O and k_H/k_D will be greater than 1.

Challis et al. [17] and Long and Bigeleisen [18] suggested that the pre-equilibrium proton transfer (cleavage of H–A or D–A bond) takes place prior to the pyranose ring rupture, whereas the second proton transfer (proton from the anomeric hydroxyl group to the base) occurs simultaneously with the ring rupture (Fig. 4.8). If this were so, then a higher concentration of conjugate acid can be expected to be found in deuterium oxide than in water, which should result in k_H/k_D ratio less than 1. However, in D$_2$O, the hydrogen atom of the anomeric hydroxyl group (as well as of all other hydroxyl groups in carbohydrate molecule) will be almost instantaneously replaced by deuterium. In the subsequent simultaneous opening of the pyranose ring, the cleavage of C1–O–H in water or C1–O–D bond in deuterium oxide, which is the rate limiting step, will be slower in deuterium oxide, because the O–D bond is stronger than OH bond, and k_H/k_D ratio will be greater than 1 (Fig. 4.9).

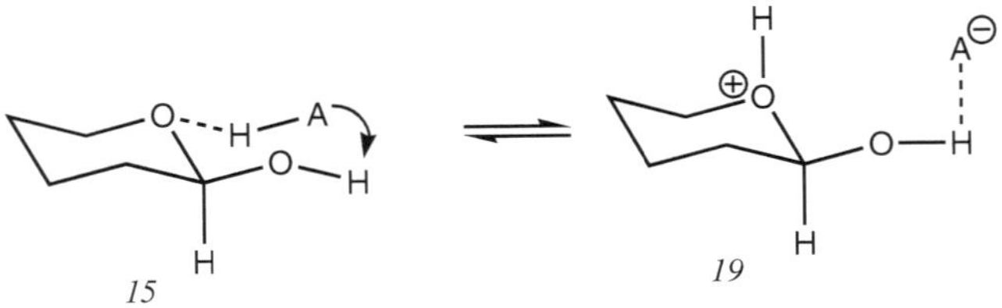

Fig. 4.8

The high value of k_H/k_D for the observed mutarotation in water was explained by postulating a *concerted mechanism* proposed by Lowry [7, 19] (Fig. 4.10) in which water acts as the general acid and the general base catalyst. Two slightly different transition state intermediates can be written for the mutarotation of sugars in water:

According to one mechanism, two molecules of monomeric water are required – one water molecule acts as the general acid and the other as the general base (Fig. 4.10) – and according to the other mechanism, one molecule of dimeric water molecule is required, in which case an eight-remembered cyclic transition state

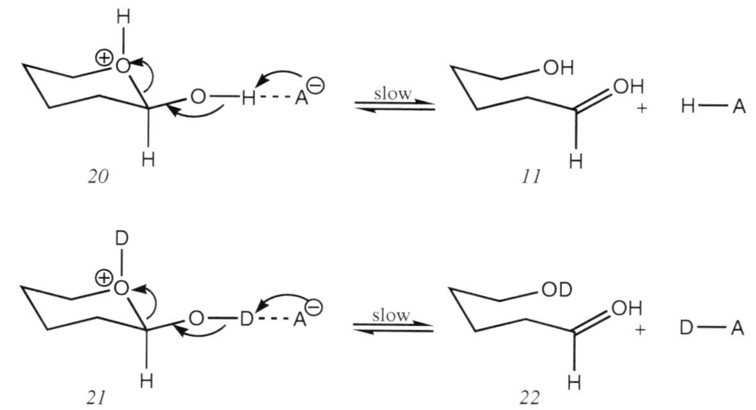

Fig. 4.9

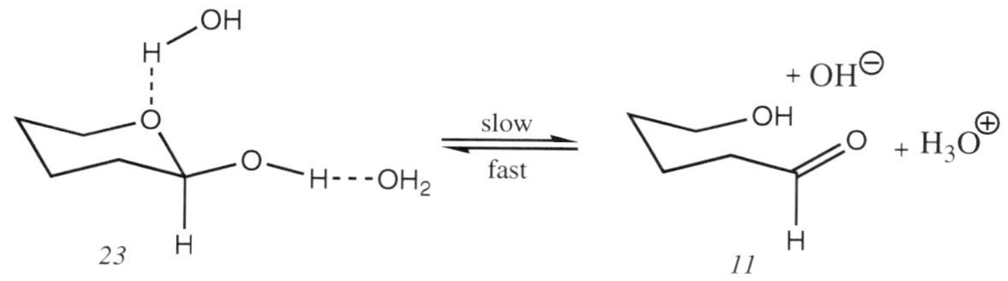

Fig. 4.10

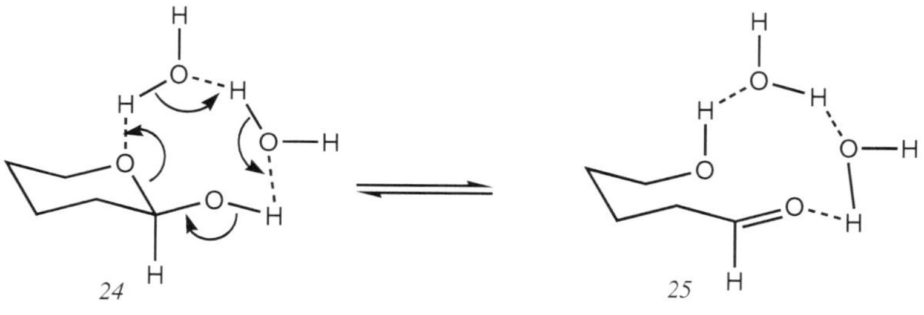

Fig. 4.11

intermediate is formed in which one part of the water dimer acts as a general acid and the other part as a general base (Fig. 4.11). Thus in the second mechanism, water dimer acts as a *bifunctional catalyst*, and the reaction is a concerted one.

The mutarotation of sugars in aqueous solutions catalyzed by acid–base catalysts involves two proton transfers (one from the acid catalyst to the sugar-ring oxygen and the second proton from the sugar anomeric hydroxyl group to the base catalyst) and the pyranose ring rupture. The reaction can start with either proton transfer.

Before we discuss these issues, let us take a brief look at how the charges are distributed in a pyranose or a furanose ring (Fig. 4.12). Zhdanov et al. [20] have calculated the electronic structures and charge distributions of nonspecified pentoses in their furanose and pyranose forms using the LCAO-MO method of inductive parameters, in the form derived by Del Re [21]. The results of their calculations are given in Tables 4.1 (for pentofuranoses) and 4.2 (for pentopyranoses).

Table 4.1 Electronic distribution in pentofuranoses

Pentofuranose							
C5	+0.0467	O5	−0.4731	(C)H5	+0.0530	(O)H5	+0.3164
C4	+0.0964	O4	−0.2642	(C)H′5	+0.0530		
C3	+0.1068	O3	−0.4748	(C)H4	+0.0507	(O)H3	+0.3161
C2	+0.1160	O2	−0.4734	(C)H3	+0.0518	(O)H2	+0.3164
C1	+0.1926	O1	−0.4627	(C)H2	+0.0529	(O)H1	+0.3184
				(C)H1	+0.0610		

Table 4.2 Electronic distribution in pentopyranoses

Pentofuranose							
C5	+0.0362	O5	−0.2631	(C)H5	+0.0518		
C4	+0.1069	O4	−0.4747	(C)H′5	+0.0518	(O)H4	+0.3161
C3	+0.1079	O3	−0.4746	(C)H4	+0.0518	(O)H3	+0.3161
C2	+0.1162	O2	−0.4734	(C)H3	+0.0520	(O)H2	+0.3164
C1	+0.1929	O1	−0.4629	(C)H2	+0.0529	(O)H1	+0.3166
				(C)H1	+0.0611		

As one can see from the results of these calculations, the electronegativities of all hydroxyl oxygens are practically identical except for the ring oxygen and the C1 oxygen of the anomeric hydroxyl group. The anomeric oxygen is slightly less electronegative (ca. −0.01) than other hydroxyl group oxygens, whereas the ring oxygen, in both pyranose and furanose rings, is significantly less electronegative (by −0.2) than other hydroxyl group oxygens. The small electronegativity of the ring oxygen (−0.2642 for pentofuranoses and −0.2631 for pentopyranoses) as compared to the

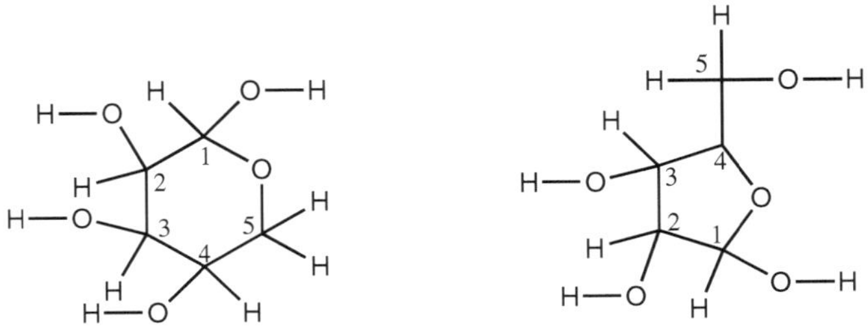

Fig. 4.12

electronegativity of the anomeric oxygen (−0.4627 for pentofuranoses and −0.4629 for pentopyranoses) immediately raises the question of why would the protonation of the ring oxygen as the first step of mutarotation be favored over the protonation of the anomeric or any other oxygen of the sugar molecules, since all of them are more electronegative.

Aside of this controversy, the mechanism for the simple mutarotation described in Fig. 4.13 is accepted as the most plausible one.

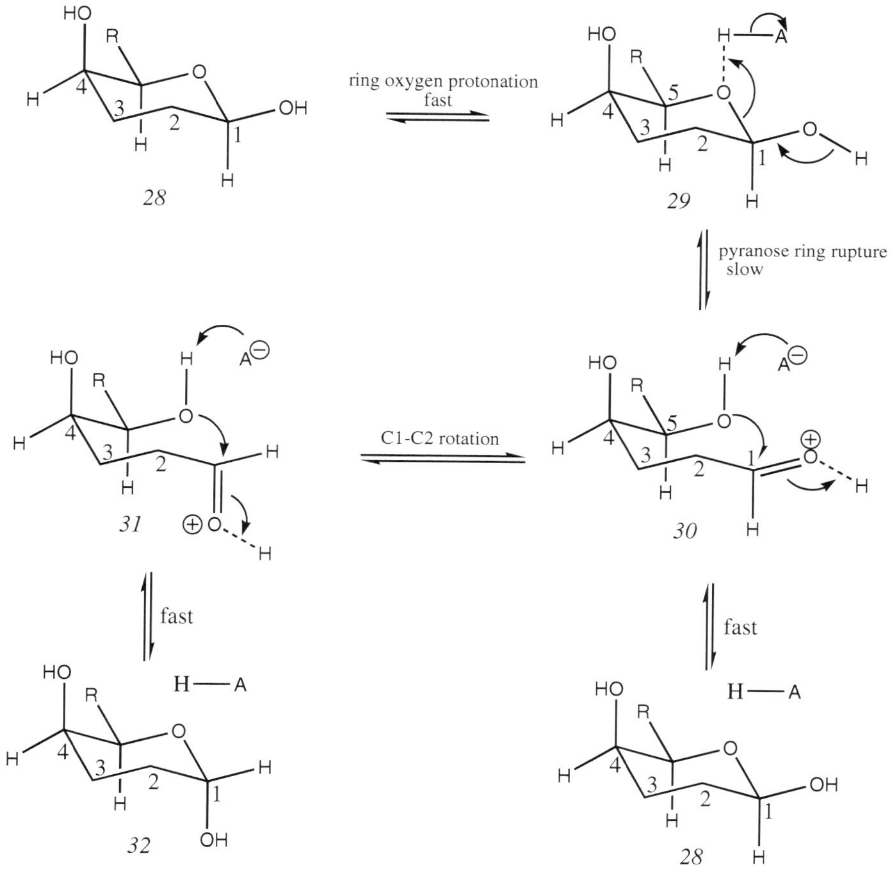

Fig. 4.13 Acid catalyzed simple mutarotation

The first step is assumed to be the fast formation of an adduct between an acid and the ring oxygen (28 → 29) which is followed by a slow (rate determining) ring rupture with the simultaneous formation of the C5–OH and a protonated carbonyl oxygen, 29 → 30. The acyclic intermediate 30 then undergoes either a fast recyclization, back to the starting β-D-pyranose, or, prior to recyclization, a rotation about the C1–C2 carbon, which takes place giving the acyclic conformer 31, which on recyclization gives the α-anomer 32.

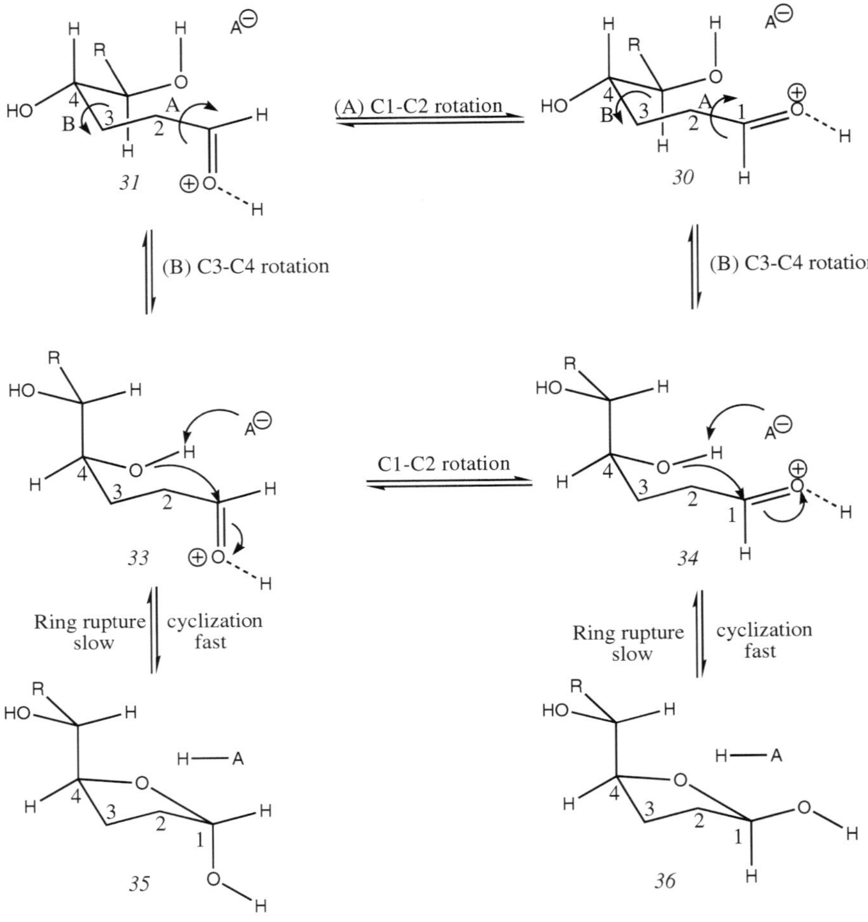

Fig. 4.14 Acid catalyzed complex mutarotation

However, an additional step can be involved in the mutarotation of sugars (Fig. 4.14). Namely, the intermediate *30* (or *31*) can, either before or after the C1–C2 rotation, undergo another rotation about the C3–C4 bond giving the conformer *33* or *34* which after recyclization is converted to the α- and β-furanoid forms of the sugar (*35* or *36*). This is the accepted mechanism for the *complex mutarotation*.

The mechanism of mutarotation of sugars catalyzed by bases differs from the one catalyzed by acids in the site of initial attack of the catalyst. Whereas the acid catalyzed mutarotation begins with the attack of an acid to the ring oxygen, in the base catalyzed mutarotation the first step is deprotonation of the anomeric hydroxyl group. The sugar anion then undergoes slow opening of the pyranose ring, with the participation of water. Closure of the ring without prior conformational change of the acyclic intermediate produces the starting material. However, the rotation about

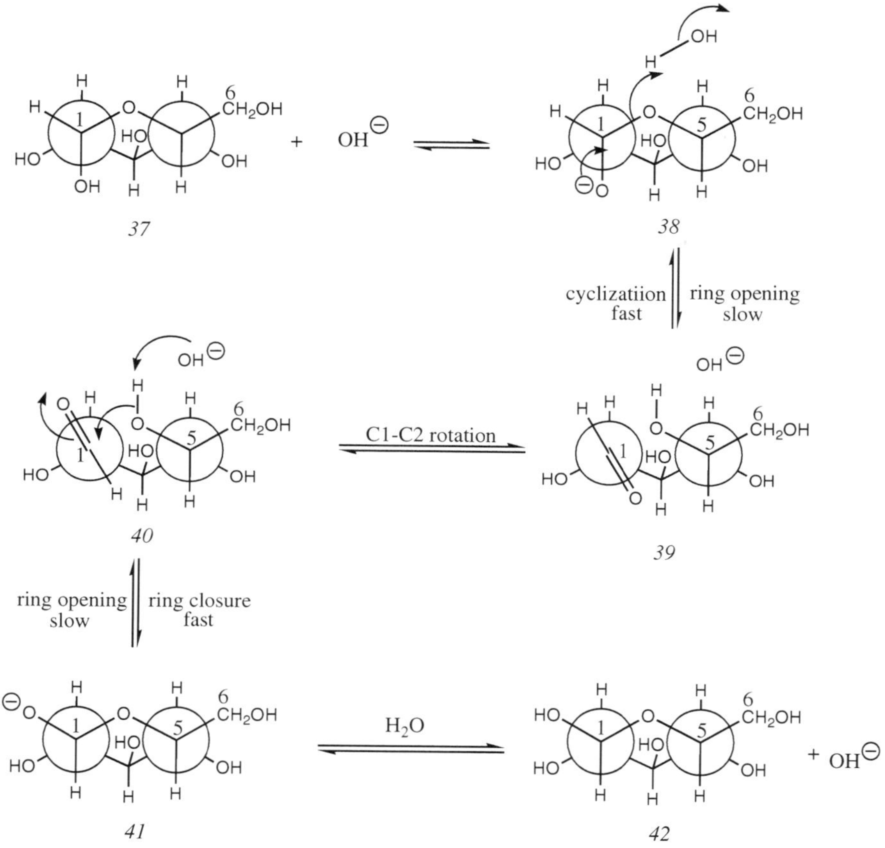

Fig. 4.15 Base catalyzed mutarotation

the C1–C2 bond prior to the ring closure gives another acyclic conformer of the parent sugar, which upon cyclization gives the anomer of starting glucopyranose (*simple mutarotation*) (Fig. 4.15).

Anomerization

The *simple* and *complex mutarotation* is typical behavior for the so-called reducing sugars, i.e., sugars with underivatized anomeric hydroxyl group. The epimerization of the C1 carbon of pento- and hexofuranoses and pento- and hexopyranoses, i.e., the α- to β-isomer conversion, and vice versa, without a ring isomerization, was named *anomerization*. An important difference between *simple mutarotation* and the *anomerization* is that the mutarotation requires presence of both an acid and a base catalyst, whereas for the anomerization only an acid catalyst is needed. In addition to that, *simple mutarotation* can take place only in protic (aqueous) solutions whereas the *anomerization* can also take place in aprotic (nonaqueous) solutions [22]. Consequently, the reaction mechanisms of *simple mutarotation* and *anomerization* are similar and different at the same time. Glycopyranosyl or glycofuranosyl derivatives that undergo anomerization are glycofuranosyl or glycopyranosyl 1-acetate, 1-amines, alkyl (or aryl) glycosides, 1-halogen derivatives, etc.

Although anomerization of glycofuranosyl or glycopyranosyl derivatives can proceed via several routes, the following four pathways are probably the most plausible.

(1) The anomerization can take place by the cleavage of C1-O5 (ring oxygen) bond when catalyzed by specific acid, followed by the C1-C2 rotation of obtained acyclic intermediate *44* and its cyclization into the C1 epimer *46* of the starting material *43* (Fig. 4.16).

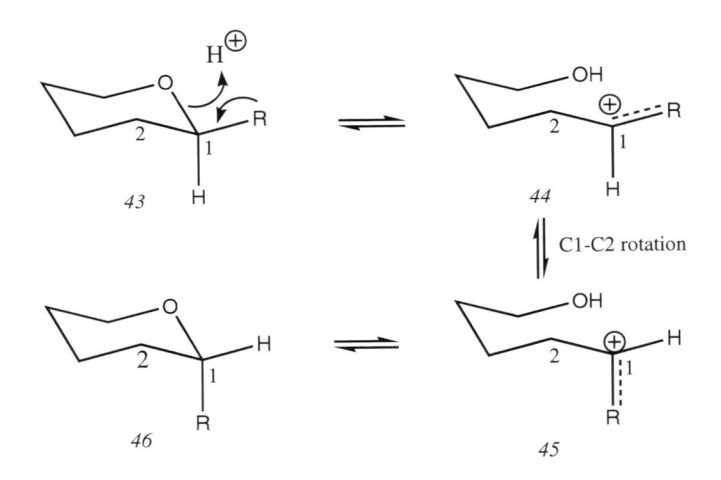

Fig. 4.16

Similar reaction pathway has been postulated for the simple mutarotation.

(2) The anomerization can take place via a complete dissociation of the C1-X bond (X = aglycon[1]) after protonation of the aglycon, followed by recombination of ions to form the starting sugar or its anomer, as shown in Fig. 4.17.

---

[1] Alkyl, aryl groups in an aglycon X are referred to as "aglycon groups." Hence the aglycon is the aglycon group and glycosidic oxygen linked together.

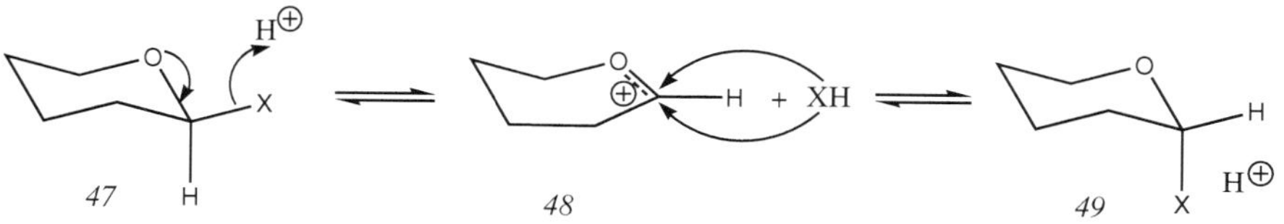

Fig. 4.17

(3) The anomerization may occur by migration (without complete dissociation) of the aglycon group (R) from one side of the ring to another, by way of an intermediate molecular complex (Fig. 4.18).

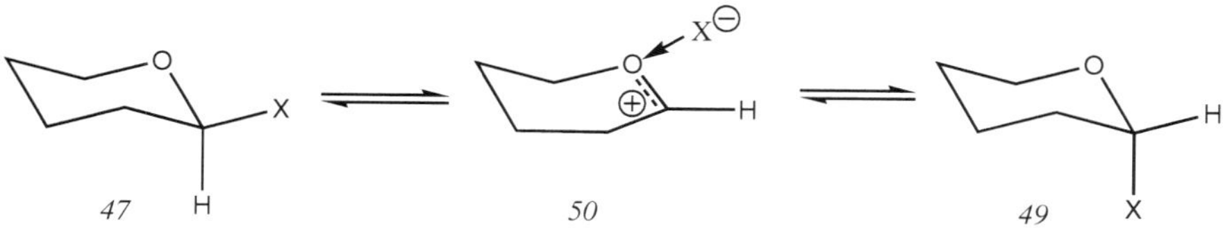

Fig. 4.18

(4) The anomerization may proceed by a way of bimolecular displacement, under the conditions where solvent supplies the aglycon group (solvolysis) (Fig. 4.19).

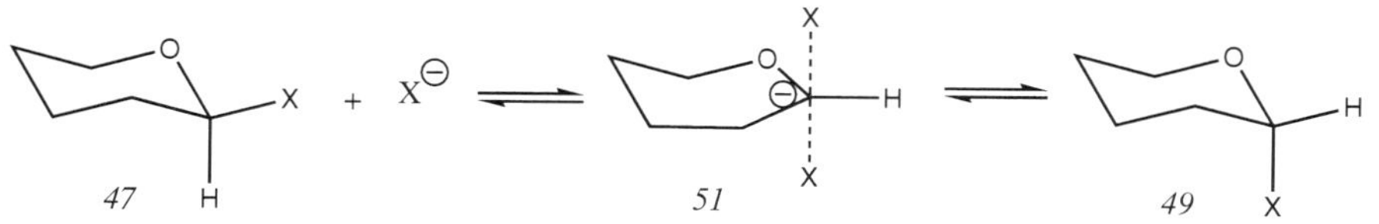

Fig. 4.19

Evidence obtained from a study of methanolysis of a number of methyl glycosides, by paper chromatography and isotope exchange, suggests that the anomerization can proceed via both acyclic (*44* and *45* in Fig. 4.16) and cyclic (*48* in Fig. 4.17) oxo-carbenium intermediate [23, 24]. Thus, from equilibration of methyl α-D-gluco- and α-D-mannopyranosides in anhydrous ^{14}C-labeled methanol containing 1% of anhydrous HCl, it was concluded that the anomerization proceeds via cyclic oxo-carbenium ion (*51* in Fig. 4.19). The same mechanism was considered to be the principal pathway for acid-catalyzed anomerization of ethyl D-xylopyranosides [25]. However, it was concluded that the anomerization of methyl β-D-glucopyranoside, methyl α- and β-D-galactopyranoside, and methyl β-D-mannopyranoside proceeds via acyclic oxo-carbenium ion (*44* and *45* in Fig. 4.16).

Some substituents of the C2 hydroxyl group of glycopyranoses, having the C1 hydroxyl group already derivatized (alkyl, acetyl, etc.), strongly favor the axial anomer in the anomeric equilibrium. Thus equilibration of methyl tetra-*O*-methyl-α and β-D-glucopyranosides in 5% methanolic HCl yields about 3:1 mixture α- to β-anomers [26].

The mechanism of anomerization has been most extensively studied with peracetylated glycopyranoses and acetylated alkyl glycopyranosides. Due to the presence of acetyl group at the C2 carbon of these substrates, the reaction mechanism is slightly different from unprotected glycopyranosides since the acetyl group is known to participate in the stabilization of oxo-carbenium ion intermediate (there will be more discussion on the neighboring group participation later).

It has been found [27] that the furanoside to pyranoside conversion proceeds predominantly with the retention of configuration at the anomeric carbon. The suggested explanation for this observation was that the acyclic ion formed on opening of the furanoside ring underwent the ring closure to the pyranoside with the same anomeric configuration at a much greater rate than that for the conformational change required to produce the other anomeric configuration (C1-C2 rotation).

Two excellent reviews on mutarotation of sugars have been published by Isbel and Pigman [28, 29].

Lobry de Bruyn–Alberda van Ekenstein Transformation

Lobry de Bruyn–Alberda van Ekenstein transformation is acid–base catalyzed aldose–aldose and ketose–ketose epimerization, and aldose–ketose isomerization [30–40] (Fig. 4.20). The reaction is essentially an enolization of an aldose or a ketose having a hydrogen at the α-carbon to the carbonyl group and proceeds via a common "enediol" 53 intermediate.

These reactions are usually base catalyzed and proceed readily in alkaline solution; however, they can also take place under acid or even neutral conditions [40]. 2-Deoxy aldoses and 2-deoxy-2-acetamido-hexoses understandably do not undergo

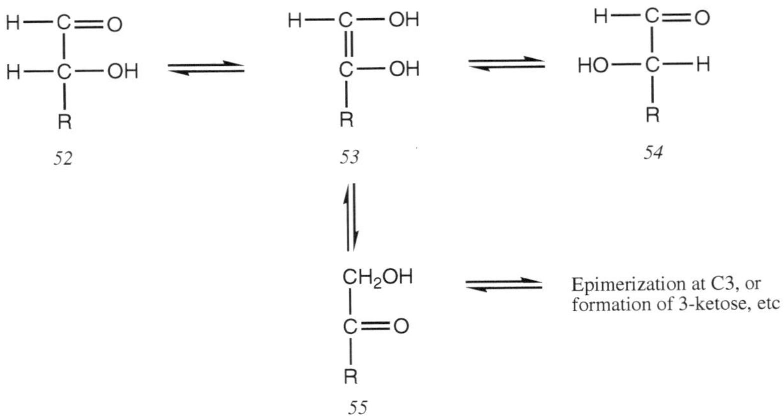

Fig. 4.20

Fig. 4.21

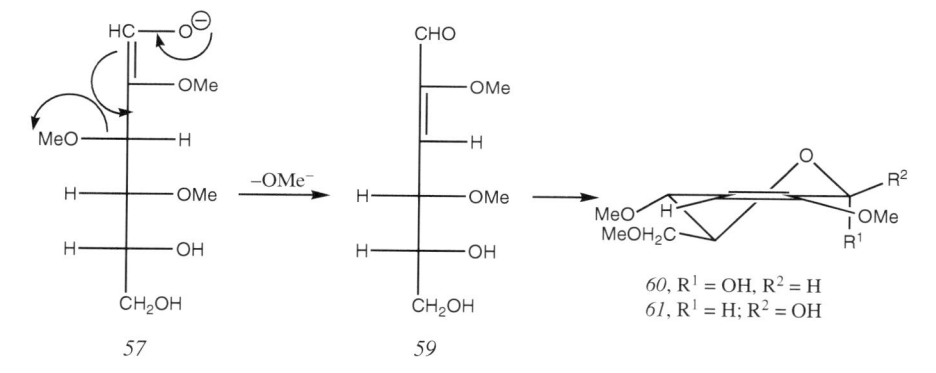

Fig. 4.22

Lobry de Bruyn–Alberta van Ekenstein transformation [41, 42]. Aldoses with their hydroxyl groups protected with alkali-stable protecting groups, such as methyl groups, on reaction with bases undergo only epimerization giving a mixture of C2 epimers (Fig. 4.20). Thus, forexample, reaction of 2,3,4,6-tetra-O-methyl-D-glucose *56* and 2,3,4,6-tetra-O-methyl-D-mannose *58* with a base (saturated lime water) gives a mixture of these two epimeric aldoses in the same proportion [43, 44] (Fig. 4.21).

The enolate *57* initially obtained by treating 2,3,4,6-tetra-O-methyl-D-glucose or 2,3,4,6-tetra-O-methyl-D-mannose with lime water slowly eliminates the C3 methoxy anion to form the unsaturated sugar *59*, which cyclizes to 3-deoxy-2,4,6-tri-O-methyl-α,β-D-*erythro*-hex-2-enopyranose *60, 61* [45, 46, 47] (Fig. 4.22).

References

1. Dubrunfaut, A. P., Compt. Rend. (1846) **23**, 38
2. Hudson, C. S., "*Multirotation of lactose*", Z. Physik. Chem. (1903) **44**, 487–494

3. Erdmann, E. D., Berichte (1880) **13**, 218
4. Urech, F., *"Zur strobometrischen Bestimmung der Invertirungsgeschwindigkeit von Rohrzucker und des Uebergangs der Birotation von Milchzucker zu seiner constanten Drehung"*, Berichte(1882) **15**, 2130–2133
5. Hudson, C. S., *"The Catalysis by Acids and Bases of the Mutarotation of Glucose"*, J. Am. Chem. Soc. (1907) **29**, 1571–1576
6. Lowry, T. M., *" Studies of dynamic isomerism. Part XVIII. The mechanism of mutarotation"*, J. Chem. Soc. (1925) **127**, 1371–1385
7. Lowry, T. M.; Richards, E. M., *"Studies of dynamic isomerism. Part XIX. Experiments on the arrest of mutarotation of tetramethylglucose"*. J. Chem. Soc. (1925) **127**, 1385–1401
8. Lowry, T. M.; Faulkner, I. J., *"Studies of dynamic isomerism. XX. Amphoteric solvents as catalysts for the mutarotation of the sugars"*, J. Chem. Soc., (1925) **127**, 2883–2887
9. Lowry, T. M.; Smith, G. F., (1930) *"Rapports sur les Hydrates de Carbone"*, 10th Conf. Intern. Union. Chem., Liege
10. Swain, C. G.; Brown, J. F.,*"Concerted Displacement Reactions. VII. The Mechanism of Acid-Base Catalysis in Non-aqueous Solvents"*, J. Am. Chem. Soc. (1952) **74**, 2534–2537
11. Hamil, W. H.; La Mer, V. K., *"The Acid-Base Catalysis of the Mutarotation of Glucose in Protium Oxide-Deuterium Oxide Mixtures"*, J. Chem. Phys. (1936) **4**, 395–401
12. Nicolle, J.; Weisbuch, F., *"Comparison of the rates of mutarotation of various sugars in water and in deuterium oxide"*, Compt. Rend. (1955) **240**, 84–85
13. Fredenhagen, H.; Bonhoeffer, K. F., *"Hexose rearrangement in heavy water"*, Z. Physik. Chem. (1938) **A181** 392–405
14. Bonhoeffer, K. F., *"Deuteron transfer in solutions"*, Trans. Faraday. Soc. (1938) **34**, 252–259
15. Bell, R. P., *"Acid-Base Catalysis"*, Oxford University Press, London, 1941, p. 82
16. Purlee, E. L., *"On the Solvent Isotope Effect of Deuterium in Aqueous Acid Solutions"*, J. Am. Chem. Soc. (1959) **81**, 263–272
17. Challis, B. C.; Long, F. A.; Pocker, Y., *"Relative rates of mutarotation of tetra- O-methyl-α-D-glucose in H_2O and D_2O and the mechanism of the reaction"*, J. Chem. Soc. (1957) 4679–4681
18. Long, F. A.; Bigeleisen, J., *"Correlations of relative rates in the solvents D2O and H2O with mechanisms of acid and base catalysis"*, Trans. Faraday Soc. (1959) **55**, 2077–2083
19. Richards, E. M.; Faulkner, I. J.; Lowry, T. M., *"Studies of dynamic isomerism. Part XXIII. Mutarotation in aqueous alcohols"*, J. Chem. Soc. (1925) **127**, 1733–1739
20. Zhdanov, Yu. A.; Minkin, V. I.; Ostroumov, Yu. A.; Dorofenko, G. N., *"Quantum chemistry of carbohydrates: Part I. The electronic structure of some pentoses"*, Carbohydr. Res. (1968) **7**, 156–160
21. Del Re, G., *"A simple M.O.-L.C.A.O. method for calculating the charge distribution in saturated organic molecules"*, J. Chem. Soc. (1958), 4031–4040
22. Lemieux, R. U., *"Some Implications in Carbohydrate Chemistry of Theories Relating to the Mechanisms of Replacement Reactions"*, Advan. Carbohydr. Chem. (1954) **9**, 1–57
23. Swiderski, J.; Temeriusz, A., *"Studies on the mechanism of methanolysis of some methyl image-glycopyranosides by the method of isotope exchange"*, Carbohydr. Res. (1966) **3**, 225–229
24. Temeriusz, A., *"Chromatographic analysis of the deacetylation products of methyl 2,3,4,6-tetra-O-acetyl-D-glucopyranosides in conditions of Fischer's methanolysis"*, Rocz. Chem. (1966) **40**, 825–829
25. Ferrier, R. J.; Hatton, L. R.; Overend, W. G., *"Studies with radioactive sugars. III. The mechanism of the anomerization of ethyl α- and β-D-xylopyranoside"*, Carbohydr. Res. (1968) 8, 56–60
26. Jungius, C. L., *"Isomeric changes of some dextrose derivatives, and the mutarotation of the sugars"*, Z. Phys.Chem. (1905) **52**, 97–108
27. Bishop, C. T.; Cooper, F. P., *"Glycosidation of Sugars: I. Formation of Methyl-D- Xylosides"*, Can. J. Chem. (1962) **40**, 224–232

References

28. Isbell, H. S.; Pigman, W., "*Mutarotation of Sugars in Solution: Part I*", Adv. Carbohydr. Chem. (1968) **23**, 11–57
29. Isbell, H. S.; Pigman, W., "*Mutarotation of Sugars in Solution: Part II*", Adv. Carbohydr. Chem. (1969) **24**, 13–65
30. Lobry de Bruyn, C. A., "*Action of dilute alkalis on the carbohydrates*" Rec. Trav. Chim. (1895) **14**, 156–165
31. Lobry de Bruyn, C. A.; van Ekenstein, W. A., "*Action of alkalis on the sugars. Reciprocal transformation of glucose, fructose, and mannose*" Rec. Trav. Chim. (1895) **14**, 201–206
32. Lobry de Bruyn, C. A.; Alberda van Ekenstein, W., "*Action of alkalis on the sugars. IV"*, Rec. Trav. Chim. (1897)**16**, 257–261
33. Lobry de Bruyn, C. A.; Alberda van Ekenstein, W., Rec. Trav. Chim. (1897)**16**, 241
34. Lobry de Bruyn, C. A.; Alberda van Ekenstein, W., "*Action of alkalis on the sugars. V. Transformation of galactose. The tagatoses and galtose*", Rec. Trav. Chim. (1897), **16**, 262–273
35. Lobry de Bruyn, C. A.; Van Alberda Ekenstein, W.,"*Action of alkalis on the sugars. VI. Glutose and ψ-fructose*", Rec. Trav. Chim. (1897) **16**, 274–281
36. Lobry de Bruyn, C. A.; Alberda Van Ekenstein, W., "*Action of boiling water on d- fructose (levulose)*", Rec. Trav. Chim. (1897) **16**, 282–283
37. Lobry de Bruyn, C. A.; Alberda van Ekenstein, W., "*Action of alkalis on the sugars. VII. Maltose, lactose, and melibiose*", Rec. Trav. Chim. (1899) **18**, 147–149
38. Lobry de Bruyn, C. A.; Alberda van Ekenstein, W., "*D-Sorbose and L-sorbose (ψ- tagatose) and their configuration*", Rec. Trav. Chim. (1900) **19**, 1–11
39. Alberda van Ekenstein, W.; Blanksma, J. J., "*The Transformation of l-Gulose and of l-Idose into l-Sorbose*", Rec. Trav. chim. (1908) **27**, 1–4
40. Speck, J. C., "*Lobry De Bruyn-Alberta Van Ekenstein Transformation*", Adv. Carbohydr. Chem. (1958) **13**, 63–103
41. Spivak, C. T.; Roseman, S., "*Preparation of N-Acetyl-D-mannosamine (2-Acetamido- 2-deoxy-D-mannose) and D-Mannosamine Hydrochloride (2-Amino-2-deoxy-D-man nose)*", J. Am. Chem. Soc. (1959) **81**, 2403–2404
42. Coxon, B.; Hough, L., "*Epimerization of 2-acetamido-2-deoxy-D-pentoses*", J. Chem. Soc. (1961) 1577–1579
43. Wolfrom, M. L.; Lewis, W. L., "*The reactivity of the methylated sugars. II. The action of dilute alkali on tetramethyl glucose*", J. Am. Chem. Soc. (1928) **50**, 837–854
44. Green, R. D.; Lewis, W. L., "*The reactivity of the methylated sugars. III. The action of diluted alkali on tetramethyl-d-mannose*", J. Am.Chem. Soc. (1928) **50**, 2813–2825
45. Anet, E. F. L. J., "*Unsaturated sugars: enols of 3-deoxy-D-"glucosone*", Chem. Ind. (London) (1963) 1035–1036
46. Anet, E. F. L. J., "*Degradation of carbohydrates. VI. Isolation and structure of some Glycos-2-enes*", Aust. J. Chem. (1965) **18**, 837–844
47. Klemer, A.; Lukowski, H.; Zerhusen, F., "*Über den alkalischen Abbau einiger D- Glucosemethyläther: 2-Methyl- und 2.4.6- Trimethyl-D-glucoseen-(2.3)*", Chem. Berichte (1963) **96**, 1515–1519

Chapter 5
Relative Reactivity of Hydroxyl Groups in Monosaccharides

Introduction

The furanoid and pyranoid cyclic structures of monosaccharides generally may have four types of chemically distinguishable hydroxyl groups: (1) the *anomeric* (*hemiacetal, lactol*) hydroxyl group and (2) three types of alcoholic hydroxyl groups: (a) the *primary hydroxyl group*, which is always *exocyclic* with regard to the carbohydrate ring and is found in both hexopyranoses and pento- or hexofuranoses, (b) the *endocyclic secondary hydroxyl* groups found in both hexopyranoses and pento- or hexofuranoses, and (c) the *exocyclic* secondary hydroxyl groups found in hexofuranoses or in higher sugars containing hexopyranose ring.

The reactivity of hydroxyl groups, excluding the anomeric hydroxyl group, is controlled by several factors. First, it depends on whether the hydroxyl group is a primary or a secondary one. In general, the *primary hydroxyl group* is *more reactive* than the secondary one suggesting that the difference in reactivity is most likely due to steric control. The steric factor is probably also responsible for the *exocyclic secondary hydroxyl groups* being more reactive than *endocyclic ones*. The difference in the reactivity among the endocyclic secondary hydroxyl groups is most likely controlled by stereoelectronic factors and by the ability of individual hydroxyl groups to form the intramolecular hydrogen bonds with neighboring hydroxyl or alkoxy oxygen. This difference is evident from dependence of reactivity of a given hydroxyl group on its position in a furanoid and/or pyranoid ring. Finally, the reactivity of a hydroxyl group depends on its configuration, i.e., on whether it is in the axial or in the equatorial orientation on a pyranoid ring.

The chemical behavior of the anomeric hydroxyl group, which is a hemiacetal or lactol hydroxyl group, is in many respects very different from all other hydroxyl groups of a monosaccharide. Thus (1) it can be easily oxidized (in both aldopyranoses and aldofuranoses) with aqueous bromine solution to a corresponding glyconolactone, whereby the other hydroxyl groups of a carbohydrate remain unchanged and (2) it can be replaced with an alkoxy group by treating a monosaccharide with an alcohol in the presence of an anhydrous mineral acid. Thus, for example, reaction of a furanoid or pyranoid form of a monosaccharide with anhydrous methanol and catalytic amounts of anhydrous HCl (gas) will result in replacement, with the methoxy group, of only the anomeric hydroxy group giving the so-called *methyl glycoside* as the product; all other hydroxyl groups will remain unchanged under the reaction conditions. The ability of monosaccharides to undergo this, so-called, *glycosidation* reaction is by far the most important property of carbohydrates and has found an extensive use in living organisms for the synthesis of biologically important carbohydrate oligomers and polymers, as well as in the glycosylation of other bio-molecules, such as proteins, lipids, that perform specific biological functions.

The relative reactivity of individual sugar hydroxyl groups depends also on the type of the reagent that is used for their transformation to corresponding derivatives. Thus in the case of acylation it often depends upon the nature of acylating agent (acyl chloride, acyl anhydride, etc.) and in case of alkylation upon the nature of alkylating agent.

Selective Acylation (Esterification)

Discussion of selectivity of acylation of hexo- and pentopyranoses will be first focused on acyl chlorides and acyl anhydrides as acylating agents because they are the most widely studied. At the end of discussion, a select number of other acylating agents will be described, which were developed to either improve the regioselectivity or to increase the yield of acylation.

The readers are referred to three very informative reviews dealing with this topic [1–3].

Our discussion on selectivity of sugar hydroxyl groups toward acylation will be limited to the following monosaccharides: α- and β-anomers of alkyl and/or aryl D-gluco- (*1*), D-manno- (*2*) and D-galactopyranosides (*3*) among hexopyranoses and α- and β-anomers of alkyl D-xylopyranoside (*4*) among pentopyranoses. The reason for this choice is that none of these glycopyranoses has either the O:O or the C:O 1,3-*syn*–axial interaction (the O:H 1,3-*syn*–axial interaction that are present in *2* and *3* are presumed to have no influence upon the final conclusions) (Fig. 5.1). In all other hexopyranoses there is at least one 1,3-*syn*–axial interaction between the axial hydroxyl groups and the presence of these interactions may adversely interfere with the interpretation of obtained results.

For example, α-anomers of D-allopyranose *5*, D-altropyranose *6*, and D-gulopyranose *7* (Fig. 5.2) all have one 1,3-*syn*–axial interaction between the axial anomeric hydroxyl group and the axial 3-OH group. The β-anomers of these hexoses have no 1,3-*syn*–axial interactions. There is one 1,3-*syn*–axial interaction between the 2-OH and the 4-OH present in both α- and β-D-talopyranose (*8* and *9*, respectively) as well as in β-D-idopyranose *11*, whereas in the α-D-idopyranose *10*, there are two *syn*–axial interactions, one between the anomeric hydroxyl group and the C3 hydroxyl group and the other between the 2-OH and the 4-OH (Fig. 5.3).

It must be emphasized that relative reactivity of hydroxyl groups in carbohydrates is, in addition to being of theoretical interest, also of great practical importance [1].

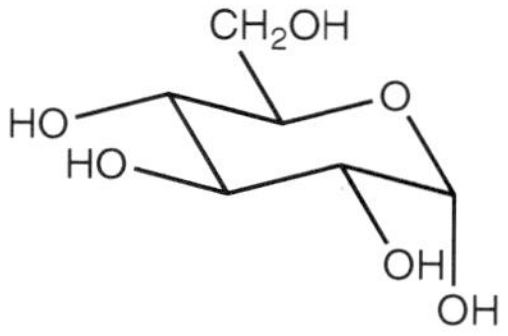

Selective Acylation (Esterification)

Fig. 5.1

α-D-glucopyranose (*1*), α-D-mannopyranose (*2*), α–D-galactopyranose (*3*), α-D-xylopyranose (*4*)

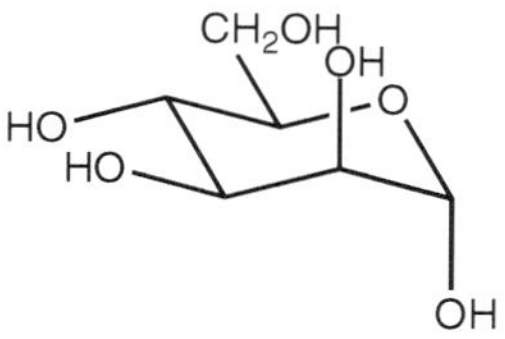

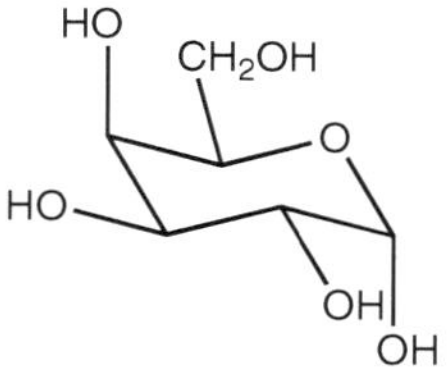

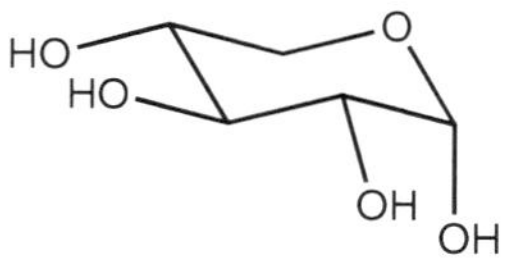

Fig. 5.2

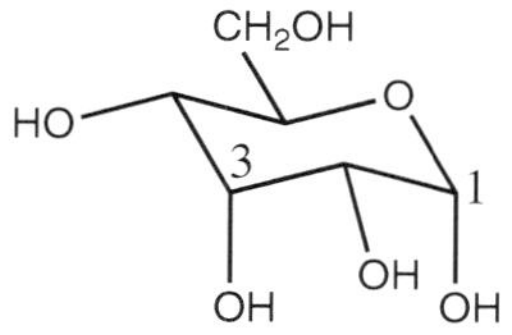

α-D-allopyranose (*5*), α-D-altropyranose (*6*), α-D-gulopyranose (*7*)

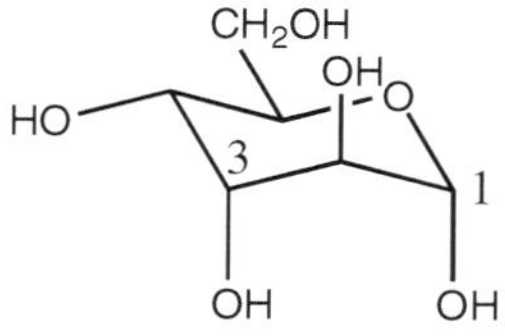

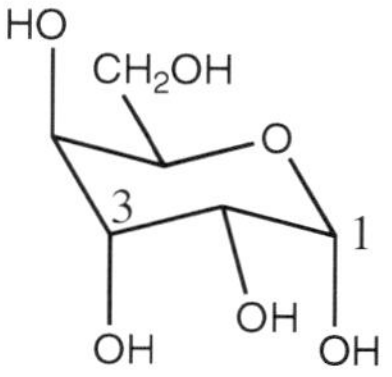

Fig. 5.3

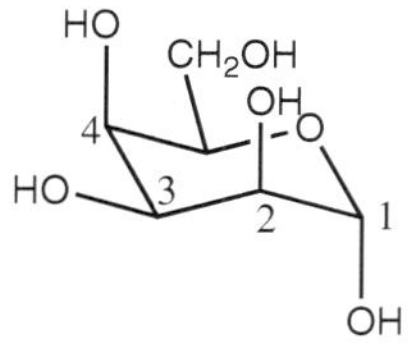

α-D-talopyranose (*8*), β-D-talopyranose (*9*), α-D-idopyranose (*10*), β-D-idopyranose (*11*)

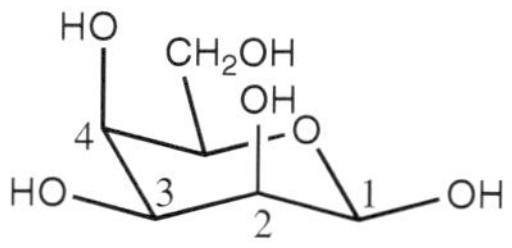

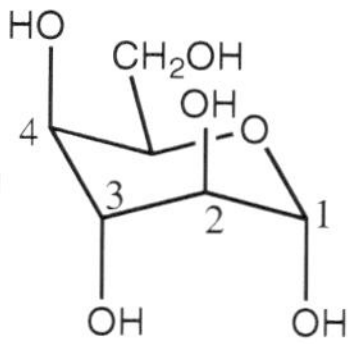

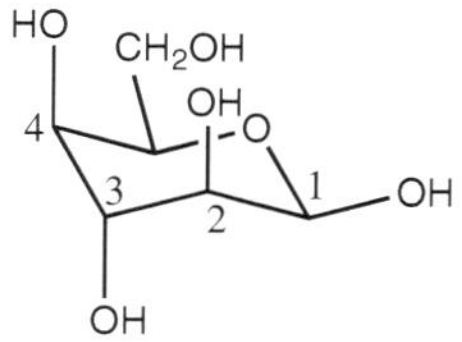

A good overall picture of relative reactivity of C2, C3, C4, and C6 hydroxyl groups in methyl α-D-glycopyranosides can be obtained by studying the partial acylation using 1, 2, and 3 mol equivalent of acylating reagent to 1 mol of methyl α- or β-D-glycopyranoside. The study of acylation of partially blocked glycopyranose derivative such as methyl 4,6-O-benzylidene-α-D-glycopyranoside permits determination of relative reactivity of the C2 and C3 hydroxyl groups.

The selective alkyl- or arylsulfonylation (methanesulfonylation – *mesylation* or *p*-toluenesulfonylation – *tosylation*) and benzoylation are the most extensively studied selective acylation reactions of monosaccharides.

Selective p-Toluenesulfonylation (Tosylation) and Methanesulfonylation (Mesylation)

Tosylation of methyl α-D-glucopyranoside *12* at 0°C with 1 mol equivalent of *p*-toluenesulfonyl (tosyl) chloride in pyridine [4] gives a mixture of tosyl esters in which the 6-O-tosylate *13* was a preponderant product (36%) (Fig. 5.4).

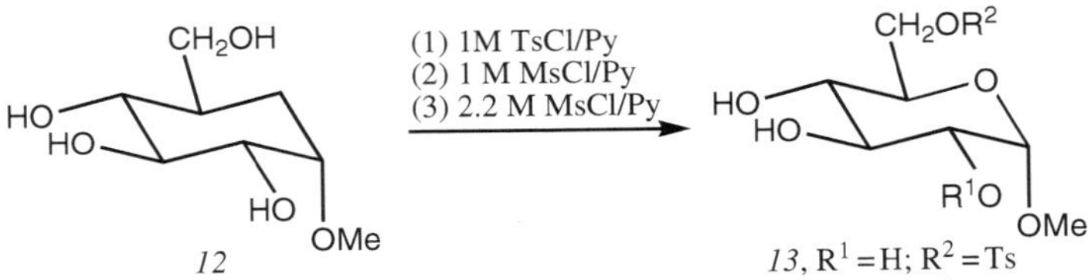

Fig. 5.4

The methanesulfonylation [5] (mesylation) of methyl α-D-glucopyranoside *12* with 1 mol equivalent of methane sulfonyl (mesyl) chloride in pyridine at −20°C gave the 6-O-mesyl derivative *14* with a considerably higher yield (67%) (Fig. 5.4).

However, in another study [6] the monomesylation of methyl α-D-glucopyranoside *12* with mesyl chloride in pyridine below −20°C was reported to give a complex mixture of products: the 6-O-mesyl ester *14* was obtained in only 20% yield, the 2,6-di-O-mesyl ester *15* was obtained in 10% yield, and 2-O-mesyl ester *15* in 2.5% yield. When mesylation was conducted with 2 mol equivalent [6] or 2.2 mol equivalent [7] of mesyl chloride in pyridine at −20°C, 2,6-di-O-mesyl derivative was obtained in 51% yield indicating that the C6 primary hydroxyl group and the C2 equatorial hydroxyl group are the most reactive hydroxyl groups in *12* (Fig. 5.4) and that the C2 hydroxyl group is more reactive toward mesylation than both the C3 and the C4 hydroxyl groups (Fig. 5.5).

The tosylation of methyl β-D-glucopyranoside [4] *17* with 1 mol equivalent of *p*-toluenesulfonyl chloride in pyridine gave 41% of 6-O-tosyl ester *18* indicating that the primary C6 hydroxyl group is the most reactive hydroxyl group toward tosylation also in β-anomers.

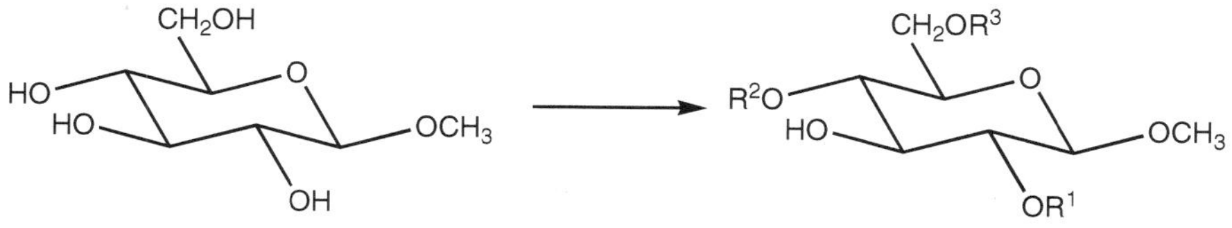

Selective Acylation (Esterification)

Fig. 5.5

17

18, $R^1 = R^2 = H; R^3 = Ts$
19, $R^2 = R^3 = Ms; R^1 = H$
20, $R^1 = R^3 = Ms; R^2 = H$
21, $R^3 = Ms; R^1 = R^2 = H$

Dimolar mesylation of methyl β-D-glucopyranoside [6] *17* was much less selective than the dimolar mesylation of α-isomer. Now the major product was methyl 4,6-di-*O*-methanesulfonyl-β-D-glucopyranoside *19* (13%), whereas the methyl 2,6-di-*O*-methanesulfonyl-β-D-glucopyranoside *20* and methyl 6-*O*-methanesulfonyl-β-D-glucopyranoside *21* were obtained with 4% yield each (Fig. 5.5).

These results show that selectivity of mesylation of endocyclic hydroxyl groups does not depend only on the type and the position of a hydroxyl group in the ring but also on the anomeric configuration. In β-D-glucopyranoside the relative reactivity of the C2 and the C4 hydroxyl groups seems to be reversed.

Tosylation of methyl 4,6-*O*-benzylidene-α-D-glucopyranoside [8] *22* in pyridine with 0.74 mol equivalent of TsCl for 24 h gives the 2-*O*-tosyl ester *23* with 64.30% yield. Tosylation of methyl 4,6-*O*-benzylidene-α-D-glucopyranoside *22* in pyridine with 1.3 mol equivalent of TsCl at room temperature for 12 h [9] gives 64.46% of 2-*O*-tosyl ester *23* (Fig. 5.6). Small amounts of starting material and the 2,3-ditosyl ester *24* were also isolated.

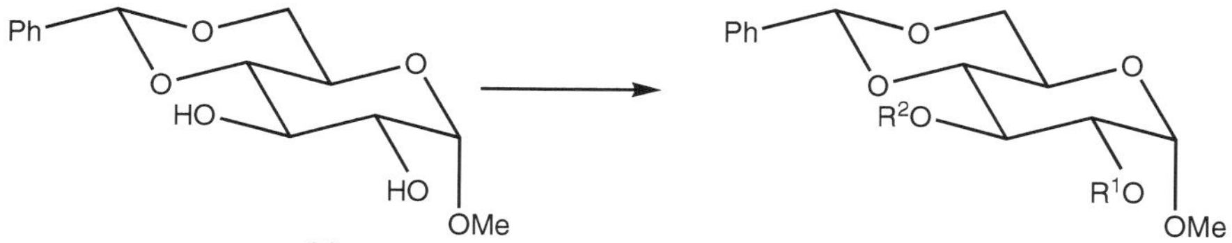

22

23, $R^1 = Ts; R^2 = H$
24, $R^1 = R^2 = Ts$
25, $R^1 = Ms; R^2 = H$
26, $R^1 = R^2 = Ms$

Fig. 5.6

The mesylation of methyl 4,6-*O*-benzylidene-α-D-glucopyranoside 22 with 1.1 mmol of methanesulfonyl chloride in pyridine gave the 2-*O*-mesyl derivative *25* with 68% yield [10] together with 16% of 2,3-dimesylate *26*.

118 5 Relative Reactivity of Hydroxyl Groups in Monosaccharides

These results are in full accord with the previous conclusion that the 2-OH is more reactive than 3-OH group in methyl α-D-glucopyranoside. This study also indicated that the selectivity of these two hydroxyl groups toward mesylation and tosylation is identical.

The tosylation of methyl 4,6-O-benzylidene-β-D-glucopyranoside [11] 27 with 1.1 mol equivalent of tosyl chloride in pyridine gave 2-O-tosyl ester 28 with 21% yield, 3-O-tosyl ester 29 with 28% yield, and 2,3-di-O-tosyl ester 30 with 4% yield (Fig. 5.7).

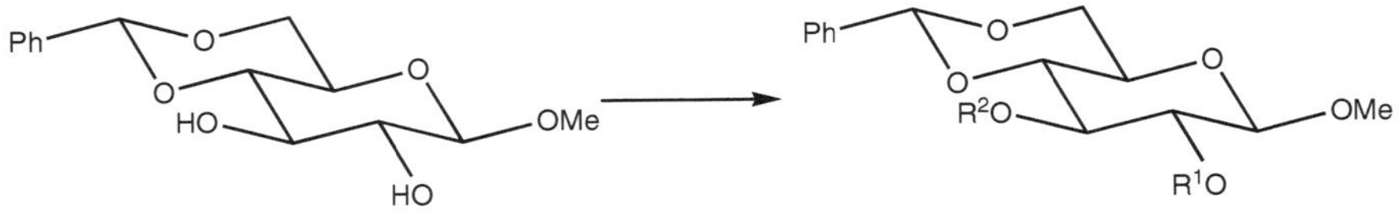

28, R^1 = Ts; R^2 = H
29, R^1 = H; R^2 = Ts
30, R^1 = R^2 = Ts
31, R^1 = R = Ms
32, R^1 = Ms; R^2 = H
33, R^1 = H; R^2 = Ms

Fig. 5.7

In another study [12] tosylation of methyl 4,6-O-benzylidene-β-D-glucopyranoside 27 with 1 mol equivalent of TsCl was conducted at 4°C for 6 days (pyridine solution of TsCl and the pyridine solution of sugar were pre-cooled with dry ice–acetone prior to mixing), 3-O-tosyl ester 29 was obtained with 8% yield, 2-O-tosyl ester 28 with 2–5% yield, and 2,3-di-O-tosyl ester 30 with 40% yield (20% of the starting material 26 was recovered after the reaction) (Fig. 5.7).

Thus, the tosylation of methyl 4,6-O-benzylidene-β-D-glucopyranoside has shown (1) a much lower selectivity from that found in the corresponding α-anomer and (2) the relative reactivity of the C2 and the C3 hydroxyl groups appears to be reversed.

Selective mesylation of methyl 4,6-O-benzylidene-β-D-glucopyranoside [12] with 1 mol equivalent of methanesulfonyl chloride in pyridine (pyridine solution of MsCl and the pyridine solution of sugar were pre-cooled with dry ice-acetone prior to mixing) by keeping the reaction mixture at 4°C overnight gave 2,3-di-O-mesylate 31 with 41% yield, 2-O-mesylate 32 with 6% yield, and 3-O-mesylate 33 with 17% yield (Fig. 5.7).

From these, as well as from other studies, it was concluded [1] that the primary hydroxyl group of both α- and β-D-glucopyranoses is more reactive toward alkyl or arylsulfonyl chlorides than any of the secondary hydroxyl groups, thus permitting the selective esterification of the terminal hydroxyl group in methyl glucopyranosides.

Kondo [13] studied selective tosylation of methyl α-D-mannopyranoside 34 using 2 mol equivalent of p-toluenesulfonyl chloride and found that two major products were 6-O-tosylate 35 (35%) and 3,6-ditosylate 36 (35%). The mixture of minor

Selective Acylation (Esterification)

products consisted of 4,6- (*37*) and 2,6-ditosylate (*38*) (14%), 2,3,6-tri-*O*-tosylate *39* (3%) and 3,4,6-tritosylate *40* (1%) (Fig. 5.8).

35, $R^1 = R^2 = R^3 = H$; $R^4 = Ts$
36, $R^2 = R^4 = Ts$; $R^1 = R^3 = H$
37, $R^1 = R^2 = H$; $R^3 = R^4 = Ts$
38, $R^1 = R^4 = Ts$; $R^2 = R^3 = H$
39, $R^1 = R^2 = R^4 = Ts$; $R^3 = H$
40, $R^1 = H$; $R^2 = R^3 = R^4 = Ts$

Fig. 5.8

Mesylation of methyl α-D-mannopyranoside with the 3 mol equivalent of MsCl gave methyl 2,3,6-tri-*O*-methanesulfonyl-α-D-mannopyranoside with 41% yield together with some other unidentified mesylation products [6].

The tosylation of methyl 4,6-*O*-ethylidene-α-D-mannopyranoside *41* with 1.0 mol equivalent of TsCl (2 days at –5°C and 1 day at 0°C) gave the 3-*O*-tosylate *42* in good yield (59.16%), whereas the 2-*O*-tosyl derivative was not detected indicating that the equatorial 3-OH is sulfonylated much more readily than the axial 2-OH group [14] (Fig. 5.9).

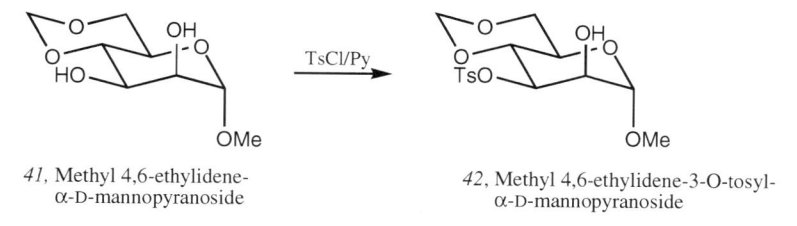

41, Methyl 4,6-ethylidene-α-D-mannopyranoside

42, Methyl 4,6-ethylidene-3-*O*-tosyl-α-D-mannopyranoside

Fig. 5.9

Tosylation of methyl 4,6-*O*-benzylidene-α-D-mannopyranoside *43* with 1.1 mol equivalent of TsCl in pyridine at 0°C for 24 h [15] gave the 3-*O*-tosylate *44* with 35.9% yield together with small amount of 2,3-di-*O*-tosyl derivative (4.3%) (Fig. 5.10).

The mesylation of methyl α-D-galactopyranoside [6] *45* with 2 mol equivalent of methane sulfonyl chloride gave 2,6-di-*O*-mesyl ester *46* with 20% yield, 2,3,6-tri-*O*-mesyl ester *47* with 10% yield, and 3,6-di-*O*-mesyl ester *48* with 4% yield. The mesylation of *45* with 3 mol equivalent of methanesulfonyl chloride [6] gave 2,3,6-tri-*O*-mesyl ester *47* with 30% yield (Fig. 5.11).

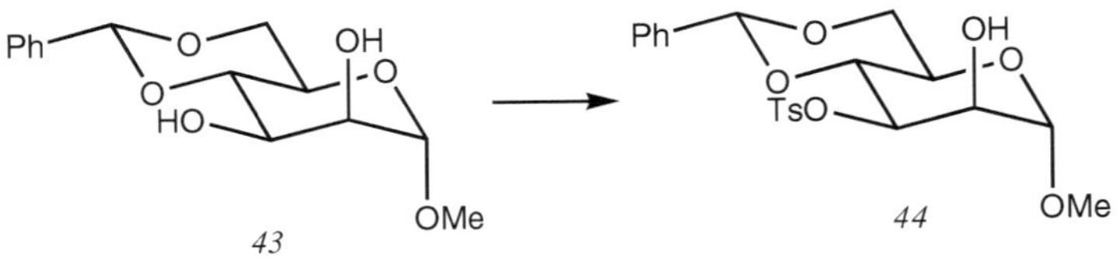

120   5 Relative Reactivity of Hydroxyl Groups in Monosaccharides

Fig. 5.10

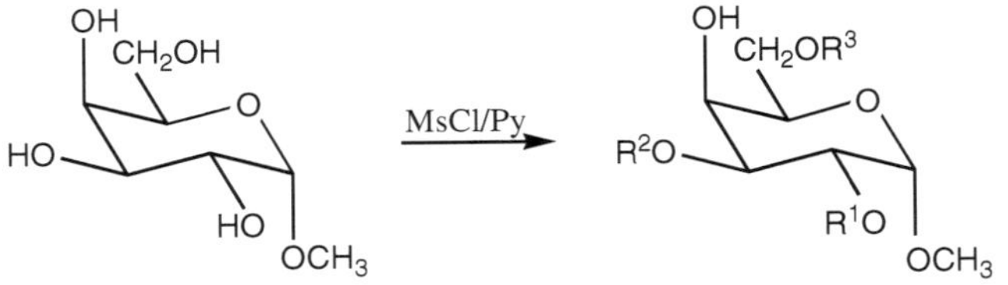

Fig. 5.11

Mesylation of methyl-β-D-galactopyranoside [6] *49* with 2 mol equivalent of methanesulfonyl chloride gave the 3,6-di-*O*-mesyl ester *50* with 26% yield (Fig. 5.11).

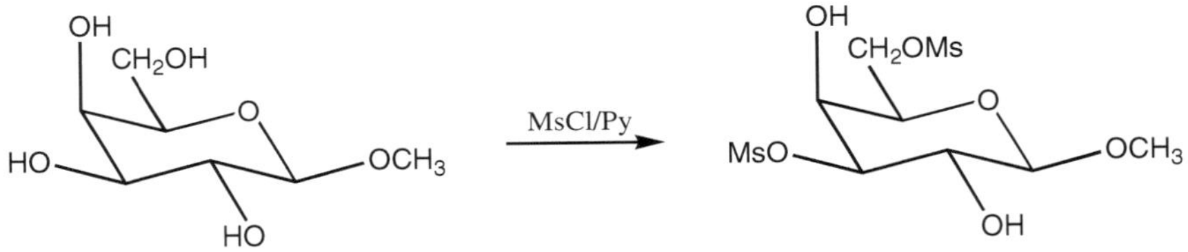

Fig. 5.12

Selective Benzoylation

Williams and Richardson [16] studied the selective benzoylation of methyl α-D-gluco-, α-D-manno-, and α-D-galactopyranoside with benzoyl chloride in pyridine. The benzoylation of methyl α-D-glucopyranoside *12* with 3.1 mol equivalent of benzoyl chloride gave a reaction mixture containing predominantly two tribenzoates: 2,3,6-tribenzoate *51*, which was the major product (67% yield), and 2,4,6-tribenzoate *52* which was the minor product (28% yield) (in pure form, *51* was isolated with 49% yield and *52* with 19% yield). Dibenzoylation of methyl α-D-glucopyranoside *12* gave 2,6-di-*O*-benzoate *53* with the 50% yield (Fig. 5.13).

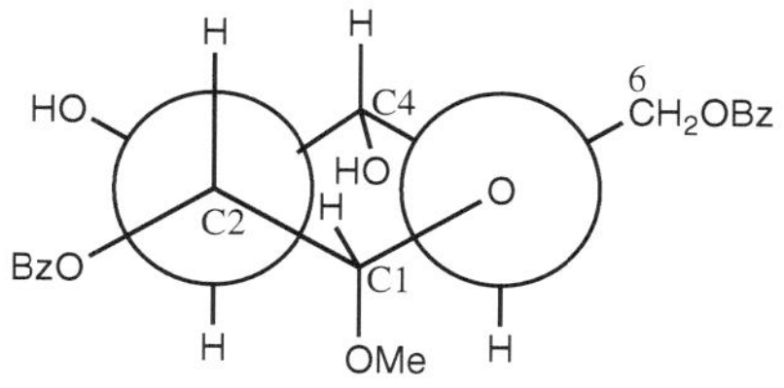

Selective Acylation (Esterification)

56, Methyl 2,6-di-O-benzoyl-α-D-glucopyranoside

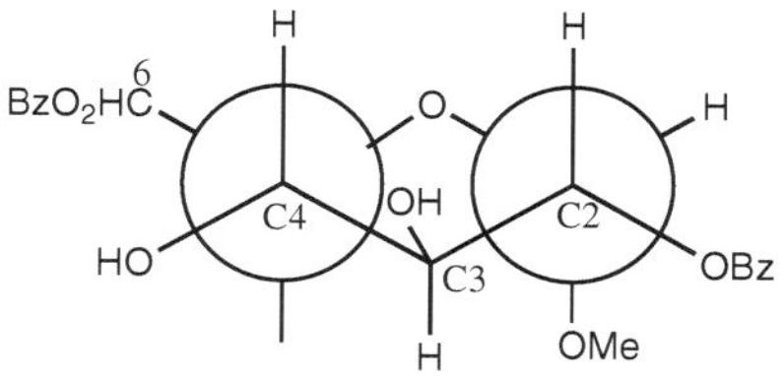

57, Methyl 2,6-di-O-benzoyl-α-D-glucopyranoside

Fig. 5.13

The lower reactivity of the C4 hydroxyl group compared to 3-OH (2,3,6-tri-O-benzoyl derivative *51* was obtained with 2.4 times greater yield than 2,4,6-tribenzoate *52*) was explained by steric hindrance (Fig. 5.14), namely it is known that the C6 primary hydroxyl and the C2 hydroxyl groups are the two most reactive hydroxyl groups in *12* toward acylation. Therefore, it can be expected that the 2,6-dibenzoate will be formed first. The tribenzoates are actually then formed by benzoylation of 2,6-dibenzoate. In 2,6-dibenzoate the C3 hydroxyl group is *gauche* to the 2-*O*-benzoate and to the C4 hydroxyl group *56* (Fig. 5.14) and the C4 hydroxyl group is *gauche* to the C6 hydroxymethyl group and to the C3 hydroxyl group *57* (Fig. 5.14). Therefore due to steric constraint of the C4 hydroxyl group, the C3 hydroxyl group is more available for acylation and thus will be benzoylated at a faster rate than the 4-OH, thus explaining the preponderance of 2,3,6-tribenzoate (67%) over the 2,4,6-tribenzoate (28%).

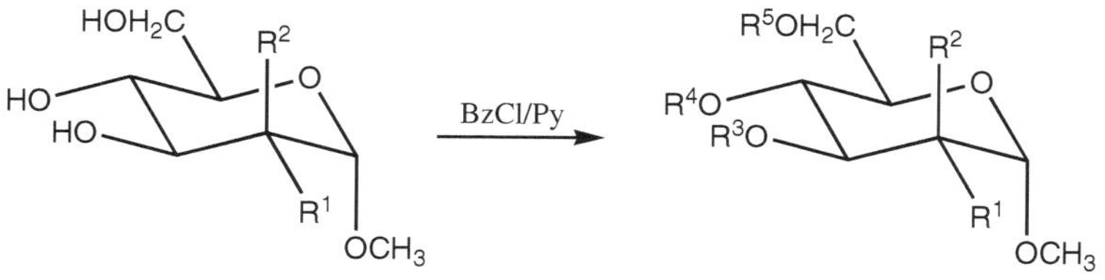

12, R^1 = OH; R^2 = H
34, R^1 = H; R^2 = OH

51, R^1 = OBz; R^3 = R^5 = Bz; R^2 = R^4 = H
52, R^1 = OBz; R^4 = R^5 = Bz; R^2 = R^3 = H
53, R^1 = OBz; R^2 = R^3 = R^4 = H; R^5 = Bz
54, R^2 = OBz; R^3 = R^5 = Bz; R^1 = R^4 = H
55, R^1 = R^4 = H; R^2 = OH; R^3 = R^5 = Bz

Fig. 5.14

Lieser and Schweizer [17, 18] obtained methyl 2,6-di-*O*-benzoyl-α-D-glucopyranoside *53* with 50% yield on benzoylation of methyl α-D-glucopyranoside *12* with 2.0 mol equivalent of benzoyl chloride in pyridine. It is interesting that the benzoylation of phenyl β-D-glucopyranoside with equimolar amount of benzoyl chloride in pyridine gave phenyl 6-*O*-benzoyl-β-D-glucopyranoside with 73% yield, indicating an unusual high selectivity of the primary hydroxyl group for benzoylation compared to mesylation or tosylation.

The order of reactivity of secondary hydroxyl groups in methyl α-D-glucopyranoside toward benzoylation with benzoyl chloride in pyridine is thus 2-OH > 3-OH > 4-OH.

Benzoylation of methyl α-D-mannopyranoside *34* (Fig. 5.13) with 3.1 M equivalent of benzoyl chloride in pyridine [16] gave 2,3,6-tri-*O*-benzoate *54* as the major product (56%) and methyl 3,6-di-*O*-benzoyl-α-D-mannopyranoside *55* as the minor product (26%).

Benzoylation of methyl α-D-mannopyranoside *34* with 2 M equivalent of benzoyl chloride in pyridine [16] gave methyl 3,6-di-*O*-benzoyl-α-D-mannopyranoside *55* with 62% yield (Fig. 5.13) suggesting that in methyl α-D-mannopyranoside the C3 hydroxyl group is the most reactive secondary hydroxyl group. The lesser reactivity of the 2-OH is probably due to its unfavorable axial orientation, since it is known [18], from conformational analysis, that axial hydroxyl groups are considerably less reactive (3.69 times) toward acylation than equatorial ones.

The fact that tribenzoylation of methyl α-D-mannopyranoside gave the 2,3,6-tribenzoate *54* as the predominant product shows that the C4 hydroxyl group is the least reactive of the three secondary hydroxyl groups. Thus the order of reactivity of secondary hydroxyl groups in methyl α-D-mannopyranoside is 3-OH > 2-OH > 4-OH. The greater reactivity of the C2 hydroxyl group compared with the C4 hydroxyl group is difficult to rationalize. Perhaps, the C2 hydroxyl group experiences less unfavorable *gauche* interactions than the 4-OH and/or is somehow activated by the anomeric group.

The benzoylation of methyl α-D-galactopyranoside *45* with 4.2 mol equivalent of benzoyl chloride in pyridine [16] at −30°C gave 2,3,6-tri-*O*-benzoate *58* with 65% yield, 3,6-di-*O*-benzoate *59* with 6% yield, and 2,3,4,6-tetra-*O*-benzoate *60* with 2% yield (Fig. 5.15).

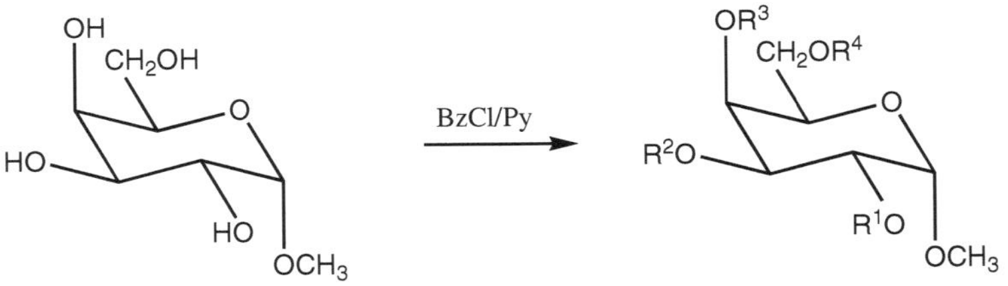

45 → BzCl/Py →

58, $R^1 = R^2 = R^4 = Bz$; $R^3 = H$
59, $R^1 = R^3 = H$; $R^2 = R^4 = Bz$
60, $R^1 = R^2 = R^3 = R^4 = Bz$

Fig. 5.15

The proposed explanation for the predominant formation of *58* (Fig. 5.15) is again that the axial C4 hydroxyl group is considerably less reactive than equatorial ones (C2 and the C3) [19].

Selective Acylation (Esterification)

Benzoylation of methyl 6-deoxy-α-L-galactopyranoside *61* (methyl α-L-fucopyranoside) with 2.1 M equivalent of benzoyl chloride in pyridine [20] at –40°C gave 2,3-di-*O*-benzoate *62* with 80% yield (Fig. 5.16).

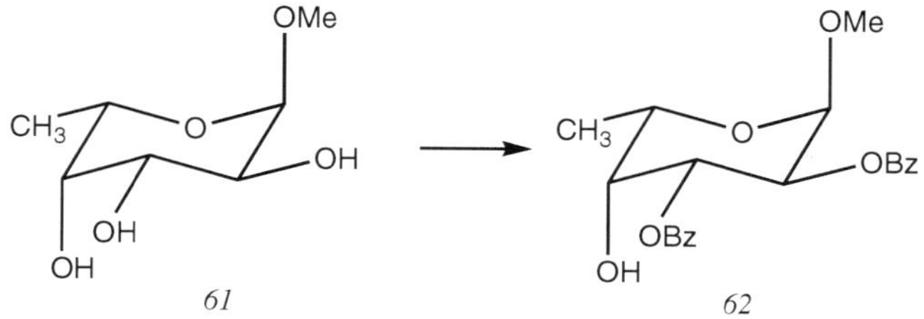

Fig. 5.16

Benzoylation of methyl 6-deoxy-α-L-mannopyranoside (methyl α-L-rhamnopyranoside) *63* with 2.0 M equivalent of benzoyl chloride at room temperature [20] gave 2,3-dibenzoate *64* with 50% yield (Fig. 5.17).

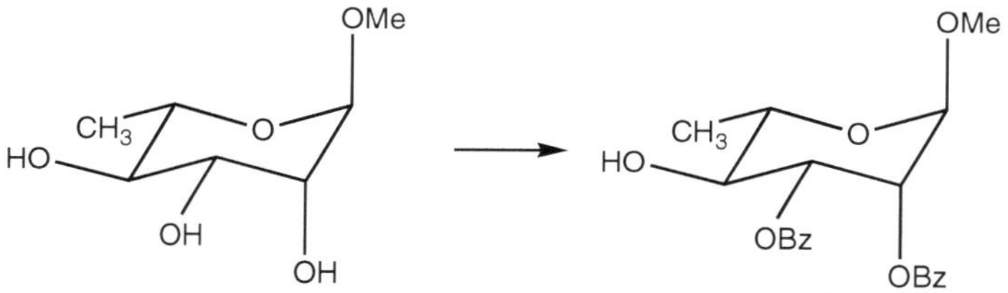

Fig. 5.17

The obtained results are similar to those obtained for benzoylation of methyl α-D-manno- and α-D-galactopyranoside. Consequently the absence of hydroxyl group at the C6 carbon appears to have little or no effect upon the reactivity of the C4 hydroxyl group [20]. With the exception of the C4 hydroxyl group, the relative reactivity of the C2 and the C3 hydroxyl groups of methyl α-D-galactopyranoside cannot be predicted from the above results. The high selectivity toward tribenzoylation only shows that the axial C4-OH is the least reactive one.

It is clear that the order of reactivity of secondary OH groups is different for each glycoside, thus it is 2-OH> 3-OH> 4-OH for the glucoside, 3-OH> 2-OH> 4-OH for the mannoside, and 2-OH, 3-OH> 4-OH for the galactoside.

In conclusion, the most reactive hydroxyl group in methyl α-D-glucopyranoside, α-D-mannopyranoside, and α-D-galactopyranoside is the primary C6 hydroxyl group whereas the least reactive hydroxyl group is the secondary C4 hydroxyl group. The reactivity of the C2 and C3 hydroxyl groups depends upon whether the C2 hydroxyl group is equatorial or axial. If it is equatorial and *cis* to the C1 methoxy group, as is the case in methyl α-D-glucopyranoside, it is more reactive than the C3 hydroxyl group; if it is axial (*trans* to C1 methoxy group) as in methyl α-D-mannopyranoside, the C3 hydroxyl group is more reactive.

Benzoylation of methyl 4,6-*O*-benzylidene-α-D-glucopyranoside 22 with 1 mol of benzoyl chloride under phase-transfer conditions [21] (dichloromethane, aqueous 40% sodium hydroxide, tetrabutylammonium chloride) in the presence of sodium iodide or perchlorate gave 2-benzoate 65 as the major product (72% after chromatography, 62% after crystallization). 3-Benzoate 66 and 2,3-dibenzoate 67 were obtained in 4% each. Benzoylation of methyl 4,6-benzylidene-α-D-mannopyranoside 43 with 1 mol of benzoyl chloride under phase-transfer conditions [21], a ~ 1:1 equilibrium mixture of the 2-(68) and 3-(69) benzoates was obtained. If acylation was performed in the presence of sodium iodide or perchlorate, 2-benzoate 65 was obtained in 52% and 3-benzoate 66 in 11% (both after chromatography (Fig. 5.18)).

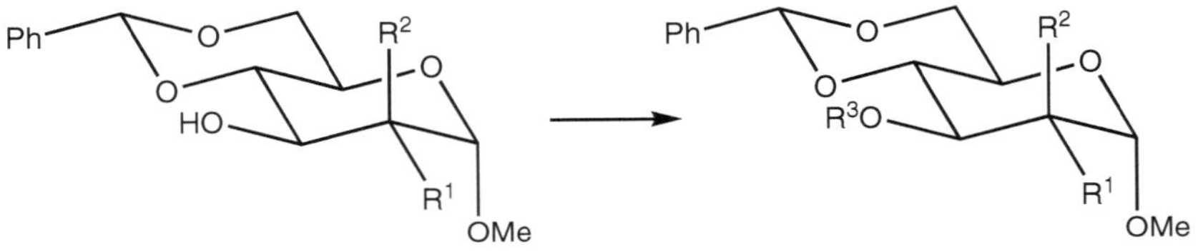

22, R^1 = OH; R^2 = H
43, R^1 = H; R^2 = OH

65, R^1 = OBz; R^2 = R^3 = H
66, R^1 = OH; R^2 = H; R^3 = Bz
67, R^1 = OBz; R^2 = H ; R^3 = Bz
68, R^1 = R^3 = H; R^2 = OBz
69, R^1 = H; R^2 = OH; R^3 = Bz

Fig. 5.18

The acylation of benzyl or methyl α-D-pentopyranosides shows different relative reactivities of secondary hydroxyl groups presumably due to the absence of exocyclic hydroxymethyl group at the C5 carbon.

Thus, dibenzoylation of benzyl α-D-xylopyranoside [22] 70 gave preponderance of the 2,4-dibenzoate 71 over the 2,3-isomer 72. The reason for this could be steric; namely the C4 hydroxyl group is *gauche* just to the C3 hydroxyl group (the C5 has

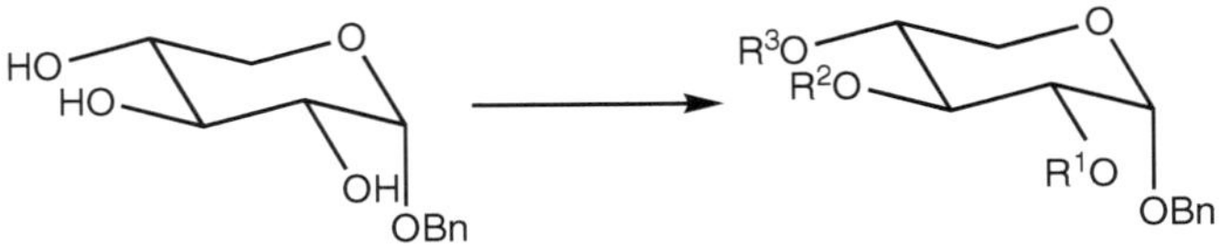

70

71, R^1 = R^3 = Bz; R^2 = H
72, R^1 = R^2 = Bz; R^3 = H
73, R^1 = R^2 = R^3 = Bz
74, R^1 = Bz; R^2 = R^3 = H

Fig. 5.19

only hydrogen atoms), whereas the C3 hydroxyl group is *gauche* to both the C4 and the C2 hydroxyl groups (Fig. 5.19).

These results were confirmed by Kondo [23] who repeated the selective benzoylation of methyl α-D-xylopyranoside *60* with 2 M equivalents of benzoyl chloride in pyridine at –40°C and obtained 2,4-dibenzoate *71* in 45%, 2,3-dibenzoate *72* in 39%, 2,3,4-tribenzoate *73* in 11%, and 2-benzoate *74* with 5% yield (Fig. 5.19).

75

76, $R^1 = R^2 = Bz; R^3 = H$
77, $R^2 = R^3 = Bz; R^1 = H$
78, $R^1 = R^3 = Bz; R^2 = H$
79, $R^1 = R^3 = R^2 = Bz$
80, $R^1 = R^3 = H; R^2 = Bz$
81, $R^1 = R^2 = H; R^3 = Bz$
82, $R^1 = Bz; R^2 = R^3 = H$

Fig. 5.20

Benzoylation of methyl β-D-xylopyranoside *75* with 2 mol equivalent of benzoyl chloride in pyridine [23] gave a mixture of 2,3-dibenzoate *76* and 3,4-dibenzoate *77* with 53% yield, 2,4-dibenzoate *78* (22%), tribenzoate *79* (17%), the 3-benzoate *80* (4%), and a mixture of 4-benzoate *81* and 2-benzoate *82* (4%) (Fig. 5.20).

Benzoylation of methyl β-D-xylopyranoside *75* with 1 M equivalent of benzoyl chloride in pyridine [23] gave 2-benzoate *82* (26%), 3-benzoate *80* (25%), 4-benzoate *81* (20%), 2,4-dibenzoate *78* (8%), a mixture of 2,3-dibenzoate *76* and 3,4-dibenzoate *77* (19%), and tribenzoate *79* (2%) (Fig. 5.20).

The higher reactivity of 2-OH group in methyl α-D-xylopyranoside was explained by intramolecular hydrogen bonding between the 2-OH group and the *cis*-oriented C1-OR substituent. The lowest reactivity of the 3-OH group in methyl α-D-xylopyranoside is in accord with the results obtained for selective benzoylation of benzyl α-D-xylopyranoside and the C4 carbon. Thus the 3-OH group has *gauche* interactions with the C2 benzoyl and the C4 hydroxyl group, whereas 4-OH has interactions with the C3 benzoyl and the C5 hydrogen atom. Therefore,

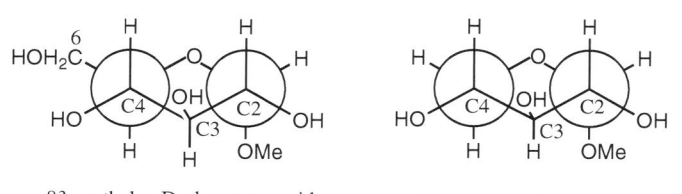

83, methyl α-D-glucopyranoside

84, methyl α-D-xylopyranoside

Fig. 5.21

4-OH group is less sterically hindered, thus causing the preponderance of the 2,4-dibenzoate over 2,3-dibenzoate. However, the preponderance of 2,3-dibenzoate over 2,4-dibenzoate in selective benzoylation of methyl β-D-xylopyranoside cannot be similarly explained [23] (Fig. 5.21).

The hypothesis that the activation of an equatorial C2 hydroxyl group toward acylation is due to the ability of this hydroxyl group to enter the hydrogen bonding with the neighboring C1 oxygen was questioned, since the IR spectra of pyridine solution of free sugars provide no evidence for intramolecular hydrogen bonding [24]. However, this may not be a valid criticism at all because the 2-OH can form the hydrogen bond with the solvent or the hydrogen bond between the 2-OH and the C1 methoxy group can be destroyed by the solvent.

Support for the rationalization that the activation of a hydroxyl group toward acylation is due to its ability to enter the hydrogen bonding with the neighboring C1 oxygen has been provided by experiments conducted by Lemieux and McInnes [25]. They have tosylated the 1,4:3,6-dianhydro-D-glucitol 85 (Fig. 5.22) with 1 mol equivalent of tosyl chloride and found that the major product was the 5-O-tosyl ester

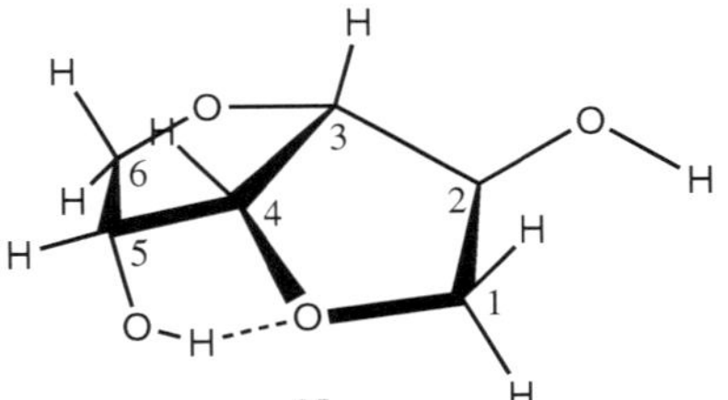

Fig. 5.22

(45%), indicating that sterically more hindered *endo*-hydroxyl group is preferably tosylated, whereas the less hindered 2-exo hydroxyl group was tosylated to a much lesser extent (12%). Although the more reactive 5-OH is comparatively shielded, it is significant that it is intramolecularly strongly hydrogen bonded to a ring oxygen atom (C1–O–C4).

Since the selectivity of acylation vary with acylating agent and the catalyst it is important to point out that the relative reactivity of secondary hydroxyl groups in hexopyranoses toward acyl chlorides (tosyl chloride, mesyl chloride, and benzoyl chloride) is not necessarily applicable to acetanhydride, benzoic anhydride, etc., or to some other acylating agents.

Comparison of acylation of methyl 4,6-O-benzylidene-α-D-glucopyranoside using acetic anhydride, benzoic anhydride, methanesulfonic anhydride, and *p*-toluene sulfonic anhydride with the acylation using acetyl chloride, benzoyl chloride, and methanesulfonyl chloride in pyridine is given in Table 5.1 [10].

As can be seen from Table 5.1 the product ratio obtained by selective acylation of methyl 4,6-O-benzylidene-α-D-glucopyranoside with 1 M equivalent of

Table 5.1 Esterification of methyl 4,6-O-benzylidene-α-D-glucopyranoside in pyridine

Reagent	Molar equivalents of reagent	2-Ester (%)	3-Ester (%)	2,3-Di-ester (%)
$(CH_3CO)_2O$	1.25	3	42	26
$(PhCO)_2O$	1.10	13	25	9
Ms_2O^a	1.1	40	–	6
Ts_2O^b	1.1	80–85	–	15
CH_3COCl	1.25	16	–	23
$PhCOCl$	1.25	24	6	35
CH_3SO_2Cl	1.1	68	–	16

[a]$(CH_3SO_2)_2O$
[b]$(p\text{-}CH_3C_6H_4SO_2)_2O$

acylating reagent in pyridine (Table 5.1) showed a marked dependence on the used reagent [10].

For example, 1 mol equivalent of acetyl chloride and benzoyl chloride in pyridine gives 2-acetyl or 2-benzoyl esters of methyl 4,6-O-benzylidene-α-D-glucopyranoside [10] with 16 and 24% yield, respectively.

The greater reactivity of the 2-OH group in α-glucopyranoside over that in β-anomer has been attributed [1] to an activating effect of anomeric oxygen. It has been suggested that the activation of the 2-OH by the axially oriented C1 methoxy group in α-anomer takes place via hydrogen bonding between the C1 methoxy oxygen and the C2 hydroxyl hydrogen (they are *cis* oriented in the α-anomer and in β-anomer they are *trans* oriented and thus less likely to form hydrogen bond).

The C3 hydroxyl group is practically in the same steric and electronic environment in both methyl α- and β-D-glucopyranosides. So the difference in the reactivity toward acylation of 2-OH and 3-OH could be due to higher reactivity of the 2-OH in the α-D-glucopyranoside compared to that in β-anomer, rather than to be due to greater reactivity of the 3-OH in the β-anomer compared to that in the α-D-glucopyranoside. Perhaps a better explanation could be that the C3 hydroxyl group is generally more reactive than the C2 hydroxyl group but in the α-anomer the equatorially oriented C2 hydroxyl group is activated via hydrogen bond by the *cis*-oriented C1 methoxy group. However, the greater reactivity of the C2 hydroxyl group over that of the C3 hydroxyl group in β-anomer cannot be rationalized.

Selective Acetylation

Unlike tosylation, mesylation, and benzoylation of carbohydrates that are usually performed by using acyl chloride and pyridine, the acetylation is usually effected by using acetic anhydride in the presence of pyridine, sodium acetate, etc. Thus, the relative reactivity of hydroxyl groups of a pyranoside toward acetylation does not have necessarily to be the same as toward acyl chlorides. For example, the acetylation of methyl 4,6-O-benzylidene-α-D-glucopyranoside *22* with 1.25 mol equivalent of acetic anhydride in pyridine at room temperature gave 42% of 3-acetate *86*, 26% of 2,3-diacetate *87*, and 3% of 2-acetate [10] *88*, whereas the acetylation of *22* with 1.1 mol equivalent of acetyl chloride in pyridine at room temperature gave 3-acetate *86* with 6% yield, 2,3-diacetate *87* with 35% yield, and 2-acetate *88* with 6% yield [10] (Fig. 5.23).

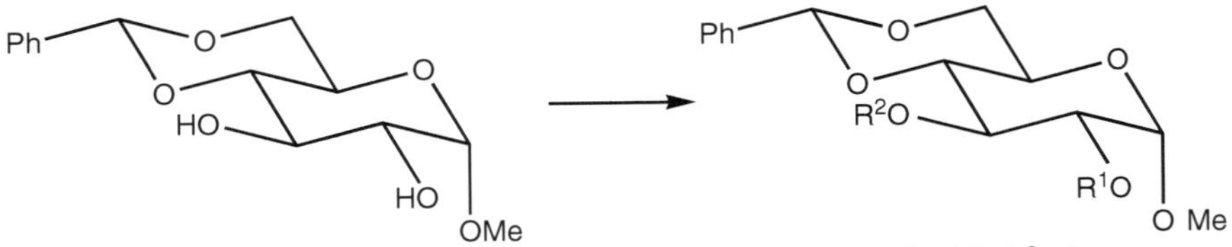

22

86, R^1 = OH; R^2 = Ac
87, R^1 = R^2 = Ac
88, R^1 = Ac; R^2 = H

Fig. 5.23

To the best of our knowledge there are no systematic studies reported on selective acetylation of alkyl or aryl α-D-gluco-, manno-, and galactopyranosides with acetic anhydride in pyridine.

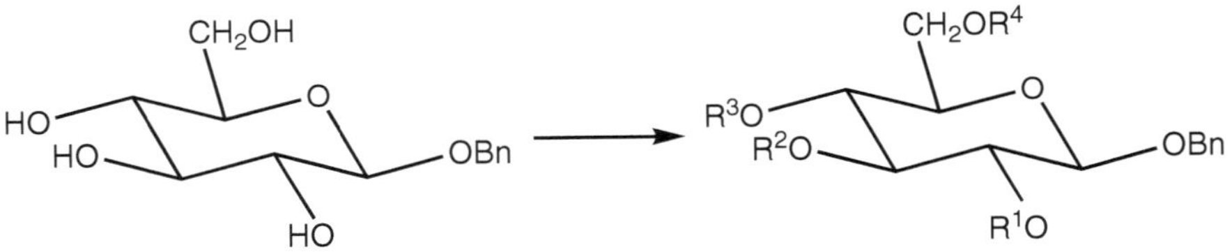

89

90, R^1= R^2 =R^3 =R^4 = Ac
91, R^1 = R^4 = R^5 = Ac; R^2 = H

Fig. 5.24

However, the selective acetylation of β-anomers of D-gluco- and D-galactopyranosides has been reported. Thus, the acetylation of benzyl β-D-glucopyranoside *89* with 8.5 mol equivalent of acetic anhydride and anhydrous sodium acetate [26], at room temperature gave a mixture of products from which the crystalline tetraacetate *90* and 2,4,6-tri-*O*-acetate *91* were isolated with 32 and 66% yield, respectively (Fig. 5.24).

Acetylation of benzyl β-D-mannopyranoside *92* with 8.5 mol equivalent of acetic anhydride in the presence of anhydrous sodium acetate at room temperature [26] gave a mixture of products consisting of tetraacetate *93* (65%) and 2,3,6-triacetate *94* (25%) (Fig. 5.25).

Acetylation of benzyl β-D-galactopyranoside *92* with 5.7 M equivalent of acetic anhydride [26] gave a very complex reaction mixture: tetraacetate *96* was the major

Selective Acylation (Esterification)

Fig. 5.25

92

93, $R^1 = R^2 = R^3 = R^4 = Ac$
94, $R^1 = R^3 = R^4 = Ac; R^2 = H$

Fig. 5.26

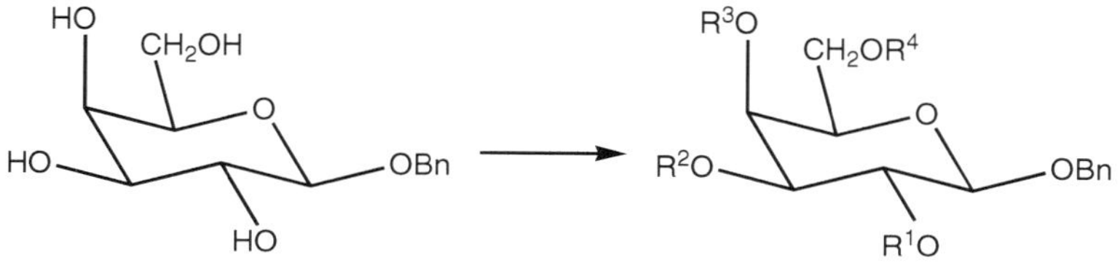

95

96, $R^1 = R^2 = R^3 = R^4 = Ac$
97, $R^1 = R^2 = R^4 = Ac; R^3 = H$
98, $R^1 = R^3 = R^4 = Ac; R^2 = H$
99, $R^2 = R^3 = R^4 = Ac; R^1 = H$

product (38%), 2,3,6-triacetate 97 was obtained with 25% yield, 2,4,6-triacetate 98 was obtained with 9% yield, and 3,4,6-triacetate 99 was obtained with 3% yield (Fig. 5.26).

The partial acetylation of benzyl 4-O-methyl-β-D-xylopyranoside [27] 100 was dependent on the reaction conditions: the 2OAc:3OAc (101:102) ratio is 1:3 with Ac_2O–$HClO_4$, 1.7:1 with Ac_2O–Py, 2:1 with Ac_2O–CH_3COONa, and 1.1:1 with AcCl–Py (Fig. 5.27).

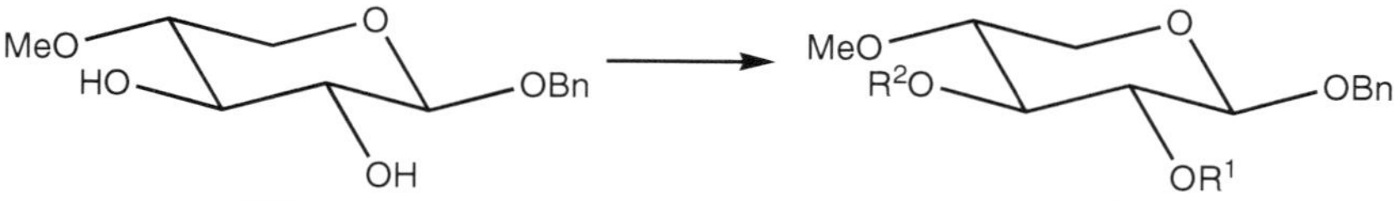

100

101, $R^1 = Ac; R^2 = H$
102, $R^1 = H; R^2 = Ac$

Fig. 5.27

Other Acylating Reagents

Selective acylation of methyl α-D-glucopyranoside with N-3,4,5-trimethoxybenzoyl imidazole [N-(tri-O-methylgalloyl) imidazol] [28] 103 (Fig. 5.28) in dioxan at 60°C gave methyl 6-O-(tri-O-methylgalloyl)-α-D-glucopyranoside with 63%

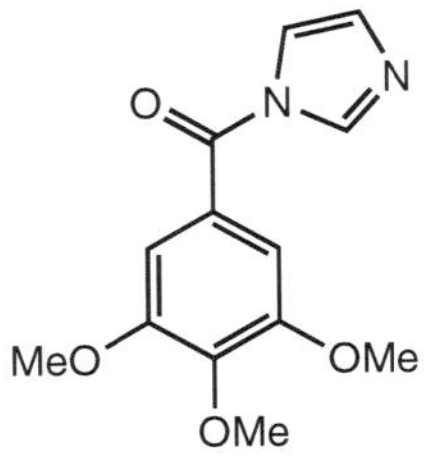

103

Fig. 5.28

yield. In a similar way methyl 2,6-di-*O*-(tri-*O*-methylgalloyl)-α-D-glucopyranoside was obtained with 31% yield and methyl 2,3,6-tri-*O*-(tri-*O*-methylgalloyl)-α-D-glucopyranoside was obtained with 65% yield.

2,4,6-Trimethylbenzenesulfonyl chloride (mesitylenesulfonyl chloride; trimsyl chloride – TmCl) *104* (Fig. 5.29) has been shown [29] to be much more selective for the monosulfonylation of vicinal secondary hydroxyl groups. Thus trimsylation of methyl 4,6-*O*-benzylidene-α-D-glucopyranoside with 1.5 M equivalent in pyridine and at room temperature (6 days) gave 2-*O*-trimsyl ester with 58% yield. The formation of 2,3-di-trimsyl ester was a much slower reaction than monotrimsylation: after treatment of methyl 4,6-*O*-benzylidene-α-D-glucopyranoside with 3 M equivalent of trimsyl chloride in pyridine, at room temperature for 4 days, only 7%

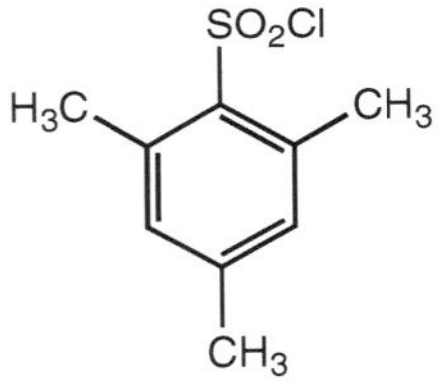

104

Fig. 5.29

of 2,3-di-trimsyl ester was obtained; by allowing reaction to proceed for 30 days, the yield was increased to 25%. Trimsylation of methyl 4,6-*O*-benzylidene-β-D-glucopyranoside with 1.5 M equivalent in pyridine at room temperature gave after 6 days 11% of 2,3-di-trimsyl ester, 23% of 2-trimsyl ester, and 33% of 3-trimsyl ester.

Benzoylation of methyl 4,6-*O*-benzylidene-α-D-glucopyranoside with benzoylimidazole [30] *105* (Fig. 5.30) in chloroform at reflux gave after 10 h 78% of 2-*O*-benzoyl ester [29].

In monobenzoylation of methyl 4,6-*O*-benzylidene-α-D-mannopyranoside using *N*-benzoylimidazole little selectivity between 2-OH and 3-OH was observed since the imidazole formed during the acylations is capable of catalyzing a facile ester migration between the two *cis*-hydroxyl groups, and the product consisted of nearly 1:1 equilibrium mixture of 2- and 3-benzoates, respectively. This contrasts the

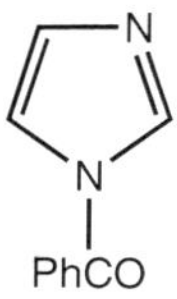

Selective Acylation (Esterification)

105, Benzoyl imidazole

Fig. 5.30

monobenzoylation of the *trans* orientation of 2-OH and 3-OH groups in a D-glucopyranose derivative in which the acyl migration is much slower under the reaction conditions, and the product is apparently formed under kinetic control.

Benzoylation of methyl 4,6-*O*-benzylidene-α-D-mannopyranoside with 1 M equivalent of benzoyl cyanide in acetonitrile in the presence of a catalytic quantity of triethyl amine [31] yielded 2- and 3-*O*-benzoates in 2.3:1 ratio. The reaction of methyl 4,6-*O*-benzylidene-α-D-glucopyranoside under similar conditions gave 2-*O*-benzoate with 62% yield and 4,6-*O*-benzylidene-β-D-galactopyranoside gave 3-*O*-benzoate with 74% yield [32].

Benzoylation of methyl 4,6-*O*-benzylidene-α-D-glucopyranoside with equimolar amounts of 1-(benzoyloxy) benzotriazole *106* and triethylamine [33] (Fig. 5.31) in methylene chloride at room temperature gave after 5 h 2-*O*-benzoyl ester with 90% yield, together with 2,3-di-*O*-benzoate (2%) and 3-*O*-benzoate (4%). Benzoylation of methyl 4,6-*O*-benzylidene-β-D-glucopyranoside was much less selective but it is not clear if the loss of selectivity was due to the solvent change, namely since methyl 4,6-*O*-benzylidene-β-D-glucopyranoside was insoluble in methylene chloride, the reaction was carried out in tetrahydrofuran; 2-*O*-benzoate was obtained with 50% yield and 3-*O*-benzoate with 43% yield [33]. Very high regioselectivity

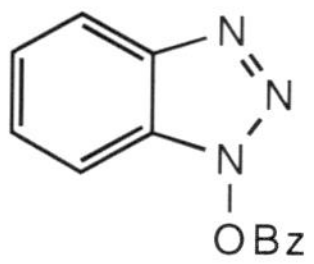

106, 1-(Benzoyloxy)benzotrizole

Fig. 5.31

was also achieved on selective benzoylation of methyl 4,6-*O*-benzylidene-α-D-altropyranoside (90% of 2-benzoate). The 2,3-dibenzoate was not formed [33]. Thus the relative reactivity of 2-OH and 3-OH in α- and β-anomers of D-glucopyranose derivatives could not be compared since they were performed in two different solvents.

Selective benzoylation of some methyl α-D-hexopyranosides was achieved using dibutylstannylene derivatives [34]. Methyl 2,3-*O*- dibutylstannylene-α-D-glucopyranoside, *107* obtained from methyl α-D-glucopyranoside and dibutyltin oxide, was benzoylated in dioxane by benzoyl chloride and triethylamine whereby 2-*O*-benzoyl ester was obtained with 70% yield, together with 2% of 2,6-dibenzoate. Benzoylation of the analogous tin derivatives of methyl β-D-glucopyranoside *110* and methyl α-D-mannopyranoside *109* failed to be selective.

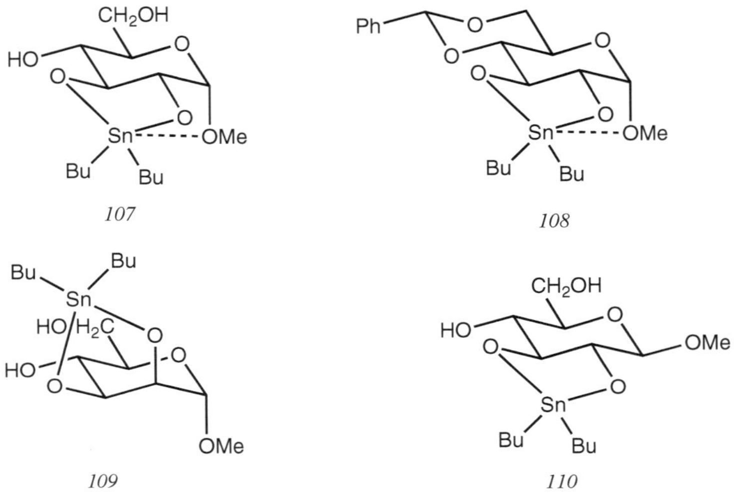

Fig. 5.32

In each case there were obtained approximately equal amounts of 2- and 3-esters in addition to a rather large amount of starting material. This difference in behavior may be due to inability of the respective tin compounds to give coordination between the metal and the α-methoxy group. This explanation is consistent with the fact that the tin compound of methyl 4,6-*O*-benzylidene-α-D-galactopyranoside gives 2-*O*-tosyl ester on treatment with tosyl chloride and triethylamine. Methyl 4,6-*O*-benzylidene-2,3-*O*-dibutylstannylene-α-D-glucopyranoside gave on benzoylation in dioxane with benzoyl chloride and triethylamine 2-*O*-benzoyl ester with 90% yield [34].

In conclusion, all factors that control the selectivity of acylation of secondary hydroxyl groups in glycopyranosides are still not fully understood. In general, the primary hydroxyl group is always the most reactive one; from the secondary endocyclic hydroxyl groups the C2 hydroxyl group is the most reactive but only if it is equatorially oriented and in α-anomers; the C4 hydroxyl group is the least reactive irrespective of the anomeric configuration and regardless whether it is equatorially or axially oriented.

Acyl Migrations

The facile acyl migration in partially acylated polyhydric alcohols was discovered by Fischer [35] who correctly proposed that this rearrangement proceeds via orthoacid intermediate. This proposal was verified by a study of acid catalyzed (0.6 N HCl) rearrangement of unlabeled glycerol-2-palmitate into glycerol-1-palmitate in the presence of 1-^{14}C-labeled glycerol that showed no incorporation of ^{14}C label into the glycerol-1-palmitate [36], thus indicating that the rearrangement is intramolecular and not intermolecular (Fig. 5.33).

Fig. 5.33

The rearrangement is both acid catalyzed and base catalyzed, thus supporting the formation of orthoacid intermediate as shown in Fig. 5.34.

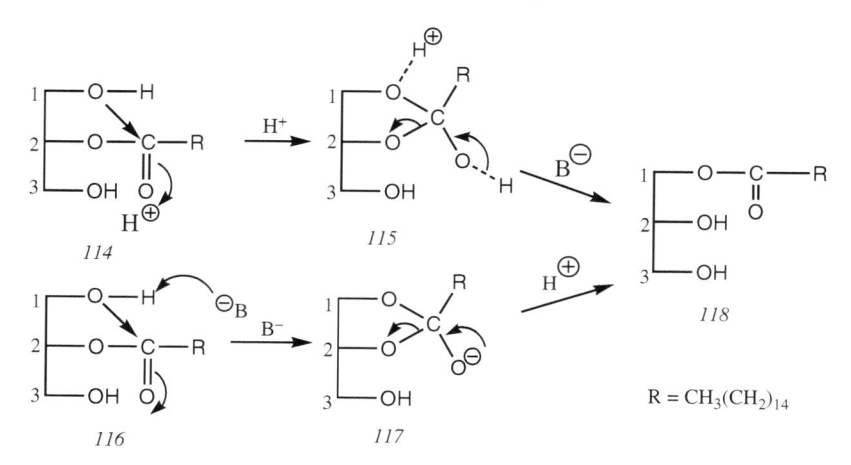

Fig. 5.34

The rate of acyl migration depends on several factors:

(a) The nature of the solvent [37, 38]
(b) The pH of the reaction solution (acidity or alkalinity) [38]
(c) Stereochemistry, i.e., the relative configurations of the two vicinal hydroxyl groups of which one is acylated [38]. The acyl migration occurs more readily in

monoacyl derivatives of vicinal diols if the two oxygen atoms are in *cis* orientation than when they are *trans* disposed, because there is less ring strain introduced into the orthoacid intermediates if they are formed from the *cis*-oriented monoacyl diols than from *trans* monoacyl diols [39]. This is more pronounced in furanose ring structures than in pyranose ring structures

(d) The acyl group generally rearranges from a secondary to the primary carbon atom, probably again due to steric factors

The acyl migration is a reversible process and the composition of equilibrium mixture is most often thermodynamically controlled. However, sometimes the composition of the equilibrium mixture is controlled by external factors, such as, for example, dramatically different solubility of starting material and the product in reaction solvent. For example, dissolution of methyl 2-*O*-benzoyl-4,6-*O*-benzylidene-α-D-glucopyranoside *65* in acetone–aqueous sodium hydroxide results

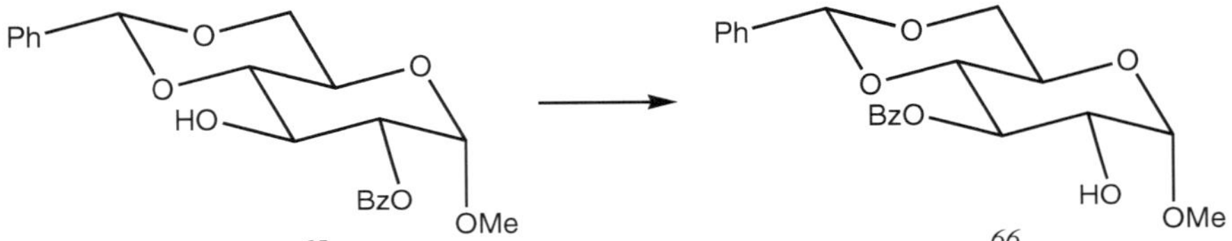

Fig. 5.35

in almost immediate crystallization of the 3-*O*-benzoate *66* with 65% yield [40] (Fig. 5.35).

Similarly, benzyl 3-*O*-benzoyl-4,6-*O*-benzylidene-β-D-galactopyranoside *119* gave 2-*O*-benzoate *120* with 81% yield even though the 2-*O*-benzoate is in homogeneous equilibrium solution, only slightly more stable than the 3-*O*-benzoate [41] (Fig. 5.36).

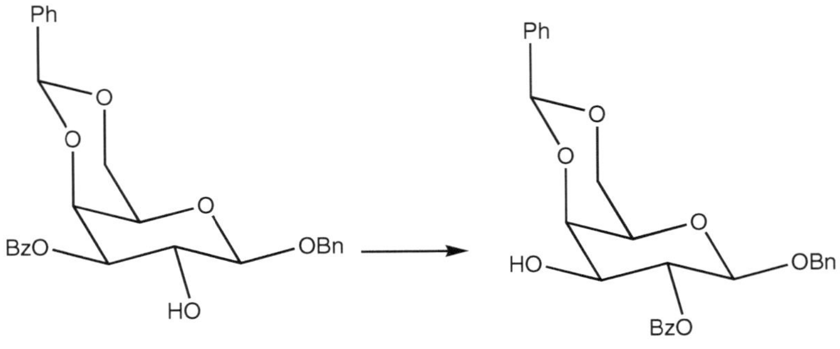

Fig. 5.36

Selective Acylation (Esterification)

Acyl migrations have often been observed during methylation of partially acylated carbohydrates. For example, methylation of methyl 2,3,4-tri-O-acetyl-α-D-glucopyranoside *121* with methyl iodide–silver oxide [42] (Purdie methylation)

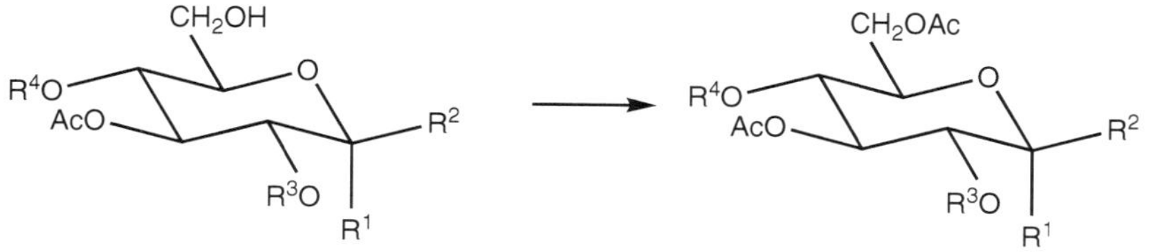

121, $R^1 = OCH_3$; $R^2 = H$; $R^3 = R^4 = Ac$
123, $R^1 = H$; $R^2 = OCH_3$; $R^3 = R^4 = Ac$

122, $R^1 = OCH_3$; $R_2 = H$; $R^3 = CH_3$; $R_4 = Ac$
124, $R^1 = H$; $R^2 = OCH_3$; $R^3 = Ac$; $R^4 = CH_3$
125, $R^1 = H$; $R^2 = OCH_3$; $R^3 = CH_3$; $R^4 = Ac$

Fig. 5.37

gave 2-O-methyl ether *122*, whereas the methylation of methyl 2,3,4-tri-O-acetyl-β-D-glucopyranoside *123* with methyl iodide–silver oxide in N,N-dimethylformamide [43] (Kuhn methylation) gave 4-O-methyl ether *124* with 45% yield. Methylation of methyl 2,3,4-tri-O-acetyl-β-D-glucopyranoside *123* under the Purdie conditions [44] gave 2-O-methyl ether *125* with 66% yield (Fig. 5.37).

Haworth et al. [45] subjected 1:2:3:4-tetraacetyl-β-D-glucopyranose *126* to methylation with methyl iodide and silver oxide and obtained methyl 2,3,4,6-tetra-O-acetyl-β-D-glucopyranoside *127* with 19.23% yield. However, methylation of 2,3,4,6-tetra-O-acetyl-β-D-glucopyranose under the same reaction conditions gave methyl 2,3,4,6-tetra-O-acetyl-β-D-glucopyranoside *127* (slightly impure) with 96% yield (Fig. 5.38), indicating that the acetyl group migration goes away from the anomeric carbon and toward the primary C6 hydroxyl group, and not the other way around.

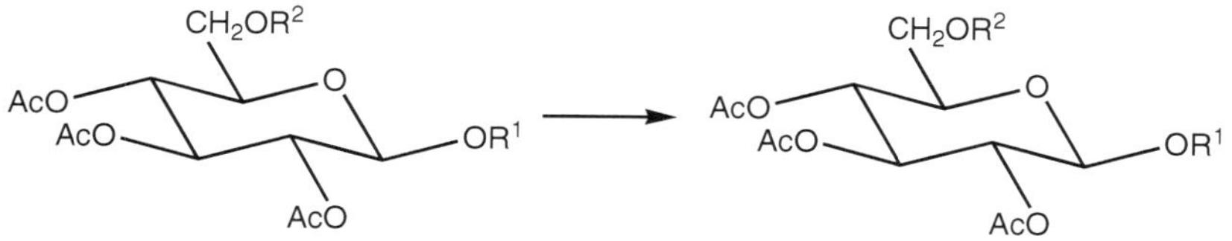

126, $R^1 = Ac$; $R^2 = H$
128, $R^1 = H$; $R^2 = Ac$

127, $R^1 = CH_3$; $R^2 = Ac$

Fig. 5.38

Selective Alkylation and/or Arylation of Glycopyranosides

Tritylation of Monosaccharides (Triphenylmethyl Ethers)

In alkylation reactions the primary hydroxyl group of hexopyranosides is favored over the secondary hydroxyl groups, especially when the alkylating agent is bulky, such as chlorotriphenylmethane (triphenylmethylchloride, trityl chloride, TrCl). Thus reaction of methyl α-D-glycopyranoside *129* with 2 mol of TrCl in pyridine at 30°C gives, after 5–6 h, 6-*O*-trityl ether *130* with 98% yield. The tritylation of any secondary hydroxyl group was not observed [46] (Fig. 5.39).

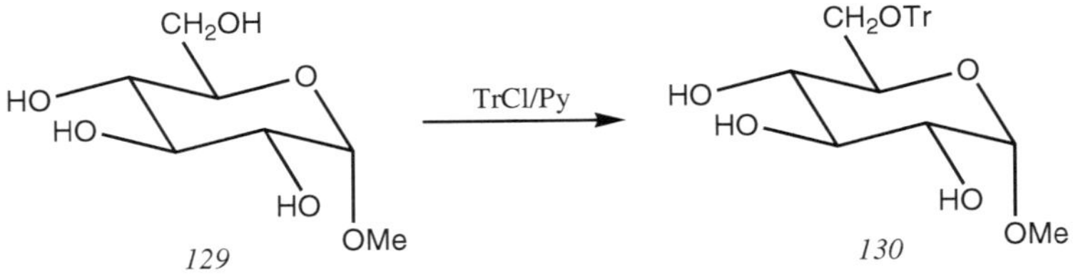

Fig. 5.39

Initially it was believed that tritylchloride, because of its size, can attack exclusively primary hydroxyl group [47–49]. However, Hockett and Hudson [50] have shown that methyl α- and β-D-xylopyranosides *131*, methyl α-D-lyxopyranosides *132*, and methyl β-D-arabinopyranosides *131* also undergo tritylation, despite the absence of an exocylic primary hydroxyl group (the C5 primary hydroxyl group in alkyl pentopyranosides is involved in the hemiacetal ring and thus is not available for tritylation) suggesting that the secondary hydroxyl groups must be the ones that are tritylated (Fig. 5.40).

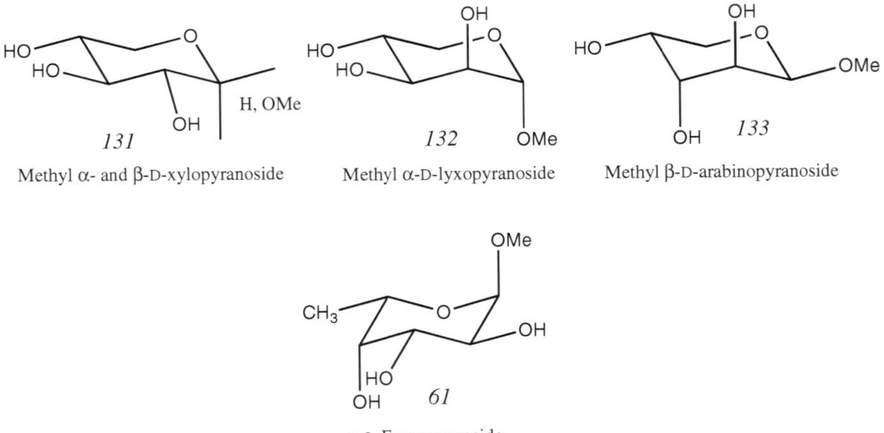

131 Methyl α- and β-D-xylopyranoside

132 Methyl α-D-lyxopyranoside

133 Methyl β-D-arabinopyranoside

61 α-L-Fucopyranoside

Fig. 5.40

Similar situation exists with methyl α-L-fucopyranoside *61* wherein the C6 carbon has no hydroxyl group. However, tritylation of methyl α-L-fucopyranoside [51] with trityl chloride in pyridine readily gave mono-trityl derivative, the structure of which was not determined; tritylation of methyl β-D-xylopyranoside in pyridine gave two isomeric di-trityl and two mono-trityl derivatives that were isolated as acetates but not identified. However it must be emphasized that all these reactions required a very long time (14 days at 20°C) [51].

Since the rates of tritylation of primary and secondary hydroxyl groups are very different (tritylation of a primary hydroxyl group is usually completed after a few hours) [52], the selective tritylation of primary hydroxyl group of hexopyranosides or hexofuranosides in the presence of free secondary hydroxyl group(s) can be accomplished in high yields.

Selective Benzylation of Monosaccharides

Partial benzylation of methyl 4,6-*O*-benzylidene-α-D-mannopyranoside *43* (Fig. 5.32) with 1.2 mol equivalent of benzyl bromide in *N,N*-dimethylformamide in the presence of barium oxide and barium hydroxide gave the 3-*O*-benzyl ether *135* as the major product (66%) along with 2,3-di-*O*-benzyl ether *136* (10%), 2-*O*-benzyl ether *134* (16%), and unreacted starting material (8%) [53] (Fig. 5.41). When

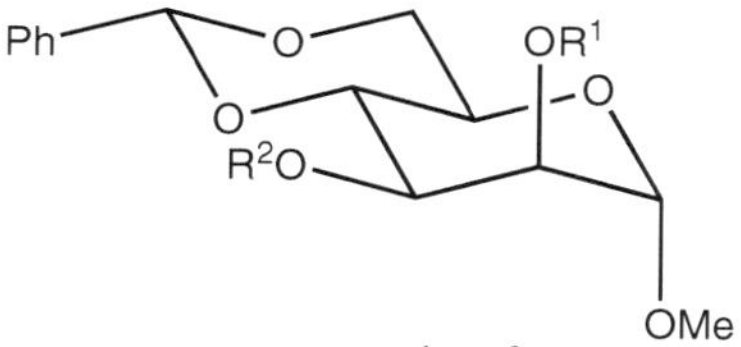

43, R^1 = R^2 = H
134, R^1 = Bn; R^2 = H
135, R^1 = H; R^2 = Bn
136, R^1 = R^2 = Bn

Fig. 5.41

benzylation of methyl 4,6-*O*-benzylidene-α-D-mannopyranoside with benzyl bromide in *N,N*-dimethylformamide was performed in the presence of silver oxide, a mixture of products was obtained in which 2-*O*-benzyl ether *134* was the predominant product (55%); 3-*O*-benzyl ether *135* was obtained in 19%, 2,3-di-*O*-benzyl ether *136* with 10% yield, and the starting material was recovered with 16% yield [53] (Fig. 5.40).

The regioselectivity of benzylation was considerably lost when benzylation was conducted in dimethylsulfoxide as the solvent and using sodium hydride as the base. Thus 2,3-di-*O*-benzyl ether *136* was obtained as the major product (65% yield), along with the 2-*O*-benzyl ether *134* (16%), 3-*O*-benzyl ether *135* (7%), and 12% of the starting material [53].

Using the same procedure Boren et al. [54] obtained methyl 2-O-benzyl-4,6-O-benzylidene-α-D-mannopyranoside *134* with 36% yield.

The results of benzylation of methyl 4,6-O-benzylidene-α-D-mannopyranoside with benzyl bromide in dimethylsulfoxide as the solvent and NaH as the base reported by Srivastava and Srivastava [55] do not agree with the results reported by Kondo [53]. Srivastava and Srivastava [55] reported that 3-O-benzyl ether *135* was obtained with 66% yield and the 2,3-di-O-benzyl ether *136* with 20% yield, together with small amounts of 2-O-benzyl ether and starting material.

From the molar ratios of reaction products it can be concluded that when benzylation of methyl 4,6-O-benzylidene-α-D-mannopyranoside is performed in the presence of barium oxide–barium hydroxide, 3-OH is more reactive than 2-OH. It was, however, shown [56] that the reactivity of hydroxyl groups of methyl 4,6-O-benzylidene-α-D-glucopyranoside under the same reaction conditions is 2-OH > 3-OH and for methyl 4,6-O-benzylidene-β-D-glucopyranoside 3-OH > 2-OH. From the foregoing findings, it seems that the *cis*-OR substituent activates the adjacent equatorial hydroxyl group also in benzylation in the presence of barium oxide, as it did in acylation of the same substrate. The order of reactivity of hydroxyl groups in benzylation in the presence of silver oxide is 2-OH > 3-OH.

The order of reactivity of the hydroxyl groups in benzylation in the presence of sodium hydride may be 2-OH > 3-OH. However, the high yield of the dibenzyl ether suggests that benzylation occurs very rapidly and that this method is not suitable for regioselective benzylation.

Selective Alkylation of Metal Complexes of Monosaccharides

The initial work of Avela et al. [57–62] showed that copper chelates of vicinal diols prepared from a sugar, sodium hydride, and methyl chloride in the molar ratio 1:2:1 are more regioselective in methylation than any other chelate they investigated.

Eby et al. [63] confirmed Avela's findings by succeeding to methylate the copper complexes of methyl 4,6-O-benzylidene derivatives of α-D-gluco-, α-D-manno-, and α-D-galactopyranosides with methyl iodide, sodium hydride, and copper chloride in the 1:2:1 molar ratio (see Table 5.2). It was found that carbohydrate derivatives having vicinal hydroxyl groups (*e, e* or *a, e*), or those having O-4 and O-5 free, were able to form 1:2:1 copper complexes. All of the complexes were soluble in tetrahydrofuran or 1,2-dimethoxyethane with formation of dark-green solutions.

Benzylation of partially stannylated methyl α-D-glucopyranoside with $(Bu_3Sn_2)_2O$ at 80–90°C gave 6-O-benzyl ether (48.6%), along with 2,6-di-O-benzyl ether (30.5%), 3,6-di-O-benzyl, ether (4.5%), and 4,6-di-O-benzyl ether (6.0%) [64]. The isolation of three dibenzyl ethers in the ratio 3:20:4 together with the 6-O-monobenzyl ether showed that the regioselectivity was only moderate, as compared with benzoylation of partially stannylated carbohydrates.

The alcohol hydroxyl groups are relatively unreactive toward diazomethane, but in the presence of certain protic acids, fluoroboric acid [65] and Lewis acids (boron

Table 5.2 Alkylation of 1:2:1 Cu complexes of methyl 4,6-O-benzylidene derivatives of α-D-gluco-, manno-, and galactopyranosides

Substrate	Solvent	Metal	Alkyl iodide	Composition (%)			
				2	3	2,3	s.m.
Methyl 4,6-O-benzylidene-α-D-glucopyranoside	THF	Cu	Methyl	20	66	–	14
	DME	Cu	Benzyl	18	74	–	8
	DME	Cu	Allyl	19	77	–	4
Methyl 4,6-O-benzylidene-α-D-mannopyranoside	DMF	Cu	Allyl	19	81	–	–
	THF	Cu	Allyl	20	80	–	–
Methyl 4,6-O-benzylidene-α-D-galactopyranoside	DME	Cu	Allyl	29	68	–	–
	DME	Cu	Benzyl	29	68	–	–

trifluoride etherate [66] and aluminum chloride [67]), the reaction is substantially facilitated. The mechanism of methylation of aliphatic alcohols with diazomethane catalyzed by boron compounds has been discussed [68].

The presence of small amounts (10–100 mM equivalents) of stannous chloride dihydrate was found to catalyze the reaction of some D-glucopyranoside derivatives with diazomethane in methanol or methanol-N,N-dimethylformamide; without this catalyst, little methylation occurred [68]. Methyl 4,6-O-benzylidene-α-D-glucopyranoside afforded 93% of the 3-O-methyl ether but alkylation of the β-anomer was much less selective giving the 2- and 3-methyl ethers with 34 and 53% yield, respectively. Reaction of methyl and phenyl α-D-glucopyranoside showed the unprecedented selectivity, yield of 74 and 81% of the respective 3- and 2-O-methyl ethers being obtained together with the minor amounts of dimethyl ethers. Methyl and phenyl β-D-glucopyranoside gave 3-methyl ethers with 54 and 47% yield, respectively, but with these β-D-glucosides, 2,3-dimethyl ethers were simultaneously obtained in much higher yields (48 and 44%, respectively) than with α-D-glucosides. Other Lewis acids (such as aluminum chloride hexahydrate, magnesium chloride hexahydrate, zinc chloride, and lead acetate trihydrate) were less active in promoting alkylation. The methylation of benzyl 4,6-O-benzylidene-β-D-galactopyranoside with diazomethane catalyzed by stannous chloride dihydrate was also highly regioselective [69], the 2-O-methyl ether being formed with 91% yield; with boron trifluoride diethyl etherate as the catalyst, 2,3-dimethyl ether was obtained with 84% yield. Involvement of 3-OH in intramolecular hydrogen bonding to 4-OH (which was suggested, is unaffected by stannous chloride, but prevented in the presence of boron trifluoride diethyl etherate) may explain these observations, as the unreactivity of strongly bonded hydroxyl groups toward diazomethane had been noted [70].

References

1. Sugihara, J. M., "*Relative reactivities of hydroxyl groups of carbohydrates*", Advan. Carbohydr. Chem. (1953) **8**, 1–44

2. Haines, A. H., "*Relative reactivities of hydroxyl groups in carbohydrates*", Advan. Carbohydr. Chem. Biochem. (1976) **33**, 11–109
3. Ball, D. H.; Parrish, F. W., "*Sulfonic esters of carbohydrates*", Adv. Carbohydr. Chem. (1968) **23**, 233–280
4. Compton, J., "*The Unimolar Tosylation of alpha- and beta-Methyl-d-glucosides*", J. Am. Chem. Soc. (1938) **60**, 395–399
5. Cramer, F.; Otterbach, H.; Springmann, H., "*Eine Synthese der 6-Desoxy-6- amino-glucose*", Chem. Ber. (1959) **92**, 384–391
6. Chalk, R. C.; Ball, D. H.; Long, L., Jr., "*Selective Mesylation of Carbohydrates. II.1a Some Mesyl Esters of Methyl α- and β-D-Glucopyranosides, Methyl α- and β-D- Galactopyranosides, and of Methyl α-D-Mannopyranoside*", J. Org. Chem. (1966) 1509–1514
7. Mitra, A. K.; Ball, D. H.; Long, L., Jr., "*Methyl 2,6-Di-O-methylsulfonyl-α-D- glucopyranoside and New Syntheses of 3,4-Di- and 3,4,6-Tri-O-methyl-D-glucose*", J. Org. Chem. (1962) **27**, 160–162
8. Robertson, G. J.; Griffith, C. F., "*The conversion of derivatives of glucose into derivatives of altrose by simple optical inversion*", J. Chem. Soc. (1935) 1193–1201
9. Bolliger, H. R.; Prins, D. A., "*2-Tosyl-4, 6-benzyliden-α-methyl-d-glucosid-(1,5) und 4,6-Benzyliden-α-methyl-d-glucosid-(1,5)-3-methyläther*", Helv. Chim. Acta (1945) **28**, 465–470
10. Jeanloz, R. W.; Jeanloz, D. A., "*Partial Esterification of Methyl 4,6-O- Benzylidene-α-D-glucopyranoside in Pyridine Solution*", J. Am. Chem. Soc. (1957) **79**, 2579–2583
11. Stirm, S.; Luederitz, O.; Westphal, O., "*Glycosides of abequose, colitose, and tyvelose*", Liebigs Ann. (1966) **696**, 180–193
12. Guthrie, R. D.; Prior, A. M.; Creasey, S. E., "*Studies on the synthesis of methyl 2,3-anhydro-4,6-O-benzylidene-α-D-allopyranoside and -mannopyranoside, and their reaction with sodium azide*", J. Chem. Soc. (C) (1970) 1961–1966
13. Kondo, Y., "*Partial tosylation of methyl α-D-mannopyranoside*", Carbohydr. Res. (1986) **154**, 305–309
14. Aspinal, G. O.; Zweifel, G., "*Selective esterification of equatorial hydroxyl groups in the synthesis of some methyl ethers of D-mannose*", J. Chem. Soc. (1957) 2271–2278
15. Buchanan, J. G.; Schwarz, J. C. P., "*Methyl 2,3-anhydro-α-D-mannoside and 3,4- anhydro-α-D-altroside and their derivatives. Part I*", J. Chem. Soc. (1962) 4770–4777
16. Williams, J. M.; Richardson, A. C., "*Selective acylation of pyranoside—I. Benzoylation of methyl α-D-glycopyranosides of mannose, glucose and galactose*", Tetrahedron (1967) **23**, 1369–1378
17. Lieser, Th.; Schweizer, R, "*Zur Kenntnis der Kohlenhydrate. V. Spezifität der Zuckerhydroxyle*", Liebigs. Ann. (1935) **519**, 271–278
18. Lieser, Th.; Schweizer, R., "*Specificity of the hydroxyls of sugars*", Naturwissen- schaften (1935) **23**, 131
19. Eliel, E. L.; Lukach, C. A., "*Conformational analysis. II. Esterification rates of Cyclohexanols*", J. Am. Chem. Soc. (1957) **79**, 5986–5992
20. Richardson, A. C.; Williams, J. M., "*Selective Acylation of Pyranosides-II. Benzoylation of Methyl 6-Deoxy-α-L-Galactopyranoside and Methyl 6-Deoxy-α-L-Manno- pyranoside*", Tetrahedron (1967) **23**, 1641–1646
21. Szeja, W., "*The selective benzoylation of methyl 4,6-O-benzylidene-α-D- glucopyranoside*", Carbohydr. Res. (1983) **115**, 240–242
22. Sivakumaran, T.; Jones, J. K. N., "*Selective benzoylation of benzyl β-l-arabino pyranoside and benzyl α-d-xylopyranoside*", Can. J. Chem. (1967) **45**, 2493–2500
23. Kondo, Y., "*Selective benzoylation of methyl α- and β-D-xylopyranoside*", Carbohydr. Res. (1982) **107**, 303–311
24. Kabayama, M. A.; Patterson, D., "*The thermodynamics of mutarotation of some sugars. II. Theoretical considerations*", Can. J. Chem. (1958) **36**, 563–573
25. Lemieux, R. U.; McInnes, A. G., "*The preferential tosylation of the endo-5-hydroxyl group of 1, 4: 3, 6- dianhydro-D-glucitol*", Can. J. Chem. (1960) **38**, 136–140

References

26. Lee, E. E.; Bruzzi, A.; O'Brien, E.; O'Colla, P. S., "*Selective acetylation of benzyl α-D-mannopyranoside, benzyl-β-D-glucopyranoside, and benzyl β-D- galactopyranoside*", Carbohydr. Res. (1974) **35**, 103–109
27. Garegg, P. J., "*Partial acetylation studies on benzyl 4-O-methyl-β-D-xylopyranoside*", Acta Chem. Scand. (1962) **16**, 1849–1857
28. Birkofer, L.; Idel, K., "*Partielle Veresterung des alpha-Methyl-D-glucopyranosids mit N-(Tri-O-methylgalloyl)imidazol1*", Liebigs Ann. (1974) 4–14
29. Creasey, S. E.; Guthrie, R. D., "*Mesitylenesulphonyl chloride: a selective sulphonylating reagent for carbohydrates*", J. Chem. Soc. Perkin Trans. I (1974) 1373–1378
30. Staab, H. A., "*Neuere Methoden der präparativen organischen Chemie IV Synthesen mit heterocyclischen Amiden (Azoliden)*", Angew. Chem. (1962) **74**, 407–423
31. Carey, F. A.; Hodgson, K. O., "*Efficient syntheses of methyl 2-O-benzoyl-4,6-O- benzylidene-α-D-glucopyranoside and methyl 2-O-benzoyl-4,6-O-benzylidene-α-D- ribo-hexopyranosid-3-ulose*", Carbohydr. Res. (1979) **12**, 463–465
32. Abbas, S. A.; Haines, A. H.; Abbas, S. A.; Haines, A. H., "*Benzoyl cyanide as a selective acylating agent*", Carbohydr. Res. (1975) **39**, 358–363
33. Kim, S.; Chang, H.; Kim, W. J, "*Selective benzoylation of diols with 1-(benzoyloxy)benzotriazole*", J. Org. Chem. (1985) **50**, 1751–1752
34. Munavu, R. M.; Szmant, H. H., "*Selective formation of 2 esters of some methyl-α-D- hexopyranosides via dibutylstannylene derivatives*", J. Org. Chem. (1976) **41**, 1832–1836
35. Fischer, E., "*Wanderung von Acyl bei den Glyceriden*", Berichte (1920) **53**, 1621–1633
36. Doerschuk, A. P., "*Acyl Migrations in Partially Acylated, Polyhydroxylic Systems*", J. Am. Chem. Soc. (1952) **74**, 4202–4203
37. Angyal, S. J.; Melrose, G. J. H, "*Cyclitols. Part XVIII. Acetyl migration: equilibrium between axial and equatorial acetates*", J. Chem. Soc. (1965) 6494–6500
38. Angyal, S. J.; Melrose, G. J. H., *Cyclitols. Part XIX. Control of acetyl migration during methylation of partially acetylated cyclitols*", J. Chem. Soc. (1965) 6501–6504
39. Angyal, S. J.; Macdonald, C. G., "*Cyclitols. Part I. isoPropylidene derivatives of inositols and quercitols. The structure of pinitol and quebrachitol*", J. Chem. Soc. (1952) 686–695
40. Bourne, E. J.; Huggard, A. J.; Tatlow, J. C., "*Studies of trifluoroacetic acid. Part VII. The synthesis of 2-benzoyl 4: 6-benzylidene methyl–D-glucopyranoside and its conversion into the isomeric 3-benzoate by an acyl migration*", J. Chem. Soc. (1953) 735–741
41. Chittenden, G. J. F.; Buchanan, J. G., "*Conversion of benzyl 3-0-benzoyl-4, 6-0- benzylidene-β-D-galactopyranoside into the 2-benzoate by acyl migration*", Carbohydr. Res. (1969) **11**, 379–385
42. Haworth, W. N.; Hirst, E. L.; Teece, E. G., "*Conversion of 2: 3: 4-triacetyl - methylglucoside into 3: 4: 6-triacetyl 2-methyl -methylglucoside*", J. Chem. Soc. (1931) 2858–2860
43. Bouveng, H.; Lindberg, B.; Theander, O., "New synthesis of 4-O-methyl-D- glucose", Acta Chem. Scand. (1957) **11**, 1788–1789
44. Finan, P. A.; Warren, C. D., "*Glycosides. Part II. The preparation of methyl 3,4,6-tri- O-acetyl-2-O-methyl–D-glucopyranoside*", J. Chem. Soc. (1962) 4214–4216
45. Haworth, W. N.; Hirst, E. L.; Teece, E. G., "*The Conversion of 1: 2: 3: 4-Tetra-acetyl β-Methylglucoside into 2: 3: 4: 6:-Tetra-acetyl β-Methylglucoside*", J. Chem. Soc. (1930) 1405–1409
46. Helferich, B.; Bigelow, N. M., "*Velocity of ether formation between α-methyl d- glucoside and triphenylmethyl chloride in pyridine*", J. Prakt. Chem. (1931) **131**, 259–265
47. Helferich, B.; Becker, J., "*Synthese eines Disacchariad-glucosids*", Annalen (1924) **440**, 1–18
48. Josephson, K., "*Über Triphenylmethyl-äther einiger Di- und Trisaccharide. Ein Bei- trag zur Kenntnis der Konstitution der Maltose, Saccharose und Raffinose*", Annalen (1929) **472**, 230–240
49. Pacsu, E., "*The constitution of Melezitose and Turanose*", J. Am. Chem. Soc. (1931) **53**, 3099–3104

50. Hockett, R. C.; Hudson, C. S., *"The Reaction of several Methylpentosides and of Al pha-Methylmannoside with Triphenylmethyl Chloride"*, J. Am. Chem. Soc. (1931) **53**, 4456–4457
51. Hockett, R. C.; Hudson, C. S., *"The Action of Triphenylmethyl Chloride on α- Methyl-l-fucoside"*, J. Am. Chem. Soc. (1934) **56**, 945–946
52. Helferich, B., *"Trityl Ethers of carbohydrates"*, Advan. Carbohydr. Chem. (1948) **3**, 79–111
53. Kondo, Y.; Noumi, K.; Kitagawa, S.; Hirano, S., *"Partial benzylation of methyl 4,6- O-benzylidene-α-D-mannopyranoside"*, Carbohydr. Res. (1983) **123**, 157–159
54. Boren, H. B.; Garegg, P. J.; Wallin, N. H., *"Synthesis of methyl 3-0-(3,6-dideoxy- D-arabino-hexopyranosyl)-D-mannopyranoside"*, Acta Chem. Scand. (1972) **26**, 1082–1086
55. Srivastava, H. C.; Srivastava, V. K., *"Use of the benzyl group during Purdie methylation: synthesis of 2-O-methyl-D-mannose"*, Carbohydr. Res. (1977) **58**, 227–229
56. Kondo, Y., *"Monomolar etherification of methyl-4,6-O-benzylidene-α- and -β-D- glucopyranosides"*, Agric. Biol. Chem. (1975) **39**, 1879–1881
57. Avela, E.; Holmbom, B., *"Reactions of metal chelates of hydroxyl compounds under anhydrous conditions. I. Selective monosubstitution at secondary hydroxyl groups of anomeric methyl 4,6-0-benzylidene-D-glucopyranosides via copper(II) chelates"*, Acta Acad. Ab. Ser. B (1971) **31**, No. 14, 14
58. Avela, E.; Melander, B.; Holmbom, B., *"Reactions of metal chelates of hydroxyl compounds under anhydrous conditions. II. Selective substitution at the primary or secondary hydroxyl group of anomeric methyl 2,3-di-O-methyl-D-glucopyranosides via copper(II) chelates"*, Acta Acad. Ab. Ser. B (1971) **31**, No. 15, 13
59. Norita, T.; Silanpaa, R.; Avela, E., Suomi Kemistil. B (1972) **45**, 188–190
60. Avela, E., *"Selective substitution of carbohydrate hydroxyl groups via metal chelates"*, Sucr. Belge (1973) **92**, 337–344
61. Avela, E.; Melander, B., Abstr. Int. Symp. Carbohydr. Chem., 6th, (1972), Madison, Wisconsin, USA
62. Avela, E., *"Selectively guiding and limiting the reactions of hydroxyl compounds"*, U.S. Pat. 3, 972, 868 8. pp.(Aug. 3, 1976)
63. Eby, R.; Webster, K. T.; Schuerch, K., *"Regioselective alkylation and acylation of carbohydrates engaged in metal complexes"*, Carbohydr. Res. (1984) **129**, 111–120
64. Ogawa, T.; Takahashi, Y.; Matsui, M., *"Regioselective alkylation via trialkylstannylation: Methyl α-D-glucopyranoside"*, Carbohydr. Res. (1982) **102**, 207–215
65. Neeman, M.; Caserio, M. C.; Roberts, J. D.; Johnson, W. S., *"Methylation of alcohols with diazomethane"*, Tetrahedron, (1959) **6**, 36–47
66. Müller, E.; Rundell, W., *"Verätherung von Alkoholen mit Diazomethan unter Bor- fluorid-Katalyse"*, Angew. Chem. (1958) **70**, 105
67. Mueller, E.; Heischkeil, R.; Bauer, M., *"Aluminiumchlorid-katalysierte Verätherung von Alkoholen mit Diazoalkanen, III"*, Liebig's Ann., (1964) **677**, 55–58
68. Aritomi, M.; Kawasaki, T., *"Partial methylation with diazomethane of the sugar moiety of some C- and O-D-glucopyranosides"*, Chem. Pharm. Bull. (1970) **18**, 677–686
69. Chittenden, G. J. F., *"Reaction of benzyl 4,6-O-benzylidene-β-D-galactopyranoside with diazomethane: synthesis of 2-O-methyl-D-galactose and some derivatives"*, Carbohydr. Res. (1975) **43**, 366–370
70. Sadekov, I. D.; Minkin, V. I.; Lutskii, A. E., *"Intramolecular hydrogen bonding and reactivity of organic compounds"*, Uspekhi Khimii (1970) **39**, 380–411

Chapter 6
Cyclic Acetals and Ketals

Cyclic and acyclic carbohydrates react with aldehydes and/or ketones, in the presence of catalysts (hard or Lewis acids), to give cyclic acetals and/or ketals (1,3-dioxolanes *3* or 1,3-dioxanes *5*, respectively) (Fig. 6.1). This reaction is routinely used in carbohydrate chemistry for the protection of hydroxyl groups in a sugar in order to prevent their interference in chemical transformation(s) of other hydroxyl group(s) of that sugar. The reaction of carbohydrates with aldehydes and ketones has been comprehensively reviewed [1–8].

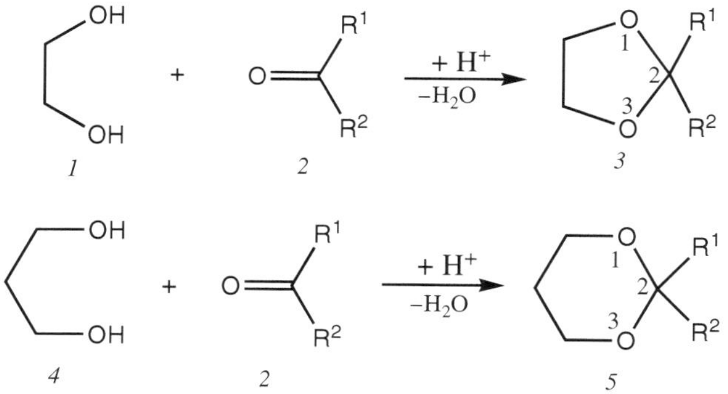

1 2 3

4 2 5

$R^1 = R^2 = CH_3$; or $R^1 = Ph, CH_3$; $R^2 = H$

Fig. 6.1

The acetal formation is believed [9] to be a stepwise process that initially involves the hemiacetal formation as represented in Fig. 6.2. The hemiacetal formation and the subsequent proton exchange are considered to be the fast reactions, whereas the formation of oxocarbenium ions *9a, b* from *8* (*8* → *9a, b*) and/or the cyclization of *9a, b* → *10* in the subsequent step are assumed to be the rate-limiting reactions. The generally accepted mechanism for the hydrolysis of most simple acyclic acetals [10] and cyclic ketals [11–15] is the S_N1cA or A1 mechanism (Ingold system for naming reaction mechanisms) involving the rate-determining heterolysis of a protonated intermediate. Thus, if the ring opening is the rate-determining step for the

M. Miljković, *Carbohydrates*, DOI 10.1007/978-0-387-92265-2_6,
© Springer Science+Business Media, LLC 2010

hydrolysis of cyclic acetals or ketals, the principle of microscopic reversibility [16] requires that the ring closure 9a, b → 10 (Fig. 6.2) be the rate-determining step for the formation of cyclic acetals or ketals.

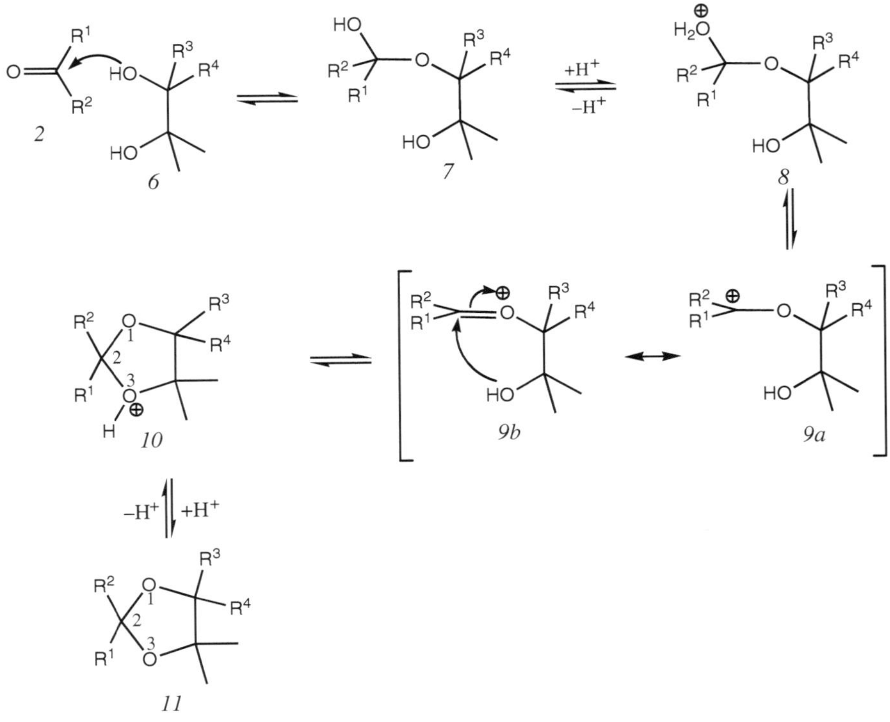

Fig. 6.2

The oxocarbenium ion *9a, b* reacts with the nearest hydroxyl group giving the (kinetic) product *11*, which, after prolonged reaction time, may rearrange to a more stable (thermodynamic) product or to an equilibrium mixture of more stable products; the composition of this mixture is determined by the relative free energies of different isomeric acetals. This is best illustrated by reaction of glycerol and benzaldehyde in anhydrous *N,N*-dimethylformamide catalyzed by *p*-toluenesulfonic acid [17]. Since benzyl protons of all four possible acetals (Fig. 6.3) have different chemical shifts, the NMR spectroscopy was ideally suited for monitoring the course of the reaction. Thus it was found that the *cis-* and *trans*-4-hydroxymethyl-2-phenyl-1,2-dioxolanes, *14* and *15*, respectively, are formed first (after less than 2 min). These initial products were then slowly rearranged into *cis-* and *trans*-5-hydroxy-2-phenyl-1,3-dioxanes, *12* and *13*, that were observable after 5 min, and after 12 min the ratio of *12*:*13*:*14*:*15* was ca. 1.1:1.0:6.2:4.0. At equilibrium that was reached after 2 days the ratio of products was 1.8:1.8:1.2:1.0, indicating that the 1,3-acetals are clearly favored products (Fig. 6.3). These results are consistent with the initial formation of hemiacetal *16* that involves primary hydroxyl group, followed by protonation and formation of oxocarbenium transition state *17* (Fig. 6.4). Subsequent cyclization involving the nearest (the C2) hydroxyl group led to *cis*- and *trans*-4-hydroxymethyl-1,2-dioxolans (*14* and *15*, respectively) (kinetic products).

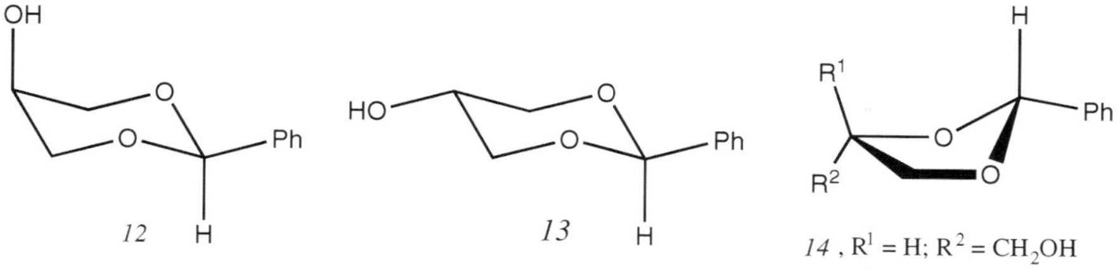

12 *13*

14 , $R^1 = H$; $R^2 = CH_2OH$
15 , $R^1 = CH_2OH$; $R^2 = H$

Fig. 6.3

This initial, kinetic phase of reaction is followed by a slow equilibration to give ultimately the thermodynamically more stable 1,3-acetal *19* (i.e., *12* and *13*) as major products (Fig. 6.4).

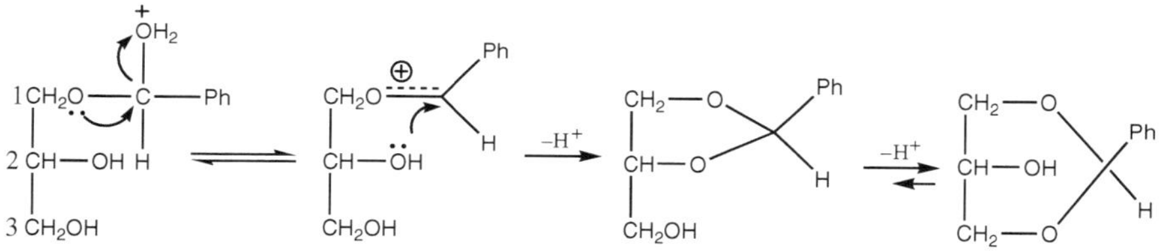

16 *17* *18* *19*

Fig. 6.4

The formation of a single stereoisomer during kinetic phase of reaction has also been demonstrated and this has important mechanistic consequences. Thus, the 1,4-anhydroerythritol *20* (Fig. 6.5) reacts with benzaldehyde giving benzylidene acetal *21* with phenyl group in the *endo*-configuration [18, 19]. Subsequently, equilibration takes place giving ultimately a near-equimolar mixture of *endo*- *21* and *exo*- *22* phenyl isomers. The stereoselective formation of one product in this reaction may

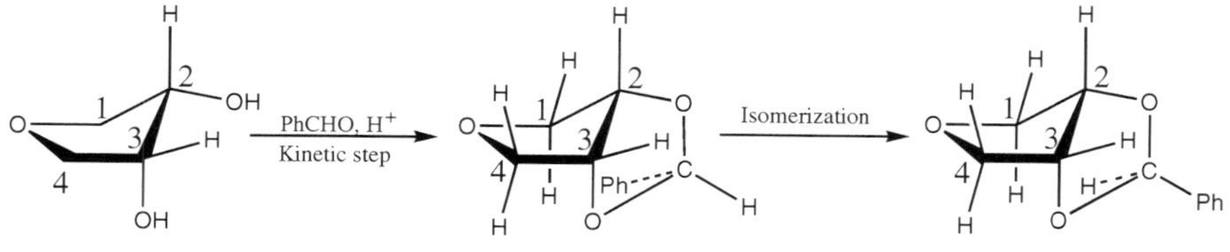

20 *21* *22*

Fig. 6.5

be rationalized [20] by assuming that the oxocarbenium ion (*9a, b*) is highly reactive and thus closely resembles the transition state. The observed relative rates of formation of particular acetals may then be explained by considering the stability or ease of formation of oxocarbenium ion. This approach is, however, valid only if the decomposition of acetal by hydrolysis or its rearrangement is negligible. This requirement is probably fulfilled in the early stages of reaction carried out in anhydrous media.

The stabilization of oxocarbenium cation (*9a, b*) in Fig. 6.2 (and Fig. 6.6) may be accomplished in two ways: (a) carbonium ion *9a* can be directly stabilized by R^1 and R^2 groups or (b) the oxonium ion *9b* can be stabilized by R^3 and R^4 groups and to a lesser extent by a substituent at the carbon atom α to the original hydroxyl group. Effect (a) will, however, not be important when determining the relative rates of formation of isomeric acetals derived from an aldehyde and a polyhydroxy alcohol. Effect (b) is involved in consideration of stabilities of oxocarbenium ions derived from the primary and secondary hydroxyl groups in a polyhydroxy alcohol. The stabilities of these ions are considered to be possible factors in the observed preferential formation of terminal five-membered ring acetals in the kinetic phase of acetal formation.

The preferential formation of a single stereoisomer during kinetic phase of the reaction of an aldehyde with a diol cannot be explained by assuming that the transition state resembles the carbonium ion (*9a*) because if it does two products should be formed from the same intermediate. It is therefore necessary to assume that the transition state resembles the oxonium ion (*9b*) in one of its rotameric forms.

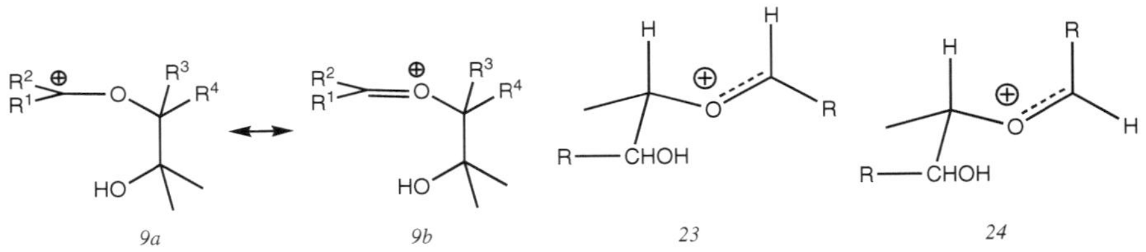

Fig. 6.6

Provided that the intermediate (*9a, b*) has considerable oxonium ion character, two distinct rotamer forms *23* and *24* (Fig. 6.6) can be recognized and designated as *E* (*23*) and *Z* (*24*), respectively. By analogy with olefins [21] the *E* arrangement should be more stable. Using Newman's projections, the structures *23* and *24* may be depicted by three gauche conformations as shown in Fig. 6.7. The conformations *25* and *28* may be dismissed as possible models for the transition state because the potential acetal carbon atom and the hydroxyl group are not sufficiently close for cyclization. If the oxocarbenium ion is next to a primary hydroxyl group, the rotamers *26* and *27* will have same stabilities. However, if the oxocarbenium ion is next to a secondary hydroxyl group, then the rotamer *30* with the group R and the carbonium carbon in the *anti*-arrangement should be more favorable than rotamer *29* wherein these two substituents are in the *syn*-arrangement. Thus it can be predicted

6 Cyclic Acetals and Ketals

that for the reaction with a secondary hydroxyl group the most likely transition state will resemble *30*, i.e., the *anti-E* conformation of oxocarbenium ion.

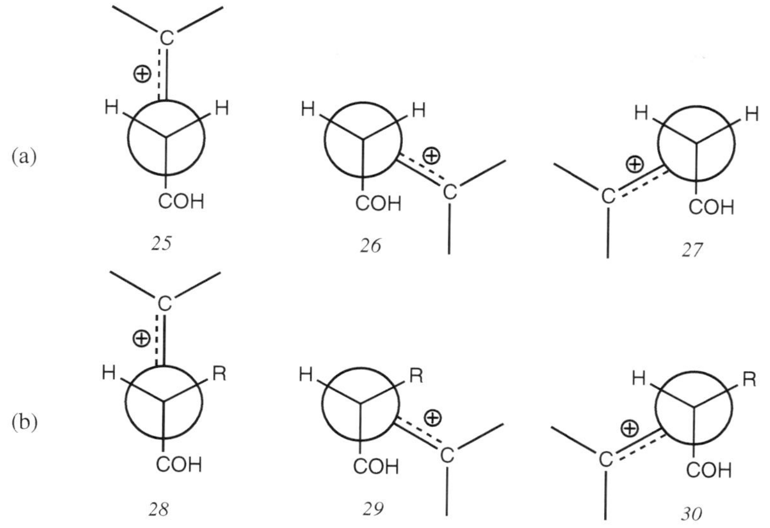

Fig. 6.7 (a) Rotamer forms of the primary oxocarbenium ion and (b) the secondary oxocarbenium ion.

For the reaction between 1,4-anhydroerythritol and benzaldehyde four conformations *31–34* (Fig. 6.8) may be drawn for the oxocarbenium ions, in which the hydroxyl group is well positioned for cyclization, and all four conformations may

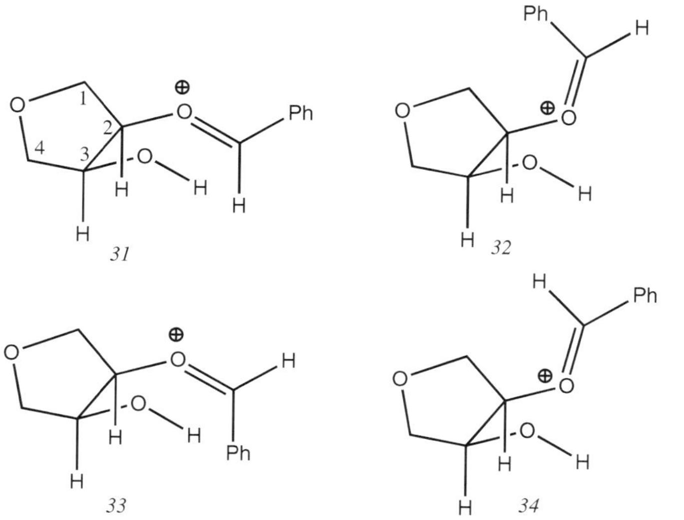

Fig. 6.8

be assumed to resemble the transition state. However, due to steric reasons, the *anti-E* structure *31* should be the most stable one, and a rapid formation of an acetal with an *endo*-phenyl group can be expected. This is in agreement with the experimental observations [18, 19]. The result is quite general and it is expected that the isomer with an *endo*-alkyl group will be the kinetic product in the acetal formation reactions.

Acetalation

Benzylidenation

Benzylidenation [22–25] of anhydrous D-glucose, *35* or *36*, with benzaldehyde in the presence of anhydrous (freshly fused) zinc chloride at room temperature gives 4,6-*O*-benzylidene-D-glucopyranose *39* or *40* with 42% yield (pure product). Similarly, methyl α- or β-D-glucopyranoside (*37* or *38*, respectively) was converted to the corresponding 4,6-*O*-benzylidene acetals [25] with 70% yield (crude product).

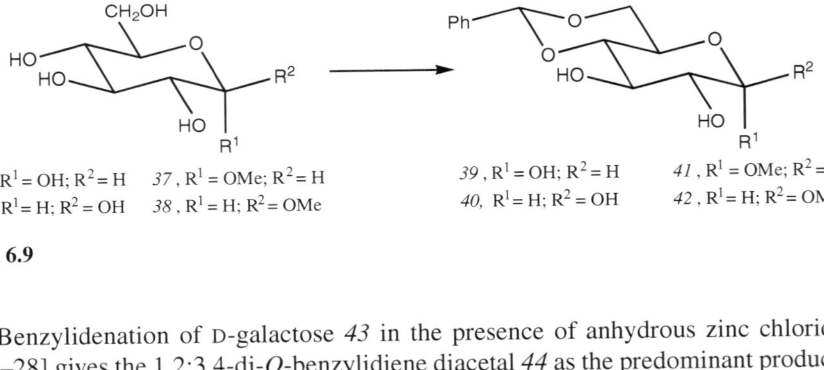

35, R^1 = OH; R^2 = H   *37*, R^1 = OMe; R^2 = H
36, R^1 = H; R^2 = OH   *38*, R^1 = H; R^2 = OMe

39, R^1 = OH; R^2 = H   *41*, R^1 = OMe; R^2 = H
40, R^1 = H; R^2 = OH   *42*, R^1 = H; R^2 = OMe

Fig. 6.9

Benzylidenation of D-galactose *43* in the presence of anhydrous zinc chloride [26–28] gives the 1,2:3,4-di-*O*-benzylidiene diacetal *44* as the predominant product, whereas the 4,6-*O*-benzylidene acetal *45* is obtained only in negligible amounts.

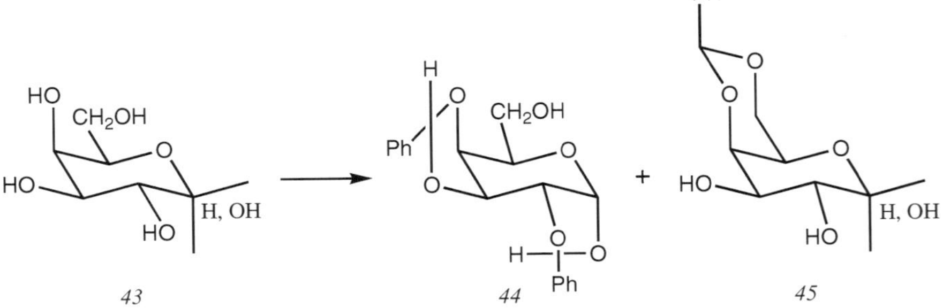

43   *44*   *45*

Fig. 6.10

Acetalation

The benzylidenation may also be carried out with sugars that already contain acetal or ketal group. For example, 1,2-*O*-isopropylidene-α-D-glucofuranose *46* gives with benzaldehyde, in the presence of zinc chloride or phosphorus pentoxide, 3,5-*O*-benzylidene-1,2-*O*-isopropylidene-α-D-glucofuranose *47* [29, 30] (Fig. 6.11). However, heating of 1,2-*O*-isopropylidene-α-D-glucofuranose *46* with

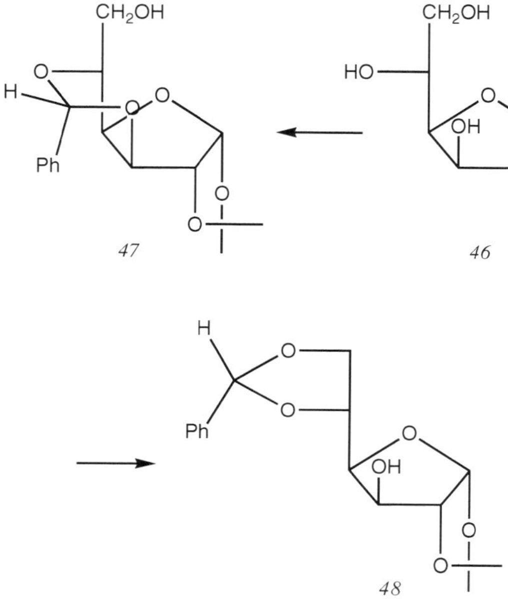

Fig. 6.11

benzaldehyde in the presence of anhydrous sodium sulfate gives 5,6-*O*-benzylidene-1,2-*O*-isopropylidene-α-D-glucofuranose *48* (Fig. 6.11). The 3,5-*O*-benzylidene-1,2-*O*-isopropylidene-α-D-glucofuranose *47* is rearranged by prolonged heating to 5,6-*O*-benzylidene-1,2-*O*-isopropylidene-α-D-glucofuranose *48* [31] indicating that *48* is more stable than *47*.

Ethylidenation

Reaction of a sugar with acetaldehyde, or paraldehyde, in the presence of sulfuric acid as the catalyst [32–36] gives 4,6-*O*-ethylidene acetal, for example, D-glucose *49* gives the 4,6-*O*-ethylidene-D-glucose *50* (Fig. 6.12).

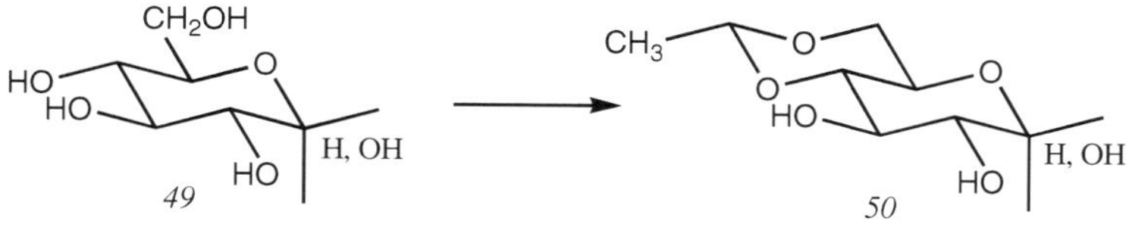

Fig. 6.12

Ketalation

Isopropylidenation (Acetonation)

Isopropylidene acetals are obtained by condensation of monosaccharides or their derivatives with acetone in the presence of mineral acids, such as concentrated sulfuric acid, as the catalysts [37]. The anhydrous acetone serves both as the solvent and as the reagent. Reaction is performed at room temperature and it is usually completed in several hours. Thus, D-glucose 49 and anhydrous acetone give in the presence of concentrated sulfuric acid 1,2:5,6-di-*O*-isopropylidene-α-D-glucofuranose 51 (46.5%) together with some 1,2-*O*-isopropylidene-α-D-glucofuranose 46 (16.8%) (Fig. 6.13).

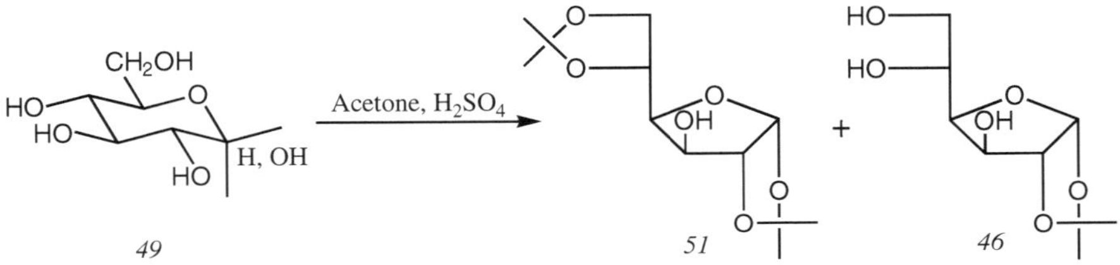

Fig. 6.13

However, when a mixture of anhydrous (freshly fused) zinc chloride and phosphoric acid (85%) is used as the catalyst 1,2:5,6-di-*O*-isopropylidene-α-D-glucofuranose 51 is obtained with 91% yield [38].

Acetonation of D-galactose 43 [39, 40] in the presence of anhydrous $ZnCl_2$ gave 1,2:3,4-di-*O*-isopropylidene-D-galactose 52 (89.9% yield of crude product) (Fig. 6.14). Mixture of concentrated sulfuric acid and anhydrous Cu(II) sulfate can also be used for the synthesis of di-*O*-isopropylidene sugars. Thus, anhydrous galactose 43 reacts with acetone, in the presence of concentrated sulfuric acid and anhydrous Cu(II) sulfate, to give 1,2:3,4-di-*O*-isopropylidene-α-D-galactopyranose 52 with 76–92% yield [41, 42] (Fig. 6.14).

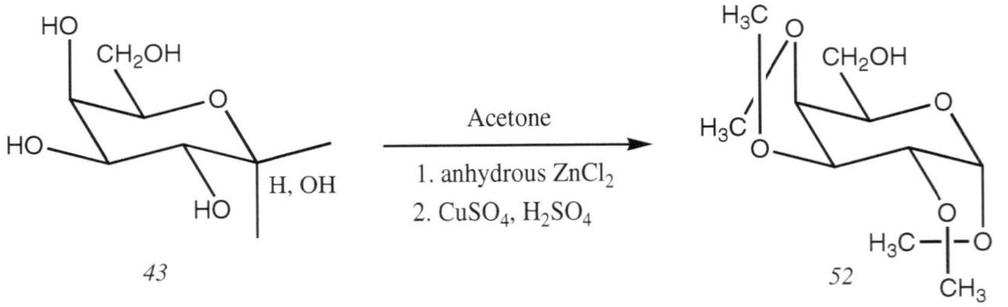

Fig. 6.14

Ketalation

When the use of strong mineral acids is not possible, as is the case, for example, with glycosides, powdered anhydrous Cu(II) sulfate alone can be used as the catalyst. However, the reaction is much slower (it takes several days for completion). Thus acetonation of methyl α-D-mannopyranoside 53 in the presence of anhydrous Cu(II) sulfate gave methyl 2,3:4,6-di-*O*-isopropylidene-α-D-mannopyranoside 54 (Fig. 6.15) with ca. 11% yield and methyl 2,3-*O*-isopropylidene-α-D-mannopyranoside 55 with ca. 6.6% yield (Fig. 6.15).

Fig. 6.15

However, methyl β-D-mannopyranoside 56 gave under the same reaction conditions, after 5 months, 45.75% of methyl 2,3:4,6-di-*O*-isopropylidene-β-D-mannopyranoside 57 and 17.86% of methyl 2,3-*O*-isopropylidene-β-D-mannopyranoside 58 [43] (Fig. 6.16).

Fig. 6.16

The nature of the catalyst may have a notable effect on acetonation reaction. Thus, for example, in the presence of zinc chloride, D-mannitol 59 affords the

59, D-mannitol

60, 1,2:5,6-di-O-isopropylidene-D-mannitol

61, 1,2:3,4:5,6-tri-O-isopropylidene-D-mannitol

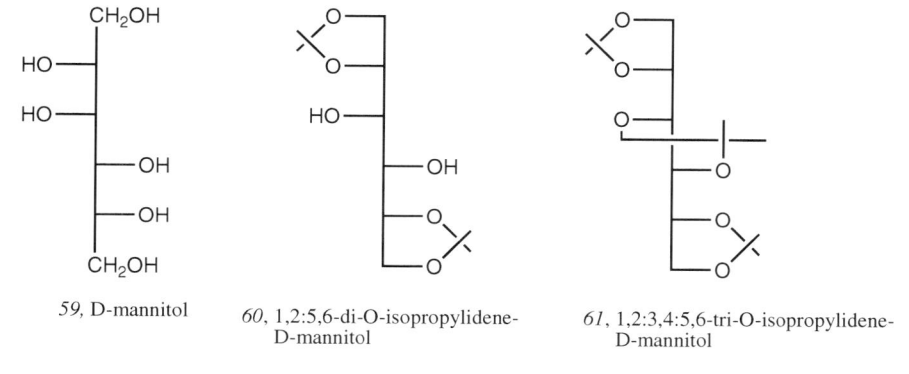

Fig. 6.17

1,2:5,6-di-*O*-isopropylidene ketal *60* [44], whereas when catalyzed by mineral acids 1,2:3,4:5,6-tri-*O*-isopropylidene ketal *61* is obtained [45] (Fig. 6.17). The pattern of acetonation of some glycosides is also influenced by the nature of catalyst [46].

Transacetalation and Transketalation

Acid-catalyzed acetal or ketal exchange, also known as *transacetalation* or *transketalation*, is widely used in carbohydrate chemistry for the synthesis of cyclic acetals and ketals. The mechanism of transacetalation is shown in Fig. 6.18. The protonation of one of the two alkoxy groups of an acetal or ketal donor *63* activates that

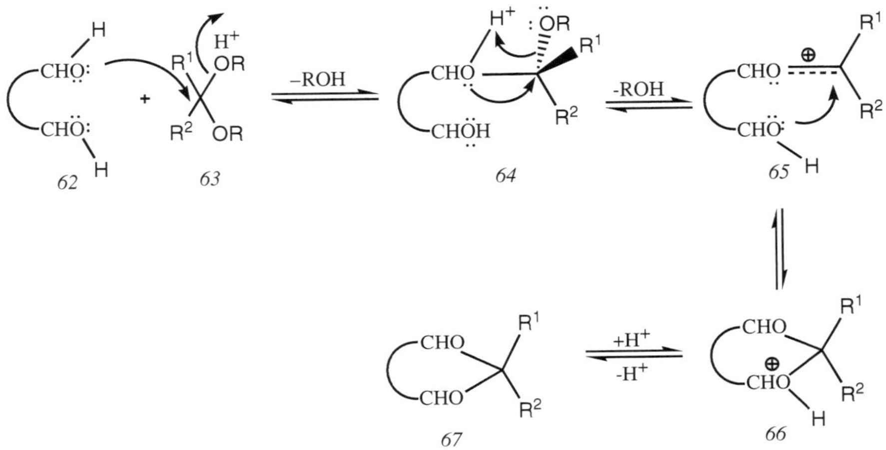

Fig. 6.18

alkoxy group so that it can be displaced by a sugar hydroxyl group and gives a mixed acetal *64*. The protonation of the second alkoxy group of acetal donor and hence the elimination of corresponding alcohol gives oxocarbenium ion *65* that reacts with the second hydroxyl group of a carbohydrate to give first the protonated acetal *66*, which, after deprotonation, yields cyclic acetal or ketal *66*.

Thus, methyl α-D-glucopyranoside *37* reacts with α,α-dimethoxytoluene in anhydrous *N,N*-dimethylformamide at 60°C in the presence of *p*-toluenesulfonic acid [47] to give 82.4% of crude (63.5% of pure) methyl 4,6-*O*-benzylidene-α-D-glucopyranoside *41* (Fig. 6.19). This is a significantly improved yield over the condensation of benzaldehyde with methyl α-D-glucopyranoside *37* in the presence of fused zinc chloride. Methyl β-D-glucopyranoside *38* gave under the same reaction conditions 58% of pure methyl 4,6-*O*-benzylidene-β-D-glucopyranoside *42* [47].

Reaction of methyl α-D-glucopyranoside *37* with 2,2-dimethoxypropane in anhydrous *N,N*-dimethylformamide and in the presence of *p*-toluenesulfonic acid [48] gave methyl 4,6-*O*-isopropylidene-α-D-glucopyranoside *68* with 79% yield, together with methyl 2,3:4,6-di-*O*-isopropylidene-α-D-glucopyranoside *69* (1.8%) and some starting material (Fig. 6.20).

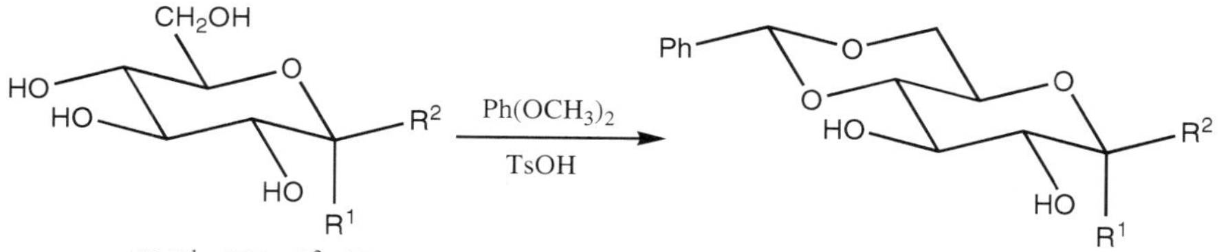

37, R^1=OMe; R^2=H
38, R^1=H; R^2=OMe

41, R^1=OMe; R^2=H
42, R^1=H; R^2=OMe

Fig. 6.19

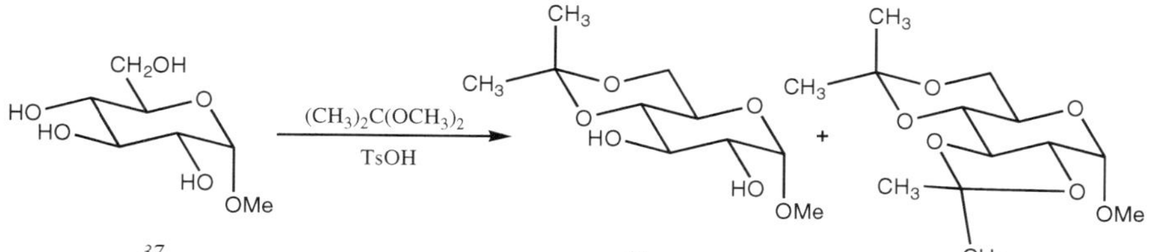

37 *68* *69*

Fig. 6.20

Similarly 2-acetamido-2-deoxy-D-glucopyranose *70* gave with 2,2-dimethoxypropane at 80–85°C in anhydrous *N,N*-dimethylformamide in the presence of *p*-toluenesulfonic acid 2-acetamido-2-deoxy-4,6-isopropylidene-D-glucopyranose *71* with 54% yield [49] (Fig. 6.21).

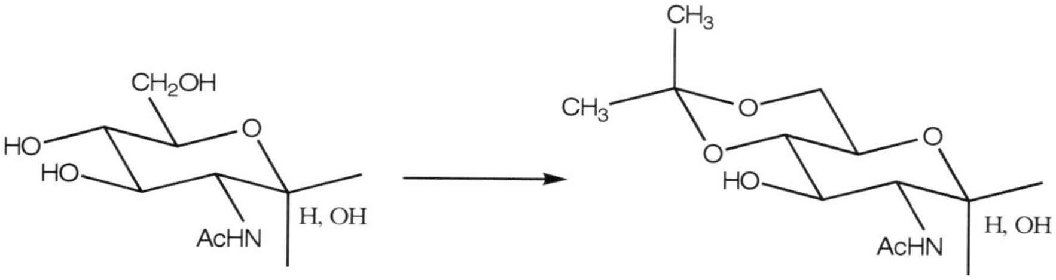

70 *71*

Fig. 6.21

Ethylidene acetals can be obtained by reacting, for example, methyl-β-D-glucopyranoside with 1,1-dimethoxyethane, in the presence of concentrated sulfuric acid, whereby the corresponding 4,6-*O*-ethylidene-acetal is obtained with 59% yield [50–52] (Fig. 6.22).

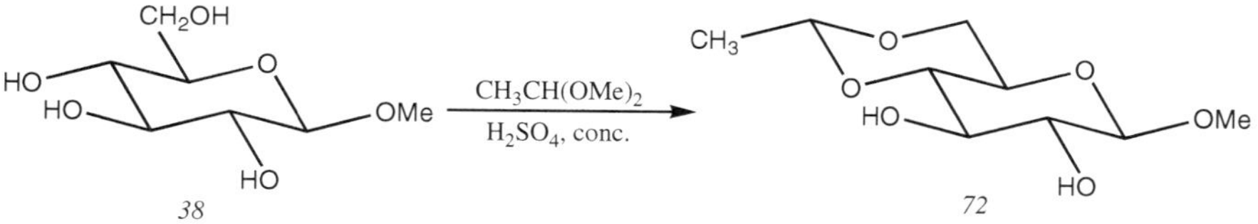

38 *72*

Fig. 6.22

The isopropylidenation of 2-acetamido-2-deoxy-D-xylose diethyl dithioacetal *73* with 1:1 (v/v) anhydrous acetone-2,2-dimethoxypropane, in the presence of concentrated sulfuric acid, gave the 3,4-*O*-isopropylidene ketal *74*, as the only product,

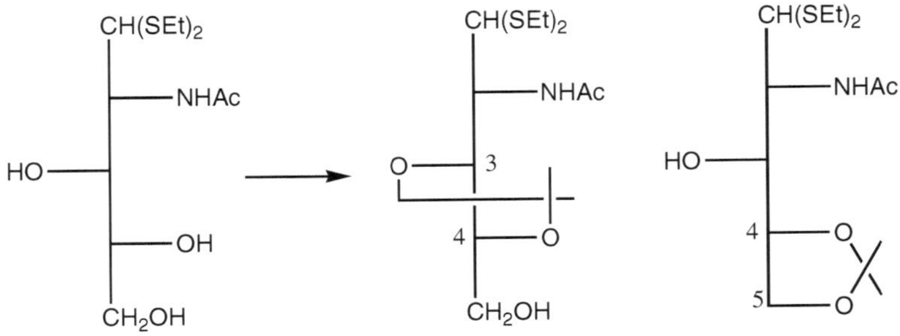

Fig. 6.23

with 90% yield. However, the isopropylidenation of *73* with 1:1 (v/v) anhydrous acetone-2,2-dimethoxypropane in the presence of copper(II) sulfate afforded the 4,5-*O*-isopropylidene ketal *75* with 74% yield [53] (Fig. 6.23).

These results suggest that isopropylidenation of alditols or acyclic sugar derivatives proceeds initially by involving the primary hydroxyl group(s) of a carbohydrate in the ketal formation (due to steric reasons) and if the catalyst is incapable of reversing the ketalation [as is the case with anhydrous copper (II) sulfate], the product is formed under kinetic control. If, however, the catalyst is capable of reversing the ketalation reaction, the end product(s) is (are) thermodynamically controlled.

The Isomerization of Cyclic Acetals and Ketals

Reaction of glycerol with acetone in the presence of acid catalyst should give the equilibrium mixture of two possible cyclic ketals, 1,3-dioxolane *78* and 1,3-dioxane *79* (Fig. 6.24). However, it has been shown that the reaction gives essentially

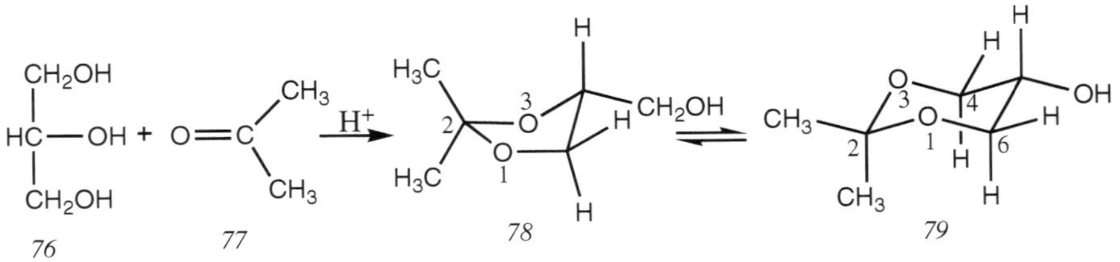

Fig. 6.24

The Isomerization of Cyclic Acetals and Ketals 155

exclusively the 1,2-O-isopropylidene glycerol (78) [54]. The formation of six-membered O-isopropylidene ketal is disfavored [55], because one methyl group must take the axial orientation in the chair conformation of the 1,3-dioxane ring 79. It has been determined that for the chair form of methylcyclohexane, the conformer with the methyl group axially oriented is about 1.9 kcal/mol less stable than the conformer with the methyl group equatorially oriented due to the *syn*–axial interactions of this methyl group with the axially oriented C3 and C5 hydrogens [56]. For the chair form of 2,2-dimethyl-1,3-dioxane 79 the axially oriented methyl group is even less favorable, since the C–O bonds are shorter (1.43 Å) than the C–C bonds (1.54 Å) and the 1,3-*syn*–axial interactions between the axially oriented C2 methyl group and the axially oriented C4 and C6 hydrogen atoms in 82 will be consequently

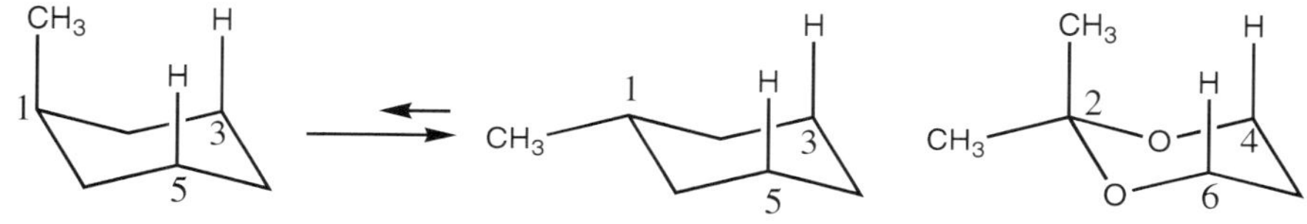

Fig. 6.25

greater than the 1,3-*syn*–axial interactions between the axially oriented C1 methyl group and the two axially oriented C3 and C5 hydrogen atoms in a cyclohexane ring 80 (Fig. 6.25).

The reason that the 2,2-dimethyl-1,3-dioxolane 78 is formed as almost the only product, however, is due not only to destabilizing 1,3-*syn*–axial interactions present in the 2,2-dimethyl-1,3-dioxane ring, but also to the fact that five-membered ring ketal 78 is relatively free from strain in the C_s conformation (shown in Fig. 6.24) wherein the hydroxymethyl group is equatorially oriented. Since the reaction of formaldehyde with glycerol favors only to a slight extent the formation of six-membered dioxane ring [57, 58] it seems that the 1,3-dioxolane ring system is inherently more stable than the 1,3-dioxane ring.

The Migration of Acetal or Ketal Group

An interesting property of isopropylidene derivatives of carbohydrates is the tendency of isopropylidene group to migrate from terminal position wherein the two carbons of isopropylidene group are the primary and secondary carbons of a sugar to a position inside the carbohydrate chain where both carbons of isopropylidene group are secondary carbons.

Thus, for example, heating of 1,2:4,5-di-O-isopropylidene-D-dulcitol 83 with pyridine hydrochloride smoothly rearranges 83 to 2,3:4,5-di-O-isopropylidene-D-dulcitol 84 [59] (Fig. 6.26).

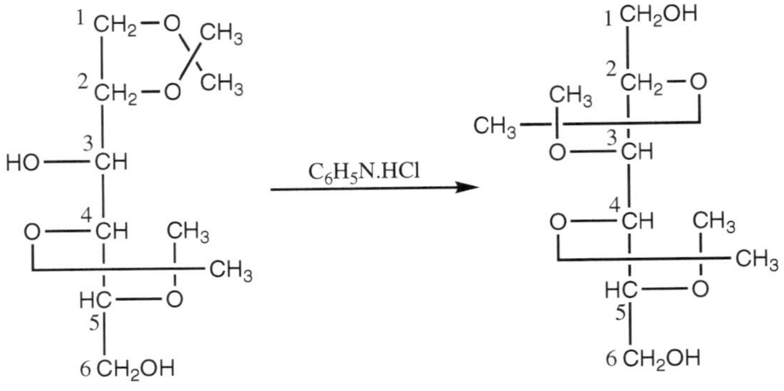

156   6 Cyclic Acetals and Ketals

Fig. 6.26

The migration of isopropylidene group takes place via an attack of a proton and an anion according to the so-called push–pull mechanism [60–62]. The protonation of the acetal oxygen linked to the primary carbon (C5) and the simultaneous attack by a base (anion) to the hydrogen atom of the secondary hydroxyl group at the neighboring carbon **C3** results in the simultaneous breakage of the O-H bond and the O1–C2' bond causing the migration of the acetal from the C1–C2 to the C2–C3 carbons.

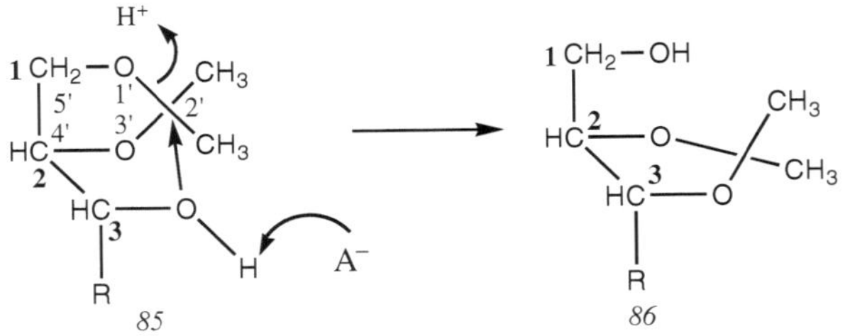

Fig. 6.27

When working with isopropylidene groups one must always consider the possibility of their migration, particularly in the presence of reagents with strong acid–base properties. Thus, for example, 1,2:5,6-di-*O*-isopropylidene-α-D-glucofuranose *51* reacts with triphenyl phosphite iodomethylate in benzene whereby the 5,6-*O*-isopropylidene group migrates to 3,5-position giving 1,2:3,5-di-*O*-isopropylidene-α-D-glucofuranose *87* [63, 64] (Fig. 6.28).

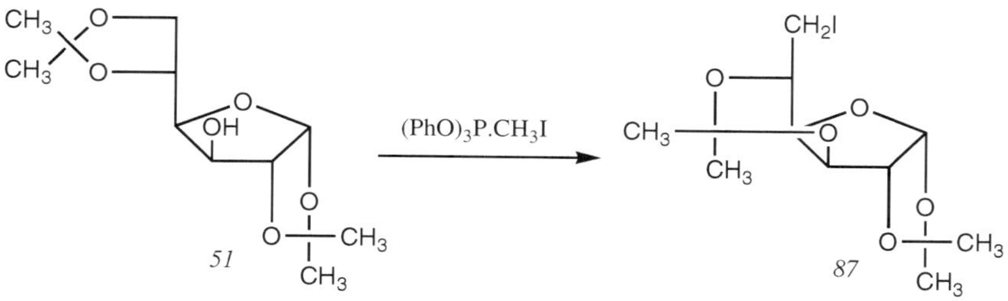

Removal of Acetal and Ketal Groups

Fig. 6.28

Removal of Acetal and Ketal Groups

Benzylidene Group

The benzylidenation is very often used for protection of 1,2- or 1,3-hydroxyl groups of carbohydrates since it is often regioselective, the yields of benzylidene acetals are high, and the benzylidene group can be easily removed in high yield.

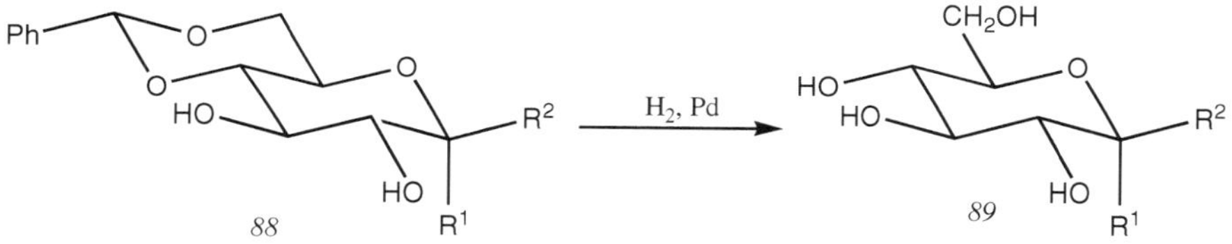

R^1 = OMe, OAc, OBz; R^2 = H
R^1 = H; R^2 = OMe, OAc, OBz

Fig. 6.29

Thus it can be removed by catalytic hydrogenation as from, for example, alkyl 4,6-*O*-benzylidene-α- or β-D-glycopyranoside or 1-*O*-acyl derivatives (acetate, benzoate) by hydrogenation using palladium black as the catalyst and methanol as the solvent [65] (Fig. 6.29). It can also be removed by acid hydrolysis using 9:1 (v/v)

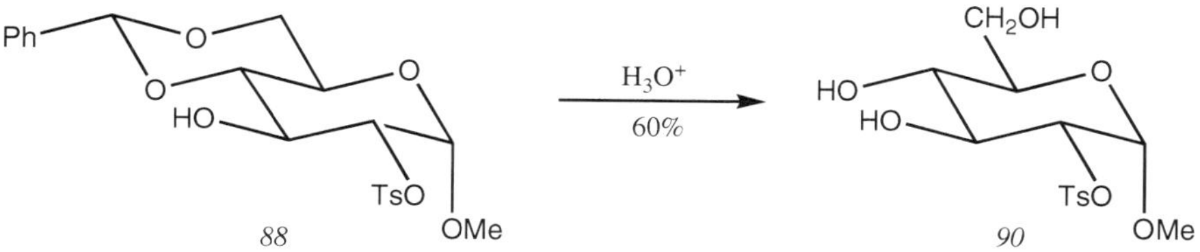

Fig. 6.30

CF$_3$COOH–water [66] by keeping the mixture for 5–10 min at room temperature (Fig. 6.30).

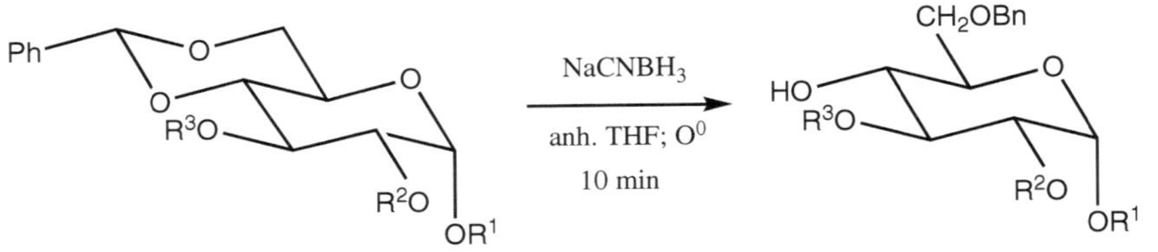

91, R^1 = OMe; R^2 = R^3 = Bn
92, R^1 = OMe; R^2 = R^3 = Bn
93, R^1 = OMe; R^2 = NHAc; R^2 = Bn

95, R^1 = OMe; R^2 = R^3 = Bn (81%)
96, R^1 = OMe; R^2 = R^3 = Bn (95%)
97, R^1 = OMe; R^2 = NHAc; R^2 = Bn (60%)

Fig. 6.31

Sodium cyanoborohydride in dry THF at 0°C converts the 4,6-*O*-benzylidene group to 6-*O*-benzyl group [67] (Fig. 6.31).

The reductive ring cleavage of carbohydrate benzylidene acetals having either the 1,3-dioxane [68–72] or 1,3-dioxolane ring [73–77] proceeds with high regioselectivity giving partially benzylated derivatives. For dioxolane-type benzylidene acetals, the direction of reaction is determined by configuration of the acetal carbon [73–77], whereas the factors that influence the reduction of the 1,3-dioxane acetals are not obvious.

Reduction of several 4,6-*O*-benzylidene derivatives of monosaccharides with LiAlH$_4$–AlCl$_3$ (1:1) [68, 69, 71, 72] gave 4-*O*-benzyl derivative as the major product, in almost every case, accompanied by various amounts of 6-*O*-benzyl ethers. For example, various 4,6-*O*-benzylidene-3-*O*-methyl-glucopyranosides gave the 6-*O*-benzyl ethers as by-products in yields of < 30%, but no 6-*O*-benzyl ethers were detected in the reduction of 3-*O*-benzyl-4,6-*O*-benzylidene glycopyranosides [68–70]. It was suggested [68, 69] that the product ratio is determined by the bulk of the C3-substituent, i.e., by steric accessibility of the acetal oxygen atoms to the chloroalane reagent [78].

A systematic study of the effect of bulkiness of the C3-substituent on the products' ratios of reductive ring cleavage of alkylated alkyl 4,6-*O*-benzylidene-D-glucopyranosides has been published [79].

Benzyl 4,6-*O*-benzylidene-β-D-glucopyranoside was alkylated to give 2,3-di-*O*-methyl (*98*), 2,3-di-*O*-ethyl (*99*), 2,3-di-*O*-propyl (*100*), and 2,3-di-*O*-benzyl (*101*) derivatives as well as 3-deoxy-2-*O*-methyl derivative (*102*), and the reductive ring cleavage of alkylated 4,6-*O*-benzylidene-glucopyranoside derivatives was performed with the LiAlH$_4$–AlCl$_3$ (1:1) reagent in ether–dichloromethane (2:1) at reflux temperature. Reactions were completed within 2 h. The structures are given in Fig. 6.32 and the results are given in Table 6.1.

Considering the product ratios in Table 6.1, it is evident that bulkier the 3-substituent, higher the proportion of the 4-*O*-benzyl ether. This increase in regioselectivity can be explained by postulating the complexation of the Lewis acid

Table 6.1 LiAlH$_4$–AlCl$_3$ reduction of 4,6-O-benzylidene-D-glucopyranosides

4,6-Acetal	3-Substituent	Product ratio[a]	Isolated yield (%)	
			4-O-Benzyl	6-O-Benzyl
98	OMe	77:23	68	14
99	OEt	91:9	75	7
100	OPr	94:6	91	3
101	OBn	93:7	91	4
102	H	53:47	43	41

[a] 4-O-Benzyl derivative/6-O-benzyl derivative.

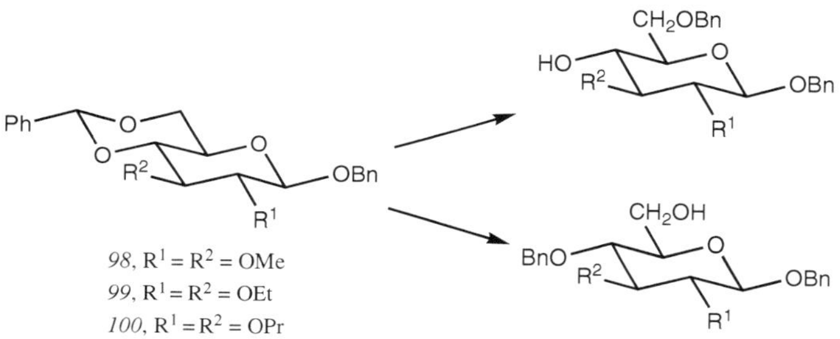

98, R^1 = R^2 = OMe
99, R^1 = R^2 = OEt
100, R^1 = R^2 = OPr
101, R^1 = R^2 = OBn
102, R^1 = OMe; R^2 = H

103, R^1 = R^2 = OMe
104, R^1 = R^2 = OEt
105, R^1 = R^2 = OPr
106, R^1 = R^2 = OBn
107, R^1 = OMe; R^2 = H

108, R^1 = R^2 = OMe
109, R^1 = R^2 = OEt
110, R^1 = R^2 = OPr
111, R^1 = R^2 = OBn
112, R^1 = OMe; R^2 = H

Fig. 6.32

chloroalane with one of the two acetal oxygens [80] (Fig. 6.33). Complex formation at O4 (Path A) is hindered by bulky substituents at the O3 whereas the complexation at O6 (Path B) is not influenced by bulkiness of O3 substituents. Hence, Path B is favored always when the O3 is substituted with bulky substituents, which will result in the greater formation of the 4-O-benzyl derivative. It is apparent from Table 6.1 that some of the 4-O-benzyl derivative was formed even in the reduction of 3-O-propyl (*100*) and 3-O-benzyl (*101*) derivatives (Fig. 6.33). These two compounds gave nearly the same product ratios of O4/O6 benzyl derivatives, indicating that the regioselectivity of this reaction is dependent not just on the length of the O-alkyl chain. Perhaps, a better insight could be obtained by comparing the product ratio of the reductive cleavage of 3-O-benzyl and 3-O-isopropyl ethers.

The proposed explanation is in agreement with earlier findings [69–71] that a higher proportion of 6-O-benzyl ethers is obtained by reductive ring cleavage of the 4,6-O-benzylidene-D-galactopyranoside derivatives than of the corresponding glucopyranoside derivatives, since steric shielding of the O4 by the 3-substituents is less effective in the former.

The reductive opening of the 4,6-O-benzylidene-D-glucopyranoside derivative was examined with various reducing agents [79]. For this purpose benzyl

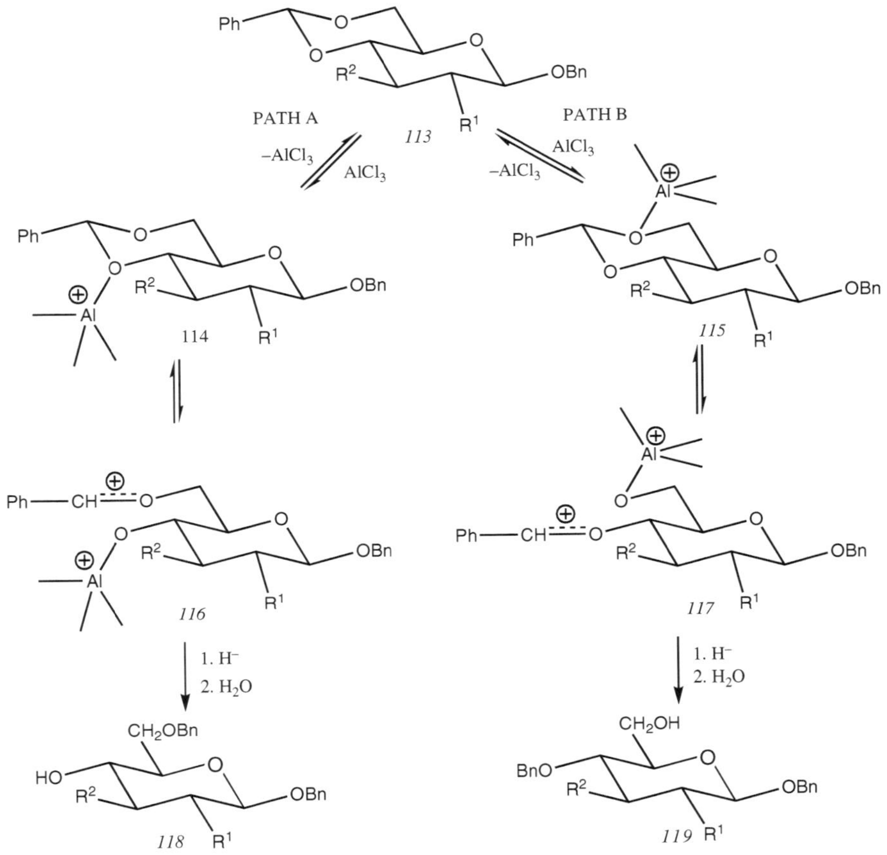

Fig. 6.33

4,6-*O*-benzylidene-2,3-di-*O*-methyl-β-D-glucopyranoside *98* (Fig. 6.32) was chosen, where the steric hindrance is relatively small so that a change in the product ratio should be maximal. The results are summarized in Table 6.2.

Table 6.2 Reductions of *98* with various reducing agents

Reagent	Solvent	Bath temperature (°C)	Reaction time (h)	Product ratios 98:103	Isolated yield (%) 98	103
LiAlH$_4$–AlCl$_3$	(2:1) Et$_2$O–CH$_2$Cl$_2$	45	2	77:23	68	14
LiAlH$_4$–AlBr$_3$	(2:1) Et$_2$O–CH$_2$Cl$_2$	45	2	84:16	73	12
(i-Bu)$_2$AlH	Benzene	0	3	46:54	34	43
(i-Bu)$_2$AlH	(9:1) Et$_2$O–CH$_2$Cl$_2$	45	8	71:29	68	25
Borane	Tetrahydrofuran	75	72	0:0	–	–
Borane	Benzene	90	120	57:11	46	7

It has been reported [81, 82] that borane in tetrahydrofuran reduces acetals and that it may have some advantages over LiAlH$_4$–AlCl$_3$ system. However, it was found that benzyl 4,6-O-benzylidene-2,3-di-O-methyl-β-D-glucopyranoside 98 did not react with an excess of borane in tetrahydrofuran after 72 h. Slow reaction also took place in benzene at elevated temperature, so that even after 120 h much of 98 was still present. The major product was the 4-O-benzyl ether, formed with a good regioselectivity, but the low reaction rate of this method makes it unsuitable for practical purposes.

Methyl 4,6-O-benzylidene-α-D-galactopyranoside 120 reacts with N-bromosuccinimide (NBS) in carbon tetrachloride at reflux and in the presence of barium carbonate for neutralizing the generated HBr giving methyl 4-O-benzoyl-6-bromo-6-deoxy-α-D-galacto pyranoside 121 with over 60% yield [83] (Fig. 6.34).

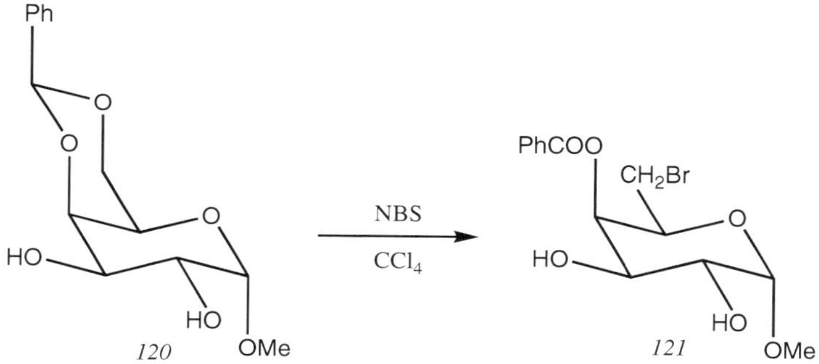

Fig. 6.34

Similarly, methyl 4,6-O-benzylidene-α-D-glucopyranoside with substituted or unsubstituted C2 and C3 hydroxyl groups gives, with N-bromosuccinimide (NBS) in carbon tetrachloride at reflux and in the presence of barium carbonate, 4-O-benzoyl-6-bromo-6-deoxy-α-D-glucopyranoside [83, 84].

Benzylidene dioxolane derivatives give, under the same reaction conditions except for the presence of water and UV irradiation (low-pressure mercury lamp), the hydroxy benzoates with an axial benzoyl group and the adjacent hydroxyl group equatorially oriented [85] with 72% yield (Fig. 6.35).

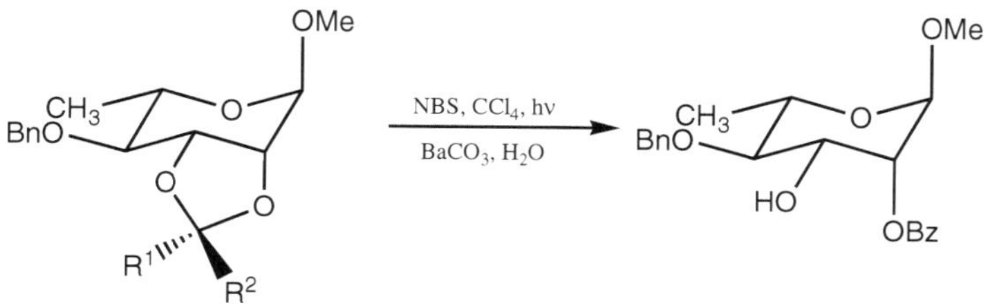

122, R^1 = Ph; R^2 = H
123, R^1 = H; R^2 = Ph

Fig. 6.35

The oxidative cleavage of acetal rings with ozone to the corresponding esters and alcohols has been studied by Deslongchamps [86, 87].

The ozonolysis of the following methyl α-D-glucopyranosides has been studied: 2,3-di-O-methyl *98*, 2,3-di-O-acetyl *125*, and 2,3-di-O-tosyl *126*. Ozonolysis of *98* in anhydrous carbon tetrachloride gave a mixture of 6-O-benzoyl derivative *127* (60%) and 4-O-benzoyl derivative *130* (40%) of methyl 2,3-di-O-methyl-α-D-glucopyranoside. The ozonolysis of 2,3-di-O-acetyl-4,6-O-benzylidene derivative *125* in glacial acetic acid gave a mixture of 6-O-benzoyl-2,3-di-O-acetyl derivative *128* (85%) and 4-O-benzoyl-2,3-di-O-acetyl derivative *131* (15%). The ozonolysis of 2,3-di-O-tosyl-4,6-O-benzylidene derivative *126* in glacial acetic acid gave methyl 4-O-benzoyl-2,3-di-O-tosyl-α-D-glucopyranoside *132* as the only product (Fig. 6.36).

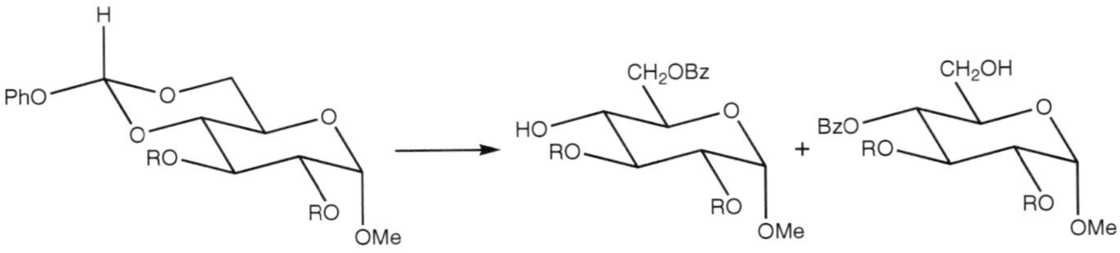

98, R = Me
125, R = Ac
126, R = Ts

127, R = Me (60%)
128, R = Ac (85%)
129, R = Ts (0%)

130, R = OMe (40%)
131, R = Ac (15%)
132, R = Ts (100%)

Fig. 6.36

It is known that methyl β-D-glucopyranoside can be oxidized by ozone [86, 87] but this reaction is slow at room temperature. Benzylidene acetals are much more reactive: the oxidation is normally completed within 2 h at −78°C. Therefore the oxidation of benzylidene acetals of alkyl 4,6-O-benzylidene-β-D-glycopyranoside can be carried out without oxidation of the anomeric carbon.

The oxidation of methyl 4,6-O-ethylidene-2,3-di-O-acetyl-α-D-glucopyranoside in acetic acid/ sodium acetate gave methyl 2,3,4,6-tetra-O-acetyl-α-D-glucopyranoside with excellent yield (Fig. 6.37). The oxidation was followed by in situ "acetylation" due to use of acetic anhydride/sodium acetate as a solvent.

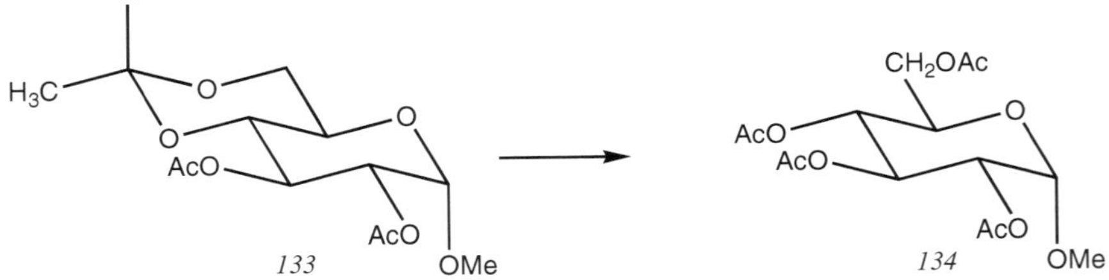

133

134

Fig. 6.37

Isopropylidene Group

There is considerable difference in stability between an exocyclic and an endocyclic isopropylidene groups (those that are fused to the furanose ring). Thus, for example, the exocyclic isopropylidene group can be selectively removed from 1,2:5,6-di-O-isopropylidene-α-D-glucofuranose *48* by acid-catalyzed hydrolysis (Fig. 6.38) giving 1,2-O-isopropylidene-α-D-glucofuranose *46* in very high yield [88, 89].

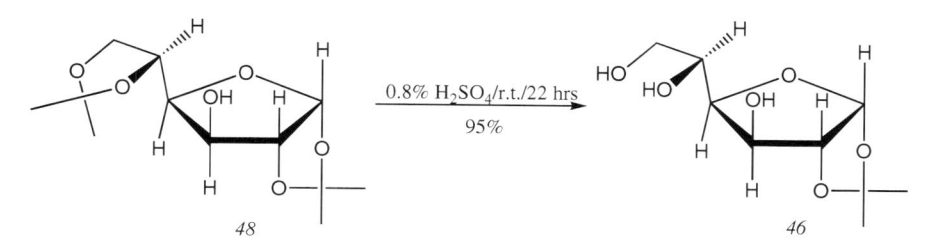

Fig. 6.38

This selective hydrolysis can also be achieved in aqueous acetic acid [90, 91] either by standing at room temperature (several days) or by heating at 100°C for 2 h. Similarly, the D-*ido*- [92] and D-*allo*- [93] isomers may be selectively hydrolyzed to the 1,2-O-isopropylidene derivatives. According to crude estimations [90] the 5,6-O-isopropylidene group hydrolyses ca. 40 times faster than 1,2-O-isopropylidene group.

Under the more vigorous conditions (mineral acid, higher acid concentration, and higher temperature) both isopropylidene groups are hydrolyzed.

References

1. Barker, S. A.; Bourne, E. J., "*Acetals and Ketals of the Tetritols, Pentitols and Hexitols*", Advan. Carbohydr. Chem. (1952) **7**, 137–207
2. de Belder, A. N., "*Cyclic acetals of the aldoses and aldosides*", Adv. Carbohydr. Chem. (1965) **20**, 219–302
3. de Belder, A. N., "*Cyclic acetals of the aldoses and aldosides*", Adv. Carbohydr. Chem. Biochem. (1977) **34**, 179–241
4. Brady, R. F., Jr., "*Cyclic acetals of Ketoses*", Adv. Carbohydr. Chem. Biochem. (1971) **26**, 197–278
5. Stoddart, J. F., *Stereochemistry of Carbohydrates*, Willey Interscience, New York (1971) pp. 186–220
6. Hough, L.; Richardson, A. C., in *Rodd's Chemistry of Carbon Compounds*, Vol. 1, Part F, Coffey, S. (Ed.), Elsevier, New York, (1967) pp. 32–38 and 351–362
7. Lemieux, R. U. in *Molecular Rearrangements*, Part II, de Mayo, P. (Ed.), Wiley-Interscience, New York (1963), pp. 723–733
8. Clode, D. M., "*Carbohydrate cyclic acetal formation and migration*", Chem. Rev. (1979) **79**, 491–513
9. Adkins, H.; Broderick, A. E., "*Hemiacetal formation and the refractive indices and densities of mixtures of certain alcohols and aldehydes*", J. Am. Chem. Soc. (1928) **50**, 499–503

10. Ingold, C. K. *Structure and Mechanism in Organic Chemistry*, 2nd ed., Cornell University Press, Ithaca, N.Y. (1969) p. 447
11. Salomaa, P.; Kankaanperä, A., *"Hydrolysis of 1,3-dioxolane and its alkyl-substituted derivatives. I. Structural factors influencing the rates of hydrolysis of a series of methyl-substituted dioxolanes"*, Acta Chem. Scand. (1961) **15**, 871–878
12. Fife, T. H.; Jao, L. K., *"Substituent effects in acetal hydrolysis"*, J. Org. Chem. (1965) **30**, 1492–1495
13. De Wolfe, R. H.; Ivanetich, K. M.; Perry, N. F., *"General acid catalysis in benzophenone ketal hydrolysis"*, J. Org. Chem. (1969) **34**, 848–854
14. Collins, P. M., *"The kinetics of the acid catalysed hydrolysis of some isopropylidene furanoses"*, Tetrahedron (1965) **21**, 1809–1815
15. Cordes, E. H.; Bull, H. G., *"Mechanism and catalysis for hydrolysis of acetals, ketals, and ortho esters"*, Chem. Rev. (1974) **74**, 581–603
16. For a treatise on this subject, see Lewis and Hammes, *"Investigation of rates and mechanism of reaction"*, 3d ed. (Vol. 6 of Weissberger, *"Techniques of Chemistry"*), 2 pts., Wiley, New York, 1974. For a monograph, see Carpenter *"Determination of organic reaction mechanisms"*, Wiley, New York, 1984
17. Baggett, N.; Duxbury, J. M.; Foster, A. B.; Webber, J. M., *"Further observations on the acid-catalysed benzaldehyde-glycerol reaction"*, Carbohydr. Res. (1966) **2**, 216–223
18. Baggett, N.; Foster, A. B.; Webber, J. M.; Lipkin, D.; Philips, B. E., *"2', 3'-O-Benzylidene nucleosides"*, Chem. Ind. (London) (1965) 136–137
19. Al-Jeboury, F. S.; Baggett, N.; Foster, A. B.; Webber, J. M., *"Observations on cyclic acetal formation and migration"*, J. Chem. Soc. Chem. Commun. (1965) 222–224
20. Clode, D. M., Ph.D. Thesis, Birmingham University, 1968
21. Turner, R. B.; Nettleton, D. E.; Perelman, M., *"Heats of Hydrogenation. VI. Heats of Hydrogenation of Some Substituted Ethylenes"*, J. Am. Chem. Soc. (1958) **80**, 1430–1433
22. Zervas, L., *"Über Benzyliden-glucose und ihre Verwendung zu Synthesen: 1-Benzoyl-glucose"*, Berichte (1931) **64**, 2289–2296
23. Wood, H. B., Jr.; Diehl, H. W.; Fletcher, H. G. Jr., *"1,2:4,6-Di-O-benzylidene-α-D-glucopyranose and Improvements in the Preparation of 4,6-O-Benzylidene-D-glucopyranose"*, J. Am. Chem. Soc. (1957) **79**, 1986–1988
24. Fletcher, H. G., Jr., *"4,6-Benzylidene Derivatives"*, in Methods in Carbohydrates Chemistry, Vol. II, Academic Press, New York (1963), pp. 307–308
25. Richtmyer, N. K., *"Methyl 4,6-O-Benzylidene-α-D-glucopyranoside"*, in Methods in Carbohydrates Chemistry, Vol. I, Academic Press, New York (1962), p. 108
26. Zinner, H.; Thielebeule,W., *"Derivatives of sugar dithioacetals. XXIII. Acyl and benzylidene derivatives of D-galacturonic acid dialkyl dithioacetals"*, Chem. Ber. (1960) **93**, 2791–2803
27. Pacák, J.; Cerny, M., *"1,2:3,4-Di-O-benzylidene-D-galactopyranose"*, Collect. Czech. Chem. Commun. (1961) **26**, 2212–2216
28. Pacák, J.; Černý, M., *"Preparation and structure of 4,6-O-benzylidene-D-galactopyranose"*, Collect Czech. Chem. Commun. (1963) **28**, 541–544
29. Brigl, P.; Grüner, H., *"Kohlenhydrate, XIII. Mitteil.: Neue Benzal- und Benzoyl- Derivate der Glucose"*, Berichte(1932) **65**, 1428–1434
30. Zervas, L.; Sessler, P., *"Synthese von d-Glucuronsäure"*, Berichte (1933) **66**, 1326–1329
31. Levene, P. A.; Raymond, A. L., *"Über die 1.2-Monoaceton-3.5-benzal- und die 1.2-Monoaceton-5.6-benzal-glucose"*, Berichte (1933) **66**, 384–386
32. Helferich, B.; Appel, H., *"Über Verbindungen von Kohlenhydraten mit Acetaldehyd: Äthylidenglucose"*, Berichte (1931) **64**, 1841–1847
33. Sutra, R., *"X-ray diffraction spectra of different varieties of starch"*, Bull. Soc. Chim. (France) (1942) **9**, 795–797
34. Hockett, R. C.; Collins, D. V.; Scattergood, A., *"The preparation of 4,6-ethylidene-D-glucopyranose from sucrose and its hydrogenation to 4,6-Ethylidene-D-sorbitol"*, J. Am. Chem. Soc. (1951) **73**, 599–601

References

35. Rappoport, D. A.; Hassid, W. Z., "*Preparation of L-Arabinose-1-C14*", J. Am. Chem. Soc. (1951) **73**, 5524–5525
36. Barker, R.; MacDonald, D. L., "*Some oxidation and reduction products of 2,4-O- Ethylidene-D-erythrose*", J. Am. Chem. Soc. (1960) **82**, 2301–2303
37. Schmidt, O. Th., "*Isopropylidene Derivatives*". in Methods in Carbohydrates Chemistry, Whistler, R. L.; Wolfrom, M. L. (Eds.), Academic Press, New York, Vol. II, (1963), pp. 318–325
38. Glen, W. L.; Myers, G. S.; Grant, G. A., "*Monoalkyl hexoses: Improved procedures for the preparation of 1- and 3-methyl ethers of fructose, and of 3-alkyl ethers of glucose*", J. Chem. Soc. (1951) 2568–2572
39. Freudenberg, K.; Hixon, R. M., "*Zur Kenntnis der Aceton-Zucker, IV.: Versuche Galaktose und Mannose*", Berichte (1923) **56**, 2119–2127
40. Tipson, R. S., "*1,2:3,4-Di-O-isopropylidene-α-D-galactopyranose*", in Methods in Carbohydrates Chemistry, Vol. II, Academic Press, New York (1963), p. 247
41. Ohle, H.; Berend, G., "*Über die Aceton-Verbindungen der Zucker und ihre Derivate, IV.: Die Konstitution der Diaceton-galaktose*", Berichte (1925) **58**, 2585–2589
42. Link, K. P.; Sell, H. M., "*D-Galacturonic acid monohydrate*", Biochem. Prep.(1953) **3**, 74–78
43. Ault, R. G.; Haworth, W. N.; Hirst, E. L., "*Acetone derivatives of methylglycosides*", J. Chem. Soc. (1935) 1012–1020
44. Baer, E.; Fischer, H. O. L., "*Studies on Acetone-Glyceraldehyde. VII. Preparation of l-Glyceraldehyde and l(-)Acetone Glycerol*", J. Am. Chem. Soc. (1939) **61**, 761–765
45. Wiggins, L. F., "*The acetone derivatives of hexahydric alcohols. Part I. Triacetone mannitol and its conversion into d-arabinose*", J. Chem. Soc. (1946) 13–14
46. Buchanan, J. G.; Saunders, R. M., "*Methyl 2,3-anhydro-D-mannoside and 3,4-anhydro-D-altroside and their derivatives. Part III*", J. Chem. Soc. (1964) 1796–1803
47. Evans, M. E., "*Methyl 4,6-O-benzylidene-α- and -β-D-glucosides*", Carbohydr. Res. (1972) **21**, 473–475
48. Evans, M. E.; Parrish, F. W.; Long, L., Jr.,, "*Acetal exchange reactions*", Carbohydr. Res. (1967) **3**, 453–462
49. Hasegawa, A.; Kiso, M., "*Acetonation of 2-(acylamino)-2-deoxy-D-glucoses*", Carbohydr. Res. (1978) **63**, 91–98
50. Bonner, T. G., "*Ethylidene Derivatives*", Methods in Carbohydrates Chemistry, Vol. II, Academic Press, New York (1963), pp. 309–313
51. Honeyman, J.; Stening, T. C., "*Ethylidene derivatives of methyl aldopyranosides*", J. Chem. Soc. (1957) 3316–3317
52. O'Meara, D.; Shepherd, D. M., "*The preparation of some glucose nitrates*", J. Chem. Soc. (1955) 4232–4235
53. Miljkovic, M.; Hagel, P., "*Regioselective isopropylidenation of 2-acetamido-2-D-xylose diethyl dithioacetal*", Carbohydr. Res.(1983) **111**, 319–324
54. Hibbert, H.; Morazain, J. G., "*Reactions relating to carbohydrates and polysaccharides, XXVIII. Structure of isopropylideneglycerol*", Can. J. Res. (1930) **2**, 214–216
55. Brown, H. C.; Brewster, J. H.; Schechter, H., "*An interpretation of the chemical behavior of five- and six-membered ring compounds*", J. Am. Chem. Soc. (1954) **76**, 467–474
56. Pitzer, K. S.; Donath, W. E., "*Conformations and strain energy of cyclopentane and its derivatives*", J. Am. Chem. Soc. (1959) **81**. 3213–3218
57. van Roon, J. D., "*Cyclic acetals*", Rec. Trav. Chim. (1929) **48**, 173–190
58. Brimacombe, J. S.; Foster, A. B.; Haines, A. H., "*Aspects of stereochemistry. Part V. Some properties of 1,2-O-methyleneglycerol and related compounds*", J. Chem. Soc. (1960) 2582–2586
59. Hann, R. M.; Maclay, W. D.; Hudson, C. S., "*The structures of the diacetone dulcitols*", J. Am. Chem. Soc. (1939) **61**, 2432–2442

60. Kochetkov, N. K.; Kudryashov, L. I.; Usov, A. I., "*Interaction between di-O-isopropylidene-α-D-glucose and the halogen complexes of triphenyl phosphite*", Dokl. Akad. Nauk. USSR (1960) **133**, 1094–1097
61. Lipták, A.; Nánási, P.; Neszmélyi, A.; Wagner, H., "*Acetal migration during Koenigs-Knorr reactions; isolation of 3-O- and 6-O-(2, 3, 4, 6-tetra-O-acetyl-β-D-glucopyranosyl) derivatives of 1, 2:5, 6- and 1, 2:3, 5-di-D-isopropylidene-α-D-glucofuranose*", Carbohydr. Res. (1980) **86**, 133–136
62. Manatt, S. L. ; Roberts, J. D., "*Small-ring compounds. XXIV. Molecular orbital calculations of the delocalization energies of some small-ring systems*", J. Org. Chem. (1959) **24**, 1336–1338
63. Breslow, R.; Kivelevich, D.; Mitchell, M. J.; Fabian, W.; Wendel, K., "*Approaches to "Push-Pull" stabilized cyclobutadienes*", J. Am. Chem. Soc. (1965) **87**, 5132 – 5139
64. Hess, B. A.; Schaad, L. J., "*Stabilization of substituted cyclobutadienes*", J. Org. Chem. (1976) **41**, 3058 – 3059
65. Fletcher, H. G., Jr., "*1-O-Benzoyl-β-D-glucopyranose*", in Methods in Carbohydrates Chemistry, Vol. II, Academic Press, New York(1963) pp. 231–233
66. Christensen, J. E.; Goodman, L., "*A mild method for the hydrolysis of acetal groups attached to sugars and nucleosides*", Carbohydr. Res.(1968) **7**, 510–512
67. Garegg, P. J.; Hultberg, H., "*A novel, reductive ring-opening of carbohydrate benzylidene acetals, with unusual regioselectivity*", Carbohydr. Res. (1981) **93**, C10-C11
68. Nánási, P.; Lipták, A., "*Carbohydrate methyl ethers. VI. Synthesis of phenyl β-D-glucopyranoside derivatives partially methylated in the sugar moiety*", Magy. Kem. Foly. (1974) **80**, 217–225
69. Lipták, A.; Jodál, I.; Nánási, P., "*Stereoselective ring-cleavage of 3-O-benzyl- and 2,3-di-O-benzyl-4,6-O-benzylidenehexopyranoside derivatives with the lithium aluminum hydride-aluminum chloride reagent*", Carbohydr. Res. (1975) **44**, 1–11
70. Lipták, A.; Jodál, I.; Nánási, P., "*Hydrogenolysis of benzylidene acetals: synthesis of benzyl 2,3,6,2', 3', 4'-hexa-O-benzyl-β-cellobioside, -maltoside, and -lactoside, benzyl 2,3,4,2', 3',4'-hexa-O-benzyl-β-allolactoside, and benzyl 2,3,6,2', 3', 6'-hexa-O-benzyl-β-lactoside*", Carbohydr. Res. (1976) **52**, 17–22
71. Lipták, A.; Fügedi, P.; Nánási, P., "*A simple method for the synthesis of benzyl 4-O-benzylhexopyranosides*", Carbohydr. Res. (1979) **68**, 151–154
72. Lipták, A.; Pekár, F.; Jánossy, L.; Jodál, I.; Fügedi, P.; Harangi, J.; Nánási, P.; Szejtli, J., "*Regioselective hydrogenolysis of 4,6-O-benzylidene derivatives of hexopyranosides. Preparation of "glyvenol"-like compounds*", Acta Chim. Acad. Sci. Hung. (1979) **99**, 201–208
73. Lipták, A.; Fügedi, P.; Nánási, P., "*Stereoselective hydrogenolysis of exo- and endo-2,3-benzylidene acetals of hexopyranosides*", Carbohydr. Res. (1976) **51**, c19–c21
74. Lipták, A., "*Hydrogenolysis of the dioxolan type exo- and endo-benzylidene derivatives of carbohydrates with the LiAlH$_4$-AlCl$_3$ reagent*", Tetrahedron. Lett. (1976) **17**. 3551–3554
75. Lipták, A.; Fügedi, P.; Nánási, P., "*Synthesis of mono- and di-benzyl ethers of benzyl α-L-rhamnopyranoside*", Carbohydr. Res. (1978) **65**, 209–217
76. Lipták, A.; Jánossy, L.; Imre, J.; Nánási, P., "*Stereoselective hydrogenolysis of dioxolane-type benzylidene acetals. Synthesis of partially substituted galactopyranoside derivatives*", Acta Chim. Acad. Sci. Hung. (1979) **101**, 81–92
77. Fügedi, P.; Lipták, A.; Nánási, P.; Neszmélyi, A., "*Synthesis of 4-O-α-D-galactopyranosyl-L-rhamnose and 4-O-α-D-galactopyranosyl-2-O-β-glucopyranosyl-L-rhamnose using dioxolane-type benzylidene acetals as temporary protecting-groups*", Carbohydr. Res. (1980) **80**, 233–239
78. Ashby, E. C.; Prather, J., "*The composition of "Mixed Hydride" reagents. A study of the Schlesinger reaction*", J. Am. Chem. Soc. (1966) **88**, 729–733
79. Fügedi, P.; Lipták, A.; Nánási, P.; Szejtli, J., "*The regioselectivity of the reductive ring-cleavage of the acetal ring of 4,6-O-benzylidenehexopyranosides*", Carbohydr. Res. (1982) **104**. 55–67

References

80. Leggetter, B. E.; Brown, R. K., "*The influence of substituents on the ease and direction of ring opening in the $LiAlH_4$–$AlCl_3$ reductive cleavage of substituted 1, 3- dioxolanes*", Can. J. Chem. (1964) **42**, 990–1004
81. Fleming, B. I.; Bolker, H. I., "*Reductive cleavage of acetals and ketals by Borane*", Can. J. Chem. (1974) **52**, 888–893
82. Bolker, H. I.; Fleming, B. I., "*Reductive cleavage of acetals and ketals by Borane. Part II. The Kinetics of the Reaction*", Can. J. Chem. (1975) **53**, 2818–2821
83. Hanessian, S., "*The reaction of O-benzylidene sugars with N-bromosuccinimide*", Carbohydr. Res. (1966) **2**, 86–88
84. Garegg, P., "*Regioselective Cleavage of O-Benzylidene Acetals to Benzyl Ethers*". in "Preparative Carbohydrates Chemistry", Hanessian, S. (Ed.), .Marcel Dekker, Inc., New York, 1997, pp. 53–67
85. Binkley, R. W.; Goewey, G. S.; Johnston, J. C., "*Regioselective ring opening of selected benzylidene acetals. A photochemically initiated reaction for partial deprotection of carbohydrates*", J. Org. Chem. (1984) **49**, 992–996
86. Deslongchamps, P.; Moreau, C., "*Ozonolysis of Acetals. (1) Ester synthesis, (2) THP ether cleavage, (3) Selective oxidation of β-Glycoside, (4) Oxidative removal of benzylidene and ethylidene protecting groups*". Can. J. Chem. (1971) **49**, 2465–2467
87. Deslongchamps, P.; Moreau, C.; Fréhel, D.; Chênevert, R., "*Oxidation of benzylidene acetals by Ozone*", Can. J. Chem. (1975) **53**, 1204–1211
88. Schmidt, O. Th., *Methods in Carbohydrates Chemistry*, Whistler, R. L.; Wolfrom, M. L.; BeMiller, J. N. (Eds.), Vol. II, Academic Press, New York, 1963, p.322
89. Blindenbacher, F.; Reichstein, T., "*Synthese des L-Glucomethylose-3-methyläthers und seine Identifizierung mit Thevetose. Desoxyzucker, 19. Mitteilung*", Helv. Chiim. Acta (1948) **31**, 1669–1676
90. Freudenberg, K.; Durr, W.; von Hochsteller, H., "*Zur Kenntnis der Aceton-Zucker, XIII: Die Hydrolyse einiger Disaccharide, Glucoside und Aceton-Zucker*", Berichte(1928) **61**, 1735–1742
91. Ohle, H.; Dickhäuser, E., "*Über die Aceton-Verbindungen der Zucker und Ihre Derivate, VI: Über Acylderivate der Monoaceton-glucose*", Chem. Ber. (1925) **58**, 2593–2606
92. Iwadare, K., "*Acetonderivate der Monosaccharide. II. Diaceton- und Monoaceton D-idose*", Bull. Chem. Soc. (Japan) (1944) **19**, 27–29
93. Theander, O., "*Chromic acid oxidation of 1,2-O-isopropylidene-α-D-glucofuranose*", Acta Chem. Scand. (1963) **17**, 1751–1760

Chapter 7
Nucleophilic Displacement and the Neighboring Group Participation

Nucleophilic Displacement

The nucleophilic substitution reaction of the type S_N2 (bimolecular nucleophilic substitution) proceeds with formation, in a single step, of the trigonal bipyramidal pentacovalent carbon transition state *2* and results in the inversion of configuration at the reacting carbon atom (the configuration of *3* vs. *1*) (Fig. 7.1); it usually follows the second-order kinetics.

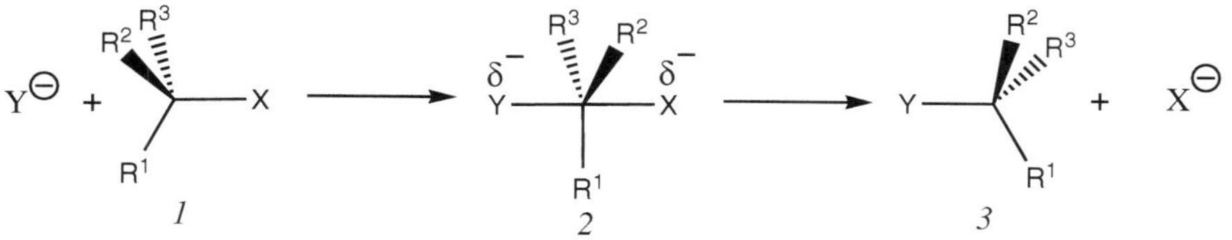

Fig. 7.1

The ligands R^1, R^2, or R^3 can be hydrogen or substituted carbon atoms. In the transition state (*2* in Fig. 7.1) the central carbon is sp^2 hybridized (Fig. 7.3), and the unhybridized *p* orbital, which is perpendicular to the planar sp^2 carbon, is used to form partial bonds with both the entering nucleophile and the leaving group.

The angles between the R groups in transition state are approximately 120° (Fig. 7.2), while the *p* orbital is oriented at an angle of 90° to these three C-R bonds (Fig. 7.3).

Thus during the course of the S_N2 reaction, the reacting carbon atom changes its hybridization from sp^3 to sp^2 and then back to sp^3; the process is accompanied by inversion of configuration.

The S_N2 reaction occurs quite readily. A large variety of nucleophiles and leaving groups are known.

The rate of an S_N2 reaction depends on the nucleophilicity of incoming nucleophile, on the nature of leaving group, on the polarity of solvent (solvation), and on steric and electronic effects. For most S_N2 reactions, steric interactions are the most

M. Miljković, *Carbohydrates*, DOI 10.1007/978-0-387-92265-2_7,
© Springer Science+Business Media, LLC 2010

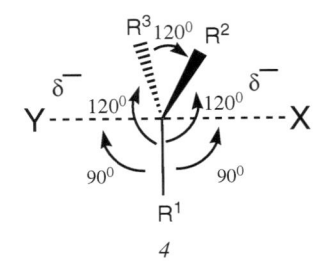

Fig. 7.2

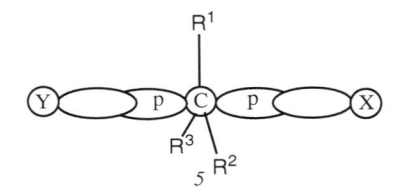

Fig. 7.3

important, because the increased crowding about the reacting carbon raises the free energy of transition state, and thus lowers the rate of reaction.

In a nucleophilic substitution, an electron pair is transferred from the nucleophile to the reacting carbon and from the latter to the leaving group. The nucleophiles could be either negatively charged (e.g., $RCOO^-$, OH^-, CN^-) or neutral (e.g., ammonia $H_3N:$, amines $-RH_2N:$, hydrazine – H_2NNH_2). Since the negatively charged nucleophiles will have their negative charge either destroyed or delocalized in the transition state, the reactants will be more highly solvated than the transition state. If a nucleophile is a neutral molecule that becomes partially positively charged in transition state, then the transition state will be more solvated than the reactants. Consequently, the polar solvent will slow down the reaction of charged nucleophiles and accelerate the reaction of neutral nucleophiles.

Nucleophilic displacement reaction has been extensively used in carbohydrate chemistry for the synthesis of a particular type of sugar, as for example, amino-, thio-, seleno-, sugar from an appropriately protected monosaccharide that is derivatized with a leaving group at the carbon where the nucleophilic displacements are to take place.

In the absence of neighboring group participation (vide infra) the nucleophilic displacement always proceeds with inversion of the configuration at the reacting carbon. However, if a carbon neighboring to the reacting carbon atom is derivatized by a group that can act as a nucleophile, such as acetate, benzoate, thioacetate, acetamido group, etc., the nucleophilic displacement may proceed with retention of configuration, if the nucleophile on the neighboring carbon is able to displace the leaving group faster than external nucleophile. The intramolecular displacement

Nucleophilic Displacement

reaction is an entropically more favored reaction than intermolecular displacement, provided that the stereochemistry of a neighboring group is favorable, i.e., it is able to form a strainless cyclic transition state.

It has already been said that steric effects exert the greatest influence upon the rate of nucleophilic displacement. For that reason the nucleophilic displacement is fastest at the primary carbon of both hexo- and pentofuranoses and of hexopyranoses.

Thus, reaction of methyl 2,6-di-O-tosyl-α-D-glucopyranoside 5 with sodium iodide in acetic anhydride gives methyl 3,4-di-O-acetyl-6-deoxy-6-iodo-2-O-tosyl-α-D-glucopyranoside 6 as the only product, indicating that the primary 6-O-tosyl group is much more reactive than the C2 secondary tosyl group [1] (Fig. 7.4).

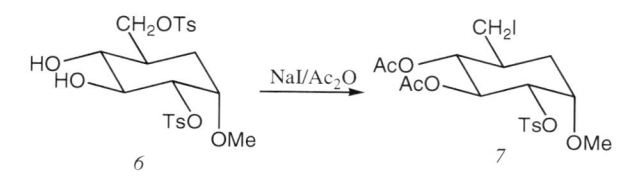

Fig. 7.4

Similarly, treatment of 3-O-acetyl-1,2-O-isopropylidene-5,6-di-O-p-toluenesulfonyl-α-D-glucofuranose 8 with sodium benzoate in N,N-dimethylformamide at 95–100°C for 7 h gives in a 64% yield [2] 3-O-acetyl-6-O-benzoyl-1,2-O-isopropylidene 5-O-p-toluenesulfonyl-α-D-glucofuranose 9 (Fig. 7.5).

Fig. 7.5

If 8 is treated with sodium benzoate in N,N-dimethylformamide at reflux for 6 h, [2] the 5,6-di-O-benzoyl-1,2-O-isopropylidene-β-L-idofuranose 10 is obtained in 50% yield (Fig. 7.6).

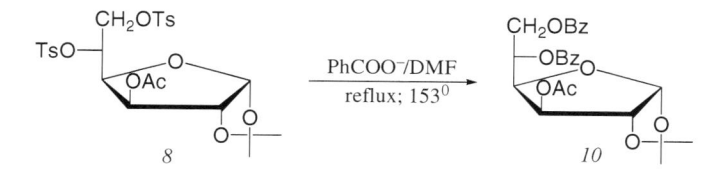

Fig. 7.6

It is interesting that not all primary sulfonates are equally reactive. Thus it is well known that the tosyl group of 1,2:3,4-di-*O*-isopropylidene-6-*O*-*p*-toluenesulfonyl-α-D-galactose *11* resists the nucleophilic displacement with iodide, fluoride, or methoxide as the nucleophile [3] (Fig. 7.7).

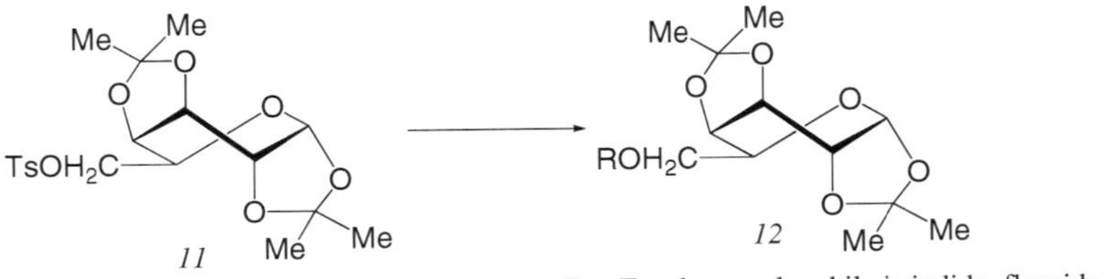

R = Ts when nucleophile is iodide, fluoride, methoxide
R = thiolacetyl, azido when nucleophiles are thiolacetate, azide

Fig. 7.7

However, this C6 tosyl group can be displaced with some other nucleophiles, such as thiolacetyl and azide.

The nucleophilic displacement of a sulfonyl group attached to a pyranoid or furanoid ring is extremely dependent on steric factors. Thus, the substitution reactions take place readily at the C4 carbon of both α- and β-D-gluco- and α-D-galactopyranosides. For example, the nucleophilic displacement, with benzoate of the axial 4-mesyl-group of methyl 2,3-di-*O*-benzoyl-4,6-di-*O*-mesyl-α-D-galactopyranoside *13* in *N,N*-dimethyl formamide at 140°C for 24 h gives methyl 2,3,4,6-tetra-*O*-benzoyl-α-D-glucopyranoside *14* in 49% yield [4] (Fig. 7.8).

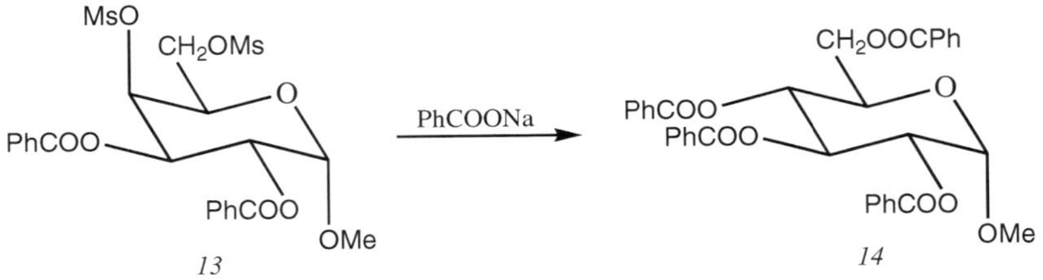

Fig. 7.8

Similarly, the nucleophilic displacement of the equatorial 4-*O*-mesyl group of methyl 2,3-di-*O*-benzoyl-4,6-di-*O*-mesyl-α-D-glucopyranoside *15* with benzoate in *N,N*-dimethylformamide at 140°C for 20 h gave methyl 2,3,4,6-tetra-*O*-benzoyl-α-D-galacto-pyranoside *16* as the only product [5] (Fig. 7.9). It should be noted, however, that an axial leaving group is displaced two to three times faster than the equatorial [4–6].

The attempted displacement with acetate of 4-*O*-mesylate, 4-*O*-tosylate, or 4-*O*-brosylate of methyl 6-deoxy-2,3-*O*-isopropylidene-α-D-mannopyranoside *17, 18,*

Nucleophilic Displacement

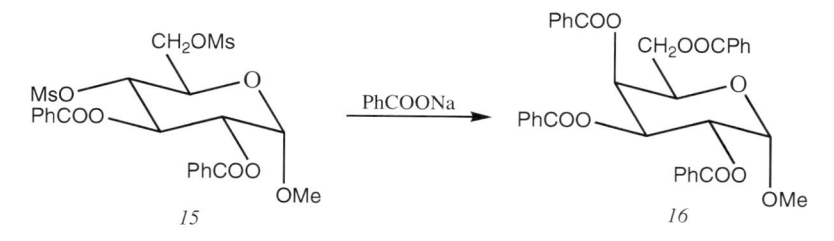

Fig. 7.9

19 resulted in ring contraction giving D-tallo- and L-allo-furanoside products *20* and *21*, respectively, in 7:1 ratio [7] (Fig. 7.10).

The formation of *20* results from inversion at the C4 and retention (or double inversion) at the C5 of *17*, while the formation of *21* is a result of inversion at both C4 and C5.

17, R = Ms
18, R = Ts
19, R = Bros

20, α-D-tallo

21, β-L-allo

Fig. 7.10

The following mechanism could be envisioned (Fig. 7.11).

Fig. 7.11

Similar rearrangement was observed by Hanessian [8] using methyl 6-deoxy-2,3-O-isopropylidene-α-L-mannopyranoside.

The nucleophilic displacement, with benzoate, of 3-*O*-mesyl group of methyl 1,2,4,6-tetra-*O*-benzoyl-3-*O*-mesyl-α- and β-D-glucopyranoses *23* and *24*, respectively, is strongly dependent on the anomeric configuration [9]. Thus, whereas the treatment of methyl 1,2,4,6-tetra-*O*-benzoyl-3-*O*-tosyl-β-D-glucopyranose *24* with sodium benzoate in *N*-methylpyrrolidone at 100°C for 16 h gave the corresponding penta-*O*-benzoyl-β-D-allopyranose *25* in 65% yield (Fig. 7.12), the α-anomer

23, R^1 = OBz; R^2 = H
24, R^1 = H; R^2 = OBz

Fig. 7.12

23 failed to undergo the displacement reaction [9]. A possible explanation [9] could be that the approach of the negatively charged nucleophile (benzoate) to the C3 carbon *26* to form the corresponding transition state *27* is impeded by unfavorable 1,3-diaxial electrostatic interaction between the approaching nucleophile (benzoate) and the axially oriented anomeric benzoate (Fig. 7.13).

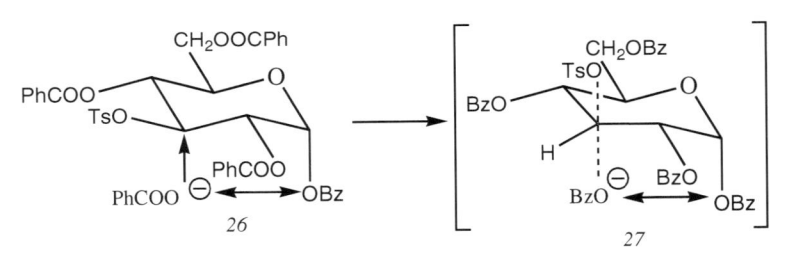

Fig. 7.13

Similarly, solvolysis of methyl 2,4,6-tri-*O*-benzoyl-3-*O*-*methanesulfonyl*-α-D-gluco-pyranoside in wet Cellosolve containing sodium acetate gave no product [10].

The 4-*O*-methanesulfonate of methyl 2,3,6-tri-*O*-benzoyl-4-*O*-methanesulfonyl-α-D-mannopyranoside *28* does not undergo direct substitution [11]. This can be again attributed to the presence of the axial 2-*O*-benzoyl group, which may hinder the approach of the nucleophile at the C4 position *28* (Fig. 7.14). This interaction can be electrostatic between the incoming nucleophile and axially oriented benzoate, or it can be the 1,3-*syn*-diaxial interaction between the C2 axial benzoate and the nucleophile in the transition state, or both [12] (Fig. 7.14).

Except for the displacement of *p*-toluenesulfonyl group of methyl 4,6-*O*-benzylidene-3-deoxy-2-*O*-*p*-toluenesulfonyl-α-D-*ribo*-hexopyranoside *30* with azide [13] (Fig. 7.15), direct displacement of the C2 sulfonyloxy group in a

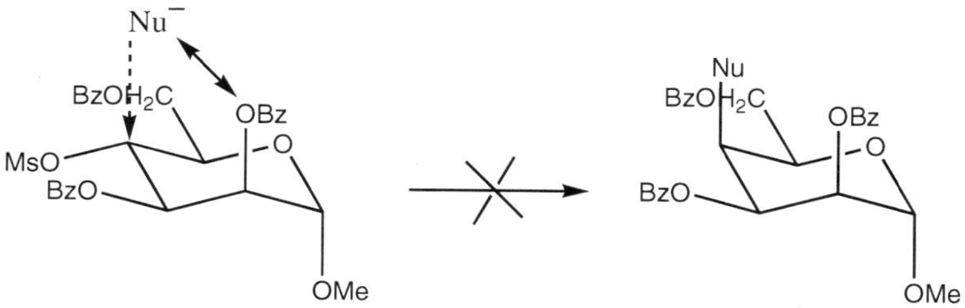

Nucleophilic Displacement

Fig. 7.14

furanoside or in a pyranoside ring with charged nucleophiles was considered, for a long time, to be very difficult if not impossible.

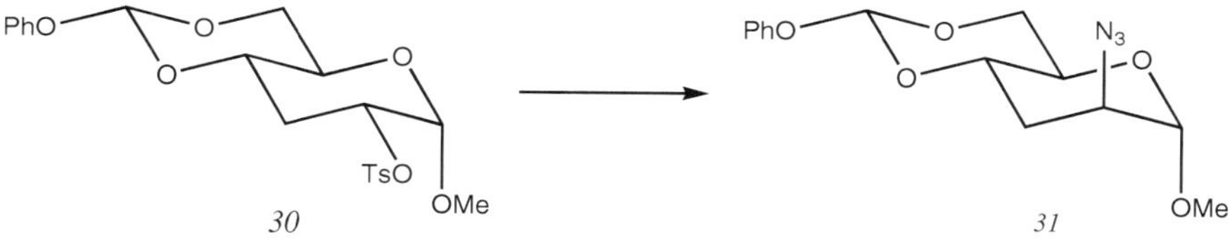

Fig. 7.15

The use of a neutral nucleophile such as hydrazine did, however, result in displacement of the C2 sulfonate in both furanose [14–16] and pyranose rings [17].

The observed unusually low reactivity of the C2 sulfonyloxy group toward displacement with charged nucleophiles was attributed to electron-withdrawing effect of the anomeric carbon and to unfavorable dipolar interaction in the transition state [18–22]. The greater reactivity of the sulfonyloxy group at the C2 carbon atom with uncharged nucleophiles was ascribed to the reversal of polarity of one of the polar bonds in the transition state resulting in a dipolar attractive force [20].

To examine the stereoelectronic interactions responsible for diminished reactivity of the C2 sulfonyloxy group toward nucleophilic displacement with charged nucleophiles Miljkovic et al. [23] studied displacement of the C2 sulfonate with benzoate in the following substrates: methyl 4,6-O-benzylidene-3-O-methyl-2-O-methylsulfonyl-α- and β-D-glucopyranosides *32* and *33* and methyl 4,6-O-benzylidene-3-O-methyl-2-O-methylsulfonyl-α- and β-D-mannopyranosides *34* and *35* (Fig. 7.16 and Table 7.1).

As can be seen from Table 7.1 the nucleophilic displacement of 2-O-sulfonate in *35* was considerably faster than displacement of 2-O-methanesulfonyl group in *33* (8 h vs. 120 h). The obtained results can be better explained if the partial structures of the C2 sulfonates *32–35* containing only the C1 and C2 carbons with their respective ligands are drawn in Newman's projection (Fig. 7.17).

Thus, in Fig. 7.17 the C1 and the C2 carbon of compounds *32–35* (structures *39–42*) as well as of the corresponding transition states (*43–46*) for the nucleophilic

Table 7.1 Displacement of 2-sulfonate of methyl 4,6-O-benzylidene-3-O-methyl-2-O-methylsulfonyl-α- and β-D-gluco- (*32* and *33*) and α- and β-D-mannopyranosides (*34* and *35*)[a]

Sugar	Reaction, t (°C)	Reaction time (h)	Yield (%)
Methyl 4,6-O-benzylidene-3-O-methyl-α-D-glucopyranoside, *32*	153	120	3.5
Methyl 4,6-O-benzylidene-3-O-methyl-β-D-glucopyranoside, *33*	153	120	62
Methyl 4,6-O-benzylidene-3-O-methyl-α-D-mannopyranoside, *34*	153	120	–
Methyl 4,6-O-benzylidene-3-O-methyl-β-D-mannopyranoside, *35*	153	8	70

[a] Displacement was effected by refluxing an N,N-dimethylformamide solution of the corresponding sugar derivative with potassium benzoate.

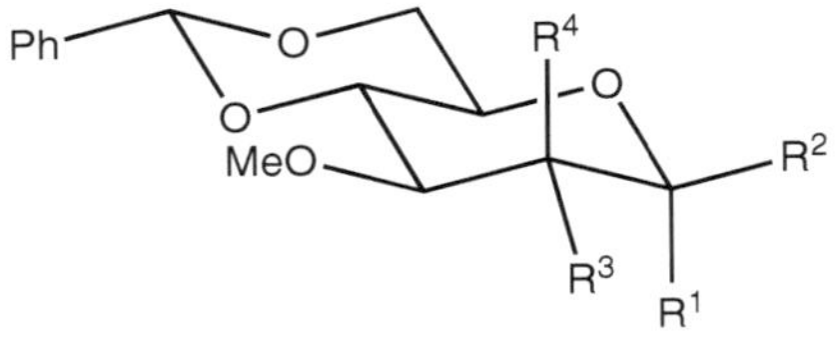

32, R^1 = OMe; R^2 = R^4 = H; R^3 = CH$_3$SO$_3$
33, R^1 = R^4 = H; R^2 = OMe; R^3 = CH$_3$SO$_3$
34, R^1 = OMe; R^2 = R^3 = H; R^4 = CH$_3$SO$_3$
35, R^1 = R^3 = H; R^2 = OMe; R^4 = CH$_3$SO$_3$
36, R^1 = OMe; R^2 = R^2 = H; PhCOO; R^4 = PhCOO
37, R^1 = R^3 = H; R^2 = OMe; R^4 = PhCOO
38, R^1 = R^4 = H; R_2 = OMe; R^3 = PhCOO

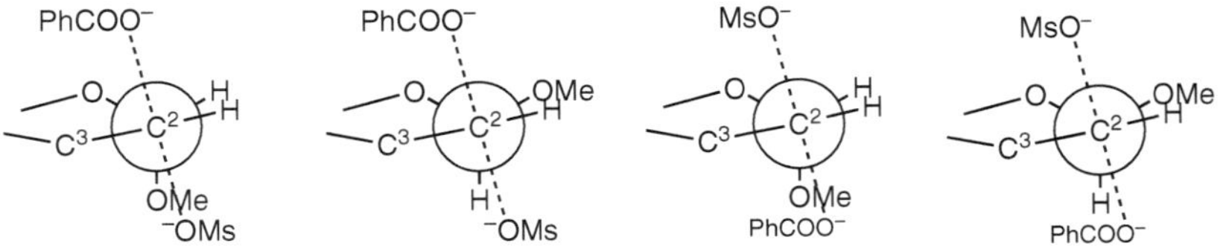

Fig. 7.16

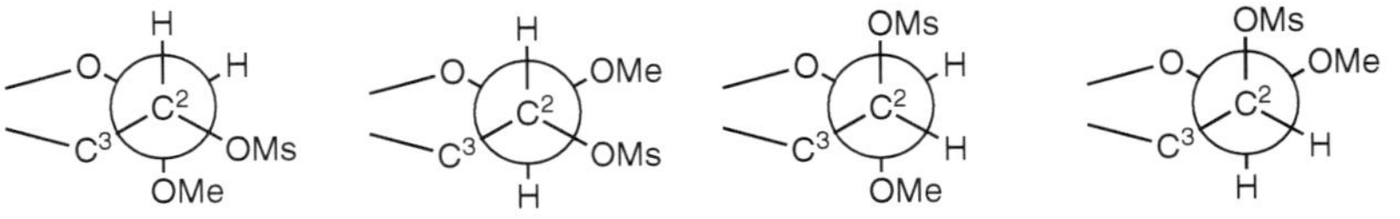

Fig. 7.17

displacements of 2-sulfonates are represented in Newman's projections. Assuming the pentacovalent bi-pyramidal transition state for the nucleophilic displacement of sulfonate, the unfavorable stereoelectronic interactions present in 43–46 are as follows: (a) in 43 there is a Pitzer strain and the electrostatic repulsion between the partially negatively charged leaving methanesulfonyl group and the electronegative C1 methoxy group (they are almost coplanar); (b) in 44 there is only one interaction between the negatively charged nucleophile that approaches the C2 carbon from a direction that bisects the resultant dipole of the β-methoxy group dipole and resultant dipole of the two nonbonding electron pairs on ring oxygen; (c) in 45 there is a Pitzer strain and electrostatic repulsion between the approaching negatively charged benzoate and the electronegative C1 methoxy group (they are almost coplanar); and (d) in 46 there is an unfavorable stereoelectronic interaction between the partially negatively charged leaving methanesulfonyloxy group and the resultant dipole of the β-methoxy group dipole and the resultant dipole of two nonbonding electron pairs on the ring oxygen.

Based on the above considerations one may conclude that the most reactive C2 sulfonate toward nucleophilic displacement will be in 42, followed by the C2 sulfonate in 40; the relative reactivity of C2 sulfonates in 39 and 41 is difficult to assess but they should be much lower than in 40 and 42.

Ishido and Sakairi [24] studied the nucleophilic displacement of 2-O-trifluoromethylsulfonyl group of methyl 4,6-O-benzylidene-3-O-methyl-2-O-trifluoromethylsulfonyl-α-D-glucopyranoside 47 with several charged nucleophiles and found that the displacements proceeded relatively fast and the yields of corresponding α-D-mannopyranosides were very good (Table 7.2 and Fig. 7.18).

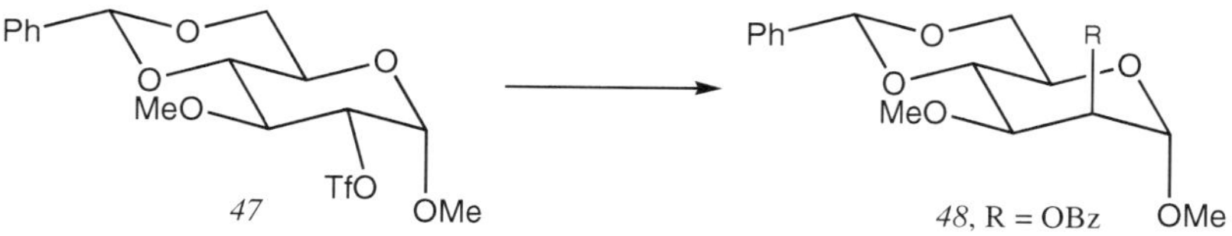

Fig. 7.18

Table 7.2 Displacement of 2-O-trifluoromethylsulfonate of methyl 4,6-O-benzylidene-3-O-methyl-2-O-trifluoromethylsulfonyl-α-D-glucopyranosides 47

Nucleophile	Temperature (°C)	Reaction time (h)	Yield (%)
PhCOO⁻	80	5	82
N_3^-	80	2	86
MeS⁻	0	1	89
PhS⁻	50	6	83

Treatment of benzyl 3,4-isopropylidene-2-O-(imidazol-1-sulfonyl)-β-L-arabino-pyranoside *52* with sodium azide in refluxing toluene [25] (110°C) gave benzyl 2-azido-2-deoxy-3,4-isopropylidene-β-L-ribopyranoside *53* in 80% yield (Fig. 7.19).

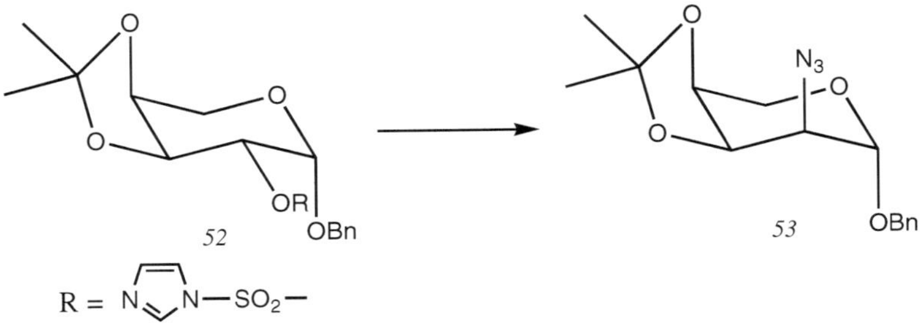

Fig. 7.19

As in the pyranoside series, the nucleophilic displacement at the C2 carbon of furanoside derivatives is also difficult [26]. Treatment of the 2-imidazylate ester of benzyl 5-deoxy-α-D-hexofuranoside derivative *54* with tetrabutylammonium azide or benzoate in refluxing toluene gave the substitution products *55* and *56* in 82 and 53% yield, respectively (Fig. 7.20).

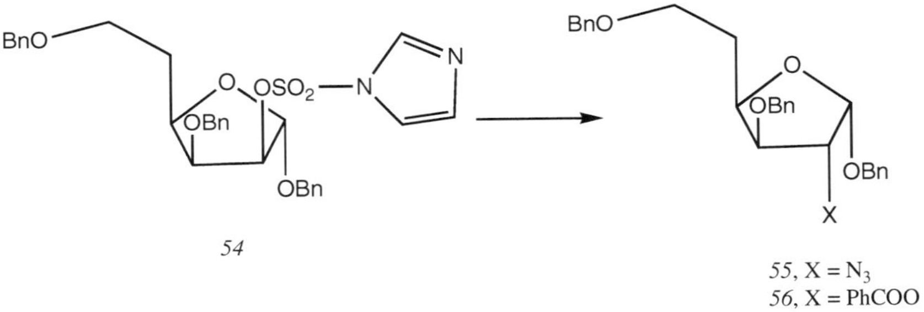

Fig. 7.20

Thus the trifluoromethanesulfonyl group and the imidazol-1-sulfonyl group seem to be very good leaving groups for the displacement of C2 sulfonate in both pyranosides and furanosides (for a good review of nucleophilic displacement of imidazol-1-sulfonates, see [27]).

In general, the nucleophilic displacement of secondary sulfonyloxy groups in a furanoid ring is reported to be slower than nucleophilic displacement of secondary sulfonyloxy groups in a pyranoside ring. The displacement, with benzoate, of the 3-O-mesyl group in methyl 2-O-benzyl-5-deoxy-3-O-mesyl-α-D-xylofuranose *57* can be used as an example where displacement proceeded very sluggishly; methyl 3-O-benzoyl-2-O-benzyl-5-deoxy-α-D-ribofuranose *58* was obtained in 53% yield,

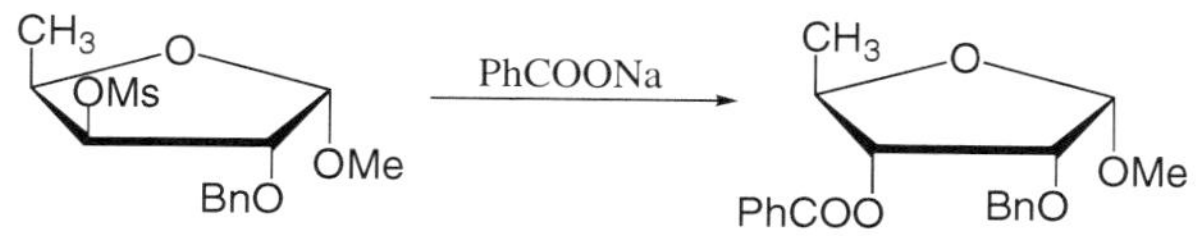

Fig. 7.21

together with several side products, after heating the reaction mixture in DMF at reflux for 40 h [28] (Fig. 7.21).

Nucleophilic Displacements with Neighboring Group Participation

When a rate of nucleophilic displacement in a molecule or the stereochemical outcome of displacement is influenced by a group that lies near the reaction site, but this influence is not a consequence of inductive, conjugative, or steric effects, it was concluded that this group participates directly in the reaction, and this influence is termed the *neighboring group participation.*

There are three types of evidence that support the concept of *neighboring group participation.*

(a) If participation occurs during the rate-determining step of a nucleophilic displacement, the reaction is usually faster than other reactions that are similar but do not involve such participation. For example, the 2-(2-chloroethyl) thioethane $ClCH_2CH_2SCH_2CH_3$ 59 is hydrolyzed over 10,000 times faster than the corresponding ether, $ClCH_2CH_2OCH_2CH_3$ 62 (1-chloro-2-ethoxy-ethane), in aqueous dioxane [29]. This rate difference is far too great to be attributed to differences in inductive, conjugative, or steric effects, but suggests rather that the hydrolysis of the sulfide (but not the ether) proceeds through the participation of sulfur atom via formation of a cyclic sulfonium ion 60. The intermediate, because of the strained three-membered ring, is readily hydrolyzed to the observed product 61 (Fig. 7.22).

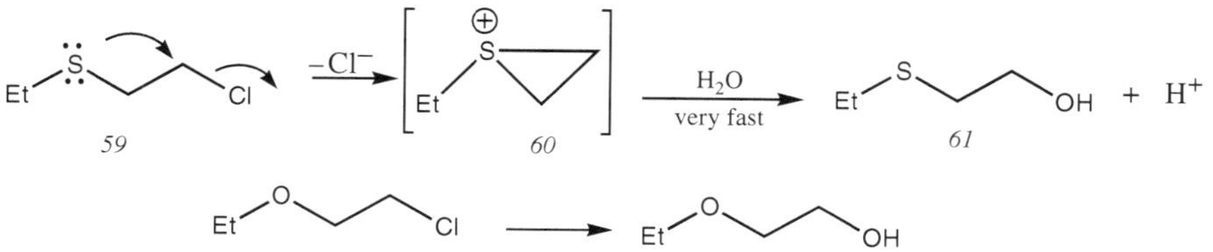

Fig. 7.22

(b) If the nucleophilic displacement occurs at the secondary chiral carbon, then stereochemistry of the reaction may suggest the involvement of a neighboring group. Since nucleophilic displacement at a chiral carbon always proceeds with the inversion of configuration, the retention of configuration would suggest the involvement of the neighboring group in the transition state. It is assumed that two displacements at the reacting carbon take place: first, the leaving group is intramolecularly displaced by a neighboring group then the cyclic transition

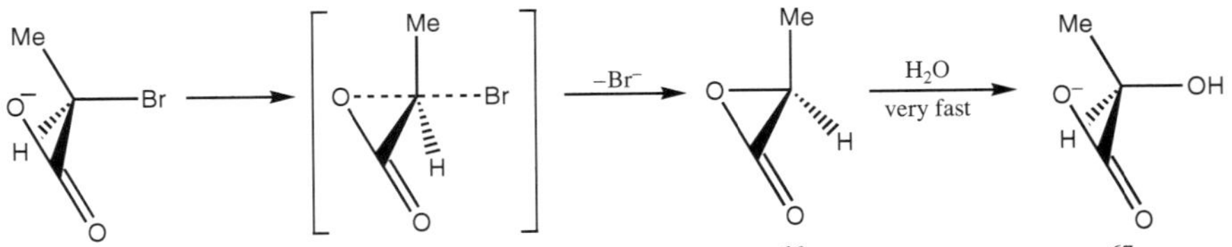

Fig. 7.23

state 65 (or cyclic transition intermediate 66) obtained by this intramolecular displacement is opened by an external nucleophile. This is illustrated by the hydrolysis of α-bromopropionic acid (Fig. 7.23). In the first step the carboxylate nucleophile displaces the bromine atom with the inversion of configuration at the α-carbon, (64→65) to form the nonisolable α-lactone (66) which is, in a very rapid step, cleaved by water giving the lactic acid 67 (66→67) that has the same configuration as the starting α-bromopropionic acid [30].

(c) Neighboring group participation may lead to a molecular rearrangement when the neighboring group remains bonded to the reaction center but breaks away from the atom to which it was originally attached in the substrate. Thus, chlorinated amine 68 yields on basic hydrolysis, the rearranged aminohydrin 70, since the diethylammonium intermediate 69 is attacked preferentially at the primary α-carbon rather than at secondary β-carbon atom [31] (Fig. 7.24).

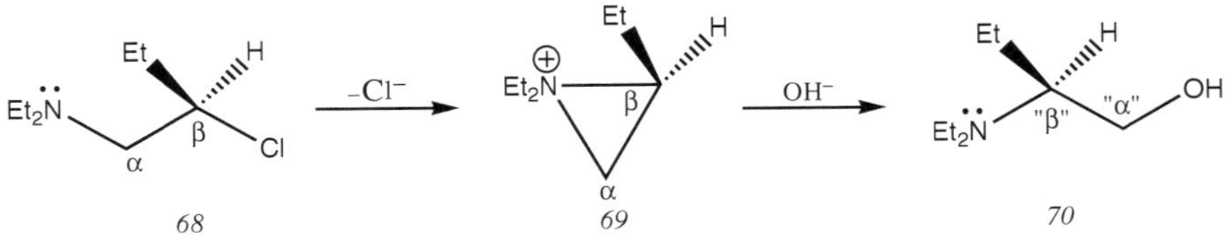

Fig. 7.24

When a transition state or a transition state intermediate is stabilized by a substituent on the same molecule by becoming chemically bonded to the reaction center due to its proximity or to proper orientation this effect is called *neighboring group participation*, and if such participation leads to an enhanced reaction rate, the group

Nucleophilic Displacements with Neighboring Group Participation 181

is said to provide *anchimeric assistance* [32, 33] (derived from the Greek: *anchi*, "adjacent"; *meros*, "part").

The most common type of neighboring group participation observed and exploited in carbohydrate chemistry is the participation of a neighboring ester group where the carbonyl oxygen acts as the nucleophile. The first step of such reactions always involves an *intramolecular nucleophilic attack* of the carbonyl oxygen, and if the carbon at which the attack occurs is sp^3 hybridized, the product is usually different from what would be expected in the absence of participation. In the course of neighboring group participation a new ring is formed that may suffer three different fates.

(1) Ring can be opened at the same carbon at which the ring closure took place (e.g., the C1 carbon in Fig. 7.25), leading to an un-rearranged product having the same configuration (β) at the reaction center as the starting material [34, 35].

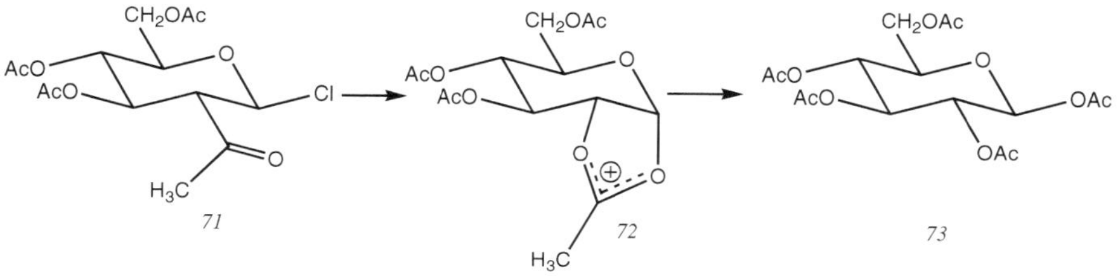

Fig. 7.25

In the absence of neighboring group participation, the direct nucleophilic displacement of C-Cl would take place giving the corresponding α-anomer *74* (Fig. 7.26). However, the direct displacement is not observed in reaction of tetra-*O*-acetyl-β-D-glucopyranosyl chloride with sodium acetate.

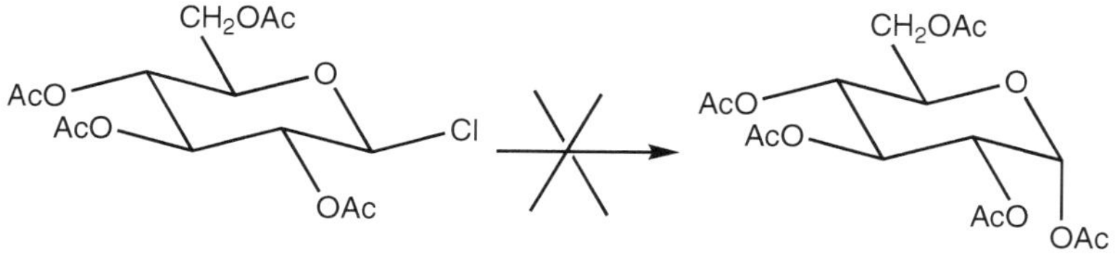

Fig. 7.26

Another example for the nucleophilic displacement with neighboring group participation that results in retention of configuration at the reacting carbon is the reaction of *N,N*-diethyl-1,2-*O*-isopropylidene-3,5-di-*O*-tosyl-α-D-glucofuranuronoamide *75* with the anhydrous Dowex 1 ion-exchange resin in

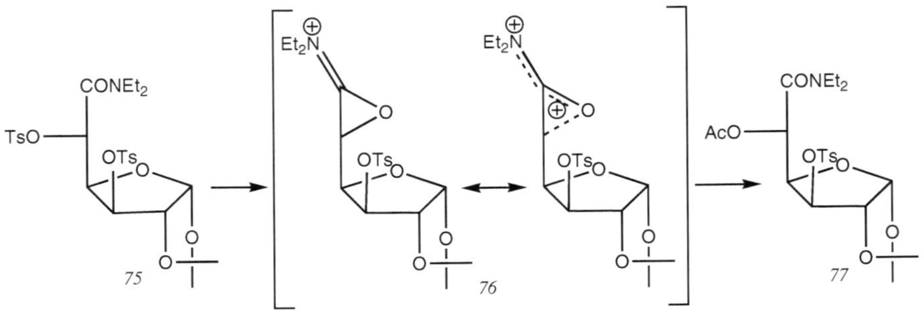

Fig. 7.27

acetate form (refluxing acetic anhydride), whereby the 5-*O*-acetyl-*N*,*N*-diethyl-1,2-*O*-isopropylidene-3-*O*-tosyl-α-D-glucofuranuronoamide 77 is obtained in 84% yield [36] (Fig. 7.27).
(2) Ring opening may take place at a different carbon from the one where the ring closure took place, leading to a rearranged product. For example, the reaction of 6-*O*-benzoyl-3,5-di-*O*-tosyl-1,2-*O*-isopropylidene-α-D-

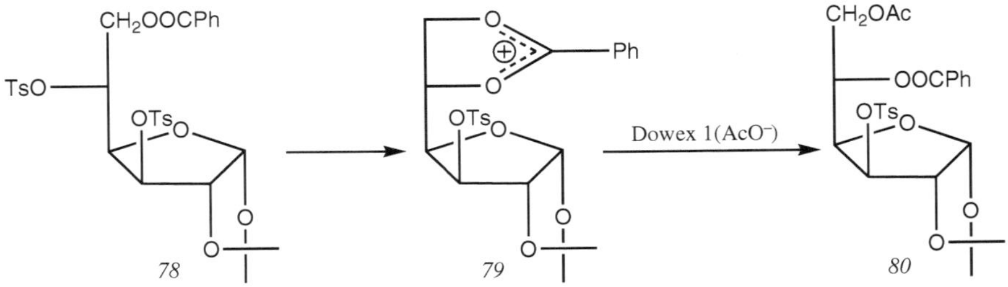

Fig. 7.28

glucofuranose 78 with anhydrous Dowex 1 (X-10, AcO⁻ form) gave 6-*O*-acetyl-5-*O*-benzoyl-3-*O*-tosyl-1,2-*O*-isopropylidene-β-L-idofuranose [37] 80 in 86% yield (Fig. 7.28). The initial attack of the carbonyl oxygen from the C6 benzoyl group gives as an intermediate the benzoyloxonium cation 79 which opens with the acetate at the C6 carbon atom (less hindered carbon) giving the 6-*O*-acetyl-5-*O*-benzoyl-L-idofuranose derivative 80.

An interesting neighboring group participation was observed [38] when methyl 4,6-di-*O*-benzoyl-3-*O*-methyl-2-*O*-methylsulfonyl-β-D-galactopyranoside 81 and methyl 2,6-di-*O*-benzoyl-3-*O*-methyl-4-*O*-methylsulfonyl-β-D-mannopyranoside 82 were treated with potassium benzoate in refluxing *N*,*N*-dimethylformamide. In both cases the only isolable products were methyl 2,4,6-tri-*O*-benzoyl-3-*O*-methyl-β-D-mannopyranoside 83 and methyl 2,4,6-tri-*O*-benzoyl-3-*O*-methyl-β-D-galactopyranoside 84 (Fig. 7.29).

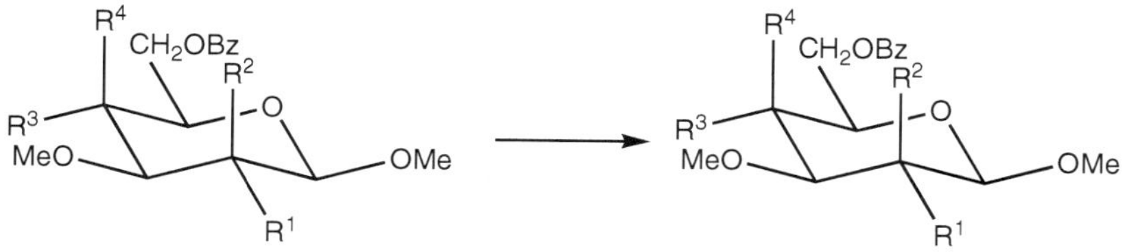

Nucleophilic Displacements with Neighboring Group Participation

Fig. 7.29

81, $R^1 = CH_3SO_3$; $R^2 = R^3 = H$; $R^4 = OBz$
82, $R^1 = R^4 = H$; $R^2 = OBz$; $R^3 = CH_3SO_3$

83, $R^1 = R^4 = H$; $R^2 = R^3 = OBz$
84, $R^1 = R^4 = OBz$; $R^2 = R^3 = H$

In the first case the reaction was complete after 120 h, and the products *83* and *84* were obtained in 54 and 18% yield (*83:84* ratio was 3:1). In the second case the reaction was complete after 10 h and the products *83* and *84* were obtained in 65 and 20% yield (*83:84* ratio was 3.25:1).

Direct displacement of the C2 or the C4 methylsulfonyl group in *81* or *82*, respectively, was excluded since in both cases there is on the β-carbon atom a *trans*-axial substituent that impedes the approach of the nucleophile to the reacting carbon. This argument was supported by a finding [38] that refluxing of an *N,N*-dimethylformamide solution of methyl 3,4,6-tri-*O*-methyl-2-*O*-methylsulfonyl-β-D-galactopyranoside with potassium benzoate for 120 h gave as the only isolable product the starting material.

Furthermore, if the direct displacement did take place the *83* and *84* would not be the obtained products but the talo derivative *85* (Fig. 7.30). Therefore, the only possible explanation for the reaction of *81* and *82* with potassium benzoate is that the reaction involves the formation of the six-membered ring acyloxonium transition state intermediate *86* (Fig. 7.31), which is then converted into products *83* and *84* by the nucleophilic attack of benzoate at either the C4 or the C2 carbon atom.

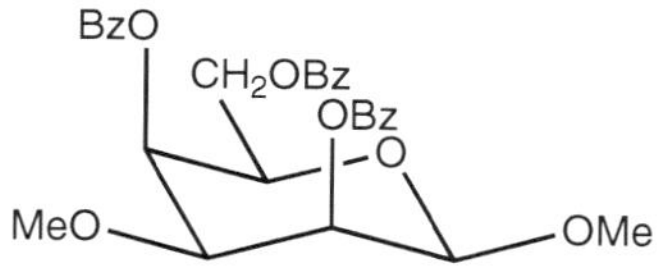

85

Fig. 7.30

The observed higher susceptibility of the C4 carbon of the cyclic six-membered benzoyloxonium intermediate *86* (Fig. 7.31) to the nucleophilic attack is in a good agreement with the observed large difference in reactivity of the C2 mesylate of *81* and the C4 mesylate of *82* (120 h vs. 10 h) and is compatible with the postulated rationalization that the electropositive character of the α-carbon to the reacting carbon atom should decrease the rate of nucleophilic displacement if the amount of

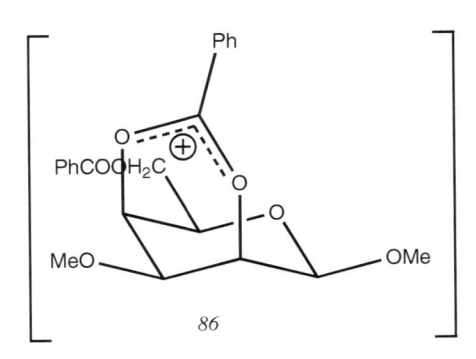

Fig. 7.31

positive charge on the reacting carbon atom in the transition state is greater than in the ground state [39].

This is also in a good agreement with previous observation [40] that direct nucleophilic displacement of C4 methanesulfonate of methyl 2,3,6-tri-*O*-methyl-4-*O*-methylsulfonyl-β-D-galactopyranoside *87* is ca. 2.7 times faster than the direct nucleophilic displacement of the C2 methylsulfonyl group of methyl 4,6-*O*-benzylidene-3-*O*-methyl-2-*O*-methanesulfonyl-β-D-mannopyranoside *88* (Fig. 7.32).

Fig. 7.32

There are other functional groups that can act as neighboring groups in nucleophilic displacements. For example, a hydroxyl or an alkoxyl group, alkyl- or arylthio groups, acylamido group (carbonyl oxygen participation or nitrogen participation), dithiocarbamoyl group.

A very nice example for the participation of hydroxyl group in the acid-catalyzed hydrolysis of dimethyl acetal of D-glucose and D-galactose was reported by Capon and Thacker [41]. The rates of hydrolysis of both acetals were substantially higher than that of glyceraldehyde dimethyl acetal where no five-membered ring transition state can be visualized. The following mechanism has been proposed for the acid hydrolysis of these acetals (in Fig. 7.33 dimethyl D-glucose acetal is used).

The above reaction is better described as intramolecular displacement by hydroxyl group of a protonated methoxy group of an acetal than as neighboring group participation by a hydroxyl group. The rate of an unassisted, two-step mechanism would be slower, it would be independent of the configuration at the C4 and the product would be free sugar.

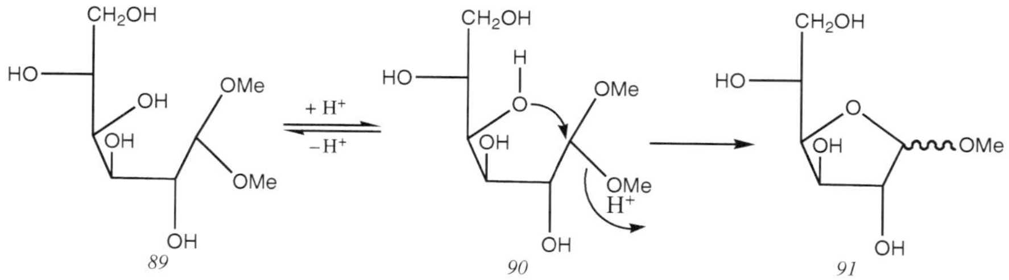

Nucleophilic Displacements with Neighboring Group Participation   185

Fig. 7.33

2,3,5-tri-*O*-Benzyl-4-*O*-*p*-toluenesulfonyl-D-ribose dimethyl acetal *92* gives with tetrabutylammonium benzoate in *N*-methylpyrrolidinone the 2,3,5-tri-*O*-benzyl-4-*O*-methyl-L-lyxose hemiacetal benzoate [42] *95*. The formation of *95* could be explained by a nucleophilic displacement of the C4 tosylate by one of

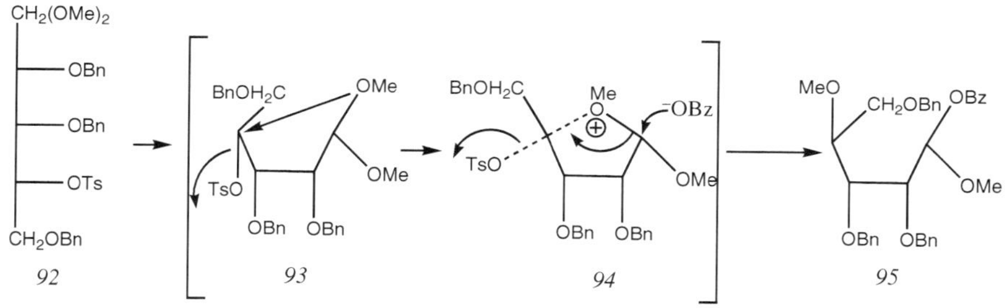

Fig. 7.34

the two methoxy groups of dimethyl acetal with formation of methoxycarbonium ion *94* which with benzoate anion then gives the L-lyxose derivative *95* (Fig. 7.34).

Similarly, 4-*O*-benzyl-1-*O*-*p*-toluenesulfonyl-1,4-pentanediol *96* undergoes a rapid solvolysis in ethanol to give 2-methyltetrahydrofuran, benzyl ethyl, and *p*-toluenesulfonic acid. The rate of the reaction and the products obtained clearly indicate anchimeric assistance from the benzyloxy group via a five-membered, cyclic, oxonium ion intermediate *97* [43] (Fig. 7.35).

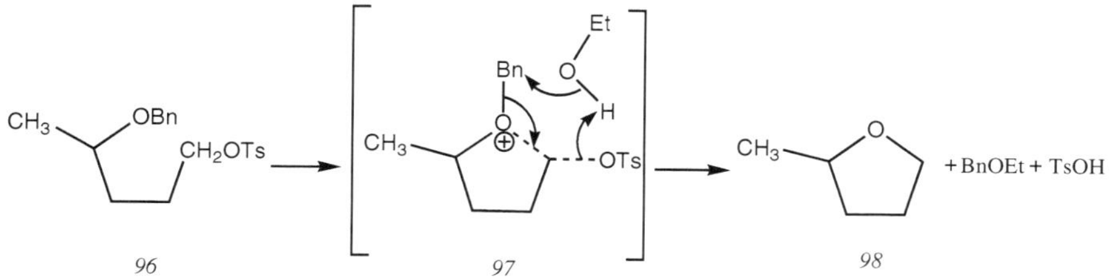

Fig. 7.35

2,3,4-tri-*O*-Benzyl-1,5-di-*O*-*p*-toluenesulfonyl-xylitol solvolyzed in ethanol [43] to give the 1,4-anhydro-2,3-di-*O*-benzyl-5-*O*-*p*-toluenesulfonyl-DL-xylitol *102* (Fig. 7.36). It should be noted that *99* is symmetrical molecule because it has a plane of symmetry passing through the C3 carbon. Consequently, if the C1 tosyl group is intramolecularly displaced, 1,4-anhydro-D-xylitol (*103*) is obtained, but if the C5 tosyl group is intramolecularly displaced the 1,4-anhydro-L-xylitol is obtained (not shown).

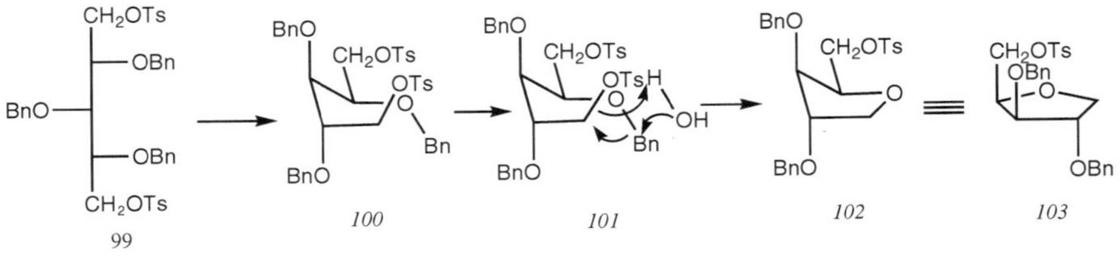

Fig. 7.36

2,3,4-tri-*O*-Benzyl-1,5-di-*O*-*p*-toluenesulfonyl derivatives of D-arabinitol and DL-ribitol underwent smooth solvolysis in ethanol, too. However, the rates of solvolysis of D-xylitol and D-arabinitol were about the same, but larger than that of D-ribitol possibly due to the presence of larger nonbonding interactions in the transition state analogous to *101*.

The acylamido group can participate in nucleophilic displacements with either the carbonyl oxygen or the amino-nitrogen of an amido group acting as a nucleophile.

An example for the first type of participation is described in Fig. 7.27. An example for the participation of nitrogen atom of an amido group is given below. Methyl 4,6-*O*-benzylidene-2-deoxy-2-benzamido-3-*O*-mesyl-α-D-altropyranoside *104* and methyl 4,6-*O*-benzylidene-3-deoxy-3-benzamido-2-*O*-mesyl-α-D-altropyranoside *106* give on treatment with basic reagents the D-manno- and D-allo-aziridines (*105* and *107*), respectively [44] (Fig. 7.37).

Aziridine formation by participation of nitrogen from various neighboring groups occurs only when the reaction conditions are sufficiently basic to convert the neighboring group into its anionic form and if the participating and the leaving groups are in the *trans* orientation (Fürst–Plattner rule [46]).

Treatment of methyl 4,6-*O*-benzylidene-3-deoxy-methyldithiocarbamoyl-2-*O*-mesyl-α-D-altropyranoside *108* (Fig. 7.38) with a hot methanolic solution of sodium methoxide gives the aziridine *109* [46].

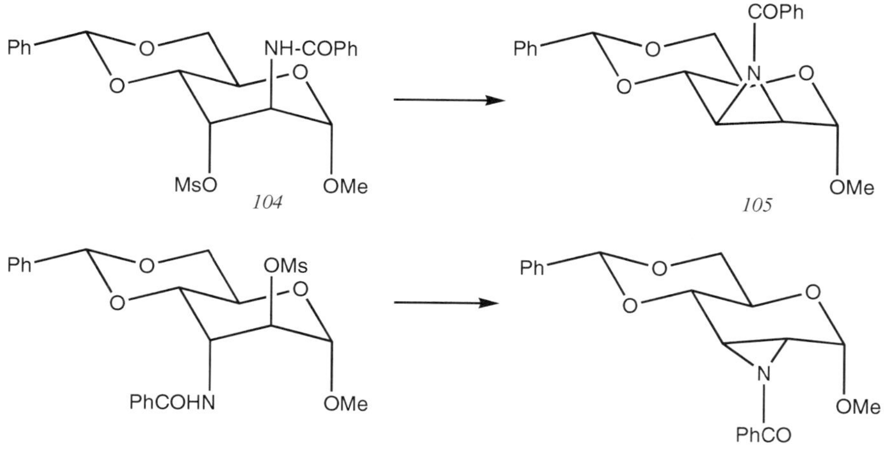

104 105

106 107

Fig. 7.37

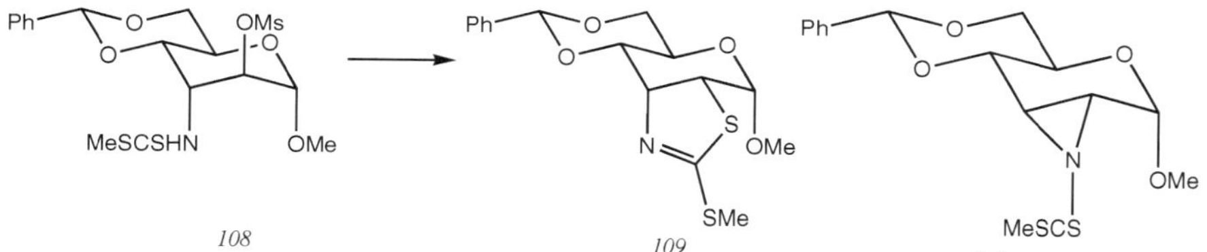

108 109 110

Fig. 7.38

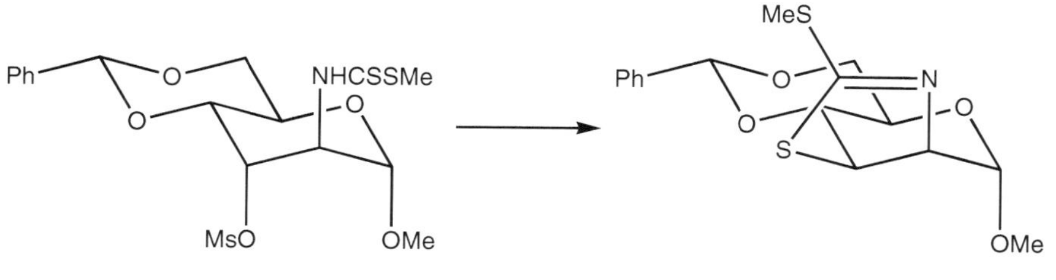

111 112

Fig. 7.39

If the 4,6-*O*-benzylidene-3-deoxy-3-methyldithiocarbamoyl-2-*O*-mesyl-α-D-altropyranoside *108* is refluxed in pyridine, the D-allo-thiazoline *109* [47] was obtained, instead of aziridine *110* (Fig. 7.39).

Heating of a pyridine solution of methyl 2-amino-4,6-*O*-benzylidene-2-deoxy-3-*O*-mesyl-2-*N*[(methylthio)thiocarbonyl]-α-D-altropyranoside *111* at 80°C for 2.5 h gave the α-D-manno thiazoline *112* in 71% yield (Fig. 7.39), which can easily be converted to methyl 2-amino-4,6-*O*-benzylidene-3-thio-α-D-mannopyranoside [48].

References

1. Jary, J.; Capek, K.; Kovar, J., "*Amino sugars. II. Preparation of derivatives of methyl 2,3-anhydro-6-deoxy- α -D-mannopyranoside*", Coll. Czech. Chem. Comm. (1964) **29**, 930–937
2. Buss, D. H.; Hall, L. D.; Hough, L., "*Some nucleophilic substitution reactions of primary and secondary sulphonate esters*", J. Chem. Soc. (1965) 1616–1619
3. Taylor, N. F.; Kent, P. W., "*168. Fluorocarbohydrates. Part I. The synthesis of 6- deoxy-6-fluoro- α -D-galactose and 5-deoxy-5-fluoro- αβ -D-ribose*", J. Chem. Soc. (1958) 872–875
4. Reist, E.; Spencer, R. R.; Baker, J. R., "*Potential anticancer agents. XXIX. Inversion of a ring carbon of a Glycoside*", J. Org. Chem. (1959) **24**, 1618–1619
5. Hill, J.; Hough, L.; Richardson, A. C. "*Replacement of methylsulfonyloxy groups: the conversion of the D-gluco into the D-galacto configuration*", Proc. Chem. Soc. (1963) 314–315
6. Hill, J.; Hough, L.; Richardson, A. C. "*Replacement of methanesulfonyloxy groups: the conversion of the D-gluco into the D-galacto configuration*", Proc. Chem. Soc. (1963) 346–347
7. Stevens, C. L.; Glinski, R. P.; Taylor, K. G.; Blumbergs, P.; Sirokman, F., "*New Rearrangement of Hexose 4- and 5-O-Sulfonates*", J. Am. Chem. Soc. (1966) **88**, 2073– 2074
8. Hanessian, S., "*Ring contraction and epimerization during a displacement reaction of Hexose Sulfonate*", Chem. Commun. (1966) 796–798
9. Hughes, N. A.; Speakman, P. R. H., "*Benzoate displacements on 3-O-toluene-p- sulphonyl-D-glucose derivatives; a new synthesis of D-allose*", J. Chem. Soc. (1965) 2236–2239
10. Jeanloz, R. W.; Jeanloz, D. A., "*The Solvolysis of Sulfonyl Esters of Methyl α-D- Glucopyranoside and Methyl α-D-Altropyranoside*", J. Am. Chem. Soc. (1958) **80**, 5692–5697
11. Richardson, A. C.; Williams, J. M., "*Selective O-acylation of pyranosides*", Chem. Commun. (1965) 104–105
12. Eliel, E. L.; Allinger, N. L.; Angyal, S. J.; Morrison, G. A., *Conformational Analysis*, p. 88, Interscience, New York (1965)
13. Nakajima, M.; Shibata, H.; Kitahara, K.; Takashi, S.; Hagesawa, A., "*Synthesis of kasuganobiosamine*", Tetrahedron Lett. (1968) **9**, 2271–2274
14. Wolfrom, M. L.; Shafizadeh, F.; Armstrong, R. K.; Shen Han, T. M., "*Synthesis of amino sugars by reduction of hydrazine derivatives; D- and L-Ribosamine, D- Lyxosamine1-3*", J. Am. Chem. Soc. (1959) **81**, 3716–3719
15. Horton, D.; Wolfrom, M. L.; Thompson, A., "*Synthesis of amino sugars by reduction of hydrazine derivatives. 2-Amino-2-deoxy-L-lyxose (L-lyxosamine) hydrochloride*", J. Org. Chem. (1961) **26**, 5069–5074
16. Roth, W.; Pigman, W., "*Methyl derivatives of D-Mannosamine*", J. Org. Chem. (1961) **26**, 2455–2458
17. Brimacombe, J. S.; How, M. J., "*Pneumococcus type V capsular polysaccharide: characterisation of pneumosamine as 2-amino-2,6-dideoxy-L-talopyranose*", J. Chem. Soc. (1962) 5037–5040
18. Hough, L.; Richardson, A. C., in "*Rodd's Chemistry of Carbon Compounds*", Vol. 1F, Coffey, S. (Ed.), 2nd ed., Elsevier, New York, N. Y., 1967, pp. 222–224 and 403–407
19. Ali, Y.; Richardson, A. C., "*Nucleophilic replacement reactions of sulphonates. Part III. The synthesis of derivatives of 2,3,4,6-tetra-amino-2,3,4,6-tetradeoxy-D-galactose and -D-idose*", J. Chem. Soc. C, (1968) 1764–1769

References

20. Richardson, A. C., "*Nucleophilic replacement reactions of sulphonates : Part VI. A summary of steric and polar factors*", Carbohydr. Res. (1969) **10**, 395–402
21. Ball, D. H.; Parrish, F. W., "*Sulfonic esters of carbohydrates: Part II*", Advan. Carbohydr. Chem. Biochem. (1969) **24**, 139–197
22. Hough, L.; Richardson, A. C., in "*The Carbohydrates, Chemistry and Biochemistry*", Vol. 1A, Pigman, W.; Horton, D. (Eds.), 2nd ed., Academic Press, Inc., New York, N. Y. (1972), p. 143
23. Richardson, A. C., "*Nucleophilic replacement reactions of sulfonates: Part VI. A Summarru of steric and polar factors*", Carbohydr. Res. (1969) **10**, 395–402
24. Miljkovic, M.; Gligorijevic, M.; Glisin, Dj., "*Steric and electrostatic interactions in reactions of carbohydrates. III. Direct displacement of the C-2 Sulfonate of Methyl 4,6-O-Benzylidene-3-O-methyl-2-O-methylsulfonyl- β -D-gluco- and mannopyranosides*", J. Org. Chem. (1974) **39**, 3223–3226
25. Ishido, Y.; Sakairi, N., "*Nucleophilic substitution-reactions at C-2 of methyl 3-O- benzoyl-4,6-O-benzylidene-2-O-(trifluoromethylsulfonyl)-α-D-glucopyranoside*", Carbohydr. Res. (1981) **97**, 151–155
26. Hashimoto, H.; Araki, K.; Saito, Y.; Kawa, M.; Yoshimura, Y., "*Preparation of 2- Azido-2-deoxypentose Derivatives*", Bull. Chem. Soc. (Japan) (1986) **59**, 3131–3136
27. Ranganathan, R., Modification of the 2′-position of purine nucleosides: syntheses of 2′-α-substituted-2′-deoxyadenosine analogs", Tetrahedron Lett. (1977) 1291–1294
28. Vatèle, J-M.; Hanessian, S., in *Preparative Carbohydrate Chemistry*, Hanessian, S. (Ed.), Marcel Dekker, Inc., New York, 1997, pp.127–149 (Chapter 7)
29. Ryan, K. J.; Arzoumanian, H.; Acton, E. M.; Goodman, L., "*Configurational Inversion within a Furanoside Ring by Anchimerically Assisted Displacement: 5-Deoxy-D- ribose from 5-Deoxy-D-xylose*", J. Am. Chem. Soc. (1964) **86**, 2497–2503
30. Bohme, H.; Sell, K., "*Die Hydrolyse halogenierter Äther und Thioäther in Dioxan- Wasser-Gemischen*", Chem. Ber. (1948) **81**, 123–130
31. Cowdrey, W. A.; Hughes, E. D.; Ingold, C. K., "*Reaction kinetics and the Walden inversion. Part III. Homogeneous hydrolysis and alcoholysis of -bromopropionic acid, its ester and anion*", J. Chem. Soc. (1937) 1208–1236
32. Ross, S. D., "*The role of neighboring nitrogen. Atom in the displacement reaction; rearrangement in the Hydrolysis of 1-Diethylamino-2-chloropropane*", J. Am. Chem. Soc. (1947) **69**, 2982–2983
33. Winstein, S.; Buckles, R. E., "*The role of neighboring groups in replacement reactions. I. Retention of configuration in the reaction of some Dihalides and Acetoxyhalides with Silver Acetate*", J. Am. Chem. Soc. (1942) **64**, 2780–2786
34. Winstein, S.; Lindegren, C. R.; Marshall, H.; Ingraham, L. L., "*Neighboring Carbon and hydrogen. XIV. Participation in solvolysis of some primary Benzenesulfonates*", J. Am. Chem. Soc. (1953) **75**, 147–155
35. Lemieux, R. U.; Brice, C., "*A comparison of the properties of pentaacetates and methyl 1,2-orthoacetates of glucose and mannose*", Can. J. Chem. (1955) **33**, 109– 119
36. Miljkovic, M.; Miljkovic, D.; Jokic, A.; Andrejevic, V.; Davidson, E. A., "*Neighboring-group participation in carbohydrate chemistry. II. Neighboring-group participation of the N,N-Diethylamido group in a nucleophilic displacement of a 5- Tosylate*", J. Org. Chem. (1971) **36**, 3218–3221
37. Miljkovic, M.; Davidson, E. A., "*An improved synthesis of 1,2-O-Isopropylidene- β - L-idofuranose*", Carbohydr. Res. (1970) **13**, 444–446
38. Miljkovic, M.; Jokic, A.; Davidson, E. A., "*Neighboring-group participation in carbohydrate chemistry. Part I. Neighboring -group participation of the 6-O- Benzoyl group in a nucleophilic displacement of a 5-p-Toluenesulfonate*", Carbohydr. Res. (1971) **17**, 155–164
39. Miljkovic, M.; Glisin, Dj.; Gligorijevic, M., "*Neighboring-group participation in carbohydrate chemistry. V. Direct evidence for the participation of the β-Trans- Axial Benzoyloxy Group in the Nucleophilic Displacement of Methanesulfonate of Methyl 4, 6-Di-O-benzoyl-3-O-methyl-2-O-methylsulfonyl-β-D-galactopyranoside and Methyl 2,6-Di-O-*

benzoyl3-O-methyl-4-O-methylsulfonyl-β-D- mannopyranoside", J. Org. Chem. (1975) **40**, 1054–1057
40. Streitwieser, A., Jr., , "*Solvolytic Displacement Reactions*", McGraw-Hill, New York, N. Y. (1962), p. 14
41. Miljkovic, M.; Glisin, Dj., unpublished results; see also Ref. 39
42. Capon, B.; Thacker, D., "*Nucleophilic assistance in the acid-catalyzed reactions of acetals and glycosides*", J. Am. Chem. Soc. (1965) **87**, 4199–4200
43. Hughes, N. A.; Speakman, P. R. H., "*1,4-Migration of a methoxy group during a benzoate displacement reaction: 4-O-methyl-L-lyxose*", Chem. Commun. (1965) 199–200
44. Gray, G. R.; Hartman, F, C.; Barker, R., "*Anchimeric assistance by Benzyloxy groups and the effect of configuration on an intramolecular displacement reaction of the pentitols*", J. Org. Chem. (1965) **30**, 2020–2024
45. Guthrie, R. D.; Murphy, D.; Buss, D. H.; Hough, L.; Richardson, A. C., "*Aziridino derivatives of carbohydrates*", Proc. Chem. Soc. (1963) 84
46. Fürst, A.; Plattner, P. A., Proc. Intern. Congr. Pure. Appl. Chem., 12th Congr., New York, 1951, Abstr. Papers, p. 409
47. Goodman, L.; Christensen, J. E., "*Potential antiradiation drugs. II. β-Aminomercap tans derived from D-Allose*", J. Am. Chem. Soc. (1961) **83**, 3823–3827
48. Miljkovic, M.; Hagel. P., "*Synthesis of Methyl 2-Acetamido-4, 6-di-O-acetyl-3-S- acetyl-2-deoxy-3-thio- α -D-mannopyranoside*", Helv. Chim. Acta (1982) **65**, 477–482

Chapter 8
Anhydrosugars

The monosaccharide derivatives obtained by intramolecular elimination of a molecule of water with simultaneous formation of a new three-, four-, five-, or six-membered heterocyclic ring are called anhydrosugars. The anhydrosugars are subdivided into two groups: (1) the anhydrosugars that involve the anomeric hydroxyl group in their formation resembling thus the intramolecular glycosides; they are called *glycosanes* and (2) the anhydrosugars that do not involve the anomeric hydroxyl group in their formation; they are simply called *anhydrosugar*s. There are several reviews of anhydrosugars [1–3].

The anhydrosugars are not found in the Nature. They are purely synthetic products, and their importance lies exclusively in their use in synthetic carbohydrate chemistry.

1,6-Anhydrosugars (Glycosanes)

In 1894 Tanret [4, 5] prepared the 1,6-anhydro-D-glucose by heating natural aromatic β-D-glucopyranosides (such as picein, salicin, and coniferin) with aqueous barium hydroxide, and he named the obtained product levoglucosan because it was levorotatory (Fig. 8.1).

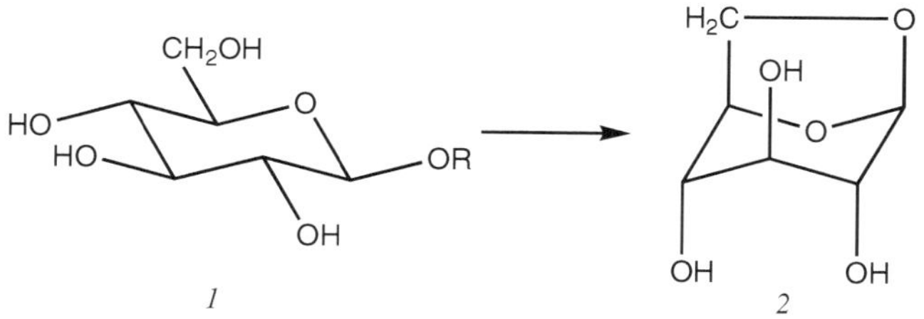

1 *2*

Fig. 8.1

Levoglucosan is a nice crystalline compound that does not reduce Fehling's solution (aqueous solution of copper (II) sulfate, sodium hydroxide, and sodium

© Springer Science+Business Media, LLC 2010

potassium tartrate), indicating the absence of free hemiacetal group. Dilute mineral acids convert the levoglucosan back to D-glucose.

The 1,6-anhydrosugars of D-galactose (β-galactosan) and D-mannose (β-mannosane) were obtained in a similar way by treating the phenyl β-D-galacto- and mannopyranoside with a base.

The fact that the phenyl α-D-glucopyranoside is very resistant to alkaline hydrolysis even under drastic conditions suggested that the hydrolysis of β-D-glucoside takes place with the participation of the C-2-oxygen atom [6] (which is not possible for the α-D-anomer). According to this explanation, the β-phenoxy group is intramolecularly displaced by the neighboring alkoxy anion formed by deprotonation of the C-2 hydroxyl group by a base giving the 1,2-anhydro-α-D-glucopyranose 5 in the first step of alkaline hydrolysis of phenyl β-D-glucopyranoside. In the second step of reaction, the C6 alkoxy anion 6 formed by deprotonation of the C6 hydroxyl group opens the 1,2-oxirane ring of 5 by attacking the C1 carbon and giving the 1,6-anhydro-D-glucose as the final product (Fig. 8.2).

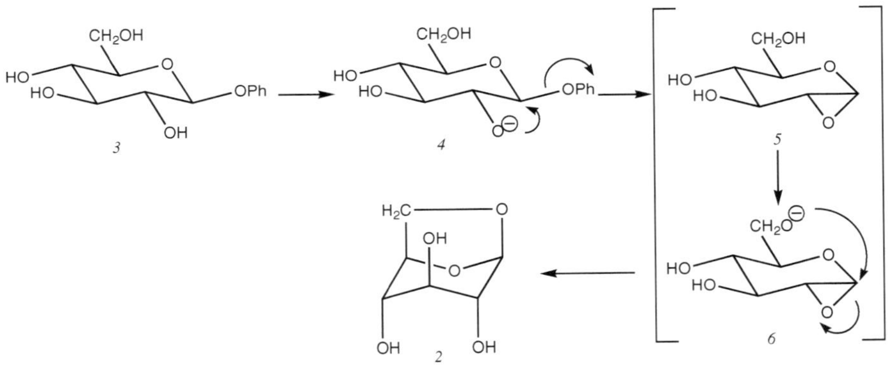

Fig. 8.2

This explanation is supported by the observation that 2-O-methyl-β-D-glucopyranoside [7] and 2,3-di-O-methyl-β-D-glucopyranoside [6] are highly resistant to alkaline hydrolysis, whereas phenyl 2,4,6-tri-O-acetyl-3-O-methyl-β-D-glucopyranoside is converted by ethanolic alkali to 1,6-anhydro-3-O-methyl-β-D-glucopyranose [7]. Finally, treatment of tri-O-acetyl-1,2-anhydro-α-D-glucopyranose with alkali gives 1,6-anhydro-β-D-glucopyranose 2 [7, 8].

The reaction mechanism has been discussed in detail by Coleman [6, 7], Lemieux [9], Ballou [10], Micheel [11], and Janson [12].

Over the years, a number of methods have been described for the synthesis of 1,6-anhydrosugars. Thus, for example, β-D-glucosan was synthesized by treating the tetraacetyl-α-D-glucopyranosyl bromide with triethylamine and then by reacting the obtained quaternary ammonium salt with barium hydroxide [13] (Fig. 8.3).

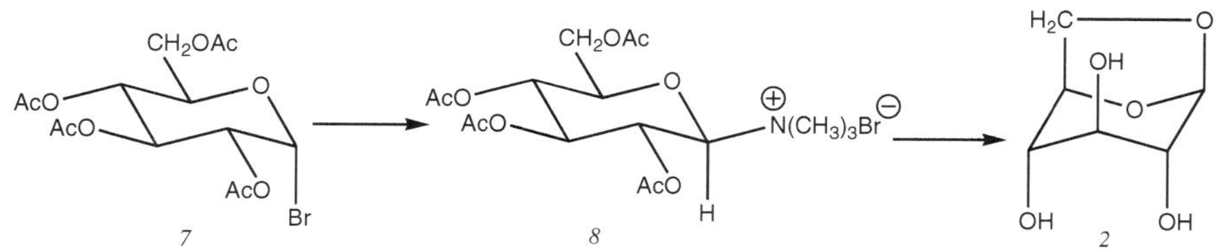

7 8 2

Fig. 8.3

This reaction was successfully used for the synthesis of galactosan [14] from acetobromo-α-D-galactose, though mannosan could not be synthesized in this way from acetobromo-α-D-mannose [15].

1,6-Anhydrosugars have been successfully prepared by action of strong bases on hexopyranosyl derivatives having a good C1 leaving group, such as fluorides [16, 17], bromides, azides [18–21], and tosylates [22] (Fig. 8.4).

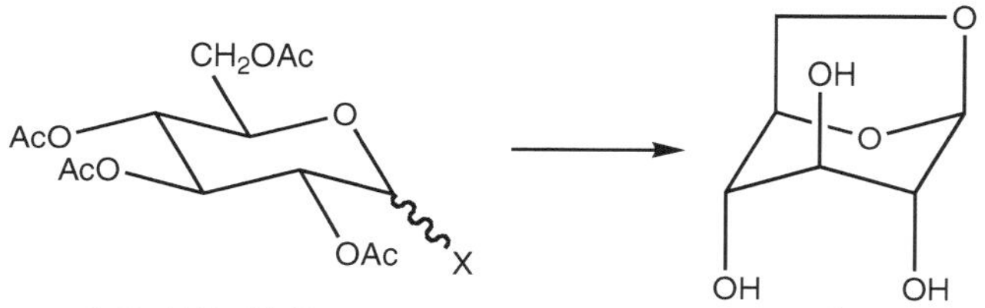

9, X = F, Br, N_3, Ts 2

Fig. 8.4

Glycosans have also been prepared by the action of bases (or the ion-exchange resins in OH^- form) upon glycosyl azides [17].

1,6-Cyclization of 6-O-tritylated [23, 24] 4- or 6-O-benzylated [25, 26] hexopyranose 1-acetate in the presence of $SnCl_4$ or $TiCl_4$ (and other Lewis acids [23]) was another approach for the synthesis of 1,6-anhydrosugars (Fig. 8.5).

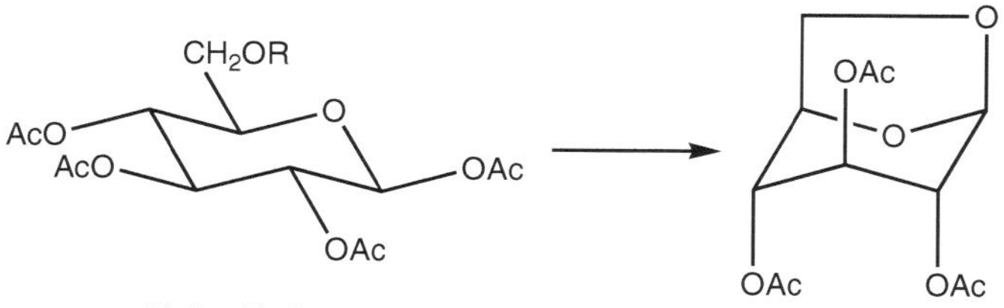

10, R = Tr, Bn *11*

Fig. 8.5

The 1,6-anhydrosugars were also prepared by a base catalyzed cyclization of 6-*O*-tosyl hexopyranoses (for example D-glucose and D-mannose) having a free or acetylated anomeric hydroxyl group [27–30] (Fig. 8.6). However, this method failed with D-galactose [31].

12, R = Ts

Fig. 8.6

There are other methods of preparation of 1,6-anhydrosugars; for a fuller discussion of this topic, the reader is referred to [1, 3, 31–33].

Figure 8.7 gives all 1,6-anhydro-β-D-hexopyranoses that were prepared: *allo* [34–36] (*13*), *altro* [37, 38] (*14*), *gluco* [23] (*2*), *manno* [27, 39–41] (*15*), *gulo* [42] (*16*), *ido* [43] (*17*), *galacto* [19] (*18*), and *talo* [44] (*19*).

13, 1,6-anhydro-β-D-allopyranose

14, 1,6-anhydro-β-D-altropyranose

2, 1,6-anhydro-β-D-glucopyranose

15, 1,6-anhydro-β-D-mannopyranose

16, 1,6-anhydro-β-D-gulopyranose

17, 1,6-anhydro-β-D-idopyranose

18, 1,6-anhydro-β-D-galactopyranose

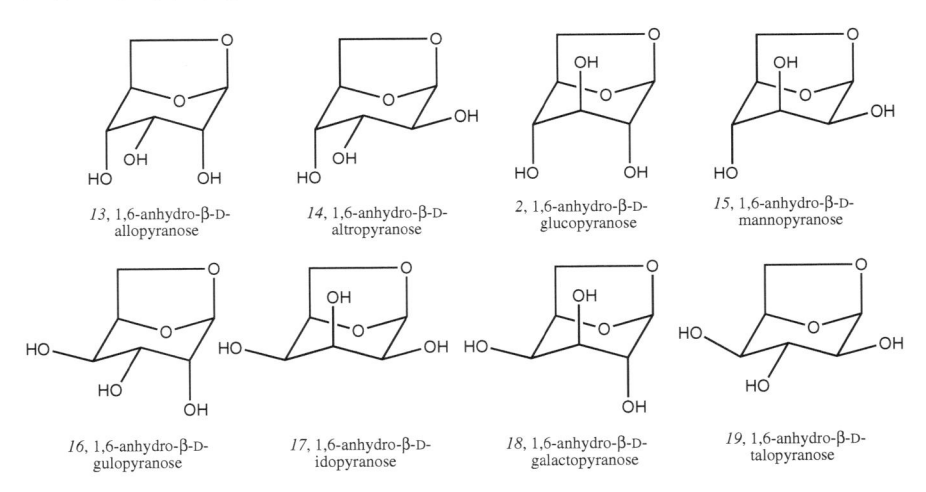

19, 1,6-anhydro-β-D-talopyranose

Fig. 8.7

The relative reactivity of individual hydroxyl groups of a number of 1,6-anhydrosugars toward sulfonylation with *p*-toluenesulfonyl chloride has been found to depend upon both the orientation of the hydroxyl group (axial or equatorial) and its position in the pyranose ring [45].

The axially oriented C3 hydroxyl group of a 1,6-anhydrohexopyranose is sterically more hindered than the axially oriented C2 and C4 hydroxyl groups and consequently the least reactive (*15* in Fig. 8.7). This permits selective modifications of axial or equatorial hydroxyl groups at the C2 or the C4 carbon, such as acylation (tosylation, benzoylation), alkylation (benzylation), oxidation [46, 47]. The equatorially oriented hydroxyl groups of 1,6-anhydrosugars are generally, with a few exceptions [44, 48], more reactive than the axial ones.

Sulfonyloxy groups, both axial and equatorial, may undergo nucleophilic displacement with external or, via neighboring group participation, with internal nucleophiles, depending on their axial or equatorial orientation. Axial or equatorial C4 sulfonyloxy groups can be replaced with inversion of configuration. Thus, for example, the equatorial trifluoromethanesulfonate of 1,6-anhydro-β-D-galactopyranose *21* can be displaced with the sulfide anion of sodium salt of 1-thio-α-D-glucopyranose *20* to give the precursor *22* of thiomaltose in 61% yield [49] (Fig. 8.8).

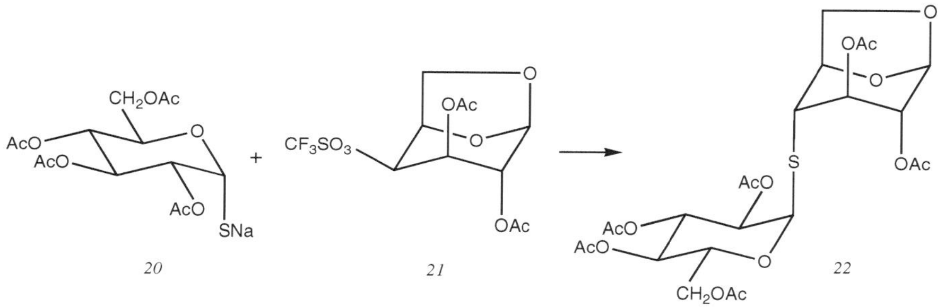

Fig. 8.8

Similarly the S-linked chitobiose has been synthesized [50].

The ammonolysis of 1,6-anhydro-2,4-di-*O*-*p*-toluenesulfonyl-β-D glucopyranose *23* with methanolic ammonia at 0°C gave 1,6-anhydro-2,4-diamino-2,4-dideoxy-β-D-glucopyranose *24* together with unidentified monoamino derivatives [51] (Fig. 8.9).

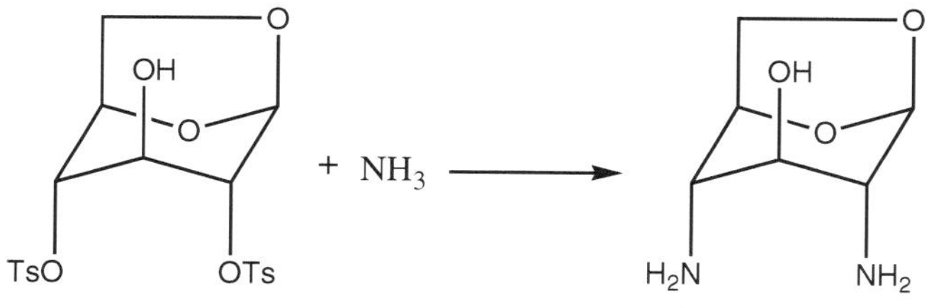

Fig. 8.9

Direct displacement of the C2 and C4 tosyl groups can be excluded since the obtained product was not 1,6-anhydro-2,4-diamino-2,4-dideoxy-β-D-talopyranose as would be expected if the direct displacement took place because this would be accompanied with the inversion of configurations at the C2 and the C4 carbons. The only plausible explanation for the formation of 1,6-anhydro-2,4-diamino-2,

4-dideoxy-β-D-glucopyranose would be that the displacement of these two leaving groups took place via the participation of the C3 hydroxyl anion with formation of the corresponding epoxides as intermediates since the C3 hydroxyl group was left unprotected. The epoxide intermediates were then opened with ammonia (as shown in the Fig. 8.10) to give *24*. There are several reasons for this reaction pathway. First, the C3 hydroxyl group or better alkoxy anion is in the *trans*-diaxial orientation to leaving groups (tosylates) at the C2 and the C4 carbon. Second, intramolecular reactions are thermodynamically more favored than intermolecular reactions, and third the pentacovalent carbon in the SN_2 transition state would place the incoming nucleophile in close proximity with either the ring oxygen or the C6 carbon and thus will be the cause of severe stereoselectronic interactions.

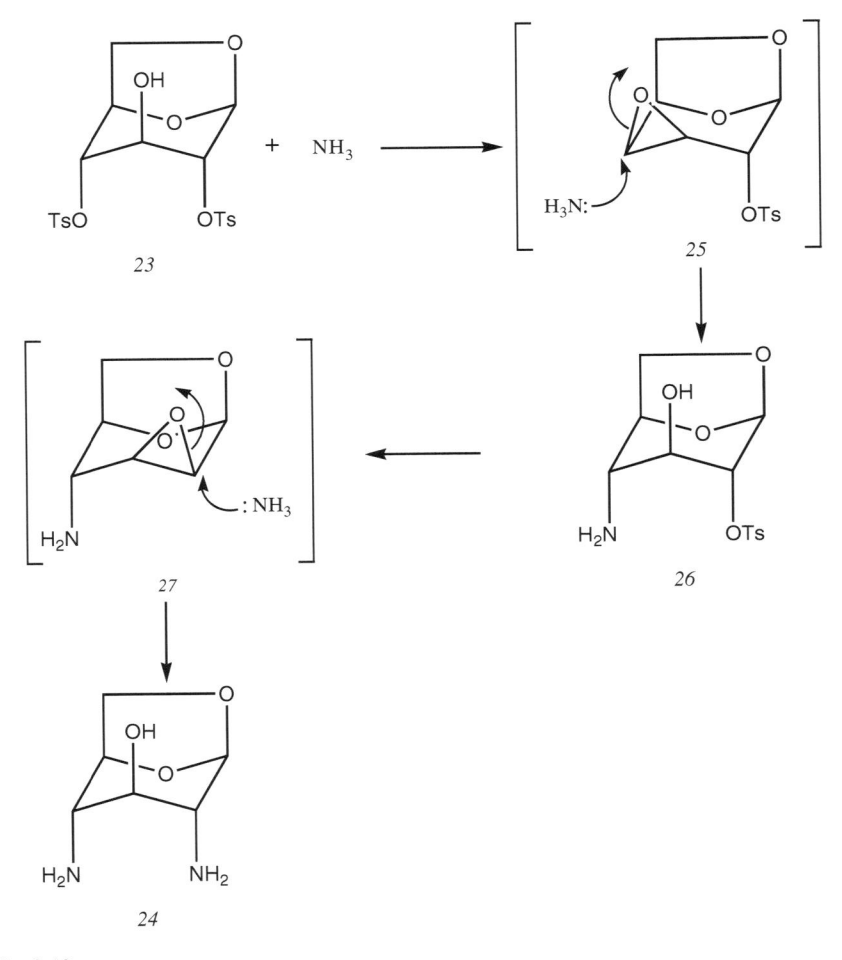

Fig. 8.10

1,6-Anhydrosugars (Glycosanes)

The axial C2 trifluoromethanesulfonylgroup can be, however, directly displaced by acetate [52], under forcing conditions and in the presence of nonparticipating C3 axial azido group (Fig. 8.11).

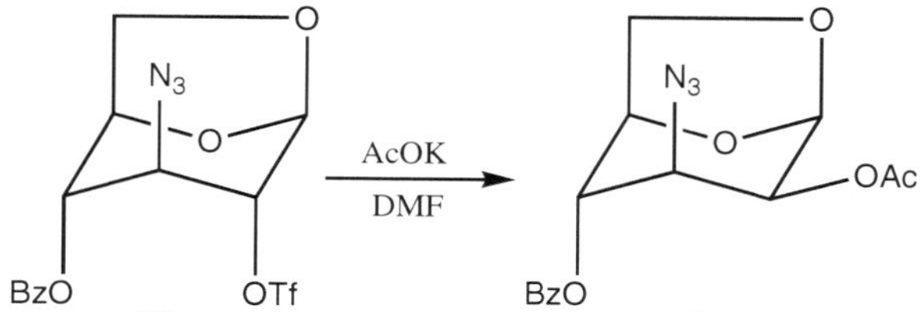

Fig. 8.11

The equatorial C2 trifluoromethanesulfonyl group can be displaced with inversion of configuration by azide [27, 53, 54] and fluoride ions [55].

The axial C3 sulfonyloxy group can be displaced by external nucleophile with inversion of configuration, providing that no axial and nonparticipating substituents are present at the C2 and C4 carbon [56].

The 1,6-anhydrosugars, being actually the intramolecular glycosides, are easily hydrolyzed by mineral acids, such as HCl or H_2SO_4 to their parent sugars (Fig. 8.12). The rate of hydrolysis is higher than with the ordinary glycosides [57, 58]. This increased rate of hydrolysis has been attributed to the ring strain, based on studies of other bicyclic acetals [59].

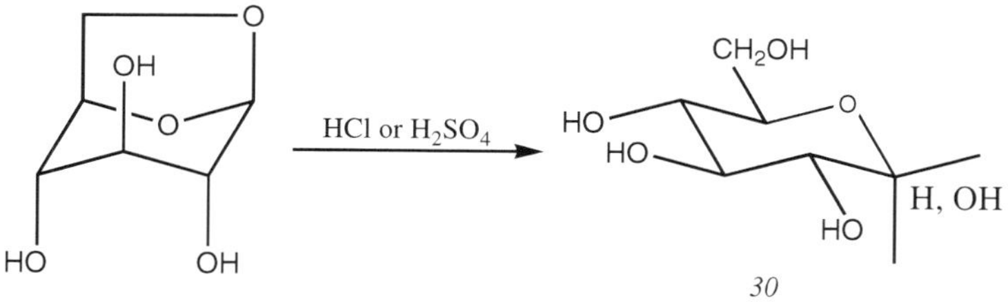

Fig. 8.12

Methanolic hydrogen chloride, at elevated temperatures, converts the 1,6-anhydrohexopyranoses (e.g., 2) into a mixture of corresponding methyl glycosides 32 [14]. Similarly, ethanethiol and zinc chloride convert the 2,3,4-tri-O-benzyl-1,6-anhydro-D-glucopyranose 32 to ethyl 2,3,4-tri-O-benzyl-1-thio-α-D-glucopyranoside 33 [60, 61] (Fig. 8.13).

Treatment of 2,3,4-tri-O-acetyl-levoglucosan 11 with hydrogen bromide in acetic acid [62–66], or with hydrogen bromide in acetic anhydride [66, 67] gave 2,3,4,6-tetra-O-acetyl-α-D-glucopyranosyl bromide 34 (Fig. 8.14).

The 1,6-anhydro ring of levoglucosan triacetate can also be cleaved by titanium tetra-chloride [68–70] or tetrabromide [68, 69] in chloroform giving the 2,3,

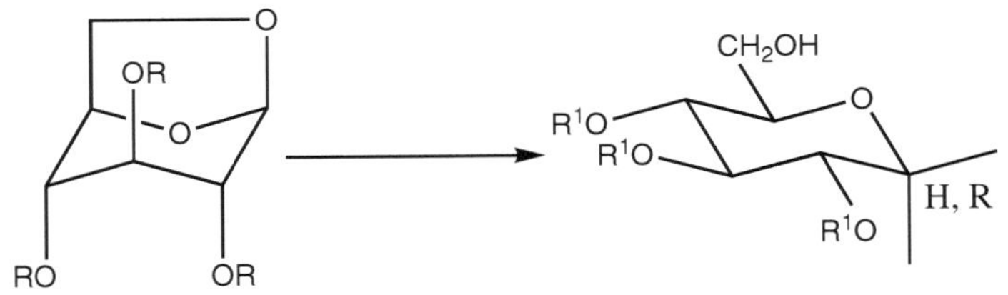

Fig. 8.13

2, R = H
31, R = Bn

32, R = OCH$_3$; R^1 = H
33, R = SC$_2$H$_5$; R^1 = Bn

Fig. 8.14

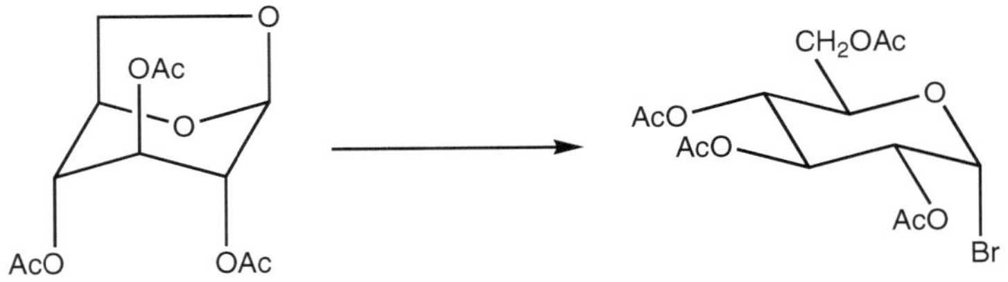

11

34

4-tri-*O*-acetyl-α-D-glucopyranosyl chloride or bromide. Titanium tetrachloride is also effective with 1,6-anhydro-β-D-galactopyranose derivatives [71, 72], but not with 1,6-anhydro-β-D-manno-pyranose derivatives [73].

The 1,6-anhydrosugars are useful synthetic intermediates. First, the anomeric and the C6 carbon atom are simultaneously protected and thus excluded from reactions involving other hydroxyl groups of a pyranose ring. Second, there may be considerable differences in reactivities of the remaining hydroxyl groups of the pyranose ring. Third, the stereo-selectivity of nucleophilic addition on the C2, C3, and C4 carbon is considerably increased.

1,4-Anhydrosugars

This class of anhydrosugars (Fig. 8.15) can be regarded as 1,4-glycosans of glyco-pyranoses, or 1,5-glycosans of glycofuranoses. However, according to the Nomenclature of Carbohydrates (IUPAC–IUBMB) recommendations 1996 [74], they should be named 1,4-anhydropyranoses and not 1,5-anhydrofuranoses because the order of preference of ring size designator is pyranose > furanose.

Treatment of 2,3,6-tri-*O*-methyl-4-*O*-*p*-toluenesulfonyl-D-glucose *37* with alkali gives 1,4-anhydro-2,3,6-tri-*O*-methyl-β-D-galactopyranose *38* [75](Fig. 8.16).

6-Azido-2,3-di-*O*-benzoyl-6-deoxy-4-*O*-methanesulfonyl-α-D-glucopyranosyl acetate *39* is converted to 6-azido-2,3-di-*O*-benzoyl-6-deoxy-1,4-anhydro-galactopyranose *44* [76–78] by treatment with sodium azide in *N*,*N*-dimethylamide

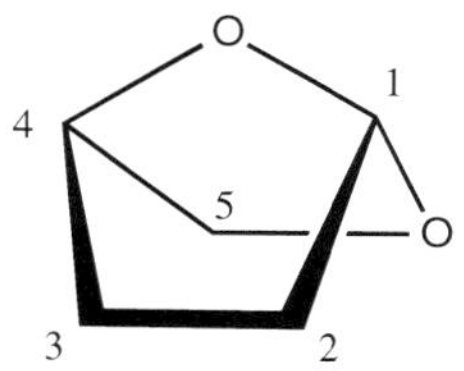

1,4-Anhydrosugars

35, 1,4-glycopyranose, or
36, 1,5-glycofuranose

Fig. 8.15

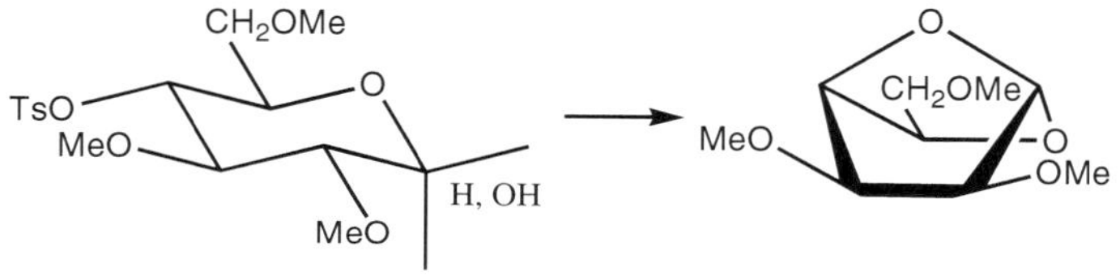

Fig. 8.16

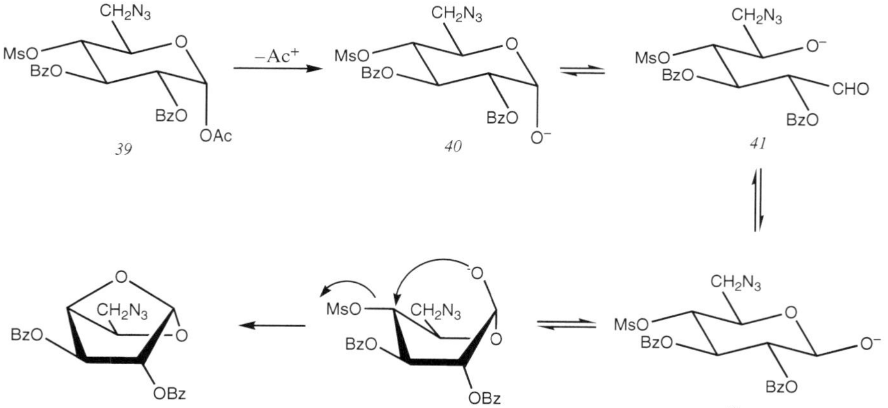

Fig. 8.17

at 140 °C (Fig. 8.17). It was surprising that the 4-O-methanesulfonate group of **39** did not undergo direct S_N2 displacement with the azide ion, but the deacetylation with the formation of C1 alkoxy anion was the preferred reaction which subsequently led to the formation of a 1,4-anhydro-D-galactose derivative.

The action of alkali on 1-O-acetyl-2,3,6-tri-O-methyl-5-O-p-toluenesulfonyl-α-D-glucofuranose **45** gave 1,4-anhydro-2,3,6-tri-O-methyl-α-L-idopyranose **46** [79] (Fig. 8.18).

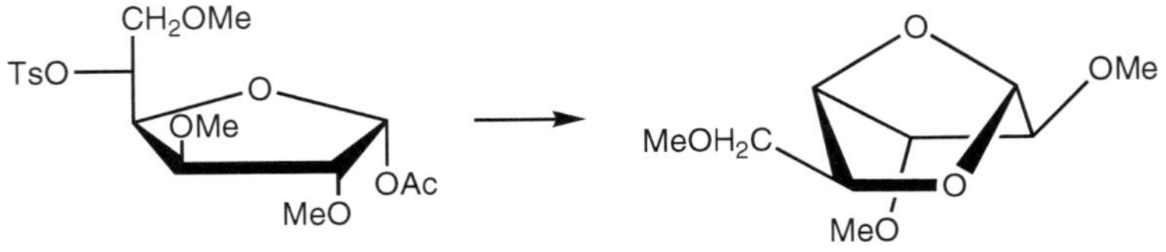

Fig. 8.18

Various anhydrosugars have been investigated as monomer units for the synthesis of synthetic polysaccharides and oligosaccharides. Nokami [80] studied the scope and limitations of 1,4-anhydrosugars as glycofuranosyl building blocks for oligofuranoside synthesis. The oligofuranosides are interesting because they constitute a major part of mycolyl-arabinogalactan (AG), a component of the mycobacterial cell wall [81]. The several 1,4-galactopyranoses were prepared by microwave irradiation of acetonitrile solution of substituted methyl α-D-galactopyranosides in the presence of $FeCl_3$ as the catalyst [82] (Fig. 8.19).

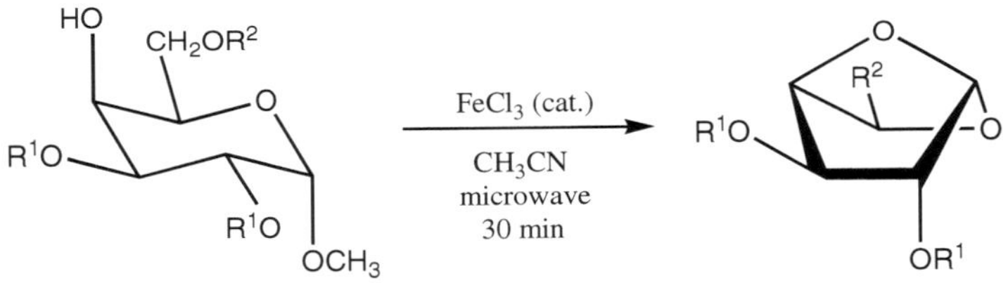

47, $R^1 = R^2 = Bn$
48, $R^1 = Bz; R^2 = Bn$
49, $R^1 = Bn; R^2$ TBS

50, $R^1 = Bn; R^2 = CH_2OBn$ (51%)
51, $R^1 = Bz; R^2 = CH_2OBn$ (12%)
52, $R^1 = Bn; R^2 = CH_2OTBS$ (25%)

Fig. 8.19

It is interesting that starting galactopyranosides are consumed within 30 min by microwave irradiation, while the same reaction takes several days by conventional heating with comparable yields. As can be seen, the yields are generally poor. As far as the usefulness of 1,4-anhydrosugars as glycofuranosyl building blocks for oligofuranoside synthesis, the results presented are too preliminary and will not be described.

1,2-Anhydrosugars (Brigl's Anhydrides)

The 1,2-anhydrosugars are synthetically very important and useful because they can be easily converted to glycosides or di-, tri-, etc., saccharides because the

1,2-Anhydrosugars (Brigl's Anhydrides)   201

1,2-epoxy ring is very reactive and always opens with the attack of a nucleophile at the anomeric (C1) carbon. The configuration of the obtained glycosidic bond is determined by the configuration of the epoxide ring.

The 1,2-anhydrosugars were first prepared by Brigl [83] from penta-*O*-acetyl-β-D-gluco-pyranose as shown in Fig. 8.20. Treatment of penta-*O*-acetyl-β-D-glucopyranose *53* with phosphorus pentachloride gave 3,4,6-tri-*O*-acetyl-2-*O*-trichloroacetyl-β-D-glucopyranosyl chloride *54* which on ammonolysis in ether at 0°C gave 3,4,6-tri-*O*-acetyl-β-D-glucopyranosyl chloride 55. Further treatment of *55* with ammonia converts *55* to 3,4,6-tri-*O*-acetyl-1,2-anhydro-α-D-glucopyranose *56*.

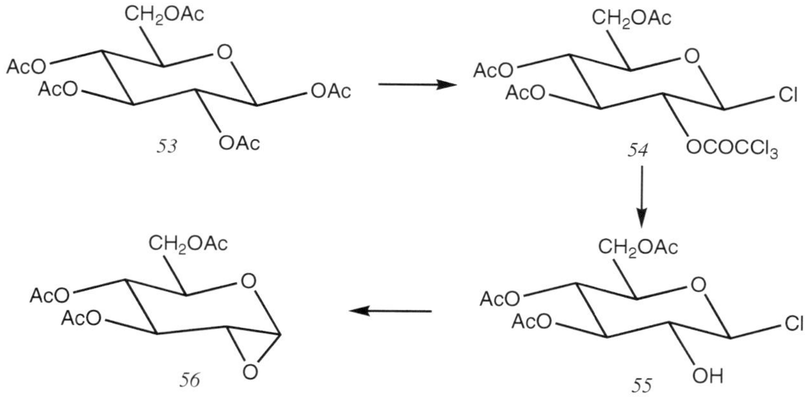

Fig. 8.20

Lemieux and Huber [84, 85] used 3,4,6-tri-*O*-acetyl-1,2-anhydro-α-D-glucopyranose *57* for the first chemical synthesis of sucrose *59*. Thus heating of *56* with 1,3,4,6-tetra-*O*-acetyl-D-fructofuranose *58* (Fig. 8.21) at 100°C in a sealed glass tube for 104 h gave sucrose derivative in 5.5% yield. The mechanism of this "abnormal" epoxide ring opening to form α-D-glucopyranoside cannot be predicted with certainty, but Lemieux considered as the most plausible route the one which involves the participation of the C6 acetate in the first stage of the reaction to yield the 1,2-diaxial carboxonium ion *57* because the C6-oxygen atom is certainly suitably positioned for such a participation. Hickinbottom et al. [86, 87] have observed that 3,4,6-tri-*O*-acetyl-1,2-anhydro-α-D-glucopyranose *56* rapidly reacts with methanol at room temperature giving the corresponding methyl β-D-glucoside excluding thus the participation of the C6 acetoxy group as proposed by Lemieux and Huber [84, 85] for the participation of the C6 acetate. However, heating of *55* with phenol at 100°C for 20 h gives exclusively the corresponding phenyl 3,4,6-tri-*O*-acetyl-α-D-glucopyranoside which is in agreement with Lemieux and Huber [84, 85] proposal for the participation of the C6 acetate. Hardegger and Pascual [88] have reported that the reaction of *56* with isopropanol (16 h at

60 °C), benzylalcohol (16 h at 110 °C), or *tert*-butyl alcohol (16 h at 60 °C) gives, in high yield, the corresponding β-glucosides, excluding thus the participation of the C6 acetoxy group. However, cholesterol (16 h at 120 °C) gave, in poor yield, the corresponding α-glucoside again in agreement with the Lemieux and Huber [84, 85] proposal for the participation of the C6 acetate. Finally, the oxygen atom of the C6 acetate group that acts as the nucleophile in Lemieux mechanism is not the carbonyl oxygen but the alkoxy oxygen of the acetate that is not very nucleophilic. The formation of the intermediate 57 thus must be considered not proven.

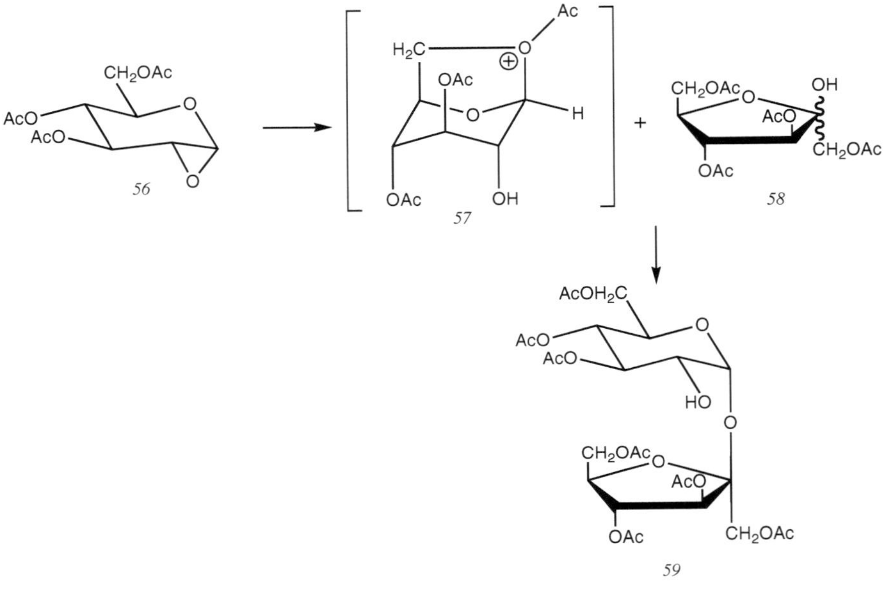

Fig. 8.21

As already stated, Brigl's anhydrides are extremely reactive. Nucleophiles open them by attacking exclusively the anomeric carbon. In case of 3,4,6-tri-*O*-acetyl-1,2-anhydro-α-D-glucopyranose 56, the Fürst–Plattner rule would be violated if the anomeric hydroxyl group of 1,3,4,6-tetra-*O*-acetyl-D-fructose directly attacked the 1,2-epoxide ring forming the β-glycosidic bond. However, the opening of the 1,2-epoxide ring of 56 takes place by the attack of the C6 oxygen on the anomeric carbon giving the five-membered glucosan 57 as an intermediate, which reaction does not violate the Fürst–Plattner rule; the anomeric hydroxyl group of 1,3,4,6-tetra-*O*-acetyl-D-fructose then opens the acetylium-1,6-anhydro ring of 57 to give the α-glycosidic bond.

Anhydrosugars Not Involving the Anomeric Carbon

Epoxides or Oxiranes

Oxiranes (epoxides) are another synthetically important group of anhydrosugars. They are very valuable intermediates for the synthesis of a wide variety of carbohydrate derivatives.

The common strategy employed for the synthesis of this class of anhydrosugars is that all hydroxyl groups of a monosaccharide, except the two hydroxyl groups that are involved in making of the epoxide, must be blocked. Then one of these two hydroxyl groups is derivatized with a good leaving group and the other is left free. Treatment with alkali initiates the intramolecular nucleophilic displacement (S_N2) with the inversion of the configuration at the carbon bearing the leaving group, as shown in Fig. 8.22.

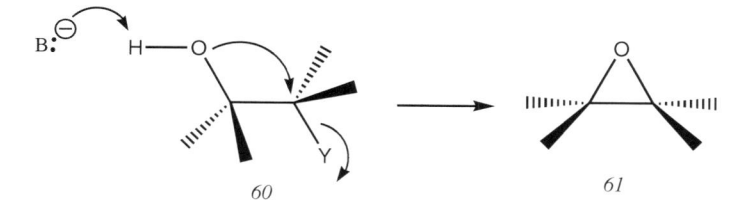

Fig. 8.22

Leaving group Y is most often sulfonic ester, such as *p*-toluenesulfonate or methanesulfonate, but it can also be a halogen (bromide or iodide), sulfate, nitrate, diazonium group, etc. The formation of an epoxide on a pyranoid ring obeys strict stereochemical requirements, i.e., the two reacting groups must be in the antiparallel *trans*-diaxial orientation relative to each other, as shown in Fig. 8.23 (*63*) but not in a *gauche* orientation such as *trans*-diequatorial *62* or in *cis* axial–equatorial *65* orientation.

Fig. 8.23

It has been shown in steroid field that where this requirement is fulfilled without conformational change, the rate of epoxide formation is considerably greater than in cases where a change to a less favorable conformation is required for this condition to be fulfilled [89]. Thus, for example, 1,6-anhydro-2-*O*-mesyl-β-D-galactopyranose *66* and 1,6-anhydro-4-*O*-tosyl-β-D-mannopyranose *68* are readily converted to

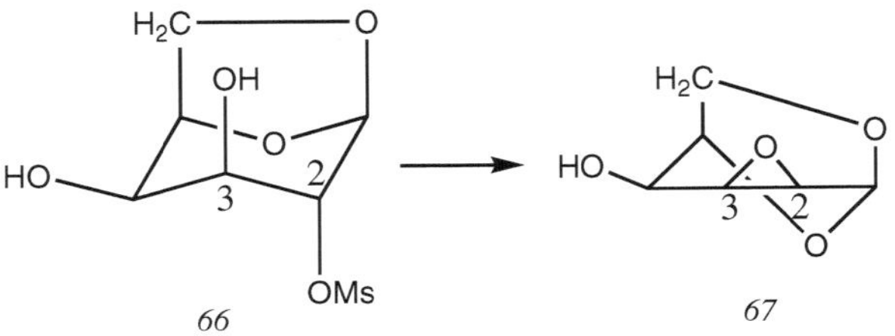

66 *67*

Fig. 8.24

2,3- and 3,4-*talo*-epoxides, *67* and *69*, respectively, since in both of them the hydroxyl and sulfonyloxy groups are in the required coplanar *trans*-diaxial orientation [90, 91] (Figs. 8.24 and 8.25).

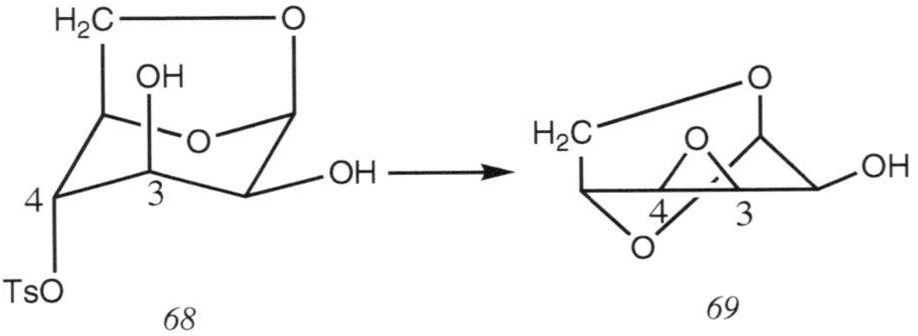

68 *69*

Fig. 8.25

However, in methyl 4,6-*O*-benzylidene-2-*O*-*p*-toluenesulfonyl-α-D-glucopyranoside *70* (Fig. 8.26) or methyl 4,6-*O*-benzylidene-3-*O*-*p*-toluenesulfonyl-α-D-glucopyranoside *73* (Fig. 8.27), the tosyl and the hydroxyl

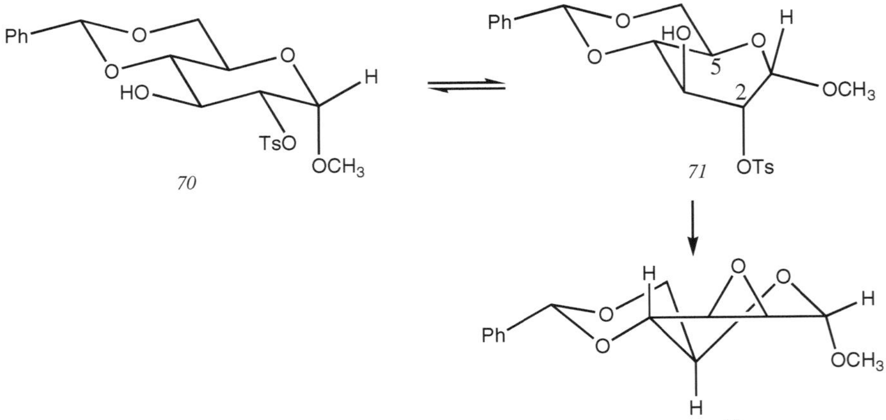

70 *71* *72*

Fig. 8.26

groups are oriented *trans*-diequatorially relative to each other in the preferred $^4C_1$ conformation. In order for these groups to adopt the coplanar *trans*-diaxial arrangement necessary for epoxide formation, the conformation of the pyranoid ring must be changed from $^4C_1$ to $B_{2,5}$ (*71* in Fig. 8.26, or *74* in Fig. 8.27).

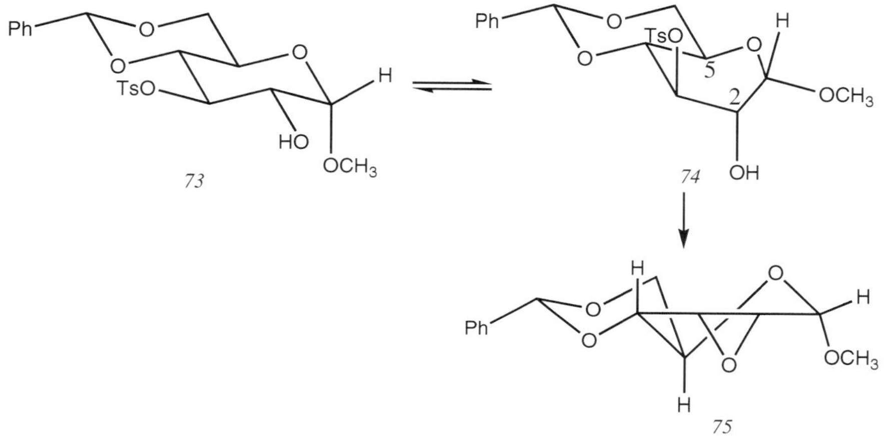

Fig. 8.27

The $^4C_1 \rightarrow B_{2,5}$ conformational change of the α-anomer requires that the 2-tosyloxy and 1-methoxy groups pass each other giving rise to a strong stereoelectronic interaction (Fig. 8.28). This stereoelectronic interaction has been used to explain why the β-anomer of methyl-4,6-*O*-benzylidene-2-*O*-*p*-toluenesulfonyl-D-galactopyranoside is more readily converted to the 2,3-oxirane than the α-anomer.

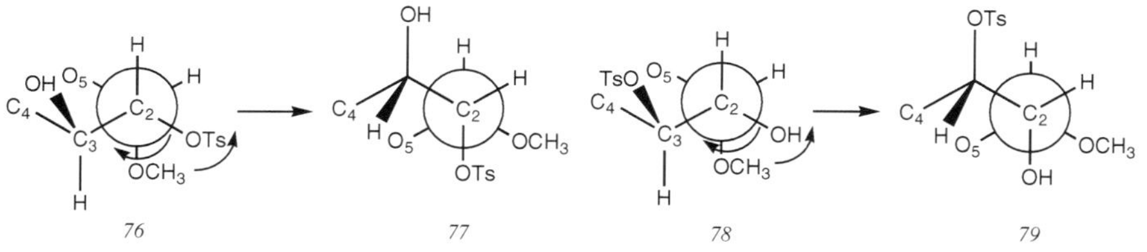

Fig. 8.28

Methyl 4,6-*O*-benzylidene-2,3-di-*O*-*p*-toluenesulfonyl-D-hexopyranosides serve as convenient precursors of 2,3-epoxides. Thus α-D-glucopyranoside *80* gives with sodium methoxide exclusively the 2,3-*allo*-epoxide *81* [92], whereas the α-D-altropyranoside *82* gives 2,3-*manno*-epoxide *83* [93] (Fig. 8.29)

It has been suggested that these reactions proceed by initial base hydrolysis of the more reactive 2-*O*-sulfonate followed by ring epoxide closure [94]. This explanation is supported by the isolation of some methyl 4,6-*O*-benzylidene-3-*O*-*p*-toluenesulfonyl-α-D-gluco-pyranoside from ditosylate by a mild treatment

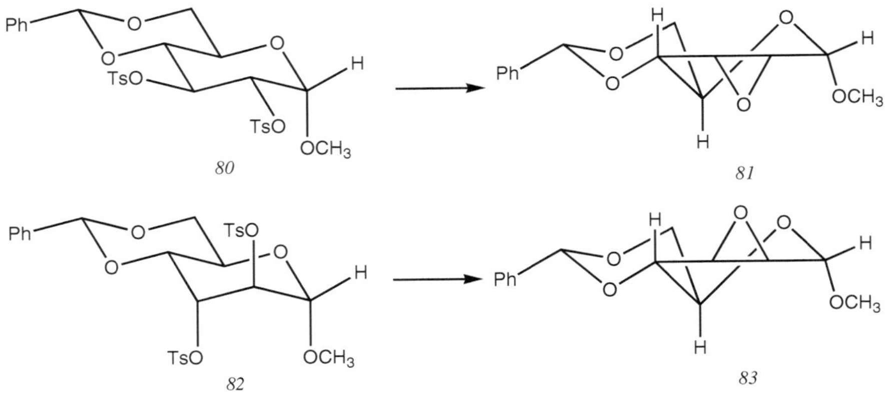

80 *81* *82* *83*

Fig. 8.29

with alkali [95]. This explanation is, however, not applicable to methyl 4,6-*O*-benzylidene-2,3-di-*O*-*p*-toluenesulfonyl-β-D-galactopyranoside *84*, since the reaction with alkali gives the corresponding *talo*-epoxide *86* (Fig. 8.30), suggesting that the 3-*O*-*p*-toluenesulfonate was hydrolyzed by a base [96] rather than 2-*O*-tosylate. However, the α-anomer *86* gave the *gulo*-epoxide *88* together with small amount of *talo*-epoxide *88* [97].

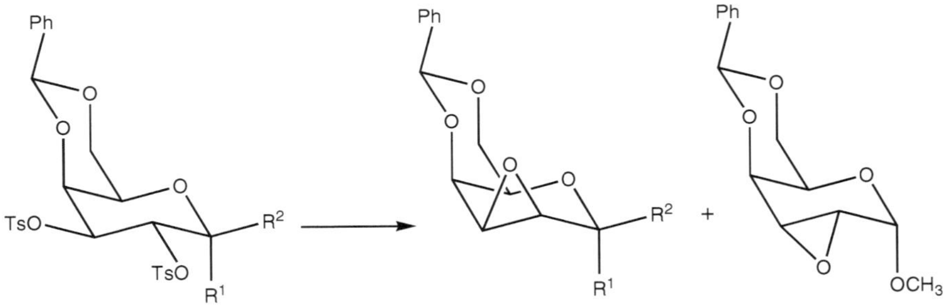

84, $R^1 = H$; $R^2 = OCH_3$
85, $R^1 = OCH_3$; $R^2 = H$

86, $R^1 = H$; $R^2 = OCH_3$
87, $R^1 = OCH_3$; $R^2 = H$

88

Fig. 8.30

It is interesting that 1,2-*O*-isopropylidene-5,6-di-*O*-*p*-toluenesulfonyl-α-D-glucofuranose *89* with alkali does not give the 5,6-epoxide *90* but the 3,6-anhydro-1,2-*O*-isopropylidene-5-*O*-*p*-toluenesulfonyl-α-D-glucofuranose *91* [98] (Fig. 8.31).

The opening of epoxide rings is highly stereoselective, making thus epoxides very versatile and valuable synthetic intermediates in synthetic carbohydrate chemistry (for a general review of epoxides see [99]; for a review of the chemistry of sugar epoxides see [100]).

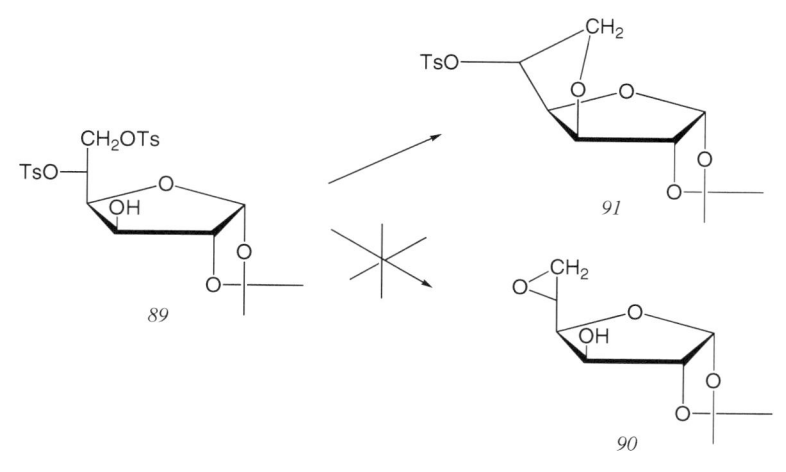

Anhydrosugars Not Involving the Anomeric Carbon

Fig. 8.31

The epoxide rings can be opened by many nucleophiles under both acidic and basic conditions as shown in Fig. 8.32, where N = HO$^-$, RO$^-$, NH$_2$, H$^-$, Cl$^-$, Br$^-$, N$_3^-$, RS$^-$, etc.

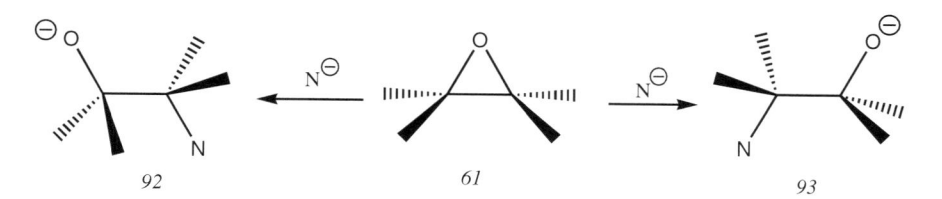

Fig. 8.32

The chiral epoxides, such as those found in carbohydrates, although able to theoretically give two products, in practice, one product is predominantly obtained depending on the site of the nucleophile attack (Fig. 8.32).

When the epoxide ring occupies the terminal position, the nucleophilic attack takes place almost exclusively at the terminal (primary) carbon (for steric reasons). Consequently the 5,6-anhydro-1,2-O-isopropylidene-α-D-glucofuranose *94* is a good intermediate for the synthesis of various six-substituted glucose derivatives *95* (Fig. 8.33).

All epoxide rings having both carbon atoms in their ring structures secondary (the epoxide ring is fused to a pyranoid ring), open with the inversion of configuration at the carbon attacked by a nucleophile, whereas the epoxides containing primary and secondary carbon atom in their structure are opened at the primary carbon and hence without inversion of the configuration of the chiral (secondary) carbon. When the epoxide ring is fused to a pyranoid ring with a rigid conformation,

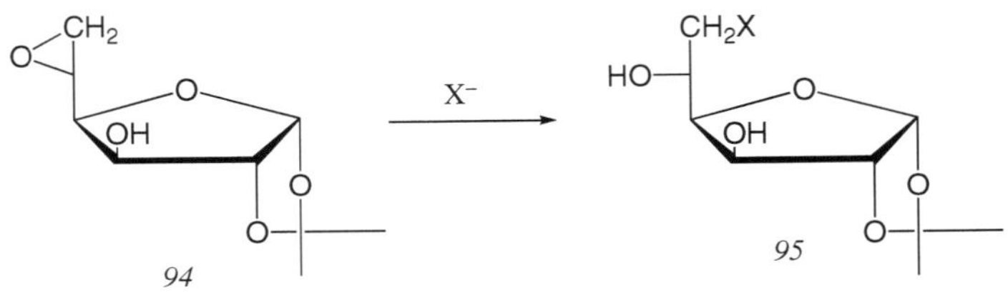

Fig. 8.33

as is the case in *trans*-fused 4,6-*O*-benzylidene-hexopyranosides, or the 1,6-anhydrohexopyranoses, the ring opening occurs with predominant formation of *trans*-diaxial products [101]. For example, 1,6:2,3-dianhydro-β-D-talopyranose *96* and 1,6:3,4-dianhydro-β-D-talopyranose *98* give 1,6-anhydro-2-substituted galactopyranose *98* and 1,6-anhydro-β-D-mannopyranose *99*, respectively, as predominant products [90] (Fig. 8.34).

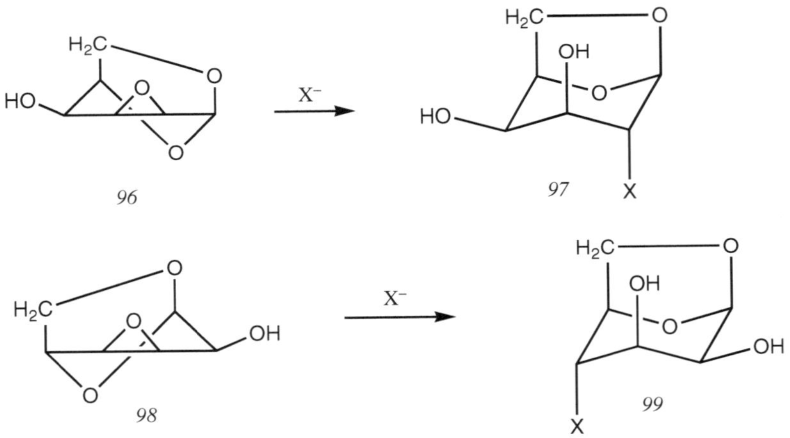

Fig. 8.34

The *trans*-diequatorial products are in both cases obtained in yields under 10%. Similarly, the epoxide rings of methyl 2,3-anhydro-4,6-*O*-benzylidene-α-D-mannopyranoside *100* and methyl 2,3-anhydro-4,6-*O*-benzylidene-α-D-allopyranoside *102* are opened with nucleophiles again diaxially giving the corresponding C3- or C2-substituted altropyranoside derivatives, *101* or *103*, respectively (Fig. 8.35).

This approach has been used for the synthesis of a wide variety of carbohydrate derivatives. Thus, by using azide, hydrazine, ammonia, and primary or secondary amines as nucleophiles many aminosugars can be synthesized.

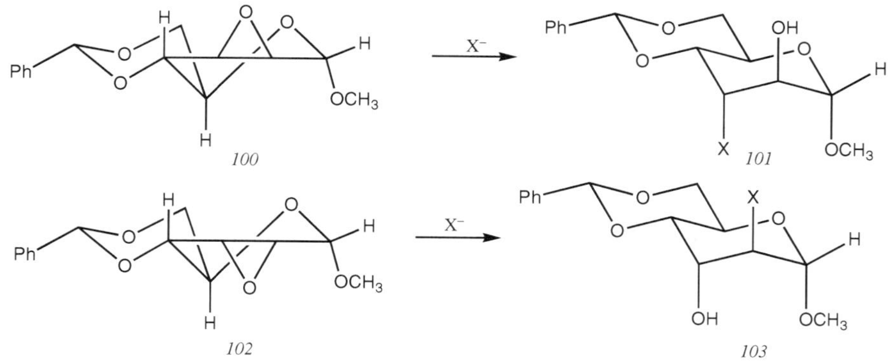

Anhydrosugars Not Involving the Anomeric Carbon

Fig. 8.35

Reduction of epoxides with lithium aluminum hydride affords deoxy-sugar derivatives in regioselective manner. However, the catalytic hydrogenation converts the epoxides to deoxy-sugars, but with much less regioselectivity.

Sodium hydrogen sulfide, sodium thiocarboxylates, sodium alkanethiolate, thiocyanates, etc., give thio-sugars.

Grignard reagents [102, 103] and magnesium halides [104] react with epoxides to give halogeno derivatives.

Diethylylmagnesium [105], sodium cyanide [106], and sodium salt of diethylmalonate [107] gave the corresponding branched-chain sugar (methyl 2-C-carbethoxymethyl-2-deoxy-α-D-altropyranoside).

The nucleophilic opening of 2,3- and 3,4-epoxides on flexible pyranoid rings has been extensively studied. Thus, for example, the epoxide ring of methyl 3,4-anhydro-β-D-galactopyranoside *105* has been opened with the following nucleophiles: H_2S [108], MeSH [109], $LiAlH_4$ [110], and CH_3O^- [111, 112], giving in all cases the 3-substituted D-gulo product (Fig. 8.36).

The methyl 3,4-anhydro-β-D-galactopyranoside can exist in two conformations: *104a* and *104b* (Fig. 8.36). The conformation *104a* is obviously more stable than conformation *104b* since *104b* is destabilized by a strong *syn*–diaxial interaction between the axial C1 methoxy group and the axial C5 hydroxymethyl group.

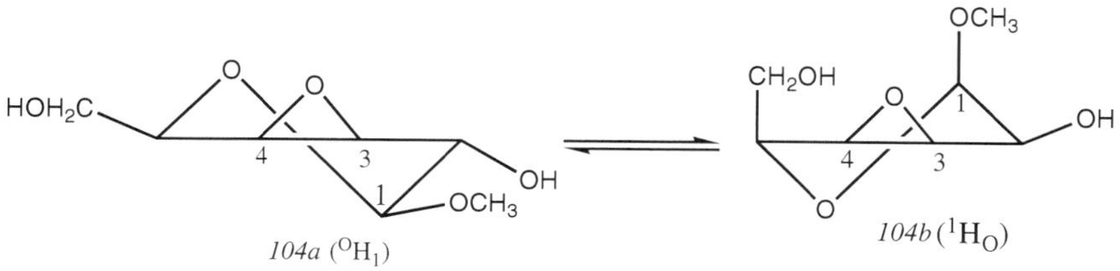

Fig. 8.36

Hence the methyl 3,4-anhydro-β-D-galactopyranoside will be attacked by a nucleophile at the C3 carbon giving the 3-substituted D-gulo product (Fig. 8.37).

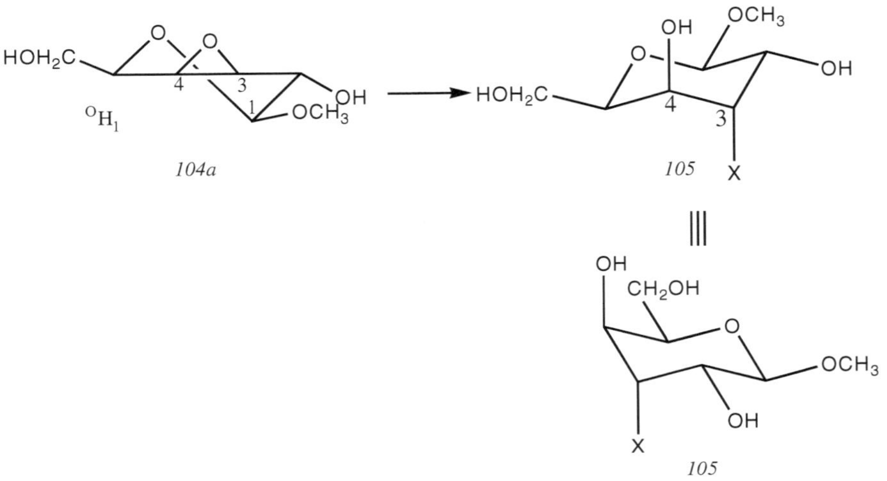

Fig. 8.37

Similarly, methyl 3,4-anhydro-β-L-ribopyranoside can also exist in two conformations: *106a* and *106b* (Fig. 8.38). The conformation *106a* is obviously more stable than the conformation *106b* because the former has fewer nonbonding steric interactions.

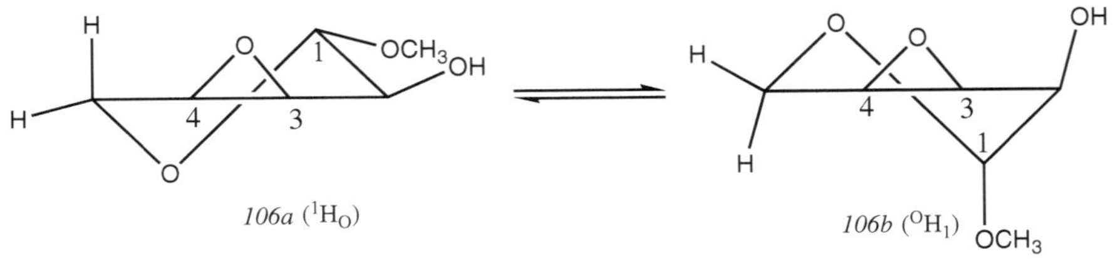

Fig. 8.38

Consequently, the reaction of methyl 3,4-anhydro-β-L-ribopyranoside *106* with hydrogen bromide [113] or various amines [108] opens the 3,4-epoxide ring exclusively at the C4 carbon atom, giving methyl 4-substituted-4-deoxy-α-D-lyxosides (Fig. 8.39).

Fig. 8.39

Rearrangements of Anhydrosugars

Epoxide Migration

Treatment of 1,6-anhydro-3-*O*-*p*-toluenesulfonyl-β-D-altropyranose *109* with a base gave, instead of 1,6:2,3-dianhydro-β-D-mannopyranose *110*, the 1,6:3,4-dianhydro-β-D-altropyranose *111* (Fig. 8.40). Obviously, the C3 tosylate of *109* cannot be displaced by the C4 alkoxy anion because these two groups are *cis* oriented. So

Fig. 8.40

the formation of *111* can only be explained by postulating the "migration of epoxide group," i.e., the first reaction product is *110* because the C2 hydroxyl and the C3 tosyl groups of *109* are *trans* oriented. The 2,3-manno derivative *110* rearranges

112, 1,6:2,3-di-anhydro-β-D-gulopyranose

113, 1,6:3,4-dianhydro-β-D-galactopyranose

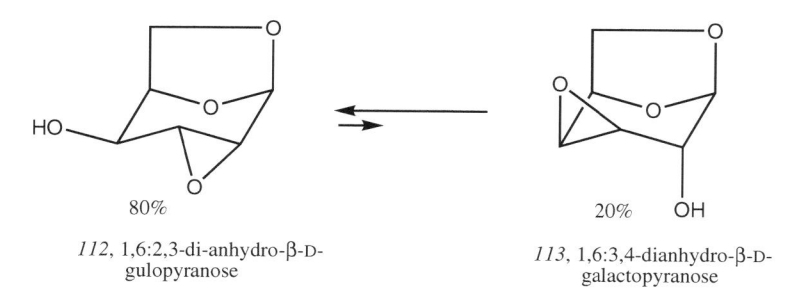

Fig. 8.41

then to the more stable 3,4-altro derivative *111* having the C2 hydroxyl group equatorially oriented [114, 115]. An independent study has shown that the *altro* isomer preponderates in this equilibrium [116]. It has also been shown [117] that in the equilibrium between 1,6:2,3-dianhydro-β-D-gulopyranose *112* and 1,6:3,4-dianhydro-β-D-galactopyranose *113*, the *gulo* isomer *112* (equatorial hydroxyl group) preponderates (Fig. 8.41).

The situation is obviously more complex with flexible oxirane derivatives, as was shown by comparison of equilibrium mixtures obtained after epoxide migration of different types of anhydro hexopyranoses [118–120], their 6-deoxy derivatives [121], and branched-chain hexoses [121, 122] (Fig. 8.42).

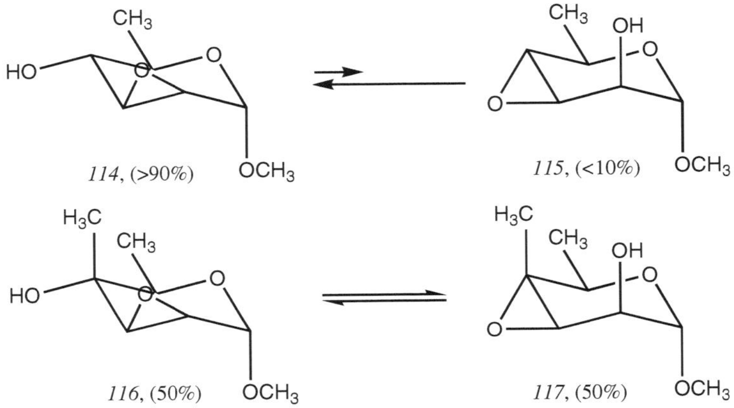

114, (>90%) *115*, (<10%) *116*, (50%) *117*, (50%)

Fig. 8.42

Other Isomerizations of Epoxides

Treatment of methyl 2,3-anhydro-β-D-ribofuranoside (*118*) with 1 N sodium hydroxide at 100°C for 18 h gave methyl 3,5-anhydro-β-D-xylofuranoside *119* in 57% yield [123, 124] (Fig. 8.43).

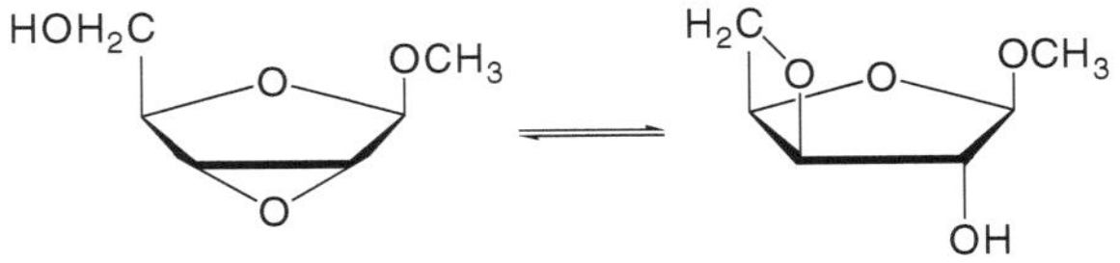

118 119

Fig. 8.43

However, treatment of methyl 2,3-anhydro-α-D-ribofuranoside *120* with sodium methoxide gave methyl 2-*O*-methyl-α-D-arabinofuranoside as the major product [123] (Fig. 8.44).

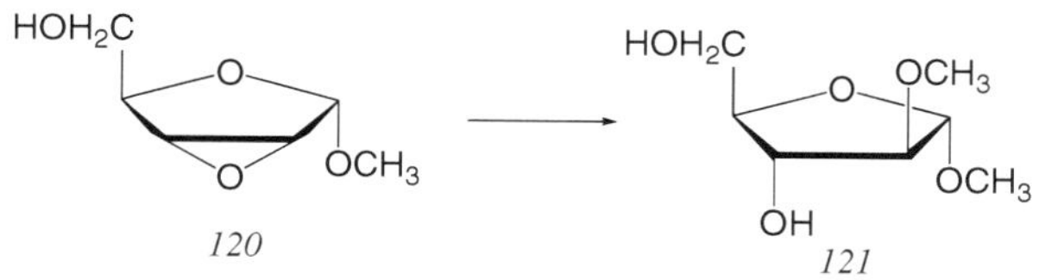

120 *121*

Fig. 8.44

Acid hydrolysis of methyl 2,3-anhydro-α-D-gulopyranoside *122* gave as the major product 3,6-anhydro-D-galactose *123* (Fig. 8.45).

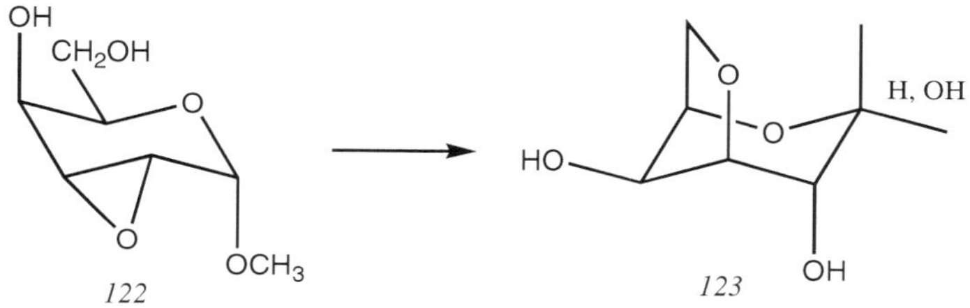

122 *123*

Fig. 8.45

The reaction may follow two courses: (A) the 2,3-anhydro ring is opened first giving initially the methyl 3,6-anhydro-α-D-galactopyranoside *124* and then the hydrolysis of glycosidic bond takes place, giving *123,* or (B) the glycosidic bond may be hydrolyzed first giving the 2,3-anhydro-D-gulose, which is subsequently converted to 3,6-anhydro-D-galactose *123* (Fig. 8.46).

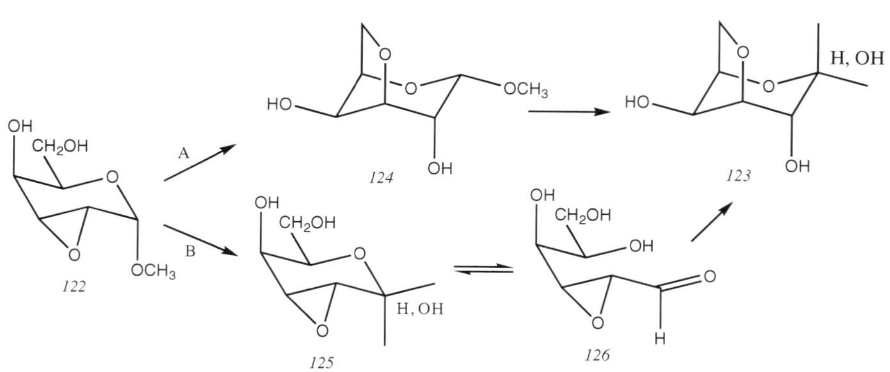

122 *124* *123* *125* *126*

Fig. 8.46

References

1. Peat, S, *"The chemistry of anhydro sugars"*, Adv. Carbohydr. Chem. (1946), **2**, 37–77
2. Schuerch, C., *"Synthesis and polymerization of anhydro sugars"*, Adv. Carbohydr. Chem. Biochem. (1981) **39**, 157–212
3. Černý, M., *"Chemistry of anhydro sugars"*, Adv. Carbohydr. Chem. Biochem. (2003) **58**, 121–198
4. Tanret, C., *"Levoglucosan"*, Compt. Rend. (1894) **119**, 158–161
5. Tanret, C., Bull. Soc. Chim. (France) (1894) **211**, 949
6. McCloskey, C. M.; Coleman, G. H., *"A proposed inversion mechanism for the formation of Levoglucosan from Phenyl β - d -Glucoside and Trimethylglucosylammoniium compounds"*, J. Org. Chem. (1945) **10**, 184–193
7. Bardolph, M. P.; Coleman, G. H., *"Mechanism of the formation of Levoglucosan"*, J. Org. Chem. (1950) **15**, 169–173
8. Dyfverman, A.; Lindberg, B., *"The alkaline hydrolysis of phenyl β-glucosides"*, Acta Chem. Scand. (1950) **4**, 878–884
9. Lemieux, R. U., *"Some implications in carbohydrate chemistry of theories relating to the mechanisms of replacement reaction"*, Adv. Carbohydr. Chem. (1954) **9**, 1–57
10. Ballou, C. E., *"Alkali-Sensitive Glycosides"*, Adv. Carbohydr. Chem. (1954) **9**, 59–95
11. Micheel, F.; Klemer, A., *"Über den Reaktionsmechanismus der Anhydridbildung bei Zuckern"*, Chem. Ber. (1958) **91**, 194–197
12. Janson, J.; Lindberg, B., *"Alkaline hydrolysis of glycosidic linkages. IV. Action of alkali on glucopyranosides"*, Acta Chem. Scand. (1959) **13**, 138–143
13. Karrer, P.; Smirnoff, A. P., *"Eine neue Methode zur Gewinnung von Anhydrozuckern"*, Helv. Chim. Acta (1921) **4**, 817–820
14. Micheel, F., *"Über das Galaktosan <α 1.5><β 1.6>. (Zuckeranhydride, I. Mitteil.)"*, Chem. Ber. (1929) **62**, 687–693
15. Zemplén, G.; Gerecs, A.; Valatin, T., *"Über Lävomannosan"*, Chem. Ber. (1940) **73B**, 575–580
16. Micheel, F.; Klemer, A., *"Eine neue Darstellungsmethode für Zuckeranhydride"*, Chem. Ber. (1952) **85**, 187–188
17. Micheel, F.; Baum, G., *"Darstellung von Zuckeranhydriden mit Hilfe von alkalischen Austauschern"*, Chem. Ber. (1955) **88**, 479–481
18. Micheel, F.; Klemer, A., *"Glycosyl Fluorides and Azides"*, Adv. Carbohydr. Chem. (1961) **16**, 85–103
19. Micheel, F.; Klemer, A.; Baum, G. (mitbearbeitet von Predrag Risticcaron und Fritz Zumbülte), *"Synthesen von Zuckeranhydriden aus 1-Fluor- und 1-Azido-zuckern"*, Chem. Ber. (1955) **88**, 475–479
20. Barnett, J. E. G., *"Acid and alkaline hydrolysis of glycopyranosyl fluorides"*, Carbohydr. Res. (1969) **9**, 21–31
21. Yamamoto, K.; Haga, M.; Tejima, S., *"Thiosugars. XVIII. Synthesis of 2-acy lamino-1,6-anhydro-2,6-dideoxy-6-thio-β- d -glucopyranose"*, Chem. Pharm. Bull. (1975) **23**, 233–236
22. Wiśniewski, A.; Madaj, J.; Skorupowa, E.; Sokolowski, J., *"1,6-Cyclization reactions of selected aldohexopyranoses via their 1-O-tosyl derivatives"*, J. Carbohydr. Chem. (1994) **13**, 873–880
23. Rao, M. V.; Nagarajan, M., *"An improved Synthesis of 2, 3, 4,-tri-O-acetyl-1,6-anhydro-β- d -glucopyranose (levoglucosan triacetate)"*, Carbohydr. Res. (1987) **162**, 141–144
24. Zàrà-Kacziàn, E.; Deàk, G.; Holly, S., *"Mechanism of the stannic chloride-catalyzed conversion of 1,2,3,4-tetra-O-acetyl-β- d -glucopyranose to triacetyllevoglucosan"*, Acta Chim. Acad. Sci. Hung. (1983) **113**, 379–391; Chem. Abstr. (1983) **99**, 176172
25. Kanie, O.; Takeda, T.; Ogihara, Y., *"Synthetic studies on oligosaccharide of a glycolipid from the spermatozoa of bivalves. Part V. A convenient synthesis of 2,3-di-O-acetyl-1,6-anhydro-β- d -glucopyranose"*, J. Carbohydr. Chem. (1990) **9**, 159–165

26. Katano, K.; Chang, P.-I.; Millar, A.; Pozsgay, V.; Minster, D. K.; Ohgi, T.; Hecht, S. M., "*Synthesis of the carbohydrate moiety of bleomycin. 1,3,4,6-Tetra-O-substituted L-gulose derivatives*", J. Org. Chem. (1985) **50**, 5807–5815
27. Kloosterman, M.; De Nijs, M. P.; Van Boom, J. H., "*Synthesis of 1,6-anhydro-2-O-(trifluoromethanesulfonyl)-β- d -mannopyranose derivatives and their conversion into the corresponding 1,6-anhydro-2-azido-2-deoxy-β- d -glucopyranoses: a convenient and efficient approach*", J. Carbohydr. Chem. (1986) **5**, 215–233
28. Sondheimer, S. J.; Eby, R.; Schuerch, C., "*A synthesis of 1,6-anhydro-2,3,4-tri-O-benzyl-β- d -mannopyranose*", Carbohydr. Res. (1978) **60**, 187–192
29. Zottola, M. A.; Alonso, R.; Vite, G. D.; Fraser-Reid, B., "*A practical, efficient large-scale synthesis of 1,6-anhydrohexopyranose*", J. Org. Chem. (1989) **54**, 6123–6125
30. Lafont, D.; Boullanger, P.; Cadas, O.; Descotes, G., "*A mild procedure for the preparation of 1,6-Anhydro-β- d -hexopyranoses and derivatives*", Synthesis (1989) 191–194
31. Guthrie, R. D. in Pigman, W.; Horton, D. (Eds.), *The Carbohydrates: The Chemistry and Biochemistry*, 2nd ed., Vol. 1A, Academic Press, New York, 1972, pp. 423–478;
32. Staněk, J.; Černý, M.; Kocourek, J.; Pacák, J., *The Monosaccharides*, Publishing House of Czechoslovak Academy of Sciences, Prague, 1963, pp. 358–383
33. Dimler, R. J., "*1,6-Anhydrohexofuranoses, a new class of hexosans*", Adv. Carbohydr. Chem. (1952) **7**, 37–52
34. Matsumoto, K.; Ebata, T.; Koseki, K.; Kawakami, H.; Matsushita, H., "*Synthesis of D-allosan from levoglucosenone*", Heterocycles (1991) **32**, 2225–2240
35. Černý, M.; Kalvoda, L.; Pacák, J., "*Syntheses with anhydro sugars. V. Preparation of 2,4-di-O-substituted 1,3-anhydro-β- d -hexopyranos-3-uloses, their isomerization, and reduction*", Collect.Czech. Chem. Commun. (1968) **33**, 1143–1156
36. Pratt, J. W.; Richtmyer, N. K., "*Transformation of D-Allose to 1,6-Anhydro-β- d - allopyranose in acid solution*", J. Am. Chem. Soc. (1955) **77**, 1906–1908
37. Matsumoto, K.; Ebata, T.; Koseki, K.; Kawakami, H.; Matsushita, H., "*Synthesis of D-altrose via D-altrosan from levoglucosenone*", Bull. Chem. Soc. Jpn. (1991) **64**, 2309–2310
38. Richmyer, N. K.; Hudson, C. S., "*The ring structure of D-Altrosan*", J. Am. Chem. Soc. (1940) **62**, 961–964
39. Furneaux, R. H.; Shafizadeh, F., "*Pyrolytic production of 1,6-anhydro-β- d -manno pyranose*", Carbohydr. Res. (1979) **74**, 354–360
40. Georges, M.; Fraser-Reid, B., "*A simple, one-flask, two-step synthesis of 1,6-anhydro-β- d -mannopyranose (D-mannosan) from D-mannose*", Carbohydr. Res. (1984) **127**, 162–164
41. Montgomery, E. M.; Richtmyer, N. K.; Hudson, C. S., "*The Alkaline Degradation of Phenylglycosides; a New Method for Determining the Configuration of Glycosides and Sugars*", J. Am. Chem. Soc. (1943) **65**, 3–7
42. Stewart, L. C.; Richtmyer, N. K., "*Transformation of D-Gulose to 1,6-Anhydro-β- d - gulopyranose in Acid Solution*", J. Am. Chem. Soc. (1955) **77**, 1021–1024
43. Sorkin, E.; Reichstein, T., "*D-Idose aus D(+)-Galaktose*", Helv. Chim. Acta (1945) **28**, 1–17
44. Heyns, K.; Weyer, J.; Paulsen, H., "*Über selektive katalytische Oxydationen, XXIV. Selektive katalytische Oxydation von 1.6-Anhydro-β- d -hexopyranosen zu 1.6-Anhydro-β- d - hexopyranos-ulosen*", Chem. Ber. (1967) **100**, 2317–2334
45. Černý, M.; Staněk, Jr., J., "*1, 6-Anhydro derivatives of Aldohexoses*", Adv. Carbohydr. Chem. (1977) **34**, 23–177
46. Černý, M.; Gut, V.; Pacák, J., "*Partial substitution of 1,6-anhydro-β- d -glucopyranose*", Collect. Czech. Chem. Commun. (1961) **26**, 2542–2550
47. Paulsen, H.; Kolář, Č.; Stenzel, W., "*Bausteine von Oligosacchariden, XI: Synthese α-glycosidisch verknüpfter Disaccharide der 2-Amino-2-desoxy- d -galactopyranose*", Chem. Ber. (1978) **111**, 2358–2369
48. McLeod, J. M.; Schroeder, L. R.; Seib, P. A., "*Selective esterification of 1,6-anhydrohexopyranoses: The possible role of intramolecular hydrogen-bonding*", Carbohydr. Res. (1973) **30**, 337–347

49. Blanc-Muesser, M.; Defaye, J.; Driguez, H., "*Stereoselective thioglycoside syntheses. Part 4. A new approach to 1,4-linked 1-thio-disaccharides and a synthesis of thiomaltose*", J. Chem. Soc. Perkin Trans. (1982) **1**, 15–18
50. Auzanneau, F.-I.; Bennis, K.; Fanton, E.; Promé, D.; Defaye, J.; Gelas, J., "*Synthesis of S-linked thiooligosaccharide analogues of Nod factors. Part 1: selectively N-protected 4-thiochitobiose precursors*", J. Chem. Soc. Perkin Trans. (1998) **1**, 3629–3636
51. Jeanloz, R. W.; Rapin, A. M. C., "*The Ammonolysis of 1,6-Anhydro-2,4-di-O-p-tolylsulfonyl-β- d -glucopyranose and the Synthesis of 2,4-Diamino-2,4-dideoxy- d -glucose*", J. Org. Chem. (1963) **28**, 2978–2983
52. Holla, E. W.; Sinnwell, V.; Klaffke, W.,"*Two syntheses of 3-Azido-3-deoxy- d -mannose*", Synlett (1992) 413–414
53. Brimacombe, J. S.; Hunedy, F.; Mather, A. M.; Tucker, L. C. N., "*Studies related to the synthesis of derivatives of 2,6-diamino-2,3,4,6-tetradeoxy- d -erythro-hexose (purpurosamine C), a component of gentamicin C $_{12}$* ", Carbohydr. Res. (1979),**68**, 231–238
54. Sakari, N.; Takahashi, S.; Wang, F.; Ueno, Y.; Kuzuhara, H., "*Facile preparation of 1,6-anhydro-2-azido-3-O-benzyl-2-deoxy-β- d -glucopyranose and its 4-O-substituted derivatives.*", Bull. Chem. Soc. (Japan) (1994) **67**, 1756–1758
55. Haradahira, T.; Maeda, M.; Kai, Y.; Kojima, M., "*A new, high yield synthesis of 2-deoxy-2-fluoro- d -glucose*", J. Chem. Soc. Chem. Commun. (1958) 364–365
56. Černý, M.; Staněk, Jr.; Pacák, J., "*Syntheses with anhydro sugars. VII. Deoxy sugars. 4. Preparation of 4-deoxy- d -ribo-hexose (4-deoxy- d -allose), 4-deoxy- d -lyxo-hexose (4-deoxy- d -mannose), and their 1,6-anhydro derivatives*", Collect. Czech. Chem. Commun. (1969) **34**, 1750–1764
57. Freudenberg, K.; Kuhn, W.; Dürr, W.; Bolz, F.; Steinbrunn, G., "*Die Hydrolyse der Polysaccharide (14. Mitteil. über Lignin und Cellulose)*", Chem. Ber. (1930) **63**, 1510–1530
58. Freudenberg, K.; Nagai, W., "*Die synthese der Cellobiose*", Chem. Ber. (1933) **66**, 27–29
59. Hall, H. ; Jr., K.; DeBlauwe, F., "*2,6- and 2,7-Dioxabicyclo[2.2.1]heptanes*", J. Am. Chem. Soc. (1975) **97**, 655–656
60. Koto, S.; Uchida, T.; Zen, S., "*Synthesis of isomaltose, isomaltotetraose, and iso maltooctaose*", Chem. Lett. (1972) 1049–1052
61. Koto, S.; Uchida, T.; Zen, S., "*Synthesis of isomaltose, isomaltotetraose, and isomaltooctaose*", Bull Chem. Soc. (Japan) (1973) **46**, 2520–2523
62. Ohle, H.; Spencker, K., "*Über die Aceton-Verbindungen der Zucker und ihre Derivate, VII.: Die Konstitution einiger Mono-acyl-derivate der Mono-aceton-glucose und die Ringstruktur der Glucose*", Chem. Ber. (1926) **59**, 1836–1848
63. Bergmann, M.; Koch, F. K. v., "*Notiz über Gewinnung gemischt-acylierter Zucker*", Berichte (1929) **62**, 311–313
64. Josephson, K., "*Neue Acylderivate der Glucose und des β-Methyl-glucosids aus Levoglucosan*", Chem. Ber. (1929) **62**, 317–321
65. Jeans, A.; Wilham, C. A.; Hilbert, G. E., "*Acetobrominolysis of Di- and Polysaccharide Acetates*", J. Am. Chem. Soc. (1953) **75**, 3667–3673
66. Csürös, Z.; Deák, G.; Haraszthy-Papp, M., "*Reaction of levoglucosan [1,6-anhydro-β- d -glucopyranose]esters with hydrogen bromide and with acid bromides in glacial acetic acid*", Acta Chim. Acad. Sci. Hung. (1961) **29**, 227–235
67. Freudenberg, K.; Soff, K., "*Über den Abbau der Stärke mit Acetylbromid*", Chem. Ber. (1936) **69**, 1252–1257
68. Zemplén, G.; Gerecs, A., "*Einwirkung von Quecksilbersalzen auf Aceto-halogen zucker, VI. Mitteil.: Synthese von Gentionbiose- und Cellobiosido-6-Glykose-Derivaten*", Chem. Ber. (1931) **64**, 1545–1554
69. Haq, S.; Whelan, W. J., "*The chemical synthesis of polysaccharides. Part I. Synthesis of gentiodextrins*", J. Chem. Soc. (1956) 4543–4549
70. Zemplén, G.; Csürös, Z, "*Aufspaltung des Laevoglykosans mit Titantetrachlorid*", Chem. Ber. (1929) **62**, 993–996

References

71. Zemplén, G.; Gerecs, A.; Flesch, H., "*Einwirkung von Quecksilbersalzen auf Actohalogen-Zucker, XI. Mitteil.: Synthese einiger Derivate der β-1- l -rhamnosido-6- d -galaktose*", Chem. Ber. (1938) **71**, 774–776
72. Thompson, A.; Wolfrom, M. L.; Inatome, M., "*Tetraacetates of D-Glucose and D-Galactose*", J. Am. Chem. Soc. (1955) **77**, 3160–3161
73. Zemplén, G.; Gerecs, Á.; Valatin, T., "*Über Lävomannosan*", Chem. Ber. (1940) **73**, 575–580
74. Carbohydrates (IUPAC-IUBMB) recommendations 1996, Carbohydr. Res. (1997) **297**, 1–92
75. Kops, J.; Schuerch, C., "*Syntheses of 1,4-Anhydro-2,3,6-tri-O-methyl- d -galactose and 1,4-Anhydro-2,3-di-O-methyl- l -arabinose*", J. Org. Chem. (1965) **30**, 3951–3953
76. Brimacombe, J. S.; Minshall, J.; Tucker, L. C. N., "*Nucleophilic displacement reactions in carbohydrates. Part XXII. Formation of 1,4-anhydropyranoses from 1-O-acetyl-6-deoxy-2,3-O-isopropylidene-4-O-methylsulphonyl-α- l -manno- and talo-pyranose with sodium azide*", J. Chem. Soc. Perkin 1 (1973) 2691–2694
77. Bullock, C.; Hough, L.; Richardson, A. C., "*A novel route to 1,4-anhydro derivatives of β-image-galactopyranose*", Carbohydr. Res. (1990) **197**, 131–138
78. Dessinges, A.; Castillon, S.; Olesker, A.; Thang, T. T.; Lukacs, G., "*Oxygen-17 NMR and oxygen-18-induced isotopic shifts in carbon-13 NMR for the elucidation of a controversial reaction mechanism in carbohydrate chemistry*", J. Am. Chem. Soc. (1984) **106**, 450–451
79. Hess, K.; Heumann, K. E., "*Über ein weiteres Anhydrid aus 2.3.6.-Trimethyl-glucose (IX. Mitteil. über synthetische Zucker-anhydride)*", Chem. Ber. (1939) **72**, 137–148
80. Nokami, T.; Werz, D. B.; Seeberger, P. H., "*Synthesis and Reactions of 1,4-Anhydro galactopyranose and 1,4-Anhydroarabinose – Steric and Electronic Limitations*", Helv. Chim. Acta (2005) **88**, 2823–2831
81. Lowary, T. L., "*Mycobacterial Cell Wall Components*", in Glycoscience III, Fraser-Reid, B.; Tatsuta, K.; Thiem, J. (Eds.), Springer, Berlin, 2001, pp. 2005–2080
82. Aberg, P.-M.; Ernst, B., "*Facile preparation of 1,6-anhydrohexoses using solvent effects and a catalytic amount of a Lewis acid*", Acta Chem. Scand. (1994) **48**, 228–233
83. Brigl, P., "*Carbohydrates. II. A new anhydride of glucose*", Hoppe-Seyler's Z. physiol. Chem. (1922) **122**, 245–262
84. Lemieux, R. U.; Huber, G., "*A chemical synthesis of sucrose*", J. Am. Chem. Soc. (1953) **75**, 4118–4118
85. Lemieux, R. U.; Huber, G., "*A chemical synthesis of sucrose. A conformational analysis of the reactions of 1,2-Anhydro-α- d -glucopyranose Triacetate*", J. Am. Chem. Soc. (1956) **78**, 4117–4119
86. Hickinbottom, W. J., "*Glucosides. Part I. The formation of glucosides from 3:4:6-triacetyl glucose 1:2-anhydride*", J. Chem. Soc. (1928) 3140–3147
87. Haworth, W. N.; Hickinbottom, W. J., "*Synthesis of a new disaccharide, neotrehalose*", J. Chem. Soc. (1931) 2847–2850
88. Hardegger, E.; De Pascual, J., "*Glucoside und β-1,3,4,6-Tetraacetyl-glucose aus Triacetyl-glucosan-α<1,2>β<1,5>*", Helv. Chim. Acta (1948) **31**, 281–286
89. Alt, G. H.; Barton, D. H. R., "*Some conformational aspects of neighbouring-group participation*", J. Chem. Soc. (1954) 4284–4294
90. James, S. P.; Smith, F.; Stacey, M.; Wiggins, L. F., "*The action of alkaline reagents on 2:3-1:6- and 3:4-1: 6-dianhydro β-talose. A constitutional synthesis of chondrosamine and other amino-sugar derivatives*", J. Chem. Soc. (1946) 625–628
91. Hann, R. M.; Hudson, C. S., "*An Anhydro derivative of D-Mannosan <1,5>β<1,6> (presumably 3,4-Anhydro- d -talosan <1,5>β<1,6>)*", J. Am. Chem. Soc. (1942) **64**, 925–928
92. Richtmyer, N. K.; Hudson, C. S., "*Crystalline α-Methyl- d -altroside and some new derivatives of D-Altrose*", J. Am. Chem. Soc. (1941) **63**, 1727–1731
93. Robertson, G. J.; Whitehead, W., "*Walden inversion in the altrose series*", J. Chem. Soc. (1940) 319–323

94. Angyal, S. J.; Gilham, P. T., "*Cyclitols. Part VII. Anhydroinositols and the epoxide migration*", J. Chem. Soc., 1957, 3691–3699
95. Honeyman, J.; Morgan, J. W. W., *Sugar nitrates. Part II. The preparation and reactions of some nitrates, sulphonates, sulphinates, and other esters of methyl 4:6-O-benzylidene-α- d -glucoside*#x201D;, J. Chem. Soc. (1955) 3660–3674
96. Sorkin, E.; Reichstein, T., "*d-Idose aus d(+)-Galaktose*", Helv. Chim. Acta (1945) **28**, 1–17
97. Gyr, M.; Reichstein, T., "*α-Methyl-d-idosid-<1,5>-monomethyläther-(2) und (3)*", Helv. Chim. Acta (1945) **28**, 226–233
98. Ohle, H.; Thiel, H., "*Über Aceton-Verbindungen der Zucker und ihre Umwandlungsprodukte, XVIII. Mitteil.: 6-p-Toluolsulfo- d -galaktose und 3.6-Anhydro- d -galaktose*", Chem. Ber. (1933) **66**, 525–532
99. Parker, R. E.; Isaacs, N. S., "*Mechanisms of epoxide reactions*", Chem. Rev. (1959) **59**, 737–799
100. Newth, F. H., "*Sugar epoxides*", Quart. Rev. (1959) **13**, 30–47
101. Mills, J. A., "*Stereochemistry of cyclic derivatives of carbohydrates*", Adv. Carbohydr. Chem.(1955) **10**, 1–53
102. Newth, F. H.; Richards, G. N.; Wiggins, L. F., "*The action of Grignard reagents on anhydro-sugars of ethylene oxide type. Part I. The behaviour of derivatives of α-methyl-2:3-anhydroalloside towards methylmagnesium iodide*", J. Chem. Soc. (1950) 2356–2364
103. Richards, G. N.; Wiggins, L. F., "*The action of Grignard reagents on anhydro-sugars of ethylene oxide type. Part II. The behaviour of 4: 6-benzylidene 2: 3-anhydro-α-methyl- d -alloside towards ethyl- and phenyl-magnesium halides*", J. Chem. Soc. (1953) 2442–2446
104. Richards, G. N.; Wiggins, L. F.; Wise, W. S., "*The reaction of magnesium halides with αβ-anhydro-sugars*", J. Chem. Soc. (1956) 496–500
105. Foster, A. B.; Overend, W. G.; Stacey, M.; Vaughn, G., "*Structure and reactivity of anhydrosugars. Part I. Branched-chain sugars. Part I. Action of diethylmagnesium on methyl 2: 3-anhydro 4: 6-O-benzylidene-α- d -mannoside*", J. Chem. Soc. (1953) 3308–3313
106. Austin, P. W.; Buchanan, J. G.; Oakes, E. M., "*Reaction of methyl 2,3-anhydro- d -ribofuranosides with nucleophiles*", Chem. Commun. (1965) 374–375
107. Kochetkov, N. K.; Kudryashov, L. I.; Klyagina, A. P., "*Monosaccharides. II. Reaction of methyl 2,3-anhydro-4,6-O-benzylidene-α- d -allopyranoside with sodium malonic ester*", Zhur. Obshch. Khim. (1962) **32**, 410–413
108. Dahlgard, M., "*Methyl 3,4-Anhydro-β- d -galactopyranoside. III. Reaction with Hydrogen Sulfide*", J. Org, Chem. (1965) **30**, 4352–4353
109. Dahlgard, M.; Chastain, B. H.; Han, R.-J. L., "*Methyl 3,4-Anhydro-β- d -galacto pyranoside. II. Reaction with Methanethiol*", J. Org. Chem. (1962) **27**, 932–934
110. Dahlgard, M.; Chastain, B. H.; Han, R.-J. L., "*Methyl 3,4-Anhydro-β- d -galactopyranoside. I. Reduction*", J. Org. Chem. (1962) **27**, 929–931
111. Müller, A., "*Über das Anhydro-β-methylhexosid aus Triacetyl-4-toluolsulfonyl-β-methylglucosid*", Berichte (1934) **67**, 421–424
112. Müller, A., "*Die Waldensche Umkehrung in der Zucker-Gruppe, I. Mitteil.: Die Aufspaltung des 3.4-Anhydro-β-methyl-hexosids*", Chem. Ber. (1935) **68**, 1094– 1097
113. Kent, P. W.; Ward, P. F. V., "*Synthesis of 4-deoxy- l -ribose from D-lyxose*", J. Chem. Soc. (1953) 416–418
114. Newth, F. H., "*O-toluene-p-sulphonyl derivatives of 1:6-anhydro-β- d -altrose and their behaviour towards alkali*", J. Chem. Soc. (1956) 441–447
115. Angyal, S. J.; Gilham, P. T., "*Cyclitols. Part VII. Anhydroinositols and the epoxide migration*", J. Chem. Soc. (1957) 3691–3699
116. Černý, M.; Pacàk, J.; Staněk, J., "*Syntheses with anhydro sugars. IV. Preparation of 1,6:2,3-dianhydro-β- d -mannopyranose and its isomerization to 1,6:3,4-dianhydro-β- d -altropyranose*", Coll. Czech. Chem. Commun. (1965) **30**, 1151–1157
117. Buben, I.; Černý, M.; Pacàk, J., "*Syntheses with anhydro sugars. III. Treatment of 1,6:3,4-dianhydro-2-O-(p-tolylsulfonyl)-β- d -galactopyranose with sodium hydroxide*", Coll. Czech. Chem. Commun. (1963) **28**, 1569–1578

References

118. Buchanan, J. G.; Clode, D. M., "*Synthesis and properties of 2,3-anhydro- d -mannose and 3,4-anhydro- d -altrose*", J. Chem. Soc. Perkin **1** (1974) 388–394
119. Jarý, J.; Čapek, K., "*Amino sugars. V. Preparation of methyl 3,4-anhydro-6-deoxy-α- d -galactopyranoside derivatives*", Coll. Czech. Chem. Commun. (1966) **31**, 315–320
120. Ataie, M.; Buchanan, J. G.; Edgar, A. R.; Kinsman, R. G.; Lyssikatou, M.; Mahon, M. F.; Welsh, P. M., "*3,4-Anhydro-1,2-O-isopropylidene-β- d -tagatopyranose and 4,5-anhydro-1,2-O-isopropylidene-β- d -fructopyranose*", Carbohydr. Res. (2000) **323**, 36–43
121. Paulsen, H.; Eberstein, K., "*Verzweigte Zucker, XI. Epoxidumlagerungen an verz-weigten Zuckern*", Chem. Ber. (1976) **109**, 3891–3906
122. Al Janabi, S. A. S.; Buchanan, J. G.; Edgar, A. R., "*Base-catalysed equilibration and conformational analysis of some methyl 2,3- and 3,4-anhydro-6-deoxy-β- d -hexopyranosides*", Carbohydr. Res. (1974) **35**, 151–164
123. Austin, P. W.; Buchanan, J. G.; Oakes, E. M., "*Reaction of methyl 2,3-anhydro- d -ribofuranosides with nucleophiles*", Chem. Commun. (1965) 374–375
124. Buchanan, J. G., "*Migration of Epoxide Rings and Stereoselctive Ring Opening of Acetoxyepoxides*" in Methods in Carbohydrates Chemistry Vol. 6, Academic Press, New York, (1972) pp. 135–141

Chapter 9
Amino Sugars

Amino sugars are monosaccharides in which one or more hydroxyl groups of a sugar chain is replaced by an amino group. The amino group(s) can be free or derivatized (it is usually acylated, but it can be alkylated, too).

The amino sugars are widespread in nature. They are building blocks of many complex saccharides such as heparin, hyaluronate, and keratan sulfate (*N*-acetylglucosamine *1*), chondroitin 4- and 6-sulfate, and dermatan sulfate (*N*-acetylgalactosamine *2*). They are found in glycoconjugates of both vertebrates and invertebrates, such as glycosaminoglycans (vertebrates) and peptidoglycans (bacteria) (*N*-acetylglucosamine *1*), glycoproteins (sialic acid *3*, structure determined by Gottschalk [1]) as well as in many natural products such as antibiotics: *streptomycin* (2-deoxy-2-methylamino-L-glucopyranoside, *4*), *erythromycin* (*desosamine* – 3,4,6-trideoxy-3-dimethylamino-D- *xylo*-hexopyranose *5*, structure determined by Bolton et al. [2, 3]), *nystatin* (*mycosamine 6*, structure determined by Walters et al. [4, 5] and the stereochemistry by von Saltza et al. [6, 7]), neomycins [*neosamine B* or *paromose* – 2,6-diamino-2,6-dideoxy-L-idose, *7* from neomycin B, structure determined by Haskell et al. [8]], *neosamine* **C** – 2,6-diamino-2,6-dideoxy-D-glucose, *8* from neomycin C, *muramic acid*, *9* [9, 10] (Fig. 9.1).

Due to their biological importance, synthesis of amino sugars has attracted a great attention of carbohydrate chemists over the years. There are several strategies employed in the synthesis of amino sugars, as for example, the ammonolysis of sugar epoxides (oxiranes), direct displacement of alkyl or arylsulfonates, or halides by nitrogen nucleophiles, such as ammonia, hydrazine, or azide (in two latter cases followed by reduction), and reduction of oximes obtained from aldosuloses.

Ammonolysis of Oxiranes

The opening of an epoxide ring always takes place in such a way that the amino and the alcohol group in the product are *trans* oriented (Fürst–Plattner rule [11]) (Fig. 9.2).

The proportion of each isomer (*11* or *12*) will depend on the structure and preferred conformation of parent sugar epoxide (i.e., on steric interactions between

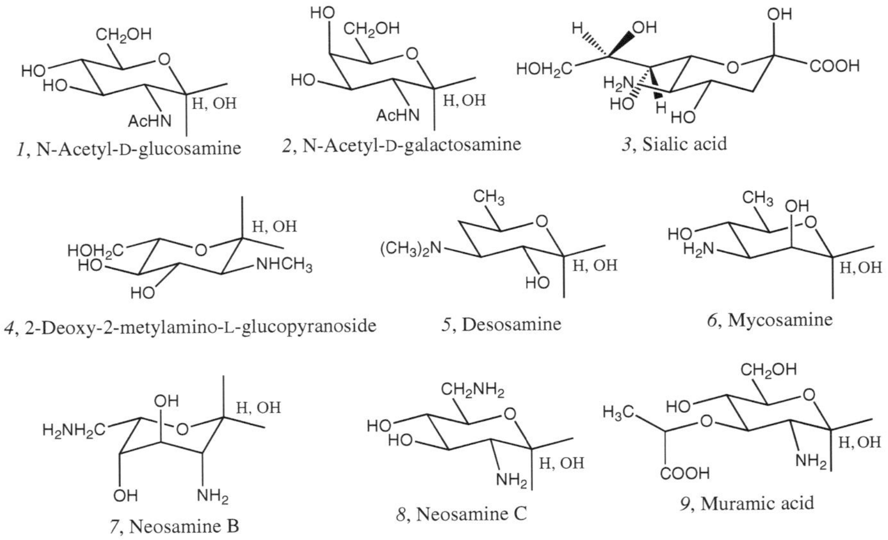

1, N-Acetyl-D-glucosamine *2*, N-Acetyl-D-galactosamine *3*, Sialic acid

4, 2-Deoxy-2-metylamino-L-glucopyranoside *5*, Desosamine *6*, Mycosamine

7, Neosamine B *8*, Neosamine C *9*, Muramic acid

Fig. 9.1

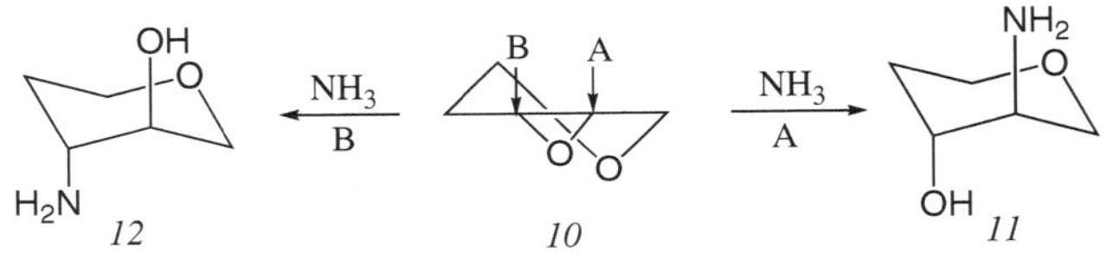

12 *10* *11*

Fig. 9.2

the incoming ammonia and various substituents on a pyranoside ring) [12–15]. Thus, for example, the ammonolysis of methyl 2,3-anhydro-4,6-di-*O*-methyl-β-D-mannopyranoside *13* gives 90% of methyl 3-amino-3-deoxy-β-D-altropyranoside *15* and 10% of methyl 2-amino-2-deoxy-β-D-glucopyranoside *14* [16] (Fig. 9.3).

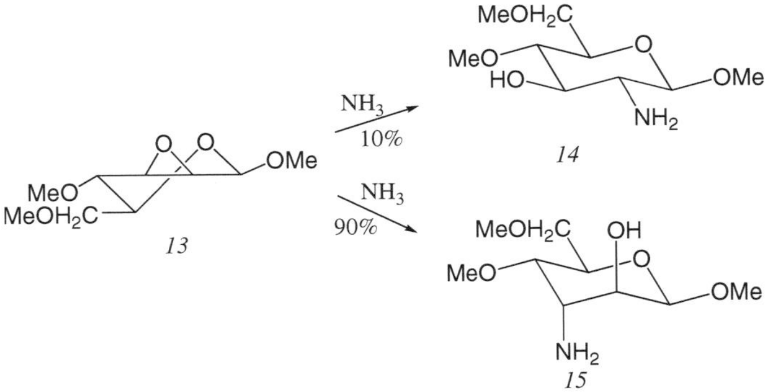

13 *14* *15*

Fig. 9.3

The ammonolysis of methyl 2,3-anhydro-4,6-*O*-benzylidene-α-D-talopyranoside *16* gives methyl 3-amino-4,6-*O*-benzylidene-3-deoxy-α-D-idopyranoside *17* [17–19] as the predominant product (Fig. 9.4). The reason for this is that the epoxides undergo exclusively *trans*-diaxial opening. When chair–chair inversion is not possible, there is a chair–twist boat inversion which then allows a *trans*-diaxial opening leading to a chair which then undergoes a final chair interconversion to the more stable chair. This results then in *trans*-diequatorial products [13, 14].

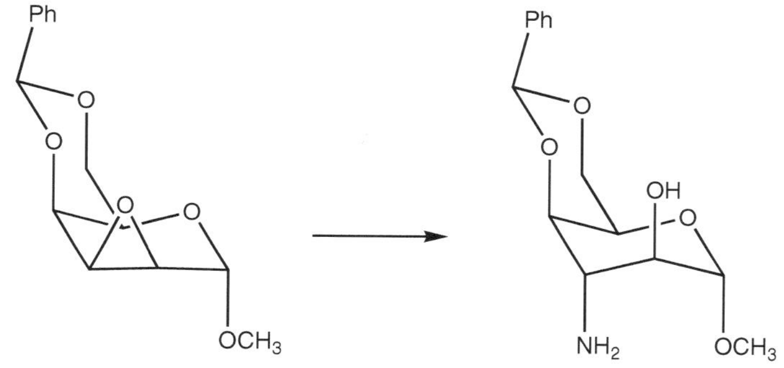

Fig. 9.4

Ammonolysis of methyl 2,3-anhydro-4,6-*O*-benzylidene-α-D-gulopyranoside *18* gives methyl 2-amino-4,6-*O*-benzylidene-2-deoxy-α-D-idopyranoside *19* as the major product [18] (Fig. 9.5). Methylamine also attacks the C2 carbon giving the corresponding methylamino derivative.

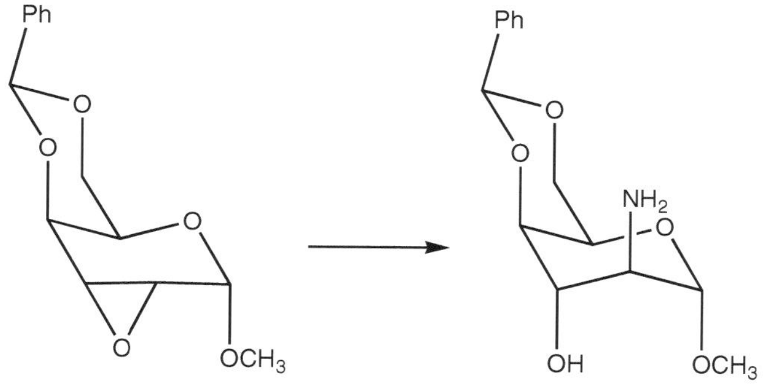

Fig. 9.5

Ammonolysis of methyl 3,4-anhydro-6-deoxy-α-L-talopyranoside *20* gave exclusively methyl 3-amino-3-deoxy-α-L-idopyranoside *21* [20] (Fig. 9.6). Similar preference for the attack of a nucleophile to C3 carbon in methyl 3,4-anhydro-6-deoxy-α-L-talopyranoside *20* was previously reported by Charalambous and

Percival [21] during their study on the opening of the 3,4-epoxide ring of methyl 3-deoxy-α-L-talopyranoside *20* with sodium methoxide (diaxial opening). They isolated two products, methyl 6-deoxy-3-*O*-methyl-L-idose *23* and methyl 6-deoxy-4-

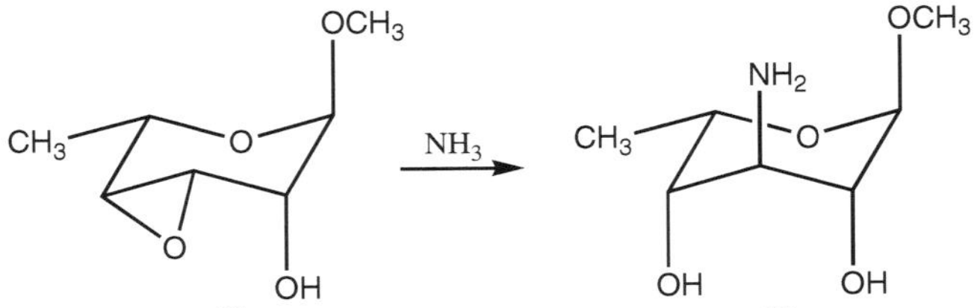

Fig. 9.6

O-methyl-L-mannose *24* in 2:1 ratio (Fig. 9.7). However, when they treated methyl 6-deoxy-3,4-anhydro-2-*O*-methyl-α-L-talopyranoside *22* with sodium methoxide under the same experimental conditions they obtained exclusively 6-deoxy-2,4-di-*O*-methyl-L-mannose *25*. The attack of methoxide anion took place exclusively at the C4 carbon (Fig. 9.7).

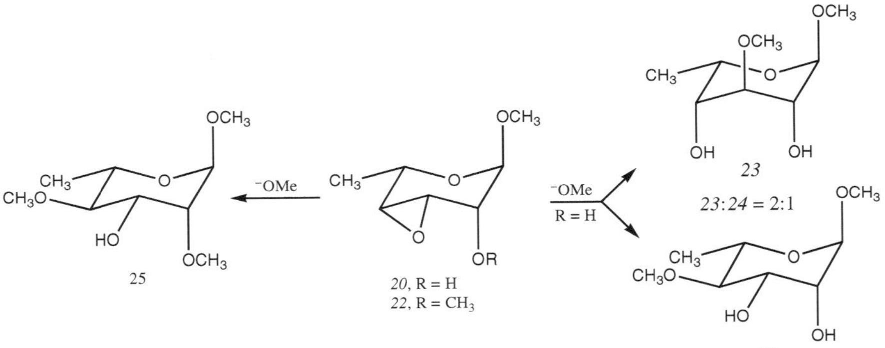

Fig. 9.7

Stevens et al. [22, 23] used these observations to prove the structure of perosamine (4-amino-4,6-dideoxy-D-mannose) and for its synthesis. Since, in the presence of a substituent at the O2 (methyl group), the attack of a nucleophile to 3,4-epoxide of *27* (Fig. 9.8) is directed away from the C3 carbon and to the C4 carbon it is reasonable to assume that the interference of the O2 substituent and the incoming nucleophile is steric in nature. If so then increase in the bulkiness of the O2 substituent should increase the overall regioselectivity. Thus instead of a methyl group as the O2 substituent Stevens et al. opted for the benzoyl group since benzoyl group is much larger than methyl group. However, since benzoate is easily removed by alkali, such as ammonia, the azide was selected as the nitrogen nucleophile.

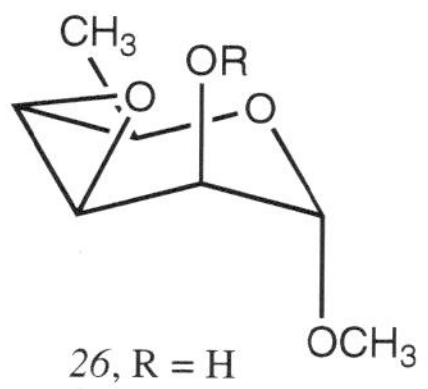

Ammonolysis of Oxiranes 225

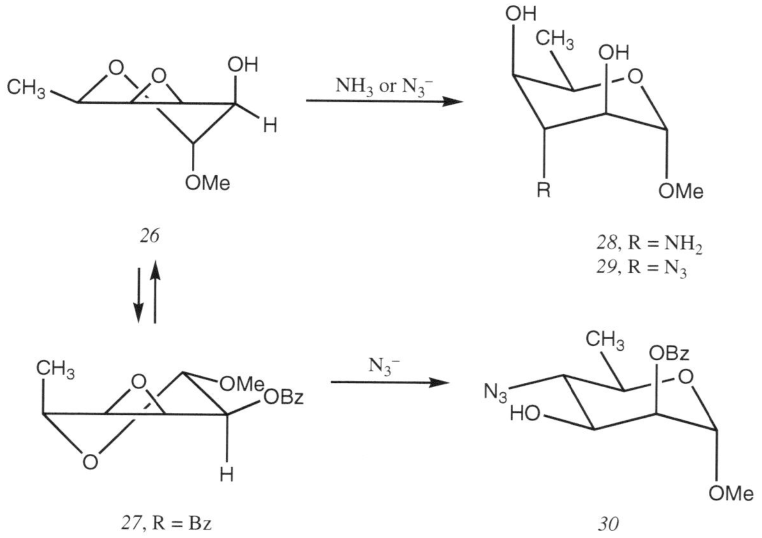

Fig. 9.8

26, R = H
27, R = Bz

Fig. 9.9

26

27, R = Bz

28, R = NH$_2$
29, R = N$_3$

30

In full agreement with Percival et al. [21] findings, the attack of azide nucleophile to the 3,4-anhydro ring of 26 (Fig. 9.8) took place predominantly at the C3 carbon (C3:C4 attack ratio was 3:1) whereas the azide nucleophile attacked predominantly the C4 carbon of methyl 3,4-anhydro-2-O-benzoyl-6-deoxy-β-D-talopyranoside 27 (Fig. 9.9) giving methyl 4-azido-4,6-dideoxy-α-D-mannopyranoside 30 [22, 23] (C3:C4 attack ratio was 1:4). The lack of higher regioselectivity was explained by relative conformational flexibility of 26 and 27 and by the steric interaction in transition state (Fig. 9.9).

In case of terminal epoxides having in their epoxide ring structures a primary and a secondary carbon atom a nitrogen nucleophile will attack exclusively the primary carbon of this terminal epoxide.

For example, reaction of 31 with liquid ammonia gives the 6-amino-6-deoxy-1,2-O-isopropylidene-3-O-methanesulfonyl-β-L-idofuranose 32 [24] (Fig. 9.10).

Sterically shielded 3-*O*-methanesulfonyl group does not react with ammonia under the reaction conditions.

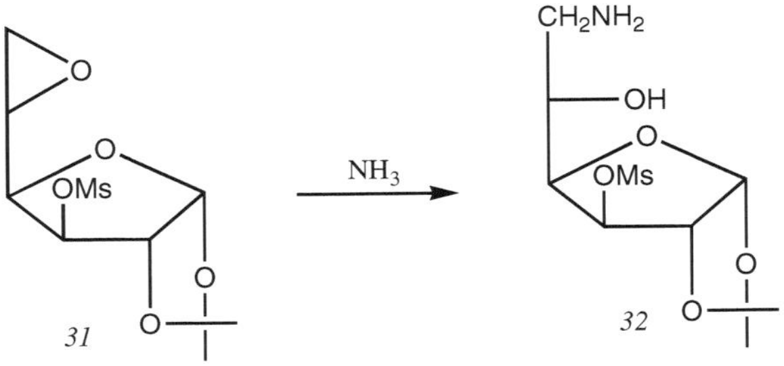

Fig. 9.10

Nucleophilic Displacement of Sulfonates (or Halides) with Nitrogen Nucleophiles

Halides (or sulfonic esters) at the primary carbon of a monosaccharide (the C6 in hexopyranoses and hexofuranoses or the C5 in pentofuranoses) can be directly displaced with ammonia to give 6-amino-6-deoxy or 5-amino-5-deoxy sugar. For example, methyl 2,3,4-tri-*O*-acetyl-6-bromo-6-deoxy-D-glucopyranoside *33* reacts with ammonia to give methyl 6-amino-6-deoxy-D-glucopyranoside *34* [25] (Fig. 9.11).

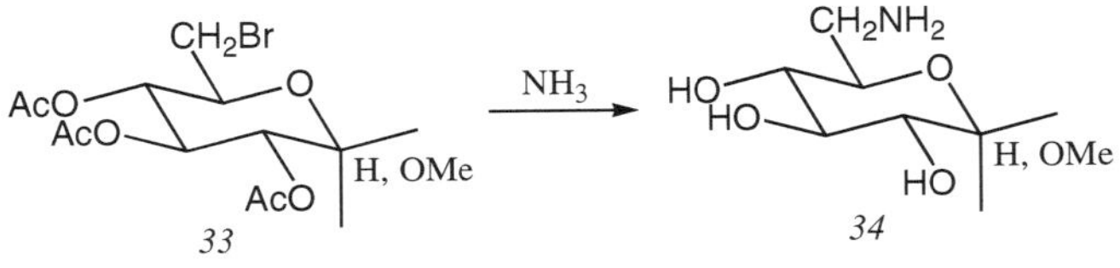

Fig. 9.11

Ammonolysis of 1,2-*O*-isopropylidene-5-*O*-*p*-toluenesulfonyl-α-D-xylofuranose *35* gave 5-amino-5-deoxy-α-D-xylofuranose *36* [26, 27] (Fig. 9.12).

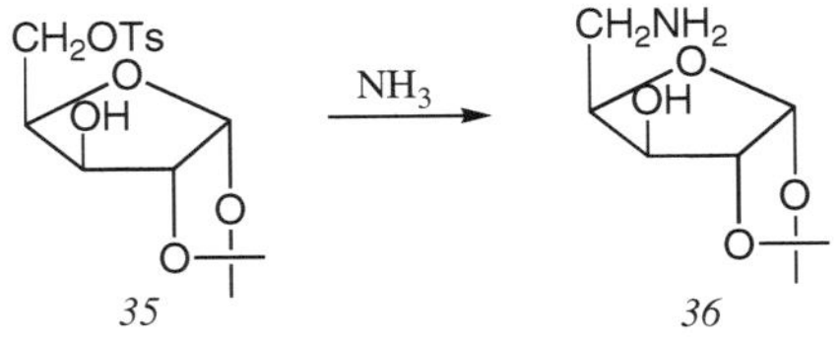

Fig. 9.12

The hydrazinolysis of *35* followed by reduction of the 5-hydrazo product also gave *36* [28].

The nucleophilic displacement of primary *p*-toluenesulfonates with azide, followed by reduction (or hydrogenation) of the obtained azido derivative to the amino group is probably the best method for obtaining the primary amino sugars [29, 30].

The direct nucleophilic displacement of sulfonic esters of secondary alcohols by ammonia is much more difficult than that of primary sulfonates. The ammonolysis of secondary sulfonates requires higher temperatures resulting, due to the harsher reaction conditions, in increased formation of side products and decreased yield of desired product. Much better results are obtained by displacing the secondary sulfonate with hydrazine, or even better with azide as nitrogen nucleophile. The resulting hydrazino or azido derivatives must then be reduced to the corresponding amino sugars.

In the absence of neighboring group participation, the nucleophilic displacement of secondary sulfonates is highly stereoselective and always takes place with the inversion of configuration at the reacting carbon. For example, 1,2:5,6-di-*O*-isopropylidene-3-*O*-*p*-toluenesulfonyl-α-D-glucofuranose *37* reacts with ammonia [31, 32] to give 3-amino-3-deoxy-1,2:5,6-di-*O*-isopropylidene-α-D-allofuranose *38* or with hydrazine to give 3-deoxy-3-hydrazino-1,2:5,6-di-*O*-isopropylidene-α-D-allofuranose *39* [32, 33] (Fig. 9.13).

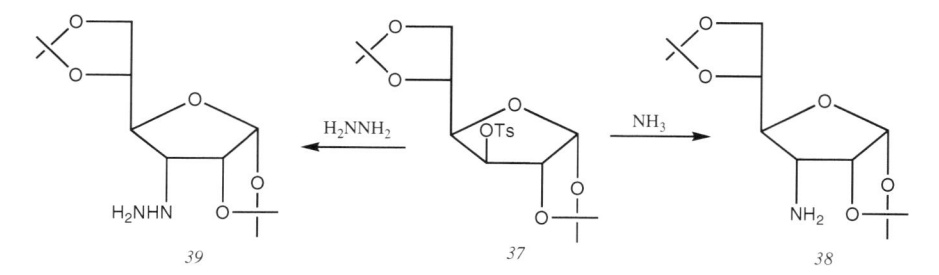

Fig. 9.13

It is interesting to note that 3-*p*-toluenesulfonate of *37* is highly resistant to direct displacement with charged nucleophiles such as acetate or benzoate; displacement with azide, however, does takes place, but only at high temperatures and in *N,N*-dimethylformamide [34] or hexamethylphosphoric triamide [35] which solvents are known to solvate well cations but not anions, thus not increasing the bulkiness of a negatively charged nucleophile via solvation.

Nucleophilic displacement of alkyl or arylsulfonates by azide is generally the preferred method for introducing the amino group (after reduction of azide), because azide is a better nucleophile than ammonia or hydrazine and it is less basic resulting in fewer side reactions. For example, methyl 2,3,6-tri-*O*-benzoyl-4-*O*-methanesulfonyl-α-D-galactopyranoside *41* was converted to methyl 4-amino-2,3,6-tri-*O*-benzoyl-4-deoxy-α-D-glucopyranoside *42* [36] (Fig. 9.14).

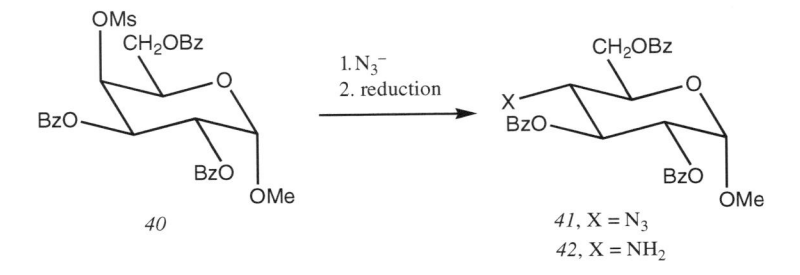

Fig. 9.14

Interestingly, methyl 2,6-di-*O*-benzoyl-3,4-di-*O*-methanesulfonyl-α, β-D-galactopyranoside *43* did not react with sodium azide after extended heating in hexamethylphosphoric triamide [37], whereas methyl 2,3,4,6-tetra-*O*-methanesulfonyl-α-D-glucopyranoside *44* gave methyl 4,6-diazido-4,6-dideoxy-2,3-di-*O*-methanesulfonyl-α-D-galactopyranoside *45* under less forcing conditions [38] (Fig. 9.15).

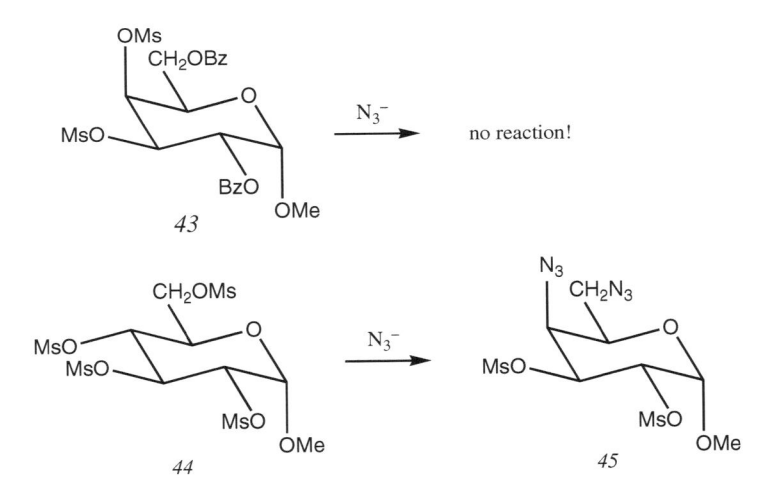

Fig. 9.15

Reaction of 6-deoxy-2,3-*O*-isopropylidene-4-*O*-methanesulfonyl-α-D-mannopyranoside *46* (Fig. 9.16) proceeds by contraction of the pyranoside to furanoside ring via the participation of the ring oxygen, giving methyl 5-azido-5,6-dideoxy-2,3-*O*-isopropylidene-α-L-talofuranoside *47* as the only product [39].

Hydrazinolysis of *46* proceeds, however, primarily with the direct displacement [40] of the C4-sulfonate, giving, after hydrogenation, methyl 4-amino-4,6-dideoxy-2,3-*O*-isopropylidene-α-L-talopyranoside *48*, together with small amounts of rearranged products *47*.

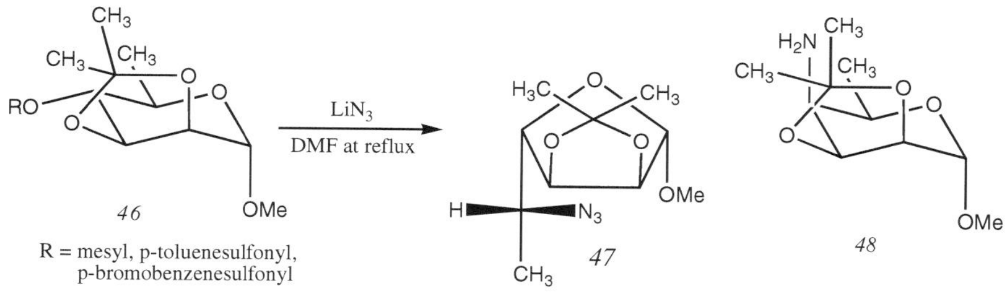

Fig. 9.16

2-Deoxy-2-amino sugars are biologically probably the most important amino sugars because they are the building blocks of many glycoconjugates (glycoproteins, glycolipids, glycosaminoglycans, peptidoglycans, etc.). Thus, for example, glycoproteins are proteins that are glycosylated at either L-serine or L-threonine (the O-linked oligosaccharides) (Fig. 9.17) or at the amido nitrogen of L-asparagine residue of a protein backbone (the N-linked oligosaccharides) (Fig. 9.18).

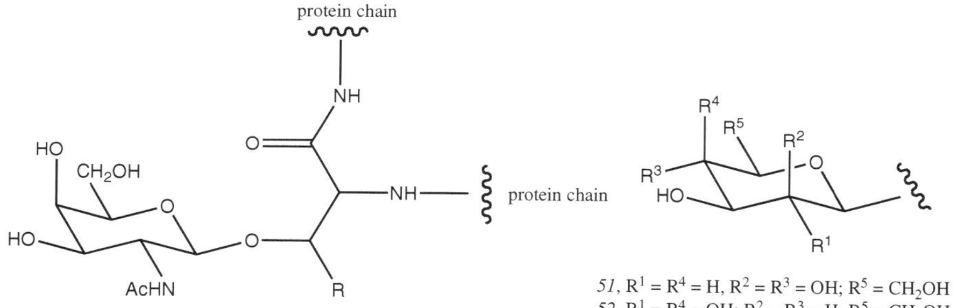

51, $R^1 = R^4 = H$, $R^2 = R^3 = OH$; $R^5 = CH_2OH$
52, $R^1 = R^4 = OH$; $R^2 = R^3 = H$; $R^5 = CH_2OH$
53, $R^1 = R^3 = OH$; $R^2 = R^4 = R^5 = H$

49, R = H, Serine
50, R = CH$_3$, Threonine

Fig. 9.17

The first monosaccharide that is glycosidically linked to an amino acid of a protein is, in both cases, a 2-deoxy-2-amino sugar: in glycoproteins having the O-linked oligosaccharide is the *N*-acetylgalactosamine; however, D-mannose *51*, D-galactose *52*, and D-xylose *53* residues are also found as the first carbohydrate glycosidically linked to the hydroxyl group of L-serine or L-threonine (Fig. 9.17). The first carbohydrate residue linked to L-asparagine (found in N-linked oligosaccharides) is always *N*-acetylglucosamine (Fig. 9.18).

Glycosyl phosphatidylinositol (GPI)-linked proteins are not in a true sense of the word glycoproteins because the role of a carbohydrate moiety is not to modify the protein's physico-chemical properties and thus its biological activity, but to anchor the protein via phosphatidylinositol to the cell membrane. Typically, phosphatidylinositol is glycosylated with *N*-acetylglucosamine which is then deacetylated. To the obtained glucosamine are then added three mannose residues. In the next step,

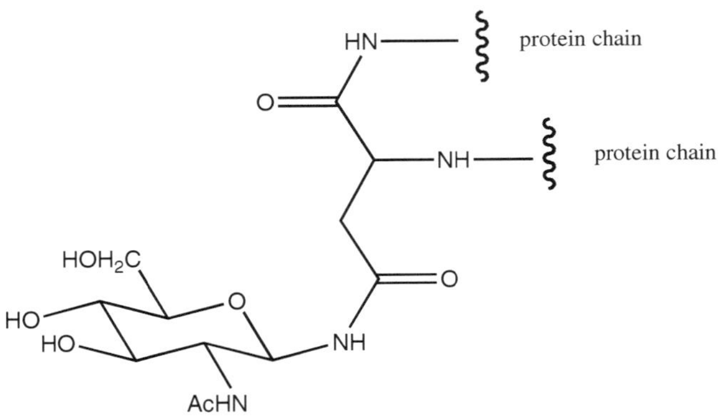

Fig. 9.18

ethanolamine is added to the terminal mannose residue via a phosphate diester bond and to the amino group of ethanolamine a protein is linked with its carboxy terminal via an amide bond 55 (Fig. 9.19).

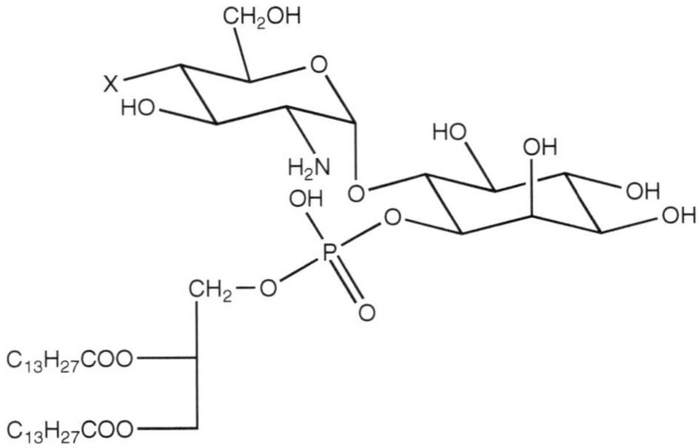

55, X = three mannose residues + phosphate + ethanol amine + protein

Fig. 9.19

This is perhaps the unique example of an amino sugar existing in nature with its amino group underivatized. The possible reason could be that the proximity of the phosphate proton to the glycosidic oxygen of D-glucosamine residue could represent a threat to the survival of the glycosidic bond of the amino sugar and phosphatidylinositol if the amino group is acetylated. Removal of the acetate from the *N*-acetamido group makes the glycosidic bond very stable and resistant to acid-catalyzed hydrolysis (vide supra). It is well established that essentially all eukaryotes add glycolipid anchor to certain proteins in order to express these proteins at the cell surface and to regulate their presence in that location. GPI-anchored proteins are also present in trypanosomes, such as *Plasmodium falciparum* (malaria parasite).

Glycosylamines and N-Glycosides

Glycosylamines are generally defined as 1-amino-1-deoxy derivatives of monosaccharides *56* or *N*-alkylamino-1-deoxy or *N*-arylamino-1-deoxy derivatives of monosaccharides *57*, as shown in Fig. 9.20.

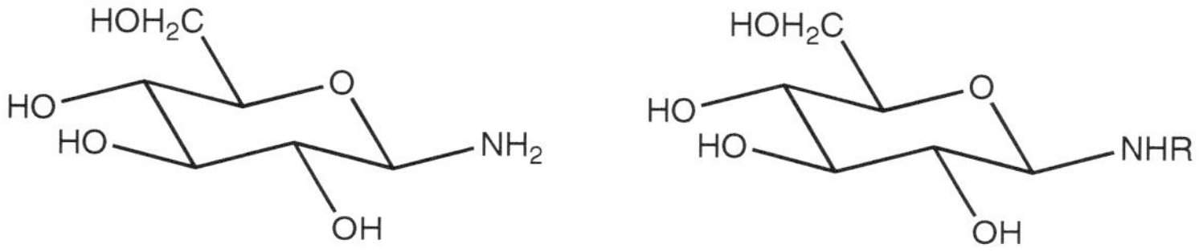

Fig. 9.20

The 1-amino sugars are discussed separately from the other amino sugars not only because they are chemically very different but also because the understanding of their chemistry has a direct bearing on the understanding of the chemical behavior of ribonucleic and deoxyribonucleic acids (RNA and DNA).

Cyclic forms of sugars with a free reducing group react with ammonia, alkyl-, and arylamines, esters of amino acids, and urea derivatives giving the corresponding glycosylamine derivatives. Amides normally do not react with reducing sugars because the amido nitrogen is insufficiently nucleophilic. However, aldoses do undergo acid-catalyzed condensation with urea and thiourea [41–46]. The great stability of *N*-glycosylureas and glycosylguanidines toward acid-catalyzed hydrolysis can be attributed to the involvement of the electron pair of the nitrogen atom at the C1 in the delocalization within the amide or guanidine group making this nitrogen unavailable for protonation. This may be the reason why in N-linked glycoproteins the oligosaccharide chain is attached to a protein via *N*-acetylglucosamine glycosidically linked to the amido nitrogen of asparagine residue of protein.

Glycosylamines are obtained by treating aldohexoses [47–50] or aldopentoses [50, 51] with cold concentrated solution of ammonia in alcohol. The product usually crystallizes when solution is kept for a prolonged time.

N-Alkylaldosamines are easily prepared by reacting aldoses with primary or secondary aliphatic amines [52–59].

Several *N*-alkyl D-glucosylamines were prepared by heating D-glucose with the corresponding amine in the presence of small amount of 0.5 N HCl [52].

N-Arylaldosamines have been known for a long time. Sorokin [60–63] synthesized *N*-phenyl D-glucosamine, D-galactosamine, and D-fructosylamine by heating aniline with free sugars at 130–135°C, or by heating the amine and the reducing sugar in boiling methanol or ethanol, containing up to 10% of water. Also small amounts of acids have been used to catalyze the reaction. N-Phenylglucosamine and N-phenylgalactosamine were in this way obtained in good yields (Fig. 9.21).

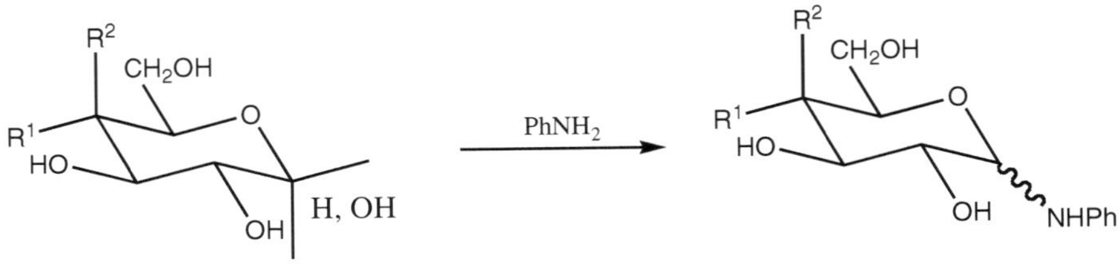

58, R^1 = OH; R^2 = H
59, R^1 = H; R^2 = OH

60, R^1 = OH; R^2 = H
61, R^1 = H; R^2 = OH

Fig. 9.21

A simple general method, suitable for large-scale preparations, was developed by Weygand [64]. It consists of dissolving the sugar in a minimum amount of hot water and heating the solution with an arylamine for a few minutes longer than required for achieving the complete miscibility; the product crystallizes on adding alcohol and cooling.

The preparation of N-alkylglycosamines is often complicated by subsequent and rapid *Amadori rearrangement* [65], whereby an N-substituted glycosylamine isomerizes to an N-substituted 1-amino-1-deoxy-ketose [66] (Fig. 9.22).

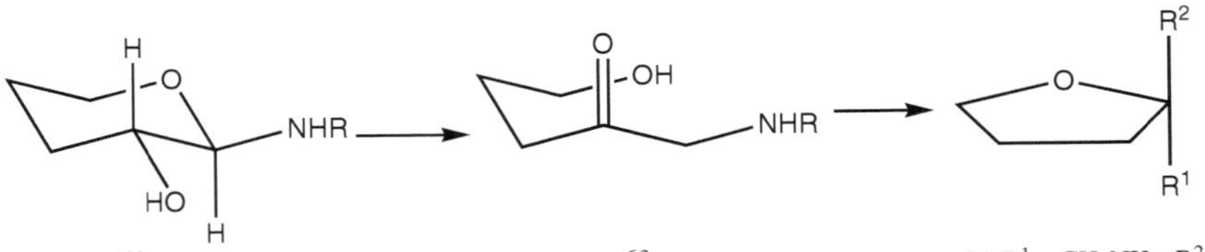

62

63

64, R^1 = CH_2NH_2; R^2 = OH
65, R^1 = OH; R^2 = CH_2NH_2

Fig. 9.22

The two mechanisms depicted in Fig. 9.23 were postulated by Kuhn et al. [67, 68] and Weygand [69] (Mechanism A) and Isbell and Frush [50] (Mechanism B).

These two reaction mechanisms proposed for this rearrangement (Fig. 9.23) differ in the site of protonation in the initial step. According to Weygand [69] (Mechanism A) the initial protonation of N-glycosylamine 66 takes place at the nitrogen atom 67. Proton is then transferred to the ring oxygen causing the opening of the pyranoside ring with the simultaneous formation of immonium transition state intermediate 69 ⇌ 70 which collapses to enolamine 71. Tautomerization to acyclic 1-substituted amino ketose 72 followed by cyclization gives 1-N-substituted ketosylamine 73 (Fig. 9.23).

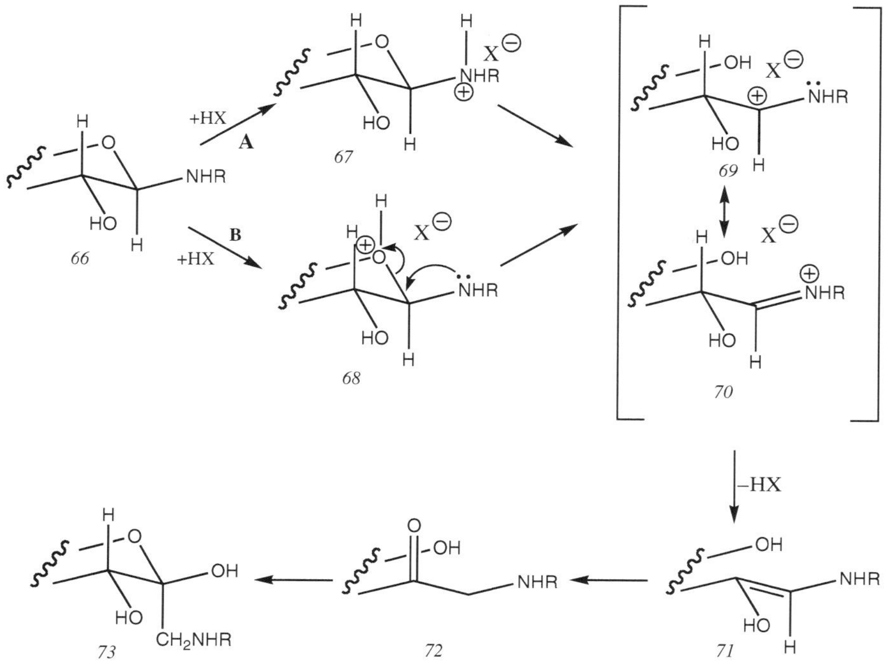

Glycosylamines and N-Glycosides

Fig. 9.23

Isbell and Frush [50] developed the currently accepted view [70] (Mechanism B) that postulates that the initial protonation takes place at the ring oxygen giving 68 which is probably more likely because of the analogy with the mechanism of mutarotation and hydrolysis of glycosylamines. The positively charged C1 carbon of immonium intermediate 69 withdraws the electrons from the C2 carbon weakening thus the C2–H bond so that it can be easily removed from the C2 by a base and gives "enolamine" 71. This intermediate rearranges first to the amino-ketone 72 and then to the corresponding cyclic form 73.

Ketosylamines undergo a rearrangement to 2-amino-2-deoxyaldoses. This rearrangement, termed *Heyns rearrangement*, can be in a way regarded as reverse Amadori rearrangement, because the 1-N-substituted ketose is converted into a 2-amino aldose. Since new asymmetric C2 carbon is produced, two epimeric rearrangement products are obtained (Fig. 9.24). Thus, for example, N-substituted D-fructosylamine 74, obtained by reaction of D-fructose in liquid ammonia or in methanolic ammonia, rearranges stereoselectively to 2-amino-2-deoxy D-glucose by acid catalysis [71–73]. D-Fructose undergoes spontaneous rearrangement to 2-alkylamino-2-deoxy-D-glucose derivatives when reacted with cyclohexyl- and isopropyl amines [74], as well as with butylamine [75] in the cold.

The mechanism of Heyns rearrangement is similar to the mechanism of Amadori rearrangement, as shown in Fig. 9.25. The protonation of the ring oxygen of 77,

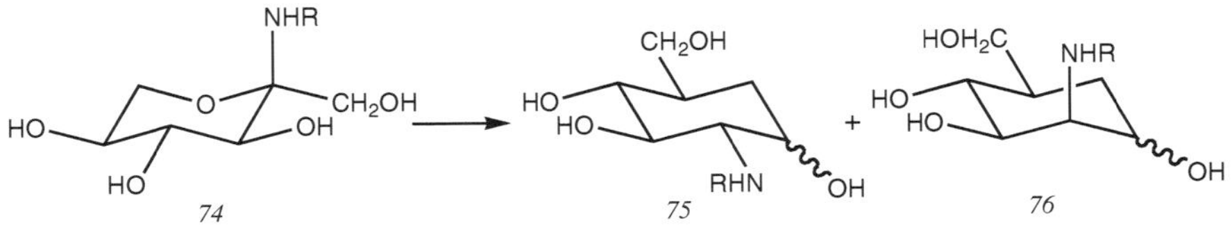

Fig. 9.24

followed by ring opening, gives immonium transition state intermediate (78 ↔ 79) that is by deprotonation of the hydroxymethyl group of 79 converted to "enolamine" 80. The isomerization of "enolamine" 80 to 2-amino-aldehyde 81 followed by cyclization gives the 2-amino-2-deoxy aldopyranose 82 (Fig. 9.25).

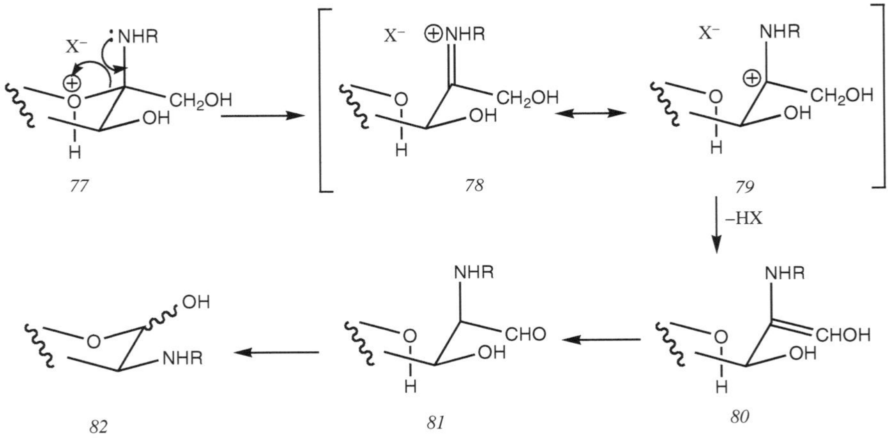

Fig. 9.25

Glycosylamines undergo simple mutarotation in solution. A kinetic study of the mutarotation of N-*p*-tolylglycosylamines in aqueous NaOH at pH 12.99 showed that the reaction is first order [76–78]. Isbell and Frush [50] proposed a mechanism for mutarotation assuming that the initial protonation takes place at the ring oxygen 83 that is followed by the ring cleavage and formation of open-chain immonium intermediate 87. Rotation about the C1–C2 bond and the re-closure of the pyranoid ring results in the formation of the other anomer 88 (Fig. 9.26).

In aqueous solutions, the acid-catalyzed hydrolysis of all glycosylamines is preceded by a rapid anomerization to a mixture of about 10% α- and 90% β-form [76]. Although the anomerization must involve an acyclic intermediate, this form cannot be present in appreciable concentrations since the reaction shows good first-order kinetics and the rate constants are equal, regardless of whether starting from α- or the β-form. Thus when studying the acid-catalyzed hydrolysis of glycosamines it is always a rapidly interconverting mixture of isomers that is being studied [76].

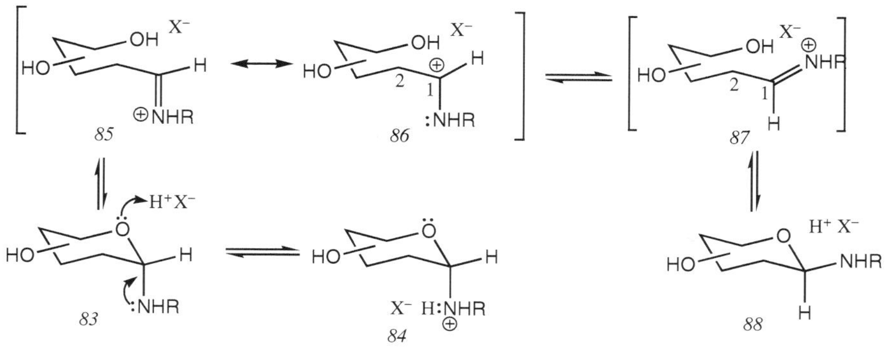

Glycosylamines and N-Glycosides

Fig. 9.26

The hydrolysis could be envisaged to proceed either via the formation of cyclic oxocarbenium ion *91*, as is the case in the acid-catalyzed hydrolysis of *O*-glycosides, or via the formation of acyclic immonium ion *85* (protonated Schiff base) (Fig. 9.27).

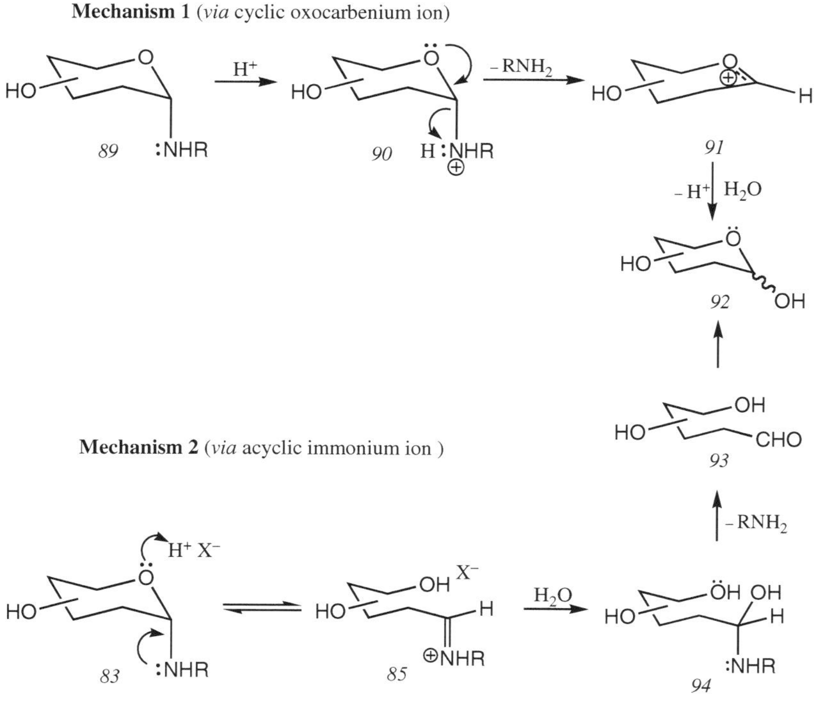

Mechanism 1 (*via* cyclic oxocarbenium ion)

Mechanism 2 (*via* acyclic immonium ion)

Fig. 9.27

The hydrolysis of the more easily hydrolyzed glycosylamines (for example, N-*p*-tolyl) was studied in buffers and was found to undergo general acid catalysis, excluding thus the Mechanism 1. However, it is consistent with Mechanism 2, since general acid catalysis mechanism in the hydrolysis of Schiff bases is well established [79, 80] and has also been shown to involve kinetically equivalent specific acid/general base catalysis [81]. The *N*-arylglycosylamines are far more resistant to hydrolysis than *N*-alkylglycosylamines, because the lone pair of electrons on nitrogen is more available to form the immonium cation in alkylamines than in arylamines where it is overlapping with the aromatic π-electrons of benzene ring.

A very important group of *N*-glycosides are nucleosides (or nucleotides) that are the building blocks of ribo- and deoxyribonucleic acids (DNA and RNA, respectively). These *N*-glycosides are chemically different from *N*-alkyl and *N*-arylglycosylamines because the glycosidic nitrogen as a part of a ring system is conjugated with a network of double bonds in pyrimidine or purine base. This electronic orbital overlap makes the *N*-glycosidic nitrogen atom of nucleosides less susceptible to protonation and thus makes nucleosides more stable to acid-catalyzed hydrolysis than *N*-alkyl and *N*-arylglycosylamines. In Figs. 9.29 and 9.31 the structures of most important nucleosides are given. As can be seen there are two classes

Pyrimidine bases

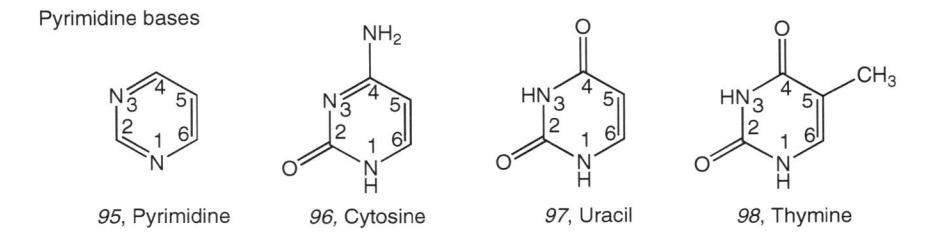

95, Pyrimidine   96, Cytosine   97, Uracil   98, Thymine

Fig. 9.28

Pyrimidine nucleosides

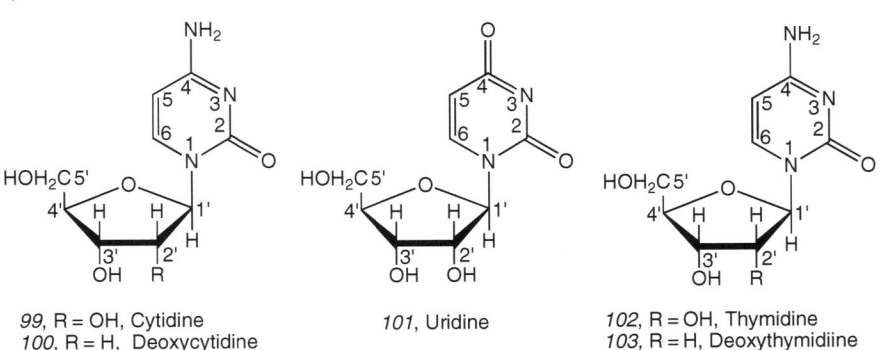

99, R = OH, Cytidine
100, R = H, Deoxycytidine

101, Uridine

102, R = OH, Thymidine
103, R = H, Deoxythymidiine

Fig. 9.29

of nucleosides: one having a pyrimidine derivative as the base (Figs. 9.28 and 9.29) and the other group of nucleosides having a purine derivative as the base (Figs. 9.30 and 9.31). In pyrimidine nucleosides the N1 nitrogen is glycosidically linked via β glycosidic bond to ribose or to 2-deoxy ribose molecule.

In purine nucleosides the N9 nitrogen is linked via a β-glycosidic bond to D-ribofuranose (in ribonucleic acids, RNA) or to 2-deoxy-D-ribofuranose (in deoxyribonucleic acids, DNA).

Pyrine bases

104, Purine   *105*, Adenine   *106*. Guanine

Fig. 9.30

Purine nucleosides

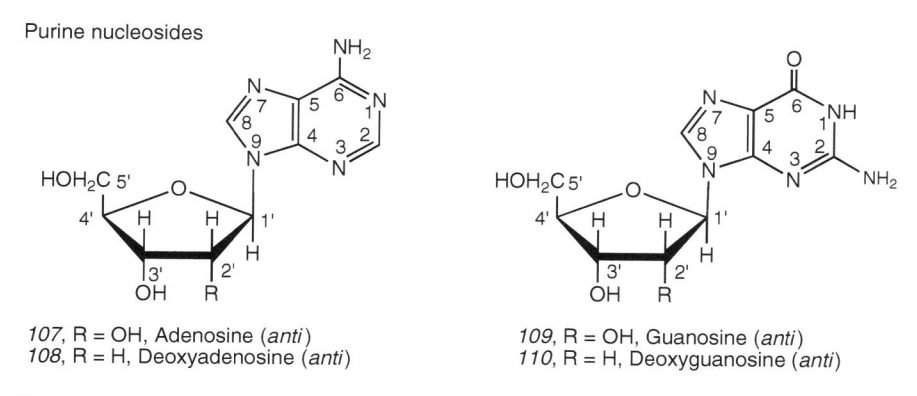

107, R = OH, Adenosine (*anti*)
108, R = H, Deoxyadenosine (*anti*)

109, R = OH, Guanosine (*anti*)
110, R = H, Deoxyguanosine (*anti*)

Fig. 9.31

Acid-Catalyzed Hydrolysis of Purine and Pyrimidine Nucleosides

Acid-catalyzed hydrolysis of purine and pyrimidine nucleosides has been extensively studied [82–96].

It was initially believed that the acid-catalyzed hydrolysis of nucleosides (and hence nucleotides) proceeds by a similar mechanism as the hydrolysis of alkyl- and arylglycosylamines, i.e., via an open-chain protonated Schiff base intermediate [82]. However, while in case of glycosylamines the sugar ring cleavage is aided by resonance stabilization of the open-chain form by the amine nitrogen forming an imine, such stabilization is less likely with nucleosides because the aromatic nitrogens at N1 (pyrimidines) and N9 (purines) cannot donate their lone pair electrons due to conjugation with the aromatic ring π electrons. Consequently, with the overwhelming support of experimental data the consensus regarding the mechanism of acid-catalyzed hydrolysis of nucleosides has shifted toward direct C1'-N fission with the formation of oxocarbenium transition state.

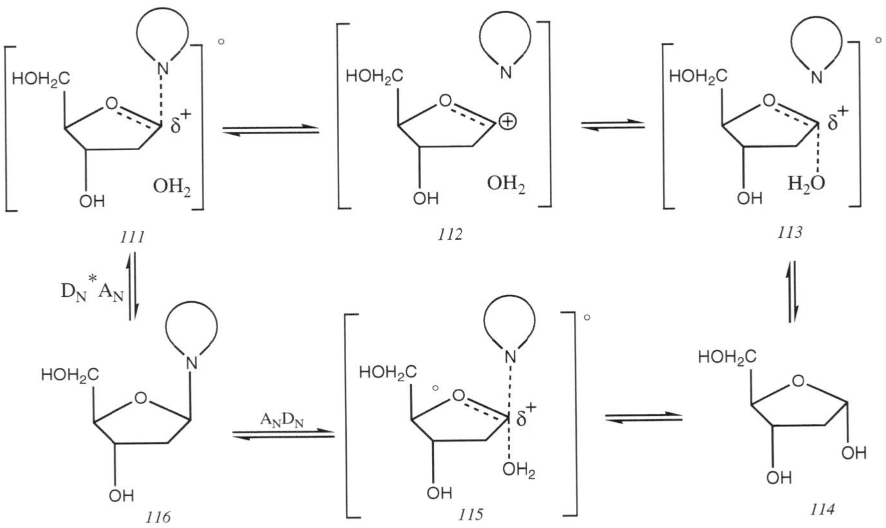

Fig. 9.32

Glycoside hydrolyses tread the borderline between the concerted and stepwise reactions and almost invariably proceed through either stepwise $D_N^*A_N$ (S_N1) reactions or highly dissociative $A_N D_N$ (S_N2) mechanisms [82, 97–99] (Fig. 9.32). Thus, these reactions form either a discrete oxocarbenium ion intermediate or an oxocarbenium ion-like transition state (highly dissociative $A_N D_N$ reactions are occasionally referred to as being "S_N1-like," which is incorrect). If both the nucleophile and the leaving group are in the reaction coordinate at the transition state, then the mechanism is $A_N D_N$ (S_N2), no matter how dissociative is the transition state. Therefore, highly dissociative $A_N D_N$ transition states should be perhaps referred to as "oxocarbenium ion-like" rather than "S_N1-like" where leaving group departure is far advanced over nucleophile approach, and positive charge accumulates on the sugar ring.

The ribofuranosyl oxocarbenium ion is expected to be more stable than glucopyranosyl oxocarbenium ion, based on higher reactivity of O-ribosides [82] and therefore more prone to proceed through $D_N^*A_N$ (S_N1) mechanism. However, acid-catalyzed hydrolysis of nucleosides can proceed by either of these two mechanisms: (AMP) [100] and (NAD$^+$) [101] proceed via $A_N D_N$ mechanism, though hydrolysis of dAMP appears to be stepwise $D_N^*A_N$ [102].

Acid-Catalyzed Hydrolysis of Purine and Pyrimidine Nucleosides 239

The stability of the glycosyl linkage toward acid-catalyzed hydrolysis varies widely. It is dependent on both the nature of the base (and its substituents) and the nature of the sugar (and its substituents). 2-Deoxyribonucleosides are more labile than ribonucleosides; purine nucleosides are more reactive than pyrimidine substrates [103].

The stability of 2-deoxyribonucleosides toward acid-catalyzed hydrolysis decreases in the order uracil (or thymine) > cytosine > adenine > guanine nucleoside. Acylation of the heterocyclic base can have a marked effect on the rate of acid-catalyzed cleavage of the glycosyl linkage [104, 105], notwithstanding implications to the contrary [106]. Esterification of sugar hydroxyl groups affects the stability of the glycosyl bond as shown in case of $2'O$-acetyldeoxyguanosines [107], thymidine phosphates [108, 109], and $2'$-O-p-toluenesulfonyl adenosine [110], and by the remarkable stability of $2',3',5'$-tri-O-(3,5-dinitrobenzoyl)-uridine [110].

DNA is <10-fold more stable to depurination than nucleosides, and the native DNA was 4-fold more stable than denatured DNA [111].

Depyrimidation is <20-fold slower than depurination in DNA [112]. It is also slower in DNA than in nucleosides and slower in native than denatured DNA by a factor of 2–6 [112].

Nonenzymatic mechanisms of purine and pyrimidine hydrolysis have been studied extensively [82–96].

Hyperconjugation between the $2'$-hydrogen atoms of (deoxy)nucleosides and the cationic center at $C1'$ in oxocarbenium ion helps to stabilize the transition state. Hyperconjugation can be described as donation of electron density from the $C2'$–$H2'$ σ-bond(s) into the empty p orbital at $C1'$, forming the $C2'$–$C1'$ π-bond. Hyperconjugation in ribosyl oxocarbenium ions is conformation dependent because there is only one H2'. In deoxyribosyl oxocarbenium ions, both H2' hydrogens can contribute, lessening thus the conformational dependence (Fig. 9.33).

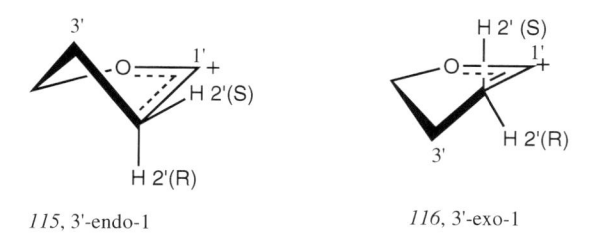

115, 3'-endo-1   *116*, 3'-exo-1

Fig. 9.33

TS analyses have shown that all nonenzymatic N-glycoside hydrolyses and most enzymatic reactions form oxocarbenium ion-like TS in the $3'$-*exo* conformation.

Forming an oxocarbenium ion requires rehybridization of the anomeric carbon ($C1'$) from sp^3 to sp^2. There is increased π-bonding between C1' and O4', the ring oxygen, and between the C1' and C2' to a lesser extent. Resonance forms imply a molecule that is somewhere between carbocation, *110a*, and an oxonium ion, *110b* (Fig. 9.34).

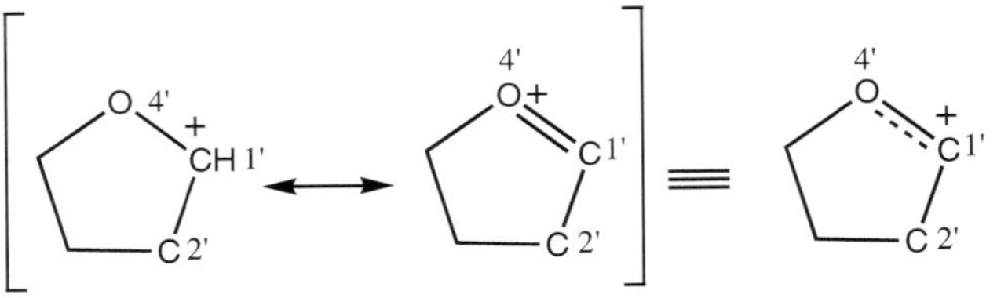

Fig. 9.34

117a ⟷ 117b ≡ 117

The +1 charge on an oxocarbenium ion is centered at the anomeric carbon, C1', but is delocalized around the furanoid ring. Although the ring oxygen is formally trivalent in the dominant oxonium resonance form, *117b*, quantum-mechanical calculations showed it still bears a negative charge in the oxocarbenium ion [97, 112].

References

1. Gottschalk, A., "*Structural relationship between Sialic Acid, Neuraminic Acid and 2-Carboxy-Pyrrole*", Nature (1955) **176**, 881–882
2. Bolton, C. H.; Foster, A. B.; Stacey, M.; Webber, J. M., "*Carbohydrate components of antibiotics. Part I. Degradation of desosamine by alkali: its absolute configuration at position 5*", J. Chem. Soc. (1961) 4831–4836
3. Bolton, C. H.; Foster, A. B.; Stacey, M.; Webber, J. M., "*The configuration of desosamine*", Chem. Ind. (London) (1962) 1945–1946
4. Walters, D. R.; Dutcher, J. D.; Winterstiner, O., "*The structure of mycosamine*", J. Am. Chem. Soc. (1957) **79**, 5076–5077
5. Dutcher, J. D.; Walters, D. R.; Winterstiner, O., "*Nystatin. III. Mycosamine: Preparation and determination of structure*", J. Org. Chem. (1963) **28**, 995–999
6. von Saltza, M. H.; Reid, J.; Dutcher, J. D.; Winterstiner, O., "*Nystatin. II. The Stereochemistry of Mycosamine*", J. Am. Chem. Soc. (1961) **83**, 2785–2785
7. von Saltza, M. H.; Dutcher, J. D.; Reid, J.; Winterstiner, O., "*Nystatin IV. The Stereochemistry of Mycosamine*", J. Org. Chem. (1963) **28**, 999–1004
8. Haskell, T. H.; Hanessian, S., "*The configuration of paromose*", J. Org. Chem. (1963) **28**, 2598–2604
9. Strange, R. E., "*The structure of an Amino Sugar present in Certain Spores and Bacterial cell walls*", Biochem. J. (1956) **64**, 23P
10. Strange, R. E.; Kent, L. H., "*The isolation, characterization and chemical synthesis of muramic acid*", Biochem. J. (1959) **71**, 333–339
11. Fürst, A.; Plattner, P. A., Proc. Intern. Congr. Pure Appl. Chem., 12th, New York, p. 405
12. Overend, W. G.; Vaughan, G., "*Sugar transformations-direction of ring opening of anhydrosugars*", Chem. Ind. (London) (1955) 995–1000
13. Newth, F. H., "*Sugar epoxides*", Quart. Rev. (London) (1959) **13**, 30–47
14. Parker, R. E.; Isaacs, N. S., "*Mechanisms of epoxide reaction*", Chem. Rev. (1959) **59**, 737–799
15. Huber, G.; Schier, O., "*Zum Verständnis der Epoxydöffnung an Pyranose-Ringen*", Helv. Chim. Acta (1960) **43**, 129–135
16. Haworth, W. N.; Lake, W. H. G.; Peat, S., "*The configuration of glucosamine (chitosamine)*", J. Chem. Soc. (1939) 271–274

17. Wiggins, L. F., "*The conversion of galactose into derivatives of d-idose*", J. Chem. Soc. (1944) 522–526
18. Buchanan, J. G.; Miller, K. J., "*The action of ammonia on methyl 2,3-anhydro-4,6-O-benzylidene- -d- guloside and -taloside*", J. Chem. Soc. (1960) 3392–3394
19. Jeanloz, R. W.; Tarasiejska-Glazer, Z.; Jeanloz, D. A., "*2-Amino-2-deoxy-d-idose (D-Idosamine) and 2-Amino-2-deoxy-d-talose (D-Talosamine)*", J. Org. Chem. (1961) **26**, 532–536
20. Jarý, J.; Čapek, K.; Kovář, J., "*Synthesis of derivatives of 3,6-dideoxy-3-amino -l- idose*", Coll. Czech. Chem. Commun. (1963) **28**, 2171–2181
21. Charalambous, G.; Percival, E., "*Products from the alkaline and reductive fission of the epoxide ring of methyl 3 : 4- and 2 : 3-anhydro-6-deoxy- α -l- taloside and of their methylated derivatives*", J. Chem. Soc. (1954) 2443 – 2448
22. Stevens, C. L.; Gupta, S. K.; Glinski, R. P.; Taylor, K. G.; Blumbergs, P.; Schaffner, C. P.; Lee, C.-H., "*Proof of structure, stereochemistry, and synthesis of perosamine (4-amino-4,6-dideoxy -d- mannose) derivatives*", Carbohydr. Res. (1968) **7**, 502–504
23. Stevens, C. L.; Glinski, R. P.; Taylor, K. G.; Blumbergs, P.; Gupta, S. K., "*Synthesis and proof of structure of perosamine (4-amino-4,6-dideoxy -d- mannose) derivatives*", J. Am. Chem. Soc. (1970) **92**, 3160–3168
24. Kovář, J.; Jarý, J., "*Amino sugars. XV. Synthesis of derivatives of 3,6-diamino-3,6- dideoxy -l- talose*", Coll. Czech. Chem. Commun. (1968) **33**, 549–555
25. Fischer, E.; Zach, K., "*Neue Synthese von Basen der Zuckergruppe*", Chem. Ber. (1911) **44**, 132–135
26. Akiya, S.; Ossawa, T., "*Nitrogen-containing sugars. III. Synthesis and deamination of 5-amino-5-deoxy-1,2-isopropylidene -d- xylofuranose*", Yakugaku Zasshi (1956) **76**, 1280–1282
27. Helferich, B.; Burgdorf, M., "*Ueber derivate der D-xylose*", Tetrahedron (1958) **3**, 274–278
28. Wolfrom, M. L; Shafizadeh, F.; Armstrong, R. K.; Shen Han, T. M., "*Synthesis of Amino Sugars by reduction of Hydrazine Derivatives; D- and L-Ribosamine, D- Lyxosamine1-3*", J. Am. Chem. Soc. (1959) **81**, 3716–3719
29. Meyer zu Reckendorf, W., "*Synthese der 2.6-Diamino-2.6-didesoxy -l- idose*", Angew. Chem. (1963) **75**, 573–573
30. Meyer zu Reckendorf, W., "*Diaminozucker, II. Die Synthese der 2.6-Didesoxy- 2.6- diamino -d- galaktose*", Chem. Ber. (1963) **96**, 2019–2023
31. Freudenberg, K.; Burkhart, O.; Braun, E., "*Zur Kenntnis der Aceton-Zucker, VIII.: Eine neue Amino-glucose*", Chem. Ber. (1926) **59**, 714–720
32. Lemieux, R. U.; Chu, P., "*1, -d- allose   2: 5, 6-Di-O-isopropylidene 3-Deoxy-3-amino-*", J. Am. Chem. Soc. (1958) **80**, 4745
33. Freudenberg, K.; Brauns, F., "*Zur Kenntnis der Aceton-Zucker, I.: Umwandlungen der Diaceton-glucose*", Chem. Ber. (1922) **55**, 3233–3238
34. Nayak, U. G.; Whistler, R. L., "*Nucleophilic displacement in 1,2:5,6-di-O-isopropylidene-3-O-(p-tolylsulfonyl)- α -d- glucofuranose*", J. Org. Chem. (1969) **34**, 3819– 3822
35. Whistler, R. L.; Doner, L. W., "*Displacement of the p-Toluenesulfonyloxy Group in 1,2:5, 6-Di-O-isopropyliedener-3-p-toluenesulfonyl- α -d- glucofuranose*", Methods Carbohydr. Chem. (1972) **6**, 215–217
36. Reist, E. J.; Spencer, R. R.; Baker, B. R.; Goodman, L., "*Sodium azide in dimethy- formamide for the preparation of amino sugars*", Chem. Ind. (London) (1962) 1794– 1795
37. Watanabe, K. A.; Goody, R. S.; Fox, J. J., "*Nucleosides—LXVIII : Synthetic studies on nucleoside antibiotics. 5. 4-amino-2,3-unsaturated sugars related to the carbohydrate moiety of blasticidin S*", Tetrahedron (1970) **26**, 3883–3903
38. Hess, K.; Stenzel, H., "*Über ein unterschiedliches Verhalten von alpha- und beta- Methylglucosid gegenüber Tosylchlorid-Pyridin*", Chem. Ber. (1935) **68**, 981–989
39. Stevens, C. L.; Glinski, R. P.; Taylor, K. G.; Sirokman, F., "*Rearrangement reactions of hexose 4-0-sulfonates in the presence of azide and phthalimide nucleophiles*", J. Org. Chem. (1970) **35**, 592–596

40. Jarý, J.; Samek, P. N., "*Aminozucker, (XXIV1) Die Reaktion von Methyl-2.3-O-isopropyliden-4-O-mesyl- α -l- rhamnosid mit Natriumazid und Hydrazin*", Liebigs Ann. (1970) **740**, 98–111
41. Goodman, I., "*Glycosyl ureids*", Adv. Carbohydr. Chem. (1958) **13**, 215–236
42. Schoorl, M. N., "*Sugar ureides*", Rec. Trav. Chim. (Pays-Bas) (1903) **22**, 31
43. Helferich, B.; Kosche, W., "*Über Verbindungen von Aldosen mit Harnstoff und ihre Verwendung zur Synthese stickstoff-haltiger Glucoside*", Chem. Ber. (1926) **59**, 69– 79
44. Benn, M. H.; Jones, A. S., "*Glycosylureas. I. Preparation and some reactions of D- glucosylureas and D-ribosylureas*", J. Chem. Soc. (1960) 3837–3841
45. Jones, A. S.; Ross, G. W., "*The structure of d-glucosylureas*", Tetrahedron (1962) **18**, 189–193
46. Jensen, W. E.; Jones, A. S.; Ross, G. W., "*Glycosylureas. Part II. The synthesis and properties of 2-deoxy -d- ribosylureas*", J. Chem. Soc. (1965) 2463–2465
47. Lobry de Bruyn, C. A.; Franchimont, A. P. N., "*Crystalline amido-derivatives of the carbohydrates*", Rec. Trav. Chim. (Pays-Bas) (1893) **12**, 286–289
48. Hodge, J. E.; Moy, B. F., "*Preparation and properties of Dialditylamines*", J. Org. Chem. (1963) **28**, 2784–2789
49. Frush, H. L.; Isbell, H. S., "*Mutarotation, hydrolysis, and structure of D-galactosylamines*", J. Res. Nat. Bur. Stand. (1951) **47**, 239–247
50. Isbell, H. S.; Frush, H. L., "*Mutarotation, hydrolysis, and rearrangement reactions of Glycosylamines*", J. Org. Chem. (1958) **23**, 1309–1319
51. Isbell, H. S.; Frush, H. L., "*Mechanisms for the mutarotation and hydrolysis of the glycosylamines and the mutarotation of the sugars*", J. Res. Nat. Bur. Stand. (1951) **46**, 132–144
52. Mitts, E.; Hixon, R. M., "*The reaction of glucose with some amines*", J. Am. Chem. Soc. (1944) **66**, 483–486
53. Hodge, J. E.; Rist, C. E., "*N-Glycosyl derivatives of secondary amines*", J. Am. Chem. Soc. (1952) **74**, 1494–1497
54. Hodge, J. E.; Rist, C. E., "*The Amadori rearrangement under new conditions and its significance for non-enzymatic browning reactions*", J. Am. Chem. Soc. (1953) **75**, 316–322
55. Micheel, F.; Hagemann, G., "*Darstellung aliphatischer Amadori-Produkte*", Chem. Ber. (1959) **92**, 2836–2840
56. Micheel, F.; Hagemann, G., "*Darstellung aliphatischer Amadori-Produkte*", Chem. Ber. (1960) **93**, 2381–2383
57. Stepanenko, B. N.; Greshnykh, R. D., "*Syntheses of some N-alkylglycosylamines*", Dokl. Akad. Nauk SSSR (1966) **170**, 121–124
58. Ames, G. R.; King, T. A., "*Long-Chain derivatives of sugars. I. Some reactions of N-Octadecyl -d- glucosylamine*", J. Org. Chem. (1962) **27**, 390–395
59. Erickson, J. G., "*Reactions of long chain Amines. V. Reactions with sugars*", J. Am. Chem. Soc. (1955) **77**, 2839–2843
60. Sorokin, B., "*Ueber Anilide der Glycose*", Chem. Ber. (1886) **19**, 513
61. Sorokin, B., J. Russ. Phys. Chem. Soc., (1887) **Pt. 1**, 377
62. Sorokin, B., Berichte (1887) **20**, (Referata), 783
63. Sorokin, B., J. prakt. Chem. (1888) **37**, 291
64. Weygand, F., "*Darstellung von N-Glykosiden des Anilins und substituierter Aniline*", Chem. Ber. (1939) **72**, 1663–1667
65. Amadori, M., Atti real. Acad. Lincei (1925) **2**, 337; (1929) **9**, 68; (1929) **9**, 226; (1931) **13**, 72
66. Hodge, J. E., "*The Amadori Rearrangement*", Adv. Carbohydr. Chem. (1955) **10**, 169–205
67. Kuhn, R.; Dansi, A., "*Über eine molekulare Umlagerung von N-Glucosiden*", Chem. Ber. (1936) **69**, 1745–1754
68. Kuhn, R.; Weygand, F., "*Die Amadori-Umlagerung*", Chem. Ber. (1937) **70**, 769– 772
69. Weygand, F., "*Über N-Glykoside, II. Mitteil.: Amadori-Umlagerungen*", Chem. Ber. (1940) **73**, 1259–1278; German Pat. 727,402 (Oct. 1, 1942); U.S. Pat. 2,356,846 (Aug. 1, 1944)
70. Simon, H.; Kraus, A., "*Mechanistische Untersuchungen über Glykosylamine, Zuck- erhydrazone, Amadori-Umlagerungsprodukte und Osazone*", Fortschr. Chem. Forsch. (1970) **14**, 430–471

71. Heyns, K.; Paulsen, H.; Eichstedt, R.; Rolle, M., "*Über die Gewinnung von 2-Amino-Aldosen Durch Umlagerung von Ketosylaminen*", Chem. Ber. (1957) **90**, 2039–2049
72. Heyns, K.; Meinecke, K.-H., "*Über Bildung und Darstellung von d-Glucosamin*", Chem. Ber. (1953) **86**, 1453–1462
73. Heyns, K.; Koch, W., "*Über die Bildung eines Aminozuckers as d-Fructose und Ammoniak*", Z. Naturforsch. (1952) **7b**, 486–488
74. Carson, J. F., "*The reaction of fructose with isopropylamine and cyclohexylamine*", J. Am. Chem. Soc. (1955) **77**, 1881–1884
75. Carson, J. F., "*The reaction of fructose with aliphatic amines*", J. Am. Chem. Soc. (1955) **77**, 5957–5960
76. Capon, B.; Connett, B. E., "*The mechanism of the hydrolysis of N-aryl -d- glucosylamines*", J. Chem. Soc, (1965) 4497–4502
77. Jasinski, T.; Smiataczowa, K., "*Mutarotation of N-(p-chlorophenyl) -d- glucosylamine in methanol-dioxane mixtures in the presence of benzoic acids*", Z. phys. Chem. (1967) **235**, 49–56
78. Jasinski, T.; Smiataczowa, K.; Sokolowski, J., "*Mutarotation of N-glycosides as a new method of acid strength studies. IV. Thermodynamic characteristics of mutarotation of N-C-glucosyl-p-chloroaniline in methanol catalyzed by benzoic acid and its derivatives*", Rocz. Chem. (1968) **42**, 107–115
79. Willi, A. V.; Robertson, R. E., "*A kinetic study of the hydrolysis of Benzalaniline*", Canad. J. Chem. (1963) **31**, 361–376
80. Willi, A. V., "*Kinetik der Hydrolyse von Benzalanilin II: Die pH-Abhängigkeit der Reaktionsgeschwindigkeit in ungepufferten Lösungen und die Rolle der Aminoalkohol-Zwischenstufe*", Helv. Chim. Acta (1956) **39**, 1193–1203
81. Cordes, E. H.; Jencks, W. P., "*The mechanism of hydrolysis of schiff bases derived from aliphatic amines*", J. Am. Chem. Soc. (1963) **85**, 2843–2848
82. Capon, B., "*Mechanism in carbohydrate chemistry*", Chem. Rev. (1969) **69**, 407–498
83. Fujii, T.; Saito, T.; Nakasaka, T, "*Purines. XXXIV. 3-Methyladenosine and 3- methyl-2'-deoxyadenosine: Their synthesis, glycosidic hydrolysis, and ring fission*", Chem. Prarm. Bull. (1989) **37**, 2601–2609
84. Lindahl, T.; Nyberg, B., "*Rate of depurination of native deoxyribonucleic acid*", Biochemistry (1972) **11**, 3610–3618
85. Gates, K. S.; Nooner, T.; Dutta, S., "*Biologically relevant chemical reactions of N7-Alkylguanine Residues in DNA*", Chem. Res. Toxicol. (2004) **17**, 839–856
86. Lindahl, T.; Karlstrom, O., "*Heat-induced depyrimidination of deoxyribonucleic acid in neutral solution*", Biochemistry (1973) **12**, 5151–5154
87. Kampf, G.; Kapinos, L. E.; Griesser, R.; Lippert, B.; Sigel, H., "*Comparison of the acid-base properties of purine derivatives in aqueous solution. Determination of intrinsic proton affinities of various basic sites*", J. Chem. Soc. Perkin Trans. 2 (2002)1320–1327
88. Zoltewicz, J. A.; Clark, D. F.; Sharpless, T. W.; Grahe, G., "*Kinetics and mechanism of the acid-catalyzed hydrolysis of some purine nucleosides*", J. Am. Chem. Soc. (1970) **92**, 1741–1750
89. Venner, H., "*Nucleic acids. IX. Stability of the N-glycosidic linkage of nucleosides*", Hoppe-Seyler's Z. Physiol. Chem. (1964) **339**, 14–27
90. Cadet, J.; Teoule, R., "*Nucleic acid hydrolysis. I. Isomerization and anomerization of pyrimidic deoxyribonucleosides in an acidic medium*", J. Am. Chem. Soc. (1974) **96**, 6517–6519
91. Venner, H., "*Nucleic acids. XII. Stability of the N-glycoside linkage in nucleotides*", Hoppe-Seyler's Z. Physiol. Chem. (1966) **344**, 189–196
92. Garrett, E. R.; Seydel, , J. K.; Sharpen, A. J., "*The Acid-Catalyzed solvolysis of Pyrimidine Nucleosides*", J. Org. Chem. (1966) **31,2**219–2227
93. Shapiro, R.; Kang, S., "*Uncatalyzed hydrolysis of deoxyuridine, thymidine, and 5- bromodeoxyuridine*", Biochemistry (1969) **8**, 180–1810

94. Shapiro, R.; Danzig, M., "*Acidic hydrolysis of deoxycytidine and deoxyuridine derivatives. General mechanism of deoxyribonucleoside hydrolysis*", Biochemistry (1972) **11,**23–29
95. Hevesi, L.; Wolfson-Davidson, E.; Nagy, J. B.; Nagy, O. B.; Bruylants, A., "*Contribution to the mechanism of the acid-catalyzed hydrolysis of purine nucleosides*", J. Am. Chem. Soc. (1972) **94**, 4715–4720
96. Bennet, A. J.; Kitos, T. E., "*Mechanisms of glycopyranosyl and 5-thioglycopyranosyl transfer reactions in solution*", J. Chem. Soc. Perkin Trans. (2002) **2**, 1207–1222
97. Berti, P. J.; Tanaka, K. S. E., "*Transition state analysis using multiple kinetic isotope effects: mechanisms of enzymatic and non-enzymatic glycoside hydrolysis and transfer*", Adv. Phys. Org. Chem. (2002) **37**, 239–314
98. BeMiller, J. N., "*Acid-catalyzed Hydrolysis of Glycosides*", Adv. Carbohydr. Chem. (1967) **22**, 25
99. Jencks, W. P., "*How does a reaction choose its mechanism?*", Chem. Soc. Rev. (1981) **10**, 345
100. Mentch, F.; Parkin, D. W.; Schramm, V. L., "*Transition-state structures for N- glycoside hydrolysis of AMP by acid and by AMP nucleosidase in the presence and absence of allosteric activator*", Biochemistry (1987) **26**, 921–930
101. Berti, P. J.; Schramm, V. L., "*The transition-State Structure of the Solvolytic Hydrolysis of NAD$^+$*", J. Am. Chem. Soc. (1997) **119**, 1206
102. McCann and Berti, unpublished results
103. Michelson, A. M., "*The Chemistry of Nucleosides and Nucleotides*", Academic Press, London and New York, 1963, pp. 26–27
104. Michelson, A. M.; Todd, A. R., "*Nucleotides. Part XXIII. Mononucleotides derived from deoxycytidine. Note on the structure of cytidylic acids a and b*", J. Chem. Soc. (1954) 34–40
105. Khorana, H. G.; Turner, A. F.; Vizsolyi, J. P., "*Studies on Polynucleotides. IX. Experiments on the Polymerization of Mononucleotides. Certain Protected Derivatives of Deoxycytidine-5' Phosphate and the Synthesis of Deoxycytidine Polynucleotides*", J. Am. Chem. Soc. (1961) **83**, 686–698
106. Gilham, P. T.; Khorana, H. G., "*Studies on Polynucleotides. I. A New and general method for the chemical synthesis of the C_5'-C_3' internucleotidic linkage. Synthesis of Deoxyribodinucleotides*", J. Am. Chem. Soc. (1958) **80**, 6212–6222
107. Michelson, A. M., "*The organic chemistry of deoxynucleosides and deoxynucleotides*", Tetrahedron (1948) **2**, 333–340
108. Shapiro, H. S.; Chargaff, E., "*Studies on the nucleoside arrangement in deoxyribonucleic acids I. The relationship between the production of pyrimidine nucleoside 3'5'-diphosphates and specific features of nucleotide sequence*", Biochim. Biophys. Acta (1957) **26**, 596–608
109. Shapiro, H. S.; Chargaff, E., "*Studies on nucleotide arrangement in deoxyribonucleic acids III. Identification of methylcytidine derivatives among the acid degradation products of rye germ DNA*", Biochim. Biophys. Acta (1960) **39**, 62–67
110. Brown, D. M.; Fasman, G., D.; Magrath, D. I.;; Todd, A. R., "*Nucleotides. Part XXVII. The structures of adenylic acids a and b*", J. Chem. Soc. (1954) 1448–1455
111. Wempen, I.; Doerr, I. L.; Kaplan, L.; Fox, J. J., "*Pyrimidine Nucleosides. VI. Nitration of Nucleosides*", J. Am. Chem. Soc. (1960) **82**, 1624–1629
112. Loverix, S.; Geerlings, P.; McNaughton, M.; Augustyns, K.; Vandemeulebroucke, A.; Steyaert, J.; Verses, W., "*Substrate-assisted leaving group activation in Enzyme-catalyzed N-Glycosidic Bond Cleavage*", J. Biol. Chem. (2005) **28**, 14799–14802

Chapter 10
Oxidation of Monosaccharides

Oxidation is a very important reaction in carbohydrate chemistry because it enables the synthesis of a great variety of monosaccharides and their derivatives from simple monosaccharides. As already mentioned in Chapter 5 the carbohydrates have three types of chemically distinct hydroxyl groups: (1) the primary hydroxyl groups which are always exocyclic and (2) two types of secondary hydroxyl groups: endo- and exocyclic hydroxyl groups.

The anomeric and the primary hydroxyl groups are the only two hydroxyl groups that can be selectively oxidized in the presence of unprotected other hydroxyl groups: the anomeric hydroxyl group probably due to electronic reasons (it is hemiacetal or lactol hydroxyl group and hence much more reactive than ordinary alcoholic hydroxyl group) and the primary hydroxyl group probably due to steric reasons.

Reactivity of all other secondary hydroxyl groups toward oxidation is practically indistinguishable and therefore the oxidation of a particular hydroxyl group in a monosaccharide requires regioselective protection of all other hydroxyl groups with protecting groups that are stable toward the used oxidation reagent.

Our discussion of oxidation of monosaccharides will be divided into two sections: (1) the selective oxidation of carbohydrates that involves the oxidation of primary and anomeric hydroxyl group and (2) the nonselective oxidation of any other secondary hydroxyl group.

We should, however, mention a third type of oxidation of carbohydrates which is accompanied by the C–C bond cleavage between the vicinal carbons bearing hydroxyl groups. This type of oxidation will be discussed at the end of this chapter.

Selective Oxidations of Monosaccharides

Catalytic Oxidation

Catalytic oxidation of carbohydrates has been extensively investigated by Heynes and coworkers and the subject has been thoroughly discussed [1–4]. The most effective catalyst is platinum black or platinum on carbon, and the oxidant is air that is passed through a vigorously stirred solution of a substrate. The solvent is usually water, but organic solvents have also been used. Yields vary from excellent to poor, but the high selectivity and the ease of isolation of the oxidation product make this method preparatively very useful. The catalytic oxidations can be carried out at room or elevated temperatures. Mechanistically, the catalytic oxidations are regarded as dehydrogenations [1, 4].

Aldoses are readily oxidized to aldonic acids. Oxidation of D-glucose at room temperature [5] or at elevated temperature [6] (55°C) in the presence of theoretical amount of alkali that is required to combine with the formed gluconic acid gives a nearly quantitative yield of gluconic acid. A palladium precipitated on calcium carbonate was used in both cases as the catalyst, and the gluconic acid was isolated as calcium salt (Fig. 10.1).

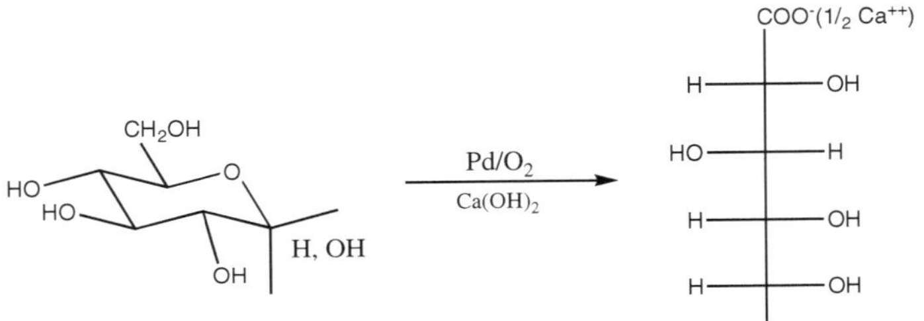

Fig. 10.1

No oxidation of primary hydroxyl group took place under these reaction conditions. This oxidation can also be performed with platinum on carbon as the catalyst and in the presence of 1 mol-equivalent of a base [7]. D-Galactose, D-mannose, D-xylose, and L-arabinose have all been converted to the corresponding aldonic acids [8] in this way. The pentoses are oxidized more rapidly than hexoses (usually in 45 min at room temperature as compared to several hours for hexoses).

If the oxidation of aldoses is performed at elevated temperatures (e.g., 50°C) with platinum-on-carbon as the catalyst, the primary C6 hydroxyl group will also be oxidized in addition to the anomeric hydroxyl group giving the corresponding glycaric acids. D-Glucose is in this way converted to D-glucaric acid *3* in 54% yield [9] (Fig. 10.2).

Hence, the anomeric carbon of free aldoses can be selectively oxidized at room temperature to carboxylic group giving aldonic acids or aldonolactones, whereas at higher temperatures, the C6 hydroxyl group is also oxidized giving aldaric acids.

If the reactive anomeric hydroxyl group is suitably blocked, as is the case in alkyl or aryl glycosides, uronic acids can be readily obtained by catalytic oxidation of the primary C6 hydroxyl group of hexopyranoses using platinum as the catalyst. Thus, for example, methyl α-D-glucopyranoside *4* is catalytically oxidized in 87% yield to

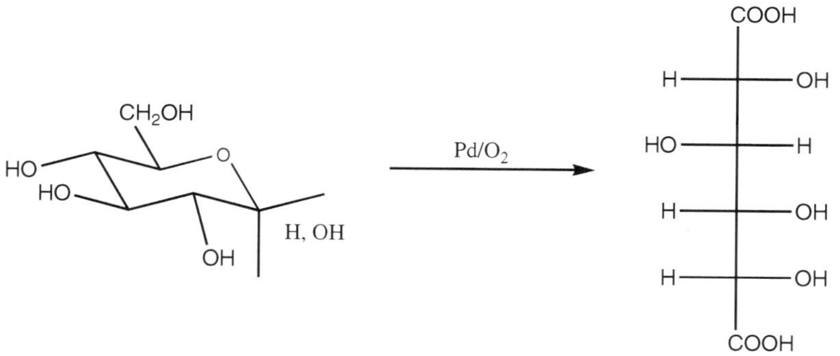

Selective Oxidations of Monosaccharides

Fig. 10.2

1, D-glucopyranose → Pd/O$_2$ → *3*, D-glucaric acid

methyl α-D-glucopyranosiduronic acid *6* [10]. D-Glucurono-6,3-lactone is obtained in only 16% yield since the acid hydrolysis of methyl glucosidic bond requires harsh conditions. The methyl β-D-glucopyranoside *5* gives on catalytic oxidation the corresponding glucuronic acid *7* in 68% yield. The oxygen is used as oxidizing agent, and Pt black, Pt/C, Pt/Al$_2$O$_3$, PtO$_2$ as the catalyst (Fig. 10.3).

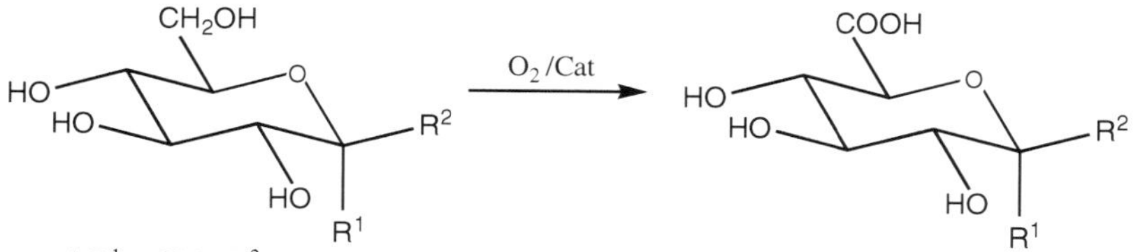

4, R^1 = OMe; R^2 = H
5, R^1 = H; R^2 = OMe

6, R^1 = OMe; R^2 = H
7, R^1 = H; R^2 = OMe

Fig. 10.3

Bromine Oxidation

Bromine reacts with free aldoses (for example, *1*) in both acid and alkaline solutions, whereas iodine reacts only in alkaline solution. Since this oxidation is a two-electron transfer reaction, 1 mol of bromine is consumed per mole of aldose giving 1 mol of glyconolactone *8* and 2 mol of hydrobromic acid (Fig. 10.4).

The accumulation of hydrobromic acid during bromine oxidation greatly lowers the rate of reaction and inhibits further oxidation. To minimize this inhibiting effect of hydrobromic acid, the oxidation is conducted in the presence of calcium or barium carbonate or calcium benzoate [11, 12] as "solid bases." It is interesting to note that although other strong acids inhibit the rate of bromine oxidation too, the

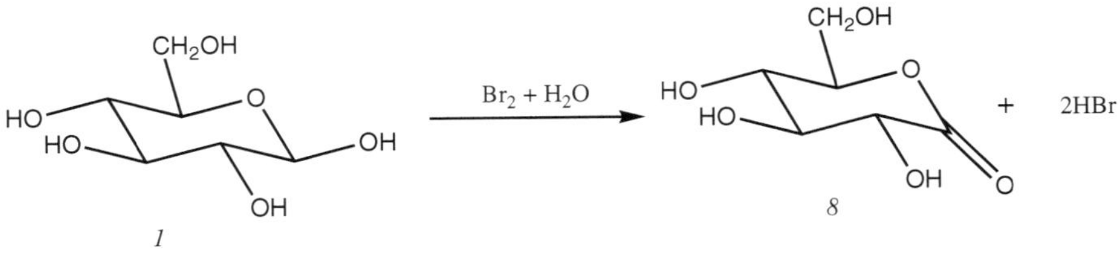

Fig. 10.4

inhibition is greatest with HBr and HCl [13]. It has been speculated that this effect may be due, in part, to complexation of 1 mol of hydrobromic acid with 1 mol of free bromine giving Br_3^-, which is ineffective as oxidant [14]. The yields of oxidation products are usually very high: D-gluconic acid 8 is obtained from D-glucose 1 in 96% yield, whereas D-xylonic acid is obtained in 90% yield from D-xylose in buffered solutions [14].

From a study of bromine oxidation of free aldoses in the presence of barium carbonate and bromides (pH about 5.4) Isbell and Pigman [15] have concluded that the active oxidant is free bromine, not hypobromous acid. Furthermore, it was found [16] that cyclic forms of a monosaccharide, not free aldehyde, are oxidized directly under these conditions. Thus pyranoses yield 1,5-lactones and furanoses, 1,4-lactones directly and in high yield.

The faster oxidation of β-anomers as opposed to α-anomers has been explained [17, 18] by using the known fact from conformational analysis that equatorial substituents are more reactive than axial ones (the hydroxyl group is in β-D-anomers equatorially oriented).

Barker et al. [17, 18] have found that the rates of bromine oxidation of α-D-anomers may be related to the rate of mutarotation into β-D-anomers so that the actual rate of oxidation of α-D-anomers is much lower. A straight line relationship was obtained for the observed rate of bromine oxidation and the rate of α- to β-mutarotation of studied sugars. Hence, the rate-determining step in the oxidation of the α-D-glucose is its anomerization into the β-D-anomer. It was found that the true rate of oxidation of α-D-glucose is about 1/250th that of the β-D-anomer.

Nonselective Oxidation of Secondary Hydroxyl Groups

Generally, the oxidation of a particular hydroxyl group in a monosaccharide requires blocking (protecting) with appropriate protective groups of all other hydroxyl groups that would also react under the used experimental conditions. Thus in nonselective oxidation of carbohydrates, a selective protection of all hydroxyl groups, except the one which is to be oxidized, has to be performed. The topic of selective protection of hydroxyl groups of glycofuranoses and glycopyranoses and their deprotection will be dealt with in a separate chapter.

Ruthenium Tetroxide (RuO$_4$) Oxidation

Ruthenium tetroxide was discovered by Claus [19] in 1860, but it took almost 100 years to realize that RuO$_4$ is rapidly reduced to RuO$_2$ by alcohol, acetaldehyde, and the like [20] and that it can be used as a very potent oxidant in neutral solution and at room temperature [21]. The most suitable solvent for RuO$_4$ oxidations is carbon tetrachloride wherein the RuO$_4$ dissolves with a dark red color. Alcohol-free chloroform and dichloromethane were also found to be satisfactory. Thus ruthenium tetroxide is stable in all these solvents. Ether, benzene, and pyridine are highly unsuitable as solvents because on contact with RuO$_4$ they either explode or catch fire [21].

Beynon et al. [22] were the first to demonstrate the usefulness of RuO$_4$ for oxidation of secondary hydroxyl groups in carbohydrates. Using the equivalent amounts of carbon tetrachloride solution of ruthenium tetroxide they were able to oxidize a number of suitably protected methyl glycosides to corresponding glycopyranosiduloses. The oxidation was conducted at room temperature and it was usually completed within 1–4 h. The structures of monosaccharides that were oxidized and the structures of obtained corresponding "uloses" are given in Figs. 10.5 and 10.6. The yields of products obtained are given in Table 10.1.

The study of RuO$_4$ oxidation of partially protected methyl glycosides has shown not only that the glycosidic linkage is unaffected by the reagent but also that benzoate, benzylidene, and isopropylidene groups can all be safely used for selective protection of hydroxyl groups in monosaccharides which are to be oxidized.

Two procedures are generally employed for the RuO$_4$ oxidation of carbohydrates [23]:

(a) Oxidation of protected monosaccharide with a slight molar excess of ruthenium tetroxide (both dissolved in carbon tetrachloride) (monophasic system); or (less satisfactorily)
(b) Oxidation of protected monosaccharide with catalytic quantity of ruthenium tetroxide [24] (substrate and RuO$_4$ dissolved in carbon tetrachloride) in the presence of aqueous solution of sodium metaperiodate (biphasic system). The reaction mixture is vigorously stirred so that the RuO$_2$ obtained after partial oxidation of the monosaccharide is reoxidized back to RuO$_4$ with sodium metaperiodate. The regenerated RuO$_4$ then redissolves in carbon tetrachloride (RuO$_4$ is much more soluble in carbon tetrachloride than in water) where it oxidizes another quantity of protected monosaccharide. This process is being repeated until all of the monosaccharide has been oxidized. Regenerated RuO$_4$ redissolves in CCl$_4$ and oxidizes there an additional amount of selectively protected carbohydrate. Sodium metaperiodate is reduced to sodium iodate in this reaction. This method requires vigorous stirring or shaking.

Although CrO$_3$–Py can oxidize a chemically resistant secondary hydroxyl group of a monosaccharide [25–27], the yields are generally better with RuO$_4$ as the oxidant [23]. For example, oxidation of 5-*O*-benzoyl-1,2-*O*-isopropylidene-α-D-xylofuranose *29* (Fig. 10.7) was not successful with CrO$_3$ in pyridine, acetone,

9, Metyl 6-deoxy-2,3-isopropylidene-α-L-mannopyranoside

10, Metyl 6-deoxy-2,3-isopropylidene-α-L-*lyxo*-4-hexulopyranoside

11, Methyl 6-deoxy-3,4-O-isopropylidene-α-L-galactopyranoside

12, Methyl 6-deoxy-3,4-O-isopropylidene-α-L-*lyxo*-hexulopyranoside

13, Methyl 3,4-O-isopropylidene-β-L-arabinopyranoside

14, Methyl 3,4-O-isopropylidene-β-L-*erythro*-pentulopyranoside

15, Methyl 4,6-O-benzylidene-2-deoxy-α-D-*arabino*-hexopyranoside

16, Methyl 4,6-O-benzylidene-2-deoxy-α-D-*erythro*-hexopyranosid-3-ulose

17, 1,2:5,6-di-O-isopropylidene-α-D-glucofuranose

18, 1,2:5,6-di-O-isopropylidene-α-D-*ribo*-hexopyranosid-3-ulose

19, Methyl 3,4,6,-tri-O-benzoyl-α-D-glucopyranoside

20, Methyl 3,4,6-tri-O-benzoyl-α-D-*arabino*-hexopyranosidulose

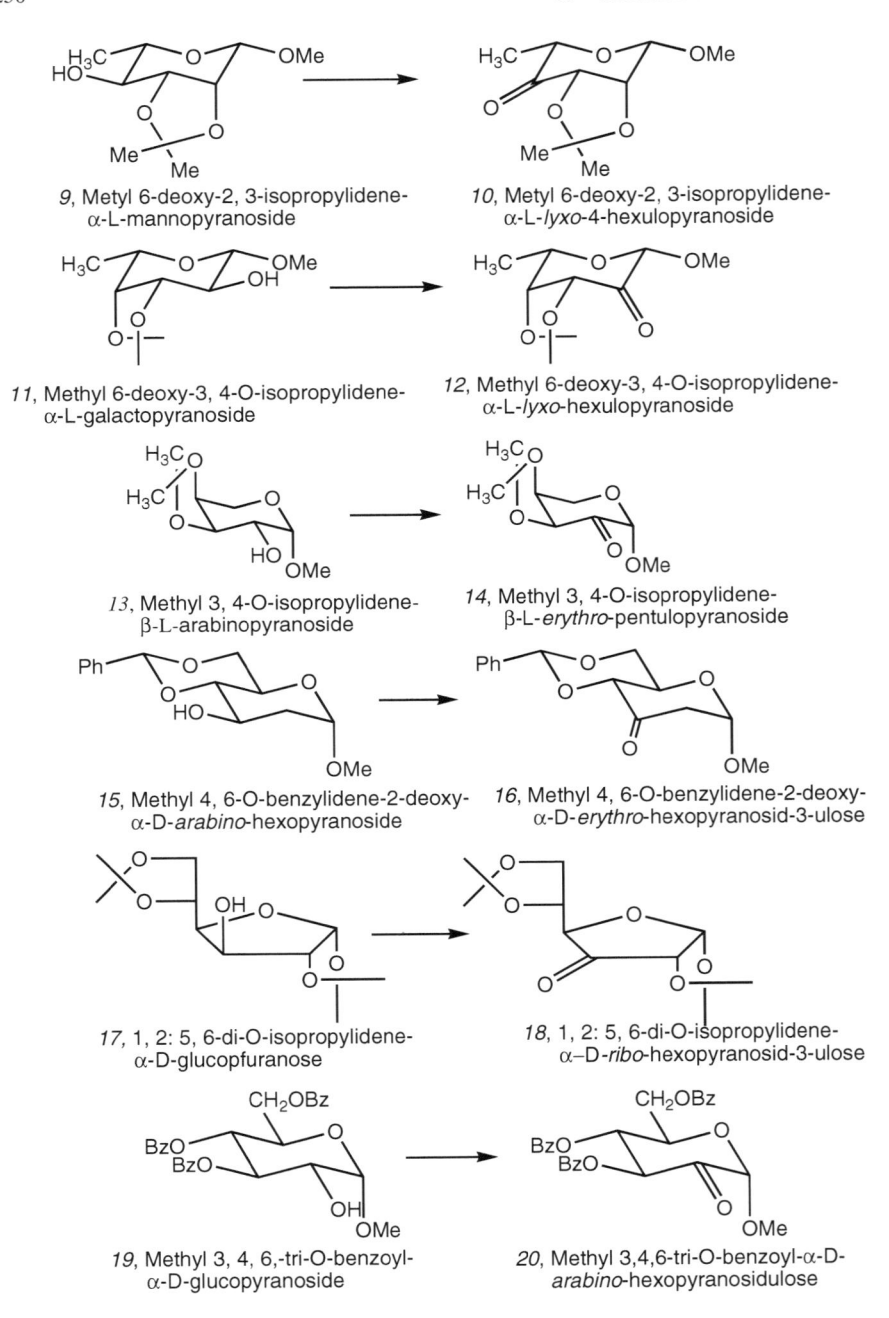

Fig. 10.5

Nonselective Oxidation of Secondary Hydroxyl Groups

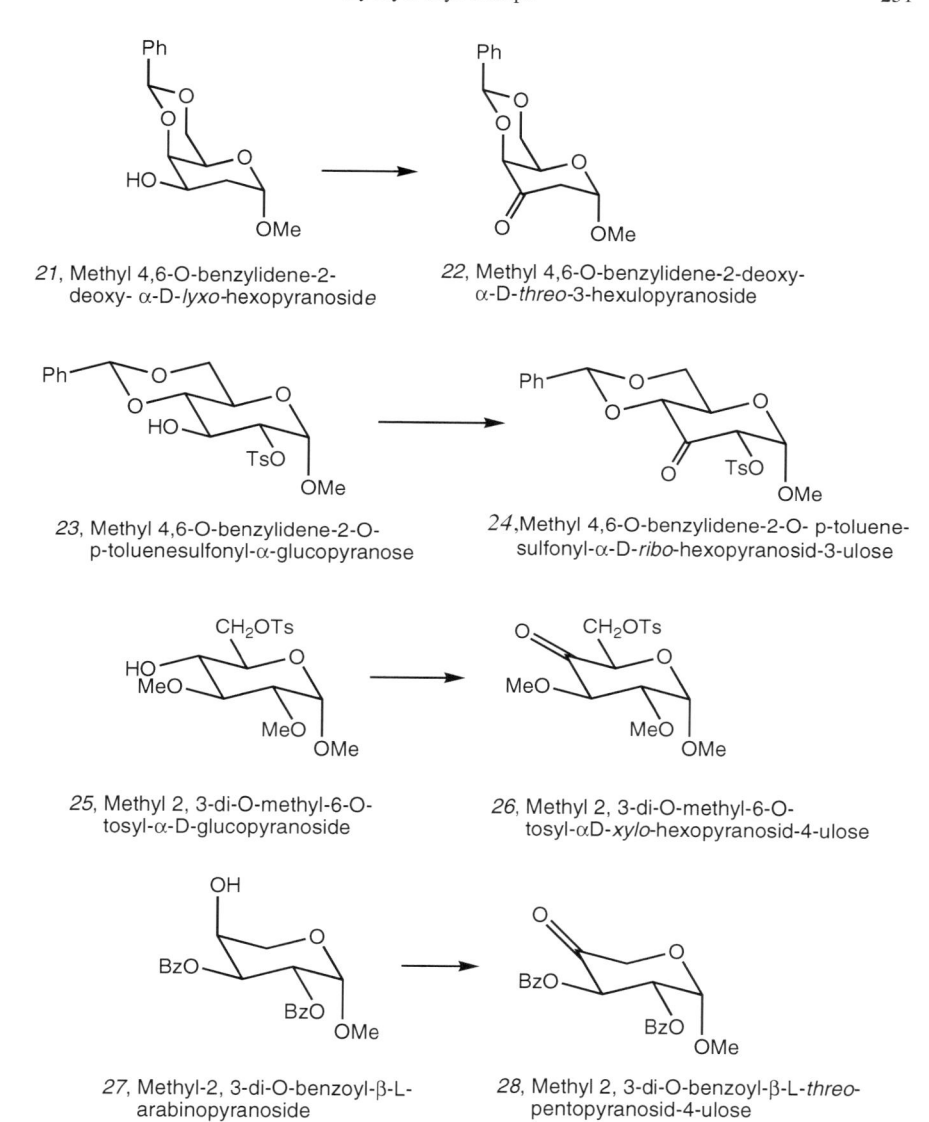

Fig. 10.6

21, Methyl 4,6-O-benzylidene-2-deoxy-α-D-*lyxo*-hexopyranoside

22, Methyl 4,6-O-benzylidene-2-deoxy-α-D-*threo*-3-hexulopyranoside

23, Methyl 4,6-O-benzylidene-2-O-p-toluenesulfonyl-α-glucopyranose

24, Methyl 4,6-O-benzylidene-2-O-p-toluenesulfonyl-α-D-*ribo*-hexopyranosid-3-ulose

25, Methyl 2,3-di-O-methyl-6-O-tosyl-α-D-glucopyranoside

26, Methyl 2,3-di-O-methyl-6-O-tosyl-αD-*xylo*-hexopyranosid-4-ulose

27, Methyl-2,3-di-O-benzoyl-β-L-arabinopyranoside

28, Methyl 2,3-di-O-benzoyl-β-L-*threo*-pentopyranosid-4-ulose

or acetic acid. It could also not be accomplished with aluminum isopropoxide, potassium permanganate, or lead tetraacetate in acetone. Chromium trioxide in *tert*-butyl alcohol did give some product, but in very low yield. The RuO$_4$, however, gave the corresponding 5-*O*-benzoyl-1,2-*O*-isopropylidene-α-D-*erythro*-pentos-3-ulose *30* in 55% yield [28, 29].

It has been reported that prolonged treatment of some aldofuranose derivatives with excess of ruthenium tetroxide [28, 29] gives lactones by an oxygen insertion

Table 10.1 Ruthenium tetroxide oxidation of various partially protected monosaccharides

Monosaccharide	Oxidation product	Yield (%)
Methyl 6-deoxy-2,3-O-isopropylidene-α-L-mannopyranoside 9	Methyl 6-deoxy-2,3-O-isopropylidene-α-L-*lyxo*-hexopyranosid-4-ulose 10	
Methyl 6-deoxy-3,4-O-isopropylidene-α-L-galactopyranoside 11	Methyl 6-deoxy-3,4-O-isopropylidene-α-L-*lyxo*-hexopyranosidulose 12	70 (crude) 45 (pure)
Methyl 3,4-O-isopropylidene-β-L-arabinopyranoside 13	Methyl 3,4-O-isopropylidene-β-L-*erythro*-pentopyranosidulose 14	80 (crude) 40 (pure)
Methyl 4,6-O-benzylidene-2-deoxy-α-D-*arabino*-hexopyranoside	Methyl 4,6-O-benzylidene-2-deoxy-α-D-*erythro*-hexopyranosid-3-ulose	47
1,2:5,6-di-O-isopropylidene-α-D-glucofuranose 17	1,2:5,6-di-O-isopropylidene-α-D-*ribo*-hexofuranos-3-ulose 18	ca. 75
Methyl 3,4,6-tri-O-benzoyl-α-D-glucopyranoside 19	Methyl 3,4,6-tri-O-benzoyl-α-D-*arabino*-hexopyranosidulose 20	50
Methyl 4,6-O-benzylidene-2-deoxy-α-D-*lyxo*-hexopyranoside 21	Methyl 4,6-O-benzylidene-2-deoxy-α-D-*threo*-3-hexopyranosid-3-ulose 22	47
Methyl 4,6-O-benzylidene-2-O-tosyl-α-D-glucopyranoside 23	Methyl 4,6-O-benzylidene-2-O-p-toluenesulfonyl-α-D-*ribo*-hexopyranosid-3-ulose	85
Methyl 4,6-O-benzylidene-2-deoxy-α-D-*lyxo*-hexopyranoside 25	Methyl 4,6-O-benzylidene-2-deoxy-α-D-*threo*-hexopyranosid-3-ulose	89
Methyl 2,3-di-O-methyl-6-O-tosyl-α-D-glucopyranoside 27	Methyl 2,3-di-O-methyl-6-O-p-toluene-sulfonyl-α-D-*xylo*-hexopyranosid-4-ulose	83
Methyl 2,3-di-O-benzoyl-β-L-arabinopyranoside 29	Methyl 2,3-di-O-benzoyl-β-L-*threo*-pentopyranosid-4-ulose	93

Note. The oxidation products of the first four monosaccharides were identical with the samples obtained when CrO_3–Py was used as the oxidant.

29, 5-O-Benzoyl-1,2-O-isopropylidene-α-D-xylofuranose

30, 5-O-Benzoyl-1,2-O-isopropylidene-α-D-*erythro*-pentos-3-ulose

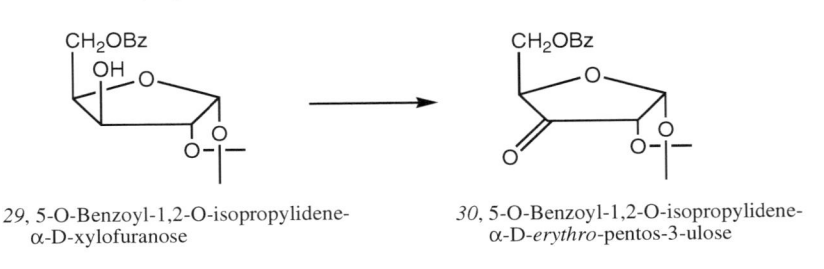

Fig. 10.7

reaction into the initially formed glycosuloses. The problem of this over oxidation is minimized if instead of sodium metaperiodate and sodium bicarbonate, potassium metaperiodate and potassium carbonate that are sparingly soluble in water are used for the reoxidation of ruthenium dioxide to ruthenium tetroxide [30].

The improved procedure for oxidation of "isolated" secondary hydroxyl groups to ketones in partially protected sugar derivatives:

A partially protected carbohydrate is dissolved in enough ethanol-free chloroform to give an approximately 15% solution, and the equal volume of water is

Nonselective Oxidation of Secondary Hydroxyl Groups

added. For each mole of substrate, 0.24 mol of anhydrous potassium carbonate, 1.3 mol of potassium periodate, and a catalytic amount (ca. 0.05 mol) of ruthenium dioxide as dihydrate are then added. The reaction mixture is vigorously stirred and the progress of oxidation is followed by tlc. At the end of reaction, the isopropyl alcohol is added to the reaction mixture to reduce the residual ruthenium tetroxide. The ruthenium dioxide is removed by filtration of the suspension, and the two layers are separated. The aqueous layer is extracted with chloroform, and the combined extract and washings dried, and evaporated in vacuo to give the oxidized product.

A few examples of monosaccharides that are successfully oxidized by using the *improved* procedure and the yields of "uloses" obtained are listed in Table 10.1 and in Fig. 10.8. Both methyl 2,3,6-tri-*O*-benzoyl-α-D-galactopyranoside *31* and 2,3,6-tri-*O*-benzoyl-α-D-glucopyranoside *33* (Fig. 10.8) were oxidized with RuO_4 (using the equivalent amount of oxidant) whereby pure, crystalline methyl 2,3,6-tri-*O*-benzoyl-α-D-*xylo*-hexopyranosid-4-ulose *32* was obtained in good yields (81% from *31* and 79% from *33*) [31] (Fig. 10.8).

The *p*-toluenesulfonyl group is stable toward the oxidant, as shown by oxidation of 4,6-*O*-benzylidene-2-*O*-*p*-toluenesulfonyl-α-D-glucopyranoside to the corresponding 3-ulose (Fig. 10.8). However, after 20 h only about 15% of the corresponding 3-ulose was obtained in addition to almost 50% of starting material [31] (the oxidation was performed with an equivalent amount of RuO_4). Thus this oxidation procedure has no advantage over the Pfitzner–Moffatt method [32] (vide infra).

The RuO_4 was also used for the oxidation of acetamidoglucosides (Fig. 10.8). For example, using the catalytic amount of RuO_4 methyl 2-acetamido-4,6-*O*-benzylidene-2-deoxy-α-D-glucopyranoside yielded the corresponding 3-ulose in 58% yield, which is not as good as the oxidation of the same substrate by Pfitzner–Moffatt reagent [32] (vide infra).

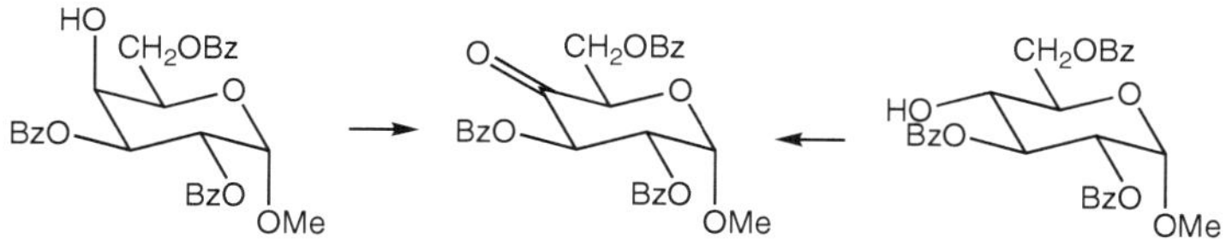

31, methyl 2, 3, 6-tri-O-benzoyl-α–D-galactopyranoside

32, methyl 2 ,3, 6-O-tri-O-benzoyl-α-D-*xylo*-hexopyranosid-4-ulose

33, Methyl 2, 3, 6-tri-O-benzoyl-α-D-glucopyranoside

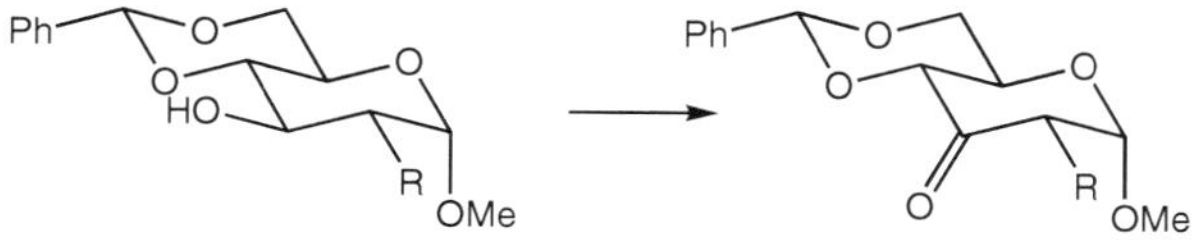

23, R = OTs, Methyl 4,6-O-Benzylidene-2-O-p-toluenesulfonyl-α-D-glucopyranoside
34, R= AcNH, Methyl 2-acetamido-4,6-benzylidene-2-deoxy-2-α-D-glucopyranoside

24, R = OTs, Methyl 4,6-O-benzylidene-2-O-p-toluenesulfonyl-α-D-*arabino*-hexopyranosid-2-ulose
35, R = AcNH, Methyl 2-acetamido-4,6-O- benzylidene-2-deoxy-α-D-*arabino*-hexopyranosid-2-ulose

Fig. 10.8

Dimethyl Sulfoxide Oxidation

The structure of dimethyl sulfoxide (DMSO) is usually represented as a resonance hybrid of the following two resonance structures (Fig. 10.9).

$$(CH_3)_2S=O \quad \longleftrightarrow \quad (CH_3)_2\overset{\oplus}{S}-\overset{\ominus}{O}$$
$$\quad\quad 36 \quad\quad\quad\quad\quad\quad\quad 37$$

Fig. 10.9

Resonance structure *36* (Fig. 10.9) owes its existence to the ability of the *3d* orbital of sulfur to accommodate an additional electron pair, in this case the *p* electrons of oxygen [33]. Although there is still debate over which hybrid best represents the structure of DMSO, or sulfoxides in general, it seems certain that the sulfur–oxygen bond can be justly characterized as being *semipolar* [34]. The molecular structure of dimethyl sulfoxide from spectroscopic and gas electron diffraction data and the force field and ab initio calculations support earlier observations emphasizing the needlessness of the hypervalency concept for describing the bonding properties in DMSO because the S:O bond is ionic and thus the octet rule is not violated [35].

The oxidizing capacity of DMSO is directly dependent on its ability to act as a nucleophile. Its basicity is slightly greater than that of water [36], and its nucleophilicity has been estimated to exceed that of ethanol toward alkylsulfonate esters [37].

Pathway A

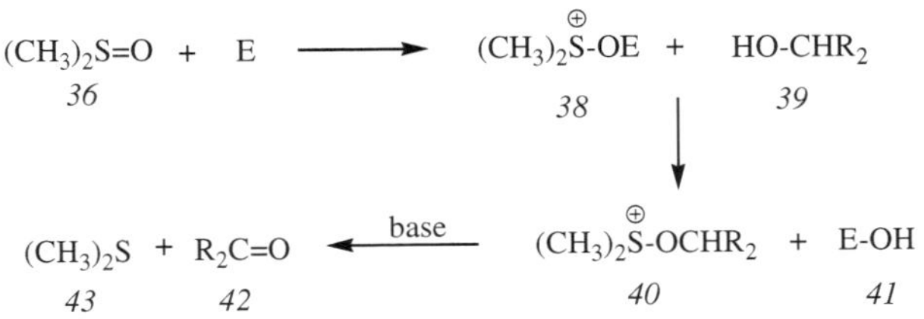

$$(CH_3)_2S=O \;+\; E \quad\longrightarrow\quad (CH_3)_2\overset{\oplus}{S}\text{-OE} \;+\; \text{HO-CHR}_2$$
$$\quad\; 36 \quad\quad\quad\quad\quad\quad\quad\quad\quad 38 \quad\quad\quad\quad 39$$

$$(CH_3)_2S \;+\; R_2C=O \quad\xleftarrow{\text{base}}\quad (CH_3)_2\overset{\oplus}{S}\text{-OCHR}_2 \;+\; \text{E-OH}$$
$$\;\; 43 \quad\quad 42 \quad\quad\quad\quad\quad\quad\quad\quad 40 \quad\quad\quad\quad 41$$

Fig. 10.10

E = electrophile.

There are two routes by which a substrate may be converted into dimethylalkoxysulfonium salt intermediate *40*; which route will be taken is determined by the structure of substrate. These two routes are illustrated in Figs. 10.10 and 10.12.

The first pathway (*Pathway A*) (Fig. 10.10) involves reaction of DMSO with an "activating" electrophilic species E forming intermediate *38*; the EO group is subsequently displaced by a substrate that is to be oxidized, usually a hydroxyl group. There is a strong indication that most of the DMSO oxidations involve the formation of the same *dimethylalkoxysulfonium salt intermediate 40* (Fig. 10.10) which subsequently reacts with a base to give the carbonyl product *42* and dimethyl sulfide *43* [(CH$_3$)$_2$S = DMS]. It has been demonstrated that dimethyl methoxysulfoxonium trifluoroborate *44* (Fig. 10.11) forms formaldehyde in the presence of a base [38, 39]. In the presence of sodium ethoxide and sodium isopropoxide it under-

$$[(CH_3)_2S\text{-}OCH_3]^{\oplus} BF_4^{\ominus}$$
44

Fig. 10.11

goes rapid alkoxide exchange, with inversion of configuration at the sulfur [40, 41] that is followed by subsequent formation of acetaldehyde and acetone, respectively [38, 39].

The oxidation of alkyl halides or sulfonates is thought to proceed via the same transition state intermediate dimethylalkoxysulfonium salt *40* which is, however, directly formed by bimolecular nucleophilic displacement of the leaving group X (X = Cl, Br, I, or a sulfonate) linked to the carbon that is to be oxidized by the oxygen of DMSO [42–45] (*Pathway B*) (Fig. 10.12).

Pathway B

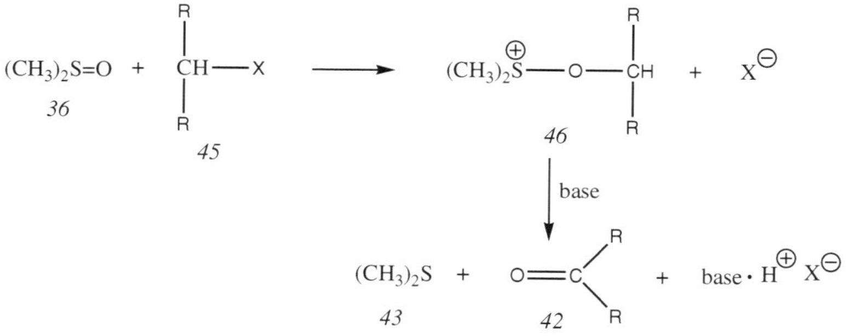

$$(CH_3)_2S=O + \underset{\underset{R}{\vert}}{\overset{\overset{R}{\vert}}{CH}}-X \longrightarrow (CH_3)_2\overset{\oplus}{S}-O-\underset{\underset{R}{\vert}}{\overset{\overset{R}{\vert}}{CH}} + X^{\ominus}$$
36   *45*   *46*

$$\downarrow \text{base}$$

$$(CH_3)_2S + O=C\underset{R}{\overset{R}{\diagup}} + \text{base} \cdot H^{\oplus} X^{\ominus}$$
     43   *42*

Fig. 10.12

Although not experimentally verified, the mechanism proposed above is generally accepted as the reaction mechanism. It should be, however, pointed out that it is quite possible that both bimolecular and unimolecular processes are operative depending on the substrate [46].

DMSO–DCC Method (Pfitzner–Moffatt Oxidation)

Pfitzner–Moffatt oxidation [47–49] involves addition of an alcohol substrate to a solution of DCC (dicyclohexyl carbodiimide) in DMSO containing phosphoric acid or pyridinium trifluoroacetate as a proton source resulting in nearly neutral reaction condition. This method is applicable for oxidation of both primary and secondary hydroxyl groups of carbohydrates. Steric effects are not important except in highly hindered structures where the oxidation of less hindered hydroxyl group will be favored. Tosylates, tertiary alcohols, olefins, and amines are unaffected by the reaction condition.

The mechanism illustrated in Fig. 10.13 was proposed by Albright and Goldman [50]. It was later proved to be correct by using ^{18}O-labeled DMSO (all of ^{18}O-label ended up in N,N′-dicyclohexylurea) and by using deuterium-labeled carbon bearing hydroxyl group which is to be oxidized (deuterium-labeled DMSO was isolated from the reaction) [51].

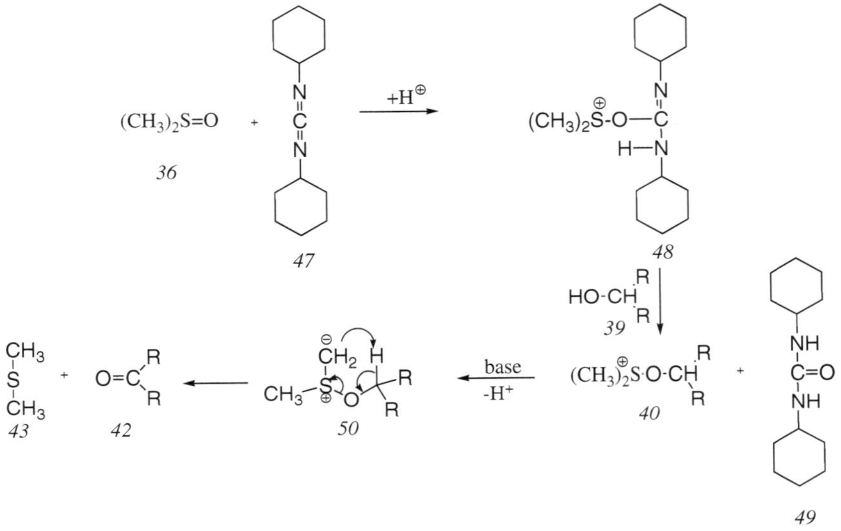

Fig. 10.13

Oxidation of 1,2:3,4-di-O-isopropylidene-L-rhamnitol (*51*, Fig. 10.14) with Pfitzner–Moffatt reagent (DMSO–DCC–anhydrous H_3PO_4) (25–30°C, 18 h) yields 1-deoxy-3,4:5,6-di-O-isopropylidene-L-fructose *52* in 49.5% yield. The oxidation of the same substrate with DMSO–P_2O_5 (25–30°C, 48 h) gave the same product in 34% yield [52].

Methyl 6-deoxy-2,3-O-isopropylidene-β-D-allofuranoside *53* (Fig. 10.15) gave on oxidation with DMSO–DCC, in the presence of pyridine and trifluoroacetic acid (30°C, 18 h), methyl 6-deoxy-2,3-O-isopropylidene-β-D-*ribo*-hexofuranosid-5-ulose *54* in 71% yield whereas oxidation of *53* with DMSO–Ac$_2$O (30°C, 24 h) gave *54* in even better yield (81%) [52].

Oxidation of 3-O-benzyl-2,4-ethylidene-D-erythritol *55* with DMSO–DCC–pyridinium trifluoroacetate (room temperature, 20 h) gave 3-O-benzyl-2,4-O-ethylidene-aldehydo-D-erythrose *56* in 84.6% yield, whereas with DMSO–Ac$_2$O the same substrate could not be oxidized into the corresponding aldehyde *56* [53] (Fig. 10.16).

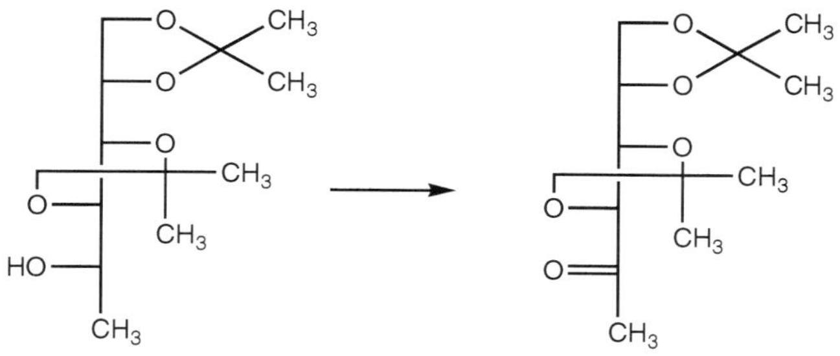

DMSO–DCC Method (Pfitzner–Moffatt Oxidation)

51, 1,2:3,4-di-O-isopropylidene-L-rhamnitol

52, 1-deoxy-3,4:5,6-di-O-isopropylidene-L-fructose

Fig. 10.14

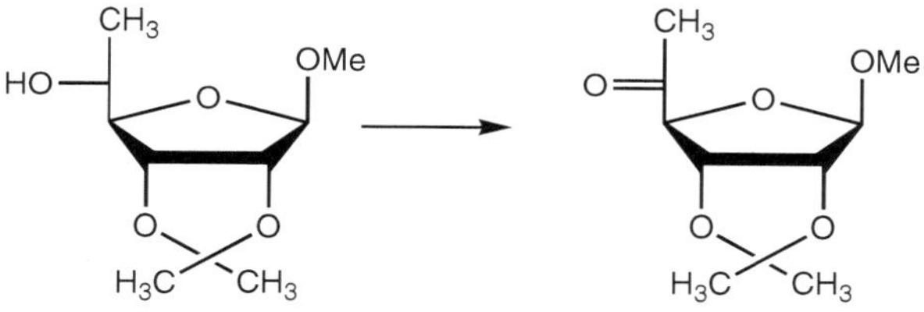

Fig. 10.15

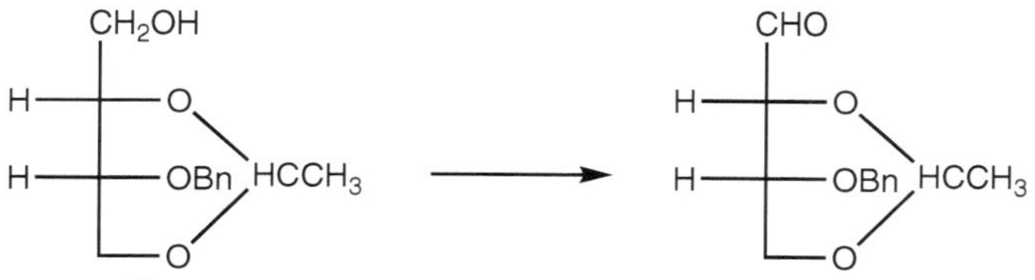

Fig. 10.16

Oxidation of 1,2:3,4-di-*O*-isopropylidene-α-D-galactopyranose *57* with DMSO–DCC–pyridinium phosphate in benzene (room temperature, 5 h) gave 1,2:3,4-di-

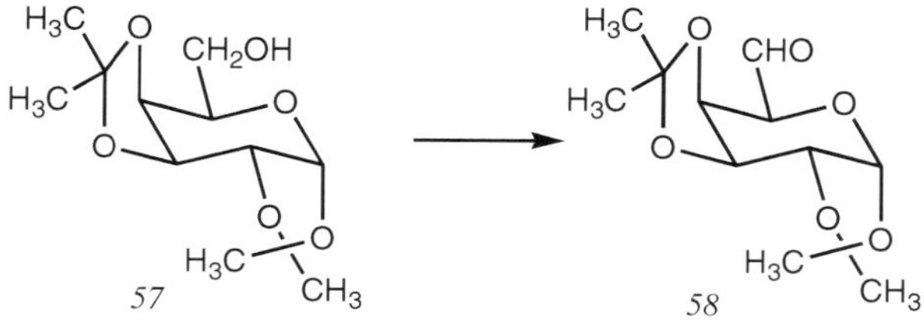

Fig. 10.17

O-isopropylidene-α-D-galacto-hexodialdo-1,5-pyranose *58* in about 80% yield [54] (Fig. 10.17).

DMSO–Acetic Anhydride Method

In this method the DMSO is activated by acetic anhydride and is used for oxidation of primary and secondary hydroxyl groups to corresponding carbonyl groups and is essentially similar to the Pfitzner–Moffatt method [55].

(a) Oxidation

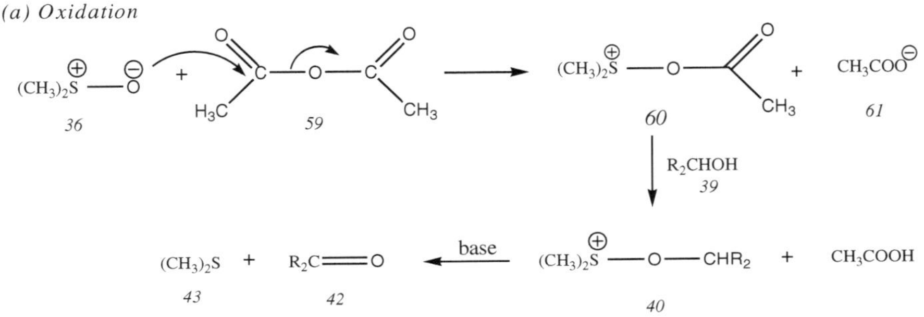

Fig. 10.18

The reaction of DMSO and organic acid anhydrides has been well studied [39, 56–58] and the intermediate *60* (Fig. 10.18), which results from nucleophilic attack of DMSO at one carbonyl carbon of acetic anhydride, may undergo two reactions: it can either react with an alkoxy group and form dimethylalkoxysulfonium salt *40* which is then transformed to carbonyl group by action of a base, or the acetate can be eliminated from *60* via an intramolecular hydrogen transfer *62* [59] giving sulfonium ylide *63* which on reaction with the hydroxyl group of a carbohydrate gives a side product methylthiomethyl ether *64* [55] (Fig. 10.19).

(b) The formation of side-product

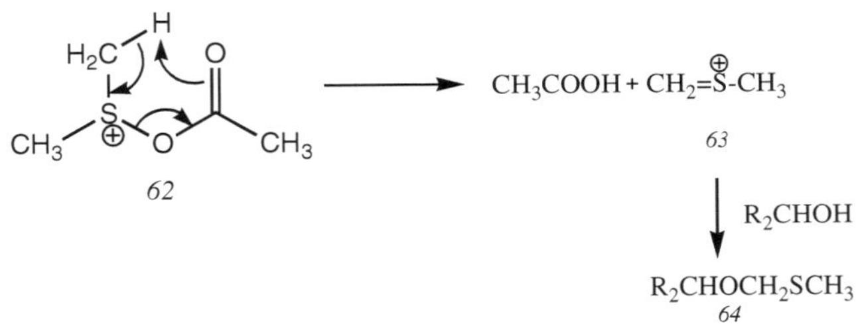

Fig. 10.19

DMSO–Acetic Anhydride Method

Compared to Pfitzner–Moffatt method this method of oxidation, in general, gives lower yields of corresponding carbonyl compounds when unhindered primary and the secondary hydroxyl groups are oxidized. Methylthiomethyl ethers and acetates are also obtained as side products. The method appears to be superior to the DMSO–DCC oxidation of hindered hydroxyl groups.

Fig. 10.20

Oxidation of 4,6-*O*-ethylidene-1,2-*O*-isopropylidene-α-D-galactopyranose *65* (Fig. 10.20) with DMSO–Ac$_2$O (25°C, 36 h) gave in 44% yield 4,6-*O*-ethylidene-1,2-*O*-isopropylidene-α-D-*xylo*-hexopyranosid-3-ulose *66* together with the 3-*O*-methylthiomethyl ether *67* (19%); oxidation of *62* with DMSO–P$_4$O$_{10}$ (65°C, 2 h) gave *66* in 63% yield [60, 61].

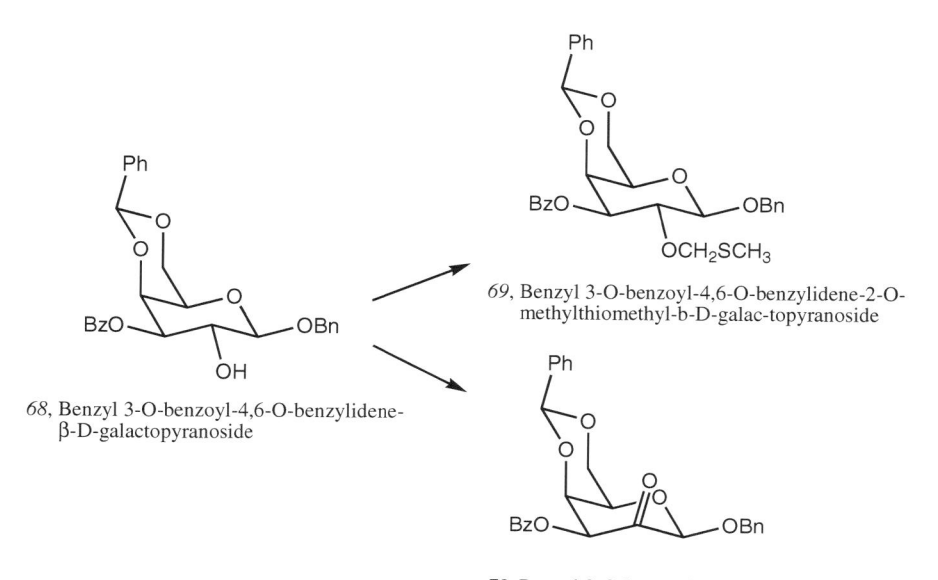

68, Benzyl 3-O-benzoyl-4,6-O-benzylidene-β-D-galactopyranoside

69, Benzyl 3-O-benzoyl-4,6-O-benzylidene-2-O-methylthiomethyl-b-D-galac-topyranoside

70, Benzyl 3-O-benzoyl-4,6-O-benzylidene-β-D-*lyxo*-2-hexulopyranoside

Fig. 10.21

The oxidation of benzyl 3-*O*-benzoyl-4,6-*O*-benzylidene-β-D-galactopyranoside *68* with DMSO–Ac$_2$O (25°C, 4 days) gave almost exclusively the 2-*O*-methylthiomethyl ether *69* (81% yield) (Fig. 10.21); the oxidation product 2-ulose *70* could be detected only in traces (tlc). However, oxidation of benzyl 3-*O*-benzoyl-4,6-*O*-benzylidene-β-D-galactopyranoside *68* with DMSO–P$_4$O$_{10}$ (60°C, 15 h) gave benzyl 3-*O*-benzoyl-4,6-*O*-benzylidene-β-D-*lyxo*-2-hexulopyranoside *70* in 73.5% [60, 61] (Fig. 10.21).

Oxidation of benzyl 2-*O*-benzoyl-4,6-*O*-benzylidene-β-D-galactopyranoside *71* with DMSO–Ac$_2$O gave a complex mixture of products (tlc), but the oxidation with DMSO–P$_4$O$_{10}$ (60–65°C, 14 h) gave benzyl 2-*O*-benzoyl-4,6-*O*-benzylidene-β-D-*xylo*-hexo-pyranosid-3-ulose *72* in 62.5% yield [60, 61] (Fig. 10.22).

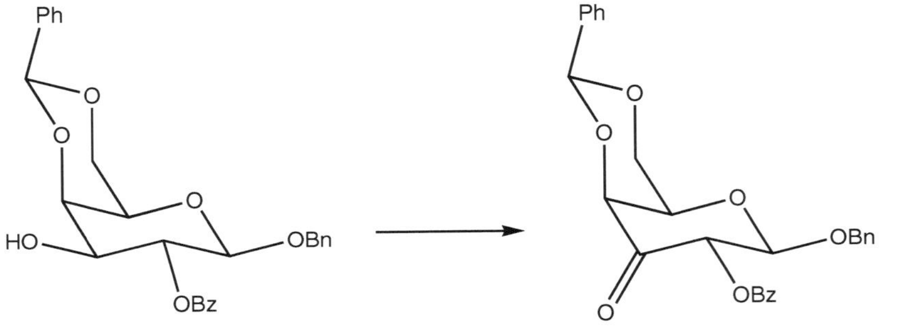

71, Benzyl 2-O-benzoyl-4,6-O-benzylidene-β-D-galactopyranoside

72, Benzyl 2-O-benzoyl-4,6-O-benzylidene-β-D-*xylo*-3-hexulopyranoside

Fig. 10.22

Oxidation of 2,3,4,6-tetra-*O*-benzyl-D-glucopyranose *73* with DMSO–Ac$_2$O (room temperature, overnight) gave 2,3,4,6-tetra-*O*-benzyl-D-glucono-1,5-lactone *74* in 84% yield [62] (Fig. 10.23).

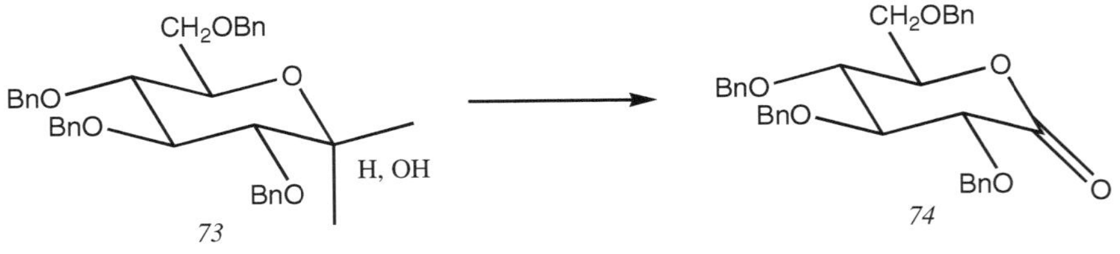

Fig. 10.23

However, the oxidation of 2,3,4,6-tetra-*O*-acetyl-D-glucopyranose *75* with DMSO–P$_4$O$_{10}$ at elevated temperature gave methyl 2,3,4,6-tetra-*O*-acetyl-D-gluconate *76* [63] (Fig. 10.24). It is highly probable that 2,3,4,6-tetra-*O*-acetyl-D-glucono-1,5-lactone was the initial product of oxidation which was converted to the corresponding methyl ester, via solvolysis with methanol during the work-up.

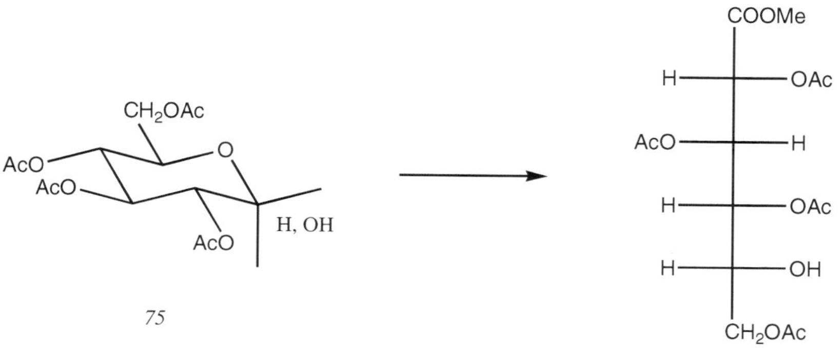

Fig. 10.24

Oxidation of 1,2:4,5-di-*O*-isopropylidene-β-D-fructopyranose *77* with DMSO–Ac$_2$O (room temperature, 48 h) gave 1,2:5,6-di-*O*-isopropylidene-β-D-*erythro*-hexopyranos-2,3-diulose *78* in 70% yield [64] (Fig. 10.25).

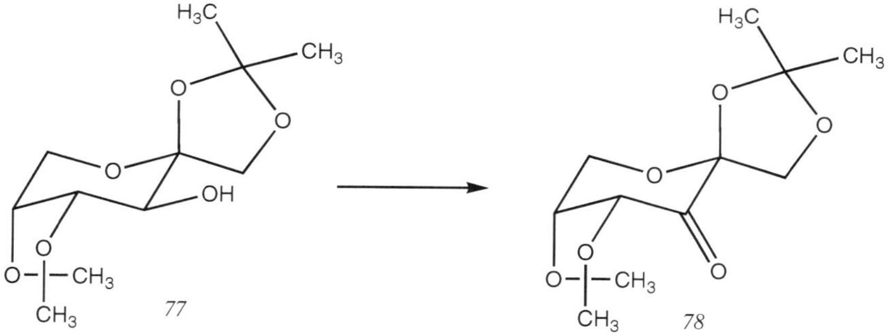

Fig. 10.25

Oxidation of 5-*O*-benzoyl-1,2-*O*-isopropylidene-α-D-xylofuranose *29* with DMSO–Ac$_2$O gave 5-*O*-benzoyl-1,2-*O*-isopropylidene-α-D-*erythro*-3-pentosulofuranose *30* in only 31% yield [65] (Fig. 10.26).

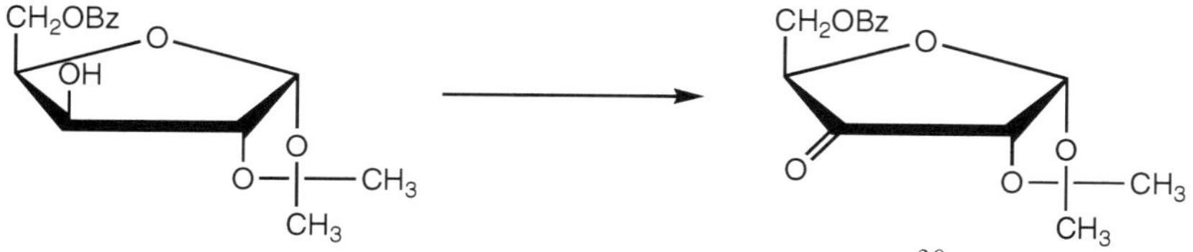

Fig. 10.26

The oxidation of 2,3,4,6-tetra-*O*-benzyl-*N,N*-dimethyl-D-gluconamide *79* with DMSO–Ac$_2$O (room temperature, overnight), gave 2,3,4,6-tetra-*O*-benzyl-*N,N*-

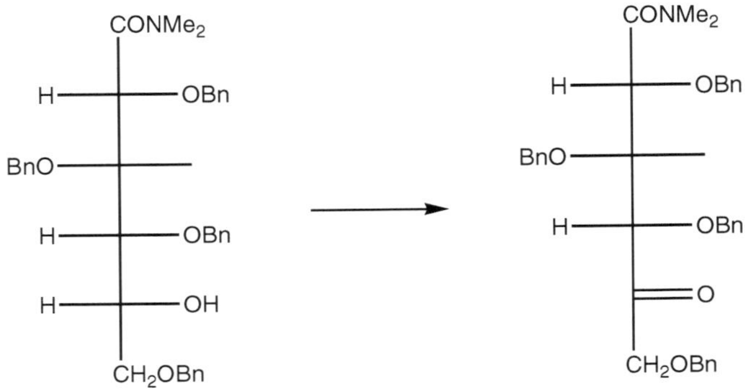

Fig. 10.27

dimethyl-D-*xylo*-5-hexulosonamide *80* (1,3,4,5-tetra-*O*-benzyl-*N,N*-dimethyl-L-sorburonamide) in almost 80% yield [66] (Fig. 10.27).

Oxidation of 2-azido-3,4,6-tri-*O*-benzyl-2-deoxy-D-altropyranose *81* with DMSO–Ac$_2$O (room temperature, 24 h) gave 2-azido-3,4,6-tri-*O*-benzyl-2-deoxy-D-allono-1,5-lactone *82* in 93% yield as a single product (Fig. 10.28). It is inter-

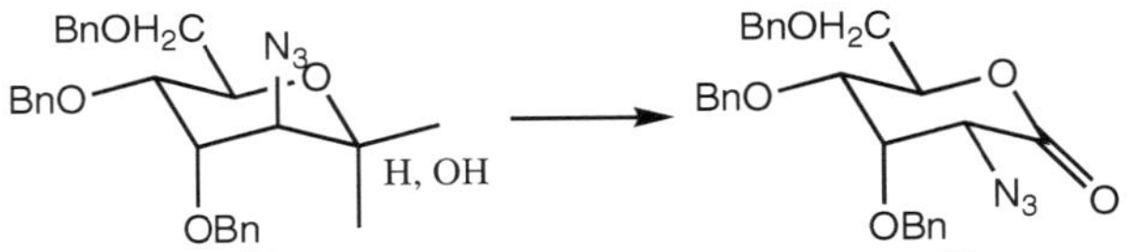

Fig. 10.28

esting that the epimerization of the C2 azido group during the oxidation reaction was almost quantitative. The oxidation of 2-azido-3,4,6-tri-*O*-benzyl-2-deoxy-D-allopyranose *83* with DMSO–Ac$_2$O (room temperature, 24 h) gave 2-azido-3,4,6-tri-*O*-benzyl-2-deoxy-D-allono-1,5-lactone *84* in 100% yield [67] (Fig. 10.29).

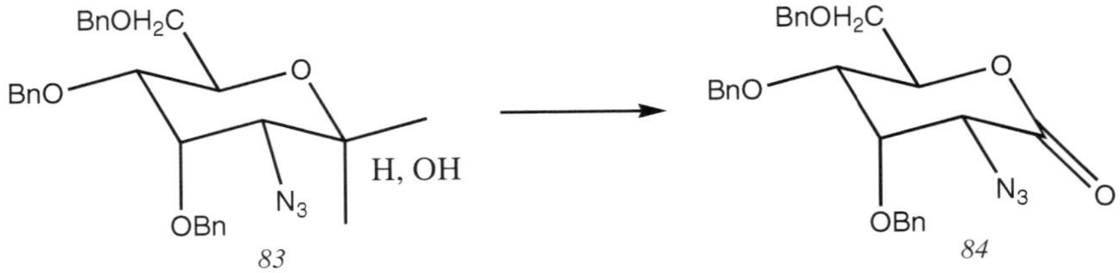

Fig. 10.29

DMSO–Phosphorus Pentoxide

Oxidation of 2-acetamido-3,4,6-tri-*O*-benzyl-2-deoxy-D-glucopyranose *85* with DMSO–Ac$_2$O (room temperature, overnight) gave the 2-acetamido-3,4,6-tri-*O*-benzyl-2-deoxy-D-glucono-1,5-lactone *86* in 92% yield [68] (Fig. 10.30).

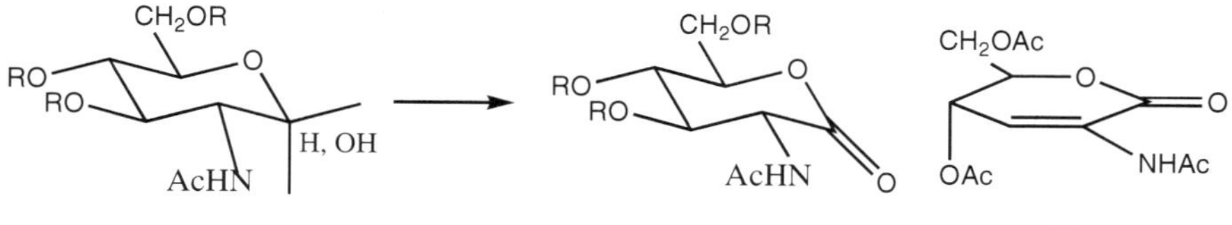

85, R = Benzyl
87, R = Ac

86, R = Benzyl
88, R = Ac

89

Fig. 10.30

However, the oxidation of 2-acetamido-3,4,6-tri-*O*-acetyl-2-deoxy-D-glucopyranose *87* with DMSO–Ac$_2$O (room temperature, overnight) gave 2-acetamido-3,4,6-tri-*O*-acetyl-2-deoxy-D-glucono-1,5-lactone *88* in only 15% yield. The major product, obtained in 50% yield, was the 2-acetamido-4,6-di-*O*-acetyl-2,3-dideoxy-D-*erythro*-hex-2-enono-1,5-lactone *89* an oxidation-elimination product [68] (Fig. 10.30).

The oxidation of 2-acetamido-3,4,6-tri-*O*-acetyl-2-deoxy-D-mannopyranose *90* with DMSO–Ac$_2$O (room temperature, overnight) gave, however, the expected 2-acetamido-3,4,6-tri-*O*-acetyl-2-deoxy-D-manno-1,5-lactone *91* as the major

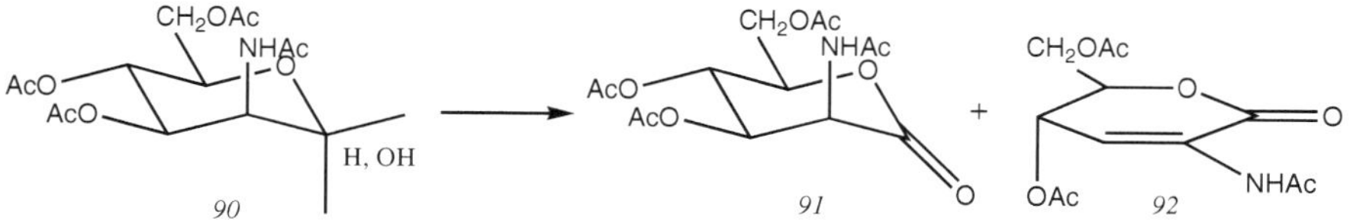

90

91

92

Fig. 10.31

product (42%); the unsaturated 1,5-lactone *92* (the oxidation-elimination product) was obtained in only 16% yield [68] (Fig. 10.31).

The above experiments suggest that DMSO–Ac$_2$O oxidation of the C1-hydroxyl group to lactone is proceeding smoothly when hexopyranose is alkylated, but when it is acylated it undergoes significant oxidation elimination giving the unwanted unsaturated sugar lactones sometimes as the major product.

DMSO–Phosphorus Pentoxide

DMSO and phosphorus pentoxide have been used for a limited number of carbohydrate oxidations [32, 63]. No mechanistic details have been elaborated, but in light of previous mechanisms, phosphorus pentoxide (P$_4$O$_{10}$), which is an anhydride, probably acts as an electrophile (E group) to activate the DMSO, resulting in oxidation via pathway A. This oxidation method like DMSO–Ac$_2$O is capable to oxidize some carbohydrates that are inert to Pfizner–Moffat oxidation [32, 63]. The formation of methylthiomethyl ether has also been reported to be the side product in this oxidation [69] as was previously observed in DMSO–Ac$_2$O oxidations.

Oxidation of 1,2:5,6-di-*O*-isopropylidene-α-D-glucofuranose *17* with DMSO–P$_2$O$_5$ (50°C, 48 h) 1,2:5,6-di-*O*-isopropylidene-α-D-ribo-hexofurano-3-ulose *18* in 45% yield [52] (Fig. 10.5). When the oxidation was performed at room temperature for 24 h 1,2:5,6-di-*O*-isopropylidene-α-D-*ribo*-hexofurano-3-ulose *18* was obtained in 65% yield [63].

Oxidation of methyl 4,6-*O*-benzylidene-2-*O*-p-toluenesulfonyl-α-D-glucopyranoside *23* with DMSO–P$_2$O$_5$ gives methyl 4,6-*O*-benzylidene-2-*O*-p-toluenesulfonyl-α-D-*ribo*-hexopyranosod-3-ulose *24* (Fig. 10.6) in 49% yield [52], whereas oxidation of benzyl 2-*O*-benzoyl-4,6-*O*-benzylidene-β-D-galactopyranoside *71* with DMSO–P$_4$O$_{10}$ (in dimethylformamide solution) (60–65°C, 14 h) gives benzyl 2-*O*-benzoyl-4,6-*O*-benzylidene-β-D-*xylo*-3-hexulopyranoside *72* in 62.5% yield [60, 61] (Fig. 10.22).

DMSO–Sulfurtrioxide Pyridine ("Parikh–Doering" Oxidation)

Primary and secondary alcohols can be oxidized to aldehydes and ketones with pyridinium–sulfurtrioxide–DMSO complex, in the presence of triethylamine [70, 71].

Methylthiomethyl ether derivative of a carbohydrate, which is often formed in significant quantities as by-product in DMSO–acetic anhydride oxidation, is formed in negligible amounts in "Parikh–Doering" oxidation and the yields of aldehydes or ketones compare favorably with those obtained in the DMSO–DCC method.

1,2:4,5-di-*O*-Isopropylidene-β-D-fructopyranose *77* was oxidized with DMSO-SO$_3$·Py-Et$_3$N to 1,2:5,6-di-*O*-isopropylidene-β-D-*erythro*-hexopyranos-2,3-diulose *78* (Fig. 10.25) in 65% yield [72]. Only a trace of methylthiomethyl ether was obtained.

Oxidation of 1,2:3,4-di-*O*-isopropylidene-α-D-galactopyranose *57* with DMSO-SO$_3$Py.-Et$_3$N gave 1,2:3,4-di-*O*-isopropylidene-α-D-galacto-hexodialdo-1,5-pyranose *58* in 85% yield [72] (Fig. 10.17).

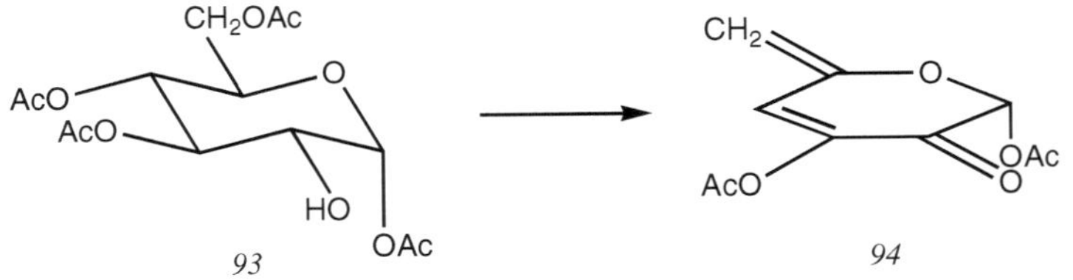

Fig. 10.32

Oxidation of 1,3,4,6-tetra-*O*-acetyl-α-D-glucopyranose *93* (Fig. 10.32) with DMSO-SO$_3$·Py-Et$_3$N gives the elimination product 3,5-diene-2-one *94* in 61% yield [72].

Oxidation of 2,3,4,6-tetra-*O*-acetyl-α-D-glucopyranose *75* gives 2,4,6-tri-*O*-acetyl-3-deoxy-D-*erythro*-hex-2-en-2-ol-1,5-lactone *95* in 81% yield, whereas the oxidation of 2,3,4,6-tetra-*O*-acetyl-α-D-mannopyranose *96* under the same reaction conditions also gives the oxidation-elimination lactone *95*, but the reaction was three times slower ($k_2 = 1/3\ k_1$) [72](Fig. 10.33).

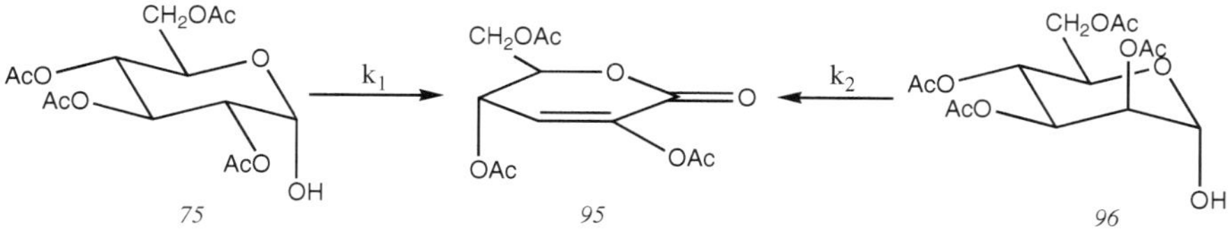

Fig. 10.33

Oxidation of methyl 2,3,4-tri-*O*-acetyl-α-D-glucopyranoside *93* with DMSO-SO$_3$·Py-Et$_3$N gives methyl 2,3-di-*O*-acetyl-6-aldo-4-deoxy-β-L-*threo*-hex-4-enopyranoside *94* in (82% yield) [72] (Fig. 10.34).

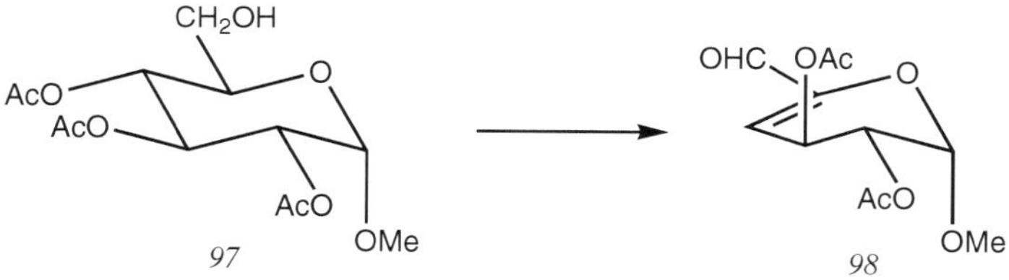

Fig. 10.34

Oxidation of methyl 2,3,4-tri-*O*-acetyl-α-D-mannopyranoside *99* with DMSO-SO$_3$·Py-Et$_3$N (room temperature, 15 min) gave methyl 2,3-di-*O*-acetyl-6-aldo-4-deoxy-β-L-*erythro*-hex-4-enopyranoside *100* in 75% yield [72] (Fig. 10.35).

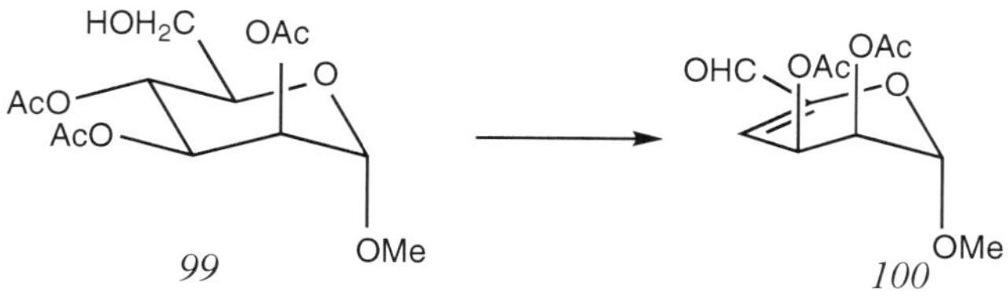

Fig. 10.35

Chromium Trioxide Oxidation

The mechanism of the chromic acid or its anhydride chromium trioxide oxidations has been very thoroughly investigated [73].

It has been postulated that the initial step in chromium trioxide oxidations is fast and reversible esterification of the hydroxyl group that is to be oxidized with chromium trioxide and the formation of a chromic acid ester. This is followed by a rate-determining deprotonation of the carbinol carbon. This deprotonation can be an

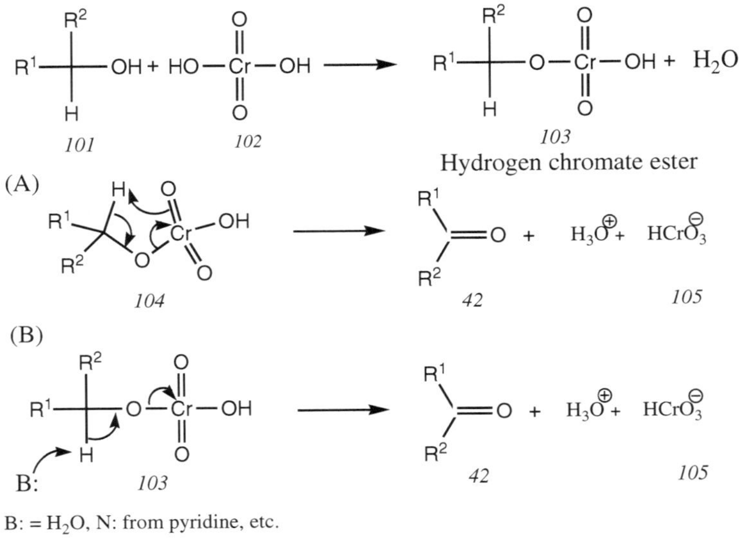

Fig. 10.36

intramolecular process involving the formation of a five-membered cyclic transition state *104* (A in Fig. 10.36) or it could be an *intermolecular* process with a water molecule acting as a general base, or pyridine molecule or any other anion serving as a base (B in Fig. 10.36). In any case, the electron pair of the carbinol carbanion formed by deprotonation of *103* is transferred to alcoholic oxygen that takes place simultaneously with the breaking of the oxygen–chromium bond and the transfer of the electron pair of the chromium oxygen bond to chromium (two-electron reduction of chromium) (Fig. 10.36).

Chromium Trioxide–Pyridine Oxidation

The chromium trioxide–pyridine complex [74–76] is a well-established reagent for the oxidation of primary and secondary alcohols to corresponding carbonyl derivatives in the steroid field. The oxidant was found to be also effective for the oxidation

Chromium Trioxide–Pyridine Oxidation

of primary and secondary hydroxyl groups of suitably protected carbohydrates. This topic together with some other oxidants has been reviewed [77].

The oxidation is generally performed in dichloromethane [78] (12.5 g of the reagent dissolves in 100 mL of dichloromethane at room temperature) and the dichloromethane solution of the oxidant is prepared in situ [79]. Using 12:1 molar ratio of oxidant to substrate the primary alcohols of carbohydrates have been smoothly oxidized to aldehydes in 53–75% yield [80].

Sugar derivatives with sterically hindered endocyclic secondary hydroxyl groups, such as the 3-OH in 1,2:5,6-di-*O*-isopropylidene-α-D-glucofuranose *17*, could not be oxidized with CrO_3–pyridine complex to the corresponding 3-ulose *18* [25–27] (Fig. 10.5).

Oxidation of methyl 6-deoxy-2,3-*O*-isopropylidene-α-L-mannopyranoside (methyl 2,3-*O*-isopropylidene-α-L-rhamnopyranoside) *106* with CrO_3–pyridine complex in pyridine gives in ca. 50% yield (34% pure) methyl 6-deoxy-2,3-*O*-isopropylidene-α-L-*lyxo*-hexopyranosid-4-ulose [81] *107* (Fig. 10.37).

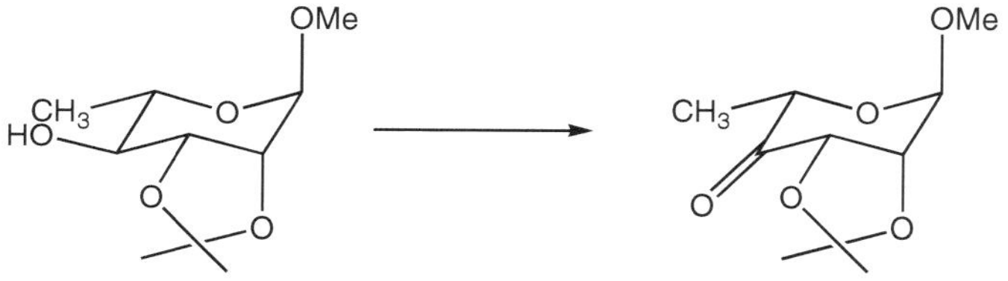

Fig. 10.37

Oxidation of methyl 6-deoxy-3,4-*O*-isopropylidene-α-L-galactopyranoside (methyl 3,4-*O*-isopropylidene-α-L-fucopyranoside) *108* with CrO_3–pyridine complex in pyridine gives methyl 6-deoxy-3,4-*O*-isopropylidene-α-L-*lyxo*-hexopyranosidulose *109* in 36% yield [82] (Fig. 10.38).

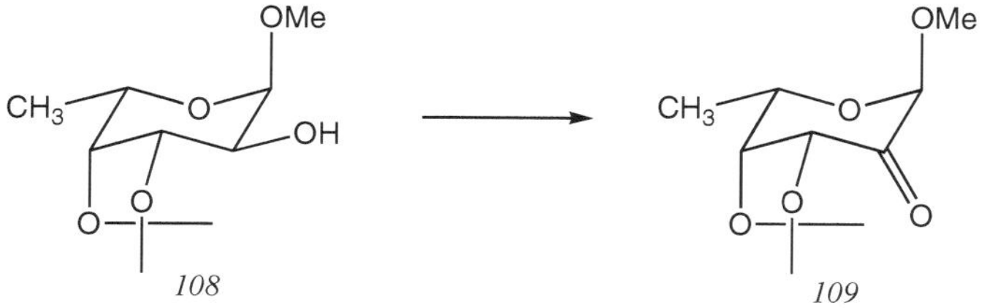

Fig. 10.38

Oxidation of methyl 4,6-*O*-benzylidene-2-deoxy-α-D-*arabino*-hexopyranoside *15* with CrO_3–pyridine complex in pyridine gives methyl 4,6-*O*-benzylidene-2-deoxy-α-D-*erythro*-hexopyranosid-3-ulose *16* in 52% yield [83] (Fig. 10.5).

Oxidation of methyl 3,4-*O*-isopropylidene-β-L-arabinopyranoside *13* with CrO_3–pyridine in pyridine gave, after 20 h, 48% of methyl 3,4-*O*-isopropylidene-β-L-*erythro*-pentopyranosidulose [84] *14* (Fig. 10.5).

Treatment of 2′-deoxyribonucleosides *110* with CrO_3–pyridine gives the corresponding uronic acids *111* (Fig. 10.39). However, the free purine or pyrimidine bases are obtained in appreciable amount as the reaction side products [85] (Fig. 10.39). The formation of free purines or pyrimidines during these oxidations was probably due to the oxidation of the 3′-hydroxyl group to the carbonyl group, thus making the 2′-hydrogen atoms more acidic and prone to deprotonation by a base which results in a base-catalyzed elimination of the purine or pyrimidine (Fig. 10.39).

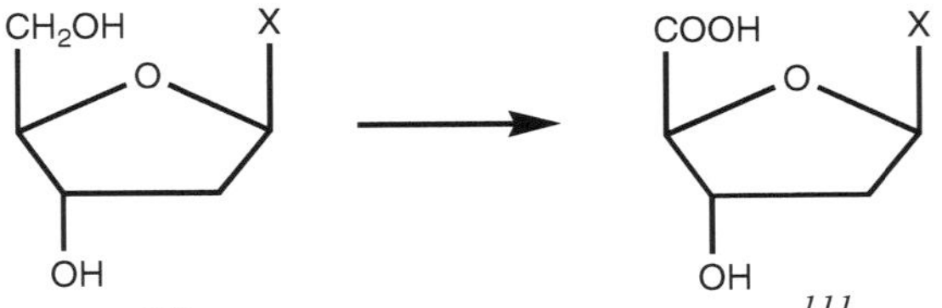

110 *111*

X = thymine, cytosine, adenine, guanine

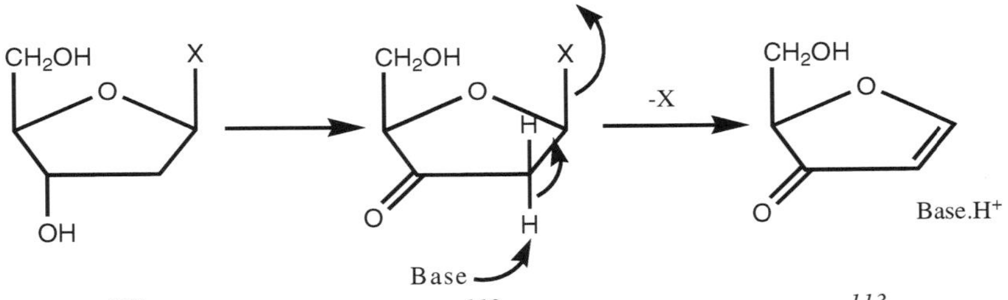

110 *112* *113*

Fig. 10.39

Garegg and Samuelson [86] studied the oxidation of carbohydrates with CrO_3–pyridine complex in dichloromethane in the presence of 1 mol of acetic anhydride per mole of oxidant. The oxidations were performed at room temperature with the excess of oxidant. The experiments with various molar ratios of reagent and hydroxyl compound demonstrated that the optimal results, with almost quantitative yields of oxidized product formed in 5–10 min, were obtained when a 4:1 ratio was used. The results are presented in Table 10.2 and some of the structures in Fig. 10.40

A strong dependence of yield upon the molar excess of used oxidant was observed for 1,2:3,4-di-*O*-isopropylidene-α-D-galactopyranose *58* only. With a molar ratio of 3:1 and 30-min reaction time, the yield of 1,2:3,4-di-*O*-isopropylidene-α-D-galacto-hexodialdo-1,5-pyranose was ca. 65%. A molar ratio

Chromium Trioxide–Pyridine Oxidation

Table 10.2 Oxidation of various partially protected carbohydrates

Starting sugar	Product	Yield (%)
1,2:5,6-di-*O*-isopropylidene-α-D-glucofuranose (*17*)	3-ulose	>90
Methyl 4,6-*O*-benzylidene-2-*O*-tosyl-α-D-glucopyranoside (*23*)	3-ulose	95
1,2:3,4-di-*O*-isopropylidene-α-D-galactopyranose (*58*)	6-aldehyde	93
2,3:5,6-di-*O*-isopropylidene-α-D-mannofuranose (*114*)	1,4-lactone	97
2,3,4-tri-*O*-benzyl-6-*O*-trityl-D-ribitol (*115*)	1-aldehyde	87

of 4:1 and a 5-min reaction time gave 93% yield. Variations in the reaction time indicated that the product is susceptible to further oxidation.

114, 2,3:5,6-di-O-isopropylidene-α-D-mannofuranose

115, 2,3,4-tri-O-benzyl-6-O-trityl-D-ribitol

Fig. 10.40

A possible reason for the high yields obtained in the oxidation when 1 M equivalent of acetic anhydride per mole of chromium trioxide–pyridine complex is added in dichloromethane may be that the acetic anhydride facilitates the reduction of chromium (VI) from the intermediate ester as shown in Fig. 10.41. In this oxidation the acetic anhydride plays a role that is analogous to the role it plays in the methyl sulfoxide–acetic anhydride oxidation (vide supra).

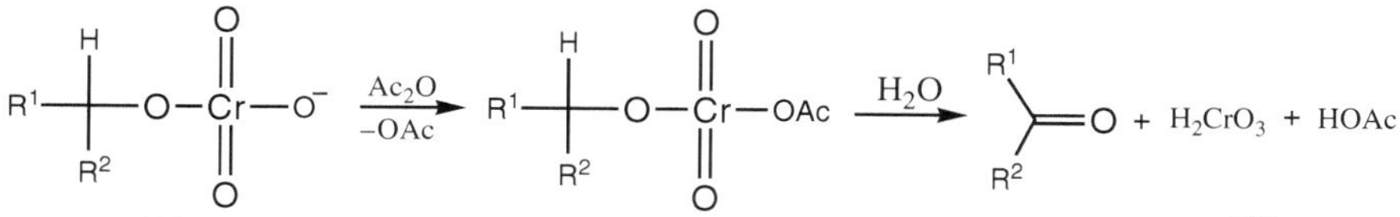

103 → *116* → *42* + H_2CrO_3 + HOAc — *105*

Fig. 10.41

The above speculation about the role of acetic anhydride in the oxidation of the primary or secondary hydroxyl groups to aldehydes or ketones (Fig. 10.41) is based on the generally accepted mechanism for the oxidation of alcohols to carbonyl derivatives with CrO_3 as proposed by Westheimer [87] (vide supra). As already stated, according to this mechanism a rapid and reversible formation of chromate ester is followed by rate-determining abstraction of proton from the carbinol carbon atom. However, the initial step (esterification) may become rate-determining if the hydroxyl group which is to be oxidized is sterically highly hindered [88, 89].

Chromium Trioxide–Acetic Acid

The oxidation of 1,5-di-*O*-benzoyl-2,3-*O*-isopropylidene-D-arabinitol *117* with CrO_3–acetic acid in benzene gave 1,5-di-*O*-benzoyl-3,4-*O*-isopropylidene-D-xylulose *118* in 50% yield [90] (Fig. 10.42).

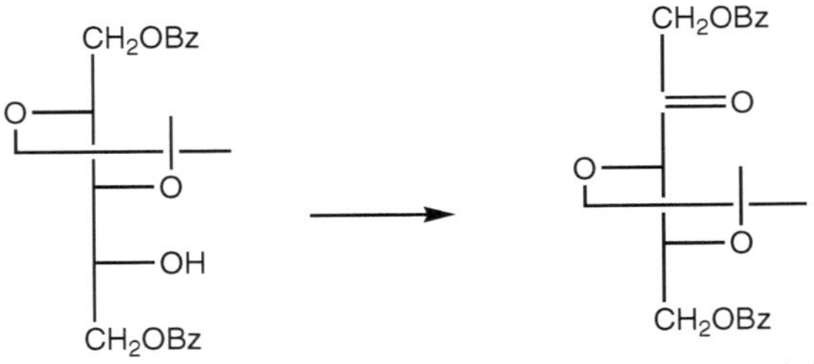

117, 1,5-di-O-benzoyl-2,3-O-isopropylidene-D-arabinitol

118, 1,5-di-O-benzoyl-3,4-O-isopropylidene-D-xylulose

Fig. 10.42

Similarly, the oxidation of 1,3:2,5-di-*O*-methylene-L-rhamnitol *119* gave 2,5:4,6-di-*O*-methylene-1-deoxy-L-*arabino*-3-hexylose *120* in 67% yield [91] (Fig. 10.43).

Stensio and Wachtmeister [92] observed that the oxidation with chromium trioxide–pyridine complex was faster when performed in acetic acid.

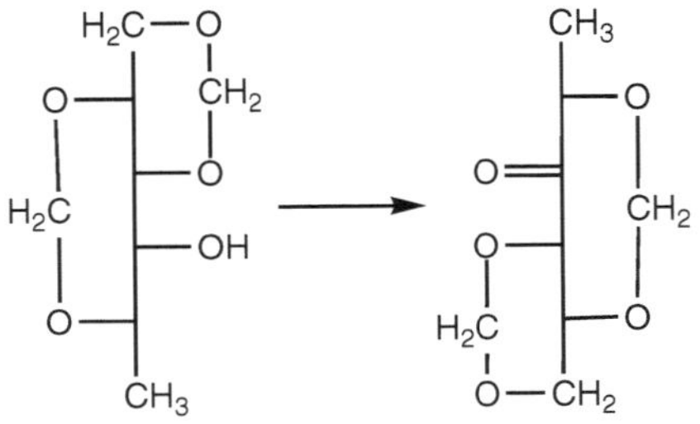

Fig. 10.43

Pyridinium Chlorochromate

Pyridinium chlorochromate (PCC) was found to be a useful reagent for oxidizing primary and secondary hydroxyl groups of partially protected carbohydrate

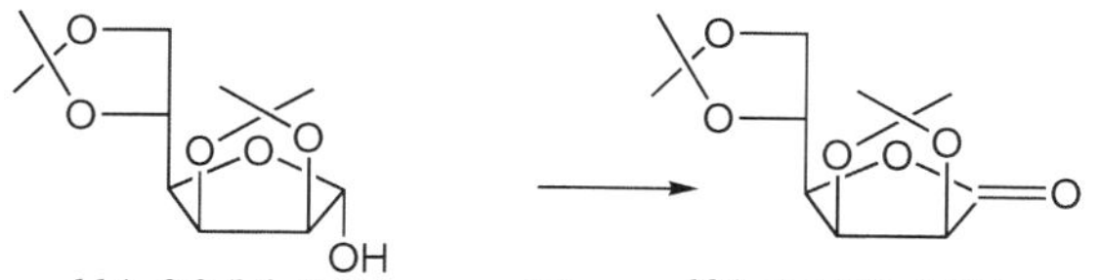

114, 2,3:5,6-di-O-isopropylidene-
α-D-mannofuranose

121, 2,3:5,6-di-O-isopropylidene-
α-D-mannofuranose

Figure 10.44

Fig. 10.44

derivatives, and it has been used for the preparation of protected aldonolactones. PCC is an inexpensive commercially available reagent which is easy to handle. Oxidation of 2,3:5,6-di-*O*-isopropylidene-D-mannofuranose *114* gave 2,3:5,6-di-*O*-isopropylidene-α-D-mannofurano-1,4-lactone *121* in 84% yield [93] (Fig. 10.44).

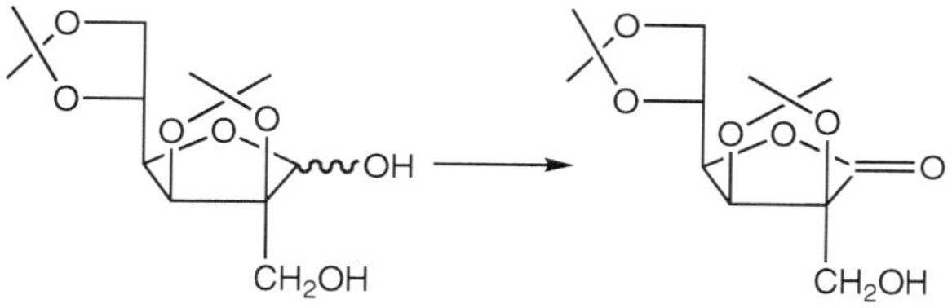

122, 2,3:5,6-di-O-isopropylidene-2-C-
hydroxymethyl-D-mannofuranose

123, 2,3:5,6-di-O-isopropylidene-2-C-
hydroxymethyl-D-mannofurano-1,4-lactone

Fig. 10.45

Oxidation of 2,3:5,6-di-*O*-isopropylidene-2-*C*-hydroxymethyl-D-mannofuranose *122* with PCC gave the corresponding lactone *123* in 71% yield [94] (Fig. 10.45).

Similarly, when 2,3-*O*-cyclohexylidene-D-ribose *124* is oxidized with PCC in dichloromethane the corresponding D-ribonolactones *125* is obtained without oxidation of the primary hydroxyl group [95] (Fig. 10.46). Pyridinium dichromate (PDC) gave the same result [95], while under the same reaction conditions the oxidation of 2,3-*O*-isopropylidene-D-ribose *126* with PCC gave a mixture of products.

In all procedures using chromium trioxide as oxidant the workups are tedious and almost always involve the chromatographic purification of product.

Pyridine chlorochromate is a good oxidant for a large-scale preparation of keto sugars [93] and its reactivity is increased when used together with a molecular sieve [96, 97].

Thus oxidation of 1,2:5,6-di-*O*-isopropylidene-α-D-glucofuranose *17* with pyridine chlorochromate gave 40–45% [98] or 63% [93] of the corresponding 3-ulose *18* (Fig. 10.5). However, the yield can be increased to up to 96% by using pyridinium dichromate–acetic anhydride [99].

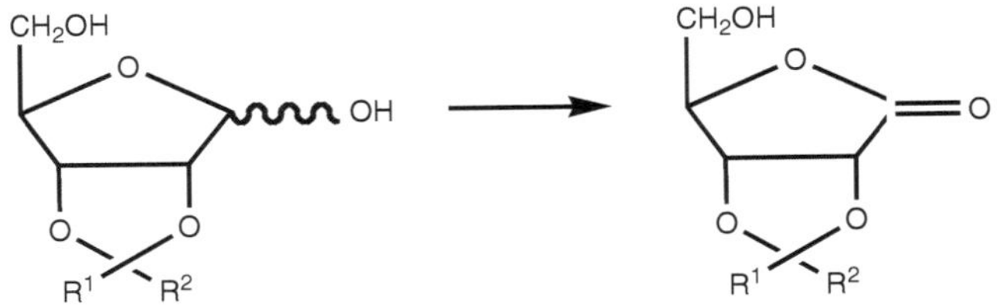

124, $R^1 = C_6H_{10}$; $R^2 = H$
2,3-O-cyclohexylidene-D-ribofuranose
126, $R^1 = R^2 = CH_3$
2,3-O-isopropylidene-D-ribofuranose

125, $R^1 = C_6H_{10}$; $R^2 = H$
2,3-O-cyclohexylidene-D-ribofuranolactone
127, $R^1 = R^2 = CH_3$
2,3-O-isopropylidene-D-ribofuranolactone

Fig. 10.46

Nicotine Dichromate

Nicotinium dichromate is another chromium (VI) oxidant that was reported to be useful in oxidations of partially protected carbohydrates [98, 100]. It was found that the rate of oxidation and the yield depend upon the solvent used. Thus the oxidation was very fast and yields were high when benzene was used as the solvent and 1:2 oxidant–pyridine mixture was used (see Table 10.3).

Table 10.3 Oxidation of Sugars with Nicotinium dichromate [100] (at 80°C; nicotine dichromate–pyridine molar ratio 1:2)

Carbohydrate	Ulose	Oxidant/sugar molar ratio	Time	Yield[a]
17	18	3:1	2.5 h	85(75)
58	59	2:1	20 min	90(75)
128	129	3:1	1 h	94(85)
130	131	3:1	30 min	92(81)

[a]The numbers in parenthesis indicate the yields of pure products (after distillation or crystallization).

The structures of compounds *128, 129, 130,* and *131* are given in Fig. 10.47.

Pyridinium Dichromate–Acetic Anhydride

Smooth and efficient oxidation of primary and secondary hydroxyl groups of partially protected carbohydrates in dichloromethane was achieved with pyridinium dichromate–acetic anhydride, whereby high yields of the corresponding uloses are obtained. Typically 0.6 M equivalent of pyridinium dichromate and 3.0 M equivalent of acetic anhydride are used. Even the unreactive hydroxyl group of 1,2:5,6-di-*O*-isopropylidene-α-D-glucofuranose *17* was oxidized efficiently (96% yield). In the following table the results of conducted studies [99] are given (Table 10.4).

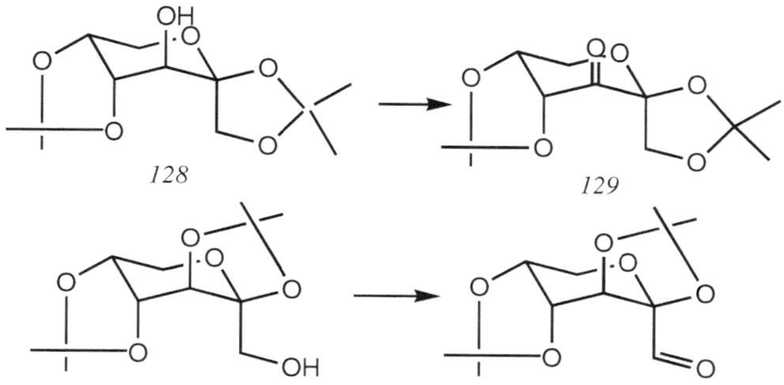

Oxidation of Carbohydrates with the Cleavage of Carbohydrate Chain

Fig. 10.47

Table 10.4

Carbohydrate	Product	Yield (%)
23 (Fig. 10.6)	*24*	93
58 (Fig. 10.17)	*59*	71
17 (Fig. 10.5)	*18*	96

Oxidations were performed at 40°C (0.5–2.0 h) with 1.0 mmol of substrate. All yields refer to isolated product.

The use of pyridinium dichromate–acetic acid-molecular sieve system has also been described [101] for small-scale oxidation of sugars.

Oxidation of Carbohydrates with the Cleavage of Carbohydrate Chain

Periodate Oxidation

Periodic acid and its salts quantitatively cleave the carbon–carbon bond of 1,2-diols [102, 103] (Fig. 10.48).

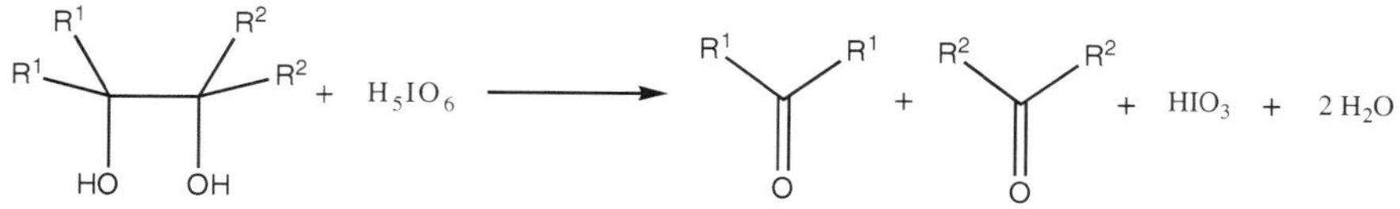

Fig. 10.48

The periodate oxidations named "Malapradian oxidations" were quantitative and reasonably fast at room temperature. They can be carried out over a wide range of pH.

The method was soon extended to the cleavage of 1,2-hydroxycarbonyl *136* and 1,2-dicarbonyl compounds [102–104] *137* and then to the oxidation of 1,2-aminoalcohols [105] *138* (Fig. 10.49).

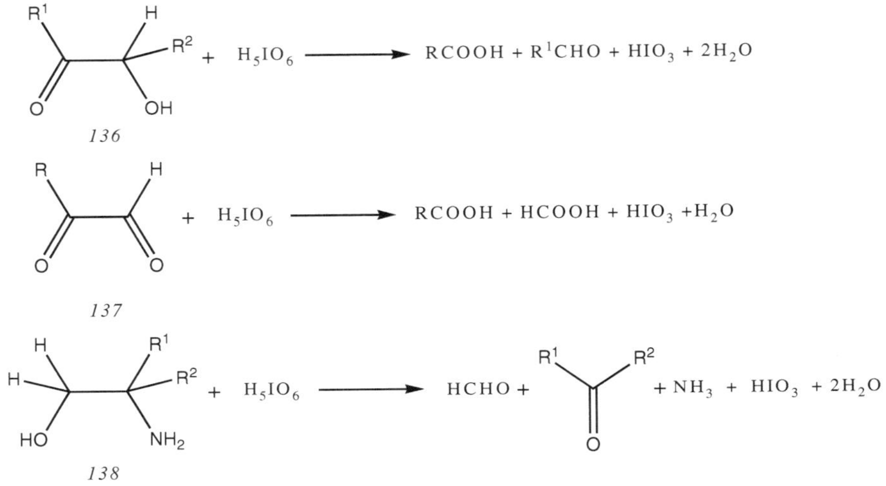

$$\underset{136}{\overset{R^1}{\underset{O}{>}}\!\!\!=\!\!\!\overset{H}{\underset{OH}{<}}\!\!\!R^2} + H_5IO_6 \longrightarrow RCOOH + R^1CHO + HIO_3 + 2H_2O$$

$$\underset{137}{\overset{R}{\underset{O}{>}}\!\!\!=\!\!\!\overset{H}{\underset{O}{<}}} + H_5IO_6 \longrightarrow RCOOH + HCOOH + HIO_3 + H_2O$$

$$\underset{138}{\overset{H}{\underset{HO}{>}}\!\!\!\overset{H}{\underset{}{<}}\!\!\!\overset{R^1}{\underset{NH_2}{<}}\!\!\!R^2} + H_5IO_6 \longrightarrow HCHO + \underset{O}{\overset{R^1}{\underset{}{>}}\!\!\!=\!\!\!\overset{R^2}{\underset{}{<}}} + NH_3 + HIO_3 + 2H_2O$$

Fig. 10.49

These periodate oxidations are also chemoselective – e.g., the monofunctional alcohols, aldehydes, and ketones are either inert or react only very slowly with the reagent. However, periodate does oxidize some other organic functional groups, such as active methylene carbons, some phenols, and thiols.

Equilibrium between periodic acid and its various anions is set up rapidly; i.e., there is rapid oxygen exchange between various periodate ions and water. There is some uncertainty as to the state of periodate monoanion in water. Considering

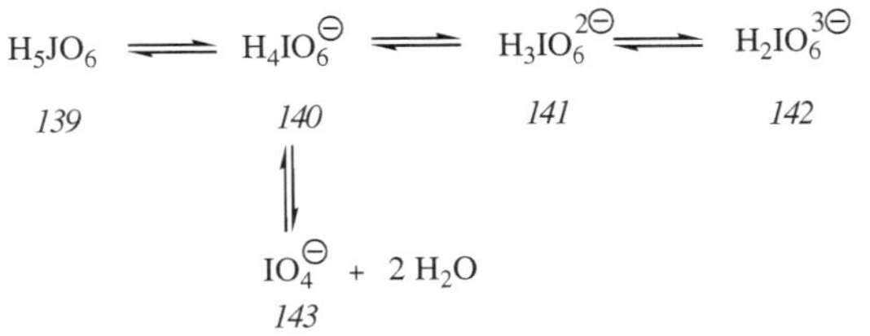

$$\underset{139}{H_5JO_6} \rightleftharpoons \underset{140}{H_4IO_6^{\ominus}} \rightleftharpoons \underset{141}{H_3IO_6^{2\ominus}} \rightleftharpoons \underset{142}{H_2IO_6^{3\ominus}}$$

$$\updownarrow$$

$$\underset{143}{IO_4^{\ominus}} + 2H_2O$$

Fig. 10.50

only monomers, the main equilibria are thought [106, 107] to be as represented in Fig. 10.50. Criegee [108] observed that periodic acid, similar to lead tetraacetate,

Oxidation of Carbohydrates with the Cleavage of Carbohydrate Chain 275

oxidized *cis*-glycols more rapidly than *trans*-isomers and suggested that a logical interpretation for this observation is that *cis*-glycols might be forming a cyclic intermediate *145*, an ester of periodic acid (Fig. 10.51).

Fig. 10.51

The *cis* isomer of cyclohexane-1,2-diol was approximately 30 times more reactive than *trans* isomer, supporting thus the proposed hypothesis for the formation of cyclic diol–periodate complex [109].

The unreactivity of some 1,2-diols to periodate provided an additional strong support for the formation of a cyclic, rather than an open-chain, intermediate [110–112], because most of these unreactive 1,2-diols have geometries that prevent the formation of a cyclic periodate ester, as is the case in *trans*-decalin-9,10-diol *147* (Fig. 10.52). Although *trans*-cyclopentane-1,2-diol *148* is cleaved by periodate

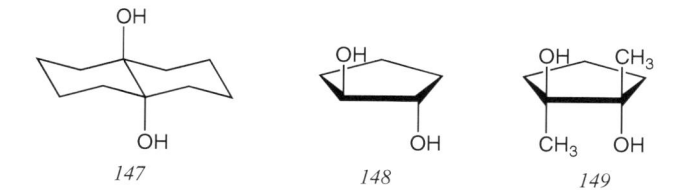

Fig. 10.52

[113, 114] the corresponding *trans*-1,2-dimethyl-cyclopentane-1,2-diol *149* is inert [115], since the formation of a cyclic intermediate would involve not only considerable distortion of the cyclopentane ring but also steric compression between nearby atoms and groups.

All these evidence suggests that a cyclic periodate ester is the key reaction intermediate, and Duke [116] showed that the kinetic form of oxidation of ethane diol by periodate could be interpreted in terms of such an intermediate *145* (Fig. 10.51), present in appreciable concentration, and decomposing slowly to product.

In full agreement with the postulate that a cyclic periodate ester is the intermediate in periodate oxidations of vicinal diols was the observation that aldohexopyranosides having *cis*-oriented hydroxyl groups in their pyranoside structures, such as the

276   10 Oxidation of Monosaccharides

3,4-hydroxyl groups in methyl α-D-galactopyranoside *150* or 2,3-hydroxyl groups in methyl α-D-mannopyranoside *151*, are oxidized faster than glycosides having their hydroxyl groups *trans*-oriented, as is the case in methyl α-D-glucopyranoside [117] *152* (Fig. 10.53).

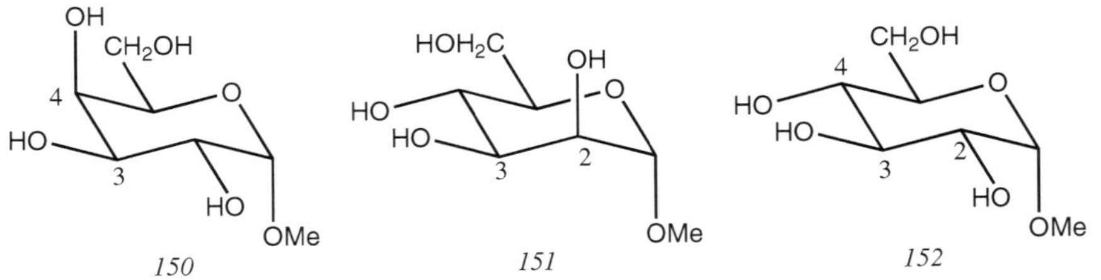

Fig. 10.53

The influence of the steric arrangements of the hydroxyl groups upon the rate of oxidation was also investigated [118–122]. Some rules for oxidation were suggested from a viewpoint of conformational analysis by Honeyman and Shaw [123, 124].

Oxidation of D-glucose was investigated in detail [125–129] (Fig. 10.54) and it was correctly concluded that the first bond cleaved was the bond between the C1 and C2 carbons, and then gradually the glycol linkages of the higher carbons are cleaved.

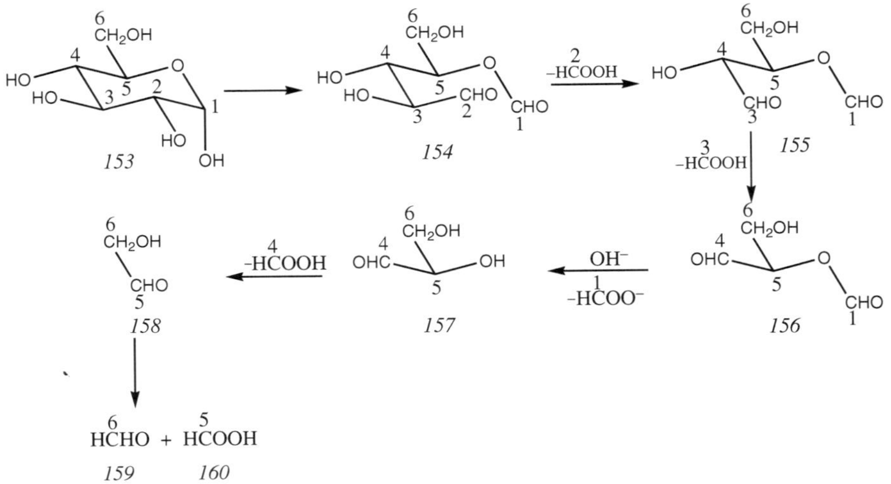

Fig. 10.54

For each 1,2-diol cleavage, there are two electrons lost from the respective carbons that are transferred to iodine of periodate reducing it to iodate, i.e., for each C–C bond cleavage 1 mol of periodate is consumed. Therefore, in the oxidation of D-glucose there will be 5 mol of periodate consumed, five formate molecules released (one by hydrolysis of formate ester), and 1 mol of formaldehyde. So, the oxidation of diols having both hydroxyl groups secondary results in the formation of formic acid, whereas oxidation of diols in which one hydroxyl group is primary and the other secondary results in the formation of formaldehyde and formic acid. Thus, formaldehyde is liberated only at the end of oxidation.

In addition to C–C bond cleavage of the diols, the C–H bond of active methylene or methine group is also oxidized. This non-typical course of oxidation, named "overoxidation" [118, 119], sometimes accompanied with the formation of elemental iodine [117, 130, 131], is encountered in treatment of some derivatives which on oxidation yield malonaldehyde or its derivatives [118]. To such substances belong, besides some disaccharides such as cellobiose, maltose, and lactose [131–135], also some hexofuranosides (e.g. methyl α-D-mannofuranoside [136]) and some deoxy sugars [130, 137] (Fig. 10.55).

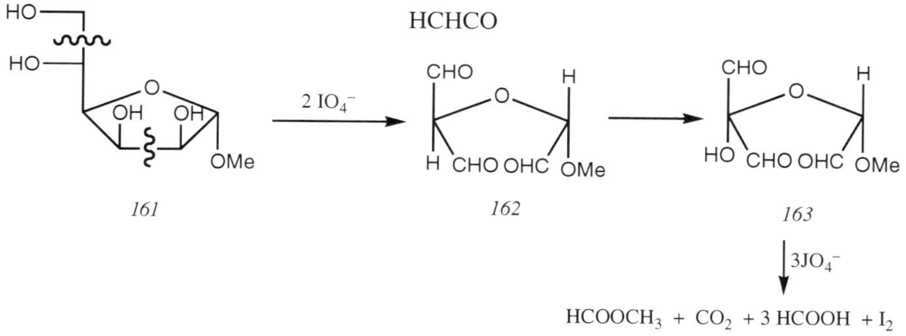

Fig. 10.55

The overoxidation depends on pH, temperature, and concentration of both sugar and periodate [138]. Complications due to overoxidation are encountered with deoxy sugars [130, 139] as well as with some nitrogen [140–149] and sulfur-containing sugars [150–152] which yield sulfoxides or even sulfones by oxidation with periodate. Phosphate esters of sugars are also complicated [153].

Lead Tetraacetate Oxidation

Lead tetraacetate is employed in carbohydrate chemistry in a similar way as the periodic acid [154–157]; however, it is a stronger oxidant because it cleaves besides 1,2-glycols also α-ketoalcohols and α-diketones, as well as α-hydroxyacids, oxalic acid, and formic acid (Fig. 10.56).

The mechanism of oxidation of 1,2-diols by lead tetraacetate was proposed by Criegee [157] and is assumed to proceed via a five-membered ring cyclic intermediate [158, 159] (Fig. 10.57).

The proposed mechanism is supported by the following facts: (1) kinetics are second order (first order in each reactant); (2) addition of acetic acid retards the

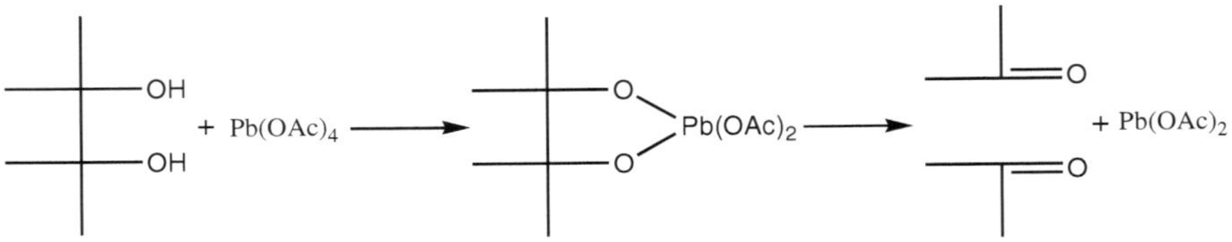

144 *164* *165*

Fig. 10.56

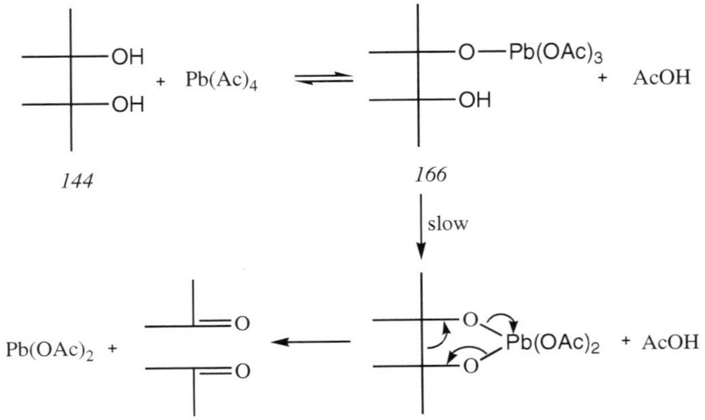

144 *166* *165* *167*

Fig. 10.57

reaction (drives the equilibrium to the left); and (3) *cis* glycols react much more rapidly than *trans* glycols [160].

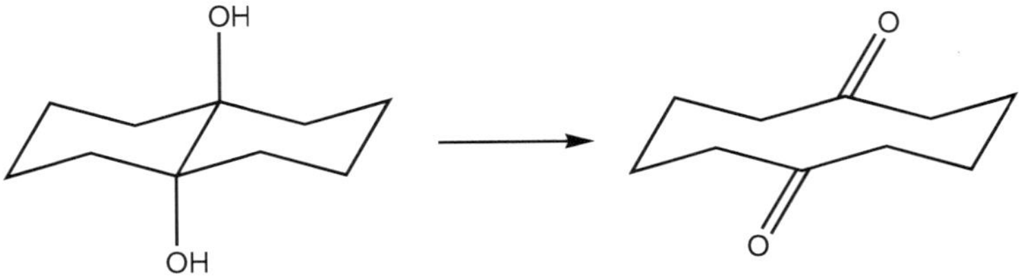

147 *168*

Fig. 10.58

Whereas the *trans*-decalin-9,10-diol *147* (Fig. 10.58) is inert to the oxidation with periodate, because of its inability to form the cyclic ester, it is cleaved by lead tetraacetate to cyclodecane-1,6-dione *168*, although other glycols that cannot form cyclic esters are not cleaved by either reagent [161].

To explain the oxidation of *147* the following transition state has been proposed [160].

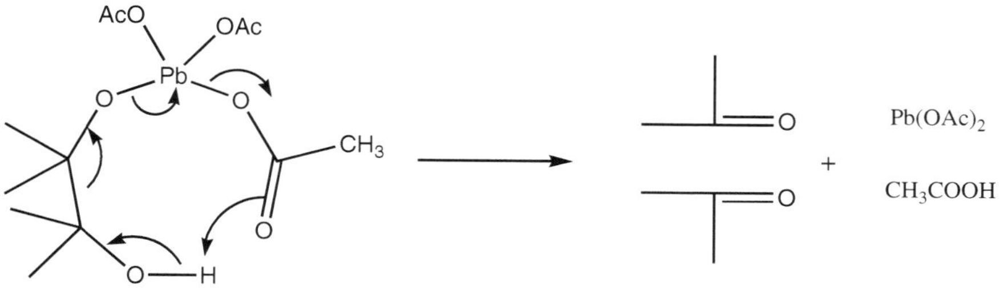

Fig. 10.59

The intermediate *169* (Fig. 10.59) might also break down with an intermolecular proton transfer to an external base [160] (*170* in Fig. 10.60), instead with an intramolecular transfer to acetate of intermediate *169*, shown in Fig. 10.59.

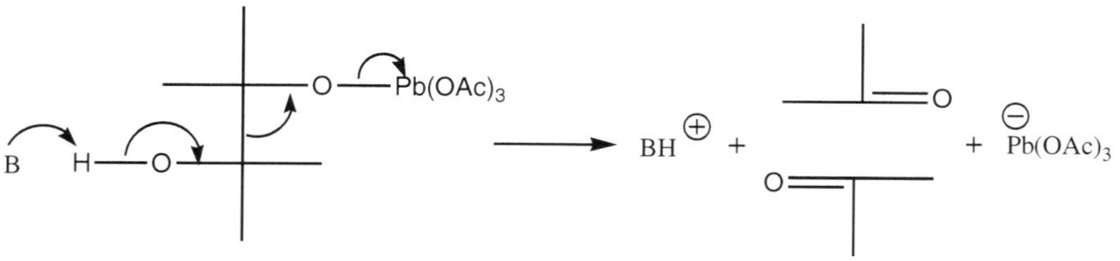

Fig. 10.60

By comparing the rates of oxidation of D-mannitol and D-glucitol (in 50% acetic acid), and D-galactitol in glacial acetic acid [162, 163] it was found that the configuration of alditol exerts little influence on the course of the reaction (Fig. 10.61).

If, however, the oxidation of hexitols was compared with the oxidation of lower alditols, the latter were found to be oxidized with greater difficulty [163, 164]. For example, in 50% acetic acid at 0°C, the percentage of the theoretical uptake of lead tetraacetate found [163] in 1-min reaction time was D-mannitol, 100; glycerol, 10; and ethylene glycol, 4. These results suggest that a *vic*-diol containing two secondary hydroxyl groups are oxidized more readily than the one containing a primary and a secondary hydroxyl group, in agreement with observations on simple aliphatic 1,2-diols [160] (Fig. 10.62).

From this it may be concluded that the oxidation occurs more rapidly toward the center of the hexitol chains than at the ends.

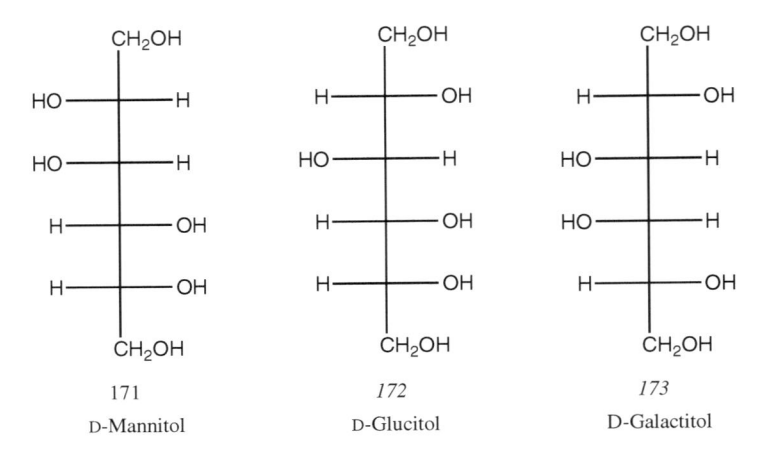

Fig. 10.61

171 D-Mannitol

```
    CH2OH
HO——H
HO——H
H——OH
H——OH
    CH2OH
```

172 D-Glucitol

```
    CH2OH
H——OH
HO——H
H——OH
H——OH
    CH2OH
```

173 D-Galactitol

```
    CH2OH
H——OH
HO——H
HO——H
H——OH
    CH2OH
```

Fig. 10.62

```
      CH2OBz
H———OH
HO∼∼∼H
H———OH
H∼∼∼OH
      CH2OBz
      174
```

$\xrightarrow{Pb(OAc)_4}$ 2 $\begin{array}{c} CHO \\ | \\ CH_2OBz \end{array}$ + 2 HCOOH

175

The use of lead tetraacetate has been particularly well illustrated by Hockett and Fletcher [165] in a study of di- and tri-benzoates of D-glucitol and D-mannitol as shown in Fig. 10.63.

On dibenzoylation of D-glucitol 15 isomeric diesters are possible. However, one major product was isolated which on oxidation with lead tetraacetate gave no trace of formaldehyde, it consumed 3 moles of oxidant and produced 2 moles of formic acid. Since only primary hydroxyl groups can be converted to formaldehyde by oxidation with lead tetraacetate, the absence of formaldehyde suggests that the two primary hydroxyl groups of D-sorbitol (D-glucitol) must be the ones benzoylated. If so then the oxidation of four internal secondary hydroxyl groups will require 3 mol of lead tetraacetate (for cleavage of each C–C bond inside D-glucitol chain 1 mol of lead tetraacetate is needed) and 2 moles of formic acid *178* will be produced in addition to 2 moles of 2-*O*-benzoyl glycolaldehyde *177*. Since this was exactly what was experimentally found it was concluded that the structure of isolated D-glucitol dibenzoate is 1,6-di-*O*-benzoyl-D-glucitol [165].

Oxidation of Carbohydrates with the Cleavage of Carbohydrate Chain   281

[Fig. 10.63 — three oxidation schemes]

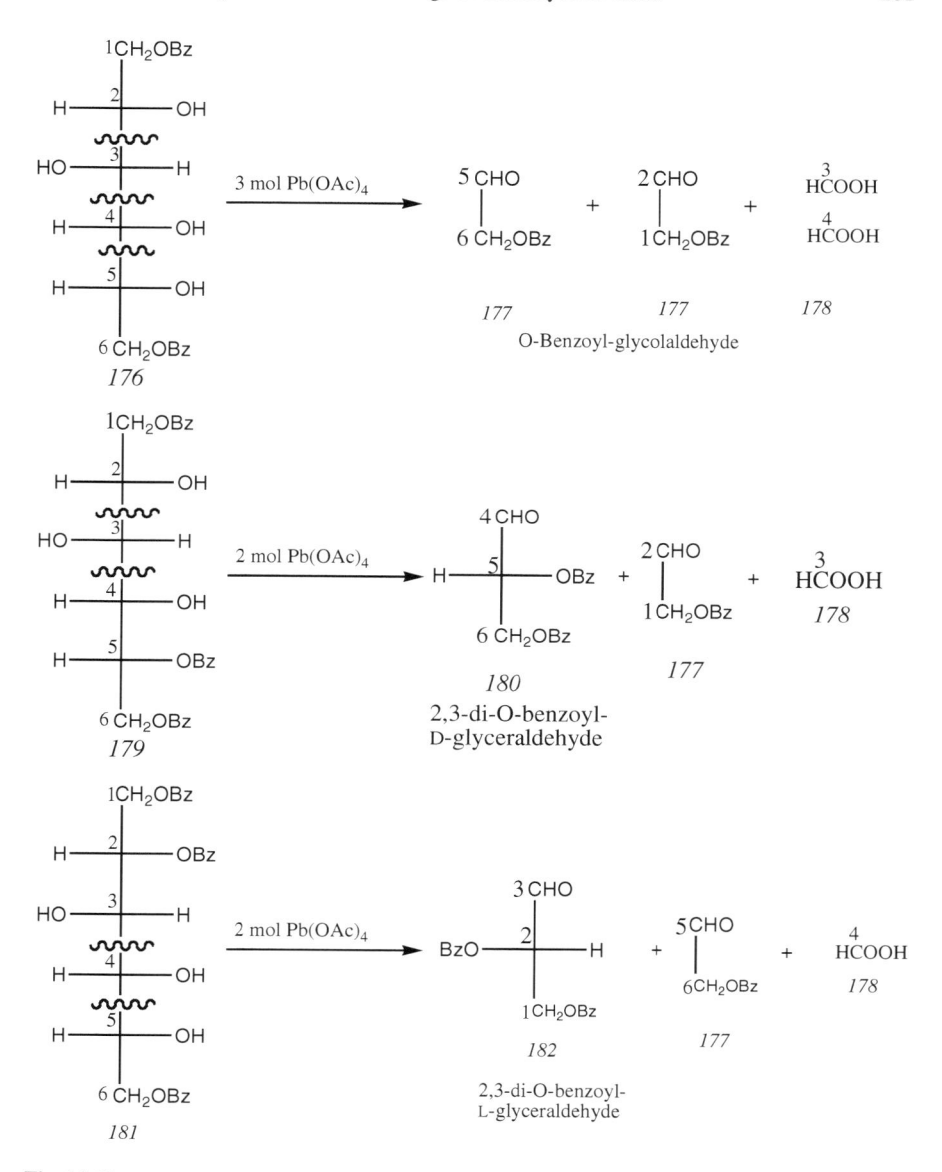

Scheme 1: Compound 176 (1,6-di-O-benzoyl-D-sorbitol with wavy bonds at C-3, C-4, C-5; 1CH₂OBz, H-C-OH (2), HO-C-H (3), H-C-OH (4), H-C-OH (5), 6CH₂OBz) —3 mol Pb(OAc)₄→ 5CHO/6CH₂OBz (177) + 2CHO/1CH₂OBz (177) O-Benzoyl-glycolaldehyde + 3HCOOH + 4HCOOH (178)

Scheme 2: Compound 179 (1,5,6-tri-O-benzoyl; 1CH₂OBz, H-C-OH (2), HO-C-H (3), H-C-OH (4), H-C-OBz (5), 6CH₂OBz) —2 mol Pb(OAc)₄→ 4CHO/H-C-OBz (5)/6CH₂OBz (180, 2,3-di-O-benzoyl-D-glyceraldehyde) + 2CHO/1CH₂OBz (177) + 3HCOOH (178)

Scheme 3: Compound 181 (1,2,6-tri-O-benzoyl; 1CH₂OBz, H-C-OBz (2), HO-C-H (3), H-C-OH (4), H-C-OH (5), 6CH₂OBz) —2 mol Pb(OAc)₄→ 3CHO/BzO-C-H (2)/1CH₂OBz (182, 2,3-di-O-benzoyl-L-glyceraldehyde) + 5CHO/6CH₂OBz (177) + 4HCOOH (178)

Fig. 10.63

A tribenzoyl derivative of D-sorbitol was obtained as a by-product in the preparation of 1,6-di-*O*-benzoyl-D-sorbitol (Fig. 10.63). The oxidation with lead tetraacetate of this D-sorbitol tribenzoate produced no formaldehyde, indicating that both primary hydroxyl groups must be benzoylated. The consumption of oxidant was 2 mol suggesting that two internal C–C bonds have been cleaved. The only two possible structures that are supported by these observations are 1,2,6-tri-*O*-benzoyl-D-sorbitol and 1,5,6-tri-*O*-benzoyl-D-sorbitol. It is obvious that these two isomers cannot be distinguished by oxidation alone. However, since 1,5,6-tri-*O*-benzoyl-D-sorbitol produces 2,3-di-*O*-benzoyl-D-glyceraldehyde *180*, whereas the 1,2,6-tri-*O*-benzoyl-D-sorbitol produces the 2,3-di-*O*-benzoyl-L-glyceraldehyde *182*, by determining the configuration of the obtained 2,3-di-*O*-benzoyl-glyceraldehyde by optical rotation, it can be determined whether D-sorbitol tribenzoate was 1,2,6- or 1,5,6-.

A tribenzoyl D-mannitol derivative obtained (Fig. 10.64) as a by-product in preparation of 1,6-di-*O*-benzoyl-D-mannitol could be 1,2,6-tri-*O*-benzoyl-D-mannitol *184*, 1,5,6-tri-*O*-benzoyl-D-mannitol *185*, 1,3,6-tri-*O*-benzoyl-D-mannitol *186*, and 1,4,6-tri-*O*-benzoyl-D-mannitol *187*, under the presumption that the primary hydroxyl groups are the most reactive ones and will be benzoylated first. The oxidation with lead tetraacetate of this tribenzoyl D-mannitol gave 2,3-di-*O*-benzoyl-D-glyceraldehyde, 2-*O*-benzoyl ethanal, and 1 mol of formic acid. Two moles of oxidant are consumed, indicating the cleavage of two C–C bonds. This result is compatible only with the structures of 1,2,6-tri-*O*-benzoyl-D-mannitol *184* or 1,5,6-tri-*O*-benzoyl-D-mannitol *185*, because the lead tetraacetate oxidation of *186* or *187* would consume only 1 mol of oxidant, and both 1,3,6-tri-*O*-benzoyl-D-mannitol *186* and 1,4,6-tri-*O*-benzoyl-D-mannitol *187* would give 2-*O*-benzoyl ethanal and 2,4-di-*O*-benzoyl-D-erythrose as oxidation products. In the case of *186* the 2,4-di-*O*-benzoyl-D-erythrose will contain the carbon atoms 1–4 and 2-*O*-benzoyl ethanal the carbon atoms 5 and 6, whereas the erythrose obtained from *187* will contain the carbon atoms 3–6, whereas the 2-*O*-benzoyl ethanal will contain the carbon atoms 1 and 2. This is because the structures *186* and *187* are

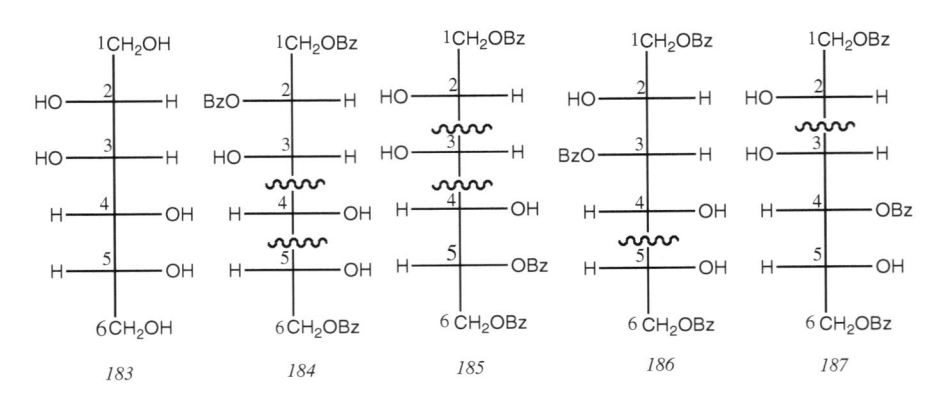

Fig. 10.64

identical (they are interconvertible via rotation in the plane of paper by 180°) (D-mannitol has a center of symmetry passing between the carbon atoms 3 and 4) (Fig. 10.65).

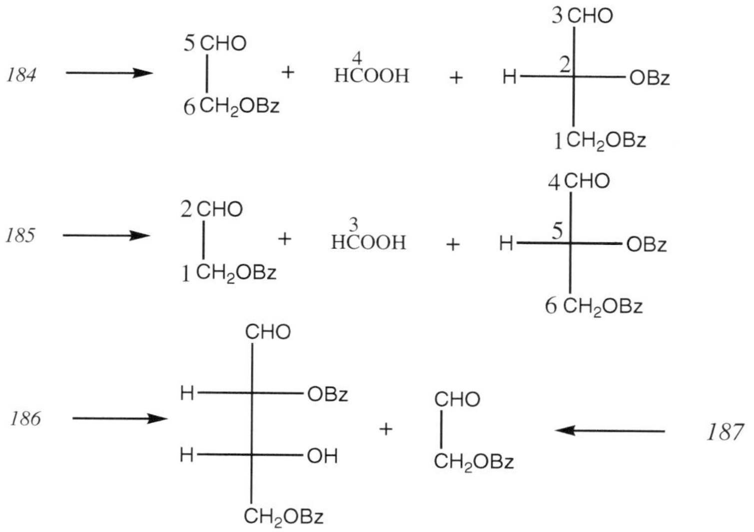

Oxidation of Carbohydrates with the Cleavage of Carbohydrate Chain

[Reaction scheme showing compounds 184, 185, 186, 187 with their oxidation products]

184 ⟶ 5 CHO | 6 CH₂OBz + 4 HCOOH + H—²—OBz (3 CHO top, 1 CH₂OBz bottom)

185 ⟶ 2 CHO | 1 CH₂OBz + 3 HCOOH + H—⁵—OBz (4 CHO top, 6 CH₂OBz bottom)

186 ⟶ CHO / H—OBz / H—OH / CH₂OBz + CHO / CH₂OBz ⟵ 187

Fig. 10.65

Pentavalent Organobismuth Reagents

Five-valent bismuth reagents, especially triphenylbismuth carbonate, show remarkable functional group selectivity, permitting alcohol oxidation even in the presence of benzenethiol, indole, and pyrrole (Fig. 10.66).

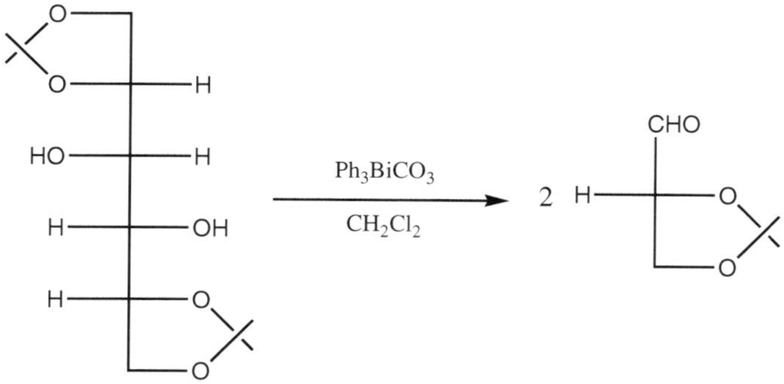

[Scheme: compound 188 (1,2:5,6-di-O-isopropylidene-D-mannitol) → Ph₃BiCO₃, CH₂Cl₂ → 2 × compound 189 (2,3-O-isopropylidene glyceraldehyde)]

Fig. 10.66

Thus, for example, 1,2:5,6-di-*O*-isopropylidene-D-mannitol *188* is oxidized (40°C, 2 h) to 2,3-*O*-isopropylidene glyceraldehyde *189* in 89% yield [166].

References

1. Heyens, K.; Paulsen, H, "*Neuere Methoden der präparativen organischen Chemie II. 8. Selektive katalytische Oxydationen mit Edelmetall-Katalysatoren*", Angew, Chem. (1957) **69**, 600–608
2. Heynes, K.; Paulsen, H. , in "*Newer Methods of Preparative Organic Chemistry*", Foerst, W. (Ed.), Academic Press, New York and London, 1963, Vol. 2, pp. 303–335
3. Heynes, K.; Paulsen, H.; Ruediger, G.; Weyer, J., "*Configuration and conformation selectivity in catalytic oxidation with oxygen on platinum catalysts*", Fortschr. Chem. Forsch. (1969) **11**, 285–374
4. Heyns, K.; Paulsen, H., "*Selective Catalytic Oxidation of carbohydrates, Employing Platinum Catalysts*", Adv. Carbohydr. Chem. (1962) **17**, 169–221
5. Busch, M., German Pat. 702, 729 (1941); Chem. Abstr. (1941) **35**, 7980
6. Chas. Pfizer & Co., Inc. (J. S. Buckley sand H. D. Embree, British Pat. 786,288 AZ(1957), Chem. Abstr. (1958) **52**, 8190
7. Heyns, K.; Heineman, R., "*Oxidation of carbohydrates. III. Catalytic oxidation of D- glucose*", Liebigs Ann. (1947) **558**, 187–192
8. Heyns, K.; Stoeckel, O., "*Oxidation of carbohydrates. IV. Catalytic oxidation of aldoses to aldonic acids*", Liebigs Ann. (1947) **558**, 192–194
9. Mehltretter, C. L.; Rist, C. E.; Alexander, B. H., U. S. Patent 2,472,168 (1949) Chem. Abstr. (1949) **43**, 7506
10. Barker, S. A.; Bourne, E. J.; Stacey, M., "*Synthesis of uronic acids*", Chem. Ind. (London) (1951) 970
11. Hudson, C. S.; Isbell, H. S., "*Relations Between Rotatory Power and Structure in the Sugar Group. XIX. Improvements in the Preparation of Aldonic Acids*", J. Am. Chem. Soc. (1929) **51**, 2225–2229
12. Hudson, C. S.; Isbell, H. S., Bur. Stand. J. Res. (1929) **3**, 57
13. Bunzel, H. H.; Mathews, A. P., "*The Mechanism of the Oxidation of Glucose by Bromine in Neutral and Acid Solution*", J. Am.; Chem. Soc. (1909) **31**, 464–479
14. Green, J. W. "Oxidative Reactions and Degradation", in *The Carbohydrates, Chemis- try and Biochemistry*, Pigman, W. W.; Horton, D. (Eds) Volume 1B, Academic Press, New York (1980), 1101–1166
15. Isbell, H. S.; Pigman, W. W., "*The oxidation of α- and β-glucose and a study of the isomeric forms of sugar in solution*", Bur. Stand. J. Research (1933) **10**, 337–356
16. Isbel, H. S.; Pigman, W. W., "*Bromine oxidation and mutarotation measurements of the α- and β-aldoses*", Bur. Stand. J. Res. (1937) **18**, 141–194
17. Barker, I. R. L.; Overend, W. G., "*Oxidation of cyclohexanol and derivatives with bromine*", Chem. Ind. (London) (1961) 558–559
18. Barker, I. R. L.; Overend, W. G.; Rees, C. W., "*Reactions at position 1 of carbohydrates. Part VI. The oxidation of - and -D-glucose with bromine*", J. Chem. Soc. (1964) 3254–3262
19. Claus, C., J. Prakt. Chem. (1860) **79**, 28
20. Martin, F. S., "*A basic trinuclear ruthenium acetate*", J. Chem. Soc. (1952) 2682–2684
21. Djerassi, C.; Engle, R. R., "*Oxidations with Ruthenium Tetroxide*", J. Am. Chem. Soc. (1953) **75**, 3838–3840
22. Beynon, P. J.; Collins, P. M.; Overend, W. G., "*The oxidation of carbohydrate derivatives with ruthenium tetroxide*", Proc. Chem. Soc. (1964) 342–343
23. Beynon, P. J.; Collins, P. M.; Doganges, P. T.; Overend, W. G., "*The oxidation of carbohydrate derivatives with ruthenium tetroxide*", J. Chem. Soc. (C) (1966) 1131–1136
24. Nakata, H., "*Oxidation reaction of steroid alcohols by ruthenium tetroxide*", Tetrahedron (1963) **19**, 1959–1963
25. Burton, J. S.; Overend, W. G.; Williams, N. R., Chem. Ind. (London) (1961) 175
26. Burton, J. S.; Overend, W. G.; Williams, N. R., "*Synthesis of L- hamamelose and its epimer*", Proc. Chem. Soc. (1962) 181

27. Burton, J. S.; Overend, W. G.; Williams, N., "*Branched-chain Sugars. Part III. The Introduction of Branching into Methyl 3,4-O-Isopropylidene-βL-arabinoside and the Synthesis of L-Hammamelose*", J. Chem. Soc.(1965) 3433–3445
28. Nutt, R. F.; Dickinson, M. J.; Holly, F. W.; Walton, E., "*Branched-chain sugar nu- cleosides. III. 3′-C-methyladenosine*", J. Org. Chem. (1968) **33**, 1789–1795
29. Nutt, R. F.; Arison, B.; Holly, F. W.; Walton, E. , "*An Oxygen Insertion Reaction of Osuloses*", J. Am. Chem. Soc. (1965) **87**, 3273–3273
30. Lawton, B. T.; Szarek, W. A.; Jones, J. K. N., "*An improved procedure for oxidation of carbohydrate derivatives with ruthenium tetraoxide*", Carbohydr. Res. (1969) **10**, 456–458
31. Collins, P. M.; Doganges, P. T.; Kolarikol, A.; Overend, W. G., "*Further studies of ruthenium tetroxide as an oxidant for carbohydrate derivatives*", Carbohydr. Res. (1969) **11**, 199–206
32. Baker, B. R.; Buss, D. H., "*Synthetic Nucleosides. LXIII.1,2 Synthesis and Reactions of Some α-Sulfonyloxy Oxo Sugars*", J. Org. Chem. (1965) **30**, 2304–2308
33. Cilento, G., "*The Expansion of the Sulfur Outer Shell*", Chem. Rev. (1960) **60**, 147–167
34. Price, C. C., "*Unraveling sulfur bonds*", Chem. Eng. News (1964) **42**, 58–63
35. Typke, V.; Dakkouri, M., "*The force field and molecular structure of dimethyl sul- foxide from spectroscopic and gas electron diffraction data and ab initio calcula- tions*", J. Mol. Struct. (2001), **599** (1–3), 177–193
36. Leake, C. D., "*Dimethyl Sulfoxide*", Science (1966) **152**, 1646
37. Smith, S. G.; Winstein, S., "*Sulfoxides as nucleophiles*", Tetrahedron (1958) **3**, 317–319
38. Johnson, C. R.; Philips, W. G., Abstracts, 149th National Meeting of the American Chemical Society, Detroit, Mich., April 1965, p. 46P
39. Johnson, C. R.; Philips, W. G., "*Reactions of alkoxides with alkoxysulfonium salts*", Tetrahedron Lett. (1965) **6**, 2101–2104
40. Johnson, C. R., "*The Inversion of Sulfoxide Configuration*", J. Am. Chem. Soc. (1963) **85**, 1020–1021
41. Johnson, C. R.; Sapp, J. B., Abstracts145th National Meeting of the American Chemical Society, New York, N.Y., Sept. 1963, p. 23Q
42. Hunsberger, I. M.; Tien, J. M., "*Preparation of ethyl glyoxylate by oxidation of ethyl bromoacetate with dimethyl sulfoxide*", Chem. Ind. (London) (1959), 88–89
43. Jarreau, F. X.; Tchoubar, B.; Goutarel, R., "*Reaction of 3β-(p-tolylsulfonyloxy) ster oids with dimethyl sulfoxide*", Bull. Soc. Chim. France (1962) 887–890
44. Jones, D. N.; Saeed, M. A., "*The reaction between steroid sulphonate esters and di- methyl sulphoxide*", J. Chem. Soc. (1963) 4657–4663
45. Nace, H. R.; Monagle, J. J., "*Reactions of Sulfoxides with Organic Halides. Prepara- tion of Aldehydes and Ketones*", J. Org. Chem. (1959) **24**, 1792–1793
46. Johnson, A. P.; Pelter, A., "*Direct oxidation of aliphatic iodides to carbonyl com- pounds*", J. Chem. Soc. (1964) 520–522
47. Pfitzner, K. E.; Moffatt, J. G., "*The Synthesis of Nucleoside-5′- Aldehydes*", J. Am. Chem. Soc. (1963) **85**, 3027–3027
48. Pfitzner, K. E.; Moffatt, J. G., "*Sulfoxide-Carbodiimide Reactions. I. A Facile Oxida- tion of Alcohols*", J. Am. Chem. Soc. (1965) **87**, 5661–5670
49. Pfitzner, K. E.; Moffatt, J. G., "*Sulfoxide-Carbodiimide Reactions. II. Scope of the Oxidation Reaction*", J. Am. Chem. Soc. (1965) **87**, 5670–5678
50. Albright, J. D.; Goldman, L., "*Dimethyl Sulfoxide-Acid Anhydride Mixtures. New Reagents for Oxidation of Alcohols*", J. Am. Chem. Soc. (1965) **87**, 4214–4216
51. Fenselau, A. H.; Moffatt, J. G., "*Sulfoxide-Carbodiimide Reactions. III.1 Mechanism of the Oxidation Reaction*", J. Am. Chem. Soc. (1966) **88**, 1762–1765
52. Brimacombe, J. S.; Bryan, J. G. H.; Husain, A.; Stacey, M.; Tolley, M. S., "*The oxidation of some carbohydrate derivatives, using acid anhydride-methyl sulphoxide mixtures and the pfitzner-moffatt reagent. Facile synthesis of 3-acetamido-3-deoxy-D-glucose and 3-amino-3-deoxy-D-xylose*", Carbohydr. Res. (1967) **3**, 318–324

53. Kampf, A.; Felsenstein, A.; Dimant, E., "*Two aldehydo-D-erythrose derivatives*", Carbohydr. Res. (1968) **6**, 220–228
54. Horton, D.; Nakadate, M.; Tronchet, J. M. J., "*1,2:3,4-di-O-isopropylidene-α-imagegalacto-hexodialdo-1,5-pyranose and its 6-aldehydrol*", Carbohydr. Res. (1968) **7**, 56–65
55. Albright, J. D.; Goldman, L., "*Indole Alkaloids. III.1 Oxidation of Secondary Alco- hols to Ketones*", J. Org. Chem. (1965) **30**, 1107–1110
56. Horner, L.; Kaiser, P., "*Studien zum Ablauf der Substitution, XVIII Über die Einwirkung von Carbonsäureanhydriden auf Sulfoxyde*", Liebigs. Ann. (1959) **626**, 19–25
57. Oae, S.; Kitao, T.; Kawamura, S.; Kitaoka, Y., "*Model pathways for enzymatic oxidate demethylation—I: The mechanism of the reaction of dimethyl sulphoxide with acetic anhydride*", Tetrahedron (1963) **19**, 817–820
58. Parham, W. E.; Groen, S. H., "*Reaction of Enol Ethers with Carbenes. VI.1 Allylic Rearrangements of Sulfur Ylids*", J. Org. Chem. (1965) **30**, 728–732
59. Sweat, F. W.; Epstein, W. W., unpublished observation in Epstein, W. W.; Sweat, F. W., "*Dimethyl Sulfoxide Oxidations*", Chem. Rev. (1967) **67**, 247–260, Ref. # 93
60. Chittenden, G. J. F., "*Synthesis of some new D-gulopyranose derivatives*", Chem. Commun. (1968) 779–780
61. Chittenden, G. J. F., "*Oxidation of some derivatives of D-galactose with methyl sulphoxideacid anhydride mixtures: a route to derivatives of D-glucose and D- talose*", Carbohydr. Res. (1970) **15**, 101–109
62. Kuzuhara, H.; Fletcher, H. G., Jr., "*Syntheses with partially benzylated sugars. VIII. Substitution at carbon-5 in aldose. The synthesis of 5-O-methyl-D-glucofuranose derivatives*", J. Org. Chem. (1967) **32**, 2531–2534
63. Onodera, K.; Hirano, S.; Kashimura, N., "*Oxidation of Carbohydrates with Dimethyl Sulfoxide Containing Phosphorus Pentoxide*", J. Am. Chem. Soc. (1965) **87**, 4651–4652
64. McDonald, E. J., "*A new synthesis of D-psicose (image-D-hexulose)*", Carbohydr. Res. (1967) **5**, 106–108
65. Tong, G. L.; Lee, W. W.; Goodman, L., "*Synthesis of some 3'-O-methylpurine ribonucleosides*", J. Org. Chem. (1967) **32**, 1984–1986
66. Kuzuhara, H.; Fletcher, H. G., "*Synthesis with partially benzylated sugars. IX. Syn- thesis of a 5-hexulosonic acid (5-oxohexonic acid) derivative and inversion of configuration at C-5 in an aldose*", J. Org. Chem. (1967) **32**, 2535–2537
67. Kuzuhara, H.; Oguchi, N.; Ohrui, H.; Emoto, S., "*Preparation of perbenzylated 2- azido-2-deoxy-D-allono-1,5-lactone and its condensation with an amino acid ester*" Carbohydr. Res.(1972) **23**, 217–222
68. Pravdic, N.; Fletcher, H. G., "*The oxidation of partially substituted 2-acetamido-2-deoxyaldoses with methyl sulfoxide—acetic anhydride. Some 2-acetamido-2-deoxyal donic acid derivatives*", Carbohydr. Res. (1971)**19**, 353–364
69. Onodera, K.; Hirano, S.; Kashimura, N.; Masuda, F.; Yajima, T.; Miyazaki, N., "*Nu- cleosides and Related Substances. V. A Synthetic Procedure for Nucleosides with Use of Phosphorus Pentoxide as Dehydrating Agent*", J. Org. Chem. (1966) **31**, 1291–1292
70. Parikh, J. R.; Doering, W. von E., "*Sulfur trioxide in the oxidation of alcohols by di- methyl sulfoxide*", J. Am. Chem. Soc. (1967) **89**, 5505–5507
71. Parikh, J. R.; Doering, W. von E., U. S. Patent 3,444,216 (May 19, 1969)
72. Cree, G. M.; Mackie, D. W.; Perlin, A. S., "*Facile elimination accompanying some methyl sulfoxide oxidations. Formation of unsaturated carbohydrates*", Can. J. Chem. (1969) **47**, 511–512
73. Wiberg, K., in "*Oxidation in Organic Chemistry*", Wiberg, K. (Ed.), Part A, (1965) Academic Press, New York and London, pp. 159–170 and references cited therein
74. Sisler, H. H.; Bush, J. D.; Accountius, O. E., "*Addition Compounds of Chromic An hydride with Some Heterocyclic Nitrogen Bases*", J. Am. Chem. Soc. (1948) **70**, 3827–3830
75. Poos, G. I.; Arth, G. E.; Beyler, R. E.; Sarret, L. H., "*Approaches to the Total Synthesis of Adrenal Steroids.1 V. 4b-Methyl-7-ethylenedioxy-1,2,3,4,4aα,4b,5,6,7,8,10,10a*

β-dodecahydrophenanthrene-4 β-ol-1-one and Related Tricyclic Derivatives", J. Am. Chem. Soc. (1953) **75**, 422–429
76. Holum, J. R., "Study of the Chromium(VI) Oxide-Pyridine Complex", J. Org. Chem. (1961) **26**, 4814–4816
77. Butterworth, R. F.; Hanessian, S., "Selected Methods of Oxidation in Carbohydrate Chemistry", Synthesis (1971) 71–88
78. Collins, J. C.; Hess, W. W.; Frank, F. J., "Dipyridine-chromium(VI) oxide oxidation of alcohols in dichloromethane", Tetrahedron. Lett. (1968) **9**, 3363–3366
79. Ratcliffe, R.; Rodehorst, R., "Improved procedure for oxidations with the chromium trioxide-pyridine complex", J. Org. Chem. (1970) **35**, 4000–4002
80. Arrick, R. E.; Baker, D. C.; Horton, D., "Chromium trioxide—dipyridine complex as an oxidant for partially protected sugars; preparation of aldehydo and certain keto sugar derivatives", Carbohydr. Res.(1973) **26**, 441–447
81. Gunner, S. W.; Overend, W. G.; Williams, N. R., "The preparation of amino sugars from methyl glycopyranosiduloses: methyl 4-acetamido-4,6-dideoxy-a-α-L- talopyranoside", Carbohydr. Res. (1967) **4**, 498–504
82. Collins, P. M.; Overend, W. G., "A synthesis of 6-deoxy-L-talose", J. Chem. Soc. (1965) 1912–1918
83. Flaherty, B.; Overend, W. G.; Williams, N. R., "Branched-chain sugars. Part VII. The synthesis of D-mycarose and D-cladinose", J. Chem Soc.(C)(1966) 398–403
84. Burton, J. S.; Overend, W. G.; Williams, N. R., "Branched-chain sugars. Part III. The introduction of branching into methyl 3,4-O-isopropylidene-β-L-arabinoside and the synthesis of L-hamamelose", J. Chem. Soc. (1965) 3433–3445
85. Jones, A. S.; Williamson, A. R.; Winkley, M., "The chromium trioxide-pyridine oxidation of deoxyribonucleosides and deoxyribonucleotides", Carbohydr. Res. (1965) **1**, 187–195
86. Garegg, P. J.; Samuelson, B., "Oxidation of primary and secondary alcohols in par- tially protected sugars with the chromium trioxide-pyridine complex in the presence of acetic anhydride", Carbohydr. Res. (1978) **67**, 267–270
87. Westheimer, F. H., "The Mechanisms of Chromic Acid Oxidations", Chem. Rev. (1949) **45**, 419–451
88. Roccaronek, J.; Westheimer, F. H.; Eschenmoser, A.; Moldovány, L.; Schreiber, J., "Chromsäureester als Zwischenprodukte bei der Oxydation von Alkoholen. Gesch- windigkeitslimitierende Veresterung eines sterisch Gehinderten Alkohols" Helv. Chim. Acta (1962) **45**, 2554–2567
89. Wu, G. Y.; Sugihara, J. M., "The effect of stereochemistry on the oxidation of substi- tuted hexitols", Carbohydr. Res. (1970) **13**, 89–95
90. Rammler, D. H.; Dekker, C. A., "The Synthesis of 1,5-Di-O-benzoyl-3,4-O- isopropylidene-D-xylulose from D-Arabitol", J. Org. Chem. (1961) **26**, 4615–4617
91. Bird, J.W.; Jones, J. K. N., "The Synthesis of 3-Hexuloses: Part II. Derivatives of 1-Deoxy-L-arabo-3-hexulose (Syn. 6-Deoxy-L-lyxo-4-hexulose)", Can. J. Chem. (1963) **41**, 1877–1881
92. Stensio, K. E.; Wachtmeister, C. A., "Rapid and selective test for alcohols by using the chromium(VI) oxide-pyridine complex in a glacial acetic acid solution", Acta Chem. Scand. (1964) **18**, 1013–1014
93. Hollenberg, D. H.; Klein, R. S.; Fox, J. J., "Pyridinium chlorochromate for the oxida- tion of carbohydrates", Carbohydr. Res. (1978) **67**, 491–494
94. Malleron, A.; David, S, "A preparation of protected 2-deoxy-2-hydroxymethyl-D- mannose and -D-glucose derivatives not involving organometallic reagents", Carbohydr. Res. (1998) **308**, 93–98
95. Liu, D.; Caperelli, C. A., "A New Synthesis of D-Ribonolactone from D-Ribose by Pyridinium Chlorochromate Oxidation", Synthesis (1991) 933–934
96. Herscovici, J.; Antonakis, K., "Molecular sieve-assisted oxidations: new methods for carbohydrate derivative oxidations", J. Chem. Soc., Chem. Commun.(1980) 561–562

97. Herscovici, J.; Egron, M.-J.; Antonakis, K., *"New oxidative systems for alcohols: mo- lecular sieves with chromium(VI) reagents"*, J. Chem. Soc., Perkin Trans.1 (1982) 1967–1973
98. Roldan, F.; Gonzalez, A.; Palomo, C., *"Nicotinium dichromate: a new cheap reagent for high-yielding large-scale oxidation of carbohydrates"*, Carbohydr. Res. (1986) **149**, C1-C4
99. Andersson, F.; Samuelson, B., *"Pyridinium dichromate-acetic anhydride: a new and highly efficient reagent for the oxidation of alcohols"*, Carbohydr. Res. (1984) **129**, C1-C3
100. Lopez, C.; Gonzalez, A.; Cosio, F. P.; Palomo, C., *"Reagents and synthetic methods. 49. 3-Carboxypyridinium dichromate (NDC) and (4-carboxypyridinium dichromate (INDC). Two new mild, stable, efficient and inexpensive chromium(VI) oxidation reagents"*, Synth. Commun. (1985) **15**, 1197–1211
101. Czernecki, S.; Georgoulis, C.; Stevens, C. L.; Vijayakumaran, K., *"Pyridinium chromate oxidation. Modifications enhancing its synthetic utility"*, Tetrahedron Lett. (1985) **26**, 1699–1702
102. Malaprade, L., *"Oxidation of some polyalcohols by periodic acid-applications"*, Compt. rend. (1928) **186**, 382–384
103. Malaprade, L., *"Action of polyalcohols on periodic acid. Analytical application"*, Bull. Soc. Chim. (France)(1928) **43** (4), 683–696
104. Clutterbuck, P. W.; Reuter, F., *"The reaction of periodic acid with -ketols, -diketones, and -ketonealdehydes"*, J. Chem. Soc. (1935) 1467–1469
105. Nicolet, B. H.; Shinn, L. A., *"The Action of Periodic Acid on α-Amino Alcohols"*, J. Am. Chem. Soc. (1939) **61**, 1615–1615
106. Anbar, M.; Guttman, S., "The Isotopic Exchange of Oxygen between Iodate Ions and Water", J. Am. Chem. Soc. (1961) **83**, 781–783
107. Brodskii, A. I.; Vysotskaya, N. A., *"Oxygen isotope exchange in solutions of acids and salts and its mechanism"*, Zh. Fiz. Khim. (1958) **32**, 1521–1531; Chem. Abstr. (1959) **53**, 1901
108. Criegee, R., Sitzber. Ges. Beförder. Ges. Naturw. Marburg (1934) **69**, 25; Chem. Abstr. (1935) **29**, 6820
109. Price, C. C.; Knell, M., *"The Kinetics of the Periodate Oxidation of 1,2-Glycols. II. Ethylene Glycol, Pinacol and cis- and trans-Cyclohexene Glycols"*, J. Am. Chem. Soc. (1942) **64**, 552–554
110. Angyal, S. J.; Young, R. J., *"Glycol Fission in Rigid Systems. II. The Cholestane- 3β,6,7-triols. Existence of a Cyclic Intermediate"*, J. Am. Chem. Soc. (1959) **81**,5251–5255,
111. Angyal, S. J.; Young, R. J., *"Glycol Fission in Rigid Systems. I. The Camphane- 2,3-diols"*, J. Am. Chem Soc. (1959) **81**,5467–5472
112. Criegee, R.; Büchner, E.; Walther, W., "Die Geschwindigkeit der Glykolspaltung mit BleiIV-acetat in Abhängigkeit von der Konstitution des Glykols", Berichte (1940) **73**, 571–575
113. Bulgrin, V. C., *"The Periodate Oxidation of cis- and trans-Cyclopentanediol-1,2"*, J. Phys. Chem. (1957) **61**, 702–704
114. Bulgrin, V. C; Dahlgren, G., *"The Effect of Methyl Substitution on the Periodate Oxidation of cis- and trans-Cyclopentanediol-1,2"*, J. Am. Chem. Soc. (1958) **80**, 3883–3887
115. Bunton, C. A.; Carr, M. D., *"The hydroxylation of cyclic olefins by iodine and silver acetate"*, J. Chem. Soc. (1963) 770–775
116. Duke, F. R., *"Theory and Kinetics of Specific Oxidation. II. The Periodate-Glycol Reaction"*, J. Am. Chem. Soc. (1947) **69**, 3054–3055
117. Halsall, T. G.; Hirst, E. L.; Jones, J. K. N., *"Oxidation of carbohydrates by the peri- odate ion"*, J. Chem. Soc. (1947) 1427–1432
118. Fleury, P., *"Some nonclassical aspects of the oxidizing action of periodic acid on organic compounds"*, Bull. Soc. Chim France (1955) 1126–1135
119. Fleury, P. F.; Courtois, J. E.; Bieder, A., *"Comparative actions of periodic acid on stereoisomeric sugars"*, Compt. rend. (1951) **233**, 1042–1044
120. Fleury, P. F.; Courtois, J. E.; Bieder, A., *"Action of periodic acid on stereoisomeric sugars and polyhydric alcohols"*, Bull. Soc. Chim. France (1952) 118–122

References

121. Viscontini, M.; Hürzeler-Jucker, E., "*Beitrag zur Struktur-Ermittlung von O- und N- Glykosiden*", Helv. Chim. Acta (1956) **39**, 1620–1631
122. Pratt, J. W.; Richtmyer, N. K.; Hudson, C. S., "*Proof of the Structure of Sedoheptu- losan as 2,7-Anhydro-β-D-altroheptulopyranose*", J. Am. Chem. Soc. (1952) **74**, 2200–2205
123. Honeyman, J.; Shaw, C. J. G., "*Periodate oxidation. Part III. The mechanism of oxi-dation of cyclic glycols*", J. Chem. Soc. (1959) 2451–2454
124. Honeyman, J.; Shaw, C. J. G., "*Periodate oxidation. Part IV. The effect of confor- mation of cyclic glycols on the rate of periodate oxidation*", J. Chem. Soc. (1959) 2454–2465
125. Karrer, P.; Pfaehler, K., "*Oxydation von Glucose und Glucosederivaten mit Perjod- säure*", Helv. Chim. Acta. (1934) **17**, 766–771
126. Hough, L.; Taylor, T. J.; Thomas, G. H. S.; Woods, B. M., "*The oxidation of mono- saccharides by periodate with reference to the formation of intermediary esters*", J. Chem. Soc. (1958) 1212–1217
127. Head, F. S. H., "*Mechanism of the periodate oxidation of D-glucose*", Chem. & Ind. (London)(1958) 360–361
128. Spencer, C. C.; McGinn, C. J., "*Indirect Method of Determining Optical Rotation of Some Monosaccharides*", Anal. Chem. (1960) **32**, 136–136
129. Warsi, S. A.; Whelan, W. J., "*Mechanism of the periodate oxidation of monosaccha- rides*", Chem. Ind. (London) (1958), 71
130. Lee, J. B., "*Periodate oxidation of deoxy-hexoses and their derivatives*", J. Chem. Soc. (1960) 1474–1479
131. Cerny, M.; Stanek, J., "*The structure of dextrans*", Monats. Chem. (1959) **90**, 157–170
132. Wolfrom, M. L.; Thompson, A.; O'Neill, A. N.; Galkowski, T. T., "*Isomaltitol*", J. Am. Chem. Soc. (1952) **74**, 1062–1064
133. Head, F. S. H.; Hughes, G., "*The oxidation of cellobiose by periodate*", J. Chem. Soc. (1954) 603–606
134. Hough, L.; Woods, B. M., "*Quantitative estimation of carbon dioxide liberated on periodate oxidation of oxygen-substituted monosaccharides via malondialdehyde derivatives*", Chem. Ind. (London) (1957) 1421–1423
135. Perlin, A. S., "*Oxidation of Carbohydrates with Periodate in the Warburg Respi- rometer*", J. Am. Chem. Soc. (1954) **76**, 4101–4103
136. Fletcher, H. G.; Diehl, H. W.; Ness, R. K., "*Methyl β-D-Gulofuranoside and Cer- tain Other Derivatives of D-Gulose*", J. Am. Chem. Soc. (1954) **76**, 3029–3031
137. Manson, L. A.; Lampen, J. O., "*Some Chemical Properties of Desoxyribose Nucleo- sides*", J. Biol. Chem. (1951)**191**, 87–93
138. Cantley, M.; Hough, L.; Pittet, A. O., "*Factors influencing the course of peroxidate oxidation of carbohydrates*", Chem. & Ind. (London) (1959) 1126–1128
139. Cleaver, A. J.; Foster, A. B.; Hedgley, E. J.; Overend, W. G., "*Periodate oxidation of deoxy-sugar derivatives*", J. Chem. Soc. (1959) 2578–2581
140. Kawashiro, I.; Tanabe, H.; Okada, T., "*Periodic acid oxidation of N-glycosides (Preliminary report)*", Yakugaku Zasshi (1953) **73**, 722–724
141. Kawashiro, I., "*Periodic oxidation of N-glucosides. I*", Yakugaku Zasshi (1953), **73**, 892–894
142. Kawashiro, I., "*Periodic acid oxidation of N-glucosides. II*", Yakugaku Zasshi (1953), **73**, 943–946
143. Kawashiro, I., "*Periodic acid oxidation of N-glucosides. III*", Yakugaku Zasshi (1954), **74**, 33–36
144. Kawashiro, I., "*Periodic acid oxidation of N-glucosides. IV*", Yakugaku Zasshi (1954), **74**, 328–330
145. Kawashiro, I., "*Periodic acid oxidation of N-glycosides. V*", Yakugaku Zasshi (1955), **75**, 97–101
146. Kawashiro, I., "*Periodic acid oxidation of N-glycosides. VI*", Yakugaku Zasshi (1955), **75**, 101–104

147. Kawashiro, I., "*Periodic acid oxidation of N-glycosides. VII*", Yakugaku Zasshi (1956), **76**, 70–73
148. Tanabe, H., J. Pharm. Soc. Japan (1956) **76**, 1023
149. Tanabe, H., J. Pharm. Soc. Japan (1957) **77**, 161
150. Bonner, W. A.; Drisko, R. W., "*Periodate Oxidations of Phenyl β-D-Thiogly copyranosides, Phenyl β-D-Glucopyranosyl Sulfones and Related Compounds*", J. Am. Chem. Soc. (1951) **73**, 3699–3701
151. Okui, S., J. Pharm. Soc. Japan (1955) **75**, 1262
152. Hough, L.; Taha, M. I., "*The periodate oxidation of some thioacetals and sul- phones*", J. Chem. Soc. (1957) 3994–3997
153. Loring, H. S.; Levy, L. W.; Moss, L. K.; Ploeser, J. Mc.T., "*Periodate Oxidation of Sugar Phosphates in Neutral Solution. I. D-Ribose 5-Phosphate*", J. Am. Chem. Soc. (1956) **78**, 3724–3727
154. Perlin, A. S., "*Action of Lead Tetraacetate on the Sugars*", Adv. Carbohydr. Chem. (1959) **14**, 9–61
155. Criegee, R., "*Neuere Untersuchungen über Oxydationen mit Bleitetraatat*", Angew. Chem. (1958) **70**, 173–179
156. Criegee, R. "*New methods in organic synthesis. III. Oxidation with lead tetraace- tate and periodic acid*", Angew. Chem. (1940) **53**, 321–326
157. Criegee, R., "*The specificity of oxidizing agents: A comparison of the oxidizing ac- tion of lead tetraacetate and periodic acid upon polyhydroxy compounds*", Sitzber. Ges. Beförder. Ges. Naturw. Marburg (1934) **69**, 25–47; Chem. Abstr. (1935) **229**, 6820
158. Criegee, R., "*Determination of the ring structure of sugars and sugar derivatives*", Annalen(1932) **495**, 211–225
159. Criegee, R.; Kraft, L.; Rank, B., "*Glycol splitting, its mechanism and its use in chemical problems*", Annalen(1933) **507**, 159–197
160. Criegeee, R.; Höger, E.; Huber, G.; Kruck, P.; Marktscheffel, F.; Schellenberger, H., "*Die Geschwindigkeit der Glykolspaltung mit Bleitetraacetat in Abhängigkeit von Konstitution und Konfiguration des Glykols. (III. Mitteilung)*", Liebigs Ann. (1956) **599**, 81–124
161. Angyal, S. J.; Young, R. J., "*Glycol Fission in Rigid Systems. II. The Cholestane- 3β,6,7-triols. Existence of a Cyclic Intermediate*", J. Am. Chem. Soc. (1959) **81**,5251–5255
162. Hocket, R. C.; Dienes, M. T.; Fletcher, H. G.Jr.; Ramsden, H. E., "*Lead Tetraace- tate Oxidations in the Sugar Group. V.1 The Rates of Oxidation of Open-Chain Polyalcohols in Dry Acetic Acid Solution*", J. Am. Chem. Soc. (1944) **66**, 467–468
163. Fleury, P. F.; Courtois, J. E.; Bieder, A., "*Action of periodic acid on stereoisomeric sugars and polyhydric alcohols*", Bull. Soc. chim. France (1952) 118–122
164. Vargha, L., "*Red lead as a selective oxidant*", Nature (1948) **162**, 927–928
165. Hockett, R. C.; Fletcher, H. G.Jr., "*Lead Tetraacetate Oxidations in the Sugar Group. VI.1 The Structures of Certain Di- and Tribenzoates of D-Sorbitol and D- Mannitol*", J. Am. Chem. Soc. (1944) **66**, 469–472
166. Barton, D. H. R.; Lester, D. J.; Motherwell, W. B.; Barros Papoula, M. T., "*Oxida- tion of organic substrates by pentavalent organobismuth reagents*", J. Chem. Soc. Chem. Commun. (1979) 705–707

Chapter 11
Addition of Nucleophiles to Glycopyranosiduloses

Two distinct types of interactions control the stereochemistry of nucleophilic addition to glycopyranosiduloses. One is the classical nonbonded steric interaction and torsional strain between the incoming nucleophile and substituents on a pyranoside ring. These types of interactions are typical for all polysubstituted six-membered cyclic compounds. The second type of interaction is the electrostatic (dipolar) or electronic interaction between the incoming nucleophile and the glycosidic and/or the ring oxygen and is typical for carbohydrates. We will illustrate these by examining the stereochemistry of addition of various nucleophiles to the carbonyl carbon of glycopyranosid-2-, 3-, and 4-ulose of both α- and β-anomers.

The Addition of a Hydride Ion (Reduction)

Reduction of methyl 4,6-*O*-benzylidene-3-*O*-methyl-α-D-*arabino*-hexopyranosid-2-ulose *1* with $LiAlH_4$ in ether gives the corresponding D-gluco-derivative *2* as the only product [1] (Fig. 11.1).

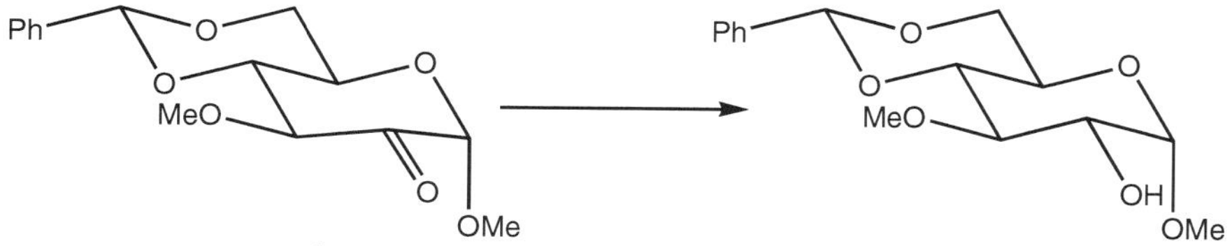

1 *2*

Fig. 11.1

However, reduction of the corresponding β-anomer *3* gives methyl 4,6-*O*-benzylidene-3-*O*-methyl-β-D-mannopyranoside *4* as the only product [1] (Fig. 11.2).

A view has been adopted [2] that the transition-state geometry for the reaction of metal hydrides (and organometallic reagents) with carbonyl groups resembles the geometry of the starting ketone and that nonbonded steric interactions, torsional strain, and electrostatic interactions (dipole–dipole repulsions) are the controlling

M. Miljković, *Carbohydrates*, DOI 10.1007/978-0-387-92265-2_11,
© Springer Science+Business Media, LLC 2010

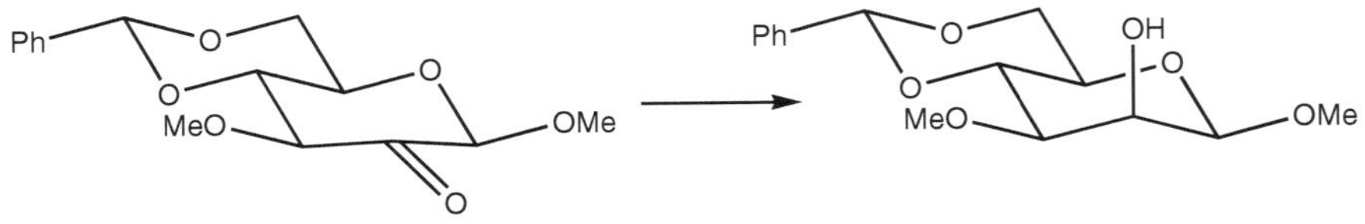

Fig. 11.2

factors in determining the direction of approach of an electronegative nucleophile to a carbonyl carbon. In the case of methyl D-*arabino*-hexopyranosid-2-ulose of the β-series, e.g., *3* (Fig. 11.2), the axial approach of metal hydride anion to the C2 carbonyl carbon, resulting in the formation of the transition state *5* (Fig. 11.3), requires that the negatively charged metal hydride ion approaches the C2 carbonyl carbon from a direction bisecting the C_1–O_1 and C_1–O_5 torsional angle. Since the

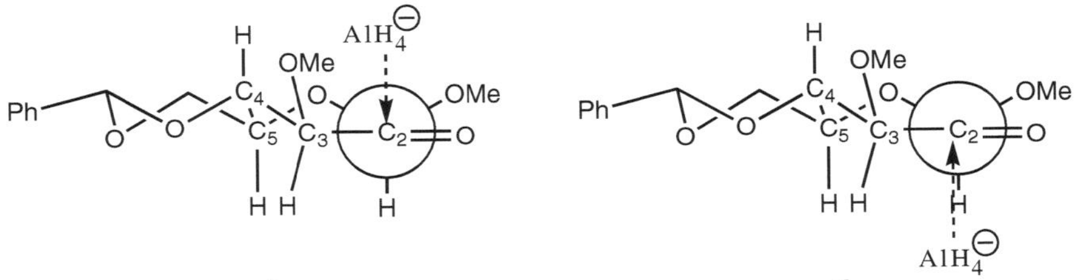

Fig. 11.3

C_1–O_1 and C_1–O_5 bonds are polarized and act as two equally oriented dipoles, with the resultant dipole bisecting this torsional angle, an approach which will appose a negatively charged ion between them should be energetically unfavorable owing to strong electrostatic interactions. An "equatorial" approach of the negatively charged metal hydride ion to the C2 carbonyl carbon of *3*, resulting in the formation of the transition state *6* (Fig. 11.3), will, however, not only be free from this electrostatic interaction, but the torsional strain and nonbonded steric interactions will be at a minimum as well.

In the transition state *7* (Fig. 11.4) which results from an "axial" approach of the negatively charged metal hydride ion to the C2 carbonyl carbon atom of methyl D-arabino-hexopyranosid-2-uloses of the α-series (*1* in Fig. 11.1), the electrostatic interactions of the type described for the transition state *5* are not present. Furthermore, there will be no torsional strain. The only interaction in *7* is one 1,3-*syn* axial steric interaction between the axially oriented C4 hydrogen atom and the incoming metal hydride anion. An "equatorial" approach of the negatively charged metal hydride ion to the C2 carbonyl carbon of *1* resulting in the formation of the transition state *8* (Fig. 11.4) should be, however, subject to a considerable torsional strain and dipolar interaction between the axially oriented C1 methoxy group and

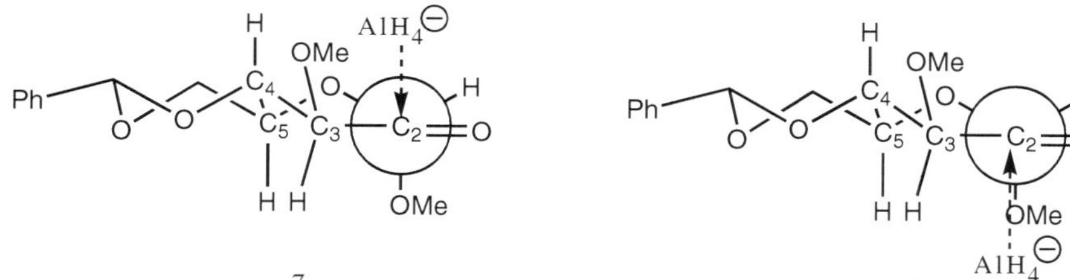

The Addition of a Hydride Ion (Reduction)

Fig. 11.4

the approaching metal hydride anion. Furthermore, in the transition state *8*, there will be two nonbonded steric interaction between the approaching metal hydride anion and the axially oriented hydrogens at the C3 and the C5 carbon.

Shaban and Jeanloz [3, 4] used this highly stereoselective reduction of β-D-*arabino*-hexopyranosid-2-uloses to β-D-mannopyranoside derivatives, to prepare a β-D-mannopyranosyl-containing oligosaccharide, from a properly protected β-D-glucopyranosyl-containing disaccharide having the C2 hydroxyl group free *9* which was oxidized to the corresponding β-D-*arabino*-hexopyranosyl-2-ulose containing disaccharide *10*. Stereoselective reduction of the hexosdiulose *10* with sodium borohydride gave then the protected β-D-mannopyranosyl-containing disaccharide *11* (Fig. 11.5).

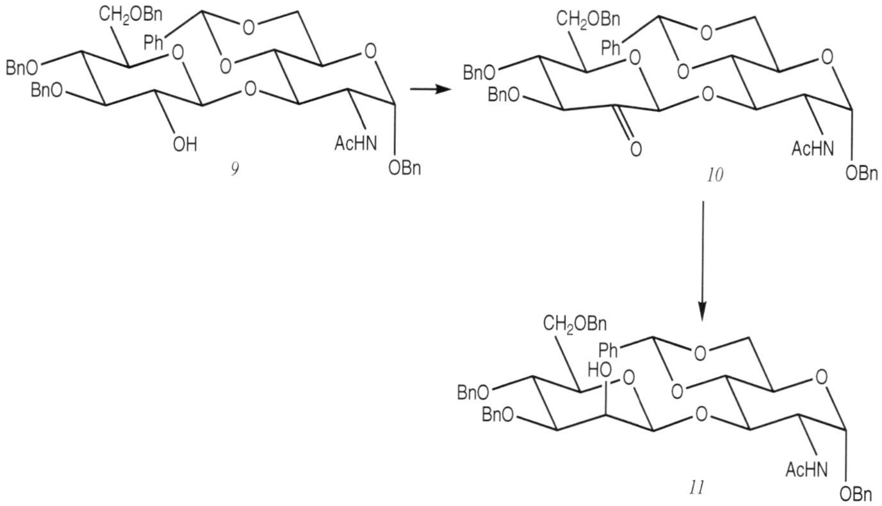

Fig. 11.5

During a study of the regioselective mono-oxidation of nonprotected or partially protected methyl glycopyranosides by bistributyltinoxide–bromine method, Tsuda et al. [5, 6] have identified the oxidation products by NMR spectroscopy and by reducing them with NaBH$_4$ in methanol. By using glc or tlc, Tsuda et al. [5, 6] identified the reduction products and determined the ratio of products obtained in cases where more than one product was obtained by a reduction of given glycopyranosidulose.

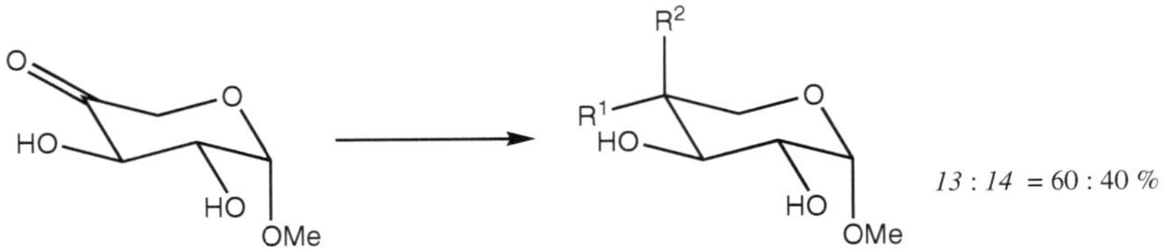

Fig. 11.6

Thus, reduction of methyl β-L-*threo*-pentopyranosid-4-ulose *12* (Fig. 11.6) gave methyl β-L-arabinopyranoside *13* and methyl α-D-xylopyranoside *14* in 60:40 ratio, whereas the reduction of methyl α-D-*xylo*-hexopyranosid-4-ulose *15* gave the mixture of methyl α-D-gluco- (*16*) and α-D-galactopyranoside (*17*) in 7:3 ratio [5].

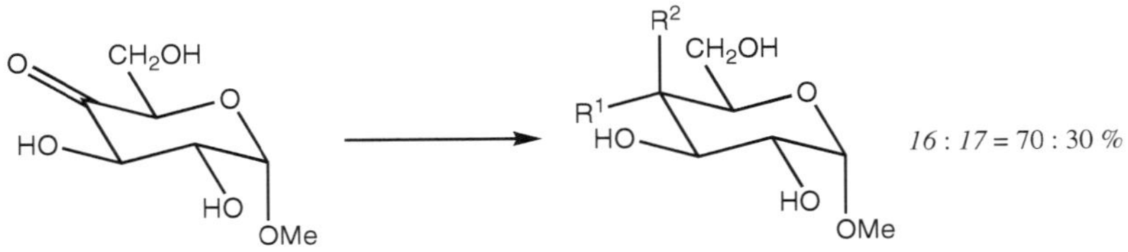

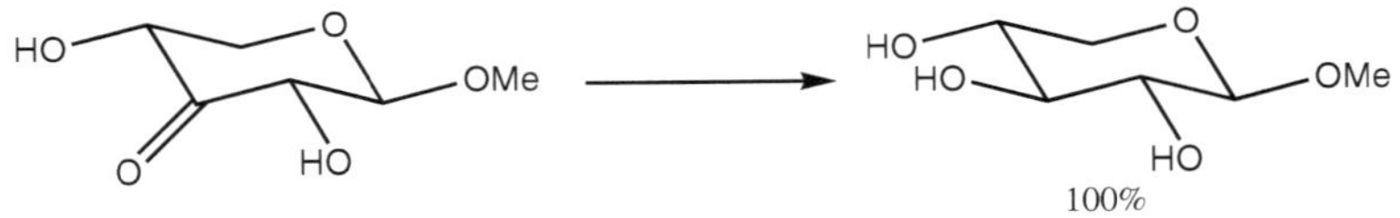

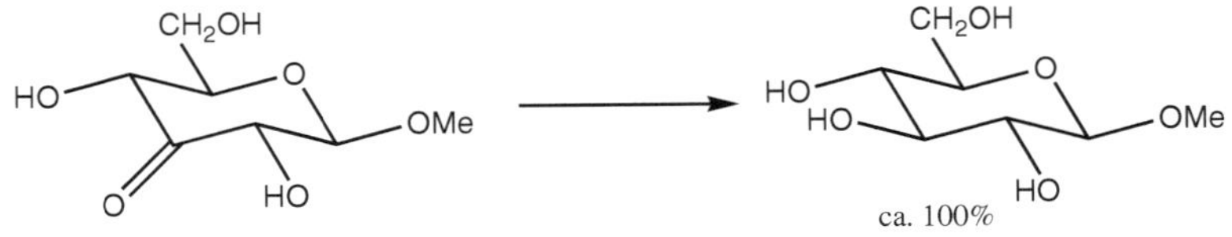

The reason for the improved stereoselectivity in the reduction of *15* over that of *12* could be due to the presence of the C6 hydroxymethyl group and stiffening of the $^4C_1$ conformation of pyranoid ring.

The Addition of a Hydride Ion (Reduction)

The reduction of both methyl β-D-*threo*-pentopyranosid-3-ulose *18* and the methyl β-D-*ribo*-hexopyranosid-3-ulose *20* proceeded with almost 100% stereoselectivity to give methyl β-D-xylopyranoside *19* in the former and methyl β-D-glucopyranoside *21* in the latter case [5].

The reduction of methyl α-D-*ribo*-hexopyranosid-3-ulose *22* (Fig. 11.7) gives highly stereoselective methyl α-D-allopyranoside *24*, with very small amounts of methyl α-D-glucopyranoside *23* (*24:23* ratio is 14:1). However, the reduction of methyl 4,6-*O*-benzylidene-β-D-*ribo*-hexopyranosid-3-ulose *25* proceeds with a poor stereoselectivity giving methyl β-D-glucopyranoside *26* and methyl β-D-allopyranoside *27* in 1:2 ratio [5, 6]. The loss of stereoselectivity observed in the reduction of *25* as compared to the reduction of *22* is probably due to the absence of electrostatic interactions between the approaching complex hydride anion and the C1–OMe dipole.

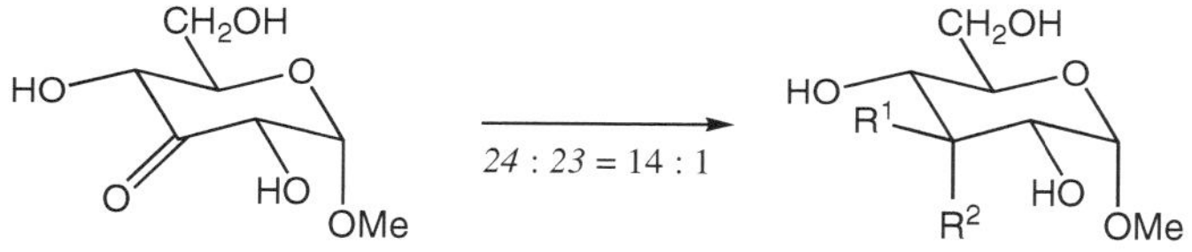

22, methyl α-D-*ribo*-hexopyranosid-3-ulose

23, R^1 = OH; R^2 = H, methyl α-D- glucopyranoside
24, R^1 = H; R^2 = OH, methyl α-D- allopyranoside

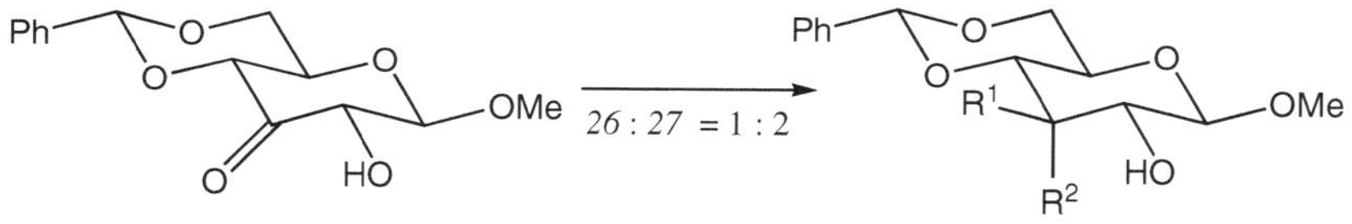

25, methyl 4,6-O-benzylidene-α-D-*ribo*-hexopyranosid-3-ulose

26, R^1 = OH; R^2 = H, methyl 4,6-O-benzylidene-α-D- glucopyranoside
27, R^1 = H; R^2 = OH, methyl 4,6-O-benzylidene-α-D- allopyranoside

Fig. 11.7

The catalytic reduction of methyl 3,4,6-tri-*O*-benzyl-β-D-*arabino*-hexopyranosid-2-ulose *28* in presence of Adams catalyst (Pt) followed by

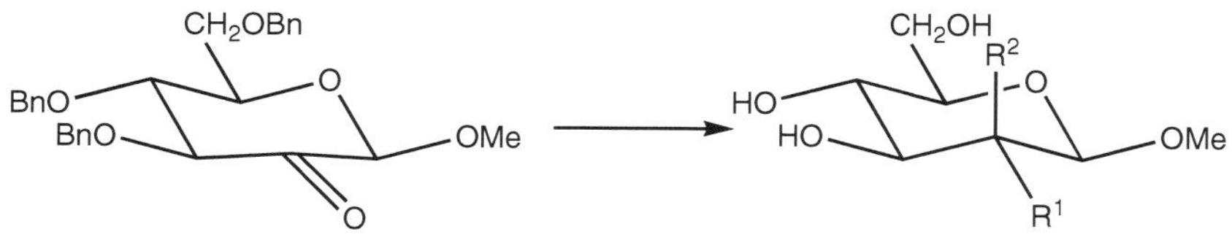

28

29, R^1 = H; R^2 = OH
30, R^1 = OH; R^2 = H

Fig. 11.8

another hydrogenation in presence of Pd–C (to remove the benzyl groups) gave in 84% yield [7] the 95:5 mixture of methyl β-D-mannopyranoside *29* and methyl β-D-glucopyranoside *30* (Fig. 11.8). It can be assumed that in this case steric interactions in the catalyst–carbonyl group transition state are the only factors that influence the stereochemical outcome of this reduction.

The comparison of the stereochemistry of catalytic hydrogenation of 2-carbonyl group of methyl α- and β-D-glycopyranosid-2-uloses and the stereochemistry of catalytic hydrogenation of 2-deoxy-2-C-methylene group of methyl α- and β-D-2-deoxy-2-C-methylene glycopyranosides (Fig. 11.9) provided interesting results. Similar to catalytic hydrogenation of the C2 carbonyl group of methyl

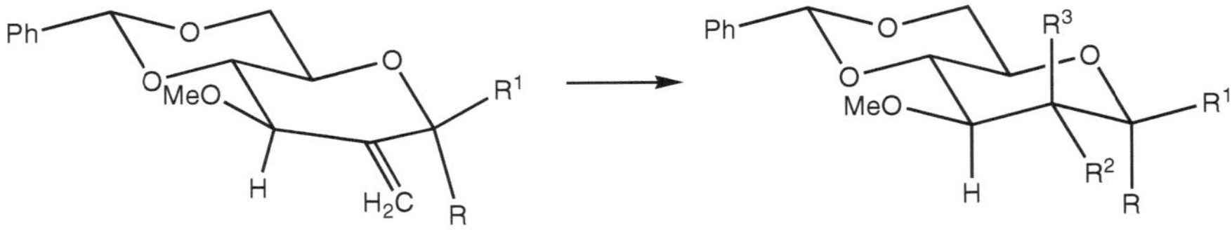

31, R = OMe; R^1 = H
32, R = H; R^1 = OMe

33, R = OMe; R^1 = R^3 = H; R^2 = CH_3
34, R = OMe; R^1 = R^2 = H; R^3 = CH_3
35, R = R^2 = H; R^1 = OMe; R^3 = CH_3

Fig. 11.9

3,4,6-tri-*O*-benzyl-β-D-*arabino*-hexopyranosid-2-ulose *28* [7], and to sodium borohydride reduction of methyl 4,6-*O*-benzylidene-3-*O*-methyl-β-D-*arabino*-hexopyranosid-2-ulose *3* where, in both cases, methyl 4,6-*O*-benzylidene-3-*O*-methyl-β-D-mannopyranoside *4* was obtained as the only product [1] (Fig. 11.2), the catalytic hydrogenation of methyl 4,6-*O*-benzylidene-2-deoxy-2-*C*-methylene-β-D-*arabino*-hexopyranoside [8] *32* (Fig. 11.9) gives, in quantitative yield, methyl 4,6-*O*-benzylidene-2-deoxy-2-*C*-methyl-β-D-mannopyranoside *35* as the only product. However, while reduction of methyl 4,6-*O*-benzylidene-3-*O*-methyl-α-D-*arabino*-hexopyranosid-2-ulose *1* with $LiAlH_4$ gives the corresponding α-D-gluco-derivative *2* as the only product [1] (Fig. 11.1) the catalytic hydrogenation of methyl 4,6-*O*-benzylidene-2-deoxy-2-*C*-methylene-3-*O*-methyl-α-D-*arabino*-hexopyranoside *31* gave a mixture of both C2 epimers, i.e., methyl 4,6-*O*-benzylidene-2-deoxy-2-*C*, 3-*O*-dimethyl-α-D-gluco- and α-D-mannopyranoside, *33* and *34* (Fig. 11.9), in the ratio ranging from 0.3 to 3.0, depending on the solvent and the nature of the catalyst.

The obtained results could be best explained by assuming that electronic interactions between the electronegative borohydride that approaches the electrophilic C2 carbonyl carbon and the electronegative anomeric methoxy group are responsible for the stereochemical outcome of the reduction of the C2 carbonyl carbon in both anomeric glycosides, whereas the stereochemical outcome of catalytic reductions of both 2-uloses and 2-deoxy-2-*C*-methylene derivatives are controlled solely by steric interactions.

The Addition of Carbon Nucleophiles: Synthesis of Branched Chain Sugars

Until 1960 the branched chain sugars were classified as rare sugars [9, 10] but discovery of numerous branched chain sugars in various antibiotics as their glycosidic components has stimulated extensive research on their synthesis, chemistry, and biochemistry.

Group 1: *Methyl-branched-chain sugars.*

36, Cladinose

37, Garosamine

38, Noviose

Group 2: *Hydroxymethyl and formyl branched-chain sugars*

39, Apiose

40, Hamamelose

41, Streptose

Group 3: *Two-carbon branched-chain sugars.*

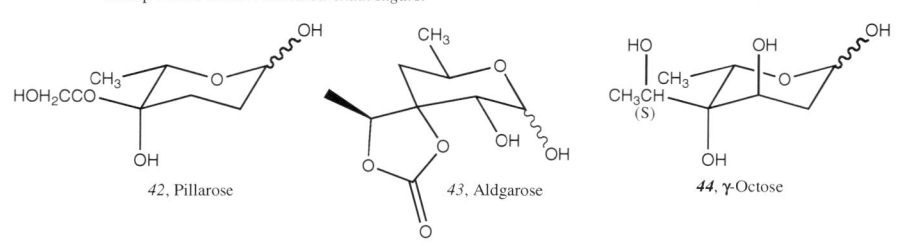

42, Pillarose

43, Aldgarose

44, γ-Octose

Group 4: *Higher branched-chain sugars.*

45

46, Blastmycinone

Fig. 11.10

The branched chain sugars have been found as glycosidic components in many antibiotics isolated from microorganisms and higher plants. Two very informative reviews have been published dealing with this topic [11, 12].

Depending on the structure of the branching group, the branched chain sugars can be classified into several groups: (1) methyl branched chain sugars, (2) hydroxymethyl- or formyl branched chain sugars, (3) two-carbon branched chain sugars, and (4) higher branched chain sugars.

In Fig. 11.10 are given a few examples of these four groups of branched chain sugars. Cladinose (*36*) is the sugar component of erythromycin, garosamine (*37*) is found in the antibiotic gentamicin, noviose (*37*) is the component of antibiotic novobiocin, apiose (*39*) and hamamelose (*40*) are found in plants, and streptose (*41*) in the antibiotic streptomycin. Pillarose (*42*), aldgarose (*43*), and γ-octose (*44*) are found in antibiotics pillaromycin A, aldgamycin E, and quinocycline A, respectively. Finally, the higher branched chain sugar *45* (unnamed) and blastmycinone (*46*) are found in loroglossin (a constituent of orchids) and in the antibiotic blastmycin.

Most branched chain sugars found in Nature have a polar group attached to the branching carbon; most often this is a hydroxyl group or its methyl ether, but an amino or a nitro group is also found attached to the tertiary branching carbon.

The addition of carbon nucleophiles to glycosiduloses has been extensively used for the synthesis of branched chain sugars in which the branching carbon has a hydroxyl group attached to it (most of the sugars in Group 1). The carbon nucleophiles used were either organometallics (such as methyl Grignard reagents and methyllithium) or diazomethane; in the latter case an extra step was required, i.e., the opening of the spiro epoxide which is initially obtained as a result of the addition of diazomethane to a glycosidulose.

The addition of phosphorus ylides (phosphoranes) to glycosiduloses (Wittig reaction) giving the corresponding *C*-alkylidene derivatives was used for the synthesis of branched chain sugars without or with the hydroxyl group attached to branching carbon.

The opening of epoxide ring of a carbohydrate oxirane derivative with a carbon nucleophile is most often used for the synthesis of the branched chain sugars of Group 2.

The Addition of Organometals

The carbon nucleophiles used for the addition to a carbonyl carbon of a glycopyranosidulose are most often methyl Grignard reagent or methyllithium. The diastereofacial selectivity is controlled by several factors and will be discussed later.

The addition of a methyl nucleophile to a glycopyranosidulose gives a branched chain sugar in which the chiral-branching carbon is a tertiary alcohol (Fig. 11.11). In order to understand the mechanism of nucleophilic addition, the absolute configuration of the newly created branched carbon must be known. However, the absence of

The Addition of Carbon Nucleophiles: Synthesis of Branched Chain Sugars

hydrogen atom at the branching carbon introduces an additional complication since the ^{1}H NMR spectroscopy cannot be used for configurational assignment.

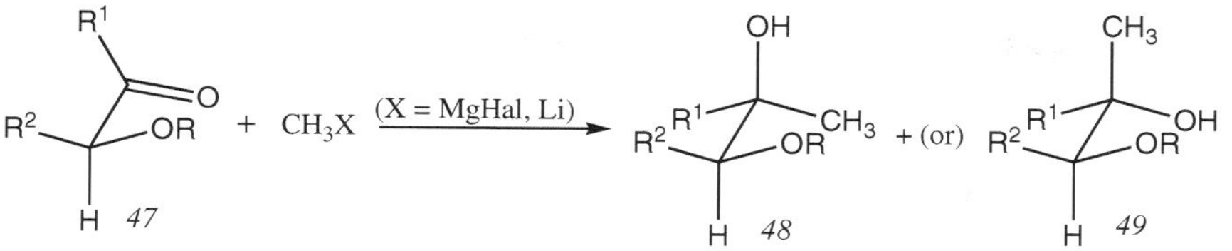

Fig. 11.11

If the R^1 has a higher priority than the CHR^2OR group according to Cahn–Ingold–Prelog convention [13], then the configuration of the branching carbon in *48* will be *S* and in *49 R*; if the R^1 has a lower priority than the CHR^2OR then the configuration of the branching carbon in *48* will be *R* and in *49 S*.

The determination of configuration of tertiary carbon in branched chain glycopyranosides was notoriously difficult, because a simple and reliable method was not available for a long time. Thus, for example, the configuration of a branching carbon has been determined from IR frequencies of the tertiary OH group [14], from chromatographic and electrophoretic mobilities in solvent systems with borate buffer [15, 16] or phenylboronic acid [14], from the kinetics of periodate oxidation [15], from the formation of cyclic carbonates [15] and bicyclic hemialdals [15, 16], and from degradation reactions [17–20]. In the case of nitroalkyl branched chain sugars or sugar alcohols, the configuration of branching carbon was deduced from ORD and CD spectra [21–24]. Finally, the configurations of tertiary alcoholic centers in branched chain sugars were also determined by NMR spectroscopy using lantanide shift reagents [25].

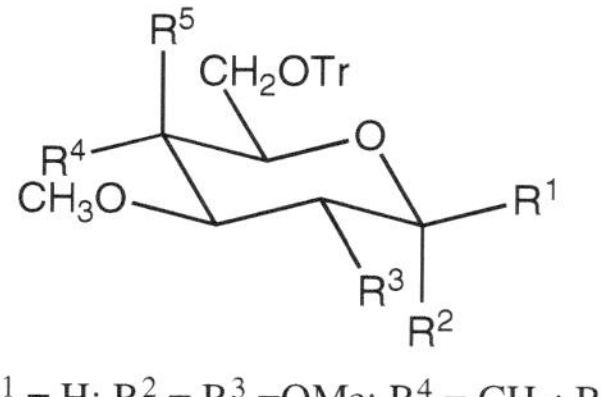

50, R^1 = H; $R^2 = R^3$ =OMe; R^4 = CH_3; R^5 = OH
51, R^1 = H; $R^2 = R^3$ =OMe; R^4 = OH; R^5 = CH_3
52, R^1 = H; R^2 = OMe; R^3 = CH_3SO_3; R^4 = CH_3; R^5 = OH
53, R^1 = H; R^2 = OMe; R^3 = CH_3SO_3; R^4 = OH; R^5 = CH_3
54, R^1 = H; R^2 = OMe; $R^3 = R^5$ = OH; R^4 = CH_3
55, R^1 = H; R^2 = OMe; $R^3 = R^4$ = OH; R^5 = CH_3
56, $R^1 = R^3$ = OMe; R^2 = H; R^4 = CH_3; R^5 = OH
57, $R^1 = R^3$ = OMe; R^2 = H; R^4 = OH; R^5 = CH_3

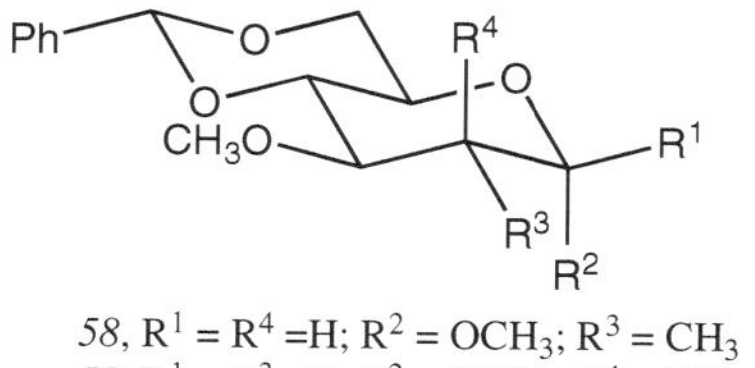

58, $R^1 = R^4$ =H; R^2 = OCH_3; R^3 = CH_3
59, $R^1 = R^3$ =H; R^2 = OCH_3; R^4 = CH_3
60, R^1 = OCH_3; $R^2 = R^3$ = H; R^4 = CH_3

Fig. 11.12

Table 11.1 C13 chemical shifts of equatorial and axial C4 methyl group in branched chain sugars

Branched chain sugar	C13 chemical shift (in ppm) of the C4 methyl group
Methyl 4-*C*-methyl-2,3-di-*O*-methyl-6-*O*-triphenylmethyl-α-D-galactopyranoside (*50*) (equatorial CH$_3$)	21.8
Methyl 4-*C*-methyl-2,3-di-*O*-methyl-6-*O*-triphenylmethyl-α-D-glucopyranoside (*51*) (axial CH$_3$)	15.6
Methyl 4-*C*-methyl-3-*O*-methyl-2-*O*-methanesulfonyl-6-*O*-triphenylmethyl-α-D-galactopyranoside (*52*) (equatorial CH$_3$)	21.7
Methyl 4-*C*-methyl-3-*O*-methyl-2-*O*-methanesulfonyl-6-*O*-triphenylmethyl-α-D-glucopyranoside (*53*) (axial CH$_3$)	15.3
Methyl 4-*C*-methyl-3-*O*-methyl-6-*O*-triphenylmethyl-α-D-galactopyranoside (*54*) (equatorial CH$_3$)	21.9
Methyl 4-*C*-methyl-3-*O*-methyl-6-*O*-triphenylmethyl-α-D-glucopyranoside (*55*) (axial CH$_3$)	15.4
Methyl 4-*C*-methyl-2,3-di-*O*-methyl-6-*O*-triphenylmethyl-β-D-galactopyranoside (*56*) (equatorial CH$_3$)	21.3
Methyl 4-*C*-methyl-2,3-di-*O*-methyl-6-*O*-triphenylmethyl-β-D-glucopyranoside (*57*) (axial CH$_3$)	16.0
Methyl 4,6-*O*-benzylidene-2-deoxy-2-*C*-methyl-3-*O*-methyl-α-D-glucopyranoside (*58*)	12.4
Methyl 4,6-*O*-benzylidene-2-deoxy-2-*C*-methyl-3-*O*-methyl-α-D-mannopyranoside (*59*) (axial CH$_3$)	11.0
Methyl 4,6-*O*-benzylidene-2-deoxy-2-*C*-methyl-3-*O*-methyl-β-D-mannopyranoside (*60*) (axial CH$_3$)	5.7

Using the observation made in the study on conformational equilibria of methyl cyclohexanes [26–28] that the carbon-13 chemical shift of an axial methyl group is ~6 ppm upfield relative to that of an equatorial methyl group, Miljkovic et al. [29] have unequivocally determined the configuration of the branching carbon atom in a number of branched chain sugars having methyl group as the branching chain (see Fig. 11.12 and Table 11.1).

The addition of methyllithium to methyl 2,3-di-*O*-methyl-6-*O*-triphenylmethyl-α-D-*xylo*-hexopyranosid-4-ulose *61* and to methyl 3-*O*-methyl-2-*O*-methanesulfonyl-6-*O*-triphenylmethyl-α-D-*xylo*-hexopyranosid-4-ulose *62* in ether at −80°C gave, in each case, only one isomer: methyl 2,3-di-*O*-methyl-4-*C*-methyl-6-*O*-triphenylmethyl-α-D-glucopyranoside *51* (70%) (from *61*) and methyl 3-*O*-methyl-2-*O*-methanesulfonyl-4c-methyl-6-*O*-triphenylmethyl-α-D-glucopyranoside *55* (53%) (from *62*) (Fig. 11.12) [30].

The branched chain sugars *50–57* were obtained by the addition of methylmagnesium iodide or methyllithium to the methyl 2,3-di-*O*-methyl-6-*O*-triphenylmethyl-α- and β-D-*xylo*-hexopyranosid-4-ulose *61* and *63* and methyl 3-*O*-methyl-2-*O*-methylsulfonyl-6-*O*-triphenylmethyl-α-D-*xylo*-hexopyranosid-4-ulose *62* in ether at −80°C (Fig. 11.13).

Reaction of oxo-sugars *61* and *62* with methylmagnesium iodide in ether at −80°C proceeded again stereospecifically, but the products obtained were the C4

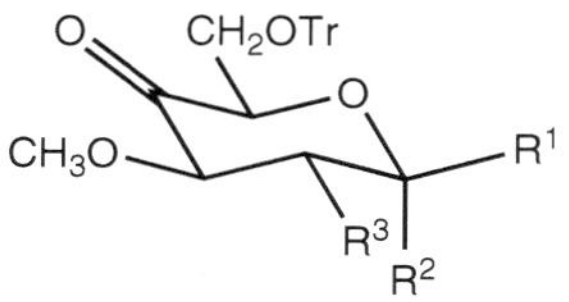

The Addition of Carbon Nucleophiles: Synthesis of Branched Chain Sugars 301

61, $R^2 = R^3 = OCH_3$; $R^1 = H$
62, $R^2 = OCH_3$; $R^1 = H$; $R^3 = CH_3SO_3$
63, $R^1 = R^3 = OCH_3$; $R^2 = H$;

Fig. 11.13

epimers of branched chain sugars *50* and *52*, i.e., *61* gave methyl 2,3-di-*O*-methyl-4-*C*-methyl-6-*O*-triphenylmethyl-α-D-galactopyranoside *51* (94%) and *62* gave methyl 3-*O*-methyl-2-*O*-methanesulfonyl-4c-methyl-6-*O*-triphenylmethyl-α-D-glucopyranoside *53* (53%) [30] (Fig. 11.12).

The stereochemistry of the addition of Grignard reagent to the glycopyranosid-4-uloses *61* and *62* was found to be dependent on the reaction temperature [30], the solvent [31, 32], and the nature of the halogen atom [30].

Thus, treatment of an ethereal solution of *61* and/or *62* with methylmagnesium iodide at −80°C afforded *50* and/or *52* as the only isolable products. At reflux, both C4 epimers, *50* and *51* (from *61*) and *52* and *54* (from *62*) were obtained, but the isomers having the methyl group in equatorial orientation (*50*, *52*, and *54*) predominated in ca. 6:1 ratio (the branched chain sugars *54* and *55* are the products of desulfonylation of branched chain sugars *52* and *53* under the given experimental conditions). The dependence of the stereochemistry of the addition reaction upon the nature of the solvent was demonstrated by refluxing a 10:1 ether–tetrahydrofuran solution of *62* with methylmagnesium chloride whereby a mixture of C4 epimers *52* and *53* was obtained in 1.3:1 ratio. The dependence of the stereochemistry of the addition reaction upon the nature of the halogen atom was demonstrated by reacting methylmagnesium iodide with *62* under the same experimental conditions whereby a mixture of C4 epimer *52* and *53* was now obtained in 2.3:1 ratio (the isomer *53* with the axial methyl group predominated) (Fig. 11.4).

The preferred equatorial addition of methyl group of methylmagnesium iodide to the C4 carbonyl group of *60* or *61* was explained [30] to be due to "chelation" of the magnesium atom of Grignard reagent with the C4 carbonyl oxygen and the C3 oxygen atom prior to the addition of methyl carbanion to the carbonyl carbon [32–34] as depicted in Fig. 11.14.

Thus the formation of a cyclic five-membered ring complex *64* will force the glycopyranosid-4-uloses *61* and *62* to adopt the $^4C_1$ conformation prior to the addition of methyl group to the C4 carbonyl carbon. The solvent dependence and to some extent the temperature dependence of stereochemistry of the addition of methyl group of Grignard reagent to the carbonyl carbon of *61* and *62* does support this view.

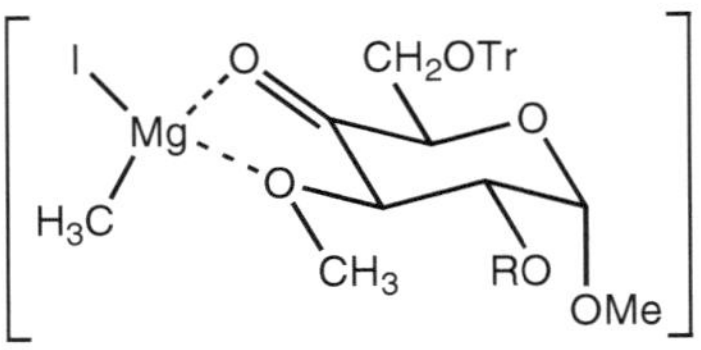

Fig. 11.14

64, R = CH$_3$, CH$_3$SO$_2$

The axial stereospecificity of methyllithium addition to the C4 carbonyl carbon of glycopyranosid-4-uloses *61* and *62* (ether and at –80°C) was rationalized as follows. Studies on conformational equilibrium of α-halocyclohexanones [35–39] have shown that conformations with the halogen atom axially oriented are strongly favored in solvents of low dielectric constant. This tendency of halogens to adopt the axial rather than equatorial orientation was attributed to the strong electrostatic

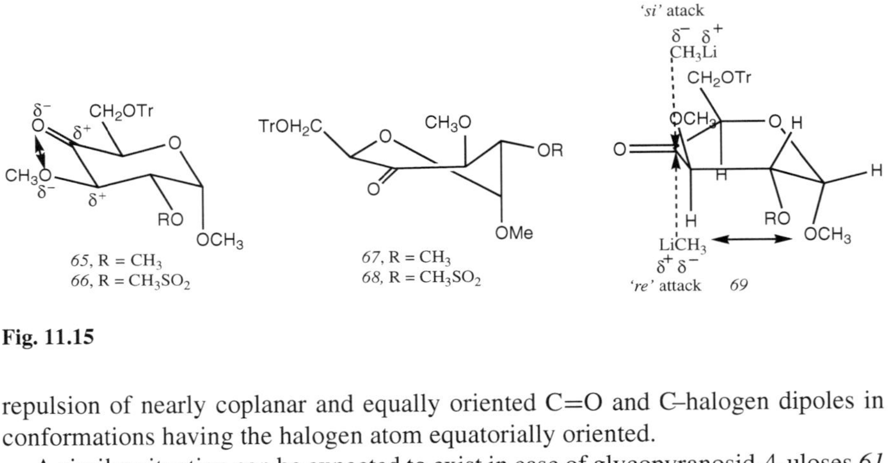

65, R = CH$_3$
66, R = CH$_3$SO$_2$

67, R = CH$_3$
68, R = CH$_3$SO$_2$

'*si*' attack

'*re*' attack 69

Fig. 11.15

repulsion of nearly coplanar and equally oriented C=O and C–halogen dipoles in conformations having the halogen atom equatorially oriented.

A similar situation can be expected to exist in case of glycopyranosid-4-uloses *61* and *62* since in the $^4C_1$ conformation the C3-methoxy group is equatorially oriented; however, due to an electrostatic repulsion of nearly coplanar and equally oriented C=O and C–OMe dipoles this conformation should be destabilized in solvents of low dielectric constant (e.g., ether). Consequently, oxo-sugars *61* and *62* will most likely adopt, in ether solution and at –80°C, either the half-chair conformation *67* or *68* (Fig. 11.15) or a conformation that is very close to the half-chair conformation *69*. The adoption of any such conformation prior to the reaction with methyllithium could then be responsible for the exclusive "*si*" attack of methyl carbanion to the C4 carbonyl carbon of *67* or *68*, since strong electrostatic and nonbonding steric interactions between the electronegative methyl group of methyllithium that approaches the C4 carbonyl carbon from the "*re*" direction and the axially oriented C1 methoxy group will completely impede the "*re*" addition of methyllithium (Fig. 11.15).

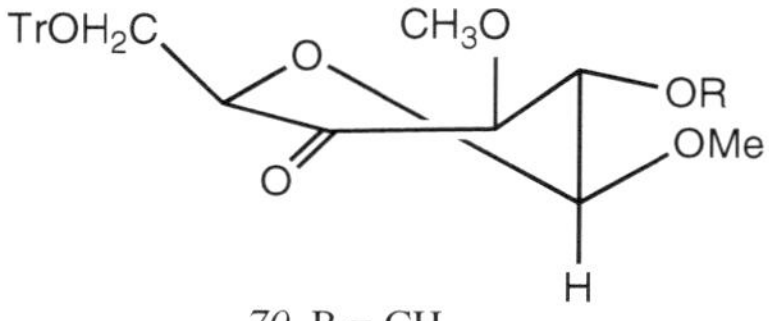

70, R = CH_3
71, R = CH_3SO_2

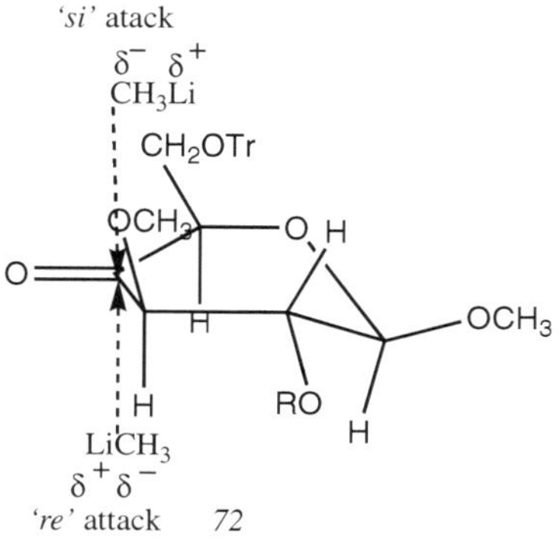

72

Fig. 11.16

This rationalization is supported by the observation that methyl 2,3-di-*O*-methyl-6-*O*-triphenylmethyl-β-D-*xylo*-hexopyranosid-4-ulose *72* (Fig. 11.16), the β-anomer of *61*, where the "1,4-*synaxial*" stereoelectronic interactions between the approaching methyl carbanion and the electronegative anomeric methoxy group does not exist when the attack comes from the "*re*" face, gave with methyllithium, in an ethereal solution at –80°C, a mixture of both C4 epimers *50* and *51* (Fig. 11.12) in which the epimer having the methyl group equatorially oriented (*50*) predominated in 3:1 ratio.

It is interesting to note that 4-*tert*-butyl-cyclohexanone reacted with methylmagnesium iodide and methyllithium in ether at –80°C considerably slower, and the addition was not stereoselective giving in each case a mixture of both C1 epimers (*74* and *75*) [30] (Fig. 11.17). Thus, methylmagnesium iodide gave a mixture of C1 isomers in 1.7:1 ratio in which the epimer with the equatorial methyl group (*74*) predominated; methyllithium gave a mixture of epimers in 3.6:1 ratio in which the epimer *74* again predominated.

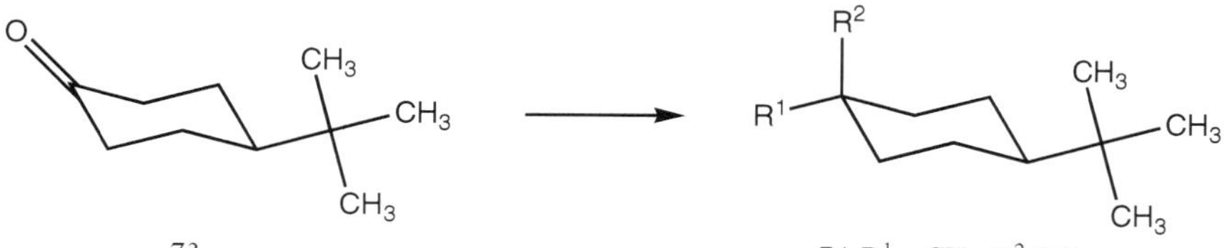

73

74, $R^1 = CH_3$; R^2 OH
75, $R^1 = OH$; $R^2 = CH_3$

Fig. 11.17

It should be noted that in reactions with both Grignard reagent and methyllithium considerable amounts of starting material were isolated: 25 and 21%, respectively.

In agreement with these findings [30], Hanessian et al. [40] reported that the addition of methyllithium to the C4 carbonyl carbon of glycopyranosid-4-ulose *76* gave, in quantitative yield, only the epimer *77* having the axially oriented methyl group at the C4 branching carbon (Fig. 11.18).

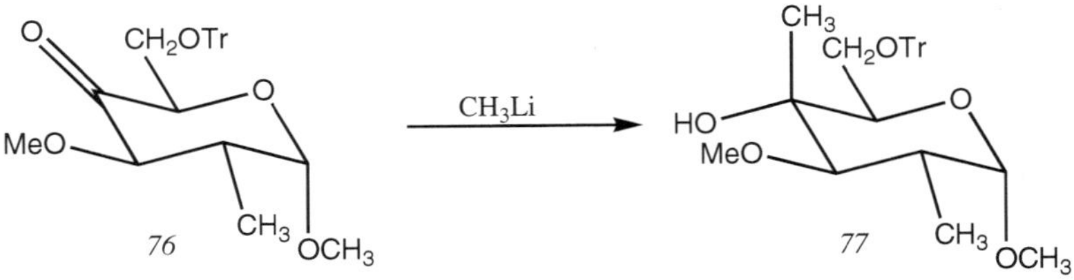

Fig. 11.18

To examine the generality of the above proposed hypothesis [30] Yoshimura et al. [41, 42] studied the stereoselectivity of addition of methyllithium and methylmagnesium iodide to a select number of hexopyranosiduloses (Figs. 11.19 and 11.20). The addition of methyllithium to methyl 6-deoxy-2,3-di-*O*-methyl-α-D-*xylo*-hexopyranosid-4-ulose *78* proceeded, as expected, highly stereoselectively giving the stereoisomer *80* as the only product (Fig. 11.19). In this study they also examined the addition of methylmagnesium iodide and methyllithium to a number of glycopyranosid-4-uloses [43] (Fig. 11.20), and the authors tried to explain the

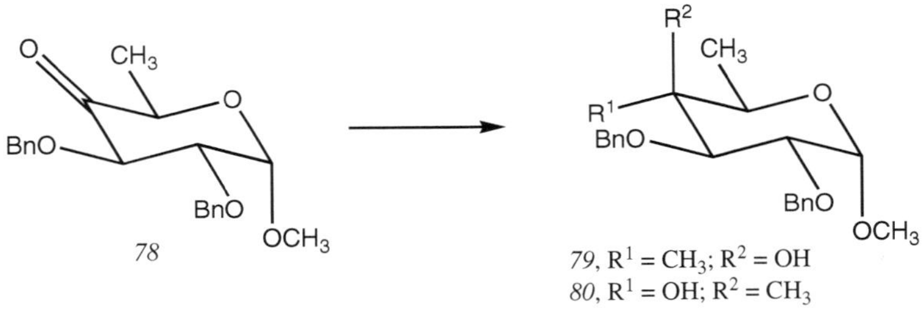

Fig. 11.19

observed stereoselectivities using Miljkovic et al. hypothesis [30]. Depending on the substrate used, the observed stereoselectivities ranged from 100% to 2.3:1.

The conclusions drawn from the obtained results are, according to the opinion of this author, flawed since the glycopyranosid-4-uloses chosen for the study had, except for the substrate (*78*), either one (*81* and *82*) or two (*83*) axial substituents on the pyranoside ring or were the bicyclic systems having the 2,3-*O*-methylene acetal five-membered ring attached to the pyranoside ring of a glycopyranosid-4-ulose (compounds *84–86*), introducing thus not only the uncertainty regarding the conformation of a sugar which reacts with methyllithium or Grignard reagent but also additional stereoelectronic interactions between the attacking nucleophile and the substituent(s) on the pyranoside ring.

The addition of methylmagnesium iodide to methyl 4,6-*O*-benzylidene-2-*O*-benzoyl-α-D-*ribo*-hexopyranosid-3-ulose *87* in ether at room temperature gave after

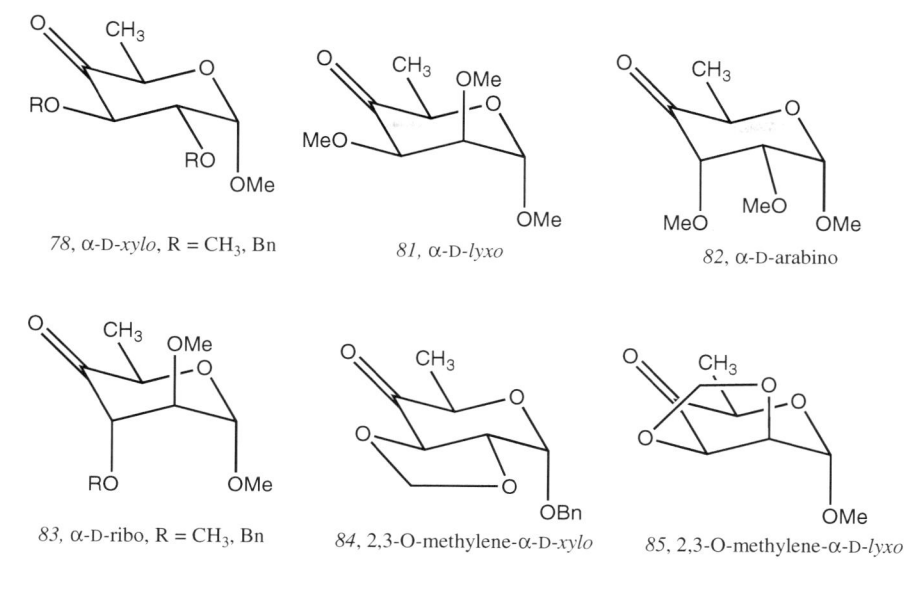

The Addition of Carbon Nucleophiles: Synthesis of Branched Chain Sugars   305

78, α-D-*xylo*, R = CH₃, Bn

81, α-D-*lyxo*

82, α-D-arabino

83, α-D-ribo, R = CH₃, Bn

84, 2,3-O-methylene-α-D-*xylo*

85, 2,3-O-methylene-α-D-*lyxo*

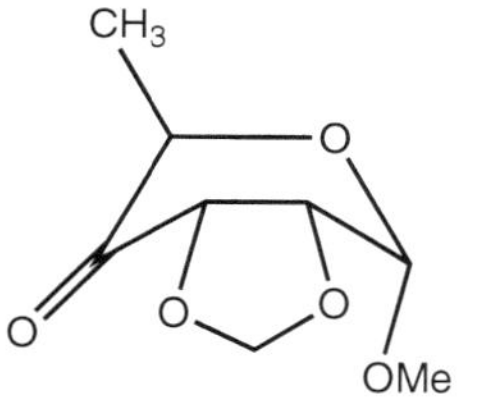

86, 2,3-O-methylene-α-D-*ribo*

Fig. 11.20

24 h the D-*talo*-derivative *88* in 40% yield (the benzoyl group was removed during the reaction) [44] (Fig. 11.21).

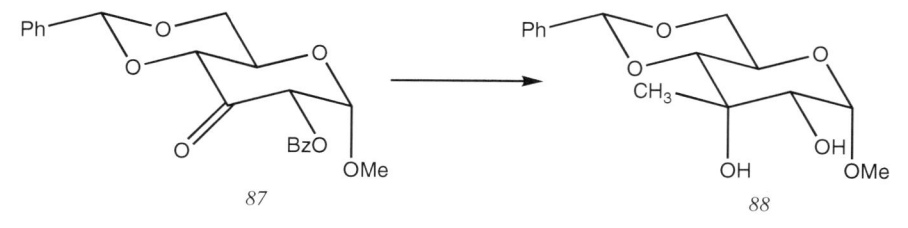

87   *88*

Fig. 11.21

Similarly, the addition of methylmagnesium iodide to methyl 4,6-*O*-benzylidene-2-deoxy-α-D-*threo*-hexopyranosid-3-ulose *89*, in ether and 0°C, gave methyl

4,6-O-benzylidene-2-deoxy-3-C-methyl-α-D-*lyxo*-hexopyranoside *90* in 91% yield together with traces of the C3 epimer methyl 4,6-O-benzylidene-2-deoxy-3-C-methyl-α-D-*xylo*-hexopyranoside [45] (Fig. 11.22).

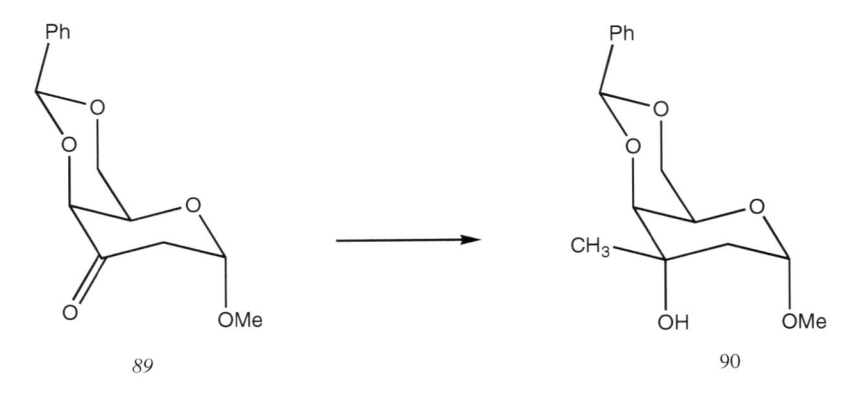

Fig. 11.22

The configuration of the branching C3 carbon was determined by converting *90* into D-arcanose *91* and then comparing it with the natural L-arcanose *92* (Fig. 11.23).

Fig. 11.23

Addition of Diazomethane

Diazomethane is a resonance hybrid of the following two canonical forms (Fig. 11.24):

Fig. 11.24

The Addition of Carbon Nucleophiles: Synthesis of Branched Chain Sugars 307

Consequently in all of its reactions diazomethane reacts as a nucleophile. Thus, it adds to the carbonyl carbon of an aldehyde or ketone giving the corresponding spiro epoxides (Fig. 11.25):

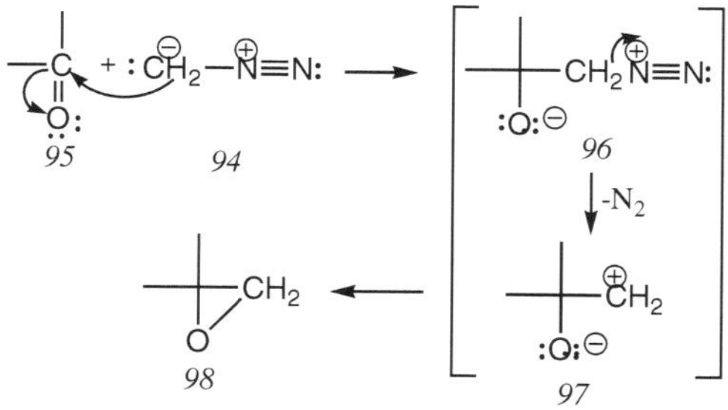

Fig. 11.25

It should be noted that intermediate *97* may undergo a *Sato rearrangement* resulting in insertion of a methylene carbon in the carbon skeleton of an aldehyde or ketone (for a review see [46]) (Fig. 11.26).

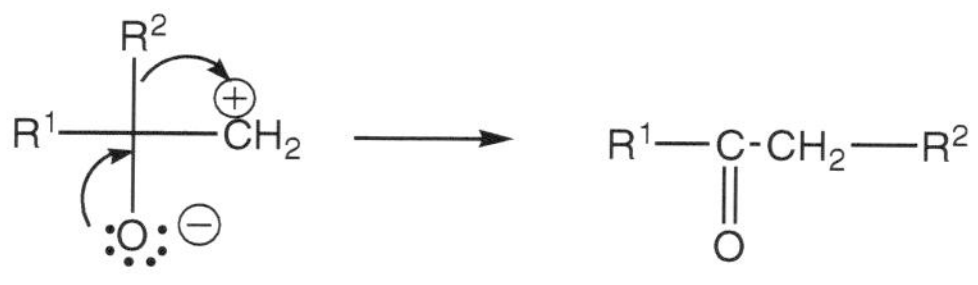

Fig. 11.26

The addition of diazomethane to methyl 4,6-*O*-benzylidene-3-*O*-methyl-α- and β-D-*arabino*-hexopyranosid-2-uloses (*1* and *3*, respectively) and methyl 4,6-*O*-benzylidene-3-*O*-methyl-α-D-*ribo*-hexopyranosid-2-ulose *99* [47] (Fig. 11.27) gave the following spiro epoxides: *1* gave a mixture of epimeric spiro epoxides *100* and *103*, *3* gave the spiro epoxide *101*, whereas *99* gave the spiro epoxide *102* as the predominant product (Fig. 8.27) (Table 11.2).

The stereochemical outcome of the addition of diazomethane to *1*, *3*, and *99* did not parallel the stereochemical outcome of the Grignard reagent addition to these glycosid-2-uloses. The authors proposed a highly speculative explanation for the observed stereoselectivity of diazomethane addition.

The addition of diazomethane to methyl 4,6-*O*-benzylidene-2-*O*-methyl-α-D-*ribo*-hexopyranosid-3-ulose *104* gave *107* in 73% yield. Methyl 4,6-*O*-benzylidene-2-*O*-methyl-β-D-*ribo*-hexopyranosid-3-ulose *105* gave diazomethane *108* and *109* (76.5 and 17.6% yield, respectively). The addition of diazomethane to methyl 4,6-*O*-benzylidene-2-*O*-methyl-α-D-*arabino*-hexopyranosid-3-ulose *106* (Fig. 11.28)

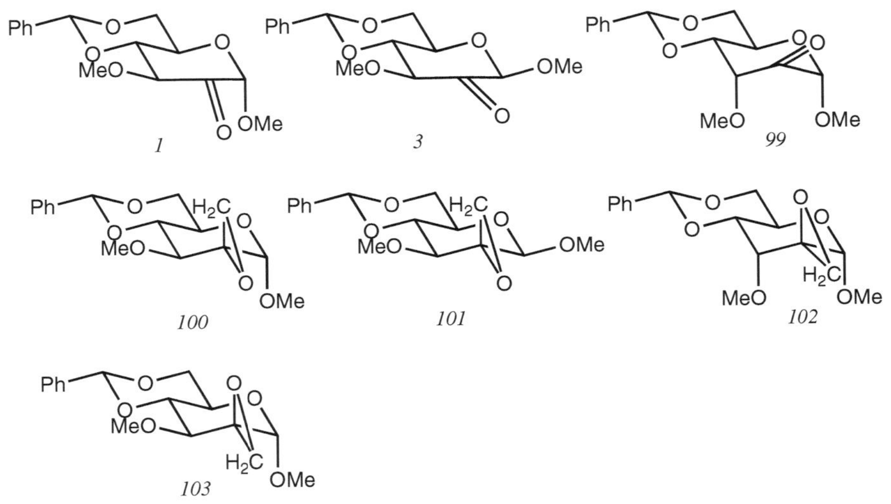

Fig. 11.27

Table 11.2 Yields of products and the direction of attack of carbon nucleophile of diazomethane to the carbonyl carbon of the methyl 4,6-*O*-benzylidene-α- and β- D-hexopyranosid-2-uloses

Glycosid-2-ulose	Yields of products (%)	
	Axial attack	Equatorial attack
α-D-*arabino*-, *1*	31.3 (*99*)	63.7 (*102*)
β-D-*arabino*-, *3*	92.5 (*100*)	–
α-D-*ribo*-, *99*	–	83.5 (*101*)

gave methyl 3,3'-anhydro-4,6-*O*-benzylidene-3-*C*-hydroxymethyl-2-*O*-methyl-α-D-altropyranoside *110* in 41% yield [48] together with a pyranoside ring expansion product.

The reduction of spiro epoxides with LiAlH$_4$ gives hydroxylated branched chain carbon with the methyl group as the branched chain, whereas the opening of spiro epoxides with a nucleophile (OH$^-$, NH$_3$, etc.) gives hydroxylated branched chain carbon with the functionalized methyl group, such as hydroxymethyl, aminomethyl, as the branched chains.

Synthesis of Branched Chain Sugars with Functionalized Branched Chain

The preparation of branched chain sugars with functionalized branched chain can be accomplished in two ways:

Fig. 11.28

(1) By opening of the spiro epoxide ring obtained by addition of diazomethane to the carbonyl carbon of a glycopyranosidulose with nucleophiles other than hydride ion
(2) By addition of a functionalized carbon nucleophile, such as (a) lithium 1,3-dithiane, (b) vinylmagnesium halide, (c) methoxyvinyllithium, (d) 1,1-dimethoxy-2-lithio-2-propene.

2-Lithio-1,3-Dithiane as the Nucleophile

2-Lithio-1,3-dithiane [49, 50] *111* (Fig. 11.29) is a stable compound and as a nucleophile was used extensively for chain extension and chain branching in synthetic carbohydrate chemistry [51]. The products obtained by the addition of 2-lithio-1,3-dithiane to glycopyranosiduloses can be converted to methyl or to formyl branched chain sugars by opening the sugar oxirane rings by catalytic hydrogenation (methyl) and via mercuric oxide–boron trifluoride hydrolysis (formyl).

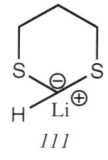

Fig. 11.29

310   11 Addition of Nucleophiles to Glycopyranosiduloses

Thus, the addition of lithium dithiane to methyl 4,6-*O*-benzylidene-2-deoxy-, 2-*O*-benzoyl, 2-acetamido, or 2-*O*-methyl-α-D-*ribo*-hexopyranosid-3-ulose gave the corresponding methyl-4,6-*O*-benzylidene-α-D-allopyranosides with dithianyl group equatorially oriented as the only product [44, 52] (see Fig. 11.30).

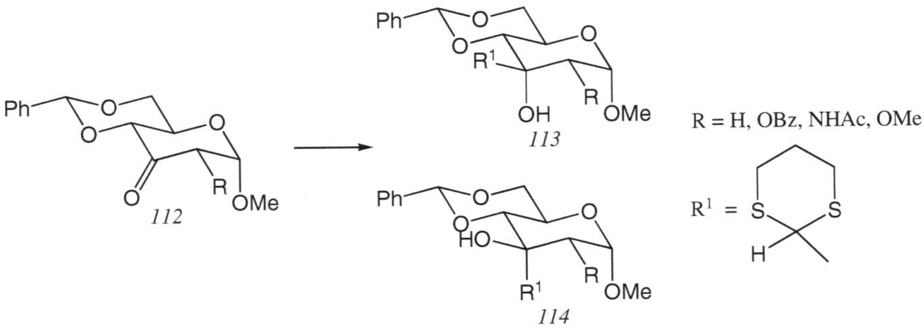

Fig. 11.30

The synthesis of branched chain sugar aldgarose *115* (Fig. 11.31) was accomplished by using dithiane carbanion as carbon nucleophile [53]. Thus the addition of 2-methyl-2-lithio-1,3-dithiane to 4,6-dideoxy-3-ulose *116* (Fig. 11.32) gave a mixture of two C3 dithiane epimers *117* and *118*, one existing in the $^4C_1$ and the other in the $^1C_4$ conformation, since in these conformations the bulkiest substituent (1,3-dithiane) is in the equatorial position. The low stereoselectivity (*117:118* = 3:2)

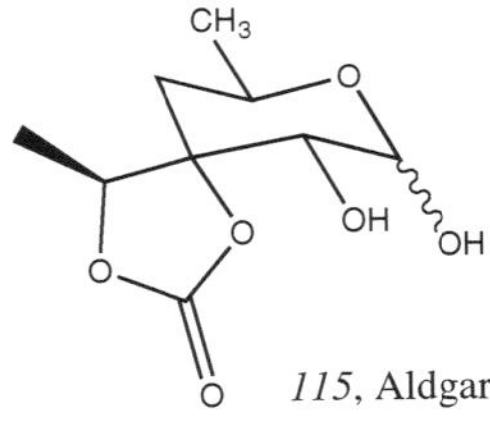

Fig. 11.31

is probably due to the fact that the anomeric methoxy group is equatorial (β) in *165*. Also, the pyranoside ring of 4-deoxy sugars may be, due to the absence of the C4 hydroxyl group, more flexible which may result in the loss of stereoselectivity. The conversion of *117* to D-aldgarose required first the conversion of dithiane group into the carbonyl group with $HgCl_2$/HgO in refluxing methanol followed by sodium borohydride reduction of the obtained acetyl group; two isomers in the branched chain *120* and *121* were obtained in 10:7.1 ratio with 85.3% overall yield. If the reduction was performed with lithium aluminum hydride in the presence of 1 mol-equivalent of *tert*-butanol the *120* and *121* were obtained in 10:4.3 ratio with overall yield of 50.3%. For the reduction in the presence of 2 mol-equivalents

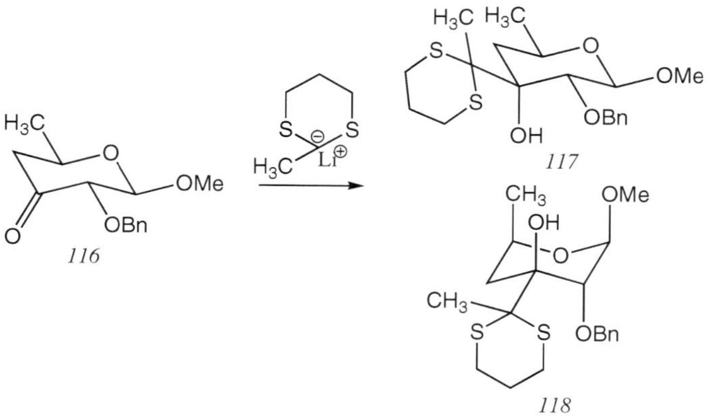

Fig. 11.32

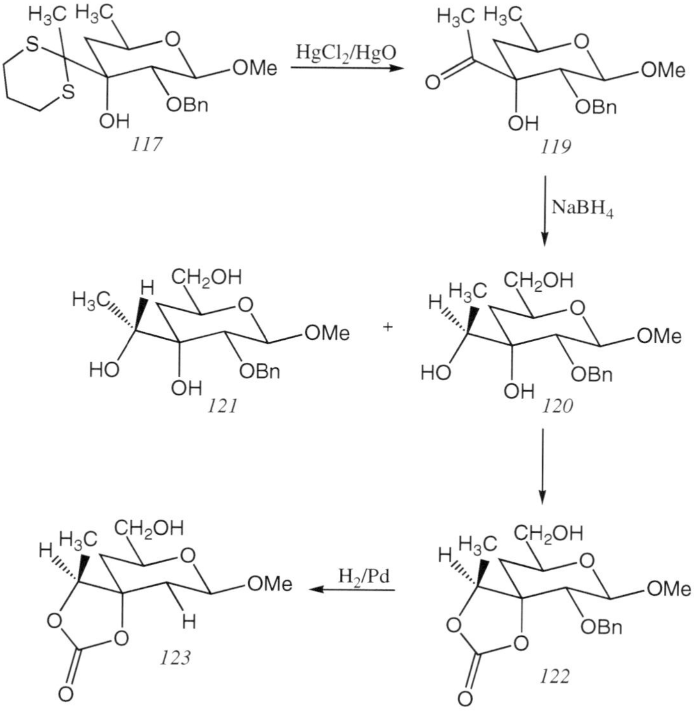

Fig. 11.33

of *tert*-butanol, *120* and *121* were obtained in 10:2.6 ratio in 61.8% overall yield (Fig. 11.33).

Vinyl Carbanion as the Nucleophile

Another method for introducing functionalized branching chain is the reaction of vinylmagnesium bromide with glycopyranosiduloses. Thus, methyl 4,6-*O*-benzylidene-2-deoxy-2-*C*-methyl-3-*C*-vinyl-α-D-allopyranoside *125* was obtained in 92% yield by addition of vinylmagnesium bromide to methyl 4,6-*O*-benzylidene-2-deoxy-2-*C*-methyl-α-D-*ribo*-hexopyranosid-3-ulose *124* (Fig. 11.34). The addition of vinylmagnesium bromide to methyl 4,6-*O*-benzylidene-2-deoxy-2-*C*-methyl-α-D-*arabino*-hexopyranosid-3-ulose *126* again resulted in the addition from the equatorial side giving in 54% yield methyl 4,6-*O*-benzylidene-2-deoxy-2-*C*-methyl-α-D-*altro*-pyranoside [54] *127*, in addition to some *allo*-branched chain sugar *125* which was formed probably by epimerization of the axially oriented C2 methyl group caused by the basic reaction conditions prior to the addition of the vinyl group.

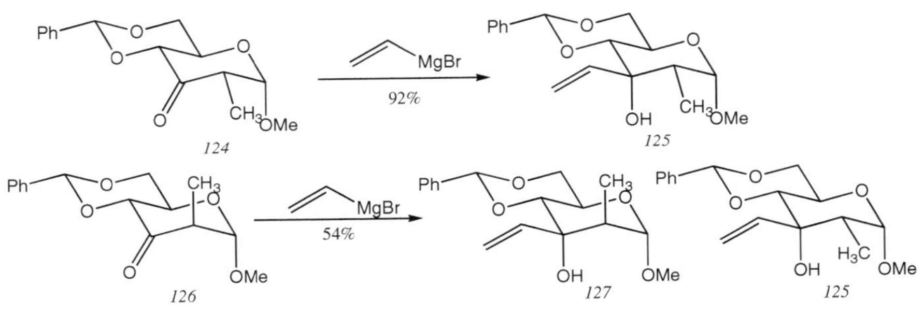

Fig. 11.34

Methoxyvinyl Lithium and 1,1-Dimethoxy-2-Lithio-2-Propene

Some other two-carbon carbanions used for the introduction of more complex branching chains are methoxyvinyl lithium *128* (introduction of acetyl group and 2-hydroxyacetyl branched chain [55, 56]) and 1,1-dimethoxy-2-lithio-2-propene *129* [57] (Fig. 11.35) which was used in the synthesis of pillarose *130* [58] (Fig. 11.35).

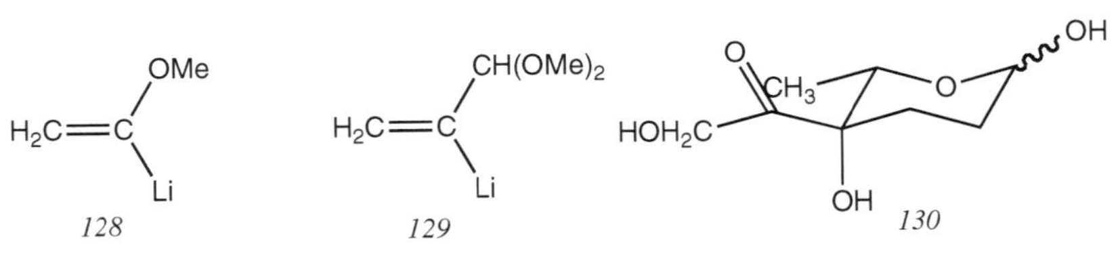

Fig. 11.35

Synthesis of Branched Chain Sugars with Functionalized Branched Chain 313

In Fig. 11.36 is shown the addition of methoxyvinyl lithium to the methyl 4,6-
O-benzylidene-3-deoxy-α-D-*erythro*-hexopyranosid-2-ulose *131*. The addition was
stereoselective, and the product *133* was obtained in 41% overall yield [58].

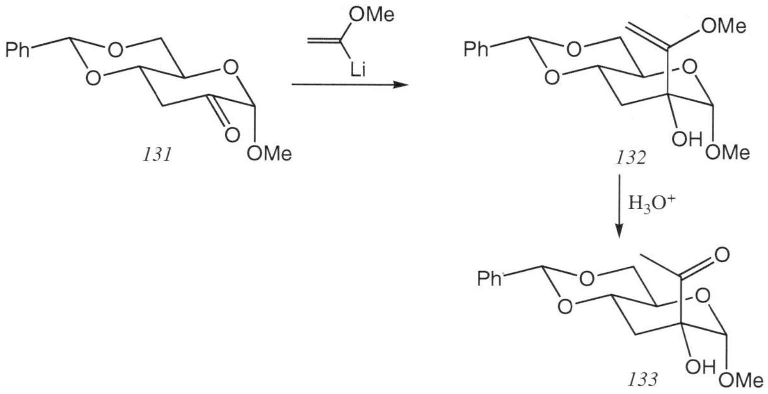

Fig. 11.36

1,1-Dimethoxy-2-lithio-2-propene *129* (Fig. 11.35) was used for the synthesis of
C-methylene branched chain sugar intermediates [57, 59] needed for the synthesis
of hamamelose G (*134*), a naturally occurring branched chain pentose (Fig. 11.37).

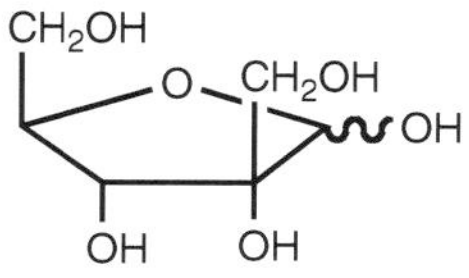

Fig. 11.37

The first step of the synthesis of hamamelose G, i.e., the addition of 1,1-
dimethoxy-2-lithio-2-propene *129* to 2,3-O-isopropylidene-D-glyceraldehyde *135*,
is depicted in Fig. 11.38.

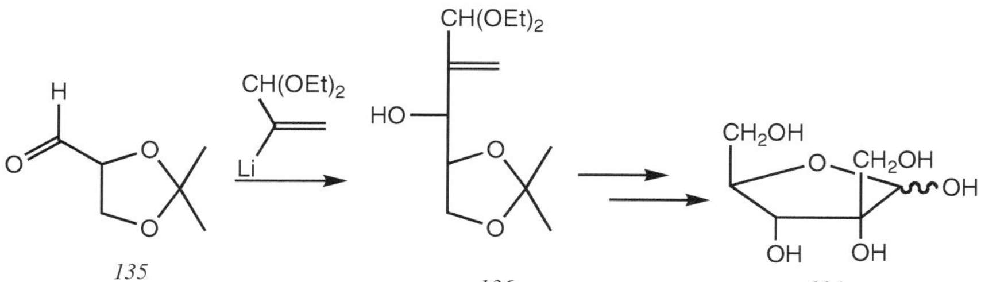

Fig. 11.38

Reformatsky Reaction

An interesting approach for the synthesis of functionalized branched chain sugars is the Reformatsky reaction [60–63]. Reformatsky reagent obtained from ethyl bromoacetate and zinc, in refluxing tetrahydrofuran, adds highly stereoselectively

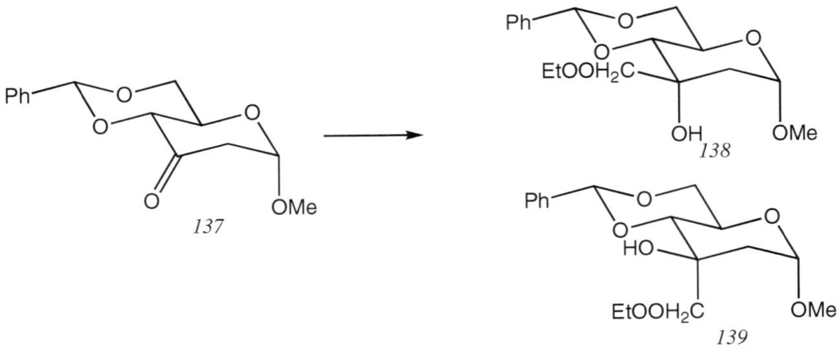

Fig. 11.39

to methyl 4,6-*O*-benzylidene-2-deoxy-α-D-*erythro*-hexopyranosid-3-ulose *137* giving in 86% overall yield the branched chain sugars *138* and *139* in 94:6 ratio [64] (Fig. 11.39). The configuration of C3 branching carbon was established by using methyl 4,6-*O*-benzylidene-2-deoxy-2,2-dideutero-α-D-*erythro*-hexopyranosid-3-ulose *140* for the Reformatsky reaction and by converting the obtained branched chain sugar to (*S*)-(−)-(1,1-^{2}H)-citric acid *142*, the absolute configuration of which is known (Fig. 11.40).

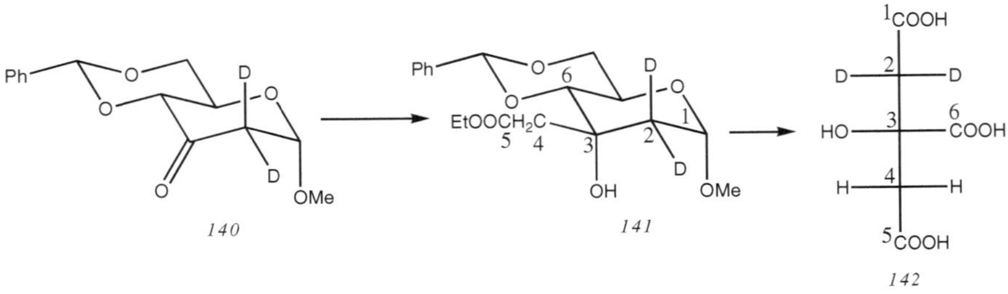

Fig. 11.40

Ethyl bromoacetate adds, at −78°C, to methyl 4,6-*O*-benzylidene-3-deoxy-α-D-*erythro*-hexopyranosid-2-ulose *143* or to methyl 4,6-*O*-benzylidene-3-deoxy-α-D-*erythro*-hexopyranosid-3-ulose *145* in the presence of zinc and laminated silver-graphite highly stereoselectively from the *re*-face to give *144* or *146*, respectively (in both products the branched chain is equatorially oriented) [65] (Fig. 11.41). The firm proof for the configuration of branching carbons was not provided.

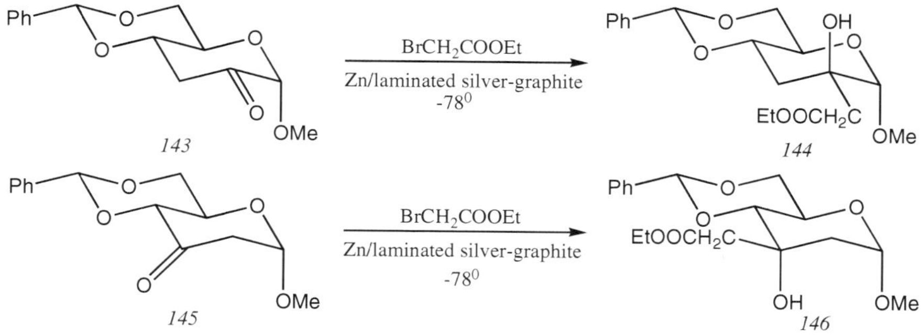

Synthesis of Branched Chain Sugars with Functionalized Branched Chain   315

Fig. 11.41

2-Bromomethyl-acrylic acid ethyl ester *147* adds to methyl 4,6-*O*-benzylidene-3-deoxy-α-D-*erythro*-hexopyranosid-2-ulose *143* in the presence of zinc, under usual conditions [66], giving, in 64% overall yield, a mixture of C2 epimers *148* and *149* in 54:10 ratio (Fig. 11.42).

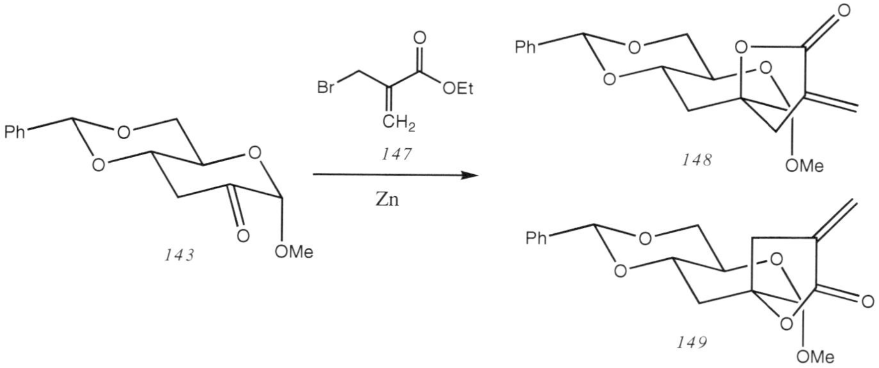

Fig. 11.42

The stereochemistry of addition of carbon nucleophiles to hexopyranosid-2-uloses in Figs. 11.41 and 11.42 seems to be less selective as compared to the stereochemistry of addition of other nucleophiles to the C2 carbonyl carbon of methyl α-D-hexopyranosid-2-uloses in which case the addition was observed to be taking place exclusively from the *si*-face due to the stereochemical control of the anomeric methoxy group (vide supra).

The addition of ethyl 2-bromomethyl-acrylate *147*, at –78°C, to methyl 4,6-*O*-benzylidene-2-deoxy-α-D-*erythro*-hexopyranosid-3-ulose *150* in the presence of laminar zinc/silver-graphite [67] gave, in 92% yield, the adduct *151* as the only product (the addition took place, as expected, from the *re*-face) (Fig. 11.43).

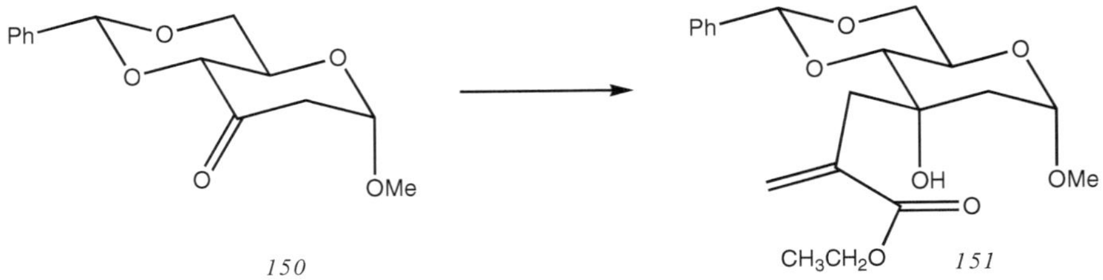

150 → *151* (CH₃CH₂O, OH, OMe)

Fig. 11.43

Unlike the adducts obtained in the addition of 2-bromomethyl-acrylate *147* to methyl hexopyranosid-2-ulose derivative *143*, the adduct *151* could not be converted to the spiro α-methylene-γ-lactone.

The authors claimed that the stereochemistry of branching carbons in *148*, *149*, and *151* were established by comparison of these products with the previously described products of similar Reformatsky reactions [65]. However, the references cited did not support these claims, and consequently the stereochemistry of branching carbons must be considered unknown.

Opening of Oxiranes with Nucleophiles

There are two types of oxiranes that were used for the synthesis of branched chain sugars: endocyclic oxiranes and spiro oxiranes. Endocyclic oxiranes consist of an epoxide ring on either the pyranose or the furanose ring (*152* and *153*) and spiro oxiranes (*154*) are obtained by the addition of diazomethane on hexo- or pentopyranosidulose or by epoxidation of *n*-deoxy-*n*-*C*-methylene sugar derivatives (Fig. 11.44).

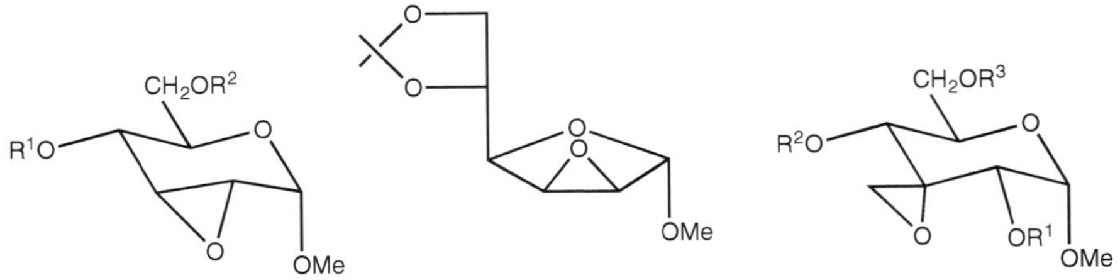

152   *153*   *154*

Fig. 11.44

The opening of an oxirane ring must always take place in the *trans*-diaxial fashion. Thus, for example, the endocyclic oxirane *155* opens with the exclusive formation of *156*, as depicted in Fig. 11.45; product *157* is not formed. This puts certain limitations on the use of this method for the synthesis of branched chain sugars, because first the branching chain will always be axially oriented and second the position of the branching carbon is predetermined by the configuration of the oxirane ring (in this case it can only be the C2 carbon).

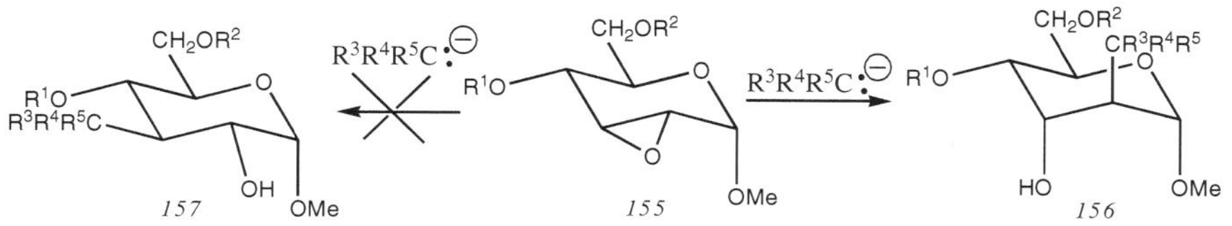

Fig. 11.45

Thus, the diaxial opening of the oxirane ring of methyl 2,3-anhydro-4,6-*O*-benzylidene-α-L-mannopyranoside *158* and methyl 2,3-anhydro-4,6-*O*-benzylidene-α-D-allopyranoside *160* (Fig. 11.46) with ethylmagnesium chloride gave 3-deoxy-3-*C*-ethyl- and 2-deoxy-2-*C*-ethyl derivatives *159* and *161*, respectively [68, 69].

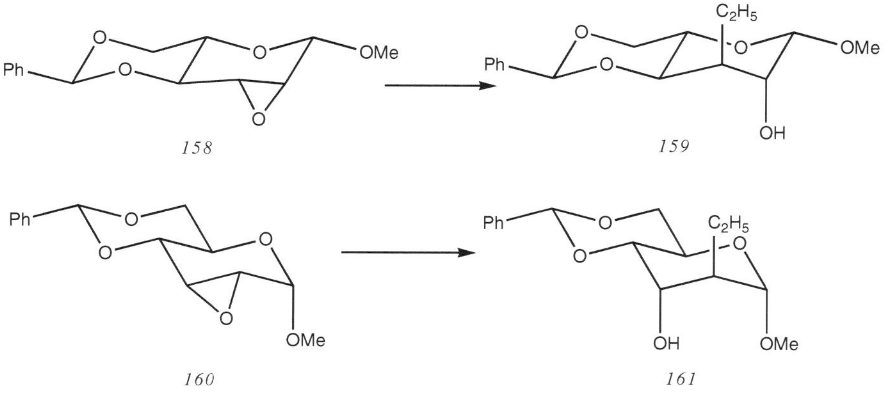

Fig. 11.46

The methyl 2,3-anhydro-5-*O*-trityl-α-D-ribofuranoside *162* reacts with the 2-lithio-1,3-dithiane regioselectively to give the 2-*C*-dithianyl derivative *164*, whereas the methyl 2,3-anhydro-5-*O*-trityl-β-D-ribofuranoside *163* reacts with methylmagnesium chloride also regioselectively to give 3-deoxy-3-*C*-methyl derivative *165* (Fig. 11.47). In this case the regioselectivity of oxirane ring opening seems to be controlled by the anomeric configuration.

Ethyl disodiomalonate [70, 71], hydrogen cyanide–triethylaluminum [72], dimethylmagnesium [73], and lithium dimethylcuprate [40, 74] are other carbon nucleophiles used for opening of the oxirane ring.

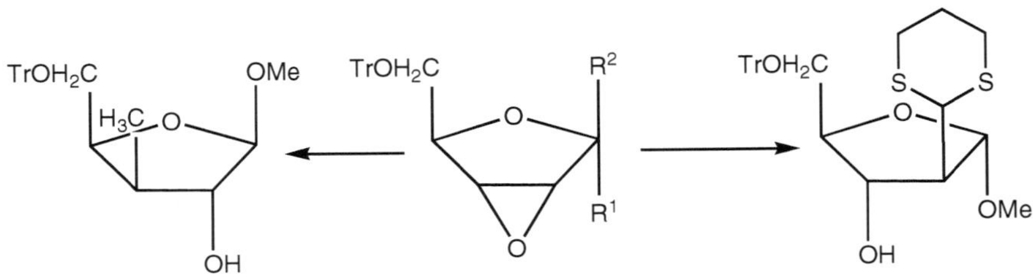

318   11   Addition of Nucleophiles to Glycopyranosiduloses

165

162, R¹ = OMe; R² = H
163, R¹ = H; R² = OMe

164

Fig. 11.47

References

1. Miljkovic, M.; Gligorijevic, M.; Miljkovic, D., *"Steric and Electrostatic Interactions in Reactions of Carbohydrates. II. Stereochemistry of Addition Reactions to the Carbonyl Group of Glycopyranosiduloses. Synthesis of Methyl 4, 6-O-Benzylidene-3-O- methyl-β-D-Mannopyranoside"*, J. Org. Chem. (1974) **39**, 2118–2120
2. House, H. O., *Modern Synthetic Reactions*, 2nd Ed. (1972), W. A. Benjamin, Menlo Park, Calif., p. 56
3. Shaban, M. A. E.; Jeanloz, R. W., *"Synthesis of 2-acetamido-2-deoxy-3-O-Ⓡ-D-mannopyranosyl-D-glucose"*, Carbohydr. Res. (1976) **52**, 103–114
4. Shaban, M. A. E.; Jeanloz, R. W., *"The synthesis of 2-acetamido-2-deoxy-4-O-Ⓡ-D-mannopyranosyl-D-glucose"*, Carbohydr. Res. (1976) **52**, 115–127
5. Tsuda, Y.; Hanajima, M.; Matsuhira, N.; Okuno, Y.; Kanemitsu, K., *"Utilization of sugars in organic synthesis. XXI. Regioselective mono-oxidation of non-protected carbohydrates by brominolysis of the tin intermediates"*, Chem. Pharm. Bull. Japan (1989) **37**, 2344–2350
6. Liu, H. M.; Sato, Y.; Tsuda, Y.,*"Utilization of sugars in organic synthesis. XXVII. Chemistry of oxo-sugars. (2). Regio- and stereo-selective synthesis of methyl D- hexopyranosiduloses and identification of their forms existing in solutions"*, Chem. Pharm. Bull. Japan (1993) **41**(3), 491–501
7. Ekborg, G.; Lindberg, B.; Lönngren, J., "Synthesis of β-D-mannopyranosides", Acta Chem. Scand. (1972) **26**, 3287–3292
8. Miljkovic, M.; Glisin, Dj., *"Synthesis of macrolide antibiotics. II. Stereoselective syn thesis of methyl 4,6-O-benzylidene-2-deoxy-2-C,3-O-dimethyl-α-D-glucopyranoside. Hydrogenation of the C-2 methylene group of methyl 4,6-O-benzylidene-2-deoxy-2-C- methylene-3-O-methyl-α- and -β-D-arabinohexopyranoside"* J. Org. Chem. (1975) **40**, 3357–3360
9. Hudson, C. S., *"Apiose and the Glycosides of the Parsley Plant"*, Adv. Carbohydr. Chem.(1949) **4**, 57–74
10. Shafizadeh, F., *"Branched-Chain Sugars of Natural Occurrence"*, Adv. Carbohydr. Chem. (1956) **11**, 263–283
11. Grisebach, H.; Schmid, R., *"Chemistry and Biochemistry of Branched-Chain Sugars"*, Angew. Chem., Int. Ed. Eng.(1972) **11**, 159–173
12. Yoshimura, J., *"Synthesis of Branched-Chain Sugars"*, Advan. Carbohydr. Chem. (1984) **42**, 69–134
13. Cahn, R. S.; Ingold, C. K.; Prelog, V., *"Specification of asymmetric configuration in organic chemistry"*, Experientia (1956) **12**, 81–94
14. Ferrier, R. J.; Overend, W. G.; Rafferty, G. A.; Wall, H. M.; Williams, N. R., *"De-termination of the configuration of branched-chain sugars"*, Proc. Chem. Soc. (London) (1963), 133

15. Hofheinz, W.; Grisebach, H.; Friebolin, H., "*Zur biogenese der makrolide—VIII: Die stereochemie der mycarose und cladinose*", Tetrahedron (1962) **18**, 1265–1274
16. Burton, J. S.; Overend, W. G.; Williams, N. R., "*Branched-chain sugars. Part III. The introduction of branching into methyl 3,4-O-isopropylidene-β-L-arabinoside and the synthesis of L-hamamelose*", J. Chem. Soc. (1965) 3433–3445
17. Keller-Schierlein, W.; Roncari, G., "*Stoffwechselprodukte von Actinomyceten. 33. Mitteilung. Hydrolyseprodukte von Lankamycin: Lankavose und 4-O-Acetyl- arcanose*", Helv. Chim. Acta (1962) **45**, 138–152
18. Keller-Schierlein, W.; Roncari, G., "*Stoffwechselprodukte von Mikroorganismen 46. Mitteilung Die Konstitution des Lankamycins*", Helv. Chim. Acta (1964) **47** 78–103
19. Lemal, D. M.; Pacht, P. D.; Woodward, R. B., "*The synthesis of L-(—)-mycarose and L-(—)-cladinose*", Tetrahedron (1962) **18**, 1275–1293
20. Roncari, G.; Keller-Schierlein, W., "*Stoffwechselprodukte von Mikroorganismen. 50. Mitteilung. Die Konfiguration der Arcanose*", Helv. Chim. Acta (1966) **49**, 705–711
21. Satoh, C.; Kiyomoto, A.; Okuda, T., *Nitrogen-containing carbohydrate derivatives: Part IV. Studies on the optical rotatory dispersion and circular dichroism of Carbohydrate C-nitro alcohols*", Carbohydr. Res. (1967) **5**, 140–148
22. Rosenthal, A.; Ong, K.-S., "*Branched-chain aminodeoxy sugars. Methyl 3-C- aminomethyl-2-deoxy-α-D-ribo-hexopyranoside and methyl 3-C-aminomethyl-2- deoxy-α-D-arabino hexopyranoside*", Can. J. Chem. (1970) **48**, 3034–3038
23. Rosenthal, A.; Ong, K.-S.; Baker, D., "*Synthesis of branched-chain nitro and amino sugars by the nitromethane route*", Carbohydr. Res. (1970) **13**, 113–125
24. Albrecht, H. P.; Moffatt, J. G., "*Synthesis of a branched chain aminosugar nucleoside*", Tetrahedron Lett. (1970) **11**, 1063–1066
25. Gero, S. D.; Horton, D.; Sepulchre, A. M.; Wander, J. D., "*Determination of the configurations of tertiary alcoholic centers in branched-chain carbohydrate derivatives: PMR spectroscopy with a lanthanide shift-reagent*", Tetrahedron (1973) **29**, 2963–2972
26. Dalling, D. K.; Grant, D. M., "*Carbon-13 magnetic resonance. IX. Methylcyclohexanes*", J. Am. Chem. Soc. (1967) **89**, 6612–6622
27. Anet, F. A. L.; Bradley, C. H.; Buchanan, G. W., "*Direct detection of the axial con- former of methylcyclohexane by 63.1 MHz carbon-13 nuclear magnetic resonance at low temperatures*", J. Am. Chem. Soc. (1971) **93**, 258–259
28. Stothers, J. B., "*Carbon-13 NMR Spectroscopy*", Academic Press, New York (1972), pp. 402 and 426
29. Miljkovic, M.; Gligorijevic, M.; Satoh, T.; Pitcher, R. G., "*Carbon-13 Nuclear Magnetic Resonance Spectra of Branched-Chain Sugars. Configuratinal Assignment of the Branching Carbon Atom of Methyl Branched-Chain Sugars*", J. Org. Chem. (1974) **39**, 3847–3850
30. Miljkovic, M.; Gligorijevic, M.; Satoh, T.; Miljkovic, D., "*Synthesis of Macrolide Antibiotics. I. Stereospecific Addition of Methyllithium and Methylmagnesium Iodide to Methyl α-D-xylo-Hexopyranosid-4-ulose Derivatives. Determination of the Configuration at the Branching Carbon Atom by Carbon-13 Nuclear Magnetic Resonance Spectroscopy*", J. Org. Chem. (1974) **39**, 1379–1384
31. Inch, T. D., "*Asymmetric synthesis: Part I. A stereoselective synthesis of benzylic centres. Derivatives of 5-C-phenyl-D-gluco-pentose and 5-C-phenyl-L-ido- pentose*", Carbohydr. Res. (1967) **5**, 45–52
32. Guillerm-dron, D.; Capmau, M.-L.; Chodkiewicz, W." *Assistance du groupe méthoxyle en α d'un carbonyle dans le cours stérique de l'addition d'or- ganométalliques insaturés*", Tetrahedron Lett. (1972) **13**, 37–40
33. Cram, D. J.; Kopecky, K. R., "*Studies in Stereochemistry. XXX. Models for Steric Control of Asymmetric Induction*", J. Am. Chem. Soc. (1959) **81**, 2748–2755
34. Yoshimura, J.; Ohgo, Y.; Ajisaka, K.; Konda, Y." *Asymmetric reactions. VI. Stereselectivities in phenyllithium and Grignard reactions with tetrahydrofurfural derivatives*", Bull Chem. Soc. Jap. (1972) **45**, 916–921

35. Corey, E. J., *"The Stereochemistry of α-Haloketones. I. The Molecular Configurations of Some Monocyclic α-Halocyclanones"*, J. Am. Chem. Soc.(1953) **75**, 2301–2304
36. Corey, E. J., *"Prediction of the stereochemistry of alpha-brominated ketosteroids"*, Experientia (1953) **9**, 329–331
37. Djerassi, C.; Geller, L. E.; Eisenbraun, E. J., *"Optical rotatory dispersion studies. XXVI. α-Haloketones. (4). Demonstration of conformational mobility in α-halocyclo hexanones"*, J. Org. Chem. (1960) **25**, 1–6
38. Allinger, N. L.; Allinger, J.; Geller, L. E.; Djerassi, C., *Conformational Analysis. VI.1a Optical Rotatory Dispersion Studies. XXVII.1b Quantitative Studies of an α-Haloketones by the Rotatory Dispersion Method"*, J. Org. Chem. (1960) **25**, 6–12
39. Eliel, E.; Allinger, N. L.; Angyal, S. J.; Morrison, G. A., *"Conformational Analysis"*, Interscience, New York (1965), p. 460
40. Hanessian, S.; Rancourt, G., *"Carbohydrates as chiral intermediates in organic synthesis. Two functionalized chemical precursors comprising eight of the ten chiral centers of erythronolide A"*, Can. J. Chem. (1977) **55**, 1111–1113
41. Yoshimura, J.; Sato, K.; Kubo, K.; Hashimoto, H., *"A facile synthesis of moenuronic acid derivatives"*, Carbohydr. Res. (1982) **99**, c1-c3
42. Sato, K.; Kubo, K.; Hong, N.; Kodama, H.; Yoshimura, J., *"Branched-chain sugars. XXIX. Synthesis of moenuronic acid (4-C-methyl-D-glucuronic acid)"*, Bull. Chem. Soc. Japan (1982) **55**, 938–942
43. Sato, K.-I.; Yoshimura, J., *"Stereoselectivities in the reactions of α-D-hexopyranosid- 4-uloses with diazomethane"*, Carbohydr. Res. (1982) **103**, 221–238
44. Carey, F. A.; Hodgson, K. O., *"Efficient syntheses of methyl 2-O-benzoyl-4,6-O- benzylidene-α-D-glucopyranoside and methyl 2-O-benzoyl-4,6-O-benzylidene-α-D- ribo-hexopyranosid-3-ulose"*, Carbohydr. Res.(1970) **12**, 463–465
45. Howarth, G. B.; Szarek, W. A.; Jones, J. K. N., *"The synthesis of D-arcanose"*, Carbohydr. Res. (1968) **7**, 284–290
46. Gutsche, C. D., Org. Reactions (1954) **8**, 364–429
47. Sato, K.-I.; Yoshimura, J., *"Stereoselectivities in the reaction of methyl 4,6-O-benzylidene-α- and β-D-hexopyranosid-2-uloses with diazomethane"*, Carbohydr. Res. (1979) **73**, 75–84
48. Sato, K.; Yoshimura, J., *"Branched-chain sugars. XII. The stereoselectivities in the reaction of methyl 4,6-O-benzylidene-α- and -β-D-hexopyranosid-3-uloses with diazomethane"*, Bull Soc. Chem. Japan (1978) **51**, 2116–2121
49. Corey, E. J.; Seebach, D., *"Carbanionen der 1,3-Dithiane, Reagentien zur C-C- Verknüpfung durch nucleophile Substitution oder Carbonyl-Addition"*, Angew. Chem., (1965) **77**, 1134–1135
50. Seebach, D., *"Nucleophile Acylierung mit 2-Lithium-1, 3-dithianen bzw. -1,3,5- trithianen"*, Synthesis (1969) 17–36
51. Wander, J. D.; Horton, D., *"Dithioacetals of Sugars"*, Adv. Carbohydr. Chem. Bio chem. (1976) **32**, 15–123
52. Flaherty, B.; Overend, W. G.; Williams, N. R., *"Branched-chain sugars. PartVII. The synthesis of D-mycarose and D-cladinose"*, J. Chem. Soc. C (1966) 398–403
53. Paulsen, H.; Redlich, H., *"Verzweigte Zucker, VI. Synthese der vier isomeren Methyl-D-aldgaroside. Strukturermittlung des Methylaldgarosids B aus Aldgamycin E"*, Chem. Berichte(1974) **107**, 2992–3012
54. López, J. C.; Lameignère, E.; Burnouf, C.; de los Angeles Laborde, M.; Ghini, A. A.; Olesker, A.; Lukacs, G., *"Efficient routes to pyranosidic homologated conjugated enals and dienes from monosaccharides"*, Tetrahedron (1993) **49**, 7701–7722
55. Brimacombe, J. S.; Mather, A. M., *"Branched-chain sugars. Part 7. A route to sugars with two-carbon branches using 1-methoxyvinyl-lithium"*, J. Chem. Soc. Perkin I (1980), 269–272
56. Brimacombe, J. S.; Mather, A. M., *"A route to branched-chain sugars using methoxyvinyl-lithium"*, Tetrahedron Lett. (1978) **19**, 1167–1170

References

57. Depezay, J-C.; Merrer, Y. L., "*Sucres branchés: Synthèse par aldolisation dirigée des désoxy-2 méthyléne-2C D-érythro et D-thréo pentoses*", Tetrahedron Lett. (1978) **19**, 2865–2868
58. Brimacombe, J. S.; Hanna, R.; Mather, A. M.; Weakley, T. J. R., "*Branched-chain sugars. Part 8. The synthesis of C-acetylpyranosides and a pillarose derivative using 1-methoxyvinyllithium*", J. Chem. Soc. Perkin 1 (1980) 273–276
59. Depezay, J.-C.; Le Merrer, Y., "*Synthèse par aldolisation dirigèe des 2-dèsoxy-2-C-mèthylene-D-erythro- ET -D-thréo-pentoses*", Carbohydr. Res. (1980) **83**, 51–62
60. Reformatsky, S."*Neue Synthese zweiatomiger einbasischer Säuren aus den Ketonen*" Berichte (1887) **20**, 1210–1211
61. Reformatsky, S., "*Action of zinc and ethyl chloroacetate on ketones and aldehydes*", J. Russ. Phys. Chem. Soc. (1890) **22**, 44–64
62. Shriner, R. L., Org. React. (1942) **1**, 1
63. Rathke, M. W., Org. React. (1975) **22**, 423
64. Brandänge, S.; Dahlman, O.; Mörch, L., "*Highly selective re-additions to a masked oxaloacetate. Absolute configurations of fluorocitric acids*", J. Am. Chem. Soc. (1981) **103**, 4452–4458
65. Csuk, R.; Fürstner, A.; Weidman, H., "*Efficient, low temperature Reformatsky reactions of extended scope*", J. Chem. Soc. Chem. Comm. (1986) 775
66. Oehler, E.; Reininger, K.; Schmidt, U. "*Einfache Synthese von alpha-Methylen- gamma-lactonen*", Angew. Chem. (1970) **82**, 480–481
67. Csuk, R.; Fürstner, A.; Sterk, H.; Weidmann, H., "*Synthesis of carbohydrate-derived α-methylene-γ-lactones by diastereoselective, low-temperature Reformatskii-type reactions*", J. Carbohydr. Chem. (1986) **5**, 459–467
68. Inch, T. D.; Lewis, G. J., "*The synthesis and degradation of benzyl 4,6-O-benzylidene-2, 3-dideoxy-3-C-ethyl-2-C-hydroxymethyl-α-D-glucopyranoside and - mannopyranoside*", Carbohydr. Res. (1972) **22**, 91–101
69. Inch, T. D.; Lewis, G. J., "*The synthesis of branched-chain, deoxy sugars by sugar epoxide-Grignard reagent reactions*". Carbohydr. Res. (1970) **15**, 1–10
70. Hanessian, S.; Dextraze, P., "*Carbanions in Carbohydrate Chemistry: Novel Methods for Chain Extension and Branching*", Can. J. Chem. (1972) **50**, 226–232
71. Hanessian, S.; Dextraze, P.; Masse, R., "*Regiospecific and asymmetric introduction of functionalized branching in carbohydrates*", Carbohydr. Res. (1973) **26**, 264–267
72. Davison, B. E.; Guthrie, R. D., "*Nitrogen-containing carbohydrate derivatives. Part XXVII. Synthesis and reactions of 3-cyano-3-deoxy-glycose derivatives*", J. Chem. Soc. Perkin I (1972), 658–662
73. Shmyrina, A. Ya.; Sviridov, A. F.; Chizov, O. S.; Shashkov, A. S.; Kochetkov, N. K., "*Synthesis of methyl-3-deoxy-3-C-methyl-4-O-benzyl-β-L-xylopyranoside*" Izv. Akad. Nauk SSSR Ser. Khim. (1977) 461–463
74. Yamamoto, H.; Sasaki, H.; Inokawa, S.," *Reaction of lithium dimethyl cuprate with methyl 2, 3-anhydro-5-deoxy-α-D-ribofuranoside. A new, convenient route for preparation of 2,5-dideoxy-2-C-methyl-D-arabinofuranose derivatives*", Carbohydr. Res. (1982) **100**, c44–c45

Chapter 12
Chemistry of the Glycosidic Bond

Introduction

Because of the importance and the role the carbohydrates play in living organisms, the formation and hydrolysis of glycosidic bond are probably the two most important reactions in carbohydrate chemistry. Just as the amino acids are the building blocks for the synthesis of peptides and proteins in living organisms, the monosaccharides are the building blocks for the synthesis of oligosaccharides, polysaccharides, and glycoconjugates (glycoproteins, glycosaminoglycans, glycolipids, and proteoglycans to name a few). The synthesis of oligo- and polysaccharides as well as glycoconjugates requires the formation of glycosidic bonds, i.e., the formation of a chemical bond between the C1 carbon of a monosaccharide and any hydroxyl oxygen of another monosaccharide or a hydroxyl oxygen of any molecule that bears hydroxyl group, such as a hydroxyamino acid (serine, threonine), a lipid (sphingosine), and phosphatidyl inositol. The glycosidic bond can also be formed between the anomeric C1 carbon of a monosaccharide and the amido nitrogen of asparagine (as, for example, in N-linked glycoproteins) or the nitrogen of a purine or a pyrimidine base (as, for example, in ribo- and deoxyribonucleosides).

The formation of glycosidic bonds takes place during the biosynthesis of various oligo- and polysaccharides or glycoconjugates whereas the hydrolysis of glycosidic bonds occurs during the processing of biosynthetic intermediates in the biosynthesis of complex saccharides or during metabolism of complex saccharides and glycoconjugates in cells.

Unlike the synthesis of peptides and proteins from amino acids, where the only issues that a synthetic chemist has to deal with are the selection of a reagent for the peptide bond formation, the selection of proper protection groups for relatively simple side chains of amino acids, and the prevention of possible racemization of amino acids during synthesis, the synthesis of oligo- and polysaccharides represents a much greater challenge to a synthetic chemist because the monosaccharides are structurally much more complex molecules than amino acids. First, they contain several hydroxyl groups (or sometimes other functional groups, such as acetamido and carboxyl) that, for the oligosaccharide or polysaccharide synthesis, require multiple regioselective protection that is often difficult to accomplish because it is difficult to distinguish them chemically. Second, there are two types of glycosidic bond that a monosaccharide can form, α- or β-glycosidic bond, so that the stereoselectivity in the glycoside bond formation is a very important issue for the synthesis of oligo- and polysaccharides. Third, one should not forget that monosaccharides exist in two ring structures, furanoses and pyranoses, and that in some instances this issue must be addressed too. Finally, the selection of protection groups in an oligosaccharide synthesis is a very important and often difficult problem, because they have to be not only regioselective but also removable under mild conditions so that the newly formed glycosidic bond will not be hydrolyzed on their removal. In peptide or protein synthesis this problem does not exist because the peptide bonds are chemically very resistant. Since the nature of the C2 substituent most often determines the stereochemistry of the newly formed glycosidic bond, much attention has been given to the protection of the C2 hydroxyl group.

Chemically the glycosidic bond is an acetal bond and generally has the same chemical properties. However, whereas in acetals or ketals formed by condensing an aldehyde or ketone with an alcohol both alkoxy groups are identical (the acetal carbon is achiral since it has a plane of symmetry) (*1* in Fig. 12.1), the glycosides are mixed acetals having as one alkoxy group the C5 or the C4 hydroxyl group of a parent monosaccharide in its pyranoid or furanoid form (*2* or *3*, respectively, in Fig. 12.1) and the other alkoxy group is an alcohol, another monosaccharide, hydroxyamino acid, lipid, etc. Consequently the anomeric carbon in glycosides is chiral (Fig. 12.1).

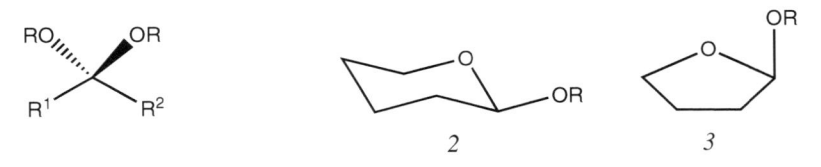

1, R = alkyl; R^1 and R^2 are H, alkyl, aryl, or carbon atoms of the same molecule

2 and *3*, R = alkyl, aryl, monosaccharide, amino acid, lipid, etc.

Fig. 12.1

Glycoside Synthesis

The first synthesis of an *O*-glycoside (to distinguish the glycosides from the more recently synthesized *C*-glycosides, which are not really glycosides, but rather C2 alkyl tetrahydrofuran or tetrahydropyran derivatives) was carried out by Michael [1–3] by reacting the 2,3,4,6-tetra-*O*-acetyl-α-D-glucopyranosyl chloride *4* with potassium phenoxide (Fig. 12.2):

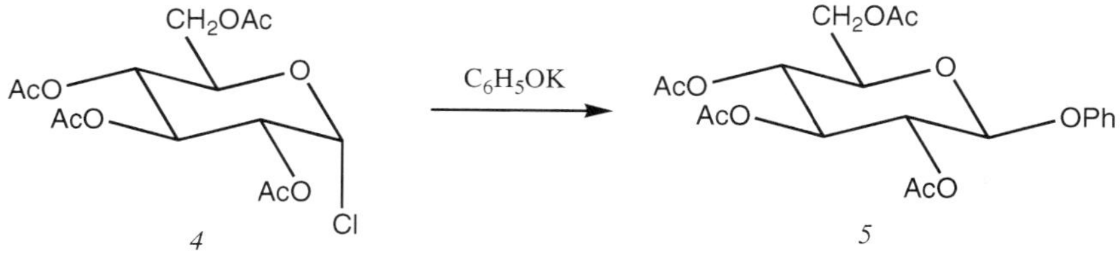

Glycoside Synthesis

Fig. 12.2

Fischer Glycosidation

In 1893 E. Fischer [4, 5] developed a synthesis of glycosides of lower alcohols by refluxing a monosaccharide (for example, D-glucose) with an alcohol in the presence of an anhydrous mineral acid (for example, HCl). If the concentration of mineral acid is several percent, α- and β-glycopyranosides are obtained (Fig. 12.3). However, if the concentration of mineral acid (HCl) is low (for example, 0.7%) and if the reaction is conducted at room temperature, the reaction of D-glucose with anhydrous methanol affords a mixture of methyl α- and β-D-glucofuranosides in good yield (Fig. 12.4).

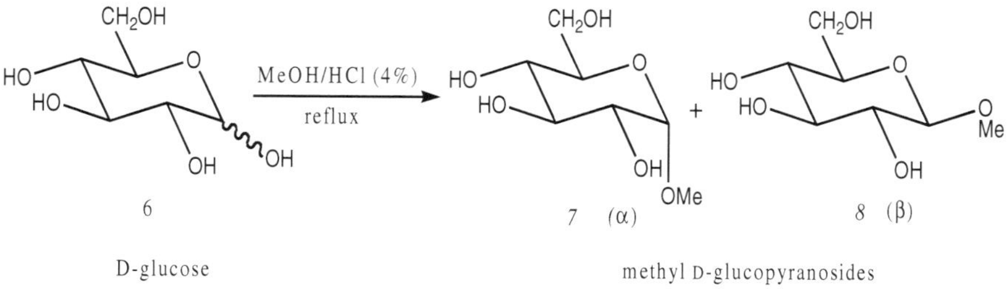

Fig. 12.3

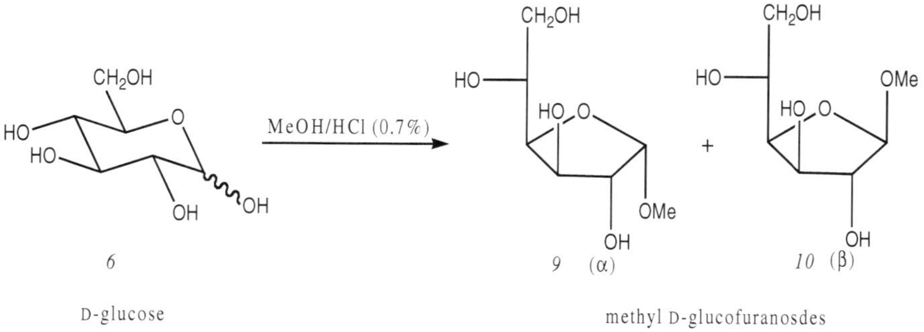

Fig. 12.4

326   12 Chemistry of the Glycosidic Bond

The above results suggest that the furanoid forms of D-glucose must be much more reactive toward glycosidation and that the obtained glucofuranosides are therefore kinetic products. Raising temperature and increasing the concentration of HCl converts the less stable methyl furanosides to more stable methyl α- and β-D-glucopyranosides suggesting that the latter are thermodynamic products. Thus, methyl glucopyranosides are most likely formed by both isomerization of methyl furanosides and direct glycosidation of glucopyranose (see Fig. 12.5).

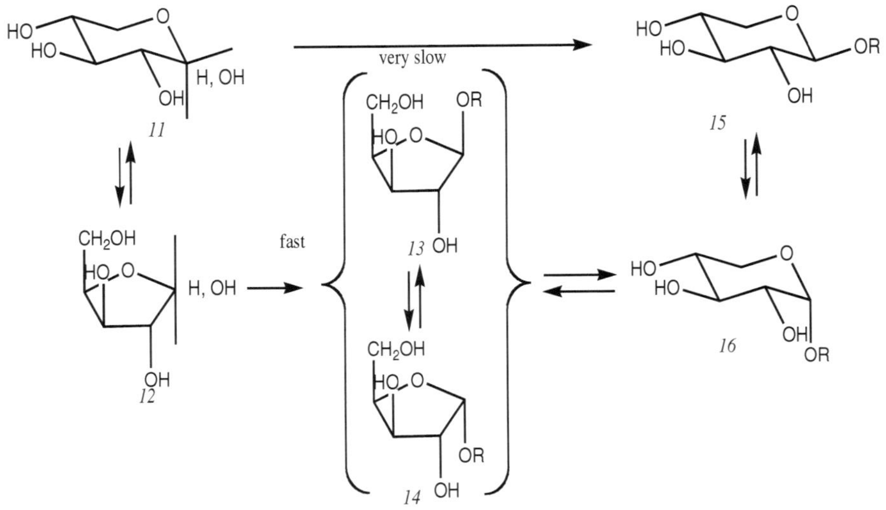

Fig. 12.5

The mechanism of Fischer glycosidation was studied in detail by using D-xylose [6, 7]. It has been shown that the formation of equilibrium mixture of glycosides (*15* and *16*) (Fig. 12.5) proceeds in four reaction steps that greatly differ in their rates of equilibrium formation (Figs. 12.5 and 12.6). In the first step, the free xylose, which is an equilibrium mixture of α- and β-D-xylopyranoses *11* and α- and β-D-xylofuranoses *12*, is converted into a mixture of α- and β-D-xylofuranosides (*13* and *14*) that, in the second step, undergoes an acid-catalyzed anomerization until an equilibrium mixture of α- and β-xylofuranosides is obtained. In the third step, the furanoside to pyranoside ring isomerization takes place, and finally in the fourth step, anomerization of α- and β-D-xylopyranosides takes place until the final equilibrium mixture of obtained pyranosides is formed (*15* and *16*) (Fig. 12.5). So the end result of these intramolecular isomerizations is the formation of an equilibrium mixture of all four methyl xylosides *13*, *14*, *15*, and *16*. It is important to note that the first two steps are considerably faster than the last two steps.

A possible reaction mechanism for the Fischer synthesis of glycosides is shown in Figs. 12.6, 12.7, and 12.8. The proposed mechanism suggests that D-xylopyranose does not undergo direct glycosidation. Instead it is first converted to xylofuranose (*11→17*) (Fig. 12.6) which then undergoes glycosidation. That is probably not

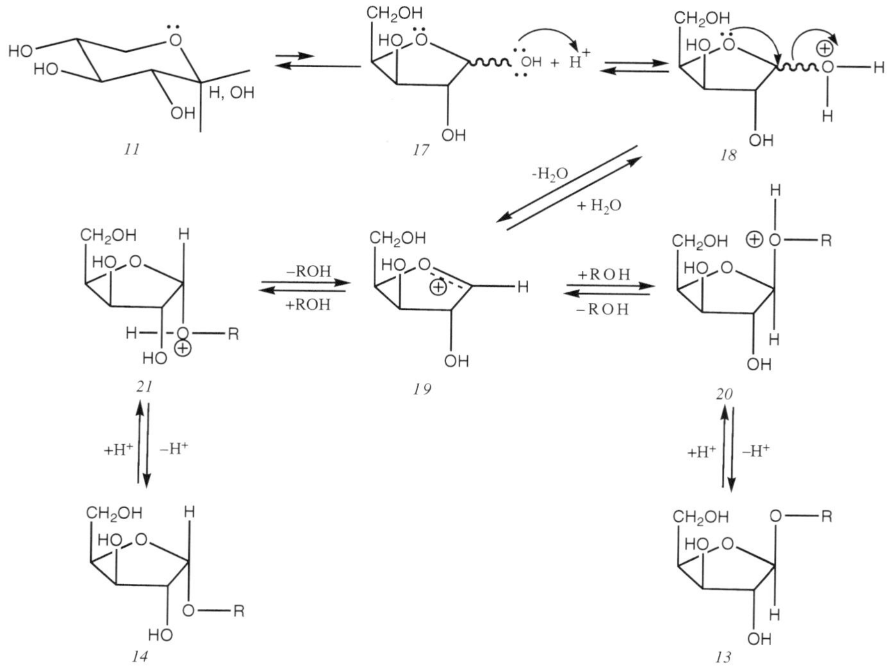

Glycoside Synthesis

Fig. 12.6

completely true, because the pyranose does, although much slowly, undergo glycosidation reaction on its own. It is assumed that protonation of glycosidic oxygen of glycofuranoses leads to the formation of oxocarbenium ion *19* which then adds an alcohol to the C1 carbon from either side of the molecule resulting in the formation of α- and β-D-xylofuranosides *13* and *14*.

The alcoholic solution of D-xylose consists predominantly of a mixture of α- and β-D-xylopyranoses *11* and a mixture of α- and β-D-xylofuranoses *12* (in Fig. 12.5). Since direct glycosidation of xylopyranoses is a very slow process and glycosidation of xylofuranoses is a fast process, the xylofuranoses will be removed from equilibrium mixture by the glycosidation resulting in the conversion of an additional amount of xylopyranoses to xylofuranoses and this process will continue until all of glycopyranose has been converted to a mixture of xylofuranose glycosides (Fig. 12.6).

The proposed reaction mechanism is in agreement with the observation that short reaction times, or small amounts of acid catalyst in the glycosylation reaction, favor the formation of furanosides as the predominant products, while higher concentrations of acid catalyst or by allowing the reaction to reach equilibrium, the predominant products are pyranosides.

The obtained mixture of α- and β-D-xylofuranosides *22* and *23* then undergoes ring isomerization via the acyclic oxocarbenium transition state *24* (Fig. 12.7). For

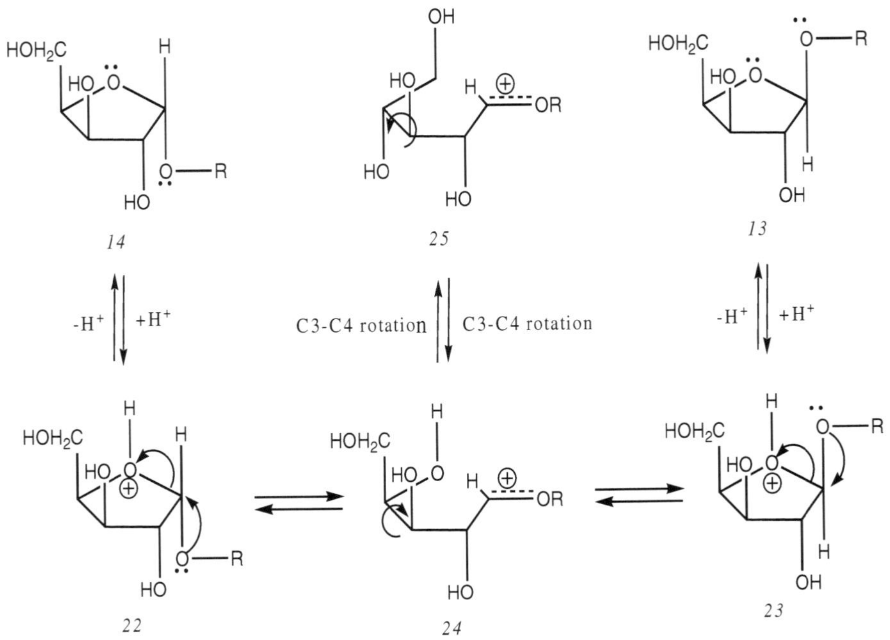

Fig. 12.7

this, the protonation of the ring oxygen must first take place followed by the C1–O4 bond rupture resulting in the formation of the acyclic oxocarbenium intermediate 24. The rotation about the C3-C4 bond prior to the ring closure (the reverse reaction of ring rupture) brings the C5 hydroxyl group in position to attack the oxocarbenium ion and thus close the pyranoside ring 25 (Fig. 12.8). The anomerization of α- and β-D-xylopyranosides takes place again via cyclic oxocarbenium ion transition state (Fig. 12.8) [8]. It should be noted that the anomerization of α- and β-D-xylopyranoses can also take place via the acyclic oxocarbenium ion 25.

Although acyclic acetals are feasible intermediates in the glycosidation of sugars, Bishop and Cooper [6, 7] did not find evidence for their presence in equilibrium mixture. However, by using radiochemical techniques, two groups of researchers have reported the presence of acyclic acetals in the glycosidation reaction mixtures. Heard and Barker [9] observed the presence of acyclic dimethyl acetal in the product mixture obtained by methanolysis of D-arabinose, and Ferrier and Hatton [10] reported the presence of acyclic dimethyl acetal in the glycosidation reaction mixture of D-xylose and D-glucose.

While still useful for the preparation of methyl glycosides, Fischer method is not at all suited for the much more complicated synthesis of oligosaccharides or for the glycosylation of other natural products (proteins, lipids, etc.).

Since Fischer synthesis often gives a complex mixture of anomeric and ring isomers it is not considered to be the preparative method of choice for the syntheses of glycosides. For example, the equilibrium mixture obtained by acid-catalyzed

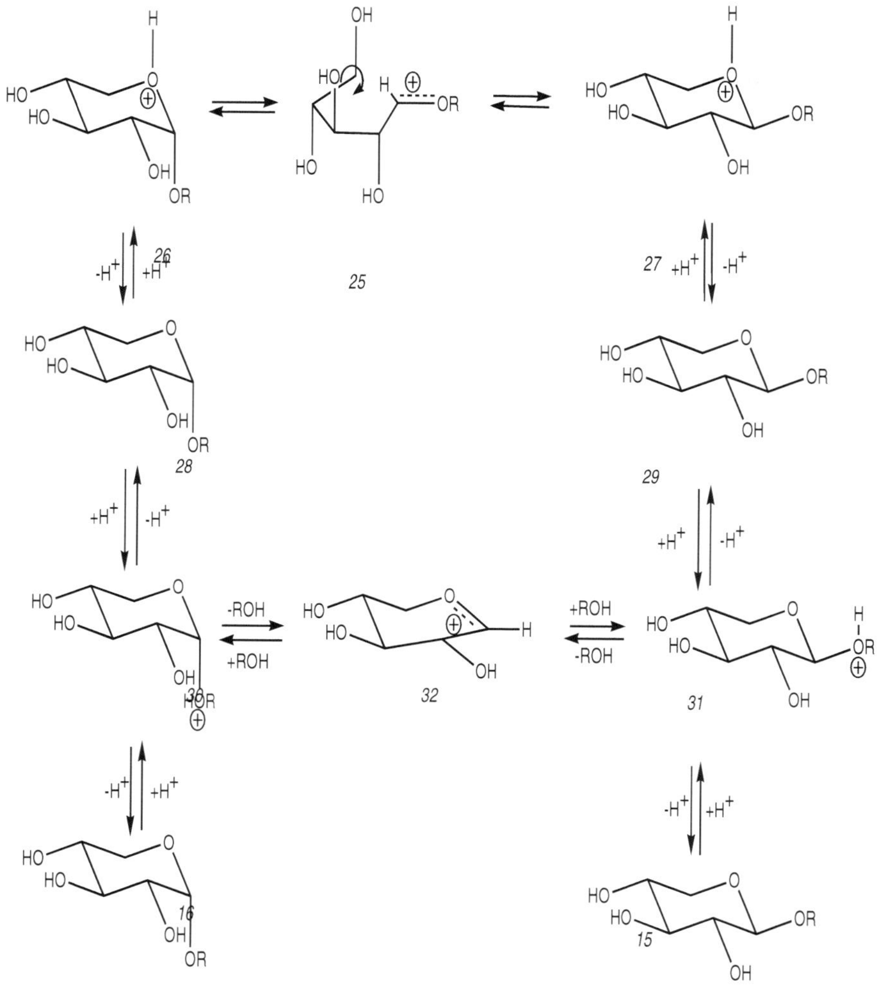

Fig. 12.8

methanolysis of D-galactose consists of methyl α-D-galactofuranoside 6.2%, methyl β-D-galactofuranoside 16.3%, methyl α-D-galactopyranoside 57.8%, and methyl β-D-galactopyranoside 19.7% [11]. Usually, the preparative application of Fischer synthesis is based on two approaches. According to one approach the Fischer reaction is terminated at a kinetically controlled stage to prepare furanosides. Thus, for example, methyl α-D-arabinofuranoside can be prepared in good yield in this way [12]. Another approach is to allow the reaction mixture to reach equilibrium and then isolate the predominant isomer. Which isomer will predominate depends on the sugar. In some cases the predominant isomer is formed in great excess, as is for example the case with methyl α-D-mannopyranoside [11] which is, at equilibrium, 94% and sometimes the desirable product crystallizes directly from equilibrium mixture, allowing thus its easy isolation, as is the case with benzyl β-L-arabinopyranoside [13] that is isolated with 75% yield.

Königs–Knorr Synthesis

In 1901 Königs and Knorr [14] developed a new method for the synthesis of glycosides which was immediately recognized as the more general and useful preparative method than Fischer method and which is still used today (modified or unmodified). The Königs–Knorr method consists of reacting a fully acetylated glycosyl halide 33 with an alcohol or with a hydroxyl group of another sugar (dissolved in a dry inert solvent) in the presence of silver carbonate or silver oxide as promoter and acid acceptor [14] (Fig. 12.9).

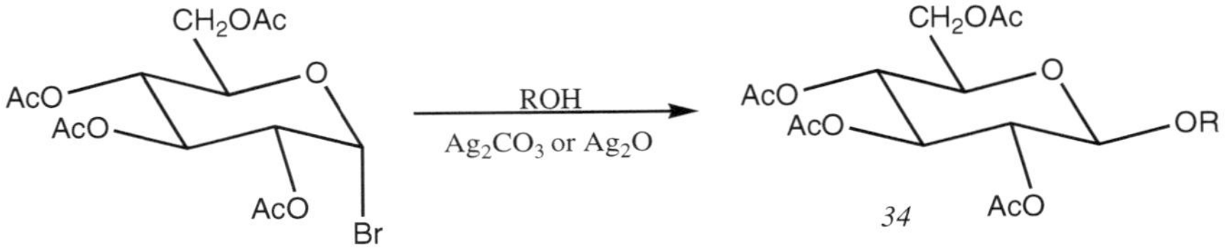

Fig. 12.9

O-Acylglycosyl halides (excluding fluorides) react with alcohols also directly (in the absence of silver carbonate or silver oxide) to yield glycosides. Thus, a variety of O-benzoylglycosyl halides react directly with methanol to give 1,2-trans-O-benzoylglycosides. For example, tri-O-benzoyl-β-D-ribopyranosyl bromide 35, in anhydrous methanol, gave methyl tri-O-benzoyl-β-D-ribopyranoside 36 with 88% yield [15] (Fig. 12.10).

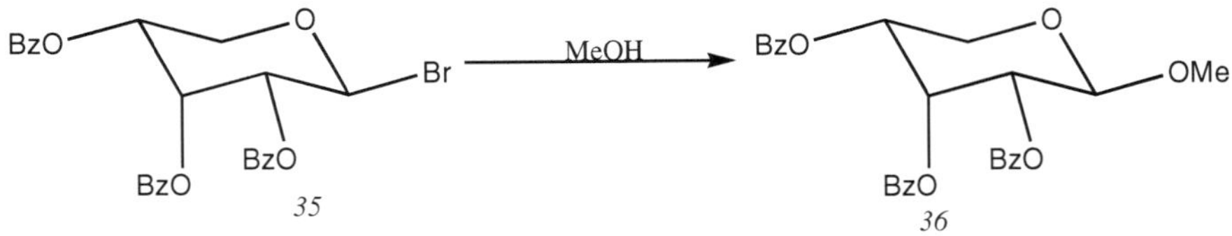

Fig. 12.10

Similarly, methyl tetra-O-benzoyl-β-D-glucopyranoside 38 was obtained with 90% yield from tetra-O-benzoyl-α-D-glucopyranosyl bromide 37 [16] (Fig. 12.11).

It has been shown that glycosyl halides having the C2 acyl substituent trans-oriented relative to the C1 halogen are much more reactive than their corresponding cis-isomers (see Table 12.1).

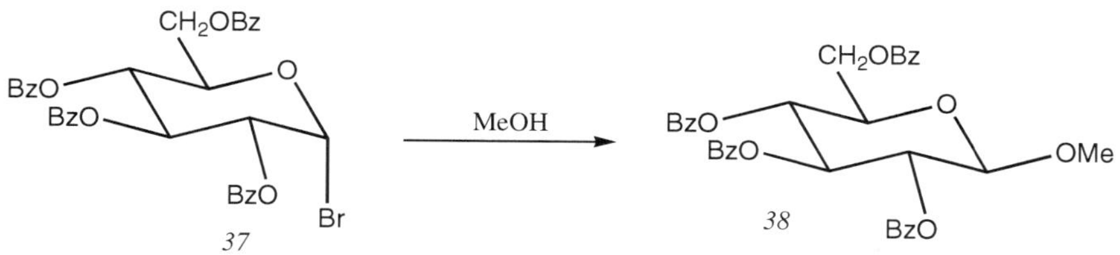

Glycoside Synthesis

Fig. 12.11

Table 12.1 Reaction of tri-O-benzoyl-D-pentopyranosyl halides with 1:9 dioxane:methanol at 20°C

Tribenzoate of	$k \times 10^4$ (min., $\log_{10}$)
1,2-*cis*-α-D-Ribosyl bromide	40
1,2-*trans*-β-D-Ribosyl bromide	760
1,2-*cis*-α-D-Ribosyl chloride	0.62
1,2-*trans*-β-D-Ribosyl chloride	53

The mechanism of Königs–Knorr reaction has been extensively studied for many years. The obtained results are often contradictory and many aspects of the reaction mechanism are still unclear [17].

The addition of hydroxyl ions does not increase the rate of solvolysis of O-acetylglycopyranosyl halides in methanol and in aqueous acetone, but the addition of water does increase the rate, thus supporting the proposed unimolecular carbonium ion mechanism [18]. The glycopyranosyl halides that bear an acyloxy group at the C2 carbon, which is *cis* to 1-halide, usually react with the inversion of anomeric configuration, whereas the glycopyranosyl halides having the C2 acyloxy group *trans* to 1-halides react with the retention of anomeric configuration. This stereochemical outcome has been explained by neighboring group participation of the C2 acyloxy group in the transition state by formation of an 1,2-cyclic carboxonium cation *41*, thus, on one side, stabilizing the transition state and, on the other side, forcing the nucleophile to approach the C1 carbon from the 1,2-*trans* direction (Fig. 12.12). The stability of the intermediary carboxonium cation has a profound influence upon the rate and the course of reaction. The formation of transition state requires flattening of the pyranoid ring and the adoption of a conformation that resembles the half-chair conformation with the C2, C1, O5, and C5 atoms lying in one plane (*40* and *41* in Fig. 12.12). It has been suggested that a large equatorial substituent at the C5 carbon hinders the formation of transition state [19] *41* and this argument was used to explain the greater rate of methanolysis of 2,3,4-tri-O-acetyl-α-D-xylopyranosyl bromide *42* than that of methyl 2,3,4,6-tetra-O-acetyl-α-D-glucopyranosyl bromide *43* (Fig. 12.13).

The slower rate of methanolysis of 2,3,4,6-tetra-O-acetyl-α-D-mannopyranosyl bromide *44* (Fig. 12.13) than that of 2,3,4,6-tetra-O-acetyl-α-D-glucopyranosyl bromide *43* has been ascribed to Pitzer strain between the C2 and the C3 acetoxy groups that increases during the formation of transition state.

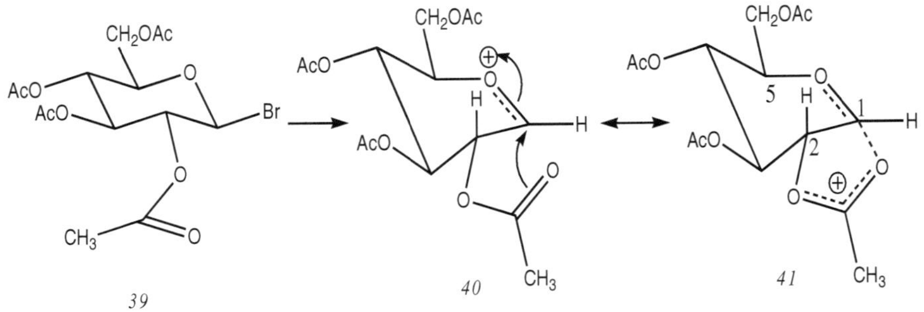

Fig. 12.12

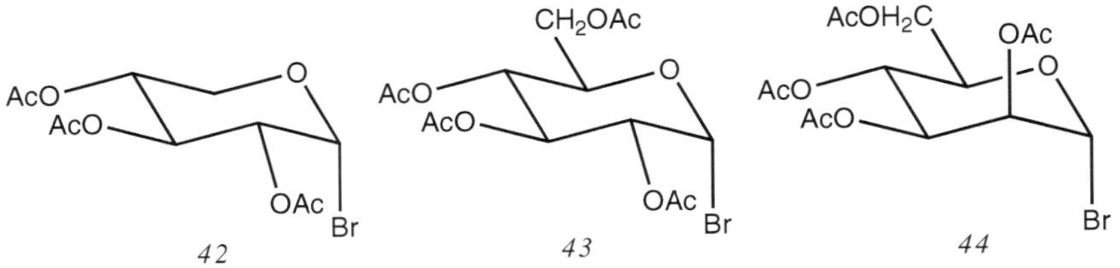

Fig. 12.13

The stereochemical consequence of stabilization of the oxocarbenium ion via neighboring group participation of the C2 acyl group in 2,3,4,6-tetra-O-acetyl-α-D-glucopyranosyl bromide 39 has been the exclusive formation of β-D-glucoside.

When, however, the nucleophilicity of carbonyl group of the C2 acetoxy group is decreased by substituting the three hydrogen atoms of methyl group with chlorine [20], as is the case in 3,4,6-tri-O-acetyl-2-O-trichloroacetyl-β-D-glucopyranosyl chloride 45, the methanolysis of 45 in pyridine, containing silver nitrate as promoter, gives methyl 3,4,6-tri-O-acetyl-2-O-trichloroacetyl-α-D-glucopyranoside 46 (Fig. 12.14).

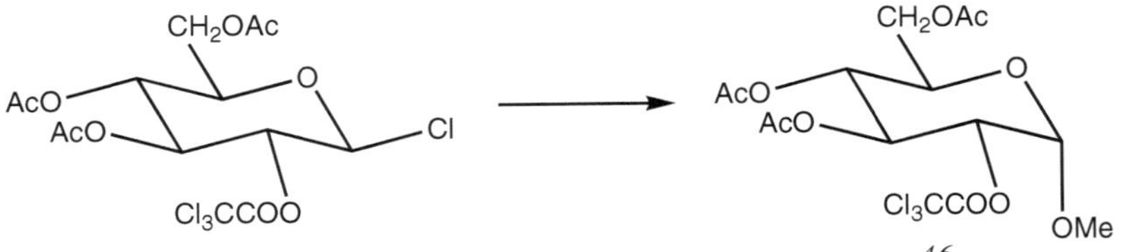

Fig. 12.14

The three chlorine atoms have removed the nucleophilicity of the carbonyl oxygen by electron-withdrawing inductive effect of these chlorine atoms and thus prevented the formation of cyclic carboxonium ion [21, 22].

Hence, in order to accomplish the synthesis of 1,2-*cis* glycosides by Königs–Knorr reaction, the protection group at the C2 carbon must be a nonparticipating group.

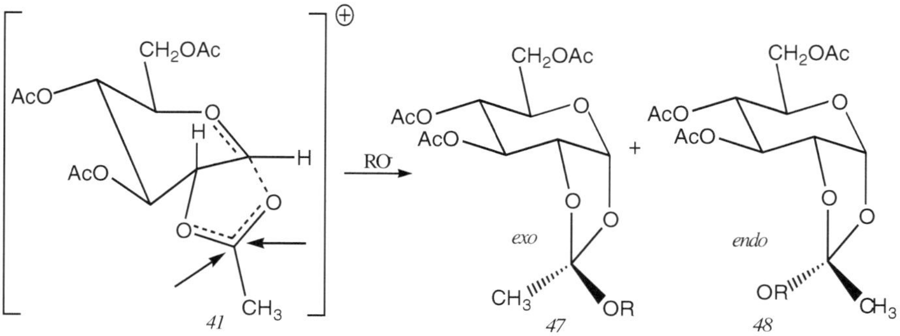

Fig. 12.15

The attack of a nucleophile on oxocarbonium ion *41* does not always have to take place at the C1 carbon; it can also attack the carbonyl carbon of the C2 acetate giving rise to one or both diastereomeric *orthoesters 47* and *48* (Fig. 12.15). The orthoester formation is often a side reaction of the Königs–Knorr reaction. By slightly changing the experimental conditions, the formation of orthoesters can be made to be the major pathway of Königs–Knorr reaction.

For example, the reaction of tri-*O*-acetyl-2-*O*-acyl-α-D-glucopyranosyl halides with a variety of alcohols in 2,4,6-trimethylpyridine containing tetraalkyl ammonium halide gives 1,2-orthoesters in high yield [23]. In each case the alkoxy group in dioxolan ring was shown (by NMR) to be in the *exo*-configuration *47* (*trans* to the pyranose ring) (Fig. 12.15). The use of 1,2-orthoesters as active intermediates for the synthesis of glycosides was developed by Kochetkov and his co-workers and will be discussed later.

Numerous modifications and improvements of Königs–Knorr reaction have been reported. For example, the use of anhydrous calcium sulfate and small amounts of iodine proved to be very useful and most often improved the yield [24, 25]. In order to avoid the need for a drying agent (drierite) for removal of water formed during Königs–Knorr reaction, silver salts of hydroxycarboxylic acids have been suggested as both promoters and acid acceptors [26–29]. The Königs–Knorr reaction was found to be strongly dependent on the solvent.

To overcome the difficulties that arise from the heterogeneity of reaction mixture that is typically associated with Königs–Knorr reaction (the catalyst and the acid acceptor are usually insoluble in the solvents used), silver salts with nonnucleophilic anion that are soluble in organic solvent were used in combination with silver carbonate or oxide. In fact silversalt in solution, for example silver perchlorate, serves as a homogenous catalyst whereas silver carbonate or oxide added to the reaction mixture serves as acid acceptor during which step the soluble catalyst is regenerated (Fig. 12.16).

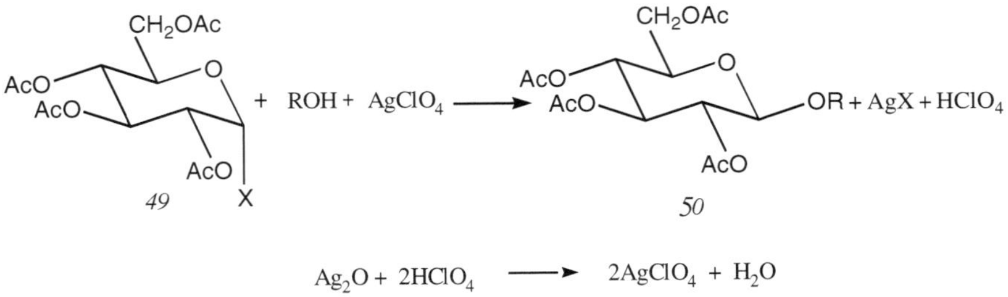

[Structure 49: tetraacetyl glycosyl halide with CH₂OAc, AcO groups and X leaving group] + ROH + AgClO₄ ⟶ [Structure 50: glycoside with OR group] + AgX + HClO₄

$$Ag_2O + 2HClO_4 \longrightarrow 2AgClO_4 + H_2O$$

Fig. 12.16

The Königs–Knorr reaction is usually faster in the presence of soluble catalysts. Various soluble silver salts were used, e.g., tetrafluoroborate, hexafluorophosphate, trifluoromethanesulfonate (triflate), or *p*-toluenesulfonate (tosylate). The yields obtained with these catalysts are almost quantitative but the stereochemical outcome was strongly dependent on reaction conditions and the structure of protecting groups in the glycosyl halide [30–32].

A solution of silver triflate catalyst in methylene chloride in combination with 1,1,3,3-tetramethylurea as a soluble acid acceptor has been shown to be an effective system for glycosidation [33, 34]. The reaction was performed under strictly anhydrous conditions in dark for 4 h. The disaccharide *53* was obtained with 47% yield. In three other experiments the respective disaccharides were obtained with yields ranging from 72 to 86% (Fig. 12.17).

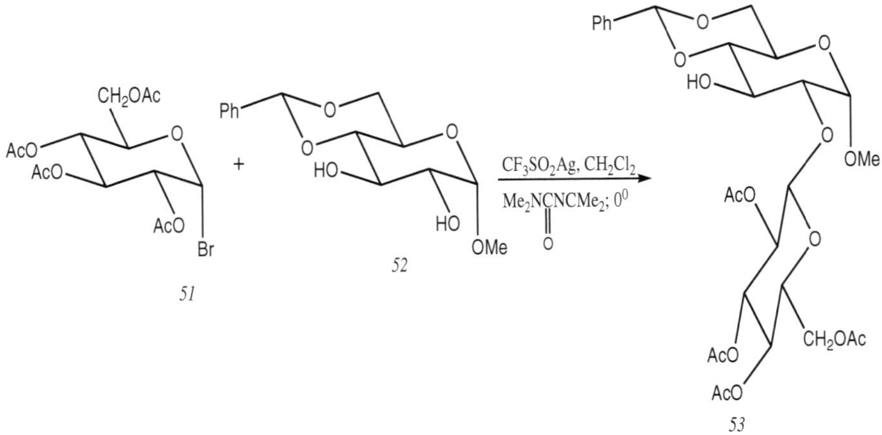

[Structure 51: acetobromoglucose] + [Structure 52: 4,6-O-benzylidene methyl glycoside with HO groups and OMe] $\xrightarrow{CF_3SO_2Ag, CH_2Cl_2 \atop Me_2NCNCMe_2; 0°}$ [Structure 53: disaccharide product]

Fig. 12.17

Helferich and Wedemeyer [35] found that mercuric cyanide in nitromethane as solvent is both the catalyst and acid acceptor in the Königs–Knorr reaction and results in significantly improved yields of glycosides.

Alkyl β-D-glucopyranoside tetra-acetates were obtained in high yield by the use of a mixture of mercuric bromide as the catalyst and yellow mercuric oxide as the acid acceptor [36]. Another modification for the synthesis of α-D-glucosides and α-D-galactosides uses a mixture of mercuric bromide and mercuric cyanide [37].

Finally, by using mercuric acetate or ferric chloride some α-D-glycosides could be synthesized in spite of the presence of a participating group at the C2 carbon [38, 39].

Synthesis of Acylated Glycosyl Chlorides and Bromides

It is now appropriate to briefly discuss the synthesis of glycosyl halides (chlorides and bromides).

Acylated glycosyl halides can be prepared by the action of hydrogen bromide (or chloride), titanium tetrachloride, or aluminum trichloride on a peracylated sugar.

Colley [40] and Königs and Knorr [14] prepared glycosyl halides from unsubstituted sugars and acyl chloride or bromide, whereby tetra-*O*-acetyl-glycosyl halides were obtained directly.

Fischer and Armstrong [41] prepared the chloride and bromide, in almost quantitative yield, by treating the β-D-glucopyranose pentaacetate with the liquid hydrogen halide (hydrogen chloride or hydrogen bromide).

A decade later, Fischer [42] developed a more convenient method for the synthesis of glycopyranosyl halides, by treating either anomer of a peracetylated sugar with the acetic acid solution of hydrogen halide instead of liquid hydrogen halide. The reaction is effected at low temperature and can be very fast [43].

The acylated glycosyl halides can be prepared from corresponding methyl glycosides by reaction with an acetic acid solution of hydrogen halide (HCl or HBr). This method is particularly useful for the synthesis of acylated glycofuranosyl halides [44].

Acetylated glycosyl chlorides can be prepared under relatively mild conditions by reacting the corresponding peracetylated sugar with titanium tetrachloride [45]. It has been reported that the anomeric configuration of C1 acetate can have a profound effect upon the reactivity of peracetylated sugar with $TiCl_4$. Thus, for example, while penta-*O*-acetyl-α-D-glucopyranose is stable toward titanium tetrachloride, at 40°C, the β-anomer reacts extremely rapidly giving the tetra-*O*-acetyl-β-D-glycopyranosyl chloride, which then slowly anomerizes to the α-D-anomer.

Aluminum chloride–phosphorus pentachloride [46, 47] and zinc chloride–thionyl chloride [48] are also convenient methods for the preparation of glycosyl chlorides.

The stability of acylated glycosyl halides depends on the configuration and inductive and neighboring group effects. Thus, both anomers of benzoylated D-ribopyranosyl chlorides [49] and D-arabinofuranosyl bromides [50] are known to be considerably stable. There is, however, a large difference in stability between the anomeric acetylated D-glycopyranosyl halides, the β-form being highly unstable and the α-form being stable. Consequently the β-anomer readily anomerizes to α-form.

The preparation of unstable tetra-*O*-acetyl-β-D-aldopyranosyl chloride can be effected by treating the tetra-*O*-acetyl-α-D-glycopyranosyl bromide with "active" silver chloride [51] or by tetraethylammonium chloride in acetonitrile [52].

The reactivity of glycosyl halides depends, first, on their configurational relationship (*cis* or *trans*) with regard to the C2 substituent and, second, on whether the substituent at the C2 carbon is capable of becoming involved in the neighboring group participation or not.

Glycosyl Fluorides in Glycosylation

Glycosyl fluorides have now been widely and effectively used for O-glycosidation reaction, because of their much higher thermal and chemical stability as compared to the low stability of other glycosyl halides, such as glycosyl chlorides and bromides. Thus, for example, the glycosyl fluorides can be purified by distillation and even by column chromatography on silica gel. There are several good reviews on glycosyl fluorides as glycosyl donors [53–55].

The use of glycosyl fluorides as glycosyl donors was first developed by Mukaiyama et al. [56]. Thus 2,3,4,6-tetra-*O*-benzyl-β-D-glucopyranosyl fluoride *54* was reacted with cyclohexanol *55* in ether, at −15°C, in the presence of 4 Å molecular sieves (to maintain the anhydrous conditions) and silver perchlorate–stannous chloride as fluorophilic activators. A mixture of anomeric cyclohexyl 2,3,4,6-tetra-*O*-benzyl-α- and β-D-glucopyranosides (*56* and *57*, respectively) was obtained with 88% yield in which the α-anomer (*60*) predominated in almost 5:1 ratio (83:17) (Fig. 12.18).

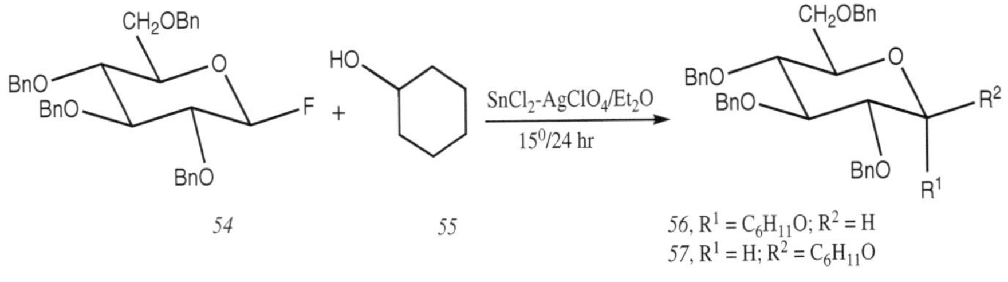

Fig. 12.18

Many activators for the O-glycosidation using glycosyl fluorides were developed over the years. For example, stannous chloride–trityl perchlorate catalyst was used for coupling 2,3,5-tri-*O*-benzyl-β-D-ribofuranosyl fluoride *58* with methyl 2,3,4-tri-*O*-benzyl-α-D-glucopyranoside *59* whereby an anomeric mixture *60* was obtained with 95% yield, with the α-anomer strongly predominating (the α/β ratio was 22:3) [57] (Fig. 12.19).

Fig. 12.19

Trimethylsilyl triflate [58] (TMSOTf) has been shown to be a very good catalyst for glycosidation of glycosyl fluorides (glycosyl donors) with trimethylsilyl ethers as glycosyl acceptors. The stereoselectivity of this glycosidation was found to be highly dependent on solvent, as illustrated in Fig. 12.20. Thus the reaction of 2,3,4-tri-*O*-benzyl-β-D-glycopyranosyl fluoride *54* and trimethylsilyl cyclohexyl ether *61* using TMSOTf as the catalyst, in acetonitrile, gave, after 2 h at 0°C, a mixture of cyclohexyl 2,3,4,6-tetra-*O*-benzyl-α- and β-D-glucopyranosides *62* and *63* with 92% yield, in which the β-anomer strongly predominated (α/β ratio was 1:6). If the glycosidation is performed in ether an anomeric mixture of cyclohexyl 2,3,4,6-

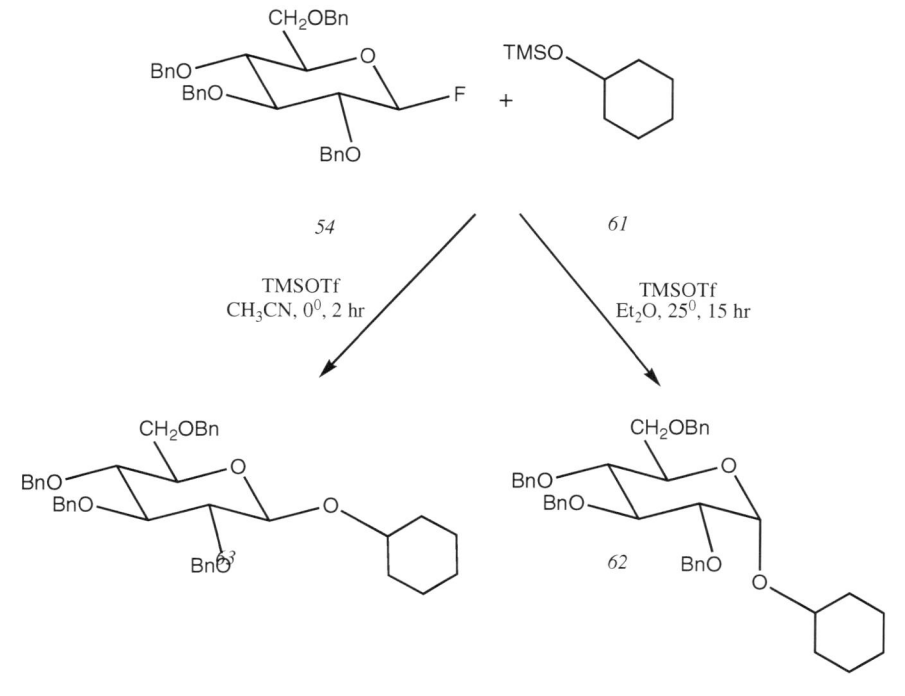

Fig. 12.20

Table 12.2 O-Glycosidation of glycosyl fluorides

Activator	X	Reference
$SnCl_2$–$AgClO_4$	H	[59, 60]
BF_3–Et_2O	H	[61–64]
Cp_2MCl_2–AgOTf (M = Zr or Hf)	H	[65–68]
Cp_2ZrCl_2–$AgBF_4$	H	[69]
$CpHfCl_2$–AgOTf	H	[69–71]
$Bu_2Sn(ClO_4)_2$	H	[72]
Me_2GaCl	H	[73]
Tf_2O	H	[74, 75]
$LiClO_4$	H	[76–78]
$Yb(OTf)_3$	H	[79]
$La(ClO_4)_3 \cdot nH_2O$ (cat.)	TMS	[80]
$La(ClO_4)_3 \cdot nH_2O$–$Sn(OTf)_2$	H	[81]
$TrB(C_6F_5)_4$ (cat.)	H	[82]

tetra-O-benzyl-α- and β-D-glucopyranoside is obtained (after 15 h at 25°C) with 81% yield, in which the α-anomer predominated (the α/β ratio was 6:1).

This procedure is operationally simple and is usable for large-scale preparations. Disaccharides are also obtainable by this method. The products do not undergo anomerization under the reaction conditions and, therefore, the observed stereochemical outcome is the result of kinetic control. The corresponding glycopyranosyl chlorides were inert to the tetrafluorosilane-promoted condensation under comparable conditions.

In Table 12.2 are listed many useful fluorophilic activators developed over the years for coupling of glycosyl fluorides with glycosyl acceptors that can be trifluorosilyl ethers or other compounds having a free hydroxyl group (for example, alcohol, sugar, hydroxyamino acid). For each promoter cited in Table 12.2 the corresponding reference is given.

Synthesis of Glycosyl Fluorides

We will now briefly describe methods for the synthesis of glycosyl fluorides (for a review see [83]).

Hydrogen fluoride–pyridine mixture (50–70%) [84] [pyridinium poly(hydrogen fluoride)(HF–Py)] converts both 1-hydroxy and 1-O-acetylated sugars to the corresponding glycosyl fluorides. Thus, 1-O-acetyl-2,3,4,6-tetra-O-benzyl-α-D-glucopyranose *64* gave on treatment with HF–Py, with 89% yield, a mixture of

Glycoside Synthesis

2,3,4,6-tetra-*O*-benzyl-α- and β-D-glycopyranosyl fluoride (*65* and *66*, respectively) with very high α-selectivity (α/β = 97:3) (Fig. 12.21).

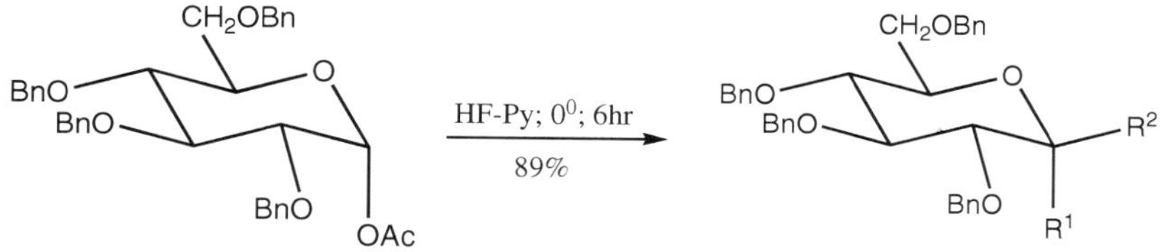

Fig. 12.21

A weaker acidic HF system, Et_3N–3HF, is suitable for preparation of kinetically favored β-glycosyl fluorides [85]. Thus, the glycopyranosyl fluorides of D-xylose *73*, L-arabinose *74*, D-glucose *75*, D-mannose *76*, and L-rhamnose derivatives as well as of galacturonic acid esters *77* and *78* were prepared from corresponding bromides (*67–72*) by bromine–fluorine exchange using this reagent (Fig. 12.22).

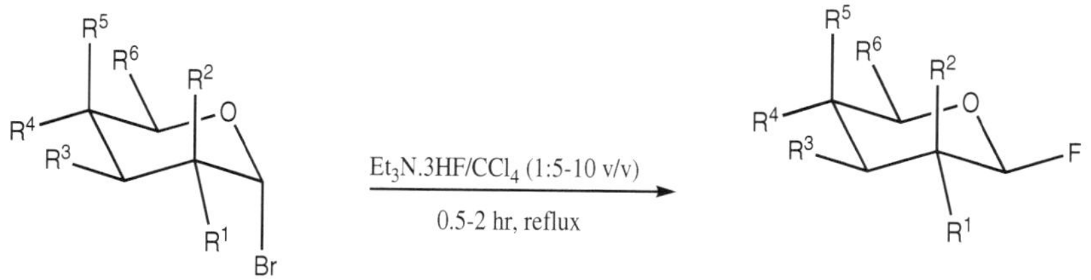

67, $R^1 = R^3 = R^4 = OAc; R^2 = R^5 = R^6 = H$
68, $R^1 = R^3 = R^5 = OAc; R^2 = R^4 = R^6 = H$
69, $R^1 = R^3 = R^4 = OMe; R^2 = R^5 = H; R^6 = CH_2OMe$
70, $R^2 = R^3 = R^4 = OBz; R^1 = R^5 = H; R^6 = CH_2OBz$
71, $R^1 = R^3 = R^5 = OAc; R^1 = R^4 = H; R^6 = COOMe$
72, $R^1 = R^3 = R^5 = OAc; R^1 = R^4 = H; R^6 = COOBn$

73, $R^1 = R^3 = R^4 = OAc; R^2 = R^5 = R^6 = H$
74, $R^1 = R^3 = R^5 = OAc; R^2 = R^4 = R^6 = H$
75, $R^1 = R^3 = R^4 = OMe; R^2 = R^5 = H; R^6 = CH_2OMe$
76, $R^2 = R^3 = R^4 = OBz; R^1 = R^5 = H; R^6 = CH_2OBz$
77, $R^1 = R^3 = R^5 = OAc; R^1 = R^4 = H; R^6 = COOMe$
78, $R^1 = R^3 = R^5 = OAc; R^1 = R^4 = H; R^6 = COOBn$

Fig. 12.22

Diethylaminosulfur trifluoride (DAST) [86, 87] was introduced into carbohydrate chemistry as fluorinating agents by Sharma and Korytnyk [88]. DAST reacts with a hydroxyl group giving an unstable and strongly electron-withdrawing group, $C-OSF_2NEt_2$, with liberation of HF; the attack of a fluoride ion (derived from HF) then forms the C–F bond (Fig. 12.23).

The first application of DAST for the replacement of anomeric hydroxyl group with fluorine was reported in 1985 [89, 90]. Since then it was widely accepted as a good procedure for the synthesis of glycosyl fluorides. For example, treatment of 2,3,5-tri-*O*-benzyl-D-ribofuranose *82* with DAST in THF gave a mixture of anomeric fluorides *83* with 94% overall yield in which the β-anomer predominated (β/α ratio = 9.9). However, if CH_2Cl_2 was used as the solvent, the β/α ratio was 2.0 [91] (Fig. 12.24).

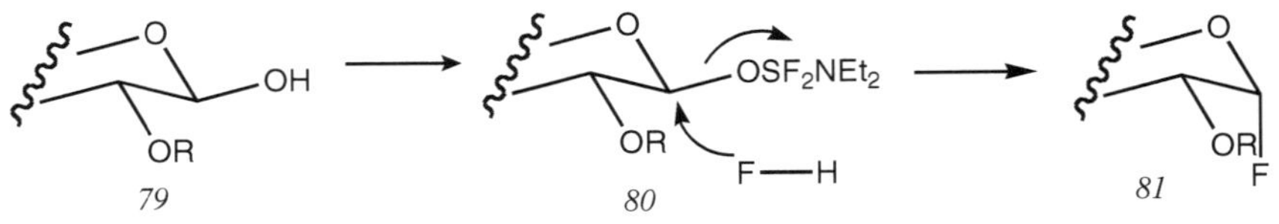

Fig. 12.23

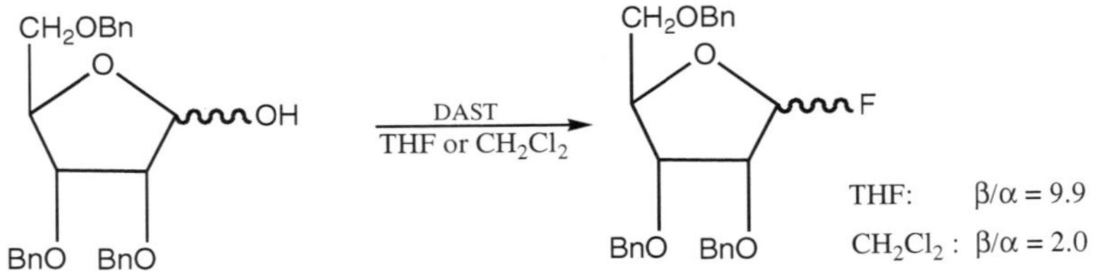

Fig. 12.24

Treatment of methyl 4,5,7-tri-*O*-benzyl-3-*O*-*t*-butyldimethylsilyl-α-D-glucohept-2-ulopyranosonate *84* with DAST gave a diastereomeric mixture of fluorides *85* and *86* with 94% overall yield [92] (the anomeric ratio was 3:1, but the assignment of anomeric configurations was not made) (Fig. 12.25).

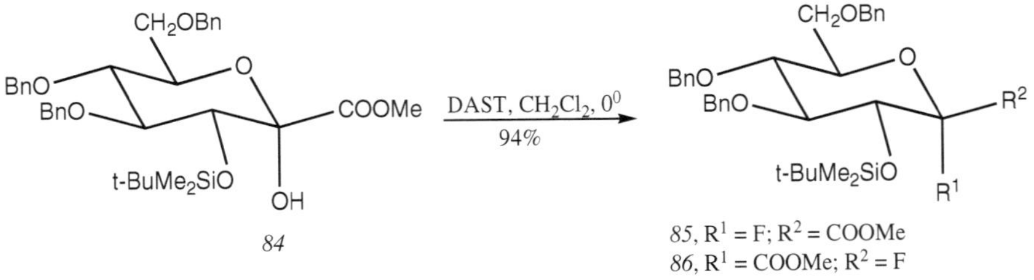

Fig. 12.25

The conversion of thioglycosides into glycosyl fluorides requires the NBS (*N*-bromosuccinimide) as sulfur activator and it is an important reaction in the synthesis of oligosaccharides. Treatment of phenyl 4-*O*-acetyl-1,6-dideoxy-2,3-di-*O*-methyl-1-thio-α- and β-D-allopyranoside *87* with DAST and NBS [61], in methylene chloride at −15°C, gave 4-*O*-acetyl-6-deoxy-2,3-di-*O*-methyl-β-D-allopyranosyl fluoride *88* with 79% yield (Fig. 12.26).

Orthoester Method of Glycosidation

It has already been mentioned that the attack of a nucleophile on oxocarbonium ion *41* (Fig. 12.27) obtained as a result of neighboring group participation of the

Glycoside Synthesis

Fig. 12.26

C2 acetoxy group in the solvolysis of glycosyl halides does not always have to take place at the C1 carbon. A nucleophile can also attack the carbonyl carbon of the C2 acetate giving one or both diastereomeric orthoesters *47* and *48*. Hence the orthoesters formation is closely related to the Königs–Knorr reaction, because by a slight change in experimental conditions the formation of orthoesters can be made to become the major pathway of Königs–Knorr reaction.

The orthoesters synthesis and Königs–Knorr reaction have the same starting materials: an acylated glycosyl halide, an alcohol, and a halogen-binding promoter. Kochetkov [93–96] has played a leading role in the mechanistic study and in the development of application of orthoester method for the synthesis of oligosaccharides [95, 97–99]. The orthoester method produces stereoselectively 1,2-*trans*-glycosides and it is generally applicable to the synthesis of both pyranosides and furanosides of pentoses and hexoses.

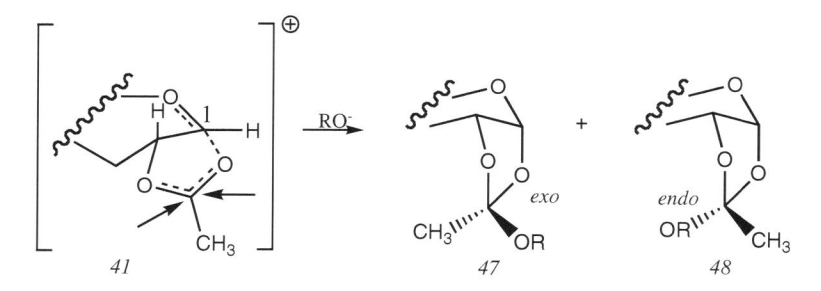

Fig. 12.27

Glycosylation of primary alcohols usually gives higher yields than that of secondary ones, although secondary alcohols are also successfully glycosylated.

The orthoester glycosylation consists typically of reacting methyl or ethyl orthoacetates (or orthobenzoates) of sugars with alcohols in nitromethane solution in the presence of catalytic amounts of mercury (II) bromide. *t*-Butyl orthoacetates of sugars in chlorobenzene with 2,6-lutidinium perchlorate catalyst give better yields [93, 100].

The synthesis of glycosidic bond using orthoester method starts by conversion of an acylated glycosyl halide into methyl, ethyl, or *tert*-butyl orthoester, which is then reacted with the alcohol that is to be glycosylated. Alternatively, the orthoester of the alcohol that is to be glycosylated may be directly prepared [95, 101–103] and then rearranged into the corresponding glycoside [95, 104] (Scheme 12.1).

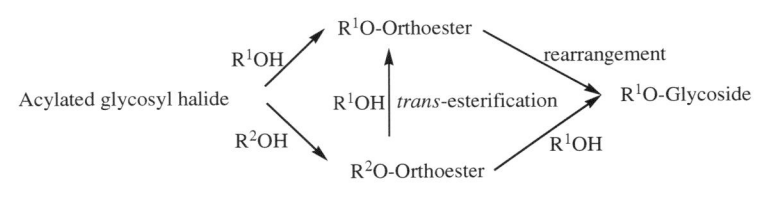

```
                              R¹O-Orthoester
                    R¹OH ↗         ↑           ↘ rearrangement
Acylated glycosyl halide      R¹OH│trans-esterification    R¹O-Glycoside
                    R²OH ↘                      ↗ R¹OH
                              R²O-Orthoester
```

R¹OH = alcohol to be glycosylated
R²OH = methanol, ethanol, tert-butanol

Scheme 12.1

Orthobenzoates of acylated sugars show significantly higher glycosylating activity in comparison with orthoacetates [94, 105].

In nonpolar solvents and in the presence of mercury(II) bromide as catalyst (the use of hard acid catalysts, such as sulfonic acids, or some other catalysts will be discussed later) an orthoester and a new alcohol react with the formation of new orthoester. In nitromethane solution reaction is catalyzed by mercury(II) bromide, and the direction of reaction depends on the amount of catalyst. Thus, in the presence of 1 mmol of mercury(II) bromide (per mol of orthoester) the preferred direction of reaction of orthoester with another alcohol is the formation of a new orthoester; however, in the presence of 20–100 mmol of catalyst the reaction proceeds in the direction of glycosidation with another alcohol and stereoselectively gives rise to 1,2-*trans*-glycosides. In most studied cases the composition of reaction mixture was carefully examined and the 1,2-*cis* isomers were not detected.

The mechanism of glycosylation via orthoesters proposed by using D-glucopyranose orthoester as the substrate [106, 107] is depicted in Fig. 12.28.

The attack of an electrophilic catalyst E (HgCl$_2$, for example) on one of the two oxygen atoms of an orthoester *89* (the exocyclic alkoxy oxygen or the C2 endocyclic oxygen, but never the endocyclic C1 oxygen) gives rise to the intermediate oxonium ions *90* or *93*, which are converted to the corresponding cyclic *91* or acyclic *94* acyloxonium ions, respectively. It was shown by ab initio calculations that the *90* → *91* or *93* → *94* conversion proceeds practically without an activation barrier [108]. The acyloxonium ions *91* and *94* are regarded as the only transition states that determine the product formation.

The cyclic oxocarbonium ion *91* can react with an alcohol in two ways. First, an alcohol (R¹OH or R²OH) can attack the carbonyl carbon of *91* giving the same orthoester if the attacking alcohol is R¹OH, or a new orthoester if the attacking alcohol is the one to be glycosylated (R²OH) (*91* → *95*) (Fig. 12.29). This step is fast. Second, the relatively slow irreversible reaction can take place if an alcohol (R¹OH or R²OH) attacks the anomeric carbon of oxocarbonium ion *91*, which gives rise to a stereoselective formation of 1,2-*trans*-glycoside (*91* → *96*) (Fig. 12.29).

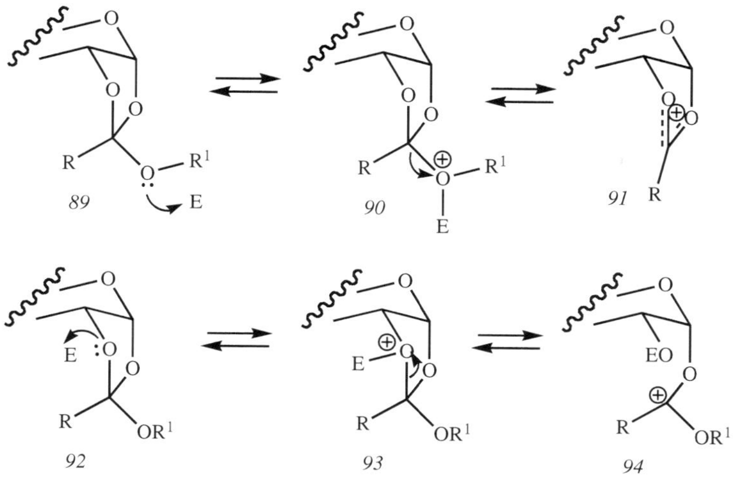

Glycoside Synthesis

Fig. 12.28

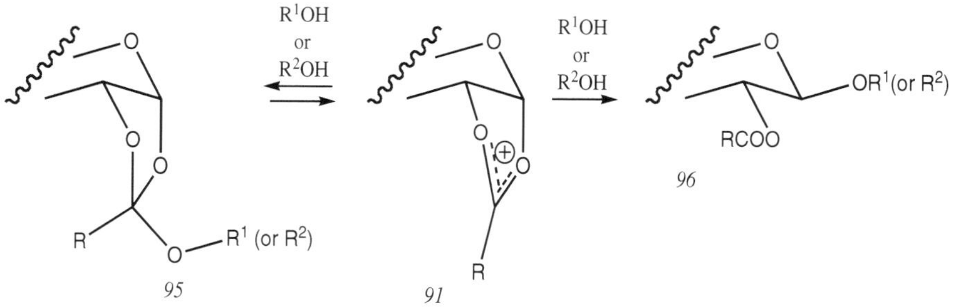

Fig. 12.29

Bochkov et al. [106] consider the attack of an alcohol on the anomeric carbon with the simultaneous breaking of the C1–O1 bond to be the rate-determining step in the glycosidation via orthoester method.

Based on experimental data the following mechanism has been proposed for the isomerization of orthoester catalyzed by mercury(II) bromide in nitromethane [107] and is depicted in Fig. 12.30. The attack of mercury(II) bromide (E in Fig. 12.28) on exocyclic oxygen of an orthoester is followed by dissociation of the complex to give an intimate ion pair *99*. The alkoxy mercury(II) bromide complex being negatively charged has relatively high nucleophilicity (certainly higher than the corresponding alcohol). The attack of this nucleophile on the C1 carbon will then result in *trans*-glycosidation (*99* → *100*) whereas the attack on the carbonyl carbon of the oxocarbonium ion will result in regeneration of orthoester (*98* → *97*) (Fig. 12.30).

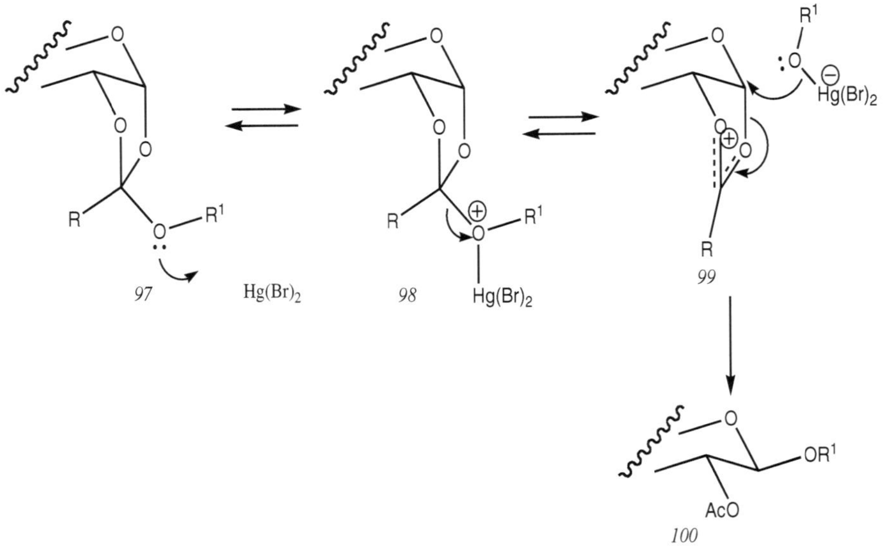

Fig. 12.30

The fate of acyclic acyloxonium ion *102* (Fig. 12.31) depends on the nature of used catalyst. When mercury(II) bromide is used as the catalyst, the initial product is the zwitter-ion *102* (Fig. 12.31), which is, due to intramolecular charge compensation, relatively unreactive.

Since this intimate ion pair is unable to dissociate, the only reaction that can take place is dissociation of the catalyst (mercuric bromide) from the oxygen–mercury complex *102* and regeneration of orthoester *95* (Fig. 12.31).

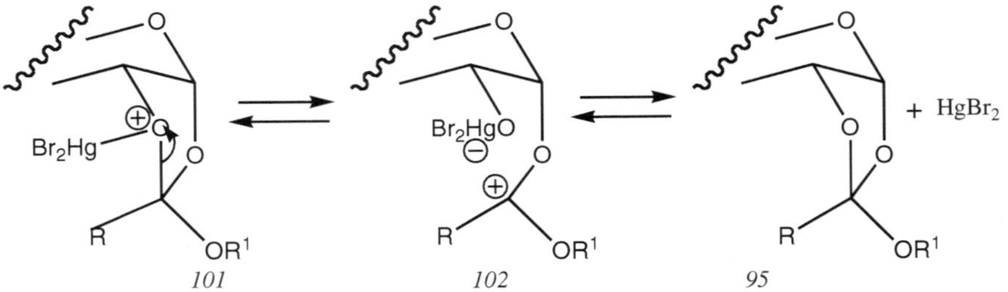

Fig. 12.31

The picture changes dramatically when the reaction is catalyzed by hard acids, such as *p*-toluenesulfonic acid (TsOH), 2,4,6-trinitrobenzenesulfonic acid $(O_2N)_3PhSO_3H$, or even picric acid $(O_2N)_3PhOH$ [95]. The 3,4,6-tri-*O*-acetyl-α-D-glucopyranosyl-orthoester *103* and 3,4,6-tri-*O*-methyl-α-D-glucopyranosyl orthoester *104* were used as model compounds for the study of acid-catalyzed

Glycoside Synthesis 345

glycosidation mechanism via orthoesters (Fig. 12.32). The composition of reaction mixtures was quantitatively analyzed for the presence of α- and β-glycosides 105→112 and cyclohexyl acetate by GLC. The conditions for the isomerization of orthoester 103 to β-glycoside 106 were optimized with regard to proton donor, counter ion (anion), solvent, and temperature.

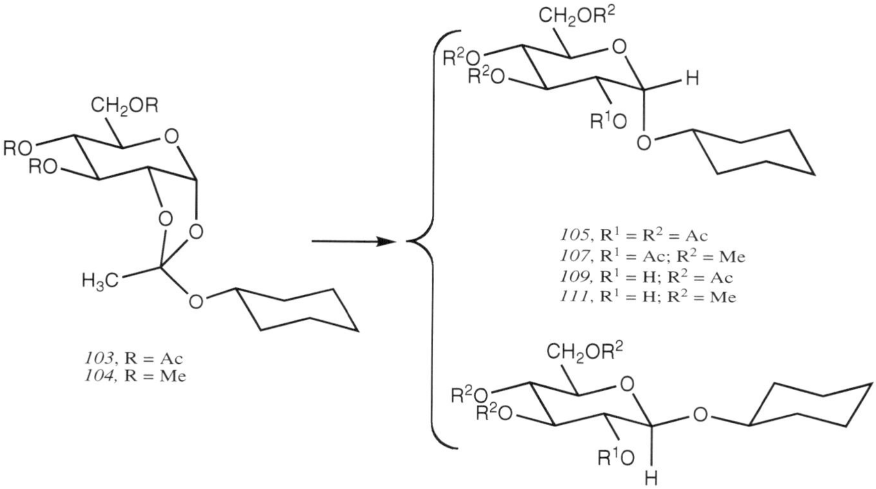

105, $R^1 = R^2 = Ac$
107, $R^1 = Ac; R^2 = Me$
109, $R^1 = H; R^2 = Ac$
111, $R^1 = H; R^2 = Me$

103, R = Ac
104, R = Me

106, $R^1 = R^2 = Ac$
108, $R^1 = Ac; R^2 = Me$
110, $R^1 = H; R^2 = Ac$
112, $R^1 = H; R^2 = Me$

Fig. 12.32

The influence of the nature of the proton donor upon the isomerization of 103→106 was studied in the presence of a large excess of tetra-n-butylammonium perchlorate with respect to acid (the orthoester 103 is stable toward this salt). From the above-mentioned hard acids [TsOH, $(NO_2)_3PhSO_3H$, $(NO_2)_3PhOH$] the best yield of isomerization product 106 was obtained with picric acid; both sulfonic acids gave unsatisfactory results. Pyridinium perchlorates as catalysts fall into two distinct groups: (1) the spatially hindered 2,6-lutidine and 2,4,6-collidine perchlorates gave equally satisfactory results; (2) the unhindered pyridinium perchlorate produced substantially lower yields probably due to N-glycosylation that is not possible with 2,6-lutidine and 2,4,6-collidine perchlorates due to steric hindrance of ortho methyl groups. Further, study of isomerization of 104→108 in the presence of variety of anions has shown that the composition of products is strongly influenced by the nature of anion. The nature of solvent has relatively small effect upon the composition of reaction mixtures. Nevertheless the best solvent for the isomerization seems to be chlorobenzene. Finally, temperature has no effect upon the composition of reaction mixture.

Under the condition studied, the proton-catalyzed isomerization of 103→106 and 104→108 (reaction 1) is accompanied by formation of deacetylated glycosides

(*110* and *112*) (reaction 2) and by formation of cyclohexyl acetate and nonglycosidic products (reaction 3).

The isomerizations *103*→*106* and *104*→*108* take place stereoselectively giving 1,2-*trans*-glycosides (reaction 1). The C2 deacetylated glycosides are formed from both cyclohexyl 3,4,6-tri-*O*-acetyl- and 3,4,6-tri-*O*-methyl-α-D-glycosyl 1,2-orthoesters (Fig. 12.33). However, whereas the 1,2-*cis* glycoside *109* is formed practically stereoselectively from *103*, *104* gives under the same experimental conditions a mixture of *111* and *112*, in which *112* slightly predominates, indicating the participation of the C3 acetoxy group in the reaction. The X-ray structural analysis of ethyl homolog of the orthoester *103* has shown [109] that the pyranose ring of this compound in crystalline state exists in the twist conformation in which the C3 ace-

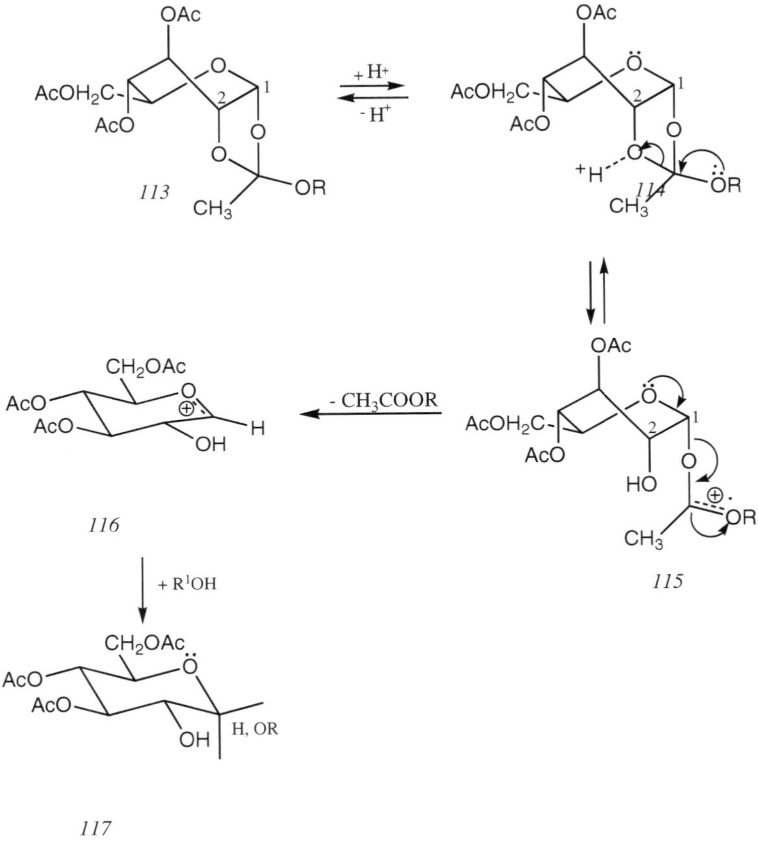

Fig. 12.33

toxy group is axially oriented and thus it is suitable for attack on the C1 whereas the CH$_2$OAc is oriented pseudo-equatorially, thus lending support to the hypothesis of participation of the C3 acetoxy group in stereoselective formation of glycosides.

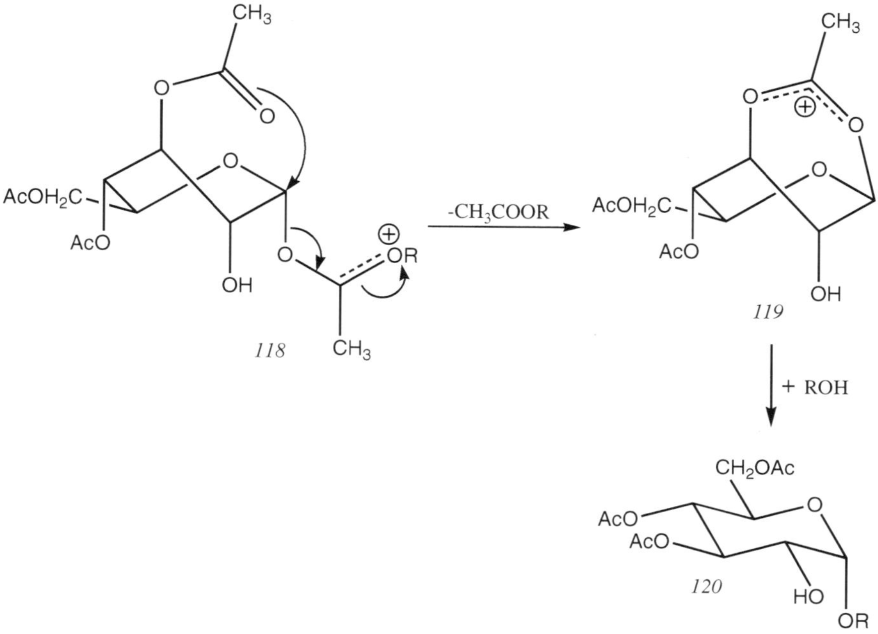

Glycoside Synthesis

Fig. 12.34

Theoretically there are three oxygen atoms in the orthoester group that can be protonated: one is the exocyclic oxygen, and the other two are the endocyclic C1 and the C2 oxygens. However, from ab initio calculations of bicyclic alkyl orthoacetates [110] the C1 oxygen atom is the least basic of all oxygen atoms of an orthoester group and thus the least likely to be protonated. Since the exocyclic and the endocyclic C2 oxygen atoms are the preferred sites for protonation these two reactions will be discussed first.

The proton attack at the C2 oxygen atom of orthoester *113* could lead to the formation of the acyclic acyloxonium ion *115* (Fig. 12.33) that can undergo cleavage of the C1 alkyl acetate without or with the participation of the C3 acetoxy group. In the first case, the cyclic oxocarbenium ion will be formed which can react with an alcohol to give a mixture of anomeric glycosides (Fig. 12.33). In the second case the axially oriented C3 acetoxy group can expel the C1 alkyl acetate (via neighboring group participation) whereby a six-membered ring cyclic acetoxonium transition state intermediate *119* will be formed. The attack of an alcohol on this intermediate will take place exclusively from the α-side, giving thus stereoselectively the α-anomer as the only product (Fig. 12.34).

The protonation of the exocyclic oxygen of orthoester *113* followed by elimination of alcohol will result in the formation of 1,2-cyclic acetoxonium ion *122*, which can be attacked by an alcohol (the same one that was part of the orthoester, or another alcohol) either at the carbonyl carbon of the acetyl group or at the C1 carbon. In the first case, the same or the isomeric orthoester will be obtained (if the attacking alcohol is the one that was part of the orthoester), or another orthoester will be obtained (if the attacking alcohol is different from the one that was part of the orthoester). In the second case, if the attack takes place at the C1 carbon, the corresponding β-glycoside *123* will be obtained in which the aglycon can be either the alcohol that was part of the orthoester or the new one (Fig. 12.35).

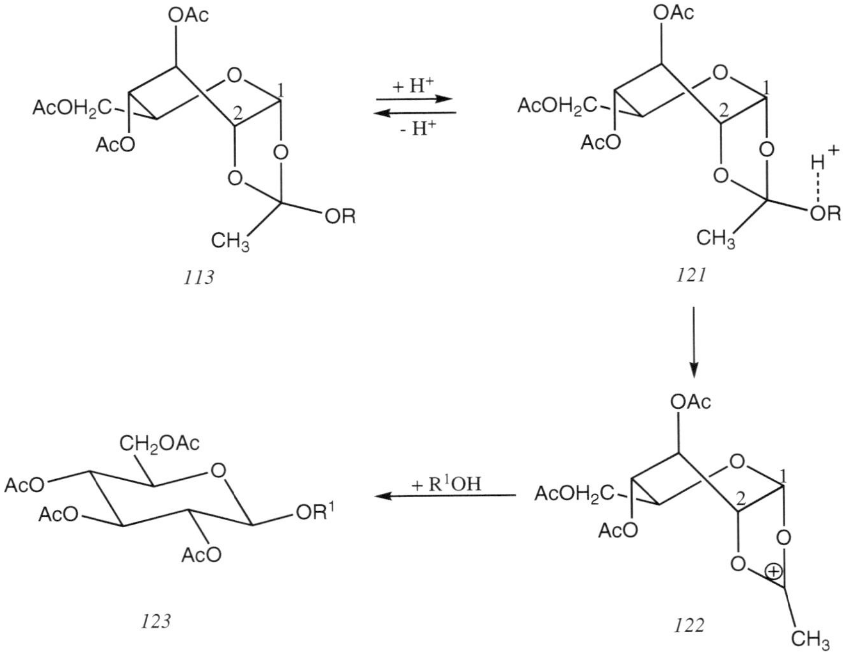

Fig. 12.35

The bicyclic orthoesters of sugars can be prepared by condensation of 1,2-*trans*-acylglycosyl halides with alcohols [111–114] in the presence of neutralizing agents. The reaction proceeds with participation of neighboring acyloxy group via orthoester cation, which reacts with alcohols to give orthoesters (Fig. 12.36) and can take place only when neutralization of liberated hydrogen halide acid is fast and efficient. Such compounds are silver oxide [111] or sterically hindered tertiary amines [112–114].

The orthoesters of sugars can also be prepared from 1,2-*cis*-glycosyl halides by condensation with alcohols in nitromethane as solvent and in the presence of 2,4,6-collidine as the base [23, 115], or in ethyl acetate as the solvent and in the presence of lead carbonate as the base [116]. In the last case the formation of orthoesters proceeds with the participation of solvent molecules (Fig. 12.37).

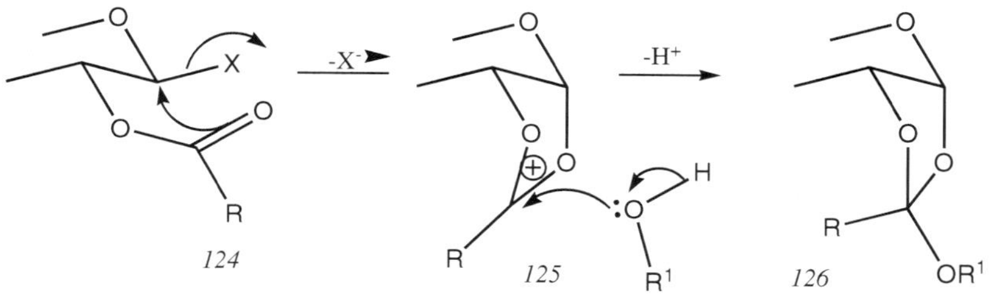

Glycoside Synthesis

Fig. 12.36

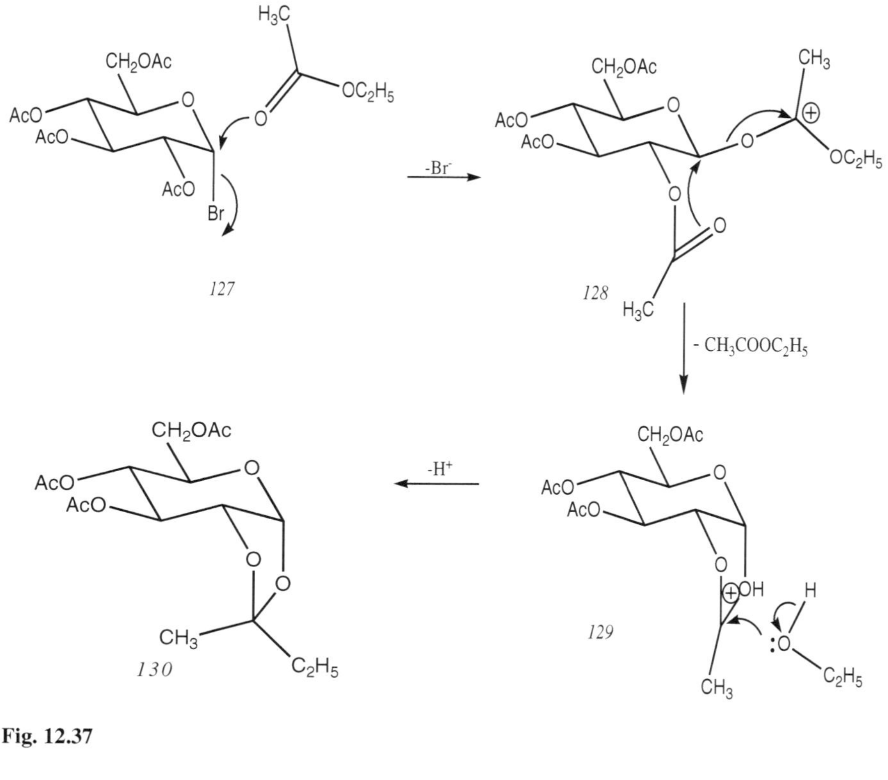

Fig. 12.37

Trichloroacetimidate Method of Glycosidation

Electron-deficient nitriles, such as trichloro- or trifluoroacetonitriles *131* (Fig. 12.38), are known to undergo direct and reversible base-catalyzed addition of alcohols giving *O*-alkyltrichloroacetimidates *132*. A detailed study of the addition of trichloroacetonitrile to 2,3,4,6-tetra-*O*-benzyl-D-glucopyranoses *133* (Fig. 12.39) has shown [117–122] that the addition of equatorial 1-oxide anion is a very rapid and reversible reaction and gives the β-trichloroacetimidate *134* as the predominant

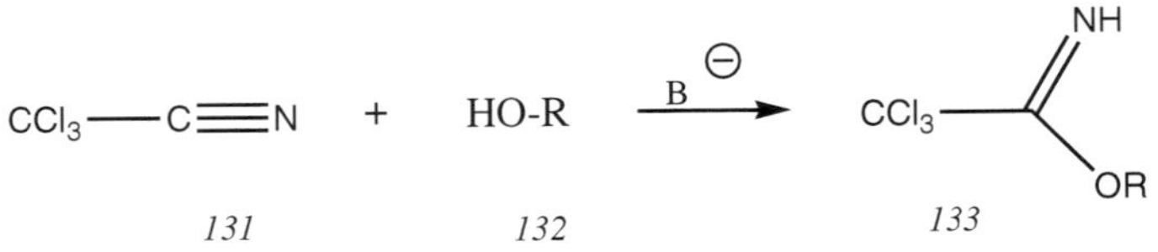

$$CCl_3-C\equiv N \;+\; HO\text{-}R \;\xrightarrow{B^{\ominus}}\; CCl_3-\underset{OR}{\overset{NH}{C}}$$

131   132   133

Fig. 12.38

or even the exclusive product. However, this product then undergoes slow, base-catalyzed anomerization (via the reverse reaction, anomerization of the 1-oxide anion, and renewed trichloroacetonitrile addition) to the α-trichloroacetimidate *135* with the electron-withdrawing 1-substituent in the axial configuration, which is, due to anomeric effect, thermodynamically a more stable anomer. Thus, depending on the base used [K$_2$CO$_3$, CsCO$_3$, and NaH or 1,8-diazabicyclo [5,4,0] undec-7-ene (DBU)] both anomeric trichloroacetimidates may be isolated in pure form and in high yield using kinetic or thermodynamic reaction control. Both anomers are thermally stable. In Table 12.3 are given several examples of synthesis of anomeric trichloroacetimidates of D-glucose.

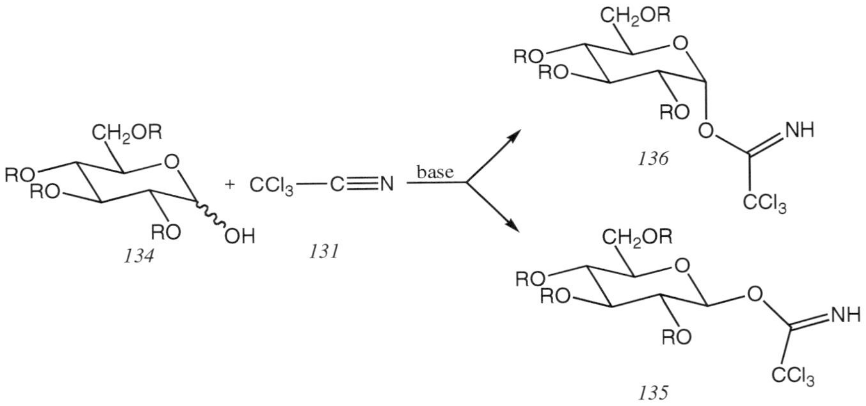

134 + CCl$_3$—C≡N $\xrightarrow{\text{base}}$ *136* / *135*

Fig. 12.39

The reaction of glycosyl acceptors (alcohols or sugars having one unprotected hydroxyl group) with *O*-glycosyl trichloroacetimidate donor requires the presence of an acid catalyst [117–120]. Boron trifluoride etherate (BF$_3$·Et$_2$O) at –40°C to room temperature (in dichloromethane or dichloromethane–*n*-hexane as solvents) or trimethylsilyl trifluoromethanesulfonate (Me$_3$SiOTf) at –80°C to room temperature (in ether or acetonitrile as solvents) proved to be very suitable acid catalysts [128, 129].

It has been shown [117, 118, 130, 131] that the solvents have an important and sometimes even dramatic influence upon the glycosidation with *O*-trichloroacetimidate glycosyl donors.

The *O*-trichloroacetimidate method for the glycoside synthesis can be illustrated with the reaction of per-*O*-acetylated α-D-glucosyl trichloroacetimidate *137* with

Table 12.3 Synthesis of trichloroacetimidates of D-glucose

Sugar	Reaction conditions	α/β ratio	yield, %	Ref.
2,3,4,6-tetra-O-benzyl-D-glucopyranosyl trichloroacetimidate (CH$_2$OBn, BnO, BnO, BnO)	CH$_2$Cl$_2$, NaH, CCl$_3$CN room temperature	1 : 0	78	123, 124
2,3,4,6-tetra-O-benzyl-D-glucopyranosyl trichloroacetimidate (CH$_2$OBn, BnO, BnO, BnO)	CH$_2$Cl$_2$, K$_2$CO$_3$, CCl$_3$CN room temperature	0 : 1	90	125-127
2,3,4,6-tetra-O-acetyl-D-glucopyranosyl trichloroacetimidate (CH$_2$OAc, AcO, AcO, AcO)	CH$_2$Cl$_2$, K$_2$CO$_3$, CCl$_3$CN 48 hr, room temperature	1 : 0	98	124
2,3,4,6-tetra-O-acetyl-D-glucopyranosyl trichloroacetimidate (CH$_2$OAc, AcO, AcO, AcO)	CH$_2$Cl$_2$, K$_2$CO$_3$, CCl$_3$CN 2 hr, room temperature	0 : 1	78	125

tetraacetyl D-glucose *138* at 0°C in the presence of boron trifluoride etherate as the catalyst whereby β,β-linked trehalose *139* is obtained in good yield (58%) (Fig. 12.40).

The glycosyl trichloroacetimidate *140* (Fig. 12.41) reacts readily at room temperature with Brønsted acids to give in high yields the corresponding glycosyl derivatives of anions of Brønsted acids *143* (Fig. 12.41).

The mechanism of this reaction is very simple. Protonation takes place at the imido-nitrogen activating the trichloroacetimidate group and making it thus a good leaving group which can easily be substituted with practically any nucleophile, as shown in Fig. 12.42.

The β-anomers that are initially formed from glycosyl-α-D-trichloroacetimidates, in the presence of strong acids, are converted, via anomerization, to thermodynamically more stable α-anomers (due to anomeric effect).

Carboxylic acids, being too weak to catalyze the anomerization, give β-*O*-acyl derivatives.

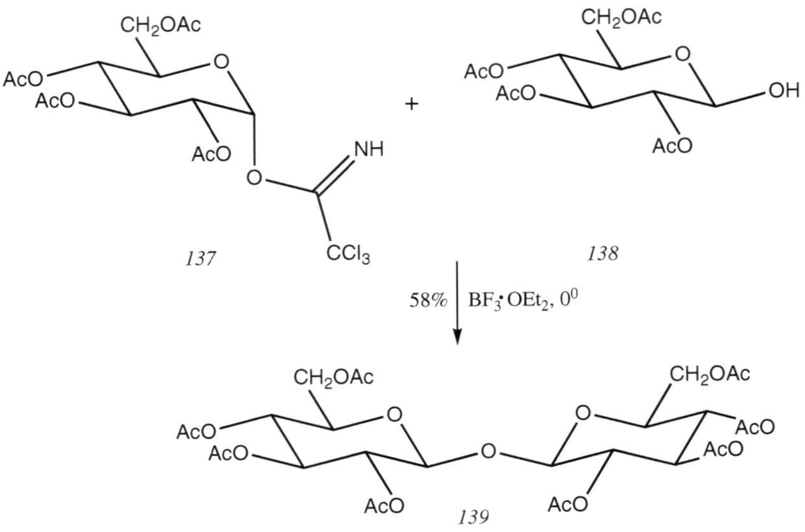

Fig. 12.40

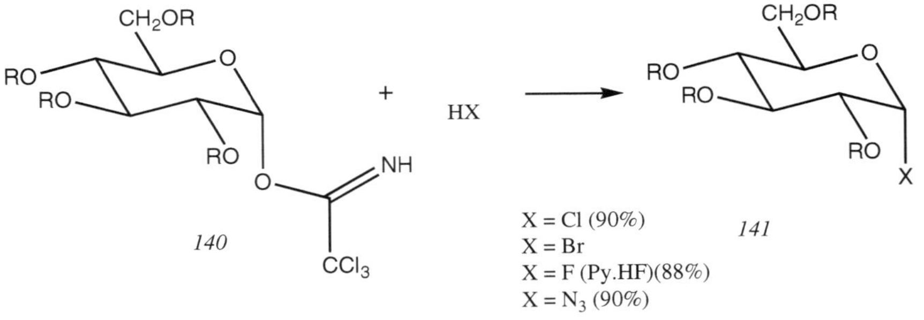

Fig. 12.41

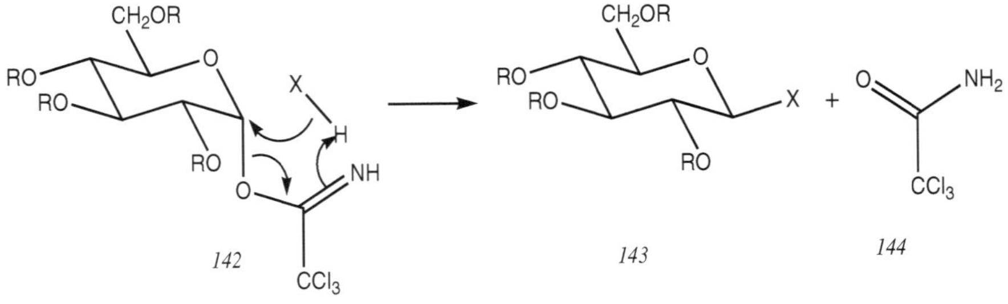

X = Cl (90%)
X = Br
X = F (Py.HF) (88%)
X = N_3 (90%)

Fig. 12.42

Glycoside Synthesis

For the reaction of alcohols (or other monosaccharides) with *O*-glycosyl trichloroacetimidates the Brønsted acids are not suitable. Hence, Lewis acid catalysts, such as boron trifluoride etherate ($BF_3 \cdot Et_2O$), at $-40°C$ to room temperature in dichloromethane or dichloromethane–*n*-hexane as solvents, or trimethylsilyltriflate at $-80°C$ to room temperature in ether or acetonitrile, respectively, are used in these instances.

Glycoside Synthesis via Remote Activation

In 1981, 2 years after R. B. Woodward's death, his group published three papers on total stereoselective synthesis of erythromycin [132–134] and in the last communication they described successful glycosidation of the C3 and C5 hydroxyl groups of erythronilide A with cladinose (α-linkage) and desosamine (β-linkage), respectively, using as glycosyl donors 2-thiopyridinyl or 2-thiopyrimidinyl glycosides (Fig. 12.43). In 1980, Hanessian published a paper [135] on a fast and efficient formation of glycosides by "remote activation" using in addition, to 2-thiopyridinyl and 2-thiopyrimidinyl glycosides, the 2-thioimidazolinyl glycosides (Fig. 12.43). It

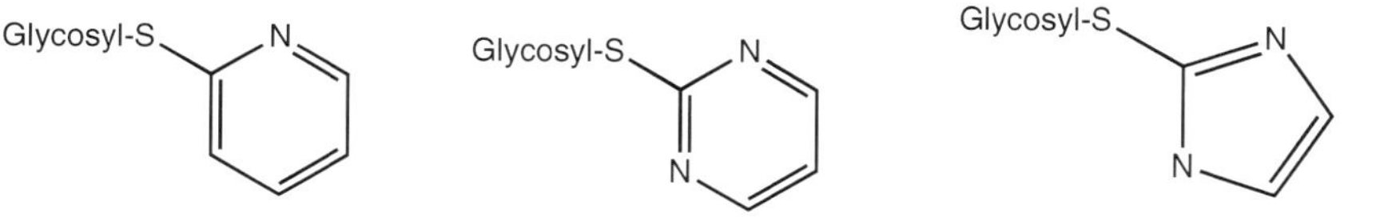

Pyridinyl-2-thioglycoside
145

Pyrimidinyl-2-thio glycoside
146

Imidazolinyl-2-thio glycoside
147

Fig. 12.43

is very difficult to say who discovered this method particularly since it is known that Woodward was pretty slow in publishing his work.

The remote activation method consists in the activation of glycosyl acceptor (an alcohol or any hydroxyl-containing substrate) by increasing the nucleophilicity of an alcohol's oxygen or of glycosyl acceptor oxygen by hydrogen bond between the alcohol's hydrogen atom and the 2-thiopyridinyl, 2-thiopyrimidinyl, or 2-thioimidazolinyl nitrogen (Fig. 12.44). The reaction requires heavy metal catalyst,

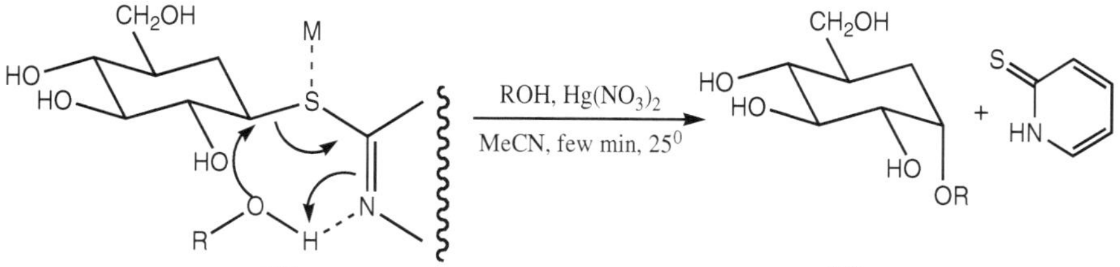

148 *149* *150*

Fig. 12.44

such as silver nitrate, or preferably mercury(II) nitrate to activate the sulfur atom of thioglycoside glycosyl donor making it thus a better leaving group. In Table 12.4 are given the yields and the α/β ratio of glucopyranosides with various alcohols and a disaccharide [135].

Table 12.4 Formation of some alkyl D-glucopyranosides and a disaccharide

Glycoside receptor	Yield (%)	α/β ratio
Methanol	95	70:30
Ethanol	85	68:32
2-Propanol	77	62:38
2,2-Dimethyl-1-propanol	47	51:49
Cyclohexanol	75	51:49
1,2:3,4-di-*O*-Isopropylidene-α-D-galactopyranose	35	55:45

n-Pentenyl Glycosides as Glycosyl Donors

n-Pentenyl glycosides (NPGs) are a special type of chemically stable glycosyl donors, in which the *n*-pentenyl group can be chemospecifically activated to provide a good leaving group, generating thus a glycosyl donor that is ready for coupling to a glycosyl acceptor [136]. The concept of this method is illustrated in Figs. 12.45 and 12.46.

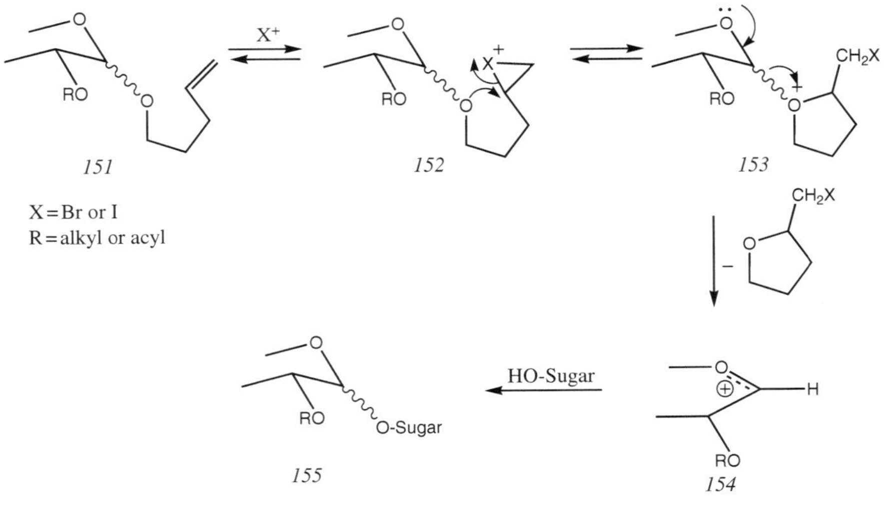

X = Br or I
R = alkyl or acyl

Fig. 12.45

Glycoside Synthesis

There is a difference in reactivity toward electrophiles between the C2 alkylated and the C2 acylated *n*-pentenyl glycosides due to the influence of the C2 substituent upon the nucleophilicity of the glycosidic oxygen. Reaction of a glycosyl donor with an electrophile produces a positively charged intermediate (*153* in Fig. 12.45), the formation of which is less favorable when there is a C2 electron-withdrawing group,

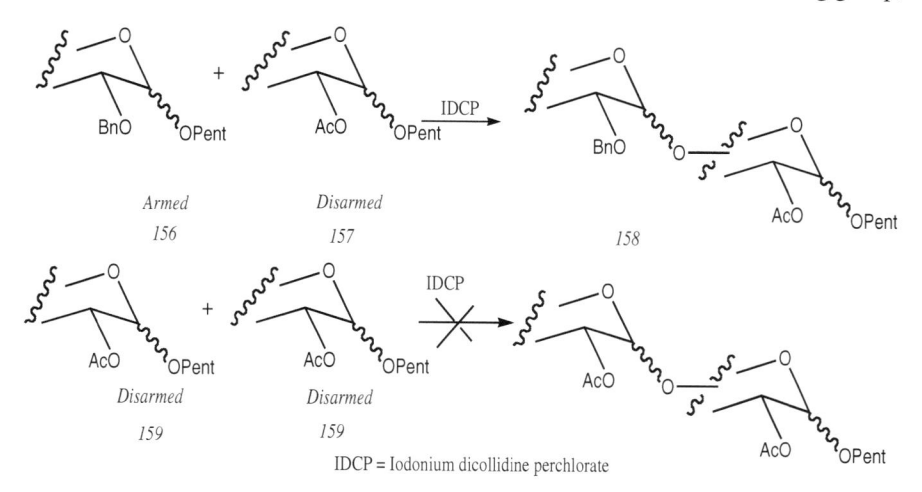

Fig. 12.46

such as an acyl group(acetyl, benzoyl, etc.). These glycosyl donors are termed *disarmed* glycosyl donors. On the other side, glycosyl donors that have an electron-donating substituent at the C2 carbon and thus favor the reaction with an electrophile that produces a positively charged intermediate are called *armed* glycosyl donors. Since the armed glycosyl donors react faster with electrophiles than the disarmed ones, in solution containing both disarmed and armed glycosyl donor molecules having one free hydroxyl group, the reaction with an electrophile will result in cross-coupling, and not self-coupling, i.e., an armed glycosyl donor will react with the disarmed one and not with an armed one, as well as a disarmed glycosyl donor will not react with a disarmed one [136], as illustrated in Fig. 12.46. The promoters for the activation of *n*-pentenyl group [137, 138] are iodonium dicollidine perchlorate (IDCP) and *N*-iodosuccinimide/triethylsilyl triflate (NIS/Et$_3$SiOTf). The mechanism of glycoside bond formation using *n*-pentenyl glycosides and IDCP is illustrated in Fig. 12.45. The electrophilic iodonium ion addition to the double bond of pentenyl moiety produces the cyclic iodonium ion *152*. Nucleophilic attack by the glycosidic oxygen results in an intermediate oxonium ion *153* that dissociates into a cyclic oxocarbenium ion *154* and an iodomethyltetrahydrofuran derivative. When the C2 hydroxyl group is protected with an electron-withdrawing ester substituent (disarmed glycoside) the glycosidic oxygen has a low nucleophilicity that will result in slow formation of *153*. On the other hand, protection of the C2 hydroxyl group with an electron-donating ether substituent (*armed glycoside*) increases the nucleophilicity and hence the substrate reacts considerably faster.

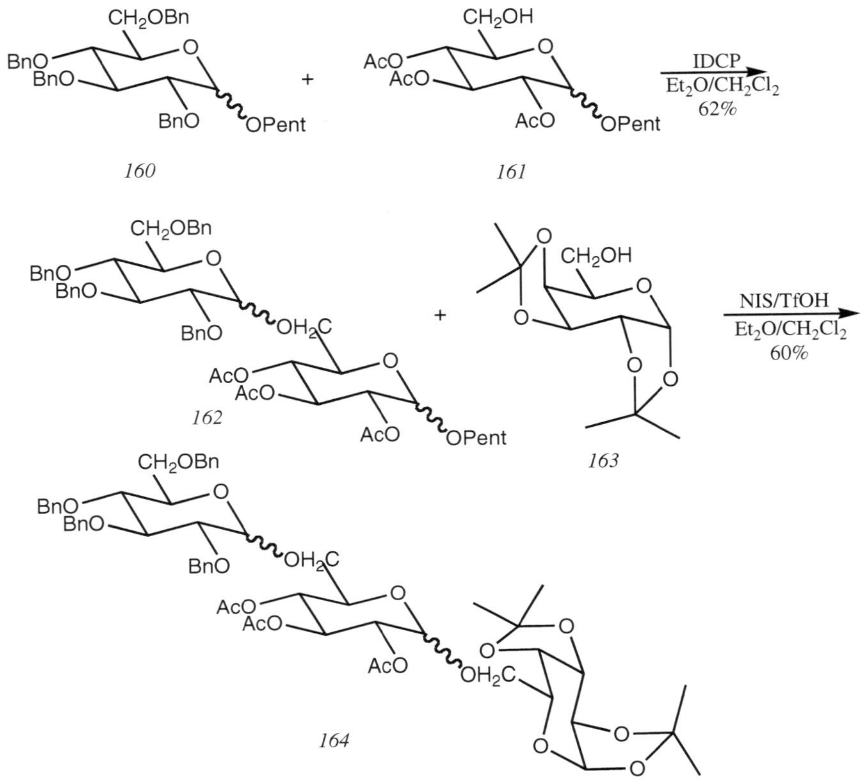

Fig. 12.47

Since the chemoselectivity relies on the fact that electron-donating C2 ether activates (*arms*) and an electron-withdrawing C2 ester deactivates (*disarms*) the anomeric carbon, coupling of an armed donor with a disarmed acceptor, in the presence of an activator, such as iodonium dicollidine perchlorate (IDCP), results in glycosidation, giving an anomeric mixture. The disarmed disaccharide could be further glycosylated with another acceptor, using the more powerful activator, *N*-iodosuccinimide/triflic acid (NIS/TfOH), giving a trisaccharide as shown in Fig. 12.47.

For reviews of this method see [139–142]. The *n*-pentenyl method of glycosidation was used successfully for the synthesis of many oligosaccharides (linear or branched) that are even fairly large (for example, nonasaccharide portion of high mannose glycoprotein is synthesized in this way [143]).

n-Pentenyl glycosides can be prepared by standard procedures that are used for the preparation of alkyl glycosides. Thus Fischer glycosidation *165→167* can be used for the preparation of *n*-pentenyl gluco-, manno-, and fucopyranoside (Fig. 12.48).

Glycoside Synthesis

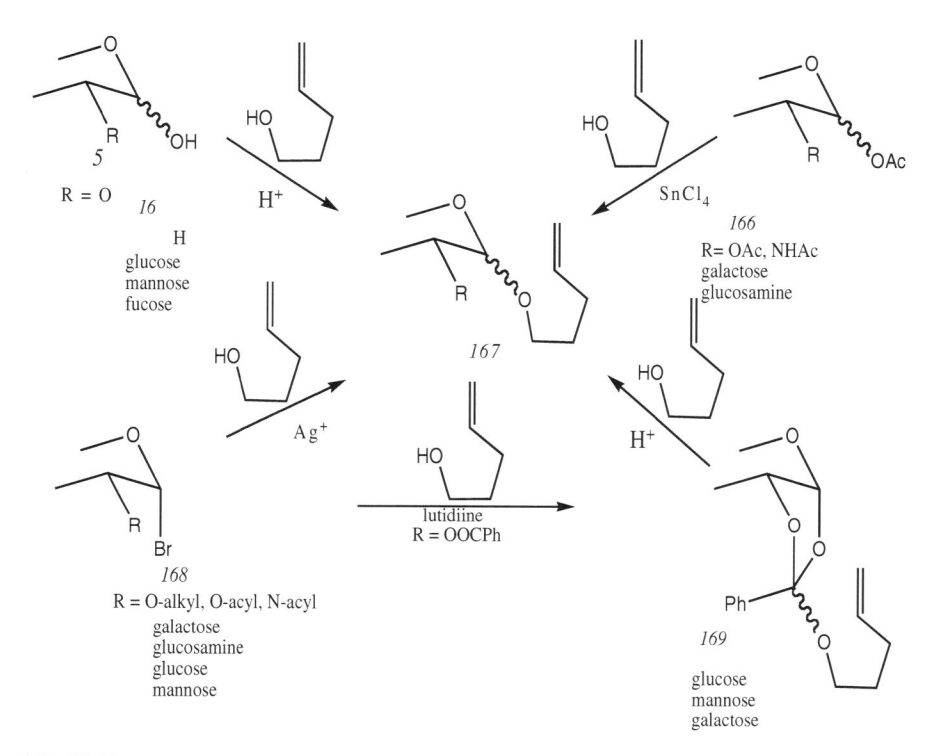

Fig. 12.48

The reaction is performed in the presence of an acid catalyst and gives a mixture of anomers; the α/β ratio for glucose is 2:1, whereas mannose gives almost exclusively α-anomer. The Fischer glycosidation with *n*-pentenol (*165→167*) gives poor yields with galactose and glucosamine. In these instances, the corresponding NPGs can be prepared using Königs–Knorr method (*168→167*) or glycosyl acetate method (*166→167*) (Fig. 12.48). The orthoesters *169* obtained by reacting perbenzoylated glycopyranosyl bromides with *n*-pentenol in the presence of lutidine (Fig. 12.48) can also serve as precursors for NPGs (*169→167*) (Fig. 12.47) or could be used directly for coupling with the glycosyl acceptors.

The promoters used for activating the NPGs for direct coupling are IDCP (intermediate potency) [144], NIS/Et$_3$SiTf [137], and even triflic acid (TfOH).

Glycals as Glycosyl Donors

The two components entering into a glycosylation reaction are differentiated according to which component contributes the anomeric carbon of the resultant glycoside. So the component that contributes the anomeric carbon is described as the glycosyl donor, and the other component is described as glycosyl acceptor. The donor reacts with glycosyl acceptor to give a glycoside. In most glycosylation reactions the acceptor is a nucleophile that supplies the oxygen of the resultant glycoside by replacing the leaving group at the anomeric carbon of the electrophilic glycosyl donor. However, with development of novel glycosylation procedures [117, 118, 145–147] the terms "glycosyl donor" and "glycosyl acceptor" should be decoupled from terms "nucleophile" and "electrophile."

Utilizing glycals as glycosyl donors in disaccharide synthesis by halonium-catalyzed coupling (iodoglycosylation) to suitably protected acceptors had been pioneered by Lemieux [148, 149] and Thiem [150–155] (Fig. 12.49). It has been shown by Thiem (vide supra) that these reactions have a high tendency for *trans*-diaxial addition thus providing an important route for the synthesis of α-linked disaccharides having an axial C2-iodo group on the nonreducing end. However, since displacement of the C2 I$^+$ iodine in such systems with a nucleophile is difficult

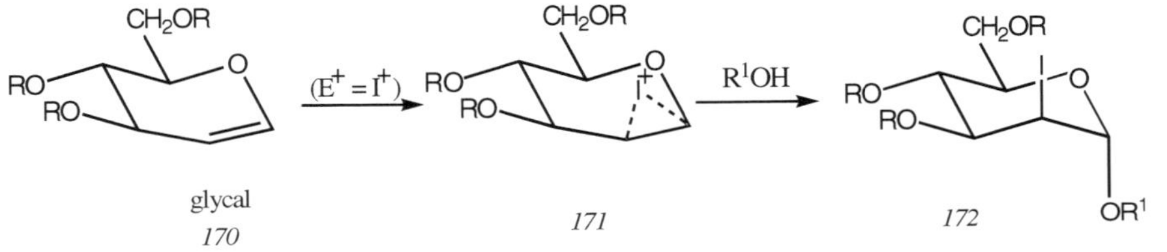

Fig. 12.49

[156] the Lemieux–Thiem method has thus far found its most useful application in the synthesis of 2-deoxyglycosides [150–155, 157].

Iodoglycosylation is carried out with glycal serving as a glycosyl donor. The glycal linkage is attacked by an "I$^+$ equivalent" reagent, such as *N*-iodosuccinimide or *sym*-collidine iodonium perchlorate, and the intermediate obtained (*171*) is attacked by a nonglycal acceptor *173* that is also present in the solution; the anomeric carbon of glycosyl acceptor has to be appropriately protected. The stereochemistry of glycosylation is controlled by *trans*-diaxial addition and the α-linked disaccharide is obtained (Fig. 12.50).

Coupling of a glycal donor with a nonglycal acceptor using iodoglycosylation approach can be used only for the synthesis of disaccharides but not for the synthesis of higher oligosaccharides. However, glycals could be used as building blocks for the synthesis of oligosaccharides higher than disaccharides if the armed–disarmed concept of Fraser-Reid could be utilized, namely if the nature of the C3 substituent does influence the rate of formation of cyclic transition state iodonium intermediate *171* (Fig. 12.51), namely if the positively charged iodonium transition state intermediate *171* is destabilized when the C3 hydroxyl group is substituted with an electron-withdrawing group, such as an acyl group (disarmed glycal), or stabilized when the C3 hydroxyl group is substituted with an electron-donating group, such as an alkyl group (armed glycal) as illustrated in Fig. 12.52.

The realization of this concept was achieved by Friesen and Danishefsky [158, 159]. Thus, the 3,4,6-tri-*O*-benzyl-glycal *177* was coupled with 3,6-di-*O*-benzoyl-

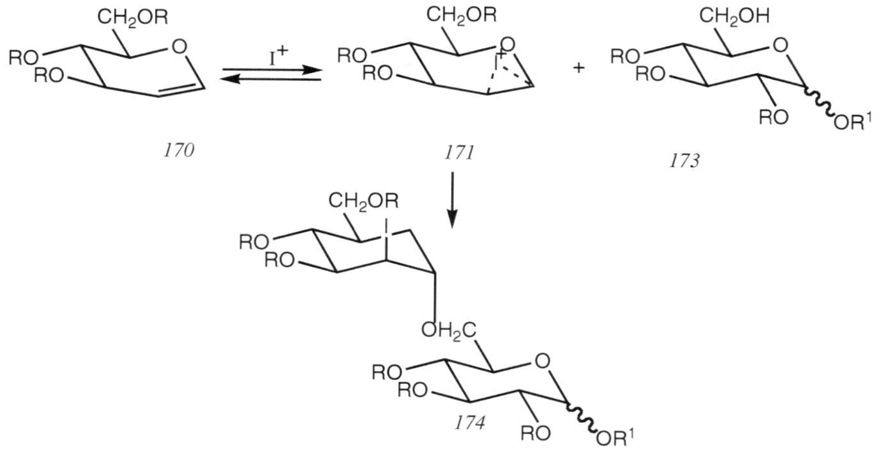

Glycoside Synthesis

Fig. 12.50

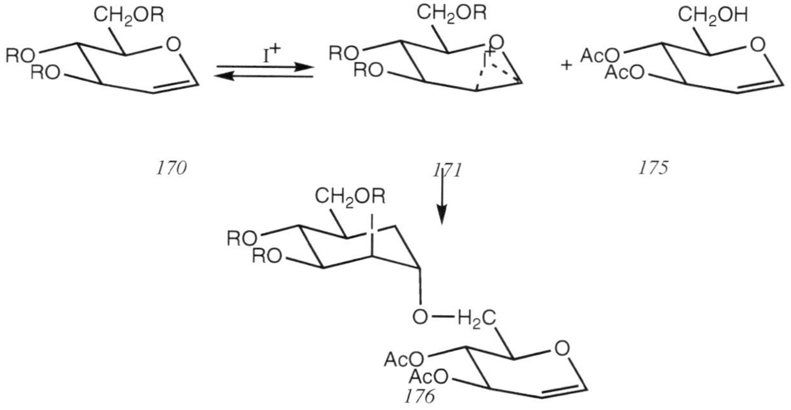

Fig. 12.51

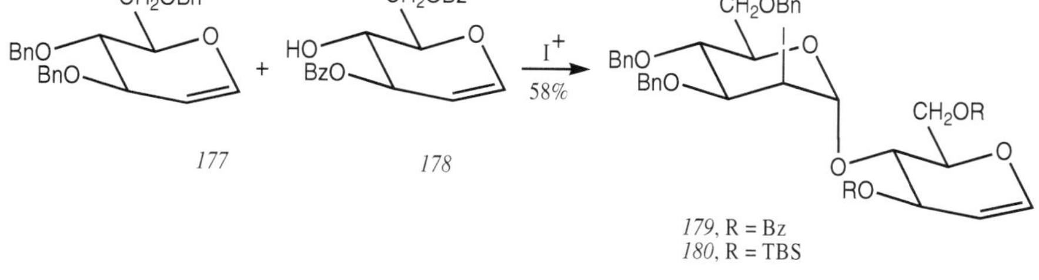

Fig. 12.52

glycal *178* to the corresponding disaccharide *179* with 58% yield (Fig. 12.52). The disaccharide *179* could be on one hand directly coupled to a nonglycal acceptor in the presence of I$^+$ with 1,2:3,4-di-*O*-isopropylidene-α-D-galactopyranose *182*, showing that glycals bearing acyl-protecting groups are competent glycosyl donors in iodoglycal portion of disaccharide iodoglycosylation reactions with nonglycal acceptors (Fig. 12.53). On the other hand, the *179* must be armed by replacing the 3- and 6-benzoyl substituents with *tert*-butyl-dimethylsilyl (TBS) groups and the modified disaccharide *183* then reacted with a glycal acceptor *178* in the presence of I$^+$ (Fig. 12.54).

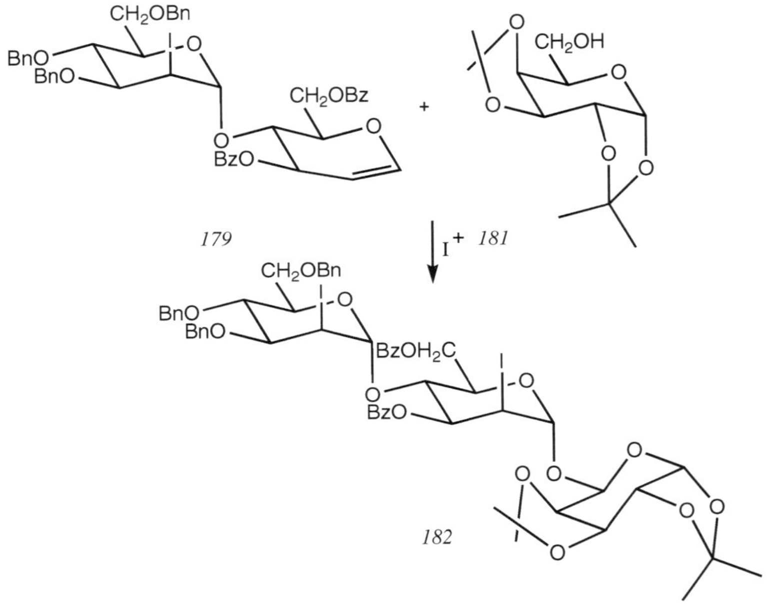

Fig. 12.53

In addition to in situ electrophilic activation of 1,2-double bond of a glycal by "I$^+$" (*N*-iodosuccinimide or *sym*-collidine iodonium perchlorate) that gives a non-isolable glycosyl donor intermediate *171* (Figs. 12.50 and 12.51) a glycal can be first converted into an isolable or identifiable glycosyl donor (for example, 1,2-anhydro sugar *185*) and then the obtained stable intermediate *185* used as actual glycosyl donor. In this approach, a glycal is only the precursor to a structurally defined glycosyl donor (*185* and *187*) (Figs. 12.55 and 12.56).

At the beginning of their investigation, Danishefsky et al. encountered two serious impediments to a broad applicability of 1,2-anhydrosugars as glycosyl donors for the synthesis of oligosaccharides. First was the actual synthesis of sugar 1,2-oxirane derivatives, and the second was the previous reports that various glycosyl acceptors add nonstereoselectively to sugar 1,2-oxiranes used as glycosyl donors.

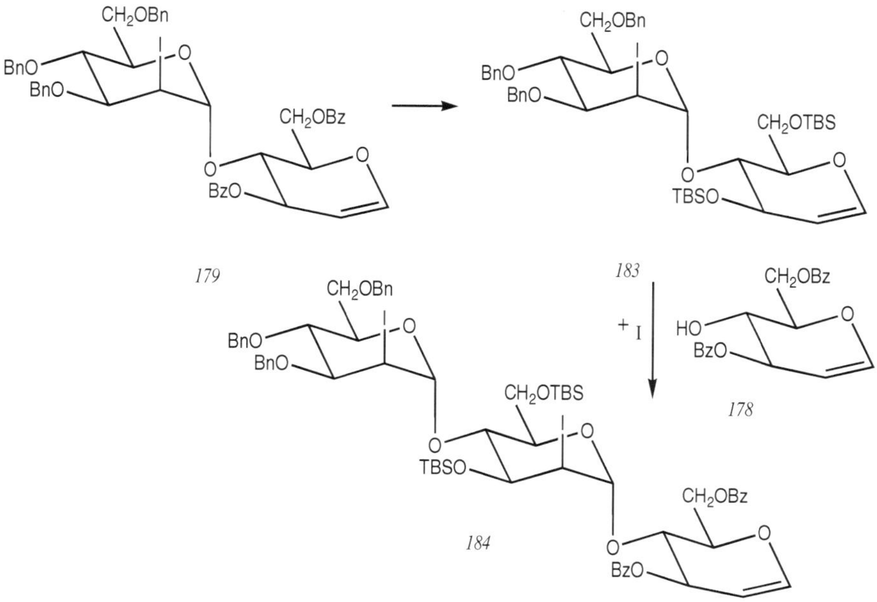

Fig. 12.54

An additional difficulty was the possibility of neighboring group participation of acyl-protecting groups in the ring opening of 1,2-epoxide.

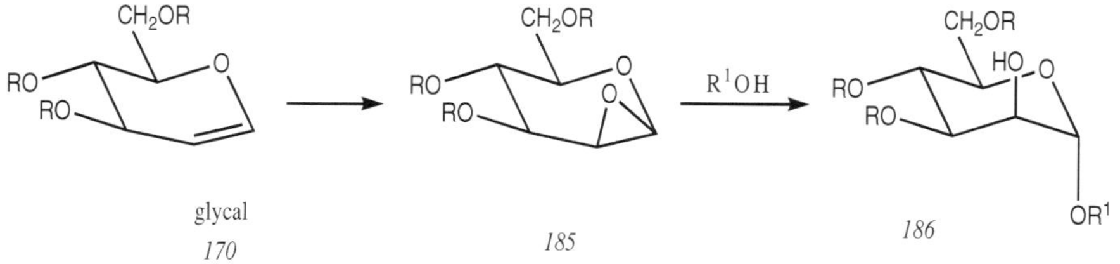

Fig. 12.55

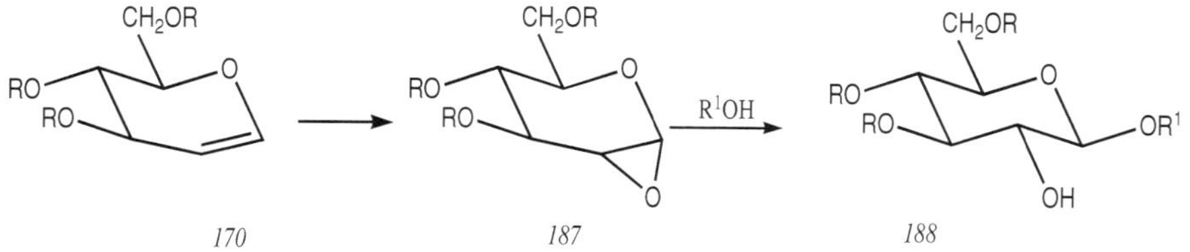

Fig. 12.56

The synthesis of sugar 1,2-oxiranes was accomplished by reacting a variety of glycals with 2,2-dimethyldioxirane [160] (DMDO) *189* in methylene chloride/acetone at 0°C [161]. The stereoselectivity of epoxidation depended upon the nature of substituents in the glycal ring. If the substituents were benzyl or *tert*-butyldimethylsilyl group (nonparticipating groups) the reaction gave highly stereoselectively the α-epoxides (*170* → *187*) (Fig. 12.56).

Solvolysis of *187* with neat methanol gave methyl β-D-glycoside *190*. However, the epoxidation of peracetylated glucal *170* gave a mixture of epoxides *191* which on solvolysis in methanol gave a mixture of methyl glycosides *192* (Fig. 12.57).

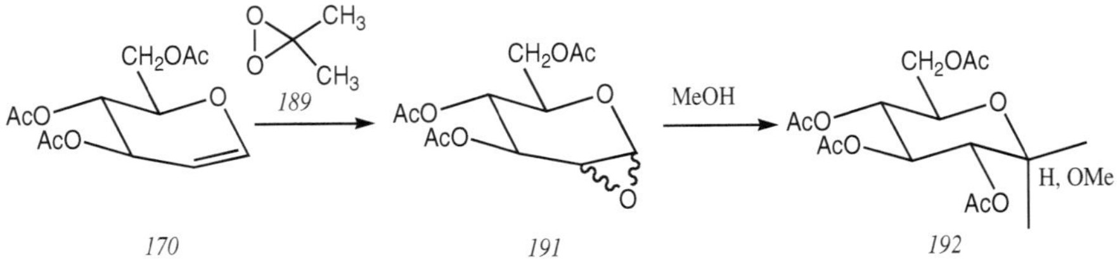

Fig. 12.57

3,4,6-Tri-*O-tert*-butyldimethylsilyl-D-galactal *193* stereoselectively gave the α-oxirane *194* (Fig. 12.58).

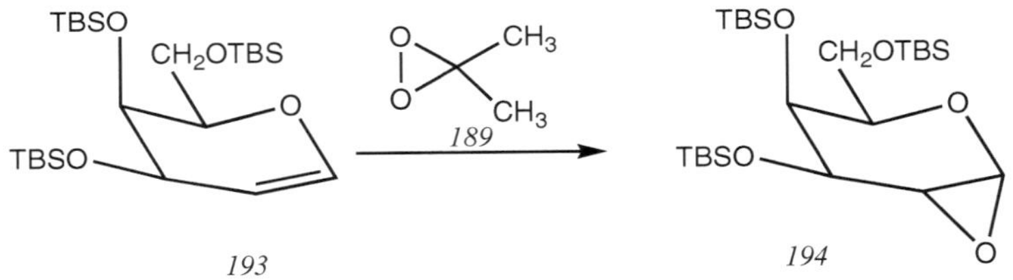

Fig. 12.58

The glycal *195* bearing the axial C3 TBSO group undergoes selective epoxidation from its β-face, giving *196* as almost the only product (Fig. 12.59). On the other hand, gulal derivative *197* having large substituents on both faces of the double bond gave the 1:1 mixture of epoxides *198* (Fig. 12.60).

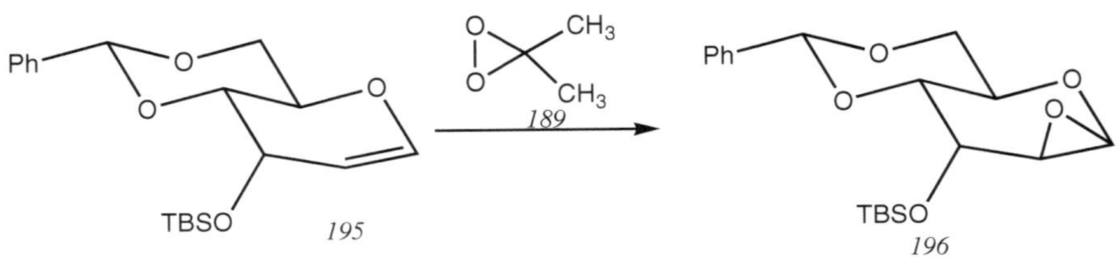

Fig. 12.59

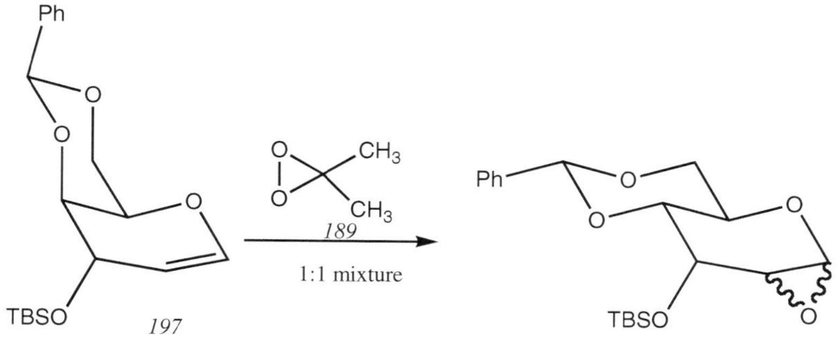

Glycoside Synthesis

Fig. 12.60

Glycosylation of acceptors more complex than methanol and present in ca. stoichiometric amount was slow and required the presence of promoters. There is no universal promoter, but with ordinary alcohol acceptors (including the hydroxyl groups of another sugar), the most widely used promoter is anhydrous zinc chloride. In some special applications stannyl derivatives generated in situ gave the best results (for example, the synthesis of gangliosides). Using this approach it was possible to glycosylate cholesterol in a relatively good yield (Fig. 12.61).

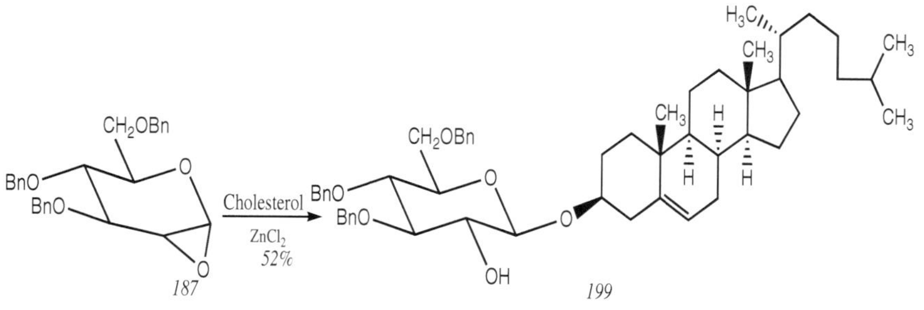

Fig. 12.61

Using 1,2-epoxy sugars as glycosyl donors Danishefsky et al. were able to synthesize many complex oligosaccharides, such as Lewis determinants, blood group determinants, and tumor antigens [162].

Thioglycosides as Glycoside Donors

It has been reported that 1-thioaldofuranosides undergo hydrolysis in aqueous solution in the presence of mercury(II) salts [163], and that acetylated aldose diethyl dithioacetals are converted into dimethyl acetals in methanol under similar conditions [164]. Further, ethyl 1-thio-α- and β-D-glucopyranosides, on treatment with bromine and silver carbonate in methanol, gave methyl β- and α-D-glucopyranosides, respectively, in high yield apparently by a way of bromosulfonium ion intermediate [165].

In 1973 Ferrier reported [166] that phenyl 1-thio-D-glucopyranosides in the presence of mercury(II) acetate are readily solvolyzed to give alkyl D-glucopyranosides with inverted anomeric configurations. Thus methanolysis of the β- and α-anomers afforded the methyl α- and β-D-glycosides with 74 and 87% yield, respectively. Furthermore, using the same procedure, they synthesized ethyl α-D-glucopyranoside with 67% yield, isopropyl 2,3,4,6-tetra-*O*-acetyl-α-D-glucopyranoside with 55% yield, cholesteryl 2,3,4,6-tetra-*O*-benzyl-α-D-glucopyranoside with 78% yield, etc.

This pioneering study attracted widespread attention for 1-thioglycosides as glycosyl donors because 1-thioglycosides are stable compounds that are readily available, the 1-sulfur atom could be activated with a wide range of electrophilic activators, and the glycosylation reaction seems to proceed with a high anomeric stereoselectivity (Fig. 12.62).

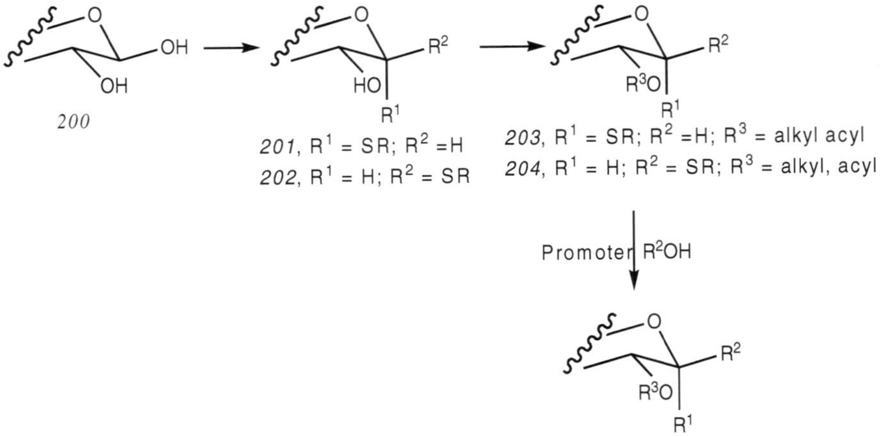

200

201, R^1 = SR; R^2 = H
202, R^1 = H; R^2 = SR

203, R^1 = SR; R^2 = H; R^3 = alkyl acyl
204, R^1 = H; R^2 = SR; R^3 = alkyl, acyl

Promoter R^2OH

205, R^1 = OR; R^2 = H; R^3 = alkyl acyl
206, R^1 = H; R^2 = OR; R^3 = alkyl, acyl

Fig. 12.62

Glycoside Synthesis

The sulfur atom in 1-thioglycosides is a "soft" nucleophile and is able to react selectively with "soft" electrophiles such as heavy (transition) metal cations, halogens, alkylating and acylating agents. This fact made 1-thioglycosides very versatile glycosyl donors in the synthesis of oligosaccharides. Additionally, the hydroxyl groups of carbohydrates are "hard" nucleophiles, which can be functionalized with "hard" reagents, without affecting the anomeric alkyl (aryl) thio group after the introduction of alkyl (aryl) thioglycoside group.

The mechanism of glycosylation using alkyl (aryl) thioglycosides is very simple. A soft electrophile activates thioglycoside by producing intermediate alkyl (aryl) sulfonium ions *208* or *212*, that dissociate leaving oxocarbenium ion *209* or *213* as actual glycosylating species that then react with an alcohol or the hydroxyl group of another molecule that can be an appropriately protected sugar giving the glycoside or a disaccharide. If the C2 carbon is protected with a nonparticipating substituent (e.g., benzyl) a mixture of anomers is obtained (*210*), whereas if the C2 substituent is a participating one (e.g., acetate) the glycoside obtained is 1,2-*trans* (β) (*214*) (Fig. 12.63).

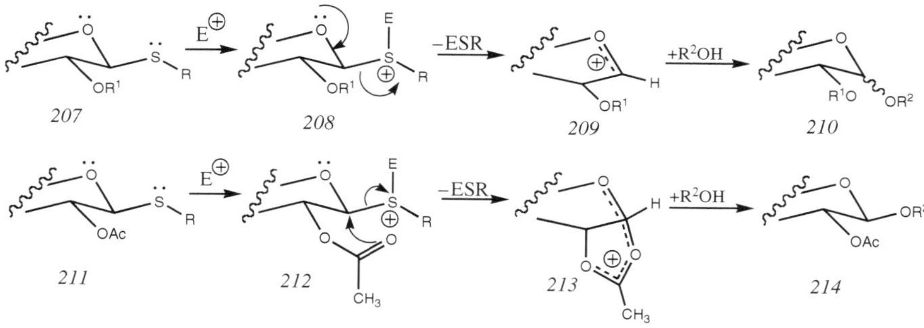

Fig. 12.63

Following these initial observations, a number of different promoters were proposed for the construction of glycosidic bonds, such as copper(II) triflate [167], mercury(II) benzoate [168], mercury(II) nitrate [135], palladium(II) perchlorate [134, 169], N-bromosuccinimide [170], phenyl mercury(II) triflate [171], and mercury(II) chloride [166, 172]. However, None of these promoters gave consistently high yields needed for the synthesis of oligosaccharides.

An interest in developing new promoters was sparked by Lönn's report that methyl triflate is an excellent thiophilic promoter for producing oxocarbenium cations from 1-thioglycosides that readily react with glycosyl acceptors (alcohols or appropriately protected sugars) to give glycosides or di- or oligosaccharides [173–176].

Unfortunately, methyl triflate had two serious disadvantages: first, it is very toxic, and second, in case of slow reacting glycosyl donors it methylates any free hydroxyl group giving thus methyl ethers in addition to glycosides. For these reasons an extensive search for new promoters has been undertaken. Most of the new promoters use not only alkyl and aryl thioglycosides as glycosyl donors, but also isothiocyanates, as well as S-pyridyl and 1-phenyl-1H-tetrazol-5-yl-thioglycosides and glycosyl 1-piperidinecarbodithioates. The list of these promoters is given in Table 12.5.

Table 12.5 Glycosyl donors and acceptors

Activator	Thioglycoside	Reference
MeOTf - methyl triflate	–SMe, –SEt, –SPh	173-176
DMTST - dimethyl(methylthio) sulfonium triflate	–SMe, –SEt, –SPh	177
NOBF$_4$ - nitrosyl tetrafluoroborate	–SMe	178
MeSOTf, MeSBr - thiomethyltriflate, Bromothiomethane	–SMe, –SEt, –SPh	179
TrClO$_4$ - trityl perchlorate	–SCN, (ROTr acceptor)	180
PhSeOTf - Selenophenyl triflate	–SMe	181
MeI - methyl iodide	–SPy	182
NIS-TfOH - N-iodosuccinimide -Triflic acid	–SMe, –SEt, –SPh	183
IDCP - iodonium dicollidine perchlorate and IDCTf - iodonium dicollidine triflate	–SEt	138
AgOTf - silver triflate	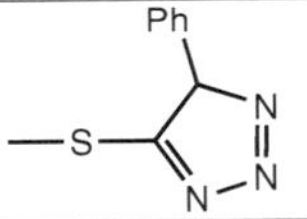 -S-(phenyl-triazole)	184
TBPA - tris (4-bromophenyl) ammoniumyl hexachloroantimonate	–SEt, –SPh	185
DMTST, AgOTf, SnCl$_4$, FeCl$_3$	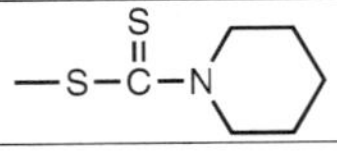 –S–C(=S)–N(piperidine)	186, 187

In Fig. 12.64 are given the structures of some of the promoters cited in Table 12.5.

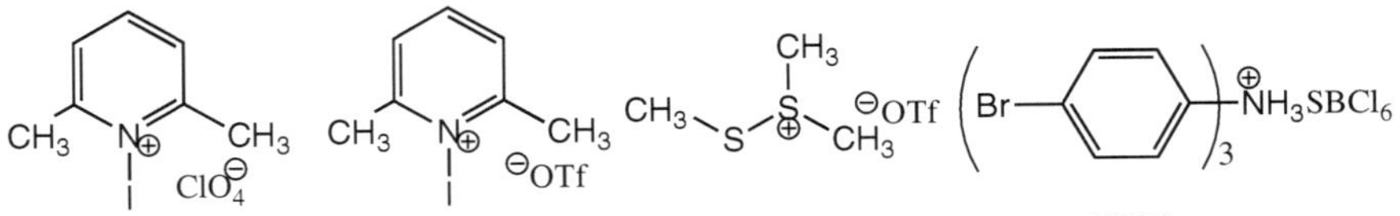

IDCP *215*   **IDCTf** *216*   **DMTST** *217*   **TBPA** *218*

Fig. 12.64

The stereoselectivity of glycosidation depends upon the nature of the C2 substituent. If the C2 substituent in a glycosyl donor, such as O-acetyl and O-benzoyl group, is capable of stabilizing the oxocarbenium cation, generated by a promoter, via neighboring group participation, 1,2-*trans*-glycosides are obtained with excellent stereoselectivity [177]. If, however, the C2 substituent in a glycosyl donor, such as O-benzyl group, is incapable of stabilizing the oxocarbenium cation via neighboring group participation, a mixture of α- and β-glycosides is obtained; the composition of the anomeric mixture is reported to be solvent dependent. Thus, for example, the proportion of 1,2-*cis* glycoside is increased if diethyl ether is used as the solvent; it was suggested that the solvent directly participates in stabilizing the intermediate oxocarbenium cation [176]. On the other hand, in acetonitrile as the solvent and with a nonparticipating C2 substituent in the glycosyl donor, 1,2-*trans*-glycosides are obtained again as a result of solvent participation [188].

The "armed–disarmed" concept developed for glycosylations with 4-pentenyl glycosides as glycosyl donors and acceptors [144] has been successfully applied for the synthesis of oligosaccharides using thioglycosides as glycosyl donors and acceptors. If the glycosyl donor is activated ("armed") by having an electron-donating group at the C2 carbon of a glycosyl donor (*219*) and an electron-withdrawing group at the C2 carbon of a glycosyl acceptor (*220*), the chemoselective activation of the glycosyl donor is possible resulting in a synthesis of an oligosaccharide (*221*) [184] as shown in Figs. 12.65 and 12.66 [189].

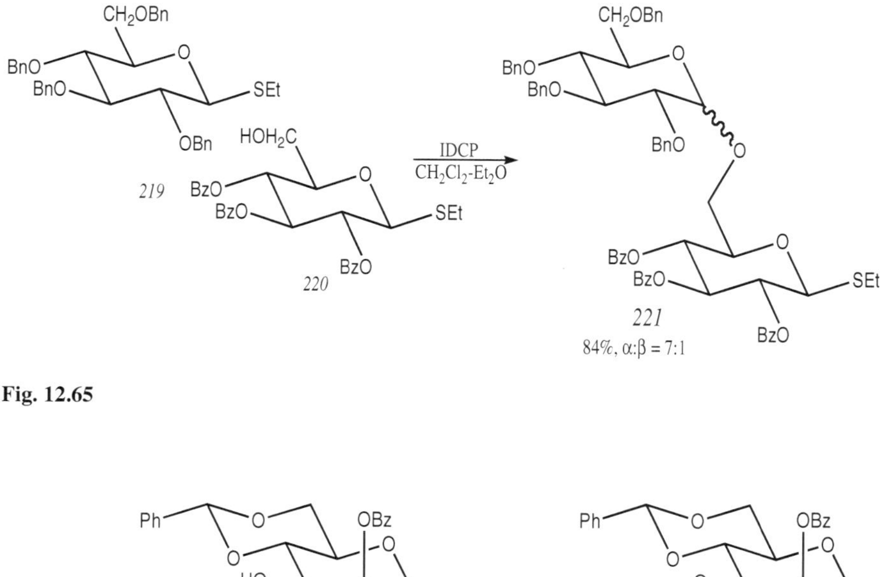

Fig. 12.65

Fig. 12.66

Thioglycosides as glycosyl donors have been used in the syntheses of a large number of oligosaccharides. Thus, for example, they have been used in synthesis of ganglioside GM_2 [190], I-active ganglioside analog [191], etc. A selection of references dealing with the use of thioglycosides as glycosyl donors can be found in the review article by Garegg [192].

Synthesis of Thioglycosides

A great variety of methods exist [193, 194] for the preparation of alkyl and aryl 1-thioglycosides. Thus they can be prepared

1. by reacting acylated aldoses with a thiol in the presence of Lewis acids [193–202]
2. by reacting acylated glycosyl halides with thiolate anion [203–211]
3. by reacting glycosyl halides with thiourea derivatives [211]
4. by partial hydrolysis of dithioacetals [193, 212]
5. by reacting 1-thioaldose derivatives with aryldiazonium salts [193, 213]
6. by decomposition of glycosyl xanthates [193, 214–216]
7. by reacting glycosyl thiocyanates with Grignard reagent [193, 217]
8. by radical addition of 1-thiols to alkenes [218]
9. using acetylated glycosyl piperidine carbodithioates [187]

Glycosyl Sulfoxides as Glycosyl Donors

The oxidation of thioglycosides to sulfinyl glycosides provided a new and powerful group of glycosyl donors – glycosyl sulfoxides (225) [219]. Glycosyl sulfoxides react with glycosyl acceptors in the presence of a promoter, to give di-, tri-, or oligosaccharides. The promoters for these sulfinyl glycosides are triflic anhydride (Tf_2O) or trimethylsilyl triflates instoichiometric amount, or triflic acid in catalytic amount. The reaction is always carried out in the presence of an acid scavenger (2,6-di-*tert*-butyl-4-methyl-pyridine – DTBMP) (Fig. 12.67).

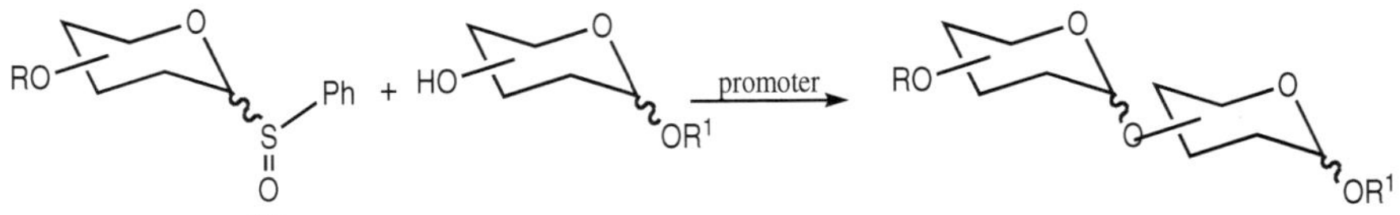

promoter: Tf_2O, TMSOTf, TfOH
acid scavenger: DTBMP

Fig. 12.67

Using this method, Kahne was able to glycosylate very unreactive hydroxyl group at the C7 carbon of a deoxycholic acid derivative [219]. The yields seem to depend on the nature of the solvent; better yields are obtained in nonpolar solvents. The stereoselectivity of glycosylation depended upon the nature of the C2 substituent. Thus glycosylation with the glycosyl sulfoxide having at the C2 carbon a participating substituent gave exclusively β-glycoside; however, glycosyl sulfoxides having at the C2 carbon a nonparticipating substituent gave a mixture of α- and β-anomers. The composition of the mixture depended again upon the solvent (Table 12.6).

Table 12.6

Glycosyl acceptor	Glycosyl donor	Solvent	α/β ratio (yield)
(steroid with OH, EtOCO, COOMe)	BnO/CH$_2$OBn glycosyl sulfoxide SPh	toluene	27:1 (86%)
		CH$_2$Cl$_2$	1:3 (80%)
		acetonitrile	1:8 (50%)
	PivO/PivOH$_2$C glycosyl sulfoxide SPh	CH$_2$Cl$_2$	all β (83%)

The glycosylation with glycosyl sulfoxides is highly efficient with rather unreactive glycosyl acceptors, it has a potential for chemoselective glycosylation, and it is applicable to the synthesis of oligosaccharides on solid supports.

One advantage of the sulfoxide method is its flexibility and wide scope. Lewis blood group antigens, namely Lewis a, Lewis b, and Lewis x (Le^a, Le^b, and Le^x), were synthesized using sulfoxides methodology [220].

Solid-Phase Synthesis of Oligosaccharides

There are two main advantages for using solid support for the synthesis of oligosaccharides over the synthesis in solution: first, there is no need for chromatographic purification of intermediates, and second, glycosyl acceptor can be used in excess raising thus the yield of reaction; the excess of glycosyl acceptor can be, after the coupling, washed out. While the automated solid-support syntheses of peptides and oligonucleotides are well-established methods for quite some time the development of solid-support oligosaccharide synthesis took much longer time.

The reasons for this is that in the oligosaccharide synthesis there are two major impediments: first, there is a requirement for the stereospecific formation of a glycosidic bond that links two monosaccharide units (α- or β-), and the second one is that the selection of protecting groups for the monosaccharide units that are to be coupled in oligosaccharide synthesis is much more complex than in peptide or oligonucleotide synthesis.

The first question that has to be addressed is how the first monosaccharide should be linked to a polymer: via the anomeric carbon or by attaching the polymer to one of the hydroxyl groups on a monosaccharide. Thus, the first monosaccharide linked to the polymer may act either as the glycosyl acceptor or as the glycosyl donor. We will illustrate both of these approaches.

In the first example [221] the polymer *228* is attached to the C6 carbon of glycal *229* giving *230*. The epoxidation, reaction of oxirane with ethane thiol, and acylation with pivaloyl chloride gave ethyl 3,4-di-*O*-benzoyl-2-*O*-pivaloyl-1-thio-β-D-glucopyranoside *231*. The reaction of *231* with 3,6-di-*O*-benzyl-D-glycal *232* in the presence of methyl triflate (Fig. 12.68) gives disaccharide *233*. Repetition of epoxidation, opening of epoxide with ethane thiol, and acylation with pivaloyl chloride, as well as coupling with glycal *232* two more times, gave a tetrasaccharide *237*.

In the second example we will describe the solid-phase synthesis of a heptasaccharide phytoalexin elicitor (HPE) [221].

Polystyrene *238* was functionalized to phenolic polystyrene *239* [222] (although only *p*-substituted polystyrene is shown, it is estimated that phenolic polystyrene contains both *p*- and *m*-hydroxyphenyl rings. The functionalization was performed to the extent 0.25–1.0 mmol/g) (Fig. 12.69).

The *o*-nitrobenzyl alcohol *242* was used for its ease of attachment to and cleavage from sugar. This aglycon was synthesized in the following way. The commercially available 5-hydroxy-2-nitrobenzaldehyde *240* (Fig. 12.70) was first reacted with 1,3-diiodopropane in the presence of Cs_2CO_3 in DMF and the obtained aldehyde *241* was reduced with $NaBH_4$ to afford iodobenzyl alcohol *242* with 92% overall yield (Fig. 12.70).

Glycosidation of *242* with phenyl 2-*O*-benzoyl-3,4-di-*O*-benzyl-6-*O*-*tert*-butyldiphenylsilyl-1-thio-β-D-glucopyranoside *243* in the presence of dimethylthiomethylsulfonium triflate (DMTST) proceeded with 95% yield to afford exclusively β-glucoside *244*. This glycoside was then attached to phenolic polystyrene *239* by using Cs_2CO_3 in DMF at 25°C to afford the conjugate *245* (Fig. 12.70) with > 90% yield (Fig. 12.70).

Cleavage of the carbohydrate fragment from the resin was effected by irradiation of *245* (Fig. 12.70) in THF at 25°C to afford the monosaccharide with 95% yield.

The building blocks for the synthesis of HPE are shown in Fig. 12.71. After several reiterative steps heptasaccharide HPE *256* (Fig. 12.72) was obtained with 20% overall yield. Figure 12.72 shows the detailed synthesis of only tetrasaccharide *254* because the remaining steps are the repetition of previous steps. In the first step, the polymer-linked sugar derivative having the C6 hydroxyl group free (*249*) was glycosylated with phenylthioglycoside *247* in the presence of dimethylthiomethylsulfonium triflate (DMTST) giving *250* with >96% yield. Removal of the fluorenylmethoxycarbonyl (Fmoc) protective group with Et_3N in CH_2Cl_2 at 25°C and glycosylation of the obtained *251* with *248* in the presence of

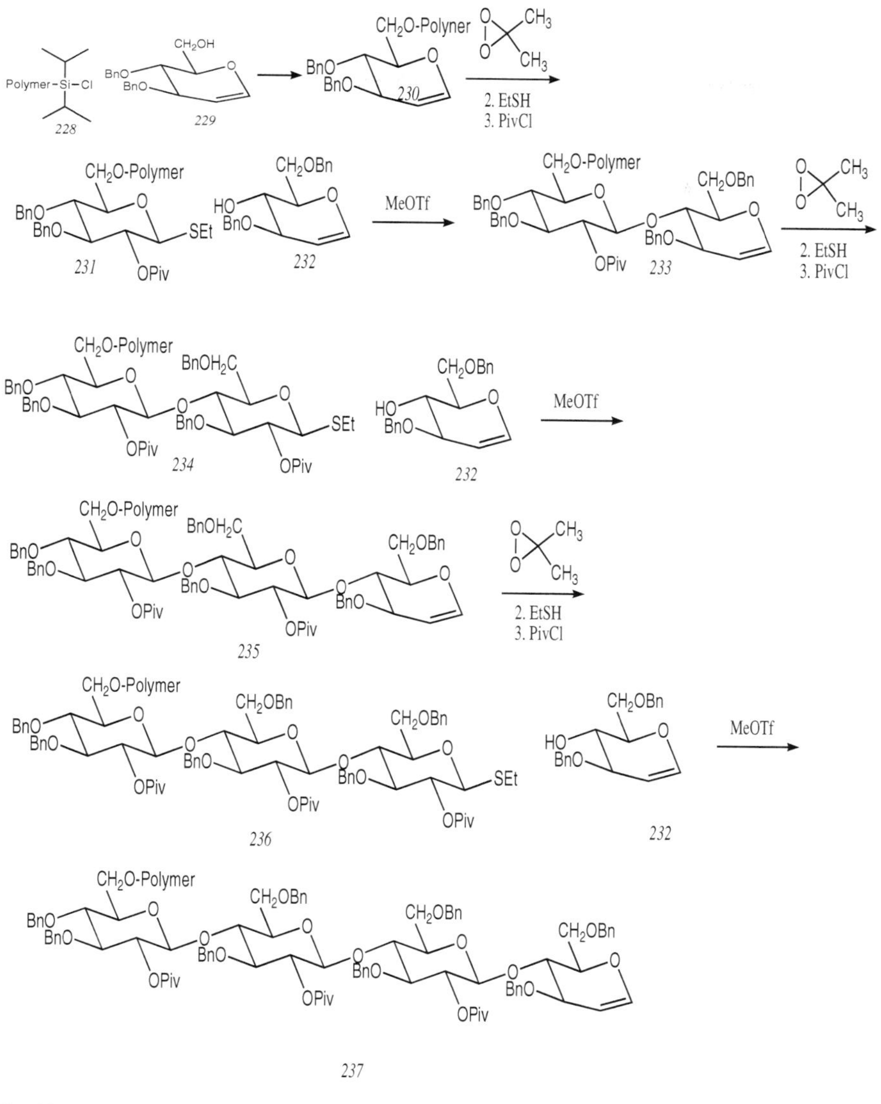

Glycoside Synthesis

Fig. 12.68

DMTST gave *252*. Removal of *tert*-butyldiphenylsilyl (TBDPS) group from *252* and glycosylation of the obtained *253* with *246* in the presence of DMTST gave tetrasaccharide *254* which was then converted into *255* by desilylation. Repeating these steps three more times, fully blocked heptasaccharide linked to the polymer was synthesized. Photolytic cleavage of heptasaccharide from the resin followed by

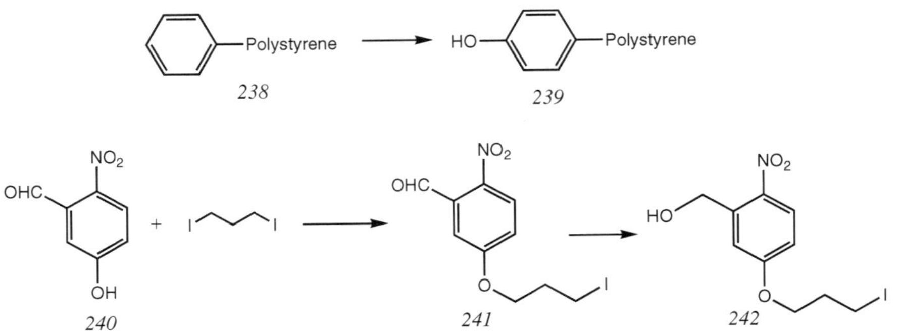

Fig. 12.69

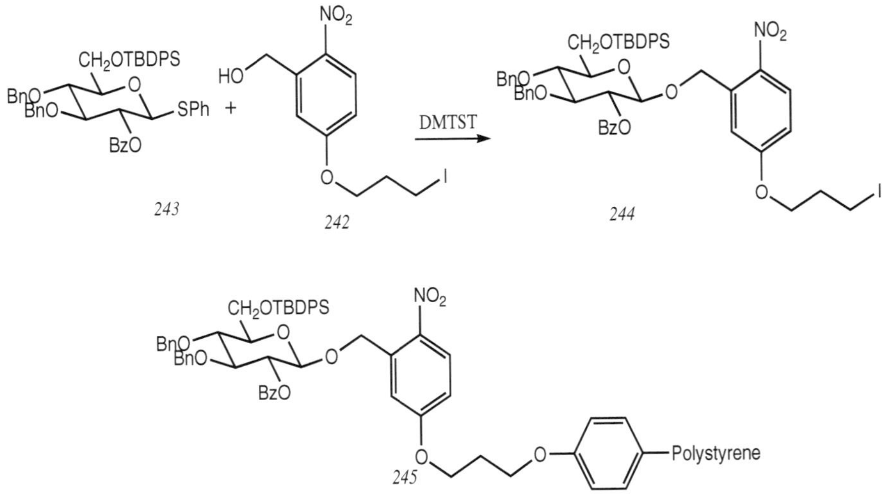

Fig. 12.70

acetylation of the anomeric hydroxyl group of monosaccharide at the reducing end gave a mixture of α- and β-anomers of fully blocked heptasaccharide HPE. Hydrogenolysis and treatment with NaOH in MeOH gave HPE *256* with 20% overall yield.

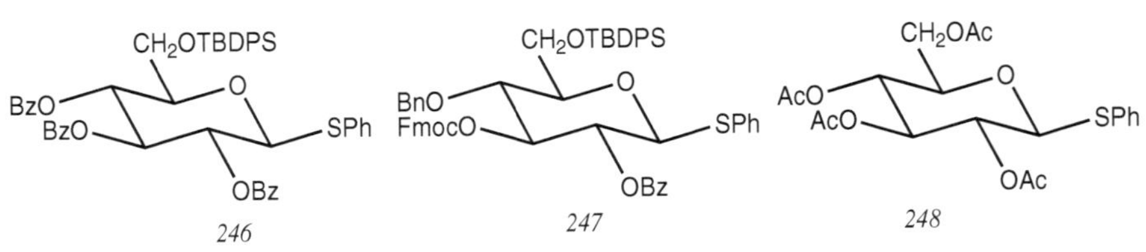

Fig. 12.71

Glycoside Synthesis

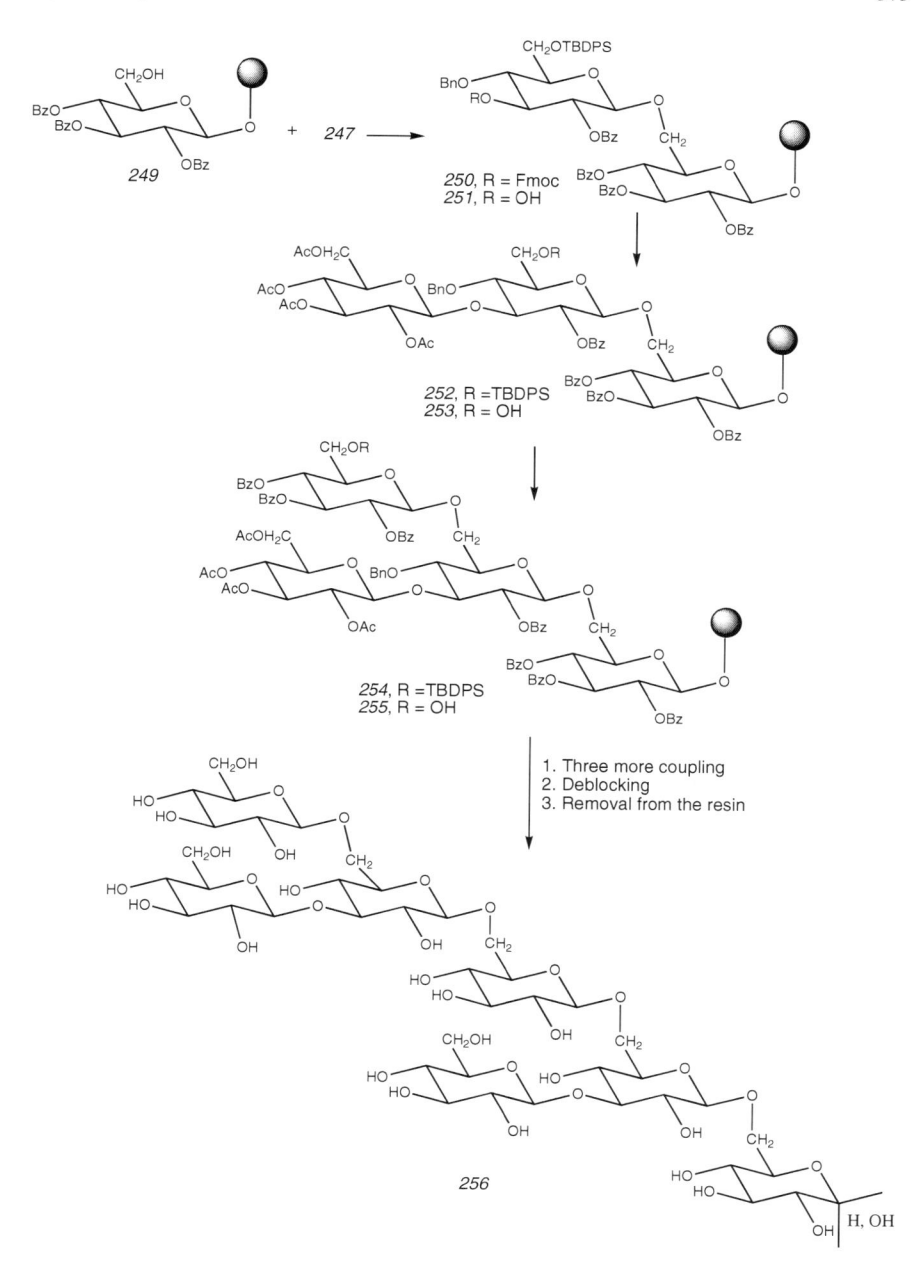

Fig. 12.72

Automated Oligosaccharide Synthesis

There is a good review article published on development of an automated oligosaccharide synthesizer by Seeberger et al. [223] and we will not discuss this topic here.

Cleavage of Glycosidic Bonds

Acid-Catalyzed Hydrolysis of Glycosides

The acid-catalyzed hydrolysis of glycosides is theoretically a reverse process of acid-catalyzed glycosidation. However, since the acid-catalyzed hydrolysis always involves a single sugar moiety, such as α- or β-anomer of a pento- or hexofuranoside or an α- or β-anomer of a pento- or hexopyranoside, whereas the acid-catalyzed glycosylation always involves a mixture of several isomeric forms of a sugar at equilibrium, such as α- and β-anomers as well as the ring isomers (the furanoid and the pyranoid forms), these two processes obviously cannot be expected to proceed via identical reaction mechanism and the same transition states. This statement is, however, not entirely true since in the acid-catalyzed glycosidation a small proportion of sugar pyranoside present in the equilibrium undergoes direct glycosidation, and only this process will be the reverse process of acid-catalyzed hydrolysis. Several good reviews have been published on this subject [224–226].

As we have already seen, the glycosides are mixed acetals wherein the aldehydo C1 carbon is on one hand linked via exocyclic oxygen atom to an alkyl, aryl, or any other molecule and, on the other hand, is linked via the endocyclic oxygen atom to the C5 (pyranosides) or the C4 (furanosides) carbon of a sugar. The initial step of an acid-catalyzed hydrolysis of a glycoside, as well as of any acetal or ketal, is the fast and reversible protonation of one of these two acetal oxygens (Fig. 12.73).

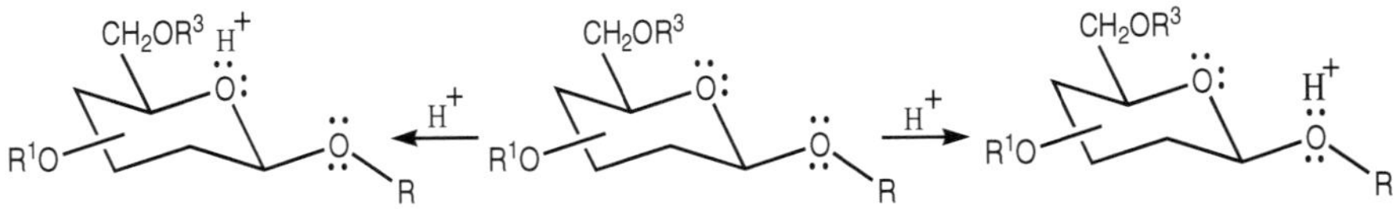

Fig. 12.73

Depending on the site of protonation two reaction mechanisms could be envisioned. If the glycosidic oxygen is protonated giving *261* (Fig. 12.74) the following step could be the unimolecular elimination of an alcohol with the assistance of the axially oriented nonbonding electron pair of the ring oxygen and the formation of the corresponding oxocarbenium ion *262*. In the presence of water in the reaction mixture (which is always the case in hydrolysis reactions) water molecules, and not alcohol molecules, will add to the positively charged carbon of oxocarbenium ion *262* giving first the hydrolyzed monosaccharide protonated at the anomeric hydroxyl

Cleavage of Glycosidic Bonds

group *263* which will, after deprotonation, give the hydrolyzed sugar *264*. Since in oxocarbenium ion *262* the C5, O5, C1, and C2 atoms all lie in one plane (the sugar molecule must assume the half-chair conformation due to the double bond character of the C1–O5 bond), the water molecules can add from either face of the oxocarbenium ion and hence a mixture of α- and β-glycopyranoses will be obtained (Fig. 12.74).

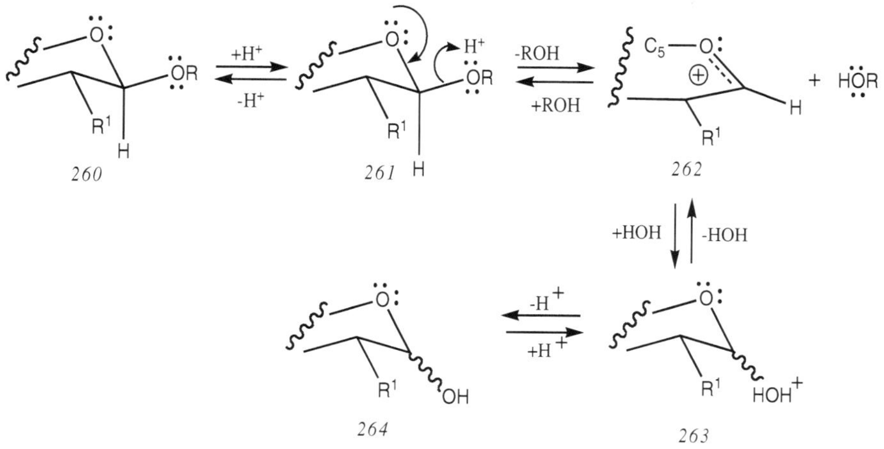

Fig. 12.74

If the ring oxygen (O5 in pyranosides, or O4 in furanosides) is protonated *265* (Fig. 12.75) the C1–O5 bond will be broken and an acyclic oxocarbenium ion *266* will be formed (this time by the participation of one nonbonded electron pair of glycosidic oxygen) (Fig. 12.75). The addition of a water molecule to *266* will form the hemiacetal *267* that is protonated at the anomeric hydroxy oxygen. This intermediate will be in equilibrium with the hemiacetal protonated at the methoxy oxygen (*268*). Now, the elimination of alcohol from *268* will result in formation of protonated aldehydo sugar *269* that by cyclization and deprotonation gives the hydrolyzed sugar *264*.

There is, however, a third possible mechanism that can be envisioned for the hydrolysis of glycosides and that is the nucleophilic (S_N2) displacement of the protonated methoxy group with water as the nucleophile, as shown in Fig. 12.76.

In order to elucidate the reaction mechanism of glycoside hydrolysis we must first fully understand the relationship between the rate of hydrolysis and steric and electronic factors present in both glycon and aglycon of a sugar glycoside. The rates of hydrolysis of many glycosides have been measured and found to be influenced by many factors such as the type of a sugar, the ring size of a sugar, the anomeric configuration of glycosidic bond, the nature of substituents on a sugar ring, the conformation of a sugar, and the size and polarity of an aglycon.

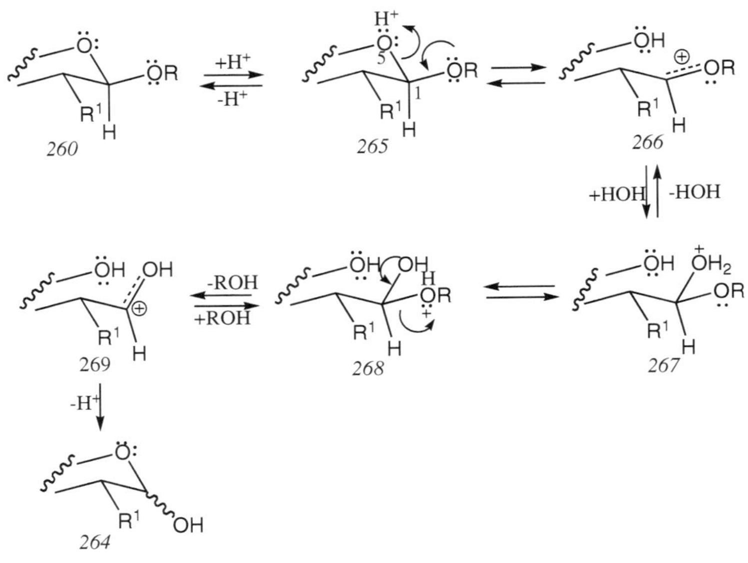

Fig. 12.75

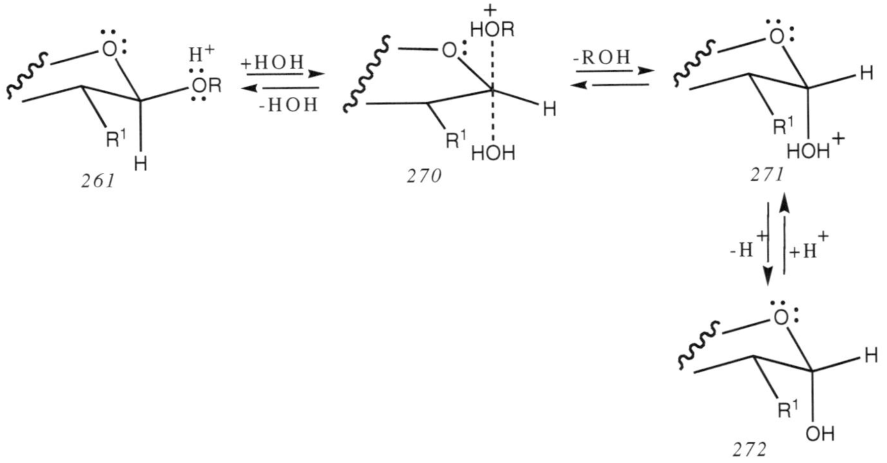

Fig. 12.76

The observations that glycofuranosides are generally hydrolyzed much faster than glycopyranosides (ca. 50–200) [227, 228] (it should be remembered that furanosides are also formed much faster than pyranosides, vide supra) and that kinetic parameters for the acid-catalyzed hydrolysis of glycofuranosides and glycopyranosides are very different (the entropies of activation for the acid-catalyzed hydrolysis of all glycofuranosides are negative, whereas they are positive for the glycopyranosides) suggest that they are probably hydrolyzed by different mechanisms and therefore we will discuss them separately.

The Acid-Catalyzed Hydrolysis of Glycopyranosides

The glycopyranosides with equatorially oriented aglycon are hydrolyzed roughly twice faster than glycopyranosides having the aglycon axially oriented and this ratio seems to be dependent neither on the structure of glycon nor on the nature of aglycon. In Table 12.7 are given the relative rates of acid-catalyzed hydrolysis of select group of methyl aldopyranosides.

Table 12.7 Relative rates of acid-catalyzed hydrolysis of methyl α- and β-aldopyranosides[19] in 0.01–0.5 M HCl or H_2SO_4 at 58–100°

Methyl D-glycopyranoside	Relative rates	α:β ratio	Orientation of 1-OMe group
α-D-Gluco-	1.0 [229]	1:1.9	Axial
β-D-Gluco	1.9 [229]		Equatorial
α-D-Galacto-	2.4 [229]	1:2.4	Axial
β-D-Galacto-	5.7 [229]		Equatorial
α-D-Manno-	5.2 [229]	1:1.8	Axial
β-D-Manno-	9.2 [229]		Equatorial
α-D-Xylo-	4.5 [229]	1:2.0	Axial
β-D-Xylo-	9.1 [229]		Equatorial
α-L-Arabino-	13.1 [229]	1.5:1	Equatorial
β-L-Arabino-	9.0 [229]		Axial
α-L-Rhamno-	8.3 [229]	1:2.3	Axial
β-L-Rhamno-	19.0 [229]		Equatorial
α-D-Glucopyranosiduronic acid	0.47 [230]	1:1.3	Axial
β-D-Glucopyranosiduronic acid	0.62 [230]		Equatorial
2-Deoxy-α-D-gluco-	2090 [231]	1:2.5	Axial
2-Deoxy-β-D-gluco-	5125 [231]		Equatorial
2,3,4,6-Tetra-O-methyl-α-D-gluco-	0.16 [232]	1:2.5	Axial
2,3,4,6-Tetra-O-methyl-β-D-gluco-	0.40 [232]		Equatorial

The removal of the hydroxyl group at either the C2 or the C6 carbon accelerates the acid-catalyzed hydrolysis. Whereas the rate acceleration of acid-catalyzed hydrolysis is enormous for the C2 deoxy glycopyranosides (ca. 2–5 × 10^3 times) the rate acceleration for the C6 deoxy glycopyranosides is much smaller (only ca. 8 times).

The acid-catalyzed hydrolysis of pentopyranosides is generally faster than that of hexopyranosides (4.5–9.0 times) but slower than acid-catalyzed hydrolysis of 6-deoxy-hexopyranosides.

The introduction of an electron-withdrawing group at the C6 carbon reduces the rate of acid-catalyzed hydrolysis (the acid-catalyzed hydrolysis of methyl glycoside of D-glucuronic acid is ca. 2 times slower).

In Table 12.8 are given the rate coefficients and kinetic parameters for the hydrolysis of select glycosides in 2.0 N HCl extrapolated to 60°C. The concentration of HCl for the hydrolysis of methyl 2-deoxy-α- and β-D-glucopyranoside was 0.1 N.

Table 12.8 Rate coefficients and kinetic parameters for the hydrolysis of select glycopyranosides [231]

Pyranoside	$10^5\ k\ (s^{-1})$	E (kcal/mol)	$\Delta S^{\neq}$ at 60°C (cal/deg mol)
Me α-D-gluco-	0.708	34.1 ± 1.0	+ 14.8
Me β-D-gluco-	1.26	34.3 ± 0.4	+ 16.5
Me α-D-galacto-	3.55	34.0 ± 0.3	+ 17.7
Me β-D-galacto-	5.13	32.3 ± 0.6	+ 13.3
Me α-D-manno-	2.09	31.9 ± 0.4	+ 10.4
Me α-D-xylo-	2.69	33.5 ± 0.9	+ 15.7
Me β-D-xylo-	5.89	33.6 ± 0.9	+ 17.5
Me 6-deoxy-α-D-galacto-	20.0	33.9 ± 0.6	+ 20.8

In order to elucidate the mechanism of acid-catalyzed hydrolysis of glycopyranosides a detailed knowledge is needed of the breakdown of conjugate acid obtained after protonation of one of the two acetal oxygens, i.e., the molecularity of the rate-determining step, which is the C1–O1 or the C1–O5 bond cleavage depending on whether the glycosidic or the ring oxygen is protonated. The experimental results suggest that the hydrolysis of glycopyranosides proceeds by an A-1 (acid-catalyzed unimolecular) mechanism (Ingold terminology [233]).

The first-order rate velocity coefficients (k_1) were found to be constant for the hydrolysis of D-glucopyranosides in perchloric acid solutions in concentrations ranging from 0.465 to 3.782 M [234]. Plots of $\log k_1$ against the Hammett acidity function, H_0, and against the pH were found to be almost linear in the first and not linear in the second instance, suggesting that analogous to the acid-catalyzed hydrolysis of acetals [235], the hydrolysis of glycopyranosides proceeds by an A-1, and not an A-2, mechanism. However, since the solvent is in large excess over the reactants, both A-1 and A-2 will follow a first-order rate law and consequently other criteria must be used to unequivocally determine the molecularity of the reaction.

The effect of a substituent on the rate of hydrolysis of a glycosidic bond is strongly dependent on its electronegativity and its size. The nature of substituent at the C2 and the C6 carbons of a pyranoside seems to have the most profound effect. Thus, for example, removal of hydroxy group from the C2 carbon dramatically increases the rate of hydrolysis of glycosidic bond. Thus, methyl 2-deoxy-β-D-arabino-hexopyranoside (*273* in Fig. 12.77) is hydrolyzed ca. 2500 times faster

Cleavage of Glycosidic Bonds

than the parent sugar, methyl β-D-glucopyranoside *274*, whereas replacement of the C2 hydroxyl group with a more electronegative group such as chlorine (*275* in Fig. 12.77) reduces the rate of hydrolysis by a factor of 35 [236] compared to the

Fig. 12.77

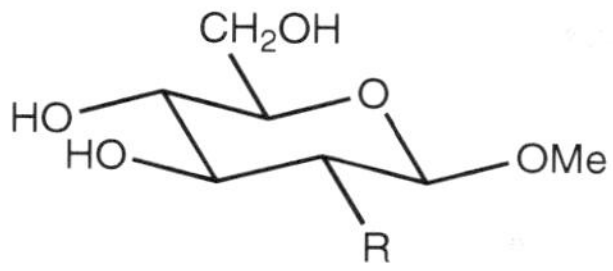

273, R = H, methyl 2-deoxy-β-D-*arabino*-hexopyranoside
274, R = OH, methyl β-D-glucopyranoside
275, R = Cl, methyl 2-chloro-2-deoxy-β-D-glucopyranoside
276, R = NH$_2$, methyl 2-amino-2-deoxy-β-D-glucopyranoside

parent sugar (see Table 12.9). The replacement of the C2 hydroxy group with amino group, which under the reaction conditions becomes positively charged, reduces dramatically the rate of hydrolysis of glycosidic bond (see Table 12.9). From these kinetic studies it can be concluded that the more electron-attracting group attached to the C2 carbon, slower the hydrolysis of glycosidic bond, supporting thus the hypothesis that the hydrolysis of glycopyranosides proceeds via oxocarbenium ion and that its formation is the rate-determining step.

Table 12.9 Rates of hydrolysis in molar acid concentration at 72.9°C for the series of methyl 2-(X-substituted) glucopyranosides for various X substituents [237]

X	Anomer	Acid concentration	t (°C)	Rate of hydrolysis k (s^{-1})	Reference
–H	α	0.10 N HCl	49.7	2.4×10^{-2}	[238]
	β	0.10 N HCl	49.7	3.5×10^{-2}	[238]
–OH	α	2.0 N HCl	71.7	2.5×10^{-5}	[239]
	β	2.0 N HCl	71.1	5.0×10^{-5}	[239]
–NHOCCH$_3$	α	pH, 0.75	78.2	5.1×10^{-6}	[237]
–NHOCCH$_3$	β	1.0 HCl	78.2	4.6×10^3	[237]
–Cl	β	2.0 N HCl	60	3.56×10^{-7}	[236]
–NH$_3^+$	β	1.0 N HCl	100	7.6×10^{-8}	[240, 241]

The influence of C5 substituent on the rate of glycoside hydrolysis was studied by comparing the rate of hydrolysis of methyl α-D-xylopyranoside *277* (Fig. 12.78) with that of methyl 6-deoxy-α-D-glucopyranoside *278*, methyl α-D-glucopyranoside *279* (the C5 substituent is hydroxymethyl group), methyl 6-*O*-methyl-α-D-glucopyranoside *280* (the C5 substituent is methoxymethyl group), methyl α-D-glucopyranosiduronic acid *281* (the C5 substituent is the carboxyl group), methyl 6-chloro-6-deoxy-α-D-glucopyranoside *282* (the C5 substituent is chloromethyl group), methyl 6-deoxy-6-iodo-α-D-glucopyranoside *283* (the C5 substituent is the iodomethyl group), and 6-amino-6-deoxy-α-D-glucopyranoside *284* (the C5 substituent is the aminomethyl group) (Fig. 12.78).

Fig. 12.78

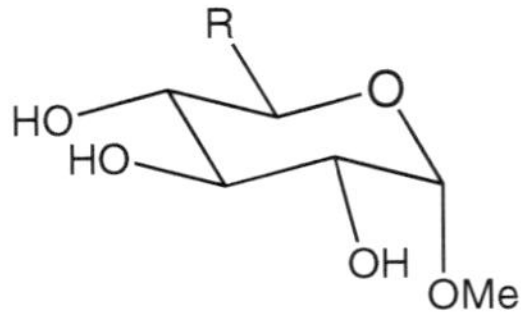

277, R = H, methyl α-D-xylopyranoside
278, R = CH$_3$, methyl 6-deoxy-α-D-glucopyranoside
279, R = CH$_2$OH, methyl α-D-glucopyranoside
280, R = CH$_2$OCH$_3$, 6-O-methyl-α-D-glucopyranoside
281, R = COOH, methyl α-D-glucopyranosiduronic acid
282, R = CH$_2$Cl, methyl 6-chloro-6-deoxy-α-D-glucopyranoside
283, R = CH$_2$I, methyl 6-deoxy-6-iodo-α-D-glucopyranoside
284, R = CH$_2$NH$_2$, methyl 6-amino-6-deoxy-α-D-glucopyranoside

The hydrolysis of alkyl glucuronopyranosides in moderately concentrated acids has been found to proceed at a lower rate than the corresponding parent glycosides. This was attributed to the inductive effect of the electron-attracting carboxyl group (Table 12.10). Support for this explanation comes from the observation that methyl 6-amino-6-deoxy-α-D-glycopyranoside is hydrolyzed more slowly than methyl α-D-glycopyranosiduronide. As one can see methyl group at C5 does not introduce any significant change of rate coefficient; the C5 hydroxymethyl group, methoxymethyl group, and carboxyl group introduce fivefold decrease in rate coefficient; chloromethyl and iodomethyl introduce another sixfold decrease in reaction rate; and finally, 6-aminomethyl introduces another twofold decrease in reaction rate. Therefore, H ≈ CH$_3$ > CH$_2$OH ≈ CH$_2$OMe ≈ COOH > CH$_2$Cl ≈ CH$_2$I > CH$_2$NH$_2$ [242].

Table 12.10 Rate coefficients for the hydrolysis in 0.5 M sulfuric acid of methyl α-D-xylopyranosides and its homologs with different substituents at C5 [231, 242]

R	$k \times 10^{-6}$ (s^{-1})		
	60°C	70°C	80°C
H	3.06	13.9	57.9
CH$_3$	3.22	14.4	61.8
CH$_2$OH	0.637	2.85	12.6
CH$_2$OMe	0.449	1.90	8.52
COOH	0.572	1.93	7.41
CH$_2$Cl	0.092	0.441	1.92
CH$_2$I	0.099	0.445	1.82
CH$_2$NH$_2$	0.065	0.284	1.04

There is practically no difference in the rate of hydrolysis when there is no C5 substituent or if the C5 substituent is methyl group.

Cleavage of Glycosidic Bonds

The removal of a hydroxyl group from various carbon atoms of a glycopyranoside has very different effect upon the rate of glycoside hydrolysis. In general, all mono-deoxy glycopyranosides are hydrolyzed faster than parent sugars, but 2-deoxy glycopyranosidesare hydrolyzed much faster than any other deoxy sugar. Thus, for

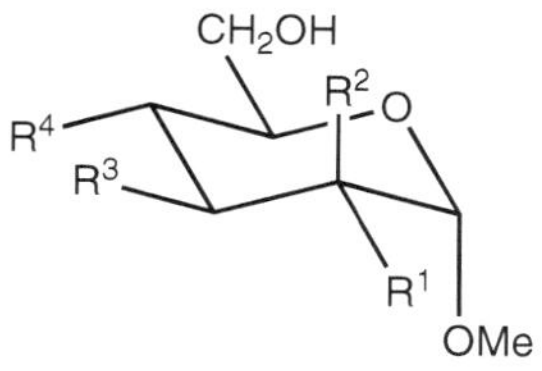

285, $R^1 = R^2 = H$; $R^3 = R^4 = OH$ methyl 2-deoxy-α-D-*arabino*-hexopyranoside
286, $R^1 = R^4 = OH$; $R^2 = R^3 = H$; methyl 3-deoxy-α-D-*ribo*-hexopyuranoside
287, $R^1 = R^3 = H$; $R^2 = R^4 = OH$; methyl 3-deoxy-α-D-*arabino*-hexopyranoside
288, $R^1 = R^3 = OH$; $R^2 = R^4 = H$; methyl 4-deoxy-α-D-*xylo*-hexopyranoside

Fig. 12.79

example, methyl 2-deoxy-α-D-arabino-hexopyranoside *285* is hydrolyzed over 2000 times faster than its parent sugar methyl α-D-glucopyranoside; methyl 3-deoxy-α-D-ribo-hexopyranoside *286* and methyl 3-deoxy-α-D-arabino-hexopyranoside *287* are hydrolyzed only 5 and 7 times faster, respectively, whereas methyl 4-deoxy-α-D-xylo-hexopyranoside *288* is hydrolyzed 40 times faster (Fig. 12.79) (Table 12.11).

Table 12.11 Relative rates of hydrolysis k/k_0 * of methyl deoxy-α-D-pyranosides relative to that of the parent sugar (see Fig. 12.75) (k_0 is the rate constant for the hydrolysis of parent glycoside under the same conditions)

Sugar	k/k_0	Conditions	Reference
2-Deoxy-(*285*)	2090	2.0 N HCl, 58°C	[231]
3-Deoxy-(*286*)	20	2.0 N HCl, 58°C	[231]
	7	1 N H_2SO_4, 100°C	[243]
3-Deoxy-(*287*)	5	1 N H_2SO_4, 100°C	[243]
4-Deoxy-(*288*)	40	2 N H_2SO_4, 58°C	[231]

Alkylation of hydroxyl group of a glycopyranoside, in general, reduces somewhat the rate of hydrolysis of the respective glycopyranoside. Although not very significant, the methylation of the C6 hydroxyl group has the largest influence on the rate of hydrolysis of glycosidic bond (the 6-*O*-methyl ether is hydrolyzed at almost half the rate of that of unsubstituted parent sugar ($k/k_0 = 0.6$); monomethyl ethers at the C2, C3, and C4 carbons of methyl β-D-glucopyranoside are hydrolyzed at 0.86, 0.99, and 0.83 of the rate of unsubstituted methyl-β-D-glucopyranoside) (Table 12.12).

Table 12.12 Rates of hydrolysis in 0.5 M sulfuric acid, at 70°C of monomethyl ethers of methyl-β-D-glucopyranoside [244]

Sugar	$k \times 10^{-6}$ (s^{-1}) 60°C	70°C	80°C	E (kcal/mol)	$\Delta S^{\neq}$ at 60°C (cal/deg mol)
2-O-Methyl	1.19	5.22	20.8	33.4	+12.9
3-O-Methyl	1.27	5.66	23.8	34.0	+41.9
4-O-Methyl	1.15	4.97	21.2	33.8	+41.1
6-O-Methyl	0.84	3.88	16.1	34.9	+16.8
Unsubstituted	1.38	–	–		

Methylation of all hydroxyl groups of a hexopyranoside has a much greater effect on the rate of glycoside hydrolysis. Thus, methyl 2,3,4,6-tetra-O-methyl-α-D-glucopyranoside is hydrolyzed more than 6 times slower than the unsubstituted sugar, whereas the β-anomer is hydrolyzed only 3 times slower. Methyl 2,3,4,6-tetra-O-methyl-α-D-mannopyranoside is hydrolyzed ca. 2.5 times slower than its parent sugar, whereas methyl 2,3,4,6-tetra-O-methyl-α-D-galactopyranoside is hydrolyzed almost 6 times slower than the parent sugar. In general, tetramethylated glycopyranosides are hydrolyzed significantly slower than monomethyl ethers (Table 12.13).

Table 12.13 Rates of hydrolysis in 0.01 N HCl at 95–100° of tetra-O-methyl ethers of α- and β-D-hexopyranosides [245]

Sugar	$k \times 10^5$ min^{-1} ($k \times 10^5$ s^{-1})
Methyl α-D-glucopyranoside	25 (0.42)
Methyl 2,3,4,6-tetra-O-methyl-α-D-glucopyranoside	4 (0.067)
Methyl β-D-glucopyranoside	30 (0.5)
Methyl 2,3,4,6-tetra-O-methyl-β-D-glucopyranoside	10 (0.17)
Methyl α-D-mannopyranoside	10 (0.17)
Methyl 2,3,4,6-tetra-O-methyl-α-D-mannopyranoside	4 (0.067)
Methyl α-D-galactopyranoside	23 (0.38)
Methyl 2,3,4,6-tetra-O-methyl-α-D-galactopyranoside	4 (0.067)

All kinetic studies thus far have strongly supported the hypothesis that hydrolysis of glycopyranosides proceeds via formation of a positively charged oxocarbenium ion and that the cleavage of C1—O1 bond is the rate-determining step in the acid-catalyzed hydrolysis, whereas the reversible protonation of one of the two acetal oxygens and the nucleophilic attack of water molecule on the oxocarbenium ion transition state intermediate are very fast processes.

The study of the rate of hydrolysis of various glycopyranoside derivatives has clearly shown that it is most sensitive to the change of electronegativity at the C2 and C5 carbons since the C2 carbon is vicinal to the C1 carbon and the C5 carbon is vicinal to the O5 ring oxygen atom, the chief players in the formation of the oxocarbenium ion. Thus, the C2 electron-attracting substituent directly inhibits the formation of a positive charge on the C1 carbon, via inductive effect (positively charged C2 carbon will prevent the formation of a positively charged C1 carbon). The C5 electron-attracting substituent reduces the rate of hydrolysis, again via inductive effect. If the hydrolysis proceeds via a cyclic oxocarbenium ion transition state the C5 electron-attracting substituent will reduce the ability of the ring (O5) oxygen to donate its nonbonding electron pair needed to stabilize the carbonium ion intermediate formed after the cleavage of the C1—O1 bond. If the acid-catalyzed hydrolysis of a glycopyranoside proceeds via the C1—O5 bond cleavage, the C5 electron-attracting substituent will lower the basicity of the ring oxygen and thus its ability to be protonated that will reduce the concentration of the reactive conjugate acid that is transformed to the acyclic oxocarbenium ion transition state.

Feather and Harris published in 1965 a paper [19] suggesting that the conversion of a $^4C_1$ or $^1C_4$ glycopyranoside conformation into the ^4H (half-chair) conformation with the C5—O5—C1—C2 atoms lying in one plane (this is presumably the conformation of oxocarbenium ion in the transition state) requires rotation about the C2—C3 and the C5—C4 bonds. Thus they proposed that there is a correlation between the ease of rotations about these bonds and the rate of hydrolysis of glycosidic bonds. The conversion of a chair conformation ($^4C_1$ or $^1C_4$) of a glycopyranoside into a half-chair (^4H or ^1H) conformation requires the counter-clockwise rotation about the C1—O5 bond (if looked along the C1—O5 bond from the direction of the C1 carbon). This rotation is accompanied by a counter-clockwise rotation about the C2—C3 bond if looked along the C2—C3 bond and from the direction of the C2 carbon in which case the substituents at the C2 and C3 carbons assume a new conformation increasing or decreasing the distance between them. Thus the rates of hydrolysis are expected to be influenced by the configurations of the C2 and C3 carbons on a pyranoside ring (Fig. 12.80). Thus counter-clockwise rotation about C2—C3 bond (if looked along the C2—C3 bond and from the direction of the C2 carbon) predicts that the acid-catalyzed hydrolysis of methyl α-D-mannopyranoside should be slower than the hydrolysis of methyl α-D-glucopyranoside since the conversion of their $^4C_1$ conformations to the respective half-chair conformation shall bring the R^2 and R^3 substituents into closer proximity, increasing thus the Pitzer strain. However, the rate coefficients of acid-catalyzed hydrolysis of methyl α-D-manno- and methyl α-D-glucopyranoside (2.0 N HCl, 60°C, Table 10.4) are 2.09×10^{-5} and 0.78×10^{-5}, respectively, i.e., methyl α-D-mannopyranoside is hydrolyzed ca. 2.7 times more rapidly than methyl α-D-glucopyranoside [219], indicating that the stereoelectronic effects are more important than Pitzer strain in the acid-catalyzed hydrolysis of methyl α-D-mannopyranoside (for example, the participation of the axial C2 oxygen in stabilization of the oxocarbenium ion).

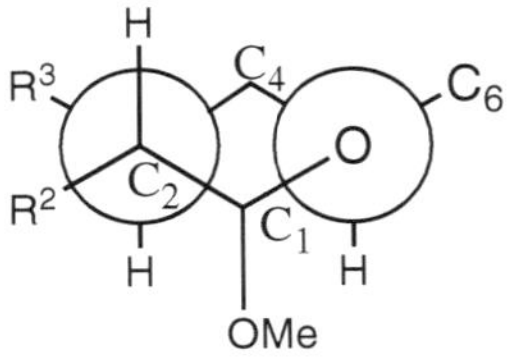

Fig. 12.80

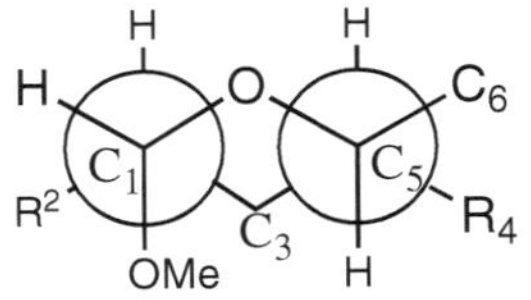

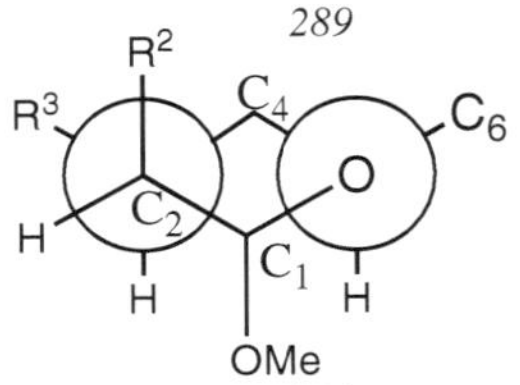

The counter-clockwise rotation about the C5—C4 bond (if looked along the C5—C4 bond from the direction of the C5 carbon) predicts that the acid-catalyzed hydrolysis of methyl α-D-galactopyranoside should be slower than that of methyl α-D-glucopyranoside. This is, however, again contrary to experimental findings. The acid-catalyzed hydrolysis (2.0 N HCl, 60°C, Table 12.14) of methyl α-D-galactopyranoside is found to be ca. 6.6 times faster than that of methyl α-D-glucopyranoside (5.13×10^{-5} and 0.78×10^{-5}) (Table 12.8) [219]. Furthermore, there is no correlation between the size of the C5 substituent and the rate of the acid-catalyzed hydrolysis of a glycopyranoside. Thus, for example, there is very little difference in the rate of acid-catalyzed hydrolysis of methyl α-D-xylopyranoside and methyl 6-deoxy-α-D-glucopyranoside (3.06×10^{-6} and 3.22×10^{-6}, respectively) (Table 12.10) although the difference in the size of C5 substituents is very large (hydrogen vs. methyl group). Also the rates of acid-catalyzed hydrolysis of methyl 6-chloro-6-deoxy- and 6-deoxy-6-iodo-α-D-glucopyranoside are very similar (0.092×10^{-6} and 0.099×10^{-6}, respectively) in spite of the very large difference in size of the C5 substituent ($-CH_2Cl$ vs. $-CH_2I$) (Fig. 12.80).

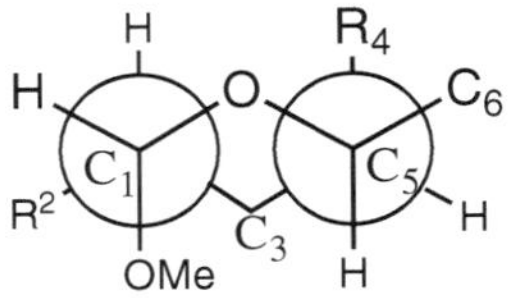

The above experimental results are in full agreement with the conclusions that can be drawn from studying molecular models. Namely, the conversion of a chair conformation of a hexopyranoside to the corresponding half-chair conformation requires counter-clockwise rotation about the C1—O5 bond and about the C2—C3 bond, but not about the C5—C4 bond. The slight increase in Pitzer strain that results from these rotations in the course of conversion of a chair to a half-chair conformation of oxocarbenium ion transition state is not sufficient to significantly influence the rate of acid-catalyzed hydrolysis of a glycopyranoside.

On the other side, the electronegativity of the C2 or the C5 substituent has a profound influence on the rate of acid-catalyzed hydrolysis of glycopyranosides as is shown in Tables 12.6–12.8.

Cleavage of Glycosidic Bonds

Acid-Catalyzed Hydrolysis of Glycofuranosides

Unlike numerous kinetic and mechanistic studies of acid-catalyzed hydrolysis of glycopyranosides [228, 231, 234, 238, 245–251] that led to the conclusion that glycopyranosides are hydrolyzed via an A-1 mechanism [the molecularity of the reaction, the entropy of activation (positive $\Delta S^{\neq}$), dissociation of methanol and the formation of oxocarbenium ion transition state intermediate] (Fig. 12.81), the acid-catalyzed hydrolysis of glycofuranosides has been much less studied [227, 231, 252–255].

A-1 Mechanism (glycopyranosides)

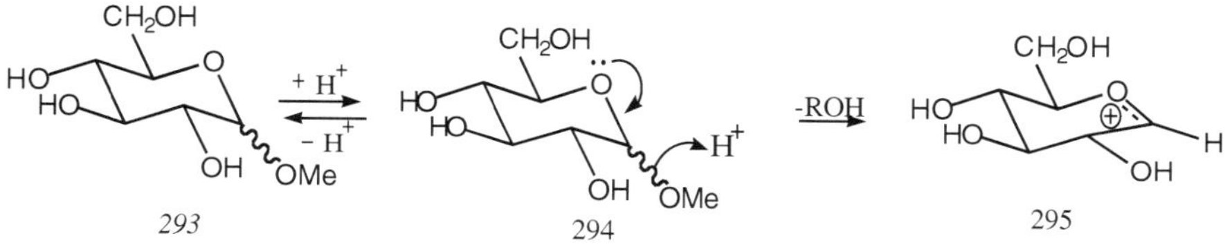

Fig. 12.81

Although there have been a number of kinetic studies of the acid-catalyzed hydrolysis of sucrose (containing a ketofuranoside) [256–259] and of methyl and benzyl fructofuranoside [260], the first systematic kinetic and mechanistic study of acid-catalyzed hydrolysis of glycofuranosides was reported by Capon and Thacker [261] (see Table 12.14).

Table 12.14 The rate coefficients and kinetic parameters for the hydrolysis of select methyl furanosides in 1 M perchloric acid [261]

Methyl furanoside	$t\,(^\circ C)$	$10^5\, k\,(s^{-1})$	E_a (kcal/mol $\pm$ 1)	$\Delta S^{\neq}$ (e.u. $\pm$ 2)
α-D-Xylo-	25.03	39.5	20.2	−8.3
	35.04	120		
β-D-Xylo-	25.01	26.3	20.3	−8.9
	35.04	79.4		
β-L-Arabino-	24.92	4.46	23.1	−2.8
	35.12	16.2		
α-D-Galacto-	25.03	3.35	21	−9.4
	34.91	10.8		
β-D-Galacto-	25.02	0.405	22.8	−8.7
	35.12	1.43		
α-D-Gluco-	24.92	59.7	19.2	−11.0
	35.12	175		
β-D-Gluco-	25.00	21.0	20.5	−9.0
	34.00	64.3		

As can be seen from Table 12.14, the entropies of activation for the hydrolysis of all glycofuranosides studied are negative. This is in strong contrast with the positive values obtained with pyranosides [231, 234, 238, 250, 251] suggesting that glycofuranosides and glycopyranosides react by different mechanism.

The solvent deuterium isotope effect for the hydrolysis of methyl α-D-xylofuranoside in 1 M hydrochloric acid at 25°C, $k_{D_2O}/k_{H_2O} = 2.5$, indicates that the first step in the hydrolysis is a rapid and reversible proton transfer with formation of a conjugate acid which can theoretically be either *296* or *297* (Fig. 12.82).

Fig. 12.82

After this initial protonation, two possible mechanisms can be postulated for the hydrolysis of glycofuranosides that are compatible with the negative entropy of activation: one that proceeds via a cyclic (protonation of the glycosidic oxygen) and the other that proceeds via an acyclic transition state intermediate (protonation of the ring oxygen).

In the first case, the large difference in entropy of activation between the acid-catalyzed hydrolysis of glycopyranosides and glycofuranosides as well as the negative sign is explained by postulating that glycofuranosides are hydrolyzed via an A-2 mechanism [231]. It is namely envisioned that the hydrolysis takes place via the protonation of glycosidic oxygen but without formation of a cyclic oxocarbenium ion transition state intermediate in the next step, as shown in the Fig. 12.83. Instead the protonated methoxy group undergoes nucleophilic displacement with a water molecule, i.e., the transition state of hydrolysis resembles the transition state of an S_N2 displacement. The A-2 mechanism is supported by the Bunnett *w* values (+1.0 to +2.4) falling in the range considered to indicate a mechanism in which water acts as a nucleophile.

An alternative explanation, however, has been proposed for the observed negative entropies of activation in the acid-catalyzed hydrolysis of glycofuranosides [261]. According to this explanation after protonation of the ring oxygen of a furanoside *303* the C1–O4 bond ruptures with the formation of acyclic oxocarbenium ion *305* in the transition state, indicating that the conjugate acid obtained by protonation of a glycofuranoside that leads to the hydrolysis of glycosidic bond is *303* and not *304* (Fig. 12.84).

The formation of the acyclic oxocarbenium ion *305* is supported by the results obtained from the study of acid-catalyzed hydrolysis of a number of 1,3-dioxolanes where it has been shown that the entropies of activation are also negative [262–265] although it is obvious that the hydrolysis must proceed with the ring opening. An

Cleavage of Glycosidic Bonds

A-2 Mechanism (glycofuranosides)

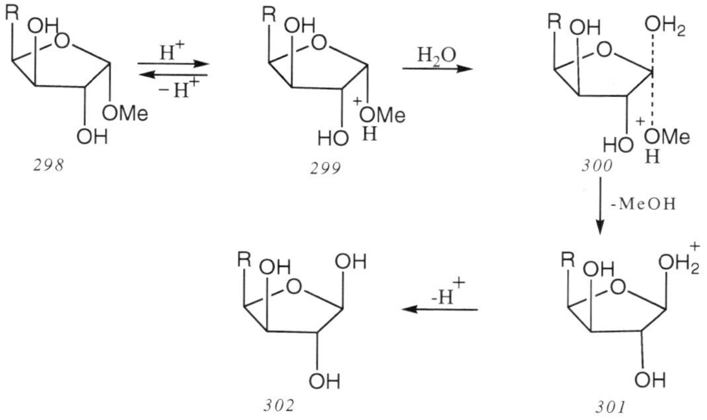

Fig. 12.83

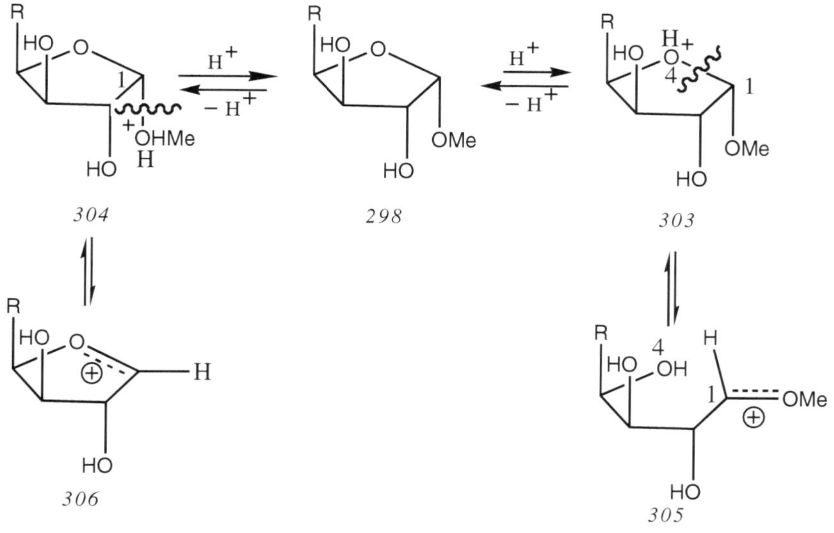

Fig. 12.84

explanation suggested by Capon and Thacker [261] is that the initial rupture of the C–O bond of the conjugate acid *308* is reversible, since the resulting hydroxyl group of *309* is part of the same molecule as the oxocarbenium ion. The reaction could be then written as described in Fig. 12.85.

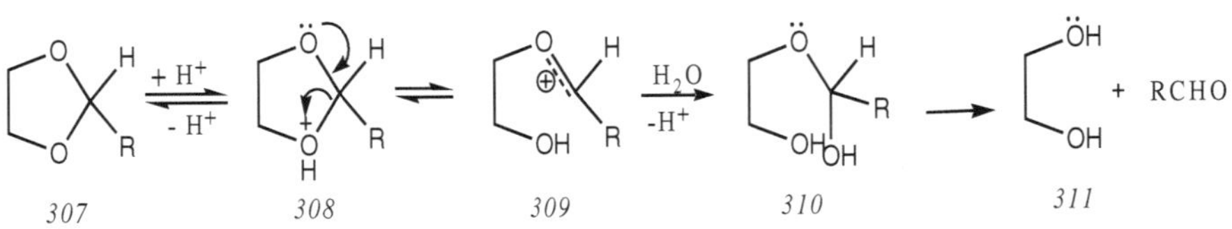

Fig. 12.85

The observed rate constant for the above reaction would then be given by $k_{obs} = k_2 K$, where K is the equilibrium constant (Fig. 12.86):

Fig. 12.86

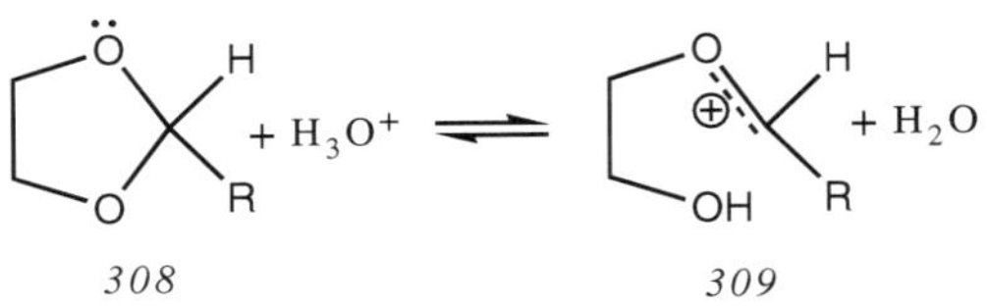

The observed entropy of activation would then be $\Delta S^{\ddagger} = \Delta S^{\circ} + \Delta S_2^{\ddagger}$, where ΔS° is the standard entropy change for the above equilibrium. This would presumably have a positive value. The value of $\Delta S_2^{\ddagger}$ would be, however, strongly negative since it is the entropy of activation for a bimolecular reaction between the oxocarbenium ion *305* and the water molecule to give an oxonium ion *312* (Fig. 12.87). The overall value for $\Delta S^{\ddagger}$ could therefore be negative.

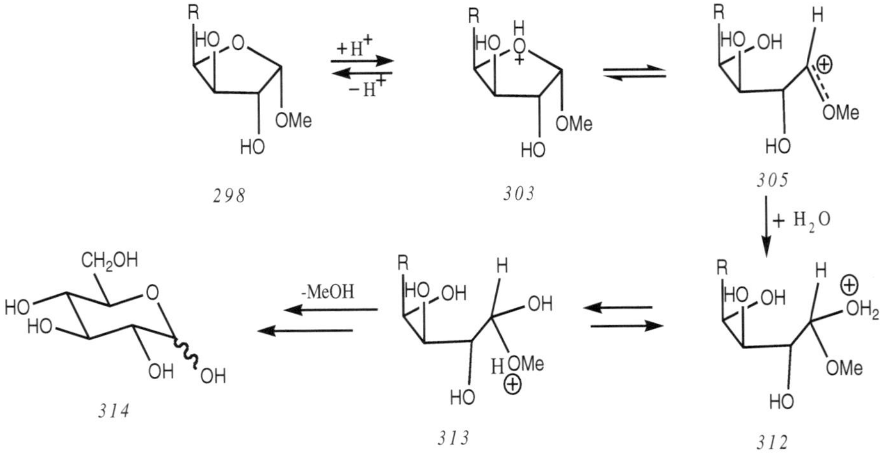

Fig. 12.87

Some Recent Developments Regarding the Mechanism of Glycoside Hydrolysis

In 1980 van Eikeren [266] undertook a study of acid-catalyzed hydrolysis of conformationally rigid methyl acetals *315* and *316* (Fig. 12.88) arguing that the rates of anomer hydrolysis may be affected by the conformation of a glycoside.

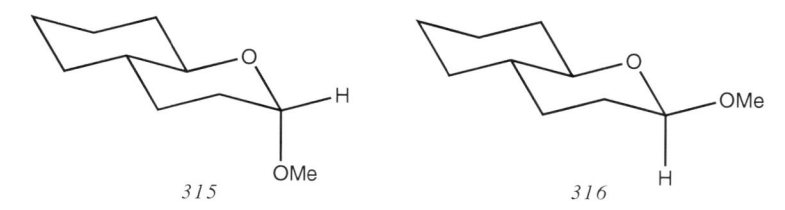

Fig. 12.88

From the composition of equilibrium mixture obtained after acid-catalyzed equilibration of *315* and *316* (68±1% of axial *315* and 32±1% of equatorial *316* anomer) it was calculated that the axial α-anomer *315α* is more stable than β-anomer *316β* by 0.45 kcal/mol which was in full agreement with the concept of anomeric effect. On the other hand, in contrast to the results for alkyl glycosides a comparison of the second-order rate constants showed that the axial anomer hydrolyses 1.51±0.22 times faster than the equatorial anomer indicating that the TS energy of the transition state in the hydrolysis of *315* is lower by 0.25 kcal/mol than the TS energy of the transition state in the hydrolysis of *316*. Thus the difference in energy between the TS of *315α* and *316β* is 0.7 kcal/mol (Fig. 12.89).

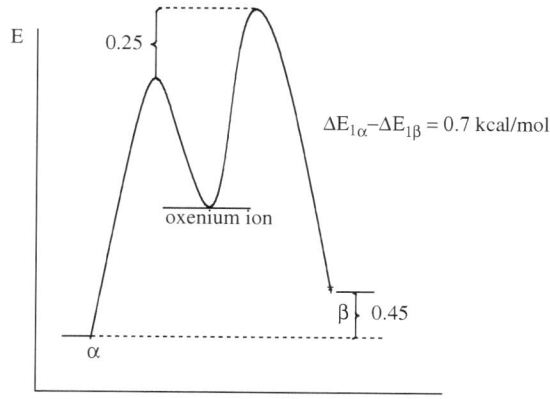

Fig. 12.89

$\Delta E_{1\alpha} - \Delta E_{1\beta} = 0.7$ kcal/mol

The acid-catalyzed hydrolyses of the axial and equatorial anomers *315* and *316* were studied in aqueous HCl–acetone solvent mixtures maintained at constant temperature in water bath. The dependence of k_{obs} on acid concentration and temperature was measured because the conclusions are justified only if the anomers show similar variations in rate with changes in the catalyzing acid and temperature.

Table 12.15 Activation parameters[a] for the hydrolysis of *315* and *316*

$[H^+]^b$ (M)	$\Delta H^{\neq 0}{}_{Ax}$ (kcal/mol)	$\Delta H^{\neq 0}{}_{Eq}$ (kcal/mol)	$\Delta S^{\neq 0}{}_{Ax}$ (cal/mol K)	$\Delta S^{\neq 0}{}_{Eq}$ (cal/mol K)
7.5×10^{-3}	+25.7	+24.6	+17	+13
2.5×10^{-2}	+26.1	+24.6	+21	+16

[a] Calculated from the slope and intercept of $\ln k_2$ vs. (temperature)$^{-1}$; temperature range 20–55°C. Ax = axial and Eq = equatorial.
[b] Aqueous HCl–acetone mixtures (1/1 v/v).

The examination of the activation parameters for the hydrolysis of *315* and *316* (Table 12.15) shows that both anomers exhibit positive enthalpies and entropies of activation as would be expected for dissociative mechanism. The observation that the axial anomer exhibits a larger positive enthalpy and entropy of activation than the equatorial anomer suggests that the rate-determining transition state of the axial anomer involves more extensive C—O bond cleavage.

Thus, van Eikeren [266] concludes that the axial and equatorial anomers *315* and *316* must hydrolyze via different transition states and that the difference in their hydrolysis rates may be explained by postulating an early transition state for the equatorial anomer with little C—O bond breakage and a late transition state for the axial anomer with more extensive C—O bond breakage.

In order to shed some new light on the mechanism of acid-catalyzed hydrolysis of α- and β-D-glucopyranosides Deslongchamps et al. [267] carried out molecular modeling study of the various endocyclic and exocyclic cleavage pathways of tetrahydropyranyl acetals *317α* and *317β* (R = H in Fig. 12.90) during acid-catalyzed hydrolysis and then compared the reached theoretical conclusions with the experimental results obtained by using the conformationally rigid bicyclic tetrahydropyranyl acetals *315* and *316* [R = (CH$_2$)$_4$ in Fig. 12.90]. The reason for selecting the bicyclic conformationally rigid tetrahydropyranyl acetals for their experimental study was again to limit the effect of conformational change of a tetrahydropyranyl acetal upon the rate of its acid-catalyzed hydrolysis (see Eikeren [266]).

Deslongchamps et al. [267] have calculated the energies of the four possible transition structures *318α*, *318β*, *319α*, *319β* and two intermediates *320* and *321* in the hydrolysis of the glycoside models *317α* and *317β* (Fig. 12.90).

For the hydrolysis of acetal *317α* → *322*, both the exocyclic (via *318α*) and the endocyclic (via *319α*) C—O bond cleavages take place via a chair-like transition state structure conformation with stereoelectronic assistance of one electron lone pair that is antiperiplanar to the leaving group. Molecular modeling indicates that the free energy of transition structure *318α* for the exocyclic mechanism is 1.72 kcal/mol lower than that of the endocyclic mechanism (*319α*). In addition, an entropy effect also favors the exocyclic mechanism (formation of two molecules: the

Cleavage of Glycosidic Bonds

oxocarbenium ion *320* and methanol). It can therefore be expected that the transition structure for the exocyclic cleavage process for α-D-glycopyranosides will be highly favored which is in agreement with published results [268–272].

For the hydrolysis of *317β* →*322*, calculations indicate that the transition structure *318β* for the exocyclic C—O bond cleavage takes place via a sofa conformation with an endocyclic oxygen lone pair periplanar to the C—O bond (*syn* or *anti*) to be cleaved [273, 274].

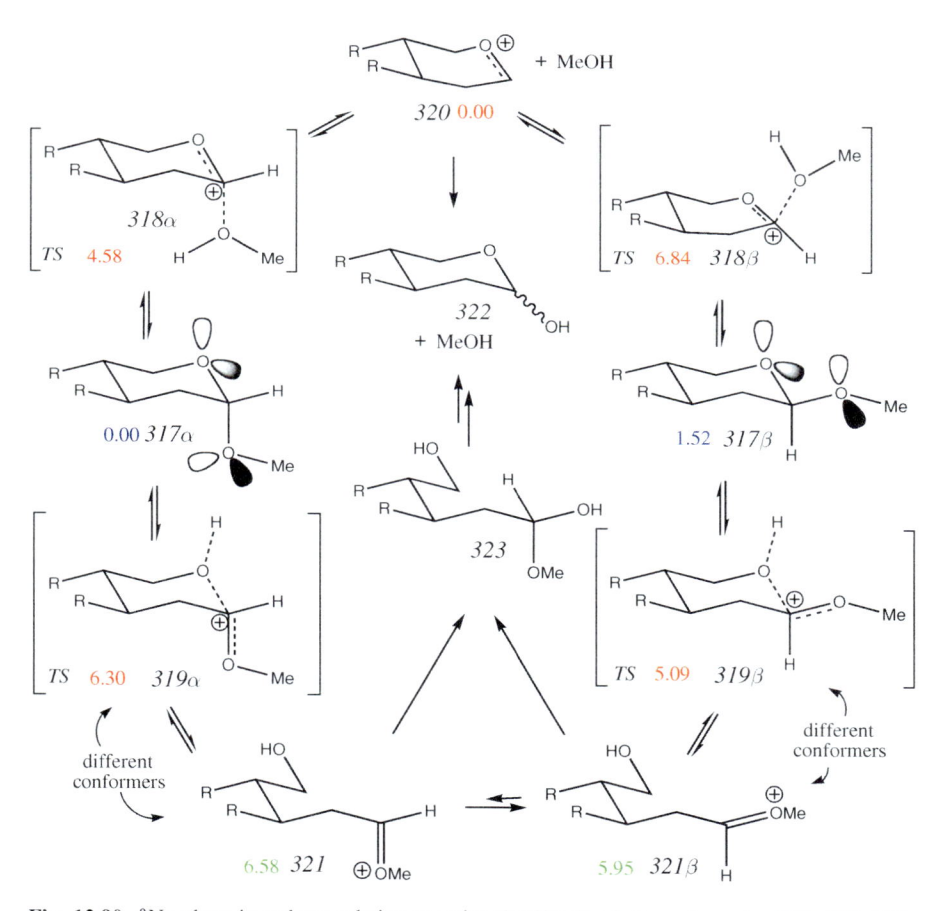

Fig. 12.90 [a]Numbers in red are relative energies (kcal/mol) of charged species 317, 320, 319, 321α, and 321β; numbers in blue are relative energies of acetals 317α and 317β, R = H (calculations), R = (CH2)4 (experimental). Numbers in green are for cations 321α and 321β in extended conformations (no interaction between cation and alcohol).

For the endocyclic C—O bond cleavage, calculations show that the transition structure geometry *319β* remains close to the chair ground-state conformation. This is the result of the participation of the exocyclic oxygen lone pair antiperiplanar to the leaving group. The enthalpy difference between the two transition state structures *318β* and *319β* is 1.75 kcal/mol, now favoring the endocyclic C—O bond cleavage. On the other hand, entropy disfavors the opening of a ring over the exocyclic C—O bond cleavage, which leads to the formation of two molecules. Since the enthalpy favors *317β→319β* process and the entropy the *317β→318β*, both processes are likely to take place concurrently which is in accord with published experimental observations [275].

The results of calculations are in full agreement with the fact that the relative rate of hydrolysis of the α-anomer in a conformationally rigid model is faster than that of the β-anomer (rate ratio 3/2) [267, 276]. The transition structure *318α* for the exocyclic C—O bond cleavage has a lower energy (4.58 kcal/mol) than the β-anomer, *318β* (6.84 kcal/mol); it has also slightly lower energy than the other competing endocyclic C—O bond cleavage of the β-isomer (*319β*, 5.09 kcal/mol). The calculations show also that there is a small energy difference (0.63 kcal/mol) between the conformers *321β* and *321α* of the corresponding oxocarbenium ion *321*. The relative energy difference of these ions increased in the corresponding transition structures *319β* and *319α*, respectively (1.21 kcal/mol favoring *319β*). During the endocyclic hydrolysis of *317β*, the hydroxy-oxocarbenium ion *321β* could undergo a rotation and recyclize via conformer *321α*, to give the more stable anomer *317α*. However, experimental results show that the isomerization of the β-anomer into the α-anomer does not take place concurrently with hydrolysis [277]. This suggests that either the exocyclic cleavage via *318β* is much more favored entropically than the endocyclic cleavage via *319β* or the recyclization barrier is too high.

The experimental and theoretical studies of acid-catalyzed hydrolysis of various conformationally rigid acetal models [266–268] such as *315* and *316* (Fig. 12.89) have shown that it takes place via late transition state.

From this study Deslongchamps et al. [267] concluded that the α-glycosides undergo hydrolysis in their ground-state chair-like conformation via an exocyclic C—O bond cleavage while following the principle of kinetic stereoelectronic control (proper orbital alignment). β-Glycosides can be, however, hydrolyzed either by an exocyclic C—O bond cleavage via distorted twist-boat or sofa conformation or by an endocyclic C—O bond cleavage in the ground-state chair-like conformation. While van Eikeren [266] suggested an early transition state for the cleavage of equatorial anomer with little C-O bond breakage and a late transition state for the cleavage of axial anomer with more extensive C-O bond breakage, Deslongchamps et al. [267] *suggested that both cleavages take place via late transition states and with stereoelectronic control* whereby the cleavage of equatorial anomer takes place somewhat earlier. The exocyclic cleavage is favored by entropy and the endocyclic cleavage might be disfavored because the resulting hydroxy-oxocarbenium ion (like *321β*) might undergo a fast cyclization to give back the β-glycoside rather than undergoing a reaction with water to produce the hydrolysis product. On that basis, the hydrolysis of β-glycosides can take place via both the exocyclic and endocyclic pathways [278], the choice depending on the specific structure of the substrate and on the reaction conditions (acid-catalyzed or enzymatic [279]).

In hydrolyses that are carried out in water or in solvents containing water, it is now generally accepted [280] that an oxocarbenium ion is too reactive to have a real life time in the presence of a nucleophile such as water [281, 282].

Consequently a glycopyranosidic bond cleavage very likely proceeds via a transient [283] *oxocarbenium-like transition state*, with an sp^2-hybridized geometry at both the C1 and O5 atoms, allowing thus a considerable double bond character between the O5 and the C1 atoms and forcing the C5, O5, C1, and C2 atoms to assume a coplanar conformation. Hence this process does not involve a discrete carbocation in a first-order reaction, but is borderline S_N1-S_N2 reaction. In other words, the C1–OMe bond breaking is taking place simultaneously with the C1–OH$_2$ bond making (S_N2-like) (Fig. 12.91).

324, α-D-glycopyranoside TS (S_N2)   *325*, β-D-glycopyranoside TS (S_N2)

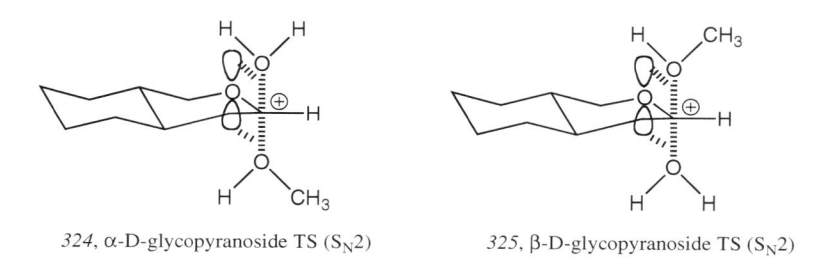

Fig. 12.91

In late transition state, both α- and β-D-glycopyranosides have essentially the same oxocarbenium ion with a CH$_3$OH group at a long distance (≥ 1.80 Å). In this way, the CH$_3$OH in α- or β-transition state will have small steric interactions with the oxocarbenium ion. As a result, it is not surprising that ΔE between TSα and TSβ is only about 0.7 kcal/mol (see Fig. 12.91).

The above rationalization is confirmed experimentally by the addition of methanol in mild acid on enol ether *326* (Fig. 12.92). Under these kinetically controlled conditions a mixture of *317α* and *317β* was obtained in the ratio 76:24. This ratio corresponds to the energy difference of 0.70 kcal/mol for the transition state favoring the formation of *317α* which is in complete agreement with the AM1 [267] and 6.31G [268] calculations. This value is in agreement with that found by van Eikeren [266]. In addition, these calculations also show that the transition states are definitely late transition states and resemble the geometry of the oxocarbenium ion.

The primary ^{13}C [284, 285] and secondary α-deuterium [286] isotope effects were consistent with this S_N2-type itinerary. The conformation of oxocarbenium ion requires the coplanarity of the C5, O5, C1, and C2 atoms of pyranosidic ring and there are four possible conformations of transition state structure in which the C5, O5, C1, and C2 will be coplanar (Fig. 12.93): (1) the $^4H_3$, (2) the $^3H_4$, (3) $B_{2,5}$, and (4) the $^{2,5}B$ conformer. The route from reactant via transition state TS into the product is called the substitution pathway.

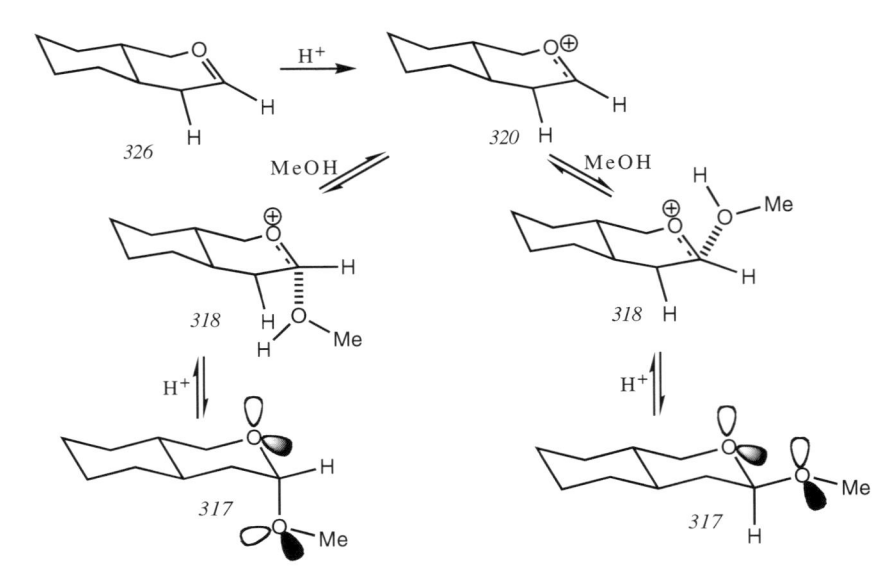

Fig. 12.92

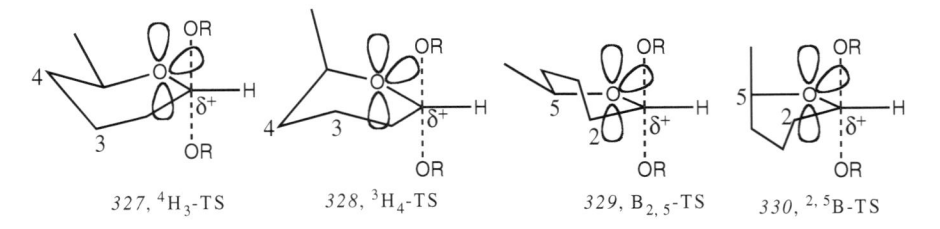

Fig. 12.93

327, $^4H_3$-TS   328, $^3H_4$-TS   329, $B_{2,5}$-TS   330, $^{2,5}B$-TS

By examining these TS geometries, Nerinckx et al. [280] have recently suggested explanation as to why there is such small difference in energy between these transition states and also as to why the rates of acid-catalyzed hydrolysis of α- and β-D-glycopyranosides are so close.

According to Antiperiplanar Lone Pair Hypothesis (ALPH) pathway the substitution of β-equatorial glycopyranosides is preceded by a conformational change from ground-state chair to a skew conformation in which the leaving group is in the axial orientation and in antiperiplanar orientation with regard to the *trans* lone electron pair of the ring oxygen. This sp^3 lone electron pair will hybridize at the TS into $2p_z$ orbital and thus allow the formation of the partial double bond toward the anomeric carbon [268, 270, 287]. The reaction then proceeds through an ALPH-compliant β-skew→$^4H_3$-TS→α-$^4C_1$ pathway (Fig. 12.94).

In the case of α-axial D-glycopyranosidic bond substitutions, the leaving group already has an ALPH-compliant orientation when the carbohydrate ring is in

Cleavage of Glycosidic Bonds

Fig. 12.94

331, β-$^4C_1^H$ *332*, β-$^1S_3$ (β-skew) *333*, $^4H_3$-TS *334*, α-$^4C_1^{Nu}$

the ground-state chair conformation. Thus the α-glycosides must hydrolyze via their ground-state conformation, as explicitly stated by Deslongchamps [268] (Fig. 12.95).

335, β-$^4C_1^L$ *336*, $^4H_3$-TS *337*, α-$^4C_1^H$

Fig. 12.95

Thus the energies of transition state intermediates for the hydrolysis of α- and β-D-glycopyranosides *338* and *339* must be very close, as shown in Fig. 12.96.

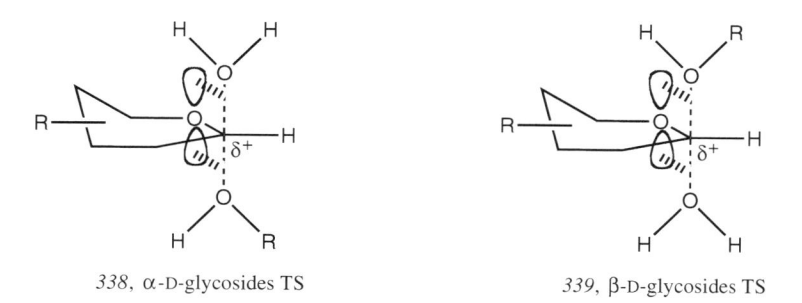

338, α-D-glycosides TS *339*, β-D-glycosides TS

Fig. 12.96

It was established that both transition states are late and have essentially the same oxocarbenium ion in which the MeOH is at a long distance from the C1 carbon (≥ 1.80 Å). What happens is while the MeOH is leaving, the H_2O is entering (S_N2-like reaction). But one should realize that TS energy is lowered by the fact that at TS, there is a *p*-orbital on O5 which assists this S_N2 reaction. Thus the anomeric effect still plays the same key role. In that case, the competing TS resemble *338* and *339* (Fig. 12.96) which again should have similar energy.

Acetolysis of Glycosides

The cleavage of a glycosidic bond by acetolysis is an alternative method to hydrolysis. Although both methods are acid catalyzed and presumably in case of glycopyranosides involve the formation of a cyclic oxocarbenium transition state they also have their differences. Thus, for example, the most important difference is that the hydrolysis is always performed in either aqueous solutions or in a solvent containing water, whereas the acetolysis is performed in nonaqueous solvents, typically acetic anhydride. In the case of acid-catalyzed hydrolysis the activation of glycosidic bond is effected by protonation of one of the two acetal oxygens (glycosidic or the ring oxygen), whereas in acetolysis the attacking species is not (H^+) but most likely the acetylium ion (Ac^+) [288–291]. Acetolysis can also be catalyzed by Lewis acids, such as ferric chloride ($FeCl_3$) [292, 293].

A review on acetolysis has been published by Guthrie and McCarthy [294].

The study of the mechanism of acid-catalyzed hydrolysis of glycosidic bonds failed to answer two very important questions: (1) Why are the β-anomers of D-glycopyranosides (having the glycosidic oxygen equatorially oriented) hydrolyzed ca. 2–3 times more rapidly than α-anomers (having the glycosidic oxygen oriented axially) irrespective of the glycopyranoside structure? (2) Why the configurations of hydroxyl groups of the pyranoid ring (e.g., D-gluco-, D-galacto-) have no or very little influence upon the rates of hydrolysis of their glycosidic bonds? The postulated mechanism also contradicts the importance of relative basicities of ring and glycosidic oxygen in a glycopyranoside upon the rate of hydrolysis of the corresponding glycoside, and thus challenges the well-documented concept of anomeric effect. The one explanation for this "anomaly" could be that perhaps the electronic effects that do exist in all glycopyranoside structures are significantly "neutralized" in aqueous solution or in solvents containing water, due to hydrogen bonding between the sugar polar groups (hydroxyl groups and ring oxygen) and water molecules. This assumption is supported by the fact that the magnitude of anomeric effect is solvent dependent [295–299]. Another possible explanation could be that the energies of transition states for the hydrolysis of both anomers are close but the ground-state energies are different due to the anomeric effect whereby the α-anomer is more stable than the β-anomer. Since both transition states are late the β-anomer should reach the transition state easier and thus sooner than α-anomer, because its ground-state energy is higher than the ground-state energy of the α-anomer (the α-anomer being more stable will reach the transition state slightly later than the β-anomer).

The observed behavior of glycosidic bonds toward acid-catalyzed hydrolysis is contrary to what one would expect from the existence of anomeric effect and ALPH. Namely, it is known that due to the anomeric effect the axially oriented oxygen should have a higher basicity than the equatorial one, due to the mixing of axially oriented nonbonding electron pair of the ring oxygen with the antibonding orbital of the C1–O1 bond. Consequently, the concentration of conjugate acid resulting from protonation of the glycosidic oxygen should be higher in solutions of glycopyranosides having the glycosidic oxygen axially oriented (α-D-anomers) than in solution of glycopyranosides having the glycosidic oxygen equatorially

Cleavage of Glycosidic Bonds

oriented (β-D-anomers). As a result, α-D-glycopyranosides should be hydrolyzed more rapidly than β-D-glycopyranosides, which is opposite to what is observed.

The mechanism of acid-catalyzed cleavage of glycosidic bonds could perhaps be better understood by studying the mechanism of acetolysis since it is performed in the absence of water (most often in acetic anhydride) and thus the electronic effects that exist in each glycopyranoside structure will not be "neutralized" by solvent and their influence upon the reactivity of glycosidic bond could be hopefully evaluated.

Dasgupta et al. [293] have studied the acetolysis of methyl α-D-glucopyranoside *340*, methyl 6-*O*-*p*-toluenesulfonyl-α-D-glucopyranoside *341*, and methyl α-D-galactopyranoside *342* in acetic anhydride at 60°C using ferric chloride as cat-

340, R = H
341, R = Ts

342

Fig. 12.97

alyst (Fig. 12.97). Methyl α-D-glucopyranoside *340* gave only two products, α- and β-penta-*O*-acetyl-D-glycopyranoses, whereas the acetolysis of methyl α-D-galactopyranoside *342* gave, under the same reaction conditions, five products of which the two major products were α- and β-penta-*O*-acetyl-D-galactopyranoses (*343* and *344*), the next two products were α- and β-D-galacto-furanose pentaacetates (*345* and *346*), and the last product was the acyclic hepta-*O*-acetyl-aldehydo-D-galactose (*347*) (Fig. 12.98). Furthermore, they observed that the acetolysis of methyl 6-*O*-*p*-toluenesulfonyl-α-D-glucopyranoside *341* was, under the same reaction conditions, ca. 4 times slower than the acetolysis of *340*. The authors suggested two reaction mechanisms: one proceeding via formation of a cyclic oxocarbenium

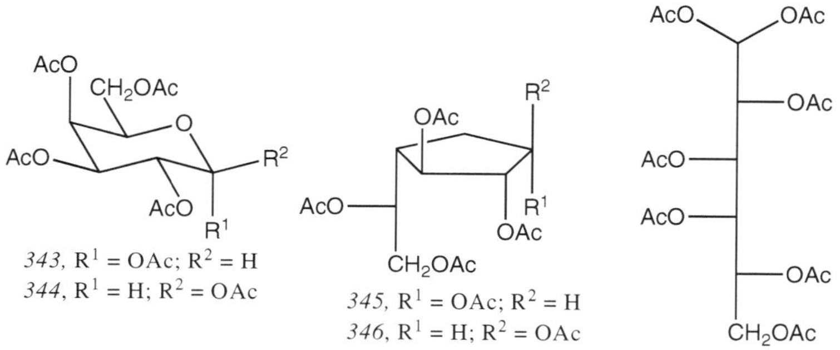

343, R^1 = OAc; R^2 = H
344, R^1 = H; R^2 = OAc

345, R^1 = OAc; R^2 = H
346, R^1 = H; R^2 = OAc

347

Fig. 12.98

ion (the initial attack of the acetylium ion taking place at the glycosidic oxygen) and the other via the formation of an acyclic oxocarbenium ion (the initial attack of the acetylium ion taking place at the Ring oxygen). Both of these pathways are actually unsubstantiated.

With a slightly modified experimental procedure McPhail et al. [300] repeated the study of Dasgupta et al. [293] on the acetolysis of methyl α- and β-D-glucopyranosides. The discussion of reaction mechanism of acetolysis based on their experiments is described below.

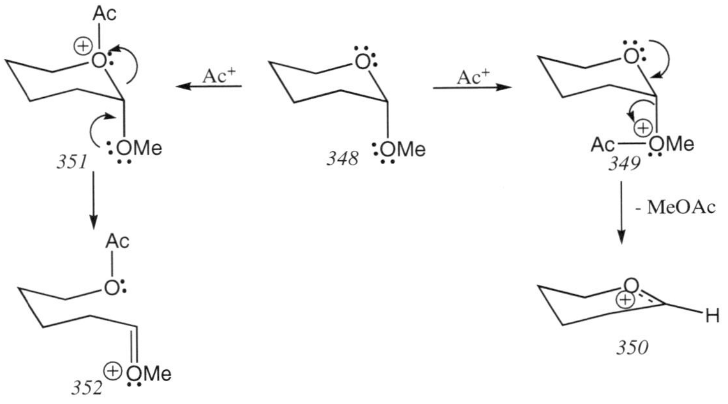

Fig. 12.99

The site of anomeric activation in glycoside cleavage has been a subject of long-standing controversy [301–306]. Early experiments were supporting the view that the activation occurs at the glycosidic oxygen (*352*), leading to the formation of cyclic oxocarbenium ion *350*, rather than at the ring oxygen (*351*) giving the acyclic counterpart *352* (Fig. 12.99).

The question of activation site is directly related to the question of relative basicities of glycosidic and ring oxygens, which is, in turn, related to anomeric effect [307, 310]. MO rationalization of anomeric effect invokes the donation of axially oriented nonbonding pair of ring oxygen to the antibonding orbital of the C1—O1 bond (n⇔σ∗ donation) (antiperiplanar orientation of these two entities) making thus the glycosidic oxygen more basic than the ring oxygen and hence the preferred site for the attack of an electrophile (Ac$^+$) [308]. An ab initio study of dimethoxymethane has provided support for this postulate by determining the proton affinities for oxygens in a *gauche* and in an *anti* orientation of methyl group and one oxygen, as shown in Fig. 12.101. As indicated by broken lines, these rotamers correspond to axial and equatorial glycosides, respectively, and as observed by Lemieux [309] the nσ∗ donations in *353* are in competition. Accordingly, Praly and Lemieux [310] found that for β-D-glycosides (*354*) (Fig. 12.100) the *exo*-anomeric effect was stronger than in α-D-glycosides. In view of these differences in oxygen basicities, a β-D-glycoside might be expected to be activated on the ring oxygen and react via

Cleavage of Glycosidic Bonds

formation of an acyclic oxocarbenium ion *352*, whereas an α-D-glycoside would be expected to be activated at both oxygens (glycosidic and the ring oxygen) and consequently react by formation of either a cyclic *350* or an acyclic oxocarbenium ion *352* (Fig. 12.99).

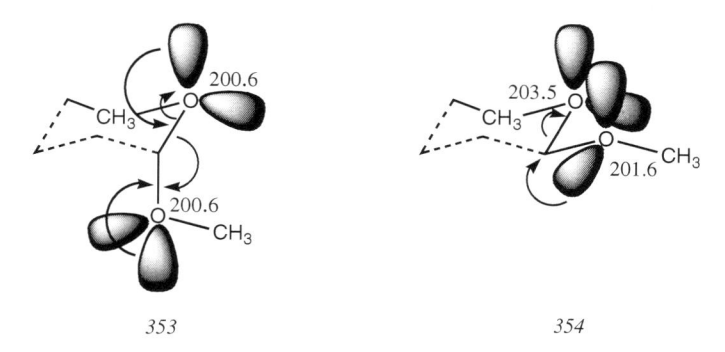

353 *354*

arrows indicate nσ* interactions
numbers represent calculated proton affinities (in kcal mol^{-1})

Fig. 12.100

355 *356*

Fig. 12.101

Their conclusion is that α- and β-D-glucopyranosides react through different mechanisms. From the fact that the ratio of the α- and β-D-glucopyranosyl acetates is 4:1 in both cases, they concluded that the cyclic oxocarbenium ion is produced from both anomers which is trapped by acetate anion. The acyclic oxocarbenium ion is responsible for the formation of both acyclic heptaacetate and the penta-*O*-acetyl furanose derivatives (Fig. 12.101).

In their study of acetolysis of methyl α-D-glucopyranoside using modified acetolysis medium (acetic anhydride, ferric chloride, and a small amount of concentrated sulfuric acid) McPhail et al. [300] found (by using gas chromatography and ^{1}NMR spectroscopy) that peracetylated α- and β-D-glucopyranoses were obtained with 73 and 18% yield, the mixture of α- and β-D-glucofuranose pentaacetates was obtained with 8% yield, and the acyclic D-glucose heptaacetate was obtained only in traces. For the β-D-glucopyranoside the same four products were obtained but in different amounts: peracetylated α- and β-D-glucopyranoses were obtained with 19 and 5% yield, the mixture of α- and β-D-glucofuranose pentaacetates with 48% yield, and the acyclic D-glucose heptaacetate with 23% yield.

From this they concluded that contrary to what might be expected on the sole consideration of oxygen basicities, it is the β- and not the α-anomer that gives rise to both cyclic and acyclic oxocarbenium ions and that therefore there are factors other than basicities of two acetal oxygens in glycosides that determine the course of glycosidic bond cleavage.

The major objection to this study is the choice of methyl α- and β-D-glucopyranosides as substrates because the acetylation is faster than acetolysis and hence their model compounds were actually the peracetylated methyl α- and β-D-glucopyranosides. The presence of acetate at the C2 carbon will enormously complicate the acetolysis reaction pathway since the acetate is known to be an excellent participating group capable of altering the reaction pathway of activated glycosides as shown in Fig. 12.102. Thus, the formation of both cyclic and acyclic oxocarbenium ions by the activation of β-glycoside via the attack of acetylium ion on either the ring or the glycosidic oxygen will most likely be assisted by the C2 equatorial acetate via neighboring group participation. *Trans*-diaxial orientation of acetylium-activated ring oxygen and the carbonyl oxygen of C2 acetate would favor the opening of the pyranoside ring with the formation of *361* over the displacement of acetylium-activated equatorially oriented glycosidic oxygen by the C2 acetate (*357→358*). Therefore it can be expected that the β-glucopyranoside would prefer a pathway that would involve the formation of acyclic oxocarbenium ion transition state.

In the case of α-D-glucopyranoside the displacement of acetylium-activated ring oxygen or the acetylium-activated glycosidic oxygen by C2 acetate via neighboring group participation is not possible due to stereochemical reasons. So the only role of the C2 acetate in the acetolysis of α-D-glucopyranosides would be the stabilization of cyclic oxocarbenium ion.

The results reported in this study are consistent with this interpretation: the predominant products of acetolysis of methyl α-D-glucopyranoside is 4:1 mixture of penta-*O*-acetyl-α- and β-D-glucopyranoses (91%) with only 8% of penta-*O*-acetyl-D-glucofuranose and traces of acyclic heptaacetate. The acetolysis of methyl β-D-glucopyranoside gave a 4:1 mixture of only 24% of penta-*O*-acetyl-α- and β-D-glucopyranoses, whereas 48 and 23% of penta-*O*-acetyl-D-glucofuranose and the acyclic heptaacetate were obtained, respectively. The 4:1 ratio of penta-*O*-acetyl-α- and β-D-glucopyranoses obtained by acetolysis of both methyl α- and β-D-glycopyranosides is due to the anomerization of the reaction mixture after the acetolysis since the authors modified the $FeCl_3$–Ac_2O original reagent of Dasgupta [293] by adding a small amount of sulfuric acid to speed up the reaction.

In 1983 Miljkovic et al. [311, 312] studied the acetolysis of permethylated methyl α- and β-D-gluco-(*362* and *363*, respectively), methyl α- and β-D-galacto-(*364* and *365*, respectively), and methyl α- and β-D-mannopyranosides (*366* and *367*, respectively) (Fig. 12.103) in acetanhydride solution at 75°C containing

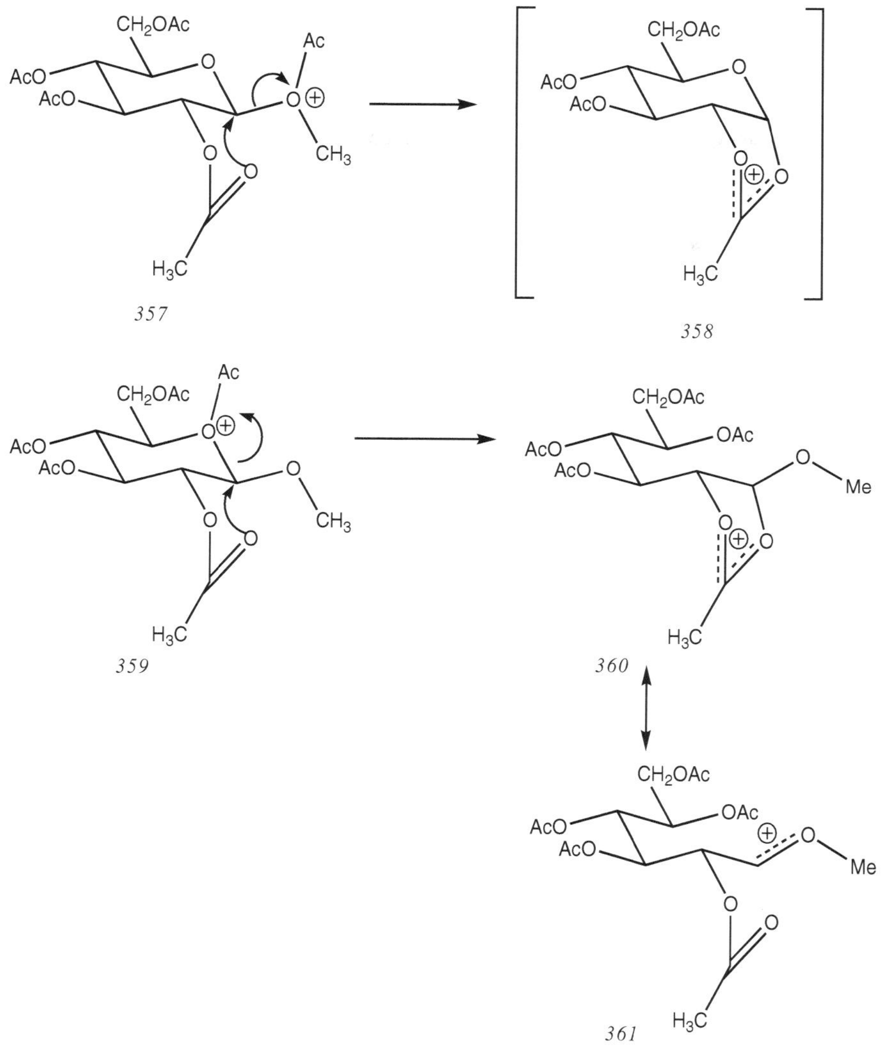

Cleavage of Glycosidic Bonds 401

Fig. 12.102

3.33% of methanesulfonic acid. Since the C2 substituent in these glycopyranosides was a nonparticipating methoxy group the acetolyses of permethylated methyl glycopyranosides of glucose, galactose, and mannose were very clean reactions, giving, in addition to starting material, a mixture of α- and β-1-acetates as the only products. In all kinetic measurements the progress of acetolysis was monitored by a change of relative concentration of starting material, using HPLC and C18 column. The kinetic data are presented in Table 12.16.

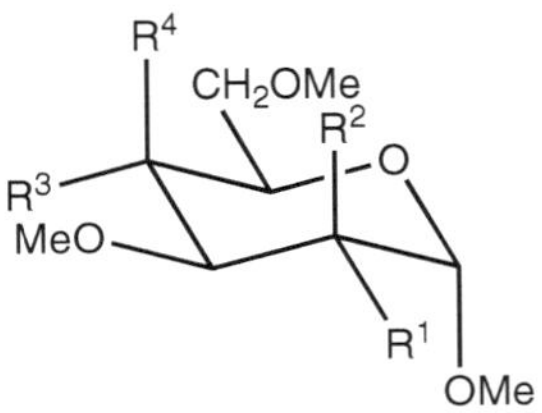

362, R¹ = R³ = OMe; R² = R⁴ = H
364, R¹ = R⁴ = OMe; R² = R³ = H
366, R¹ = R⁴ = H; R² = R³ = OMe

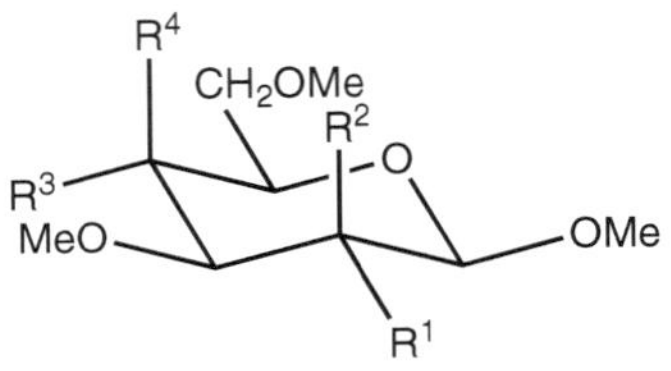

363, R¹ = R³ = OMe; R² = R⁴ = H
365, R¹ = R⁴ = OMe; R² = R³ = H
367, R¹ = R⁴ = H; R² = R³ = OMe

Fig. 12.103

Table 12.16 Kinetic data for acetolysis of permethylated methyl glycopyranosides of D-glycose, D-galactose, and D-mannose (*362–367*) with acetic anhydride–methanesulfonic acid (30:1 v/v) at 75°C

Sugar	$10^3 k_1$ (s^{-1})	$10^3 \times$ standard deviation	α/β ratio of 1-acetates
α-D-Gluco	1.87	0.0734	3.17
β-D-Gluco	0.12	0.00462	3.36
α-D-Galacto	37.10	1.43	2.97
β-D-Galacto	0.84	0.0709	3.22
α-D-Manno	1.08	0.0456	Only α-acetate
β-D-Manno	3.06	0.0631	Only α-acetate

From Table 12.16 it can be seen that the α-anomers of permethylated methyl D-gluco- and D-galactopyranosides are acetolyzed considerably faster than the corresponding β-anomers (15.58 and 44.17 times, respectively) which is in contradiction to the results obtained for acid-catalyzed hydrolysis of these two compounds where the β-anomer is hydrolyzed ca 1.9 times more rapidly than the α-anomer. The second observation is that permethylated methyl α- and β-D-galactopyranosides *364* and *365* are acetolyzed much faster than permethylated methyl α- and β-D-glucopyranosides *362* and *363* (α-gal/α-glc ≈ 20, and β-gal/β-glc ≈ 7). The permethylated methyl β-D-mannopyranoside *367* is, however, acetolyzed 2.83 times faster than the α-anomer *366* (Fig. 12.103).

It is reasonable to assume that stereoelectronic interactions that are characteristic for a given alkyl- or aryl glycopyranoside structure must play an important role in determining the overall chemical behavior of its glycosidic bond. In protic and polar solvents (e.g., water) these electronic interactions must be neutralized by intermolecular interactions with solvent dipoles (solvation), whereas in aprotic solvents particularly those having relatively low dielectric constant (e.g., acetic anhydride $\varepsilon = 20.7$, as opposed to water $\varepsilon = 84.2$) the intramolecular electronic interactions must be fully operative and could be expected to influence the chemical behavior of the anomeric carbon of a glycopyranoside. The basicity of the glycosidic oxygen, which is directly related to the anomeric effect, in nonpolar solvents with low dielectric constant can be expected to be higher than in polar protic solvents [312–314] and consequently the difference in relative basicities between the ring and glycosidic oxygens could be expected to be considerably larger. Thus the attack of acetylium cation could be expected to occur at the glycosidic rather than at the ring oxygen of a glycopyranoside. This activation will then be followed by formation of the oxo-carbenium ion that will, in the presence of acetate anion, give a mixture of α- and β-1-acetates. Due to the presence of methanesulfonic acid in acetolyzing solution, the anomerization of 1-acetates will take place until the equilibrium is reached (α:β ratio ≈ 3:1, except for the mannose). This rationale is supported by findings that an acetic anhydride solution of D-glucose pentaacetate contains, after equilibration with sulfuric acid, approximately 87% of the α- and 13% of the β-anomer (α:β ratio ≈ 6.7), whereas in aqueous solution of D-glucose there is 36% of α- and 64% of β-anomer (β:α ratio ≈ 1.78). It is clear that the magnitude of anomeric effect that favors the α-anomer in the anomeric mixture is decreased in water.

The observed higher acetolysis rate of permethylated methyl β-D-mannopyranoside as compared to the α-anomer (β:α ratio 2.83) may seem to contradict the above explanation. However, there may be other reasons for such a behavior. First, it could be the much higher ground-state energy of the β-anomer as compared to α-anomer due to unfavorable electrostatic interactions of the axially oriented C2 methoxy group with the C–OMe and C1–O5 dipoles (Δ2 effect) and due to unfavorable torsional strain between the ring, glycosidic, and C2 oxygens. Second, the n* orbital mixing in permethylated methyl-α-D-mannopyranoside could be expected to be less favored because this orbital mixing forces the flattening of the pyranoside ring and that will, in the case of mannopyranoside, result in increased torsional strain between the C2 and the C3 methoxy groups.

In order to understand why permethylated methyl α- and β-D-galactopyranosides acetolyzed 20 and 7 times more rapidly than the corresponding D-gluco derivatives, the rates of acetolysis of methyl 4-O-methyl- (*364*), 4-O-acetyl- (*368*) and 4-deoxy-4-acetamido-2,3,6-tri-O-methyl-α-D-galacto- (*370*), and the corresponding α-D-glucopyranosides (*362, 369, 371*) were compared (Fig. 12.104) and the results are given in Table 12.17 [311].

The data in Table 12.17 suggest that for 4-X-derivatives of methyl 4-X-2,3,6-tri-O-methyl-α-D-galactopyranoside the greater electronegativity of the C4 substituent the faster the rate of acetolysis. Thus 4-O-methyl derivatives acetolyze ca. 10 times faster than 4-O-acetyl derivative and the 4-O-acetyl derivatives acetolyze ca. 5 times faster than 1-acetamido derivative. When C4 substituent is equatorial, as is the case in D-glucopyranoside series, the influence of the change in electronegativity of the C4 substituent on acetolysis rates is much smaller. Thus the 4-O-methyl derivative acetolyzes only 2.9 and 3.3 times faster than 4-O-acetyl- and 4-O-acetamido derivatives, respectively. In the D-gluco series, the dependence of acetolysis rates upon the electronegativity of the C4 substituent can only be explained as a "through-bond" electronic interaction (inductive effect) with the ring oxygen, which is apparently rather small. However, in the D-galactoseries, the very large influence of the

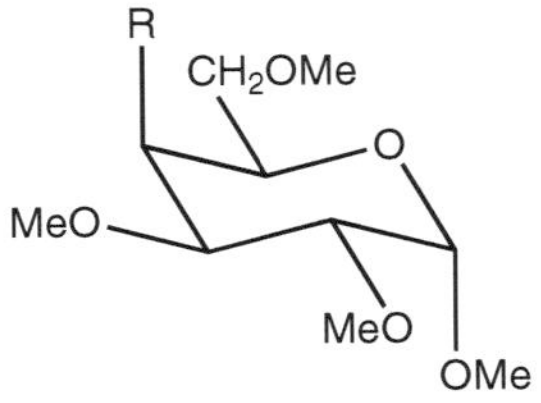

364, R = OCH₃
368, R = OAc
370, R = NHAC

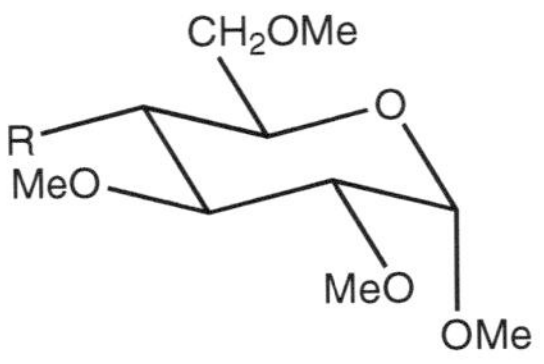

362, R = OCH₃
369, R = OAc
371, R = NHAc

Fig. 12.104

Table 12.17 Kinetic data for the acetolysis of methyl 4-O-methyl, 4-O-acetyl-, and 4-acetamido-4-deoxy derivatives of methyl 2,3,6-tri-O-methyl-α-D-galacto- (*362*, *368*, and *370*, respectively) and α-D-glucopyranosides (*362*, *369*, *371*, respectively)

Sugar	$10^3 \ k \ (s^{-1})$
Methyl	22.18
2,3,4,6-tetra-O-methyl-α-D-galactopyranoside (*364*)	25.79
Methyl 4-O-acetyl-2,3,6-tri-O-methyl-α-D-	2.44
galactopyranoside (*368*)	2.39
Methyl 4-acetamido-4-deoxy-2,3,6-tri-O-methyl-α-D-	0.44
galactopyranoside (*370*)	0.50
Methyl 2,3,4,6-tetra-O-methyl-α-D-glucopyranoside (*362*)	1.67
	1.63
Methyl 4-O-acetyl-2,3,6-tri-O-methyl-α-D-	0.63
glucopyranoside (*369*)	0.61
Methyl 4-acetamido-4-deoxy-2,3,6-tri-O-methyl-α-D-	0.56
glucopyranoside (*371*)	0.47

electronegativity of the axially oriented C4 substituent on the acetolysis rate cannot be ascribed to this small through-bond interaction. The only possible explanation for the unusually large kinetic effect observed in the D-galacto series is a strong through-space electron donation of the axially oriented electronegative substituent at C4 into the oxocarbenium ion under formation. This effect, which is destabilizing in the neutral galactopyranoside due to electrostatic repulsion, becomes very stabilizing as the oxocarbenium ion appears (Fig. 12.105).

The ab initio calculations at the 6–31G* level of theory fully supported the above conclusions [312].

Calculations were conducted on model oxocarbenium ions corresponding to D-gluco- and D-galacto series (*375–380*) (Fig. 12.106) and olefin analogs (*381–386*) stereochemically identical to oxocarbenium models of these two series. Finally, the 4-substituted 2-alkoxy-tetrahydropyrans (*387–392*) were used to calculate difference in ground-stateenergies of 4-axially substituted and 4-equatorially substituted tetrahydropyrans. When considering only the low-energy conformers of acetals

Cleavage of Glycosidic Bonds

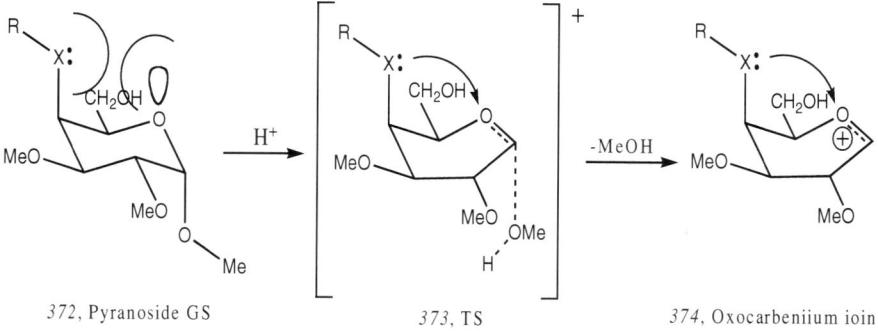

Fig. 12.105

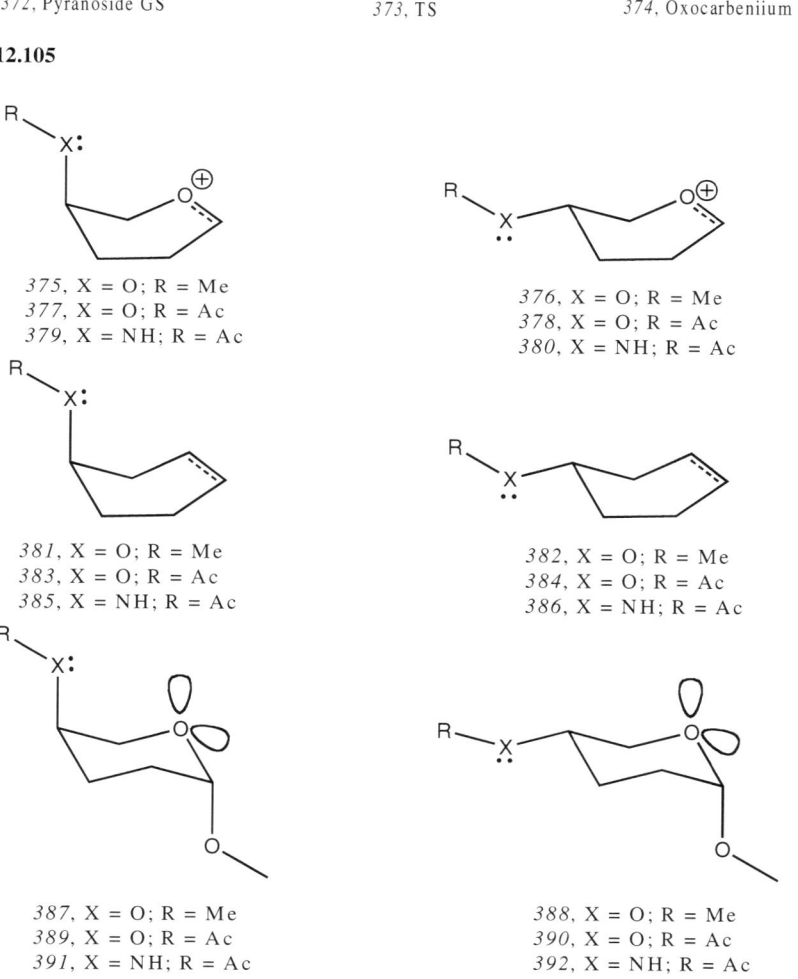

372, Pyranoside GS
373, TS
374, Oxocarbenium ion

375, X = O; R = Me
377, X = O; R = Ac
379, X = NH; R = Ac

376, X = O; R = Me
378, X = O; R = Ac
380, X = NH; R = Ac

381, X = O; R = Me
383, X = O; R = Ac
385, X = NH; R = Ac

382, X = O; R = Me
384, X = O; R = Ac
386, X = NH; R = Ac

387, X = O; R = Me
389, X = O; R = Ac
391, X = NH; R = Ac

388, X = O; R = Me
390, X = O; R = Ac
392, X = NH; R = Ac

Fig. 12.106

387–392, it was found that the compound in which the C4 methoxy group is equatorially oriented (*388*) is 0.88 kcal/mol more stable than the compound *387* having the C4 methoxy group axial. This trend is reversed for the C4 acetamido derivatives: axial isomer *391* is now more stable, by 0.55 kcal/mol, than the equatorial isomer *392*. The C4 acetates fall between these two extremes since the axial and equatorial acetates (*389* and *390*, respectively) are almost equally stable. These observations are in complete agreement with the idea of repulsion between an electron-rich C4 axial substituent and the axially oriented sp^3-hybridized lone pair of electrons on the ring oxygen. In the case of acetamido acetal there exists an electronic attraction between the axially oriented sp^3-hybridized lone pair of electrons on the ring oxygen and the axially oriented C4 acetamido nitrogen since the axial isomer is preferred. The most likely reason for this is that the nitrogen atom has a partial positive charge due to donation of its nonbonding pair of electrons to the carbonyl oxygen of the acetamido group through delocalization involving carbonyl carbon and carbonyl oxygen. While the electronic interactions are rather small in neutral acetals, they are much more significant in corresponding oxocarbenium ions which are both geometrically and energetically very similar to the acetolysis transition structures. Thus, the oxocarbenium ion having the C4 methoxy group axially oriented (*375*) was found to be 4.06 kcal/mol more stable than the oxocarbenium ion with the C4 methoxy group equatorially oriented (*376*). The same electronic interaction can be seen in the oxocarbenium ion having the C4 acetoxy group axially oriented (*377*); however, it is somewhat smaller: the axial isomer is now favored by 2.89 kcal/mol over the equatorial isomer (*378*). In the case of acetamido derivatives, this interaction seems to no longer exist. Thus the oxocarbenium ion having the C4 acetamido group axially or equatorially oriented (*379* and *380*, respectively) has very similar energies: the axial isomer, however, is again favored, but only by 0.20 kcal/mol. It is interesting that while the nitrogen atom is sp^2 hybridized in all acetamido compounds (*380, 385, 386, 391, 392*) as expected, in the two rotamers of the oxocarbenium ion *381* around the C4-X bond in which the C4 acetamido group is axially oriented the nitrogen atom exhibits a high degree of sp^3 hybridization and the nitrogen lone pair of electrons points toward the oxocarbenium ion.

References

1. Michael, A., Am. Chem. J. (1879) **1**, 305
2. Michael, A., Am. Chem. J. (1885) **6**, 336
3. Michael, A., Compt. Rend. (1879) **89**, 355
4. Fischer, E., "*Ueber die Glucoside der Alkohole*", Berichte (1893) **26**, 2400–2412
5. Fischer, E., "*Ueber die Verbindungen der Zucker mit den Alkoholen und Ketonen*", Ber. (1895) **28**, 1145–1167
6. Bishop, C. T.; Cooper, F. P., "*Glycosidation of Sugars: I. Formation of Methyl-D-Xylosides*", Can. J. Chem. (1962) **40**, 224–232
7. Bishop, C. T.; Cooper, F. P., "*Glycosidation of Sugars: II. Methanolysis of D-Xylose, D-Arabinose, D-Lyxose, and D-Ribose*", Can. J. Chem. (1963) **41**, 2743–2758

References

8. Hough, L.; Richardson, A. C., in Rodd's "*Chemistry of Carbon Compounds*", Coffey, S. (Ed.) 2nd Ed, Vol. I, Part F, Elsevier Publishing Co., Amsterdam-London- New York, 1967, p. 328
9. Heard, D. D.; Barker, R., "*Investigation of the role of dimethyl acetals in the formation of methyl glycosides*", J. Org. Chem. (1968) **33**, 740–746
10. Ferrier, R.J.; Hatton, L. R., "*Studies with radioactive sugars: Part I. Aspects of the alcoholysis of D-xylose and D-glucose; the role of the acyclic acetals*", Carbohydr. Res. (1968) **6**, 75–86
11. Smirnyagin, V.; Bishop, C. T., "*Glycosidation of sugars. IV. Methanolysis of D- glucose, D-galactose, and D-mannose*", Can. J. Chem. (1968) **46**, 3085–3090
12. Fletcher, H., Jr., in Whistler, R. L.; Wolfrom, M. L. (Eds.), *Methods in Carbohydr. Chem.* Vol. 2, Academic Press, New York and London, 1963, p. 228
13. McCormick, J. E., "*Benzyl β-L-arabinopyranoside*", Carbohydr. Res. (1967) **4**, 262–263
14. Königs, W.; Knorr, E., "*Ueber einige Derivate des Traubenzuckers und der Galactose*", Berichte (1901) **34**, 957–981
15. Jeanloz, R.; Fletcher, H. G., Jr.; Hudson, C. S., "*Some Reactions of 2,3,4-Tribenzoyl- β-D-ribopyranosyl Bromide*", J. Am. Chem. Soc. (1948) **70**, 4055–4057
16. Ness, R. K.; Fletcher, H. G.; Hudson, C. S., "*The Reaction of 2,3,4,6-Tetrabenzoyl-α-D-glucopyranosyl Bromide and 2,3,4,6-Tetrabenzoyl-α-D-mannopyranosyl Bromide with Methanol. Certain Benzoylated Derivatives of D-Glucose and D-Mannose*", J. Am. Chem. Soc. (1950) **72**, 2200–2205
17. Bochkov, A. F.; Zaikov, G. E., "*Chemistry of the O-Glycosidic Bond (Formation and Cleavage)*", Pergamon Press, Oxford-New York, 1979, pp. 16–19
18. Mattlock, G. L.; Phillips, G. O., "*The reactivity of O-acylglycosyl halides. Part VI. Steric effects of neighbouring groups*", J. Chem. Soc. (1958) 130–135
19. Feather, M. S.; Harris, J. F., "*The Acid-Catalyzed Hydrolysis of Glycopyranosides*", J. Org. Chem. (1965) **30**, 153–157
20. Hickinbottom, W. J., "*Glucosides. Part I. The formation of glucosides from 3: 4: 6-triacetyl glucose 1: 2-anhydride*", J. Chem. Soc. (1928) 3140–3147
21. Haynes, L. J.; Newth, F. H., "*The Glycosyl Halides and Their Derivatives*", Advan. Carbohydr. Chem. (1955) **10**, 207–256
22. Lemieux, R. U.; Brice, C.; Huber, G., "*The Effect of Chlorine Substitutions at the C-2 Acetoxy Group on Some Properties of the Glucose Pentaacetates*", Canad. J. Chem. (1955) **33**, 134–147
23. Lemieux, R. U.; Morgan, A. R., "*The Preparation and Configurations of Tri-O- Acetyl-α-D-Glucopyranose 1, 2-(Ortho-Esters)*", Can. J. Chem.(1965) **43**, 2198–2204
24. Reynolds, D. D.; Evans, W. L., "The Preparation of α- and β-Gentiobiose Octaacetates", J. Am. Chem. Soc. (1938) **60**, 2559–2561
25. Helferich, B.; Bohn, E.; Winkler, S., "*Ungesättigte Derivate von Gentiobiose und Cellobiose*", Ber. (1930) **63**, 989–998
26. Wulff, G.; Röhle, G., "*Results and Problems of O-Glycoside Synthesis*", Angew. Chem. Int. Ed. Engl. (1974) **13**, 157–170
27. Wulff, G.; Röhle, G.: Krüger, W., "*Untersuchungen zur Glykosidsynthese, IV. Neuar tige Silbersalze in der Glykosidsynthese*", Chem. Berichte (1972) **105**, 1097–1110
28. Wulff, G.; Röhle, G.; Schmidt, U., "*Untersuchungen zur Glykosidsynthese, V. Reaktionsprodukte und Stereospezifität der Glucosylierung in Gegenwart unlöslicher Silbersalze in Diäthyläther*", Chem. Berichte (1972) **105**, 1111–1121
29. Wulff, G.; Röhle, G., "*Untersuchungen zur Glykosidsynthese, VI. Kinetische Untersuchungen zum Mechanismus der Koenigs-Knorr-Reaktion*", Chem. Berichte (1972) **105**, 1122–1132
30. Kronzer, F. J.; Schuerch, C., "The methanolysis of some derivatives of 2,3,4-tri-O-benzyl-α-image-glucopyranosyl bromide in the presence and absence of silver salts", Carbohydr. Res. (1973) **27**, 379–390

31. Eby, R.; Schuerch, C., *The use of 1-O-tosyl-D-glucopyranose derivatives in α-D-glucoside synthesis*", Carbohydr. Res. (1974) **34**, 79–90
32. Lucas, T. J.; Schuerch, C., *"Methanolysis as a model reaction for oligosaccharide synthesis of some 6-substituted 2,3,4-tri-O-benzyl-D-galactopyranosyl derivatives"*, Carbohydr. Res. (1975) **39**, 39–45
33. Hanessian, S.; Banoub, J., *Synthetic Methods for Carbohydrates*, Am. Chem. Soc. Symposium Ser. No. 39, (1976) 36
34. Hanessian, S.; Banoub, J., *"Chemistry of the glycosidic linkage. An efficient synthesis of 1,2-trans-di-saccharides"*, Carbohydr. Res. (1977) **53**, C13–C16
35. Helferich, B.; Wedemeyer, K.-F., *"Zur Darstellung von Glucosiden aus Acetobrom- glucose"*, Ann. (1949) **563**, 139–145
36. Schroeder, L. R.; Green, J. W., *"Koenigs–Knorr syntheses with mercuric salt"*, J. Chem. Soc. (1966) 530–531
37. Helferich, B.; Zirner, J., *"Zur Synthese von Tetraacetyl-hexosen mit freiem 2-Hydroxyl. Synthese einiger Disaccharide"*, Chem. Berichte (1962) **95**, 2604–2611
38. Zemplén, G.; Csürös, Z., *"Synthesen in der Kohlenhydrat-Gruppe mit Hilfe von sublimiertem Eisenchlorid, II. Mitteil.: Darstellung der Cellobioside der alpha-Reihe"*, Berichte (1931) **64**, 993–1000
39. Zemplén, G., *"Recent results in carbohydrate research"*, Ber. (1941) **74A**, 75–92
40. Colley, A., Ann. Chim. Phys. (1870) **21**, 363
41. Fischer, E.; Armstrong, E. F., *"Ueber die isomeren Acetohalogen-Derivate des Traubenzuckers und die Synthese der Glucoside"*, Ber. (1901) **34**, 2885–2900
42. Fischer, E., *"Notiz über die Acetohalogen-glucosen und die p-Bromphenylosazone von Maltose und Melibiose"*, Berichte (1911) **44**, 1898–1904
43. Wolfrom, M. L.; Fields, D. L., *"A polymer-homologous series of beta-D-acetates from cellulose"*, Tappi (1957) **40**, 335–337
44. Glaudemans, C. P. J.; Fletcher, H. G., Jr., *"Synthesis of the Two 2-O-Nitro-3,5-di-O-p-nitrobenzoyl-D-arabinofuranosyl Chlorides, an Anomeric Pair of Crystalline Pentofuranosyl Halides Having a Nonparticipating Group at C-2"*, J. Org. Chem. (1964) **29**, 3286–3290
45. Pacsu, E., *"Über die Einwirkung von Titan (IV)-chlorid auf Zucker-Derivate, I.: Neue Methode zur Darstellung der α-Aceto-chlor-zucker und Umlagerung des β-Methyl-glucosids in seine α-Form"*, Ber. (1928) **61**, 1508–1513
46. Arlt, v. F., *"Zur Kenntnis der Glycose"*, Monatsh. Chem. (1901) **22**, 144–150
47. Skraup, Zd. H.; Kremann, R., *"Über Acetochlorglucose, -Galactose und - Milchzucker"*, Monatsh. Chem. (1901) **22**, 375–384
48. Egan, L. P.; Squires, T. G.; Vercellotti, J. R., *"Acetylated aldosyl chlorides by reaction of aldose peracetates with zinc chloride-thionyl chloride"*, Carbohydr. Res. (1970) **14**, 263–266
49. Ness, R. K.; Fletcher, H. G., Jr.; Hudson, C. S., *"New Tribenzoyl-D-ribopyranosyl Halides and Their Reactions with Methanol"*, J. Am. Chem. Soc. (1951) **73**, 959–963
50. Ness, R. K.; Fletcher, H. G., Jr.,, *"The Anomeric 2,3,5-Tri-O-benzoyl-D-arabinosyl Bromides and Other D-Arabinofuranose Derivatives"*, J. Am. Chem. Soc. (1958) **80**, 2007–2010
51. Schlubach, H. H., *"Über die isomere, linksdrehende Aceto-chlor-glucose"*, Ber. (1926) **59**, 840–844
52. Lemieux, R. U.; Hayami, J.-I., *The Mechanism of the Anomerization of the Tetra-O-Acetyl-D-Glucopyranosyl Chlorides"*, Can. J. Chem. (1965) **43**, 2162
53. Toshima, K., *"Glycosyl fluorides in glycosidations"*, Carbohydr. Res. (2000) **327**, 15–26
54. Tsuchiya, T., *"Chemistry and Developments of Fluorinated Carbohydrates"*. Adv. Carbohydr. Chem. Biochem. (1990) **48**, 91–277
55. Shimizu, M.; Togo, H.; Yokoyama, M., *"Chemistry of glycosyl fluorides"*, Synthesis (1998) **54**, 799–822
56. Mukaiyama, T.; Murai, Y.; Shoda, S., *"An efficient method for glucosylation of hydroxy compounds using glucopyranosyl fluoride"*, Chem. Lett. (1981) 431–432

57. Mukaiyama, T.; Hashimoto, Y.; Shoda, S., "*Stereoselective synthesis of 1,2-cis-glycofuranosides using glycofuranosyl fluorides*", Chem. Lett. (1983) 935–938
58. Hashimoto, S.; Hayashi, M.; Noyori, R., "*Glycosylation using glucopyranosyl fluorides and silicon-based catalysts. Solvent dependency of the stereoselection*", Tetrahedron Lett. (1984) **25**, 1379–1382
59. Ogawa, T.; Takahashi, Y., "*Total synthesis of α-cyclodextrin*", Carbohydr. Res. (1985) **138**, C5-C9
60. Takahashi, Y.; Ogawa, T., "*Total synthesis of cyclomaltohexaose*", Carbohydr. Res. (1987) **164**, 277–296
61. Nicolaou, K. C.; Chucholowski, A.; Dolle, R. E.; Randall, J. L., "*Reactions of glycosyl fluorides. Synthesis of O-, S-, and N-glycosides*", J. Chem. Soc. Chem. Commun. (1984), 1155–1156
62. Kunz, H.; Sager, W., "*Stereoselective glycosylation of Alcohols and Silyl Ethers Using Glycosyl Fluorides and Boron Trifluoride*", Helv. Chim. Acta (1985) **68**, 283–287
63. Kunz, H.; Waldmann, H., "*Directed stereoselective synthesis of - and -N-acetyl neuraminic acid–galactose disaccharides using 2-choro and 2-fluoro derivatives of neuraminic acid allyl ester*", J. Chem. Soc., Chem. Commun. (1985), 638–640
64. Vozny, Ya. V.; Galoyan, A. A.; Chizhov, O. S., "*Novel method for O-glycoside bond formation. Reaction of glycosyl fluorides with trimethylsilyl ethers*", Bioorg. Khim. (1985) **11**, 276–278
65. Matsumoto, T.; Maeta, H.; Suzuki, K.; Tsuchihashi, G., "*New glycosidation reaction 1: Combinational use of Cp_2ZrCl_2-$AgClO_4$ for activation of glycosyl fluorides and application to highly β-selective glycosidation of D-mycinos*", Tetrahedron Lett. (1988) **29**, 3567–3570
66. Suzuki, K.; Maeta, H.; Matsumoto, T.; Tsuchihashi, G., "*New glycosidation reaction 2. preparation of 1-fluoro-d-desosamine derivative and its efficient glycosidation by the use of Cp_2HfCl_2-$AgClO_4$ as the activator*", Tetrahedron Lett. (1988) **29**, 3571–3574
67. Matsumoto, T.; Maeta, H.; Suzuki, K.; Tsuchihashi, G., *First total synthesis of mycinamicin IV and VII.: Successful application of new glycosidation reaction*", Tetrahedron Lett. (1988) **29**, 3575–3578
68. Matsumoto, T.; Katsuki, M.; Suzuki, K., Chem. Lett. "*Rapid O-glycosidation of phenols with glycosyl fluoride by using the combinational activator, Cp_2HfCl_2 -$AgClO_4$*" Chem. Lett. (1989) 437–440
69. Suzuki, K.; Maeta, H.; Suzuki, T.; Matsumoto, T.,"*Cp_2ZrCl_2—$AgBF_4$ in Benzene: A new reagent system for rapid and highly selective α-mannoside synthesis from tetra-O-benzyl-image-mannosyl fluoride*", Tetrahedron Lett. (1989) **30**, 6879–6882
70. Nicolaou, K. C.; Caulfield, T. J.; Kataoka, H.; Stylianides, N. A., "*Total synthesis of the tumor-associated Lex family of glycosphingolipids*", J. Am. Chem. Soc. (1990) **112**, 3693–3695
71. Nicolaou, K. C.; Hummel, C. W.; Iwabuchi, Y., "*Total synthesis of sialyl dimeric Lex*", J. Am. Chem. Soc. (1992) **114**, 3126–3128
72. Maeta, H.; Matsumoto, T.; Suzuki, K., "*Dibutyltin diperchlorate" for activation of glycosyl fluoride*", Carbohydr. Res. (1993) **249**, 49–56
73. Kobayashi, S.; Koide, K.; Ohno, M., *Gallium reagents in organic synthesis: Dimethylgallium chloride and triflate as activators in glycosidation using glycopyranosyl fluorides*", Tetrahedron Lett. (1990) **31**, 2435–2438
74. Wessel, H. P., "*Comparison of catalysts in α-glucosylation reactions and identification of triflic anhydride as a new reactive promoter*", Tetrahedron Lett. (1990) **31**, 6863–6866
75. Wessel, H. P.; Ruiz, N.,"*α-Glucosylation reactions with 2,3,4,6-tetra-O-benzyl-β-D-glucopyranosyl fluoride and triflic anhydride as promoter*", J. Carbohydr. Chem. (1991) **10**, 901–910
76. Böhm, G.; Waldmann, H., "*Synthesis of glycosides of fucose under neutral conditions in solutions of $LiClO_4$ in organic solvents*", Tetrahedron Lett. (1995) **36**, 3843–3846

77. Böhm, G.; Waldmann, H., *"O-Glycoside Synthesis under Neutral Conditions in Concentrated Solutions of LiClO4 in Organic Solvents Employing Benzyl-Protected Glycosyl Donors"*, Liebigs Ann. Chem. (1996) 613–619
78. Böhm, G.; Waldmann, H., *"O-Glycoside Synthesis under Neutral Conditions in Concentrated Solutions of LiClO4 in Organic Solvents Employing O-Acyl-Protected Glycosyl Donors"*, Liebigs Ann. Chem. (1996) 621–625
79. Hosono, S.; Kim, W.-S.; Sasai, H.; Shibasaki, M., *"A New Glycosidation Procedure Utilizing Rare Earth Salts and Glycosyl Fluorides, with or without the Requirement of Lewis Acid"*, J. Org. Chem. (1995) **60**, 4–5
80. Kim, W.-S.; Hosono, S.; Sasai, H.; Shibasaki, M., *"Rare earth perchlorate catalyzed glycosidation of glycosyl fluorides with trimethylsilyl ethers"*, Tetrahedron Lett. (1995) **36**, 4443–4446
81. Kim, W.-S.; Sasai, H.; Shibasaki, M., *"β-Selective glycosylation with α-mannosyl fluorides using tin(II) triflate and lanthanum perchlorate"*, Tetrahedron Lett. (1996) **37**, 7797–7800
82. Takeuchi, K.; Mukaiyama, T., *"Trityl tetrakis(pentafluorophenyl)borate catalyzed stereoselective glycosylation using glycopyranosyl fluoride as a glycosyl donor"*, Chem. Lett. (1998) 555–556
83. Yokoyama, M., *"Methods of synthesis of glycosyl fluorides"*, Carbohydr. Res. (2000) **327**, 5–14
84. Hayashi, M.; Hashimoto, S.; Noyori, R., *"Simple synthesis of glycosyl fluorides"*, Chem. Lett. (1984) 1747–1750
85. Miethchen, R.; Kolp, G., *"Reactions with and in anhydrous hydrogen fluoride systems. Part 8. Triethylamine trishydrofluoride – a convenient reagent for the stereoselective synthesis of glycosyl fluorides"*, J. Fluorine Chem. (1993) **60**, 49–55
86. Markovski, L. N.; Pashinnik, V. E.; Kirsanov, A. V., *"Application of dialkylamino-sulfur trifluorides in the synthesis of fluoroorganic compounds"*, Synthesis (1973) 787–789
87. Middleton, W. J., *"New fluorinating reagents. Dialkylaminosulfur fluorides"*, J. Org. Chem. (1975) **40**, 574–578
88. Sharma, M.; Korytnyk, W., *"A general and convenient method for synthesis of 6-fluoro-6-deoxyhexoses"*, Tetrahedron Lett. (1977) **18**, 573–576
89. Card, P. J., *"Synthesis of fluorinated carbohydrates"*, J. Carbohydr. Chem. (1985) **4**, 451–487
90. Rosenbrook, W., Jr., ; Riley, D. A.; Lartey, P. A., *"A new method for the synthesis of glycosyl fluorides"*, Tetrahedron Lett.(1985) **26**, 3–4
91. Posner, G. H.; Haines, S. R., *"A convenient, one-step, high-yield replacement of an anomeric hydroxyl group by a fluorine atom using DAST. Preparation of glycosyl fluorides"*, Tetrahedron Lett. (1985) **26**, 5–8
92. Crich, D.; Lim, L. B. L., *"Diastereoselective free-radical reactions. Part 3. The methyl glucopyranos-1-yl and the 1,2-O-isopropylideneglucopyranos-1-yl radicals: conformational effects on diastereoselectivity"*, J. Chem. Soc. Perkin Trans. I (1991) 2209–2214
93. Kochetkov, N. K.; Bochkov, A. F., in R. Bognar, V. Bruckner, and Cs. Szántay, *Recent Developments in the Chemistry of Natural Carbon Compounds*, Akadémiai Kiadó, Budapest (1971), Vol. IV, p.77
94. Kochetkov, N. K.; Khorlin, A. Ya.; Bochkov, A. F., *"A new method of glycosylation"*, Tetrahedron (1967) **23**, 693–707
95. Kochetkov, N. K.; Bochkov, A. F.; Sokolovskaya, T. A.; Snyatkova, V. J., *"Modifications of the orthoester method of glycosylation"*, Carbohydr. Res. (1971) **16**, 17–27
96. Kochetkov, N. K.; Bochkov, A. F., *"Synthesis of Oligosaccharides by the Orthoester Method"*, Methods in Carbohydr. Chem. (1972) **6**, 480–486
97. Kochetkov, N. K.; Khorlin, A. Ya.; Bochkov, A. F. *"New synthesis of glycosides"*, Tetrahedron Lett. (1964) **5**, 289–293
98. Kochetkov, N. K.; Khorlin, A. Ya.; Bochkov, A. F., *"Synthesis of disaccharides"*, Dokl. Akad. Nauk SSSR (1965) **162**, 104–107

99. Kochetkov, N. K.; Khorlin, A. Ya.; Bochkov, A. F.; Demushkina, L. B.; Zolotuchin, I. O., "Ortho ester method for synthesizing trisaccharides", Zh. Obshch. Khim. (1967) **37**, 1272–1277
100. Pacsu, E., "*Carbohydrate Orthoester*", Advan. Carbohydr. Chem. (1945) **1**, 77–127
101. Wulf, G.; Krüger, W., "*Untersuchungen Glykosidsynthese: III. Teil. Eine neue dar stellungsmethode für 1,2-D-glucose-orthoester*", Carbohydr. Res. (1971) **19**, 139–142
102. Kochetkov, N. K.; Klimov, E. M.; Derevitskaya, V. A., "*Synthesis of glycosides by glycosylation of tert-butyl ethers of alcohols*", Dokl. Akad. Nauk. SSSR(1970) **192**, 336–338
103. Kochetkov, N. K.; Derevitskaya, V. A.; Klimov, E. M., "*Synthesis of glycosides via tert.-butyl ethers of alcohols*", Tetrahedron Lett. (1969) **10**, 4769–4772
104. Schulz, M.; Boeden, H.-F.; Berlin, P., "*Zuckerperoxide, III. Ein neuer Zuckerabbau durch Fragmentierung acylierter Peroxyglykoside*", Liebigs Ann. Chem. (1967) **703**, 190–201
105. Kochetkov, N. K.; Khorlin, A. Ya.; Bochkov, A. F., "*Structure and glycoside formation ability of sugar ortho esters*", Synthesis of 6-0-(α-L-arabinosyl)-D-glucoses, Zh. Obshch. Khim. (1967) **37**, 338–343
106. Bochkov, A. F.; Betaneli, V. I.; Kochetkov, N. K., "*Sugar orthoesters. XV. Relation to conditions and mechanism of proton-catalyzed reactions of protected α-D-glucopyranose 1,2-alkylorthoacetates in media of low polarity*", Bioorganicheskaya Khimiya (1976) **2**, 927–941
107. Bochkov, A. F.; Betaneli, V. I.; Kochetkov, N. K., "*Sugar orthoesters. 16. Mechanism of the mercuric bromide-catalyzed isomerization of protected α-D-glucopyranose 1,2-orthoacetates in nitromethane*", Bioorganicheskaya Khimia (1977) **3**, 39–45
108. Burshtein, K. Ya.; Fundiler, I. N.; Bochkov, A. F., "*Molecular orbital calculations relating to the mechanism of the reactions of sugar orthoesters: 1,2,4-orthoacetyl-α-D-xylopyranose*", Tetrahedron (1975) **31**, 1303–1306
109. Heitman, J. A.; Richards, G. F.; Schroeder, L. R., "*Carbohydrate orthoesters. III. The crystal and molecular structure of 3,4,6-tri-O-acetyl-1,2-O-[1-(exo-ethoxy) ethylidene]-α-D-glucopyranose*", Acta Crystallogr. (1974) **B30**, 2322–2328
110. Farkas, I.; Dinya, Z.; Szabo, I. F.; Bognar, R., "*Cleavage of sugar 1,2-(ortho esters) with dichloromethyl methyl ether*", Carbohydr. Res. (1972) **21**, 331–333
111. Lemieux, R. U.; Cipera, J. D. T., "*The Preparation and Properties of α-D-Glucopyranose 1, 2-(Ethyl Orthoacetate) Triacetate*", Can. J. Chem. (1956) **34**, 906–910
112. Fischer, E.; Bergmann, M.; Rabe, A., "*Über Acetobrom-rhamnose und ihre Verwendung zur Synthese von Rhamnosiden*", Ber. (1920) **53**, 2362–2388
113. Korytnyk, W.; Mills, J. A., "*Preparation and properties of some poly-O-acetylglycosyl chlorides of the unstable series*" J. Chem. Soc. (1959) 636–649
114. Mazurek, M.; Perlin, A. S., "*Synthesis of β-D-Mannose 1, 2-Orthoacetates*", Can. J. Chem. (1965) **43**, 1918–1923
115. Helferich, B.; Weiss, K., "*Zur Synthese von Glucosiden und von nicht- reduzierenden Disacchariden*", Chem. Berichte (1956) **89**, 314–321
116. Khorlin, A. Ya.; Bochkov, A. F.; Kochetkov, N. K., "*A new synthesis of sugar ortho esters*", Izv. Akad. Nauk SSSR, Ser. Khim. (1964) 2214–2216
117. Schmidt, R. R., "*Neue Methoden zur Glycosid- und Oligosaccharidsynthese-gibt es Alternativen zur Koenigs-Knorr-Methode?*", Angew. Chem. (1986) **98**, 213–236
118. Schmidt, R. R., "*New Methods for the Synthesis of Glycosides and Oligosaccharides. Are There Alternatives to the Koenigs-Knorr Method?*" Angew. Chem. Int. Ed. Engl. (1986) **25**, 212–235
119. Schmidt, R. R. in Bartmann, W. Sharpless, K. B.(Eds), "*Stereochemistry of Organic and Bioorganic Transformations*", Workshop Conferences Hoechst, Vol. 17, pp. 169–189. VCH Verlaggesellschaft GmbH, Weinheim, 1987
120. Schmidt, R. R., "*Recent developments in the synthesis of glycoconjugates*", Pure Appl. Chem. (1989) **61**, 1257–1270

121. Trost, B. M.; Flemming, I.; Winterfeldt, E. (Eds.), "*Comprehensive Organic Synthesis*", Vol. 6, pp. 33–64, Pergamon Press, Oxford, 1991
122. Schmidt, R. R.; Michel, J., "*Direct o-glycosyl trichloroacetimidate formation, nucleophilicity of the anomeric oxygen atom*", Tetrahedron Lett.(1984) **25**, 821–824
123. Schmidt, R. R.; Michel, J., "*Synthese von linearen und verzweigten Cellotetraosen*", Angew. Chem. (1982) **94**, 77–78; Angew. Chem. Int. Ed. Engl. (1982) **21**, 77–78; Angew. Chem. Suppl. (1982) 78–84
124. Schmidt, R. R.; Stumpp, M., "*Glycosylimidate, 8. Synthese von 1-Thioglycosiden*", Liebigs Ann. Chem. (1983) 1249–1256
125. Schmidt, R. R.; Michel, J.; Roos, M., "*Glycosylimidate, 12 Direkte Synthese von O-α- und O-β-Glycosyl-imidat*en", Liebigs Ann. Chem. (1984) 1343–1357
126. Schmidt, R. R.; Esswein, A., "*Einfache Synthese von KDO-α-Glycosiden durch anomer selektive O-Alkylierung*", Angew. Chem. (1988) **100**, 1234–1236
127. Schuhmacher, M., Dissertation, University of Constance, 1985
128. Schmidt, R. R.; Michel, J., Angew. Chem. (1980) **92**, 783–784; "*Facile Synthesis of alpha- and beta-O-Glycosyl Imidates; Preparation of Glycosides and Disaccharides*", Angew. Chem. Int. Ed. Engl. (1980) **19**, 731–732
129. Schmidt, R. R.; Behrendt, M.; Toepfer, A., "*Nitriles as Solvents in Glycosylation Reactions: Highly Selective β-Glycoside Synthesis*", Synlett (1990) 694–696
130. Paulsen, H., "*Fortschritte bei der selektiven chemischen Synthese komplexer Oligo saccharide*", Angew. Chem. (1982) **94**, 184–201; "*Advances in Selective Chemical Syntheses of Complex Oligosaccharides*", Angew. Chem. Int. Ed. Engl. (1982) **21**, 155–173
131. Wulff, G.; Röhle, G., "*Ergebnisse und Probleme der O-Glykosidsynthese*", Angew. Chem. (1974) **86**, 173–187; "*Results and Problems of O-Glycoside Synthesis*", Angew. Chem. Int. Ed. Engl. (1974) **13**, 157–170; Angew. Chem. Int. Ed. Engl. (1988) **27**, 1178–1180
132. Woodward, R. B.; Logusch, E.; Nambiar, K. P.; Sakan, K.; Ward, D. E.; Au-Yeung, B.W.; Balaram, P.; Browne, L. J.; Card, P. J.; Chen, C. H., "*Asymmetric total synthesis of erythromycin. 1. Synthesis of an erythronolide A secoacid derivative via asymmetric induction*", J. Am. Chem. Soc. (1981) **103**, 3210–3213
133. Woodward, R. B.; Au-Yeung, B. W.; Balaram, P.; Browne, L. J.; Ward, D. E.; Card, P. J.; Chen, C. H., "*Asymmetric total synthesis of erythromycin. 2. Synthesis of an erythronolide A lactone system*", J. Am. Chem Soc. (1981) **103**, 3213–3215
134. Woodward, R. B.; Logusch, E.; Nambiar, K. P.; Sakan, K.; Ward, D. E.; Au-Yeung, B.-W.; Balaram, P.; Browne, L. J.; Card, P. J.; Chen, C. H.; Chênevert, R. B.; Fliri, A.; Frobel, K.; Gais, H.-J.; Garratt, D. G.; Hayakawa, K.; Heggie, W.; Hesson, D. P.; Hoppe, D.; Hoppe, I.; Hyatt, J. A.; Ikeda, D.; Jacobi, P. A.; Kim, K. S.; Kobuke, Y.; Kojima, K.; Krowicki, K.; Lee, V. J.; Leutert, T.; Malchenko, S.; Martens, J.; Mathews, R. S.; Ong, B. S.; Press, J. B.; Rajan Babu, T. V.; Rousseau, G.; Sauter, H. M.; Suzuki, M.; Tatsuta, K.; Tolbert, L. M.; Truesdale, E. A.; Uchida, I.; Ueda, Y.; Uyehara, T.; Vasella, A. T.; Vladuchik, W. C.; Wade, P. A.; Williams, R. M.; Wong, N.-C., "*Asymmetric total synthesis of erythromycin. 3. Total synthesis of erythromycin*", J. Am. Chem. Soc. (1981) **103**, 3215–3217
135. Hanessian, S.; Bacquet, C.; Lehong, N., "*Chemistry of the glycosidic linkage. Exceptionally fast and efficient formation of glycosides by remote activation*", Carbohydr. Res. (1980) **80**, C17–C22
136. Fraser-Reid, B.; Wu, Z.; Udodong, U.E.; Ottosson, H., "*Armed/disarmed effects in glycosyl donors: rationalization and sidetracking*", J. Org. Chem. (1990) **55**, 6068–6070
137. Konradsson, P.; Mootoo, D. R.; McDevitt, R. E.; Fraser-Reid, B., "*Iodonium ion generated in situ from N-iodosuccinimide and trifluoromethanesulphonic acid promotes direct linkage of disarmed pent-4-enyl glycosides*", J. Chem. Soc. Chem. Commun. (1990) 270–272
138. Konradsson, P.; Udodong, U. E.; Fraser-Reid, B., "*Iodonium promoted reactions of disarmed thioglycosides*", Tetrahedron Lett. (1990) **31**, 4313–4316
139. Fraser-Reid, B.; Madsen, R., "*Oligosaccharide Synthesis by n-Pentenyl Glycosides*" in *Preparative Carbohydrate Chemistry*, Hanessian, S. Ed., Marcel Dekker Inc., New York, 1996, pp. 339–356

140. Fraser-Reid, B.; Udodong, U. E.; Wu., Z.; Ottosson, H.; Merritt, J. R.; Rao, C. S.; Roberts, C.; Madsen, R., "*n-Pentenyl Glycosides in Organic Chemistry: A Contemporary Example of Serendipity*", Synlett (1992) 927–942
141. Madsen, R.; Fraser-Reid, B., "*Modern Methods in Carbohydrate Synthesis*" Kahan, S.; O'Neill, R. A. (Eds.), Harwood Academic Publishers, Amsterdam, 1996, Chapt. 7
142. Boons, G.-J., "*Glycosides as Donors*", in "*Glycoscience: Chemistry and Chemical Biology*", Vol. I, Fraser-Reid, B.; Kuniaki, T.; Thiem, J., Eds., Springer Verlag, Berlin, 2001, pp. 551–581
143. Merritt, J. R.; Naisang, E.; Fraser-Reid, B., "*n-Pentenyl Mannoside Precursors for Synthesis of the Nonamannan Component of High Mannose Glycoproteins*", J. Org. Chem. (1994) **59**, 4443–4449
144. Mootoo, D. R.; Konradsson, P.; Udodong, U. E.; Fraser-Reid, B., "*Armed and dis-armed n-pentenyl glycosides in saccharide couplings leading to oligosaccharides*", J. Am. Chem. Soc. (1988) **110**, 5583–5584
145. David, S.; Lubineau, A.; Vatèle, J.-M., "*Chemical synthesis of 2-O-(-L-fucopyranosyl)-3-O-(2-acetamido-2-deoxy-α-D-galactopyranosyl)-D-galactose, the terminal structure in the blood-group A antigenic determinant*" J. Chem. Soc. Chem. Commun. (1978) 535–537
146. Vasella, A., "*New reactions and intermediates involving the anomeric center*", Pure Appl. Chem. (1991) **63**, 507–518
147. Kahne, D.; Yang, D.; Lim, J. J.; Miller, R.; Paguaga, E., "The use of alkoxy-substituted anomeric radicals for the construction of beta-glycosides" J. Am. Chem. Soc. (1988) **110**, 8716–8717
148. Lemieux, R. U.; Levine, S., "*Synthesis of Alkyl 2-Deoxy-α-D-Glycopyranosides and Their 2-Deuterio Derivatives*", Can. J. Chem. (1964) **42**, 1473–1480
149. Lemieux, R. U.; Morgan, A. R., "*The Synthesis of β-D-Glucopyranosyl 2-Deoxy-α-D-Arabino-Hexopyranoside*", Can. J. Chem. (1965) **43**, 2190–2197
150. Thiem, J.; Karl, H.; Schwentner, J., "*Synthese α-verknüpfter 2'-Deoxy-2'-iododisaccharide*" Synthesis (1978) 696–698
151. Thiem, J.; Karl, H., "*Syntheses of methyl 3-O-(α-D-olivosyl)-α-D-olivoside*", Tetrahedron Lett. (1978) **19**, 4999–5002
152. Thiem, J.; Ossowski, P., "*Studies of hexuronic acid ester glycals and the synthesis of 2-deoxy-β-glycoside precursors*", J. Carbohydr. Chem. (1984) **3**, 287–313
153. Thiem, J.; Prahst, A.; Lundt, I., "*Untersuchungen zur beta-Glycosylierung nach dem N-Iodsuccinimid-Verfahren: Synthese der terminalen Disaccharideinheit von Orthosomycinen*", Liebigs Ann. Chem. (1986) 1044–1056
154. Thiem, J.; Klaffke, W., "*Facile stereospecific synthesis of deoxyfucosyl disaccharide units of anthracyclines*", J. Org. Chem. (1989) **54**, 2006–2009
155. Thiem, J., ACS Symp. Ser. (1989) 396, Kap.8
156. Miljkovic, M.; Gligorijevic, M.; Glisin, Dj., "*Steric and Electrostatic Interactions in Reactions of Carbohydrates. III. Direct Displacement of the C-2 Sulfonate of Methyl 4, 6-O-Benzylidene-3-O-methyl-2-O-methylsulfonyl-β-D-gluco- and mannopyranosides*", J. Org. Chem. (1974) **39**, 3223–3226
157. Lemieux, R. U.; Fraser-Reid, B., "*The Mechanisms of the Halogenations and Halogenomethoxylations of D-Glucal Triacetate, D-Galactal Triacetate, and 3, 4-Dihydropyran*", Can. J. Chem. (1965) **43**, 1460–1475
158. Friesen, R. W.; Danishefsky, S. J., "*On the controlled oxidative coupling of glycals: a new strategy for the rapid assembly of oligosaccharides*", J. Am. Chem. Soc. (1989) **111**, 6656–6660
159. Friesen, R. W.; Danishefsky, S. J., "*On the use of the haloetherification method to synthesize fully functionalized disaccharides*", Tetrahedron (1990) **46**, 103–112
160. Murray, R. W.; Jeyaraman, R., "*Dioxiranes: synthesis and reactions of methyldioxirane*s", J. Org. Chem. (1985) **50**, 2847–2853

161. Halcomb, R. L.; Danishefsky, S. J., "*On the direct epoxidation of glycals: application of a reiterative strategy for the synthesis of β-linked oligosaccharides*", J. Am. Chem. Soc. (1989) **111**, 6661–6666
162. Danishefsky, S. J.; Bilodeau, M. T., "*Glycals in Organic Synthesis: The Evolution of Comprehensive Strategies for the Assembly of Oligosaccharides and Glycoconjugates of Biological Consequence*", Angew. Chem. Int. Ed. Engl. (1996) **35**, 1380–1419
163. Wolfrom, M. L.; Anno, K., "*D-Xylosamine*", J. Am. Chem. Soc. (1953) **75**, 1038–1039
164. Wolfrom, M. L.; Tanghe, L. J.; George, R. W.; Waisbrot, S. W., "*Acetals of Galactose and of Dibenzylideneglucose*", J. Am. Chem. Soc.(1938) **60**, 132–134
165. Weygand, F.; Ziemann, H., "*Glykosylbromide aus Äthylthioglykosiden, II*", Liebigs Ann. Chem. (1962) **657**, 179–198
166. Ferrier, R. J.; Hay, R. W.; Vethaviyasar, N., "*A potentially versatile synthesis of glycosides*", Carbohydr. Res. (1973)**27**, 55–61
167. Mukaiyama, T.; Nakatsuka, T.; Shoda, S., "*An efficient glucosylation of alcohol using 1-thioglucoside derivative*" Chem. Lett. (1979) 487–490
168. van Cleve, J. W., "*Reinvestigation of the preparation of cholesteryl 2,3,4,6-tetra-O-benzyl-α-image-glucopyranoside*", Carbohydr. Res. (1979) **70**, 161–164
169. Wuts, P. G. M.; Bigelow, S. S., "*Total synthesis of oleandrose and the avermectin disaccharide, benzyl.alpha.-L-oleandrosyl-.alpha.-L-4-acetoxyoleandroside*", J. Org. Chem. (1983) **48**, 3489–3493
170. Nicolaou, K. C.; Seitz, S. P.; Papahatjis, D. P., "*A mild and general method for the synthesis of O-glycosides*", J. Am. Chem. Soc. (1983) **105**, 2430–2434
171. Garegg, P. J.; Henrichson, C.; Norberg, T., "*A reinvestigation of glycosidation reactions using 1-thioglycosides as glycosyl donors and thiophilic cations as promoters*", Carbohydr. Res. (1983) **116**, 162–165
172. Tsai, T. Y. R.; Jin, H.; Wiesner, K., "*A stereoselective synthesis of digitoxin. On cardioactive steroids. XIII*", Can. J. Chem. (1984) **62**, 1403–1405
173. Lönn, H., Chem. Commun. Stockholm Univ., No. 2 (1984) 1–30
174. Lönn, H., "*Synthesis of a tri- and a hepta-saccharide which contain α-L-fucopyranosyl groups and are part of the complex type of carbohydrate moiety of glycoproteins*", Carbohydr. Res. (1985) **139**, 105–113
175. Lönn, H., "*Synthesis of a tetra- and a nona-saccharide which contain α-L-fucopyranosyl groups and are part of the complex type of carbohydrate moiety of glycoproteins*", Carbohydr. Res. (1985) **139**, 115–121
176. Lönn, H., "*Glycosylation using a thioglycoside and methyl trifluoromethanesulfonate. A new and efficient method for cis and trans glycoside formation*", J. Carbohydr. Chem. (1987) **6**, 301–306
177. Fügedi, P.; Garegg, P. J., "*A novel promoter for the efficient construction of 1,2-trans linkages in glycoside synthesis, using thioglycosides as glycosyl donors*", Carbohydr. Res. (1986) **149**, C9–C12
178. Poszgay, V.; Jennings, H. J., "*A new method for the synthesis of O-glycosides from S-glycosides*", J. Org. Chem. (1987) **52**, 4635–4637; Poszgay, V.; Jennings, H. J., "*Synthetic oligosaccharides related to group B streptococcal polysaccharides. 3. Synthesis of oligosaccharides corresponding to the common polysaccharide antigen of group B streptococci*", J. Org. Chem. (1988) **53**, 4042–4052
179. Dasgupta, F.; Garegg, P. J., "*Alkyl sulfenyl triflate as activator in the thioglycoside- mediated formation of β-glycosidic linkages during oligosaccharide synthesis*", Carbohydr. Res. (1988) **177**, C13–C17
180. Kochetkov, N. K.; Klimov, E. M.; Malysheva, N. N., "*Novel highly stereospecific method of 1,2-cis-glycosylation. Synthesis of α-D-glucosyl-D-glucoses*", Tetrahedron Lett. (1989) **30**, 5459–5462
181. Ito, Y.; Ogawa, T., "*Benzeneselenenyl triflate as a promoter of thioglycosides: A new method for O-glycosylation using thioglycosides*", Tetrahedron Lett. (1988) **29**, 1061–1064

References

182. Reddy, G. V.; Kulkarni, V. R.; Mereyalla, H. B., "*A mild general method for the synthesis of α-linked disaccharides*", Tetrahedron Lett. (1989) **30**, 4283–4286
183. Veenemann, G. H.; van Leeuwen, S. H.; van Boom, J. H., "*Iodonium ion promoted reactions at the anomeric centre. II An efficient thioglycoside mediated approach toward the formation of 1,2-trans linked glycosides and glycosidic esters*", Tetrahedron Lett. (1990) **31**, 1331–1334
184. Veenemann, G. H.; van Boom, J. H., "*An efficient thioglycoside-mediated formation of α-glycosidic linkages promoted by iodonium dicollidine perchlorate*", Tetrahedron Lett. (1990) **31**, 275–278
185. Tsuboyama, K.; Takeda, K.; Torii, K.; Ebihara, M.; Shimizu, J.; Suzuki, A.; Sato, N.; Furuhata, K.; Ogura, H., "*A convenient synthesis of S-glycosyl donors of D-glucose and O-glycosylations involving the new reagent*", Chem. Pharm. Bull. (1990) **38**, 636–638
186. Marra, A.; Mallet, J.-M.; Amatore, C.; Sinaÿ, P., "*Glycosylation Using a One-Electron-Transfer Homogeneous Reagent: A Novel and Efficient Synthesis of β-Linked Disaccharides*", Synlett (1990) 572–574
187. Fügedi, P.; Garegg, P. J.; Oscarson, S.; Rosén, G.; Silwanis, B. A., "Glycosyl 1-piperidinecarbodithioates in the synthesis of glycosides", Carbohydr. Res. (1991) **211**, 157–162
188. Braccini, I.; Derouet, C.; Esnault, J.; Hervé de Penhoat, C.; Mallet, J.-M.; Michon, V.; Sinaÿ, P., "*Conformational analysis of nitrilium intermediates in glycosylation reactions*", Carbohydr. Res. (1993) **246**, 23–41
189. Garegg, P. J.; Hällgren, C., "*Synthesis of 2-(p-trifluoroacetamidophenyl)ethyl O-β-D-mannopyranosyl-(1→2)-O-α-D-mannopyranosyl-(1→2)-O-[α-D-glucopyranosyl-(1→3)]-O-α-D-mannopyranosyl-(1→2)-O-β-D-mannopyranosyl-(1→3)-2-acetamido-2-deoxy-β-D-glucopyranoside, corresponding to the repeating unit of the Salmonella thompson, serogroup C1 O-antigen lipopolysaccharide, and of a pentasaccharide fragment thereof*", J. Carbohydr. Chem. (1992) **11**, 425–443
190. Hasegawa, A.; Nagahama, T.; Okhi, H.; Kiso, M., "*Synthetic studies on sialoglycoconjugates 41: a facile total synthesis of ganglioside GM2*", J. Carbohydr. Chem. (1992) **11**, 699–714
191. Hotta, K.; Ishida, H.; Kiso, M.; Hasegawa, A., "*Synthetic studies on sialoglycoconjugates. 54: Synthesis of I-active ganglioside analog*", J. Carbohydr. Chem.(1994) **13**, 175–191
192. Garegg, P. J., "*Thioglycosides as Glycosyl Donors in Oligosaccharide Synthesis*", Advan. Carbohydr. Chem. Biochem. (1997) **52**, 179–205, references 97–183; Garegg, P. J., "*Synthesis and Reactions of Glycosides*", Adv. Carbohydr. Chem. Biochem. (2004) **59**, 69–134
193. Horton, D.; Hutson, D. H., "*Developments in the Chemistry of Thiosugars*", Adv. Carbohydr. Chem.(1963) **18**., 123–199
194. Norberg, T., in Khan, S. H.; O'Neill, R. A. (Eds.), *Modern Methods in Carbohydrate Synthesis*, pp. 82–106, Harwood Academic Publishers, New York, 1995, and references cited therein
195. Contour, M.-O.; Defaye, J.; Little, M.; Wong, E., "*Zirconium(IV) chloride-catalyzed synthesis of 1,2-trans-1-thioglycopyranosides*", Carbohydr. Res. (1989) **193**, 283–287
196. Dasgupta, F.; Garegg, P. J., "*Synthesis of ethyl and phenyl 1-thio-1,2-trans-D-glycopyranosides from the corresponding per-O-acetylated glycopyranoses having a 1,2-trans-configuration using anhydrous ferric chloride as a promoter*", Acta Chem. Scand.(1989) **43**, 471–475
197. Pozsgay, V.; Jennings, H. J., "*A new, stereoselective synthesis of methyl 1,2-image-1-thioglycosides*", Tetrahedron Lett. (1987) **28**, 1375–1376
198. Ogawa, T.; Matsui, M., "*A new approach to 1-thioglycosides by lowering the nucleophilicity of sulfur through trialkylstannylation*", Carbohydr. Res. (1977) **54**, C17–C21
199. Ferrier, R.; Furneaux, R., "*1, 2-trans-1-Thioglycosides*", Methods Carbohydr Chem. (1980) **8**, 251–253
200. Ferrier, R. J.; Furneaux, R. H., "Synthesis of 1,2-trans-related 1-thioglycoside esters", Carbohydr. Res. (1976) **52**, 63–68

201. Lemieux, R. U., "*The Mercaptolysis of Glucose and Galactose Pentaacetate*", Can. J. Chem. (1951) **29**, 1079–1091
202. Lemieux, R. U.; Brice, C., "*A Comparison of the Properties of Pentaacetates and Methyl 1,2-Orthoacetates of Glucose and Mannose*", Can. J. Chem. (1955) **33**, 109–119
203. Horton, D., "*1-Thioglycosides*", Methods Carbohydr. Chem. (1963) **2**, 368–373, and references cited therein
204. Fischer, E.; Delbrück, K., "*Über Thiophenol-glucoside*", Chem. Berichte (1909) **42**, 1476–1482
205. Schneider, W.; Sepp, J.; Stiehler, O., "*Synthese zweier isomerer Reihen von Alkyl-thioglucosiden*", Chem. Berichte (1918) **51**, 220–234
206. Helferich, B.; Grünewald, H.; Langenhoff, F., "*Notiz über die Darstellung von Methyl-α-l-thio-arabinosid und von Methyl-β-d-thio-galaktosid*", Chem. Berichte (1953) **86**, 873–875
207. Yde, M.; De Bruyne, C. K., "*Synthesis of para-substituted phenyl 1-thio-β-D-galactopyranosides*", Carbohydr. Res. (1973) **26**, 227–229
208. Pedretti, V.; Veyriéres, A.; Sinaÿ, P., "*A novel 13 O→C silyl rearrangement in carbohydrate chemistry: Synthesis of α-D-glycopyranosyltrimethylsilanes*" Tetrahedron (1990) **46**, 77–88
209. Apparu, M.; Blanc-Muesser, M.; Defaye, J.; Driguez, H., "*Stereoselective syntheses of O- and S-nitrophenyl glycosides. Part III. Syntheses in the α-D-galactopyranose and α-maltose series*", Can. J. Chem. (1981) **59**, 314–320
210. Tropper, F. D.; Andersson, F. O.; Grand-Maître, C.; Roy, R., "*Stereospecific Synthesis of 1,2-trans-1-Phenylthio-β-D-Disaccharides Under Phase Transfer Catalysis*", Synthesis (1991) 734–736
211. Horton, D., "*1-Thio-D-glucose*", Methods Carbohydr. Chem. (1963) **2**, 433–437, and references cited therein
212. Pacsu, E., "*Preparation of Glycosides from Dithioacetals*", Methods Carbohydr. Chem. (1963) **2**, 354–367, and references therein
213. Cerny, M.; Zachystalova, D.; Pacak, J., "*Preparation of acetylated aromatic 1-thio-β-D-glucopyranosides from 2,3,4,6-tetra-O-acetyl-1-thio-β-D-glucopyranose and diazonium salts*", Coll. Czech. Chem. Commun. (1961) **26**, 2206–2211
214. Sakata, M.; Haga, M.; Tejima, S., "*Synthesis and reactions of glycosyl methyl- and benzyl-xanthates: A facile synthesis of 1-thioglycosides*", Carbohydr. Res. (1970) **13**, 379–390
215. Tropper, F. D.; Andersson, F. O.; Cao, S.; Roy, R., "*Synthesis of S-glycosyl xanthates by phase transfer catalyzed substitution of glycosyl halides*", J. Carbohydr. Chem. (1992) **11**, 741–750
216. Szeja, W.; Bogusiak, J., "*Synthesis of glycosyl xanthates from reducing sugar derivatives under phase-transfer conditions*", Carbohydr. Res. (1987) **170**, 235–239
217. Pakulski, Z.; Pierozynski, D.; Zamojski, A., "*Reaction of sugar thiocyanates with Grignard reagents. New synthesis of thioglycosides*", Tetrahedron (1994) **50**, 2975–2992
218. Lacombe, J. M.; Rakotomanomana, N.; Pavia, A. A., "*Free-radical addition of 1-thiosugars to alkenes a new general approach to the synthesis of 1-thioglycosides*", Tetrahedron Lett. (1988) **29**, 4293–4296
219. Kahne, D.; Walker, S.; Cheng, Y.; Van Engen, D. J., "*Glycosylation of unreactive substrates*", J. Am. Chem. Soc. (1989) **111**, 6881–6882
220. Yan, L.; Kahne, D., "*Generalizing Glycosylation: Synthesis of the Blood Group Antigens Le^a, Le^b, and Le^x Using a Standard Set of Reaction Conditions*", J. Am. Chem. Soc. (1996) **118**, 9239–9248
221. Nicolaou, K. C.; Winssinger, N.; Pastor, J.; DeRoose, F., "*A General and Highly Efficient Solid Phase Synthesis of Oligosaccharides. Total Synthesis of a Heptasaccharide Phytoalexin Elicitor (HPE)*", J. Am. Chem. Soc. (1997) **119**, 449–450
222. Fréchet, J. M. J.; De Smet, M. D.; Farral, M. J., "*Functionalization of crosslinked polystyrene resins. 2. Preparation of nucleophilic resins containing hydroxyl or thiol functionalities*", Polymer (1979) **20**, 675–680

223. Plante, O. J.; Palmacci, E. R.; Seeberger, P. H., "*Development of an Automated Oligosaccharide Synthesizer*", Adv. Carbohydr. Chem. Biochem. (2003) **58**, 35–54
224. Shafizadeh, F., "*Formation and Cleavage of the Oxygen Ring in Sugars*" Adv. Carbohydr. Chem. (1958) **13**, 9–61
225. Capon, B.; Overend, W. G., "*Constitution and Physicochemical Properties of Carbohydrates*", Adv. Carbohydr. Chem. (1960) **15**, 11–51
226. BeMiller, J. N., "*Acid-Catalyzed Hydrolysis of Glycosides*", Adv. Carbohydr. Chem. (1967) **22**, 25–108
227. Haworth, W. N., "*The constitution of some carbohydrates*", Chem. Berichte 1932) **65A**, 43–65
228. Heidt, L. J.; Purves, C. B., "*Thermal Rates and Activation Energies for the Aqueous Acid Hydrolysis of α- and β-Methyl, Phenyl and Benzyl-D-glucopyranosides, α- and β-Methyl and β-Benzyl-D-fructopyranosides, and α-Methyl-D-fructofuranoside*", J. Am. Chem. Soc. (1944) **66**, 1385–1389
229. Isbell, H. S.; Frush, H. L., "*α- and β-Methyl lyxosides, mannosides, gulosides and heptosides of like configuration*", J. Res. Nat. Bur. Stand. (1940) **24**, 125–151
230. Nakano, J.; Rånby, B. G., "*Acid hydrolysis of methyl glucosides and methyl glucuronosides*", Svensk Papperstid. (1962) **65**, 29–33
231. Overend, W. G.; Rees, C. W.; Sequeira, J. S., "*Reactions at position 1 of carbohydrates. Part III. The acid-catalysed hydrolysis of glycosides*", J. Chem. Soc. (1962) 3429–3440
232. Haworth, W. N.; Hirst, E. L., "*The structure of carbohydrates and their optical rotatory power. Part I. General introduction*", J. Chem. Soc. (1930) 2615–2635
233. Day, J. N. E.; Ingold, C. K., "*Mechanism and kinetics of carboxylic ester hydrolysis and carboxyl esterification*", Trans. Faraday Soc. (1941) **37**, 686–705
234. Bunton, C. A.; Lewis, T. A.; Llewellyn, D. R.; Vernon, C. A., "*Mechanisms of reactions in the sugar series. Part I. The acid-catalysed hydrolysis of α- and β-methyl and α- and β-phenyl D-glucopyranosides*", J. Chem. Soc. (1955) 4419–4423
235. McIntyre, D.; Long, F. A., "*Acid-catalyzed Hydrolysis of Methylal. I. Influence of Strong Acids and Correlation with Hammett Acidity Function*", J. Am. Chem. Soc. (1954) **76**, 3240–3242
236. Buncell, E.; Bradley, P. R., "*The acid-catalyzed hydrolysis of methyl 2-chloro-2-deoxy-β-D-glucopyranoside*", Can. J. Chem. (1967) **45**, 515–519
237. Marshall, R. D., "*Rates of Acid Hydrolysis of 2-substituted Methyl Glucopyranosides*", Nature (1963) **199**, 998–999
238. Armour, C.; Bunton, C. A.; Patai, S.; Selman, L. H.; Vernon, C. A., "*Mechanisms of reactions in the sugar series. Part III. The acid-catalysed hydrolysis of t-butyl β-D-glucopyranoside and other glycosides*", J. Chem. Soc. (1961) 412–416
239. Moelwyn-Hughes, E. A., "*The kinetics of the hydrolysis of certain glucosides, part III.; β-methylglucoside, cellobiose, melibiose, and turanose*", Trans. Faraday Soc. (1929) **25**, 503–520
240. Moggridge, R. C. G.; Neuberger, A., "*Methylglucosaminide: Its Structure, and the Kinetics of its Hydrolysis by Acids*", J. Chem. Soc. (1938) 745–750
241. Foster, A. B.; Horton, D.; Stacey, M., "*Amino-sugars and Related Compounds. Part II. Observations on the Acidic Hydrolysis of Derivatives of 2-Amino-2-deoxy-D-glucose (D-Glucosamine)*", J. Chem. Soc. (1957) 81–85
242. Timell, T. E.; Enterman, W.; Spencer, F.; Soltes, E. J., "*The Acid hydrolysis of glycosides. II. Effect of substituents at C-5*", Can. J. Chem. (1965) **43**, 2296–2305
243. Richards, G. N., "*Hydrolysis of glycosides and cyclic acetal*", Chem. Ind. (London) (1955) 228
244. Dee, K. K.; Timell, T. E., "*The acid hydrolysis of glycosides: III. Hydrolysis of O-methylated glucosides and disaccharides*", Carbohydr. Res. (1967) **4**, 72–77
245. Riiber, C. N.; Sørensen, N. A., "*Anomeric sugars*", Kgl. Norske Videnskab. Sel skabs, Skrifter (1938), (No. 1), 38 pp

246. Moelwyn-Hughes, E. A., "*The kinetics of the hydrolysis of certain glucosides (Salicin, arbutin and phloridzin)*", Trans. Faraday Soc. (1928) **24**, 309–321;
247. Moelwyn-Hughes, E. A., "*The kinetics of the hydrolysis of certain glucosides, part II.:- Trehalose, a methylglucoside and tetramethyl-a-methylglucoside*", Trans. Faraday Soc., (1929) **25**, 81–92
248. Nath, R. L.; Rydon, H. N., "*The influence of structure on the hydrolysis of substituted phenyl β-d-glucosides by emulsin*", Biochem. J. (1954) **57**, 1–10
249. Hall, A. N.; Hollingshead, S.; Rydon, H. N., "*The acid and alkaline hydrolysis of some substituted phenyl α-D-glucosides*", J. Chem. Soc. (1961) 4290–4295
250. Banks, B. E. C.; Meinwald, Y.; Rhind-Tutt, A. J.; Sheft, I.; Vernon, C. A., "*Mechanism of reactions in the sugar series. Part IV. The structure of the carbonium ions formed in the acid-catalysed solvolysis of glucopyranosides*", J. Chem. Soc. (1961) 3240–3246
251. Timell, T. E., "*The Acid Hydrolysis of Glycosides: I. General Conditions and the Effect of the Nature of the Aglycone*", Canad. J. Chem. (1964) **42**, 1456–1472
252. Augestad, I.; Berner, E.; Weigner, E., "*Chromatographic separations of anomeric glycosides*", Chem. Ind. (London) (1953) 376–377
253. Augestad, I.; Berner, E., "*Chromatographic separation of anomeric glycosides. II. New crystalline methylfuranosides of galactose, arabinose, and xylose*", Acta Chem Scand. (1954) **8**, 251–256
254. Augestad, I.; Berner, E., "*Chromatographic separation of anomeric glycosides. III. Crystalline methylfuranosides of L-fucose, D-ribose, and L-rhamnose*", Acta Chem. Scand. (1956) **10**, 911–916
255. Blom, J., "*Ein Beitrag zur Kenntnis der Konfiguration und der Konformation ano merer Aldosen und deren Glykoside*", Acta Chem. Scand. (1961) **15**, 1667–1675
256. Krieble, V. K., "*Activities and the Hydrolysis of Sucrose with Concentrated Ac ids*", J. Am. Chem. Soc. (1935) **57**, 15–19
257. Krieble, V. K.; Holst, K. A., "*Amide hydrolysis with high concentrations of mineral acids*", J. Am. Chem. Soc. (1938) **60**, 2976–2980
258. Leiniger, P. M.; Kilpatrick, M., "*The Inversion of Sucrose*", J. Am. Chem. Soc. (1938) **60**, 2891–2899
259. Moelwyn-Hughes, E. A., "*The temperature coefficient of the inversion of cane sugar*", Z. Physik. Chem. (1934) **B26** 281–287
260. Heidt, L. J.; Purves, C. B., "*The Unimolecular Rates of Hydrolysis of 0.01 Molar Methyl- and Benzylfructofuranosides and -Pyranosides and of Sucrose in 0.00965 Molar Hydrochloric Acid at 20 to 60°*", J. Am. Chem. Soc. (1938) **60**, 1206–1210
261. Capon, B.; Thacker, D., "*The mechanism of the hydrolysis of glycofuranosides*", J. Chem. Soc. (B), Phys. Org. (1967) 185–189
262. Ceder, O., "*A kinetic study of the acid hydrolysis of cyclic acetals*", Arkiv. Kemi (1954) **6**, 523–535
263. Salomaa, P.; Kankaanperä, A., "*Hydrolysis of 1,3-dioxolane and its alkyl-substituted derivatives. I. Structural factors influencing the rates of hydrolysis of a series of methyl-substituted dioxolanes*", Acta Chem. Scand. (1961) **15**, 871–878
264. Fife, T. H.; Jao, L. K., "*Substituent Effects in Acetal Hydrolysis*", J. Org. Chem. (1965) **30**, 1492–1495
265. Fife, T. H.; Hagopian, L., "*Steric Effects in Ketal Hydrolysis*", J. Org. Chem. (1966) **31**, 1772–1775
266. van Eikeren, P., "*Models for Glycoside Hydrolysis. Synthesis and Hydrolytic Studies of the Anomers of a Conformationally Rigid Acetal*", J. Org. Chem. (1980) **45**, 4641–4645
267. Deslongchamps, P.; Li, S.; Dory, Y. L., "*Hydrolysis of α- and β-Glycosides. New Experimental Data and Modeling of Reaction Pathways*", Org. Lett. (2004) **6**, 505–508
268. Deslongchamps, P., *Stereoelectronic Effects in Organic Chemistry*, Pergamon Press, Oxford, England, 1983;
269. Kirby, A. J., "*Stereoelectronic effects on acetal hydrolysis*", Acc. Chem. Res. (1985) **17**, 305

References

270. Kirby, A. J., *The Anomeric Effect and Related Stereoelectronic Effects at Oxygen*, Springer-Verlag, New York, 1983
271. Cordes, E. H.; Bull, H. G., *Transition State in Biochemical Processes*, Gandour, D. R.; Showen, R. L. (Eds.), Plenum Press, New York, 1978
272. BeMiller, J. N.; Doyle, E. R., "*Acid-catalyzed hydrolysis of alkyl α-D- glucopyranosides*", Carbohydr. Res. (1971) **20**, 23–30
273. Li, S.; Kirby, A. J.; Deslongchamps, P., "*First experimental evidence for a synperiplanar stereoelectronic effect in the acid hydrolysis of acetal*", Tetrahedron Lett. (1993) **34**, 7757–7758
274. Ratcliffe, A. J.; Mootoo, D. R.; Webster, C.; Fraser-Reid, B., "*Concerning the antiperiplanar lone pair hypothesis: oxidative hydrolysis of conformationally restrained 4-pentenyl glycosides*", J. Am. Chem. Soc. (1989) **111**, 7661–7662
275. Gupta, R. B.; Franck, R. W., "*Direct experimental evidence for cleavage of both exo- and endo-cyclic carbon-oxygen bonds in the acid-catalyzed reaction of alkyl β-tetrahydropyranyl acetals*", J. Am. Chem. Soc. (1987) **109**, 6554–6556
276. Deslongchamps, P.; Dory, Y. L.; Li, S., "*1994 R.U. Lemieux Award Lecture Hydrolysis of acetals and ketals. Position of transition states along the reaction coordinates, and stereoelectronic effects*", Can. J. Chem. (1994) **72**, 2021–2027
277. Ishij, T.; Ishizu, A.; Nakano, J., "*Acid hydrolysis of methyl chlorodeoxyglycosides*", Carbohydr. Res. (1976) **48**, 33–40
278. Liras, J. L.; Anslyn, E. V., "*Exocyclic and Endocyclic Cleavage of Pyranosides in Both Methanol and Water Detected by a Novel Probe*", J. Am. Chem. Soc. (1994) **116**, 2645–2646
279. Frank, R. W., "*The mechanism of β-glycosidases: A reassessment of some seminal papers*", Bioorg. Chem. (1992) **20**, 77–88
280. Nerinckx, W.; Desmet, T.; Claessens, M., "*Itineraries of enzymatically and non-enzymatically catalyzed substitutions at O-glycopyranosidic bonds*", ARKIVOC (2006) XIII, 90–116
281. Sinnot, M. L.; Jencks, W. P., "*Solvolysis of D-Glucopyranosyl Derivatives in Mixtures of Ethanol and 2, 2, 2-Trifluoroethanol*", J. Am. Chem Soc. (1980) **102**, 2026–2032
282. Amyes, T. L.; Jenks, W. P., "*Concerted Bimolecular Substitution Reactions of Acetal Derivatives of Propionaldehyde and Benzaldehyde*", J. Am. Chem. Soc. (1989) **111**, 7900–7909
283. Kurzynski, M., "*Enzymic catalysis as a process controlled by protein conformational relaxation*", FEBS Lett. (1993), **328**, 221–224
284. Huang, X.; Tanaka, K. S. E.; Bennet, A. J., " *Glucosidase-Catalyzed Hydrolysis of α-D-Glucopyranosyl Pyridinium Salts: Kinetic Evidence for Nucleophilic Involvement at the Glucosidation Transition State*", J. Am. Chem. Soc. (1997) **119**, 11147–11154
285. Berti, P. J.; Tanaka, K. SA. E., "*Transition state analysis using multiple kinetic isotope effects: mechanisms of enzymatic and non- enzymatic glycoside hydrolysis and transfer*", Adv. Phys. Org. Chem. (2002) **37**, 239–314
286. Vocadlo, D. J.; Wicki, J.; Rupitz, K.; Withers, S. G., "*Mechanism of Thermoanaerobacterium saccharolyticum -Xylosidase: Kinetic Studies*", Biochemistry (2002) **41**, 9727–9735
287. Lorthiois, E.; Meyyappan, M.; Vasela, A., "β *-Glycosidase inhibitors mimicking the pyranoside boat conformation*", Chem. Commun. (2000) 1829–1830
288. Murray, T. F.; Kenyon, W. O., "*The Rates of Formation of Sulfoaliphatic Acids*", J. Am. Chem. Soc. (1940) **62**, 1230–1233
289. Jeffery, E. A.; Satchell, D. P. N., "*The mechanism of sulphoacetic acid formation in the system $H_2SO_4 - Ac_2O - AcOH$*", J. Chem. Soc. (1962) 1913–1917
290. Germain, A.; Commeyras, A., "*Mechanism of the C-acylation of aromatic and ethylenic compounds. XV. Kinetic study of the formation of acetylium ion in acetic anhydride solutions in the presence of trifluoromethanesulfonic and fluorosulfonic acids*", Bull Soc. Chim. (France) (1973) 2532–2537

291. Germain, A.; Commeyras, A.; Casadevall, A., "*Mechanism of the C-acylation of aromatic and ethylenic compounds. XVI. Kinetic study of the acetylation of aromatic compounds by acetic anhydride in the presence of strongly protonating ac ids*", Bull. Soc. Chim. (France) (1973) 2537–2543
292. Dasgupta, F.; Singh, P. P.; Srivastava, H. C., "*Acetylation of carbohydrates using ferric chloride in acetic anhydride*", Carbohydr. Res. (1980) **80**, 346–349
293. Dasgupta, F.; Singh, P.; Srivastava, H. C., "*Use of ferric chloride in carbohydrate reactions. Part V. Acetolysis of methyl hexopyranosides using ferric chloride in acetic anhydride*", Indian J. Chem. (1988) **27B**, 527–529
294. Guthrie, R. D.; McCarthy, J. F., "*Acetolysis*", Advan. Carbohydr. Chem. (1967) **22**, 11–23
295. Zaccari, d. G.; Snyder, J. P.; Peralta, J. E; Taurian, O. E.; Coutreras, R. H.; Barone, V., "*Natural J Coupling (NJC) analysis of the electron lone pair effect on NMR couplings. Part 2. The anomeric effects on 1 J (C, H) couplings and its dependence on solvent*", Mol. Phys. (2002) **100**, 705–715
296. Pinto, B. M.; Johnston, B. D.; Nagelkerke, R., "*Solvent and temperature dependence of the anomeric effect in 2[(4-methoxyphenyl)seleno]-1, 3-dithianes. Dominance of the orbital interaction component*", J. Org. Chem. (1988) **53**, 5668–5672
297. Franks, F.; Lillford, P. J.; Robinson, G., "*Isomeric equilibration of monosaccharides in solution: influence of solvent and temperature*", J. Chem. Soc. Faraday Trans 1 (1989) **85**, 2417–2426
298. Paulsen, H.; Friedmann, M., "*Conformational analysis. I. Dependence of the syn-1,3-diaxial interaction on the substituents and solvents. Conformational equilibria of D-idopyranose derivatives*", Chem. Berichte (1972) **105**, 705–717
299. Bailey, W. F.; Eliel, E. L., "*Conformational Analysis. XXIX. 2-Substituted and 2, 3-disubstituted 1, 3-dioxanes. Generalized and reverse anomeric effects*", J. Am. Chem. Soc. (1974) **96**, 1798–1806
300. McPhail, D. R.; Lee, J. R.; Fraser-Reid, B., "*Exo and endo activation in glycoside cleavage: acetolysis of methyl alpha- and beta-glucopyranosides*", J. Am. Chem. Soc. (1992) **114**, 1905–1906
301. Lemieux, R. U., "*Some Implications in Carbohydrate Chemistry of Theories Relating to the Mechanisms of Replacement Reactions*", Adv. Carbohydr. Chem. (1954) **9**, 1–57
302. Capon, B., "*Mechanism in carbohydrate chemistry*", Chem. Rev. (1969) **69**, 407–498
303. Lindberg, B., "*Action of strong acids on acetylated glucosides. III. Strong acids and aliphatic glucoside tetraacetates in acetic anhydride-acetic acid solutions*", Acta Chem. Scand. (1949) **3**, 1153–1169
304. Asp, L.; Lindberg, B., "*Action of strong acids on acetylated glycosides. VII. Transglycosidation of xylosides*", Acta Chem Scand. (1950) **4**, 1446–1449
305. Lonnberg, H.; Kankaanperä, A.; Haapakka, K., "*The acid-catalyzed hydrolysis of β-D-xylofuranosides*", Carbohydr. Res. (1977) **56**, 277–287
306. Lonnberg, H.; Kulonpaa, A., "*Mechanisms for the acid-catalyzed hydrolysis of some alkyl aldofuranosides with the trans-1,2-configuration*", Acta. Chem. Scand. (1977) **A31**, 306–312
307. Lemieux, R. U. "*Rearrangements and Isomerizations in Carbohydrate Chemistry*". in Molecular Rearrangements, Part Two; de Mayo, P., Ed.; Interscience, New York, 1964; p. 709–769
309. Altona, C., Ph.D., *Thesis*, University of London, London, England, 1964
309. Lemieux, R. U., Personal communication to H. Booth in 1983, as quoted by Booth, H.; Khedhair, K. A. in "*Endo-Anomeric and exo-anomeric effects in 2-substituted tetrahydropyrans*", J. Chem. Soc. Chem. Commun. (1985) 467–468, Reference 13
310. Praly, J.-P.; Lemieux, R. U., "*Influence of solvent on the magnitude of the anomeric effect*", Can. J. Chem. (1987) **65**, 213–223
311. Miljkovic, M.; Habash-Marino, M., "*Acetolysis of permethylated O-alkyl glycopyranosides: kinetics and mechanism*", J. Org. Chem. (1983) **48**, 855–860

312. Miljkovic, M.; Yeagley, D.; Deslongchamps, P.; Dory, Y. L., "*Experimental and Theoretical Evidence of Through-Space Electrostatic Stabilization of the Incipient Oxocarbenium Ion by an Axially Oriented Electronegative Substituent During Glycopyranoside Acetolysis*", J. Org. Chem. (1997) **62**, 7597–7604
313. Tvaroška, I.; Bleha, T., "*Anomeric and Exo-Anomeric Effects in Carbohydrate Chemistry*", Adv. Carbohydr. Chem. Biochem. (1989) **47**, 45–123
314. Fuchs, B.; Ellencweig, A.; Tartakovsky, E.; Aped, P., "*Solvent Polarity and the Anomeric Effect*", Angew. Chem. Int. Eng. (1986) **25**, 287–289

Chapter 13
Synthesis of Polychiral Natural Products from Carbohydrates

Macrolide Antibiotics: Erythronolides A and B

Stereoselective synthesis of polychiral natural products is the most challenging problem for a synthetic organic chemist. The stereoselective synthesis of macrolide antibiotics represents one such difficult problem. They consist of macrocyclic lactone rings with many hydroxylated and methylated chiral carbons. In addition to that the macrocyclic lactones (macrolides) are usually glycosylated with amino sugars.

The striking resemblance of macrocyclic ring structure of macrolide antibiotics to "giant" branched chain sugars [3] inspired Woodward to describe magnamycin (carbomycin) as a *giant sugar* having at the same time the properties of a long-chain aliphatic acid. A realization that some 12- and 14-membered macrocyclic lactone rings can be dissected into two carbohydrate-like structural fragments prompted Miljkovic et al. [4–8] in 1972 to investigate the possibility of using carbohydrates for stereoselective synthesis of these stereochemically highly complex natural products because the chemical transformations of sugar molecules were known to often proceed highly stereoselectively.

Dissection of the macrocyclic lactone rings of methymycin, erythromycins A and B, picromycin, and narbomycin, as depicted in Fig. 13.1, afforded for methymycin one seven-carbon atom fragment (C1–C7) (Segment A) and one five-carbon atom fragment (C9–C13) (segment B) and for erythromycins A and B, picromycin, and Narbomycin two seven-carbon atom fragments (C1–C7) (Segment A) and (C9–C15) (Segment B) [7, 8].

Consequently, the construction of carbon skeleton of macrolides *1–5* (Fig. 13.1) from corresponding fragments would require that the C8 carbon atom be introduced either immediately before or during the coupling of the two fragments into the open-chain precursor of a given macrolide aglycone. An important advantage of dissecting the macrocyclic lactone rings as depicted in Fig. 14.1 is that it produces structurally similar fragments. This becomes particularly evident if Segments A and B of methynolide, erythronolides A and B, picronolide, and narbonolide are represented in the form of carbohydrate pyranoside-like structures (Figs. 13.2 and 13.3).

There are a couple of review articles dealing with this subject [1, 2].

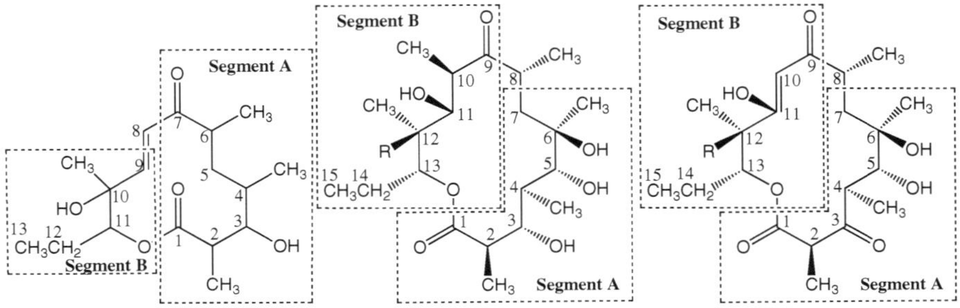

424 13 Synthesis of Polychiral Natural Products from Carbohydrates

Fig. 13.1

1, Methynolide
2, R = OH, Erythronolide A
3, R = H, Erythronolide B
4, R = OH, Picronolide
5, R = H, Narbonolide

Segments A

6 / *7* Methynolide

8 / *9* Erythronilides A and B

10 / *11* Picronolide and Narbonolide

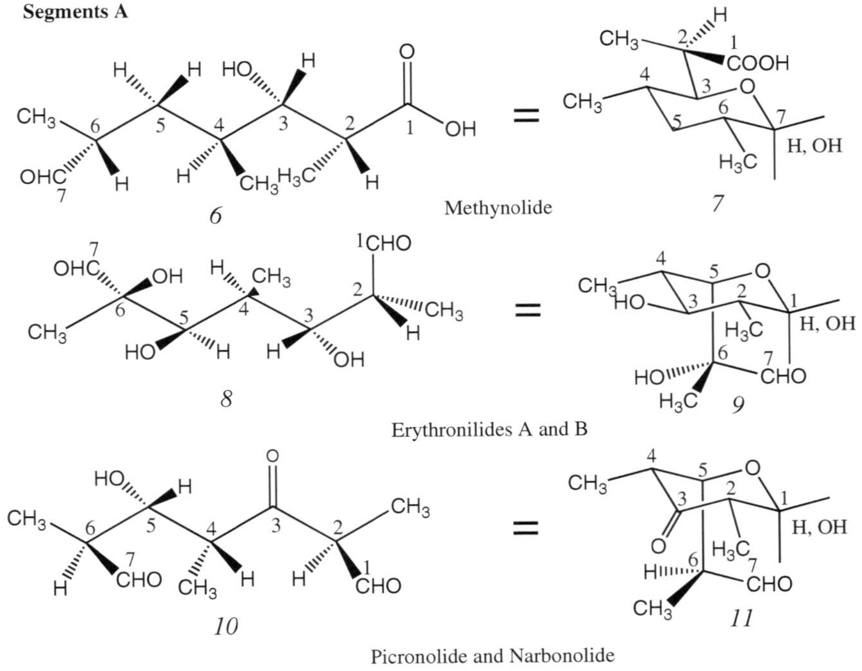

Fig. 13.2

Segment A of all five macrolide aglycones (*7*, *9*, and *11* in Fig. 13.2) has two structurally identical carbon atoms: the C2 and the C4 carbons in *9* and *11* and the C4 and the C6 carbons in *7*. All these carbon atoms have their methyl groups equatorially oriented when represented in the $^4C_1$ conformation of a pyranoside-like structure. The C3 carbon atom in *9* and *11* is oxygenated, whereas in *7* it is not linked

Macrolide Antibiotics: Erythronolides A and B   425

to oxygen. Finally, the side chain in 9 and *11*, consisting of C6 and the C7 carbon atoms of macrolide aglycones *2–5*, is axially oriented and is in *cis* configuration with respect to the C4 methyl group; however, in *7*, the side chain consisting of C1 and the C2 carbon atoms of methynolide is equatorially oriented and is in *trans* configuration with respect to the C4 methyl group.

Segment B of all five macrolide aglycones *12–21* (Fig. 13.3) has as common structural features the same side chain (ethyl group) and one configurationally identical carbon atom: the C13 carbon in *14–21* and the C11 carbon in *12* or *13*. Further, the C12 carbon in *16* and *20*, the C10 carbon in *13*, as well as the C12 carbon in *17* and *21* are structurally identical. It is important to note that the axially oriented C10

Segments B

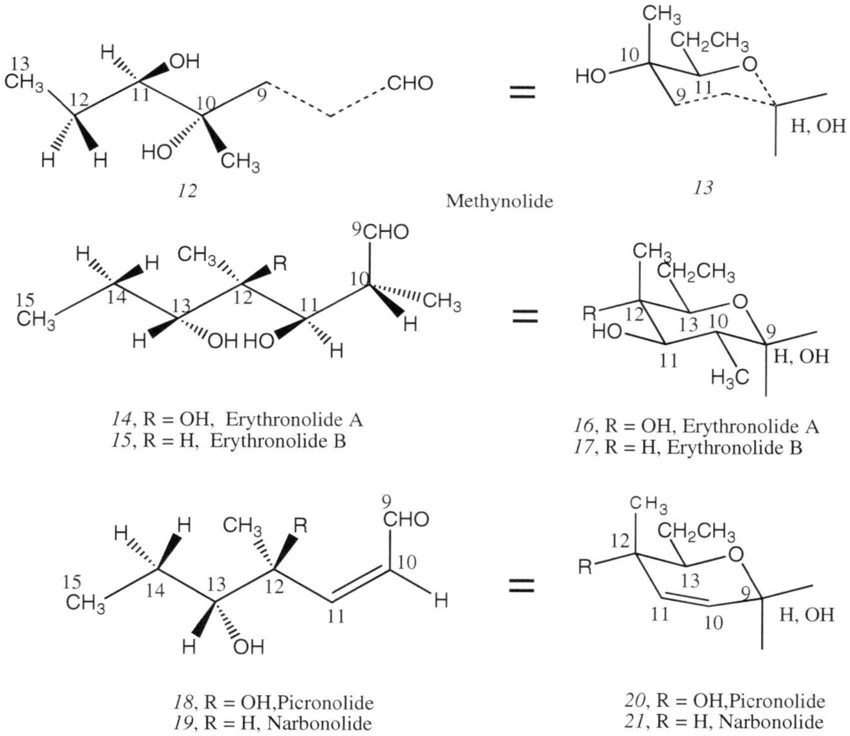

12
Methynolide

13

14, R = OH, Erythronolide A
15, R = H, Erythronolide B

16, R = OH, Erythronolide A
17, R = H, Erythronolide B

18, R = OH, Picronolide
19, R = H, Narbonolide

20, R = OH, Picronolide
21, R = H, Narbonolide

Fig. 13.3

methyl group in *13* and the C12 methyl group in *16–21* are in the *cis* configuration with respect to the equatorially oriented C11 ethyl group in *13* or with the C13 ethyl group in *16–21*.

If one compares the structure of D-glucopyranose with structures of segments A and B of erythronolides A and B, represented as glycopyranosides (synthons *9* and *16*, respectively), it becomes apparent that the stereoselective conversion of D-glucose into synthons *9* and *16* requires the following transformations:

(1) The conversion of the C5 hydroxymethyl group of a D-glucopyranoside derivative to the C5 ethyl group, i.e., the synthesis of the 6-deoxy-6-*C*-methyl homolog of D-glucopyranoside. This represents the synthesis of the side chain (ethyl group) of synthon *16*, which will later become the C14 and the C15 carbons of erythronolides A and B.
(2) Introduction of an axial methyl group at the C4 carbon atom of a D-glucopyranoside derivative whereby a branched chain sugar will be obtained in which the C4 quaternary carbon has the (*S*) configuration (this represents the synthesis of the C12 carbon of erythronolide A).
(3) Replacement of the equatorially oriented C2 hydroxyl group of a D-glucopyranoside derivative with an equatorially oriented methyl group [synthesis of the C2 and the C10 carbon atoms of erythronolide A, both having the (*R*) configuration].
(4) Inversion of the configuration of the C5 carbon of a D-glucopyranoside derivative, resulting in the formation of the corresponding L-idopyranoside derivative (synthesis of the C4 carbon of erythronolide A).
(5) Replacement of the equatorial C4 hydroxyl group of a D-glucopyranoside derivative, with an axial methyl group, resulting in a 4-deoxy-4-*C*-methyl branched chain sugar (synthesis of the C4 carbon of erythronolide B).
(6) The stereoselective addition of an alkyl group to the exocyclic C6 carbonyl carbon of 7-deoxy-L-*ido*-heptopyranoside-6-ulose derivative, resulting in a chiral C6 tertiary alcohol with a (*R*) configuration (synthesis of the C6 carbon of erythronolide A).

Except for the replacement of the primary C6 hydroxyl group with a methyl group, all other chemical transformations of D-glucopyranoside derivatives required finding a way to efficiently control the stereochemistry of reactions 2–6.

At the time Miljkovic et al. started this investigation in 1972, the configurational determination of quaternary carbon of branched chain sugars and the stereoselective synthesis of quaternary C12 carbon of erythronolide A (having the C4 methyl group in synthon *16* axially oriented) seemed to be two problems that have to be dealt with first.

Configurational determination of a quaternary carbon of branched chain sugars posed, at the time of these pioneering studies, a serious problem, since there was no single physico-chemical method available by which one could make a quick, reliable, and unequivocal assignment of the configuration of the quaternary branching carbon. In search for such a method, published studies on conformational equilibrium of methylcyclohexane by C13 NMR spectroscopy turned out to be very helpful [9–11]. In these publications it was reported that the C13 chemical shift of an axial methyl group is shifted by ca. 6 ppm toward the higher magnetic field, as compared to the C13 chemical shift of an equatorial methyl group. This observation prompted Miljkovic et al. [5] to investigate whether the C13 chemical shifts of axial and equatorial methyl groups linked to a quaternary carbon atom of branched chain sugars could be used for configurational assignments. The study which followed established that it can and that the axial methyl group linked to the quaternary carbon is shifted by ca. 6.4 ppm upfield for the α-anomers and by ca. 5.2 ppm for the β-anomers relative to an equatorial methyl group.

Simultaneously with the C13 NMR studies, a study was undertaken on the addition of methyl lithium and methyl magnesium bromide to the C4 carbon of an appropriately protected D-glucopyranosid-4-ulose derivative. From previous studies it was known that the additions of Grignard reagents and organolithium compounds to the carbonyl group in carbohydrates were highly stereoselective [12] and that in certain cases branched chain sugars epimeric at the branching carbon [13, 14] were obtained, and in other instances branched chain sugars with the same configuration at the branching carbon [15] were obtained. It was, however, not known what, if anything, controls the stereochemistry of these additions and consequently it was concluded that the addition of Grignard reagents and/or alkyl- or aryllithium to oxo-sugars cannot be reliably predicted [16].

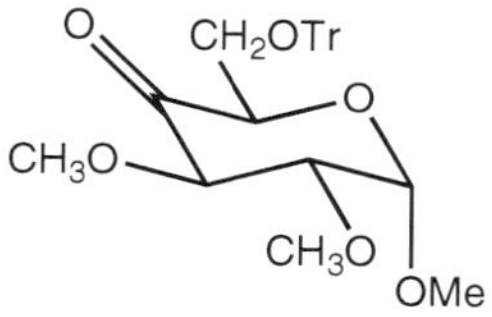

22

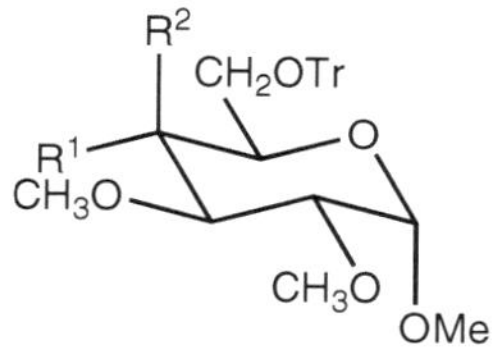

23, R^1 = OH; R^2 = CH_3
24, R^1 = CH_3; R^2 = OH

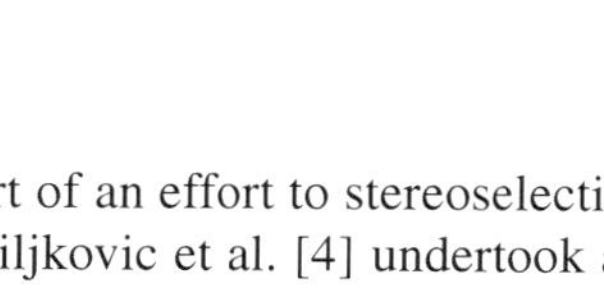

25

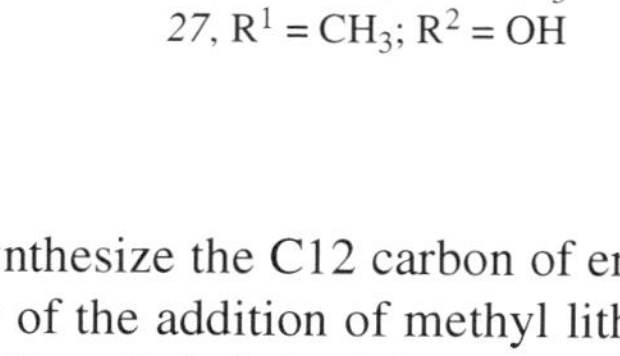

26, R^1 = OH; R^2 = CH_3
27, R^1 = CH_3; R^2 = OH

Fig. 13.4

As part of an effort to stereoselectively synthesize the C12 carbon of erythronolide A Miljkovic et al. [4] undertook a study of the addition of methyl lithium and methyl magnesium halides to methyl 2,3-di-*O*-methyl-6-*O*-triphenylmethyl-α- and β-D-*xylo*-hexo-pyranosid-4-ulose 22 and 25, respectively, in ether at −80°C (Fig. 13.4) and found that the reaction of 22 with methyl lithium (LiBr-free) afforded methyl 2,3-di-*O*-methyl-4-*C*-methyl-6-*O*-triphenylmethyl-α-D-glucopyranoside 23 as the only product in 70% yield. The reaction of 22 with methylmagnesium iodide proceeded again stereospecifically giving exclusively methyl 2,3-di-*O*-methyl-4-*C*-methyl-6-*O*-triphenylmethyl-α-D-galacto-pyranoside 24 (the C4 epimer of 22) in 94% yield (Fig. 13.4).

Contrary to the above results, methylmagnesium iodide and methyl lithium added nonstereoselectively and at a considerably slower rate to 4-*tert*-butylcyclohexanone **28** in ether and at –80°C, yielding in each case a mixture of both C1 epimers: *cis*-4-*tert*-butyl-1-methyl-cyclohexan-*r*-ol **29** and *trans*-4-*tert*-butyl-1-methyl-cyclohexan-*l*-ol **30** (Fig. 13.5). The isomer with equatorial methyl group was the preponderant product in both reactions.

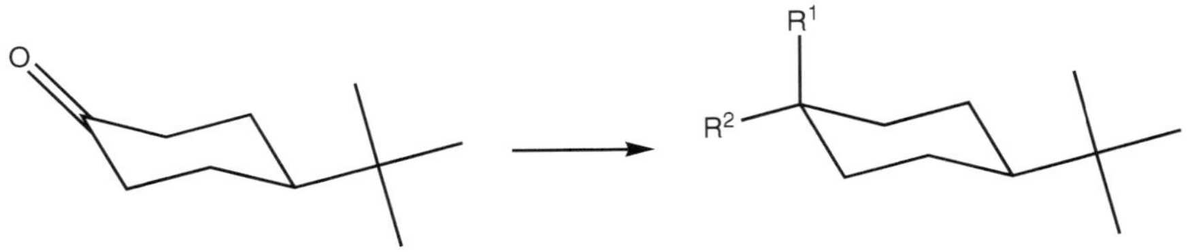

28

29, R^1 = OH; R^2 = CH_3
30, R^1 = CH_3; R^2 = OH

Fig. 13.5

The stereochemistry of the addition of Grignard reagent to the methyl α-D-*xylo*-hexo-pyranosid-4-ulose **22** was shown to depend on the reaction temperature, solvent [17, 18], and the nature of the halogen atom. Thus, treating an ethereal solution of **22** at reflux gave a mixture of both C4 epimers **23** and **24** in which the isomer with the methyl group equatorially oriented predominated in the 6:1 ratio. The dependence of stereochemistry of the addition of Grignard reagent upon the nature of the halogen atom and of the solvent was demonstrated in the following way: refluxing a 10:1 ether–tetrahydrofuran solution of **22** with methylmagnesium chloride gave a 1:1 mixture of C4 epimers **23** and **24**, whereas methylmagnesium iodide under the same reaction conditions gave a mixture of C4 epimers **23** and **24**, in which the axial isomer predominated in 2.3:1 ratio.

Dependence of the stereochemistry of addition of methyl lithium upon the anomeric configuration was discovered after the observation that the addition of methyl lithium to the methyl 2,3-di-*O*-methyl-6-*O*-triphenylmethyl-β-D-*xylo*-hexopyranosid-4-ulose **25** in ether at –80°C proceeded with considerable loss of stereoselectivity giving both C4 epimers, **26** and **27**, respectively, in ca. 3:1 ratio, with the axial epimer being the predominant product. However, the addition of Grignard reagent was unaffected by the anomeric configuration because the methylmagnesium iodide added to methyl 2,3-di-*O*-methyl-6-*O*-triphenylmethyl-β-D-*xylo*-hexopyranosid-4-ulose **25** in ether at –80°C again stereoselectively gave methyl 2,3-di-*O*-methyl-4-*C*-methyl-6-*O*-triphenylmethyl-β-D-galactopyranoside **27** as the only product.

The observed dependence of the stereochemistry of addition of methyl lithium to methyl 2,3-di-*O*-methyl-α- or β-D-*xylo*-hexopyranosid-4-ulose **22** and **25** upon the anomeric configuration and the lack of dependence of the stereochemistry of addition of methylmagnesium iodide upon **22** and **25** are discussed in Chapter 11.

Macrolide Antibiotics: Erythronolides A and B   429

The high stereoselectivity observed in the reduction of C2 keto group of methyl α- and β-D-*arabino*-hexopyranosid-2-ulose with sodium borohydride [19] and its dependence upon the anomeric configuration prompted a study of catalytic hydrogenation of methyl 4,6-*O*-benzylidene-2-deoxy-2-*C*-methylene-3-*O*-methyl-α- and β-D-*arabino*-hexopyranoside *31* and *32* (Fig. 13.6) in order to determine if the stereochemistry of catalytic hydrogenation is perhaps also controlled by the anomeric configuration [6].

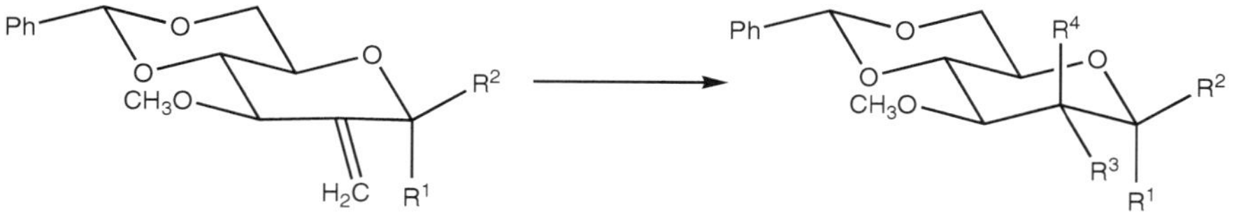

31, R^1 = OCH_3; R_2 = H
32, R^1 = H; R^2 = OCH_3

33, R^1 = R^3 = H; R^2 = OCH_3; R^4 = CH_3
34, R^1 = OCH_3; R^2 = R^3 = H; R^4 = CH_3
35, R^1 = OCH_3; R^2 = R^4 = H; R^3 = CH_3

Fig. 13.6

Unfortunately it has been found that whereas the catalytic hydrogenation of β-anomer proceeded highly stereoselectively (methyl 4,6-*O*-benzylidene-2-deoxy-2-*C*-methylene-3-*O*-methyl-β-D-*arabino*-hexopyranoside gave 2-deoxy-2-*C*-methyl-β-D-mannopyranoside *33* as the only product in 84% yield), the stereoselectivity of hydrogenation of the α-anomer *31* was not high and depended on the nature of the catalyst and the solvent used. Raney nickel catalyst and nonpolar solvents favored the isomer with the equatorial C2 methyl group (*35*) (*35*:*34* ratio was 2.9:1), whereas platinum and polar solvents favored the formation of the axial C2 methyl group (*34*) (*35*:*34* ratio 1:3.1).

With these key steps solved, the segments A and B of erythronolide A have been synthesized [7, 8] (Figs. 13.7 and 13.8).

A couple of years after Miljkovic et al. published their first studies on the synthesis of erythronolide A from D-glucose, Hanessian et al. [20–22] approached the synthesis of erythronolide A in an essentially identical way, dissecting the erythronolide 14-membered lactone ring in exactly the same way and synthesizing the segments A and B in a very similar manner. For that reason we will not attempt to describe their efforts.

In 1981, Kochetkov et al. reported the synthesis of the C1–C6 segment [23] of a number of 14-membered macrolide antibiotics and the synthesis of the C9–C13 segments [24] of erythronolides A and B and oleandonolide. In 1984 they revised the synthesis of the C9–C13 segment of erythronolide A [25] and finally in 1989, they published the full paper on stereoselective synthesis of erythronolides A and B from 1,6-anhydro-β-D-glucopyranose (levoglucosan) [26].

The synthetic strategy employed by Kochetkov's group is based on retrosynthetic considerations and envisaged the assembly of the erythronolides A and B skeleton

Synthesis of the segment B of Erythronolide A

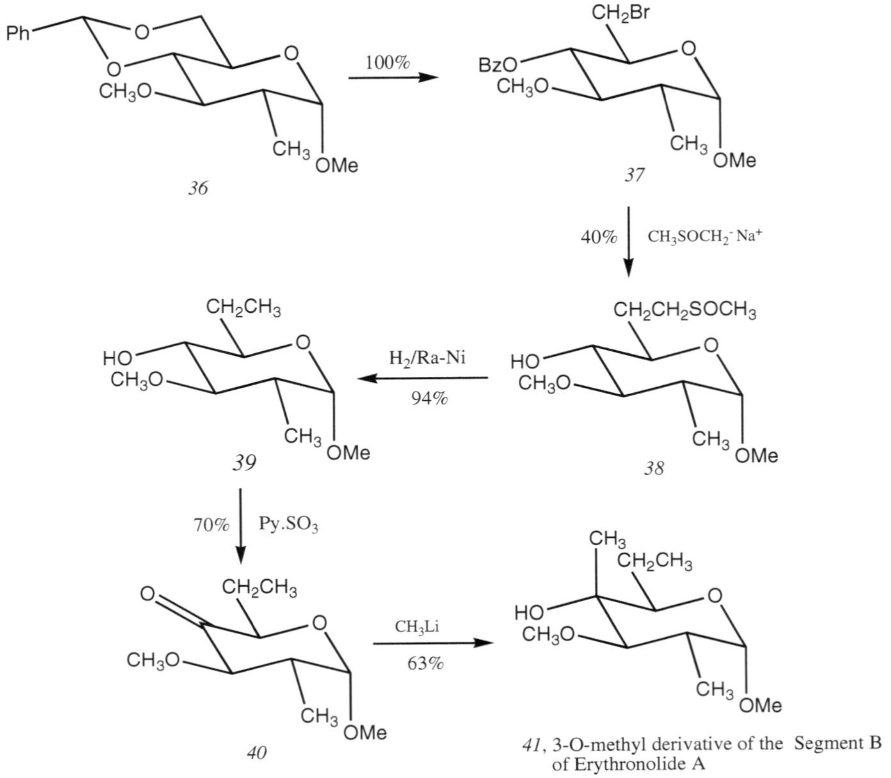

Fig. 13.7

in the C9–C13 + C7–C8 = C7–C13; C7–C13 + C1–C6 = C1–C13 sequence. The structures of C1–C6 and C9–C13 segments are shown in Fig. 13.9.

Bicyclic derivatives *59, 60,* and *61* were synthesized from 1,6-anhydro-β-D-glucopyranose (levoglucosan) *62* [25, 26, 27, 28].

Synthesis of the C1–C6 segment of erythronolides A and B is shown in Fig. 13.10. The mercaptolysis of *59* with 1,2-ethanedithiol in dichloromethane in the presence of borontrifluoro etherate at room temperature gave ethylene dithioacetal that was not isolated, but the reaction solution was cooled to –40°C and pyridine and acetic anhydride were added and the solution was kept for 2 h at –10°C. The acetate was again not isolated but the crude reaction mixture was treated with 1:1 mixture of acetone/2,2-dimethoxy propane in the presence of *p*-toluenesulfonic acid monohydrate giving *63* in 56.6% overall yield. Dethioacetalation of *63* with mercury (II) chloride–calcium carbonate mixture in acetonitrile–water gave the corresponding aldehyde *64* in 77% yield. Reaction of *64* with methylenetriphenylphosphorane

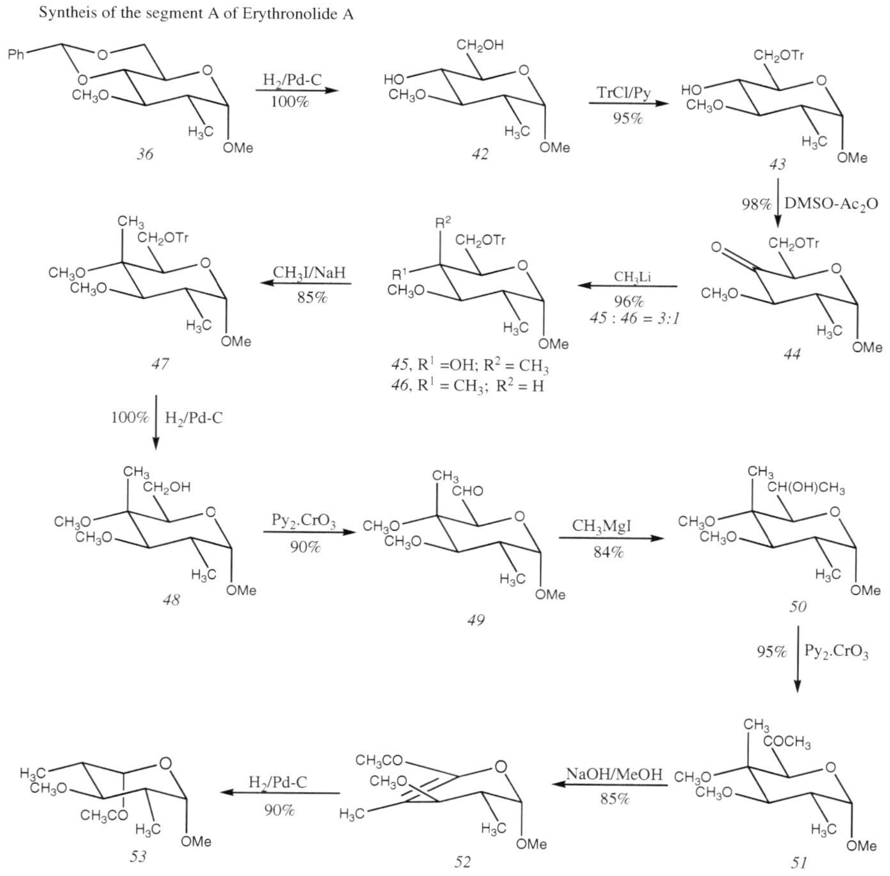

Fig. 13.8

gave a mixture of 6-*O*-acetyl 65 and 6-*O*-deacetylated olefin 66 which was treated with sodium methoxide in methanol to give 66 in 77.6% yield. Swern oxidation [29] of alcohol 66 followed by the addition of methylmagnesium chloride to the resulting C6 aldehyde and subsequent oxidation of the intermediate C6 hydroxyl group again by using the Swern procedure [loc. cit.] gave methyl ketone 67. Mild alkaline isomerization of 67 at the C5 proceeded in almost quantitative yield giving the thermodynamically more stable ketone 58 in which the C3 and the C5 hydroxyl groups are in the *cis* orientation.

Starting material for the synthesis of aldehyde 68, which is the C9–C15 segment of erythronolide B, was the bicyclic derivative 61 (Figs. 13.9 and 13.12) which has been previously synthesized [28] from 62 (Fig. 13.11).

Synthesis of the segment B of erythronolide B is shown in Fig. 13.12. Mercaptolysis of 61 and subsequent acetonation gave a dioxolan derivative 70 in 64% yield;

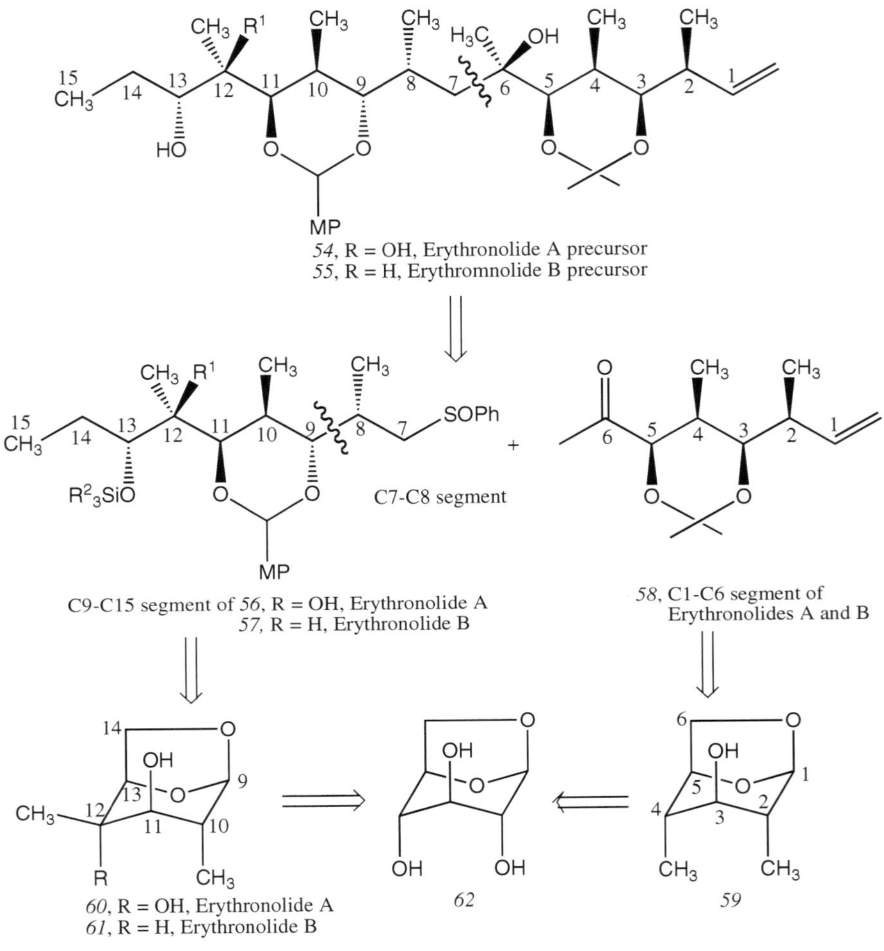

Fig. 13.9

the free C11 hydroxyl group in *70* was then protected as *p*-methoxybenzyl (PMB) ether yielding *71*. Mild acid hydrolysis of *71* gave the C13, C14 diol *72* that was converted [30] to hydroxy derivative *73*, by selective tosylation of the primary C14 hydroxyl group of *72*, the formation of the α-oxide, and opening of the latter with CH_3MgCl in the presence of a copper (I) salt. Since all of the above intermediates are extremely labile the reactions were performed as quickly as possible and without isolation of individual compounds. The protection of C13 hydroxyl group of *73* was accomplished by treating *73* with *tert*-butyldiphenylsilyl chloride, and the resulting thioacetal *74* was treated with mercuric (II) chloride–cadmium carbonate yielding the C9–C15 segment of erythronolide B (*75*).

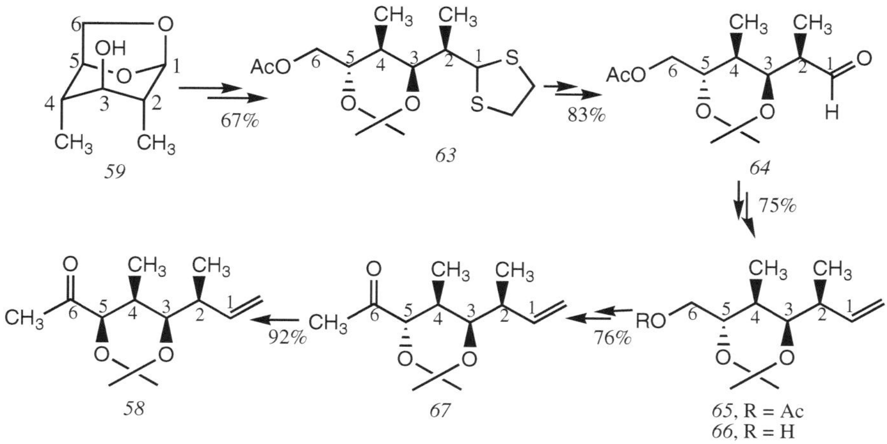

Macrolide Antibiotics: Erythronolides A and B   433

Fig. 13.10

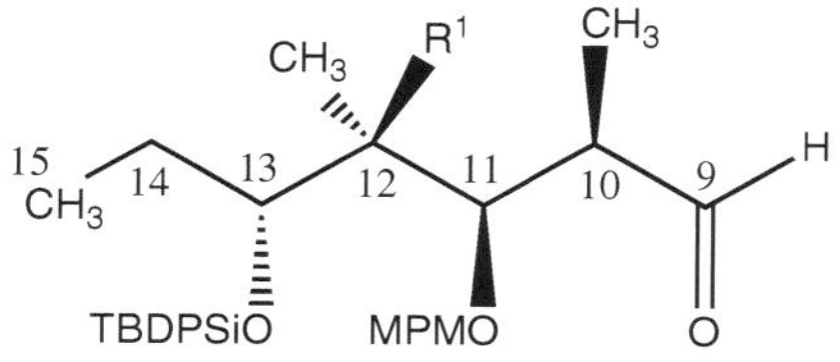

Fig. 13.11

The C9–C15 segment of erythronolide A *82* was prepared from the olefin *76* which was via stereoselective hydroxylation of the C4 methylene group with OsO$_4$-*N*-oxide-*N*-methyl-morpholine [31] converted to *60* (Fig. 13.13). The opening of the 1,6-anhydro ring of the di-*O*-benzyl derivative *77* by methanolysis (20% HCl/CH$_3$OH), whereby a mixture of methyl α- and β-D-glycopyranosides was obtained in ca. 3:1 ratio in 78% yield, was followed by oxidation of the primary hydroxyl group with DMSO/COCl$_2$/Et$_3$N in dichloromethane [29] giving in 83% yield the corresponding C14 aldehyde *79*, which with methylenetriphenylphosphorane gave the corresponding olefin *80* (80%). The C14–C15 double bond was then reduced into the ethyl group by LiAlH$_4$–CoCl$_2$ (80%) [32] giving *81*. Debenzylation of *81* by hydrogenation with Raney nickel in ethanol followed by transglycosylation of *82* with allyl alcohol in the presence of catalytic amounts of pyridinium *p*-toluenesulfonate (PPTS) gave a mixture of anomeric allyl glycosides *83*. Selective acetylation with acetic anhydride and pyridine gave the C11 monoacetyl derivative *84* in quantitative yield. Conversion of the monoacetate *84* to methoxymethyl (MOM) ether *85* followed by deacetylation proceeded also in quantitative yield to

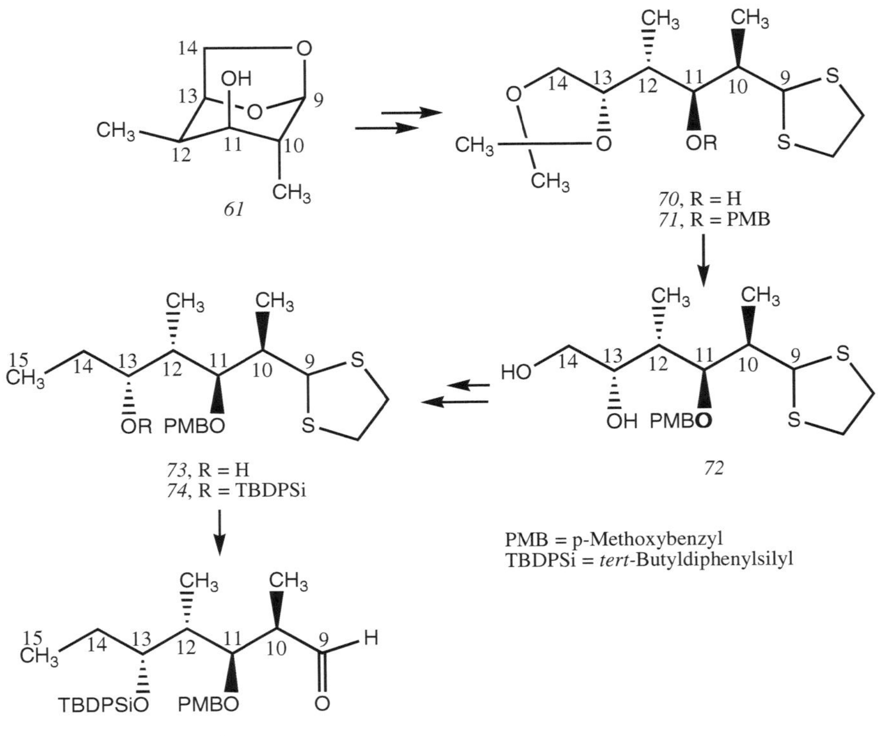

434   13 Synthesis of Polychiral Natural Products from Carbohydrates

Fig. 13.12

give C11 alcohol *86*. Alkylation of *86* with *p*-methoxybenzyl chloride (PMB chloride) gave the corresponding C11 PMB–ether *87*. The selectively protected allyl glycoside *87* was converted by the known method [33] to the free monosaccharides *88*, which was reduced with NaBH$_4$ in aqueous ethanol to give, in quantitative yield, the *89*. The selective monobenzoylation of the primary hydroxyl group gave the C9 benzoate *90*. Mild acid hydrolysis removed the MOM protection group from the C12 hydroxyl group of benzoate *90* giving diol *91*. Silylation of the C12 and C13 hydroxyl groups with *tert*-butyldimethylsilyl triflate (TBSOTf) gave the bis TBS–ether *92* which was then debenzoylated giving the C9 primary alcohol *93*. Swern oxidation [29] of *93* gave the aldehyde *94* representing the C9-C15 segment of erythronolide A.

Next step in the synthesis of erythronolides A and B was the stereoselective addition of the two-carbon fragment, that will be the C7–C8 carbon segment of erythronolides A and B, to the C9–C15 segments of erythronolides A and B *94* and *75*, to give the C7–C15 segment that will be coupled with the C1–C6 segment to give the seco-acids of erythronolides A and B. This step of synthesis is particularly challenging since it involves creation of two chiral carbons, C8 and C9, in the acyclic

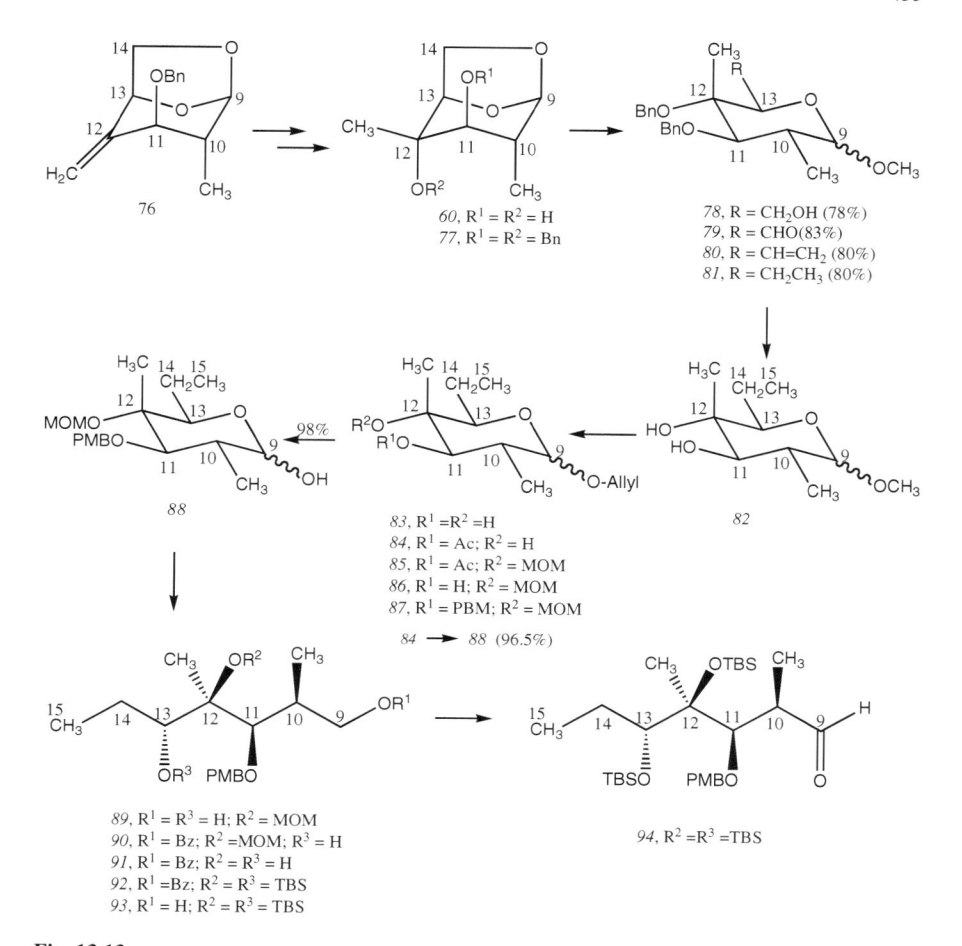

Fig. 13.13

substrates, eliminating thus the convenience of the stereocontrol of chemical transformations in pyranoside six-membered ring which was so useful in the synthesis of *94* and *75*.

Syn-selective aldol condensation of ethyl trityl ketone with aldehydes *76* and *95* gave the desired (8,9-*syn*-9,10-*anti*)-aldol *95* and *96* as the sole products (Fig. 13.14). Treatment of aldol *95* or *96* with DDQ in dichloromethane solution in the presence of molecular sieves (3 Å) [34] gave in 82–85% yield 9,11-*O*-*p*-methoxybenzylidene derivatives *97* or *98* as a single isomer at the methine carbon.

Reductive cleavage of the trityl ketones *97* or *98* with LiBHEt$_2$ [35] gave alcohols *99* and *100*, which were with Ph$_2$S$_2$–Bu$_3$P [36] readily converted to phenyl sulfides *101* and *102*. Oxidation of *101* and *102* with MCPBA (*m*-chloro-*p*-methylbenzoic

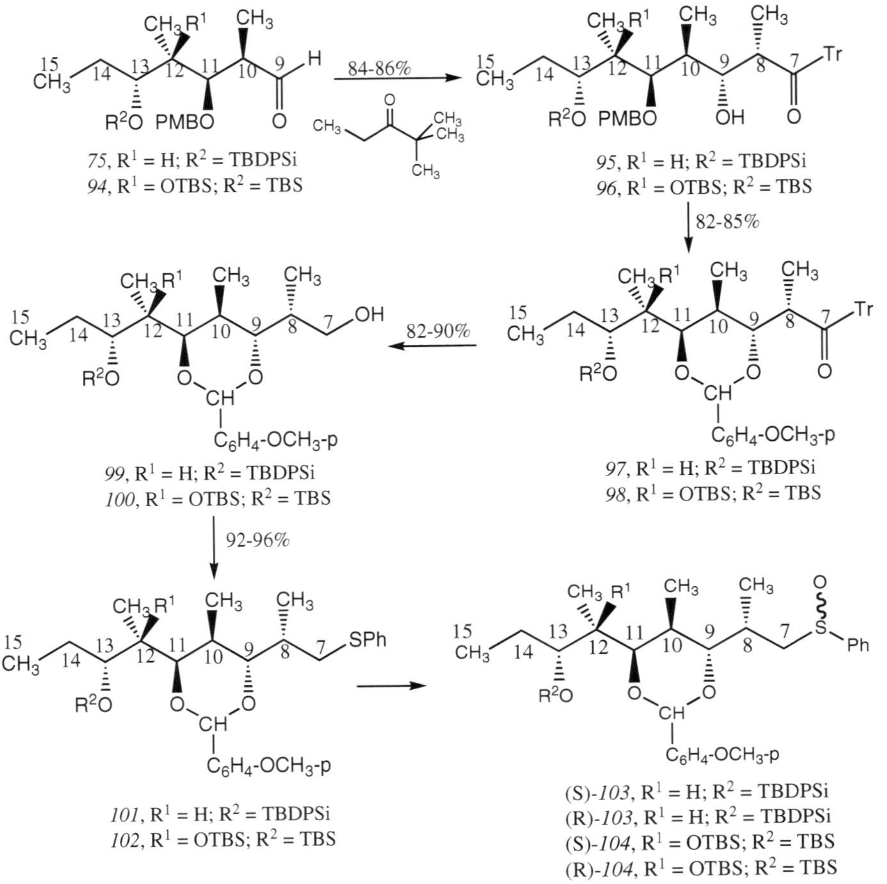

Fig. 13.14

acid) gave a mixture of epimeric sulfoxides: the (S)-epimers *103* and *104* and the (R)-epimers *103* and *104*, which can be easily separated by chromatography. Thus the C7–C15 segments of erythronolides A and B have been synthesized. The configuration at the sulfur atom has been determined on the basis of specific rotation of individual compounds [37].

The oxidation of phenyl sulfides *101* and *102* gave the (S)-isomer as the predominant product (65%) and the (R)-isomer as the minor product (33%). As it turned out, the (S)-isomer did not couple with the C1–C6 segment, whereas the (R)-isomer did give a mixture of two products in 7:1 ratio.

Since all attempts to change the selectivity of oxidation of sulfide *101* in favor of the desired (R)-isomer *103* failed, a smooth and convenient method of isomerization of sulfoxide (S)-isomer *103* was developed. Thus treatment of (S)-*103* with TFA (trifluoroacetic acid) in tetrahydrofuran-2,4,6-collidine at –78°C for 20 min followed by addition of water gave a mixture of (R)-*103* and (S)-*103* isomers in 77:23% ratio.

In this way, oxidation, separation, and isomerization of the (S)-isomer of the sulfide *101* was converted in high overall yield to (R)-sulfoxide *103*. The reaction of (R)-*103* with ketone *58* gave two products with high selectivity. The main coupling product *107* was obtained in 88% yield. By similar treatment the isomeric sulfoxides (R)-*104* and (S)-*104* were separated and (S)-sulfoxide was isomerized into mixture of (R)- and (S)-*105* in 75:22% ratio (Fig. 13.15).

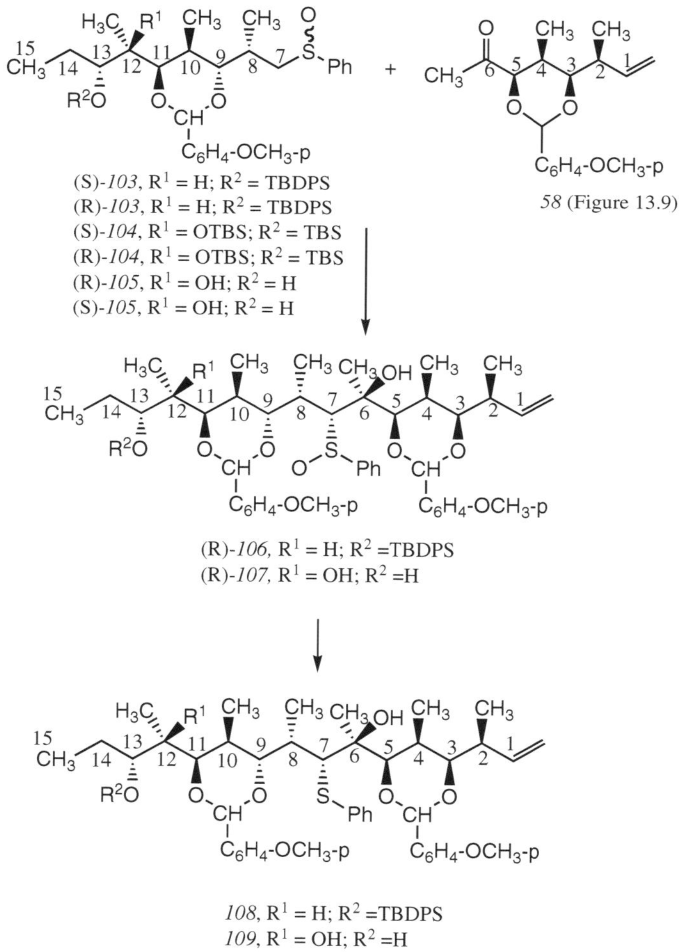

(S)-*103*, R^1 = H; R^2 = TBDPS
(R)-*103*, R^1 = H; R^2 = TBDPS
(S)-*104*, R^1 = OTBS; R^2 = TBS
(R)-*104*, R^1 = OTBS; R^2 = TBS
(R)-*105*, R^1 = OH; R^2 = H
(S)-*105*, R^1 = OH; R^2 = H

58 (Figure 13.9)

(R)-*106*, R^1 = H; R^2 = TBDPS
(R)-*107*, R^1 = OH; R^2 = H

108, R^1 = H; R^2 = TBDPS
109, R^1 = OH; R^2 = H

Fig. 13.15

The sulfoxide (R)-*104* was coupled to methyl ketone *58* in dry tetrahydrofuran at –60°C in the presence of lithium diisopropylamide (LDA). The coupling product

(R)-106 was obtained, after chromatography, in 41% yield. The sulfoxide group in the coupling product (R)-106 was immediately reduced with NaI/Na$_2$S$_2$O$_3$ in acetone since the sulfoxide is extremely labile. In this way the sulfide 108 was obtained in 84% yield.

An attempt to couple the anion of (R)-104 sulfoxide (the C7–C15 segment of erythronolide A) with the methyl ketone 58 (the C1–C6 segment of erythronolides A and B) under the same reaction conditions failed. The conformational analysis of the (R)-104 by using molecular space-filling models led to the conclusion that the absence of reaction may be due to the steric overcrowding caused on the one hand by the C8 methyl group and on the other hand by the 12-O-TBS group. Therefore, the (R)-104 sulfoxide was desilylated with (n-Bu)$_4$NF·3H$_2$O, and the desilylated product (R)-105 reacted with the methyl ketone 58 giving the adduct 107 in 45% yield. Since the reaction product was a complicated mixture containing unreacted sulfoxide 105, excess methyl ketone 58, and traces of two more products, the pure compound 107 was obtained in only 23% yield. The low yield was explained by instability of 107. Deoxygenation [38] of 107 gave the corresponding sulfide 109, which was desulfurated to 111 (Fig. 13.16), the precursor of the (9S)-dihydroerythronolide A seco-acid 113. The olefin 111 was finally converted to the (9S)-dihydroerythronolide A seco-acid 113 by ozonization and oxidation with m-chloroperbenzoic acid (Fig. 13.16).

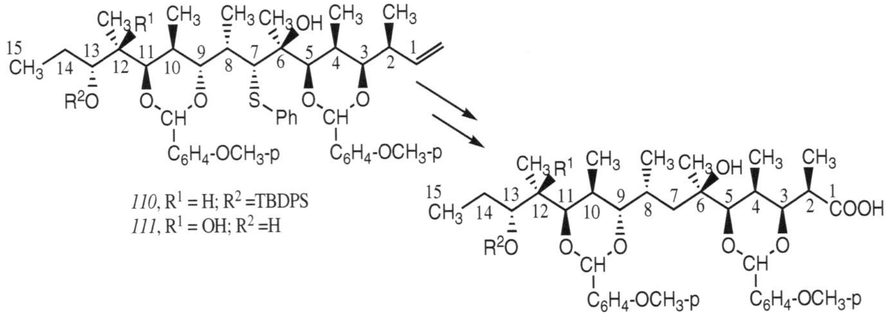

110, R^1 = H; R^2 = TBDPS
111, R^1 = OH; R^2 = H

112, R^1 = H; R^2 = TBDPS
113, R^1 = OH; R^2 = H

Fig. 13.16

Thromboxane B$_2$

Thromboxanes are members of the family of lipids known as eicosanoids. The two major thromboxanes are thromboxane A$_2$ (114) and thromboxane B$_2$ (115) (Fig. 13.17).

There are two reports dealing with stereoselective synthesis of thromboxane B$_2$ from D-glucose, one by Corey et al. [39] and the other by Hanessian et al. [40, 41]. Corey's approach [39] is shown in Fig. 13.18.

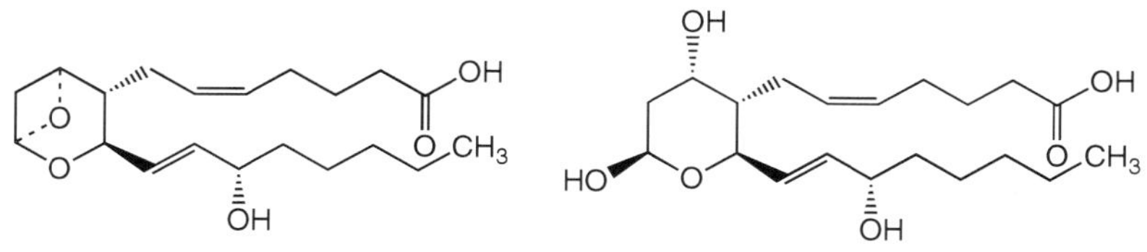

Thromboxane B$_2$

114, Tromboxane A$_2$

115, Tromboxane B$_2$

Fig. 13.17

The key intermediate in the thromboxane B$_2$ synthesis, the lactone *120*, was stereoselectively synthesized from D-glucose. Two key intermediates in this synthesis were the 3,4-unsaturated sugar *117*, obtained from D-glucose in three steps, and the stereospecific conversion of allylic alcohol *117* to the dimethylamide *118* by Claisen rearrangement [42, 43] in >75% yield. Treatment of *118* with iodine in tetrahydrofuran at 0°C afforded *119* in ca. 80% yield, which with tributyltin hydride [44] afforded, in quantitative yield, the hydroxy lactone *120* which was converted by standard methodology [44] to thromboxane B$_2$ *115* and its C15 epimer.

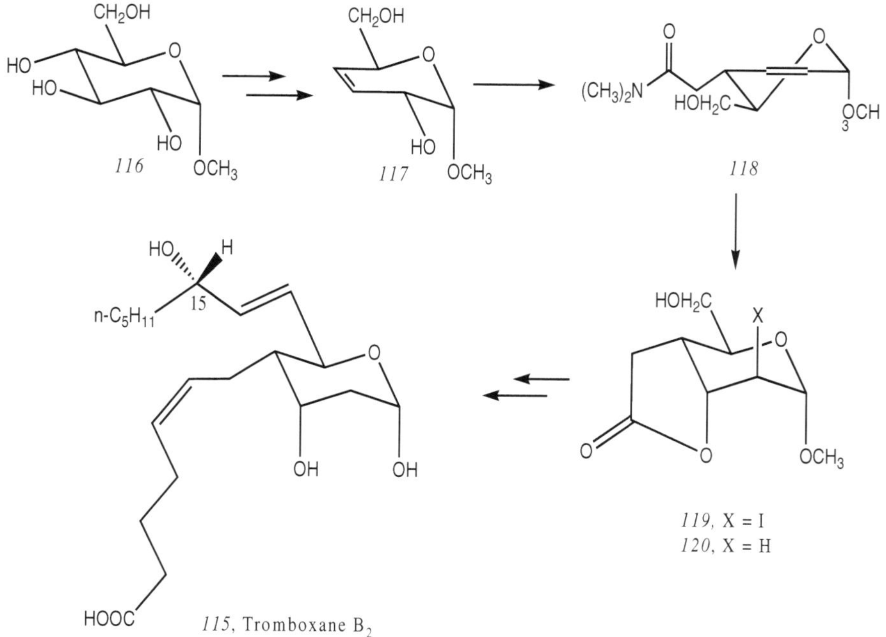

119, X = I
120, X = H

115, Tromboxane B$_2$

Fig. 13.18

Hanessian [40, 41] used D-glucose to make methyl 4,6-*O*-benzylidine-3-*O*-benzoyl-2-deoxy-α-D-*ribo*-hexopyranoside *121* which was the actual starting material for the synthesis of thromboxane B_2 (Fig. 13.19). Debenzylidenation of *121*, blocking of the primary hydroxyl group with *t*-butyldiphenylsilyl chloride, followed by oxidation with DMSO in the presence of 1-ethyl-3-(3-dimethylaminopropyl)carbodiimide hydrochloride and pyridinium trifluoroacetate afforded the C4 ulose *122* in 75% overall yield. Reaction of *122* with dimethyl (methoxycarbonyl) methylphosphonate in the presence of *tert*-butoxide gave a 1:1 mixture of *E*- and *Z*-isomers of methyl 3-*O*-benzoyl-6-*O*-*tert*-butyldiphenylsilyl)-2,4-dideoxy-*erythro*-hexopyranoside *123* and *124*. Hydrogenation of the *123* and *124* mixture with 20% palladium hydroxide on charcoal gave methyl 3-*O*-benzoyl-6-*O*-*tert*-butyldiphenyl–2,4-dideoxy-4-*C*-[(methoxycarbonyl) methyl]-α-D-*ribo*-hexopyranoside *125* in 70% yield. Although the authors do not discuss this, it is reasonable to assume that the catalytic hydrogenation gave the epimeric mixture of C4 methoxycarbonyl derivatives, since the next step which is lactonization, by treating the *125* first with potassium carbonate for 60 h at room temperature and then with Rexyn 102 (H^+), gave lactone *126* in only 51% yield. For the rest of the synthesis of thromboxane B_2 which is the synthesis of two side-chains the reader is referred to the cited reference since for their synthesis carbohydrates were not used.

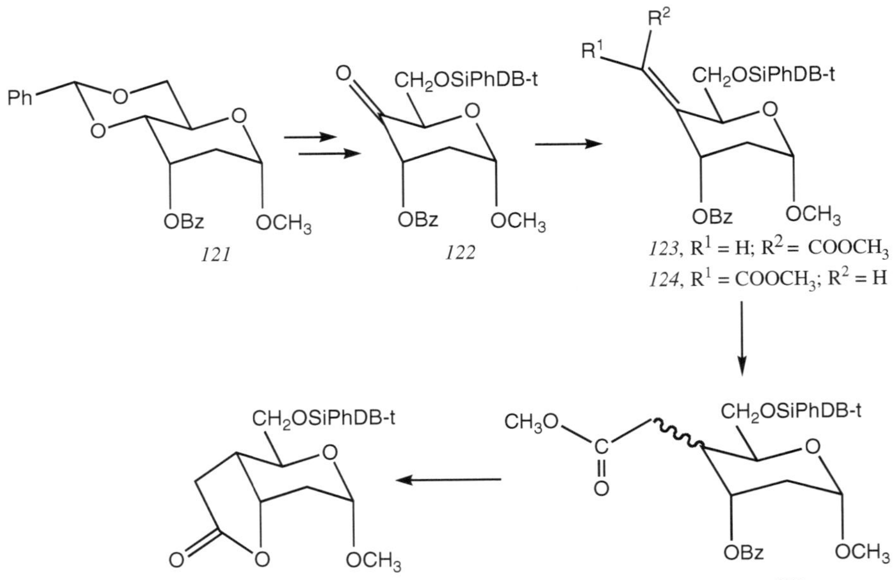

121

122

123, $R^1 = H$; $R^2 = COOCH_3$

124, $R^1 = COOCH_3$; $R^2 = H$

126

125

Fig. 13.19

Swainsonine

Swainsonine *139* (Fig. 13.20) is a very potent and specific α-mannosidase inhibitor and disrupts the processing of glycoproteins. This alkaloid has been originally isolated from the legume *Swainsona canescens* [45] but it has also been shown to be present in plants and other micro-organisms [46–49].

There are two syntheses of Swainsonine from carbohydrates published: one from 3-amino-3-deoxy-D-mannose derivative and the other from D-mannose. Synthesis of Swainsonine from 3-amino-3-deoxy-D-mannopyranose [50] was the first synthesis from a carbohydrate and starts from methyl 3-amino-3-deoxy-α-D-mannopyranoside hydrochloride *127* and is shown in Fig. 13.20.

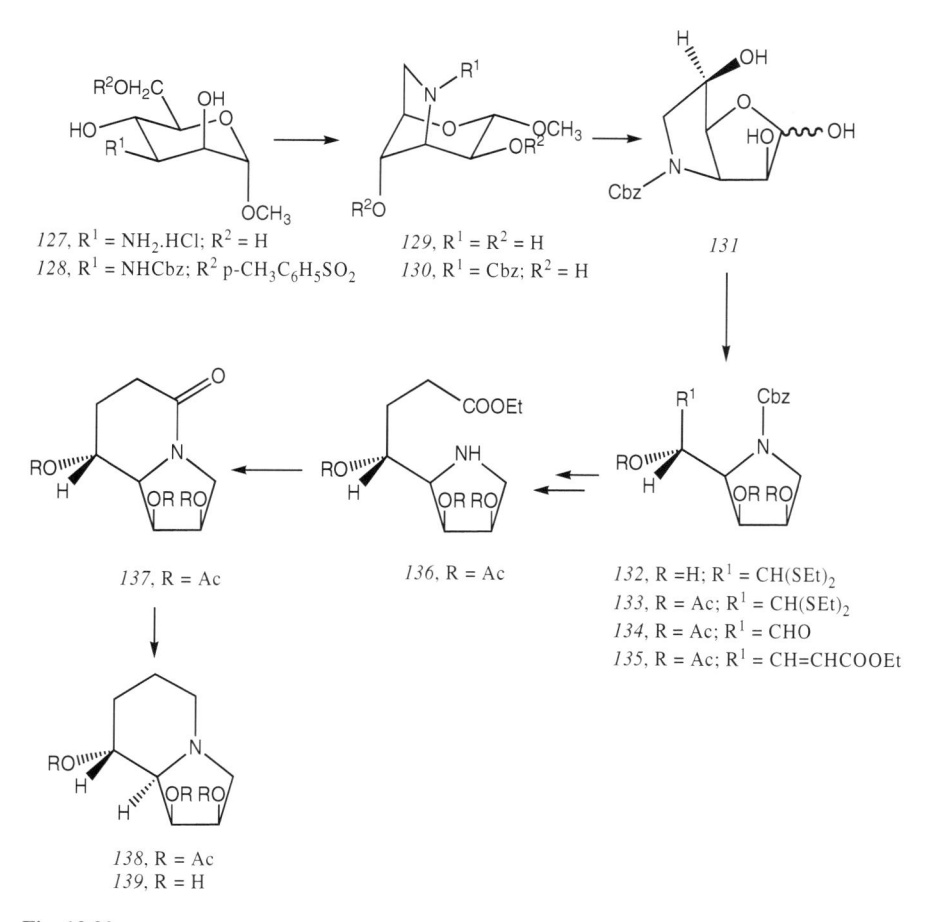

127, R^1 = NH_2·HCl; R^2 = H
128, R^1 = NHCbz; R^2 p-$CH_3C_6H_5SO_2$
129, R^1 = R^2 = H
130, R^1 = Cbz; R^2 = H
131

137, R = Ac
136, R = Ac
132, R = H; R^1 = CH(SEt)$_2$
133, R = Ac; R^1 = CH(SEt)$_2$
134, R = Ac; R^1 = CHO
135, R = Ac; R^1 = CH=CHCOOEt

138, R = Ac
139, R = H

Fig. 13.20

N-Benzyloxycarbonylation of methyl 3-amino-3-deoxy-α-D-mannopyranoside hydrochloride *127* (obtained from D-glucose in 20–25% yield [51]) and selective tosylation gave the 6-*O*-*p*-toluenesulfonate *128* in 82% overall yield. Removal of carbobenzoxy group by catalytic hydrogenation followed by refluxing of the obtained free amine in ethanol containing sodium acetate gave the 3,6-epimine *129* in >52% overall yield (from *127*). *N*-Benzyloxycarbonylation of *129* followed by acid hydrolysis of *130* afforded the free 3,6-dideoxy-3,6-iminohexofuranose *131* in 52% yield.

The 3,6-iminohexose *131* was reacted with ethanethiol in the presence of hydrochloric acid to give the diethyl dithioacetal *132* in 74% yield. Acetylation of *132* followed by dethioacetalation of *133* with mercury (II) chloride–cadmium carbonate gave the *aldehydo*-hexose *134*. Reaction of *134* with ethoxycarbonylmethylenetriphenylphosphorane gave an olefin *135* that on hydrogenation with palladium on charcoal as the catalyst hydrogenated not only the double bond but also removed the carbobenzoxy group, giving initially the free amine *136* which reacted with ethoxycarbonyl group to give a 1:1 mixture of two products, one of which was the desired cyclic lactam *137*.

Reduction of the cyclic lactam *137* with borane·dimethylsulfide complex gave a mixture of two products, one of which was tri-*O*-acetylswainsonine *138*, which on deacetylation was converted to swainsonine *139*. This was the first reported total synthesis of swainsonine which was achieved in an overall yield of 2.7% from *127*.

The synthesis from D-mannose derivative *140* [52] is shown in Fig. 13.21. Benzyl α-D-mannopyranoside *140* was treated with *tert*-butyldiphenylsilyl chloride in the presence of imidazole and the product was acetonated with 9:1 acetone/2,2-dimethoxypropane in the presence of a trace of camphor sulfonic acid to give *141* in 86% yield. The oxidation of the C4 hydroxyl group of *141* with pyridinium chlorochromate (PCC) yielded a ketone which was reduced by sodium borohydride to give benzyl 6-*tert*-butyldiphenylsilyl-2,3-*O*-isopropylidene-α-D-talopyranoside *142* in 88% overall yield. Esterification of *142* with trifluoromethanesulfonic anhydride/pyridine gave the D-talotriflate ester *143* which on treatment with sodium azide in *N,N*-dimethylformamide at room temperature gave the C4 azidomannose derivative *144* (second inversion of configuration at the C4 carbon) in 67% yield. Removal of the *tert*-butyldiphenylsilyl protection group with fluoride ion followed by oxidation of the C6 primary hydroxyl group with pyridinium chlorochromate gave an unstable aldehyde which was immediately reacted with formylmethylene triphenylphosphorane to give the azidoaldehyde *145*. Hydrogenation of azidoenal *145* in the presence of 10% palladium on charcoal in methanol led to the formation of the secondary amine *146*. Removal of the anomeric benzyl group in *146* is accomplished by hydrogenation in acetic acid using palladium black as the catalyst until all the secondary amine is consumed. This hydrogenation causes the hydrogenolysis of the benzyl group to form a lactol which is in equilibrium with an open-chain aminoaldehyde which undergoes reductive amination to form the isopropylidene swainsonine *147*. Finally, the isopropylidene group was removed with trifluoroacetic acid–D$_2$O giving swainsonine *139*.

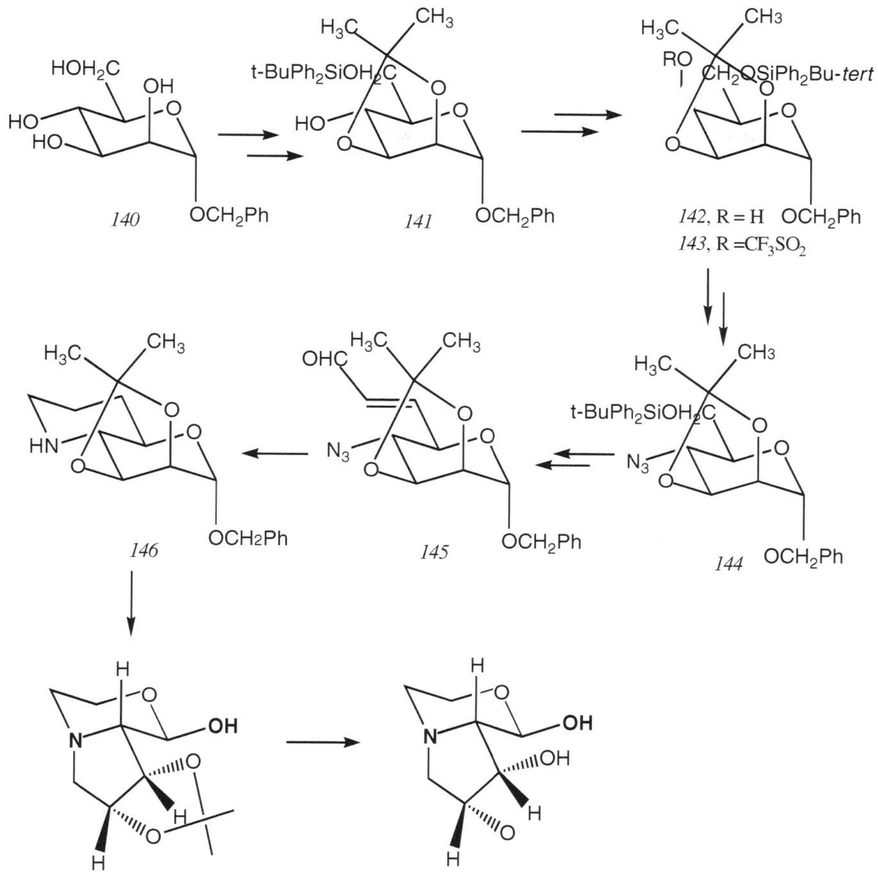

140 141 142, R = H 143, R =CF$_3$SO$_2$ 144 145 146 147 139

Fig. 13.21

Biotin

Biotin, also known as vitamin H or B7, is important in the catalysis of essential metabolic reactions. It is used in cell growth, the synthesis of fatty acids, and metabolism of fats and amino acids in gluconeogenesis. It also plays a role in the citric acid cycle and is helpful in maintaining a steady blood sugar level.

There are two stereoselective synthesis of biotin from carbohydrates reported in the literature. The first one uses D-mannose as the starting material [53] and the second one the 1,6-anhydro-β-D-glucose [54]. Use of the conformationally rigid 1,6-anhydro-β-D-glucose *62* was essential for the successful, stereoselective synthesis of biotin. The synthetic scheme is shown in Fig. 13.22.

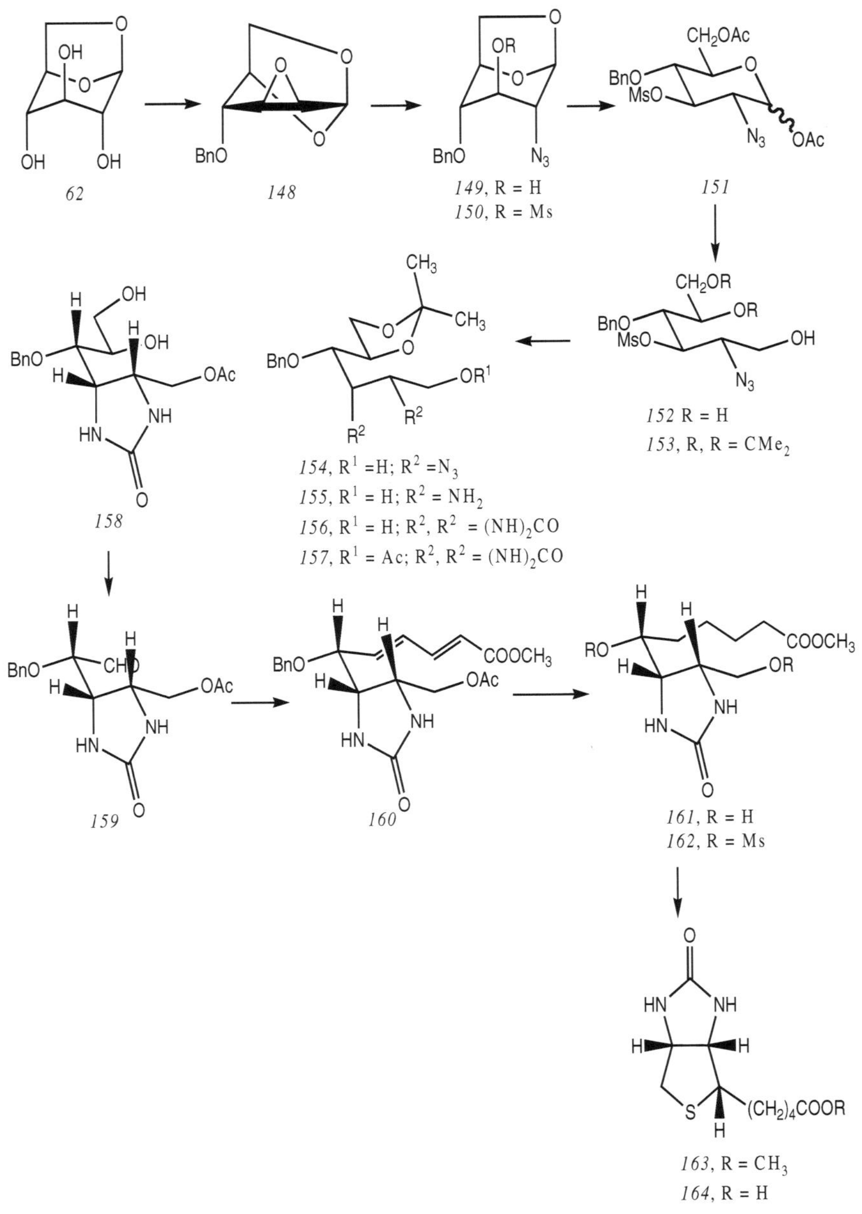

62
148
149, R = H
150, R = Ms
151
152 R = H
153, R, R = CMe2
154, R1 =H; R2 =N3
155, R1 = H; R2 = NH2
156, R1 = H; R2, R2 = (NH)2CO
157, R1 = Ac; R2, R2 = (NH)2CO
158
159
160
161, R = H
162, R = Ms
163, R = CH3
164, R = H

Fig. 13.22

Opening of the oxirane ring of 1,6:2,3-dianhydro-4-*O*-benzyl-β-D-mannopyranose *148* with sodium azide and ammonium chloride in 4:1 2-methoxyethanol–water for 22 h at 120°C gave, in 85% yield, 1,6-anhydro-2-azido-4-*O*-benzyl-2-deoxy-β-D-glucopyranose *149*. Esterification of the C3 hydroxyl group with methanesulfonyl anhydride in pyridine gave the corresponding C3 mesylate *150*. The acetolysis of *150* with acetic anhydride–4% BF_3 etherate for 3 h at room temperature gave, in 95% yield, an anomeric mixture of diacetates *151* in the α:β ratio 8:3. Solvolysis of *151* with 1% HCl in methanol for 16 h at room temperature followed by reduction with $NaBH_4$ in the presence of boric acid in ethanol at 0–5°C gave the triol *152*. Treatment of the triol *152* with 2,2-dimethoxypropane in *N,N*-dimethylformamide in the presence of catalytic amount of *p*-toluenesulfonic acid monohydrate for 15 h at room temperature gave the 5,6-isopropylidene derivative *153*. The C3 azido group was introduced by treating the *153* with sodium azide in DMF. The selective hydrogenation [55] of the diazide *154* in the presence of benzyl group was effected by using Lindlar catalyst in ethanol whereby the diamine *155* was obtained in quantitative yield. Reaction of *155* with phosgene ($COCl_2$) in carbon tetrachloride and aqueous Na_2CO_3 at 0–5°C gave ureide *156*. Acetylation of *156* with acetic anhydride–pyridine followed by deisopropylidenation of monoacetate *157* in 80% aqueous acetic acid for 3.5 h at 70°C gave 5,6-diol *158*. $NaIO_4$ oxidation of *158* in 50% aqueous ethanol for 1 h at room temperature gave aldehyde *159* and subsequent reaction of *159* with [3-(carbomethoxy)-2-propen-1-ylidene]triphenylphosphorane [53, 56] in dichloromethane at room temperature gave unsaturated ester *160*. Hydrogenation of *160* using 10% Pd–C in methanol and subsequent deacetylation by CH_3ONa in methanol gave diol ester *161*. Methanesulfonylation of *161* with 15 equivalents of methanesulfonyl chloride in pyridine-dichloroethane for 15 h at –10°C gave dimesylate *162*, which was treated without purification with large excess of Na_2S in *N,N*-dimethylformamide for 3 h at 100°C to give the biotin methyl ester *163*, which was hydrolyzed to (+)-biotin *164*.

Pseudomonic Acid C

Pseudomonic acids are a small group of antibiotics of unique structure. Three representatives of this family are Pseudomonic acid A (*165*), Pseudomonic acid B (*166*), and Pseudomonic acid C (*167*) (Fig. 13.23). They are produced by the strain *Pseudomonas fluorescens* [57–59].

The synthesis of (+)-methyl pseudomonate C from D-xylose [60] *169* is shown in Figs. 13.24 and 13.25. The D-xylopyranose *169* was converted into cyanide *171* as previously described [61] by acetylation of *168*, conversion of the tetraacetate *169* into tri-*O*-acetyl-α-D-xylopyranosyl chloride *170*, and by displacing the obtained chloride with potassium cyanide. The base-catalyzed hydrolysis of 2,3,4-tri-*O*-acetyl-β-D-xylopyranosyl cyanide *171* followed by esterification of the obtained carboxylic acid with methanol in the presence of HCl (70%) gave the corresponding

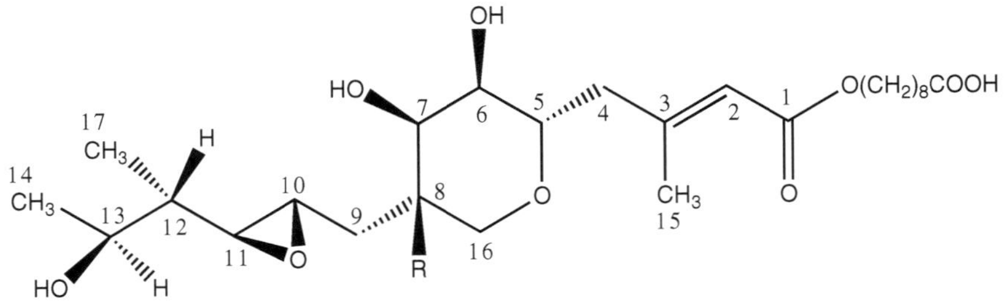

446   13 Synthesis of Polychiral Natural Products from Carbohydrates

165, R = H, Pseudomonic acid A
166, R = OH, Pseudomonic acid B

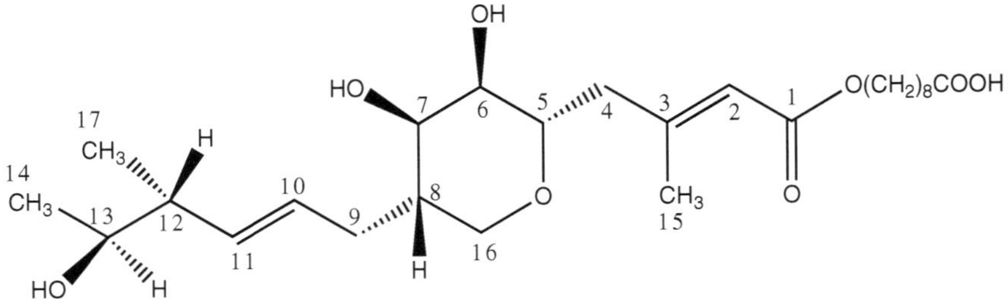

167, Pseudomonic acid C

Fig. 13.23

methyl ester *172*. By reduction of *172* with lithium aluminum hydride the methoxycarbonyl group was converted to hydroxymethyl group and the reaction of *173* with PhCH(OCH$_3$)$_2$ in the presence of *p*-toluenesulfonic acid gave the benzylidene acetal *174* (85%). Tosylation of *174* with *p*-toluenesulfonyl chloride in pyridine (70%) and the treatment of the 3,4-di-tosyl derivative *175* with sodium methoxide in chloroform at room temperature gave in quantitative yield the 3,4-epoxide *176* which represented the precursor of the central ring structure of pseudomonic acid C.

The basic assumption in this synthetic approach was that the rigid tricyclic epoxide *176* will undergo a regiospecific oxirane ring opening with a suitable allylic anion at the C4 carbon. Hence the synthesis of the left side chain of pseudomonic acid C is undertaken starting from D-glucose and is described in Scheme 2 of Fig. 13.24. D-Glucose *177* is converted to methyl 4,6-*O*-benzylidene-3-deoxy-3-*C*-methyl-α-D-altropyranoside *178* as described elsewhere [62]. The inversion of the configuration at the C2 carbon of *178* was effected by oxidation of *178* with DCCI–DMSO–TFA–pyridine at room temperature, and then by reduction of the corresponding 2-ulose with lithium aluminum hydride in ether at 0°C whereby methyl 4,6-*O*-benzylidene-3-deoxy-3-*C*-methyl-α-D-allopyranoside *179* was obtained in 95% yield. The C2 hydroxyl group was then blocked with *tert*-butyldimethylsilyl protection group by reacting *179* with *tert*-butyldimethylsilyl chloride in

Pseudomonic Acid C

Scheme 1

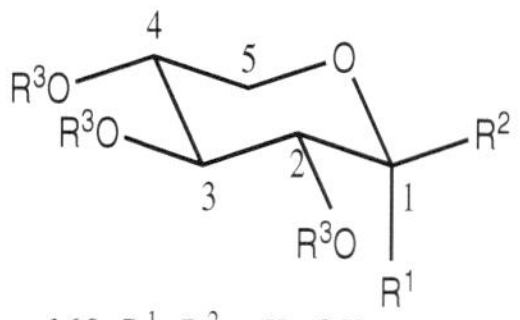

168, R¹, R² = H, OH
169, R¹, R² = H, OAc; R³ = Ac
170, R¹ = Cl; R² = H; R³ = Ac

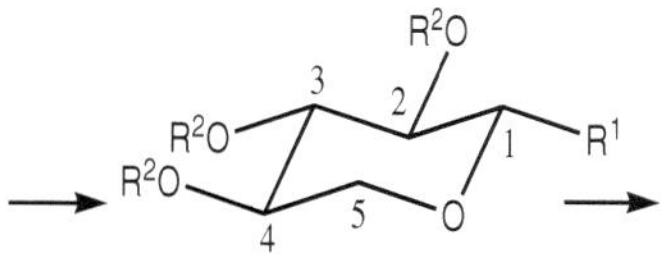

171, R¹ = CN; R² = Ac
172, R¹ = COOCH₃; R² = H
173, R¹ = CH₂OH; R² = H

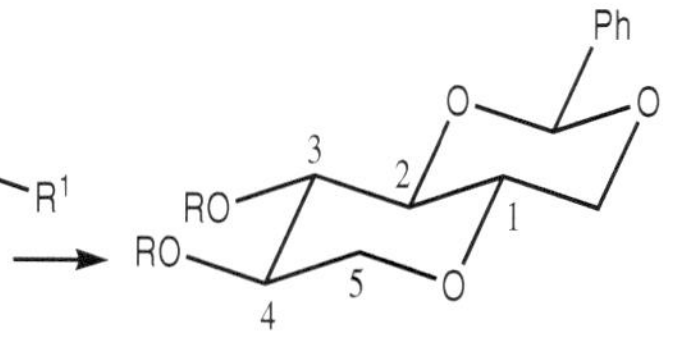

174, R = H
175, R = Ts

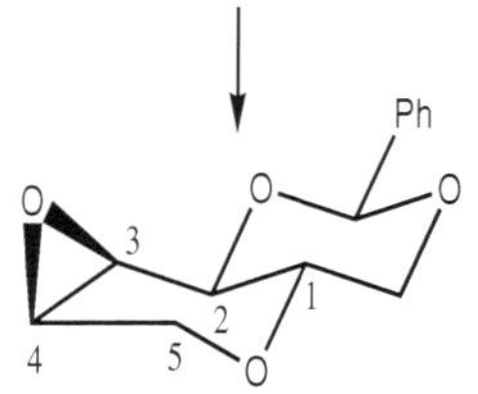

176

Scheme 2

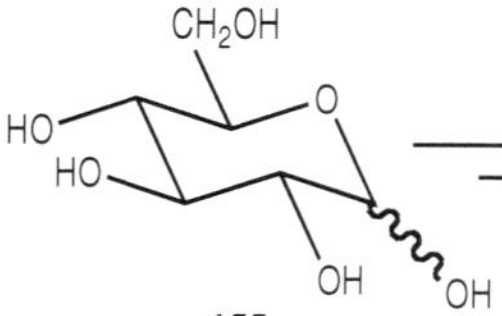

177

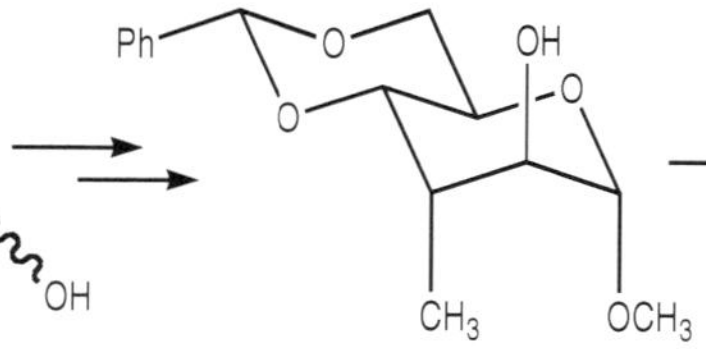

178

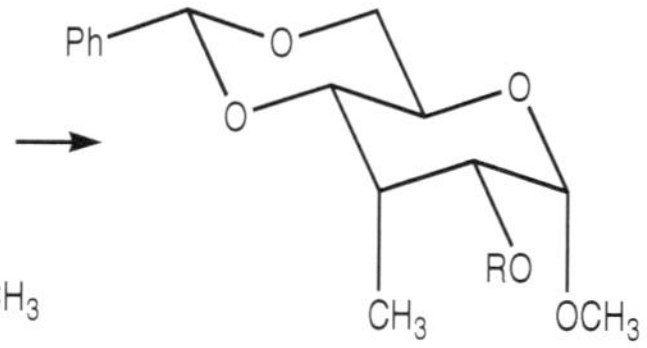

179, R = H
180, R = TBDMS

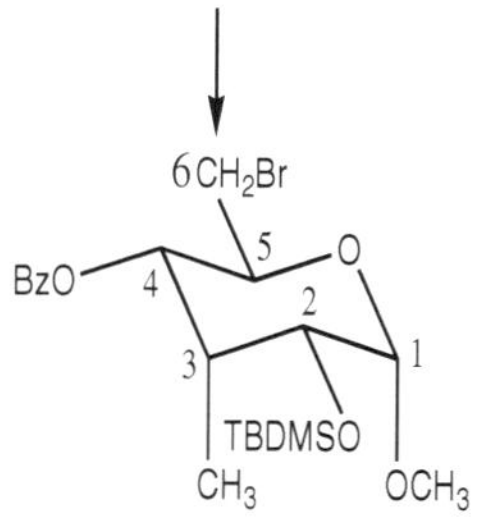

188

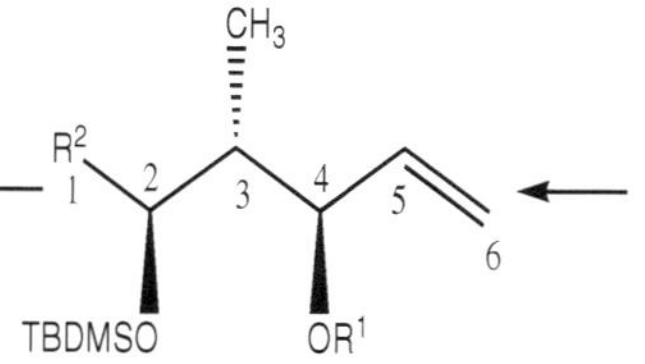

182, R¹ = Bz; R² = CHO
183, R¹ = Bz; R² = CH₂OH
184, R¹ = Bz; R² = CH₂OTs
185, R¹ = Bz; R² = I
186, R¹ = Bz; R² = CH₃
187, R¹ = H; R² = CH₃

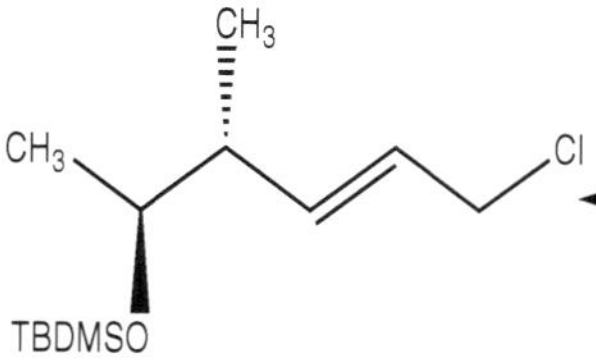

181

Fig. 13.24

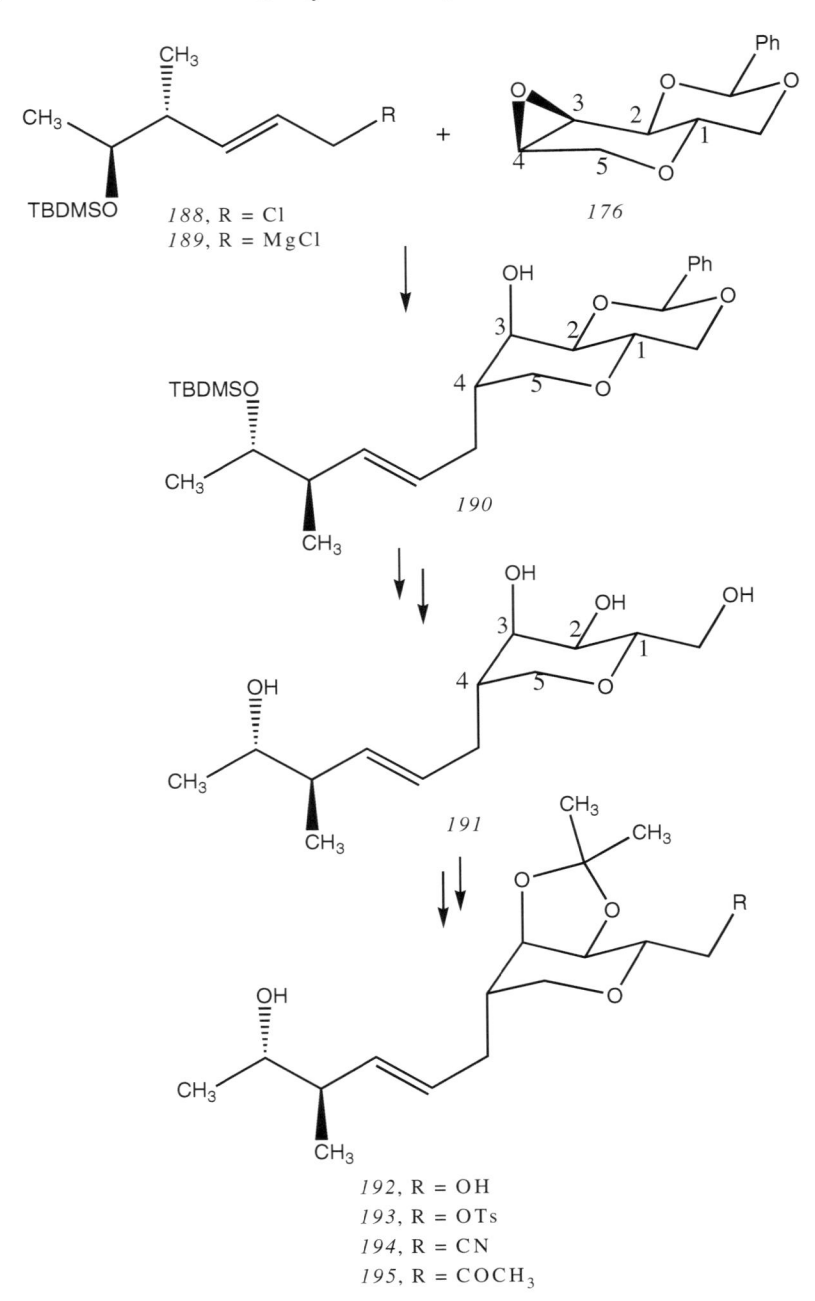

Fig. 13.25

188, R = Cl
189, R = MgCl

176

190

191

192, R = OH
193, R = OTs
194, R = CN
195, R = COCH$_3$

N,N-dimethylformamide in the presence of imidazole giving *180* in 83% yield. Debenzylidenation of *180* with N-bromosuccinimide in CCl$_4$ at 80°C in the presence of barium carbonate to neutralize the released HBr gave methyl 4-O-benzoyl-6-bromo-3,6-dideoxy-2-*tert*-butyldimethylsilyl-3-C-methyl-α-D-allopyranoside *181* in 75% yield. Reductive β-elimination of bromide *181* with activated zinc [63, 64] [acid-washed zinc (100 equivalents) in 9:1 propanol–water at 80°C for 30 min] gave *182*. This was the key reaction in this synthesis.

The copper-catalyzed (CuI) oxirane ring opening [65] of *176* with Grignard reagent *189* (Fig. 13.25) which was made from *188* (two equivalents, THF, −30°C, 10 min) afforded the expected product *190* in 43% yield. The important aspect of this reaction is that it involves the "nonrearranged" allylic Grignard reagent and gives the E double bond. These two features were critical but not obvious (for discussion of these features, see [66–69]).

Selective tosylation of *191* followed by isopropylidenation gave *193* which on treatment with potassium cyanide produced *194*. The ketone *195* was obtained from cyanide in 84% yield by treatment of the cyanide *194* with trimethylaluminum in the presence of Ni(acac)$_2$ [70] followed by acid-catalyzed hydrolysis (Fig. 13.25).

The epoxidation of the double bond of *195* with MCPBA in dichloromethane at room temperature gave a 2:3 mixture of epimeric epoxides *196* and *197*, one of which was identical with the natural product (Fig. 13.26).

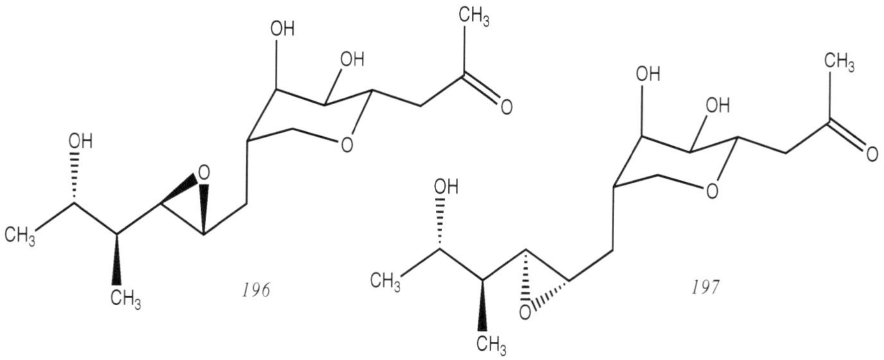

Fig. 13.26

Elongation of the right side chain was essentially performed as already described [71, 72].

A different approach for the synthesis of pseudomonic acid C was taken by Keck et al. [73]. The adopted strategy was to synthesize the pseudomonic acid C from three fragments: the lower left appendage *198*, the central tetrahydropyran segment *199*, and the upper right appendage. Stereochemically there were many challenges: the synthesis of the C2–C3 and the C10–C11 double bonds (pseudomonic acid C numbering) with E-stereochemistry and the stereoselective C1 and C4 alkylation (lyxose numbering). The stereochemistry of the C2 and the C3 hydroxyl groups (lyxose numbering) is identical with the C6 and the C7 hydroxyl groups of

Fig. 13.27

pseudomonic acid A. The key to the strategy adopted was the recognition of the structure *199* (Fig. 13.27) as that of a highly modified L-lyxose *200* (Fig. 13.28). The authors assumed that the incorporation of a latent acetaldehyde at C4 (lyxose numbering) could be accomplished by free radical allylation with allyltri-*N*-butylstannane [74] and the incorporation of a latent acetone moiety at C1 (lyxose numbering) could be similarly accomplished with either allyl- or methallyl-tri-*n*-butylstannane.

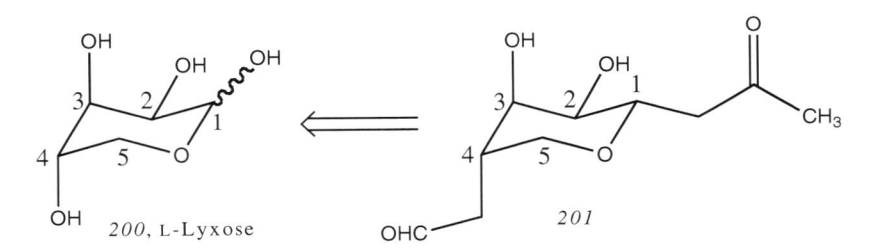

Fig. 13.28

The first task was to differentiate the C2 and C3 hydroxyl groups from C1 and C4 hydroxyl group of L-lyxose *200* and then to differentiate the C1 and the C4 hydroxyl groups from each other. Furthermore, it was desirable to make the β face of the molecule sterically as crowded as possible so that the free radical C-C bond formation at the C4 carbon takes place from the α face. Differentiation of the C2 and the C3 hydroxyl groups from the C4 hydroxyl group could very well be accomplished by acetonation because the former are in *cis* orientation, whereas the C3 and the C4 hydroxyl groups are *trans* oriented (diaxial). However, it was well known that the free lyxose preferentially forms 1,2-acetonide from its furanoside form [75]. Therefore, the anomeric hydroxyl group of L-lyxose had to be blocked first. That was accomplished by reacting L-lyxose *200* with benzyl alcohol (benzyl alcohol was used as the solvent and reagent and *p*-toluenesulfonic acid was used as the catalyst). The anomeric mixture of benzyl glycosides *201* was then treated with dimethoxypropane and *p*-toluenesulfonic acid in acetone giving the desired acetonide *202* in 85% overall yield from L-lyxose (Fig. 13.29).

The introduction of allyl residue (the latent acetaldehyde) at the C4 carbon was accomplished by using free radical methodology reported earlier [74], namely the C4 hydroxyl group was first converted to the phenyl thionocarbonate *203* by treating *202* with methyl lithium and phenyl chlorothionocarbonate in anhydrous ether at –80°C (90% yield). This was then subjected to photolysis, in toluene solution, with 450-W Hanovia lamp ($\lambda > 300$ nm) in the presence of allyl-tri-*n*-butylstannane for 65 h at 23°C, whereby the desired C4-allyl adduct *204* was obtained in 80% yield. Stereochemical control of this free radical reaction was the result of approach of allyl-tri-*n*-butylstannane to the less hindered α face of a C4 free radical derived from *203*, leading to the desired α allyl derivative *204* (the β face of the molecule was inaccessible due to the bulky 2,3-isopropylidene group). The C4-allyl derivative *204* was converted to the C4 acetaldehydo compound *205* by treating the *204* with OsO_4 and $NaIO_4$. The acetaldehyde was then reduced with sodium borohydride in ethanol to primary alcohol *206*, which was with methanesulfonyl chloride in pyridine converted to the mesylate *207*. Treatment of *207* first with sodium thiophenoxide (PhSNa) in *N,N*-dimethylformamide and then the oxidation of the phenylthio derivative *208* with peracetic acid in dichloromethane and buffered with $NaHCO_3$ gave the desired phenyl sulfone *209* (Fig. 13.29).

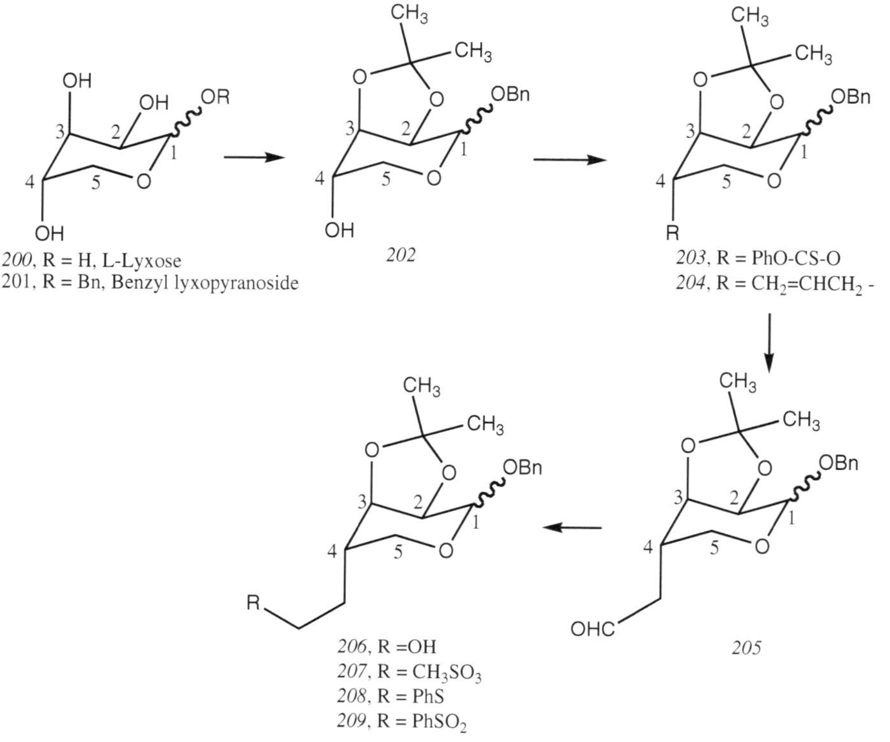

200, R = H, L-Lyxose
201, R = Bn, Benzyl lyxopyranoside

202

203, R = PhO-CS-O
204, R = $CH_2=CHCH_2$ -

206, R = OH
207, R = CH_3SO_3
208, R = PhS
209, R = $PhSO_2$

205

Fig. 13.29

452   13 Synthesis of Polychiral Natural Products from Carbohydrates

The synthesis of the lower left appendage was accomplished from commercially available (S)–(+)-ethyl-3-hydroxybutanoate *210* as shown in Fig. 13.30 (the numbering is pseudomonic acid C numbering). The methylation of dianion obtained by treating *211* with lithium diisopropylamide and methyl iodide according to the gen-

Fig. 13.30

eral protocol by Frater [76, 77] gave stereoselectively the C12 methyl derivative *211*. After blocking the C13 hydroxyl group of *211* with *tert*-butyldimethylsilyl group, the ester group of *212* was reduced with diisobutylaluminum hydride in toluene to aldehyde *213* in 87% yield.

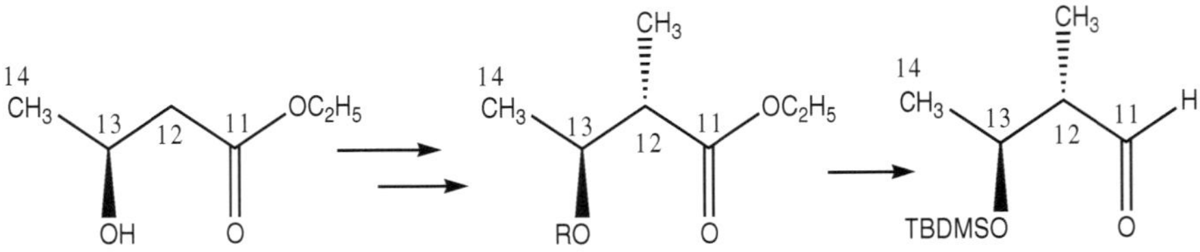

The condensation of phenyl sulfone *209* (Fig. 13.29) with the aldehyde *213* (Fig. 13.30) proceeded smoothly to afford *214* (Fig. 13.31). The C12 hydroxyl group was activated by mesylation and the obtained *215* was subjected to reductive elimination which resulted in *trans* C11–C12 double bond formation and simultaneous debenzylation giving *216* (Fig. 13.31).

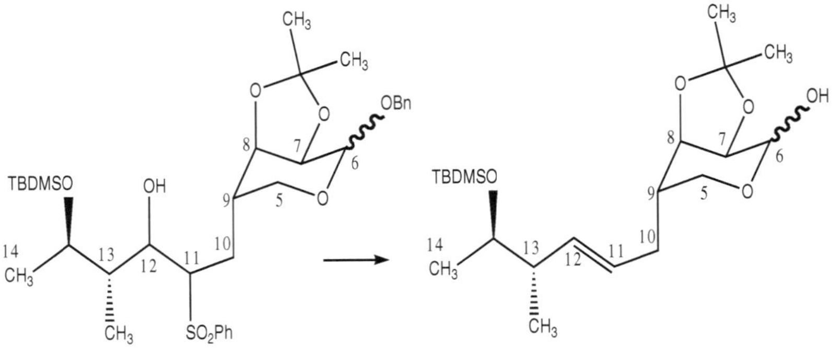

Fig. 13.31

The upper right appendage was then introduced by using Wittig reaction with stabilized ylide $(Ph)_3P=CHCOCH_3$ which was shown with free furanoses to produce *C*-glycoside in high yields [71, 78]. Thus methyl ketone *217* was obtained (Fig. 13.32).

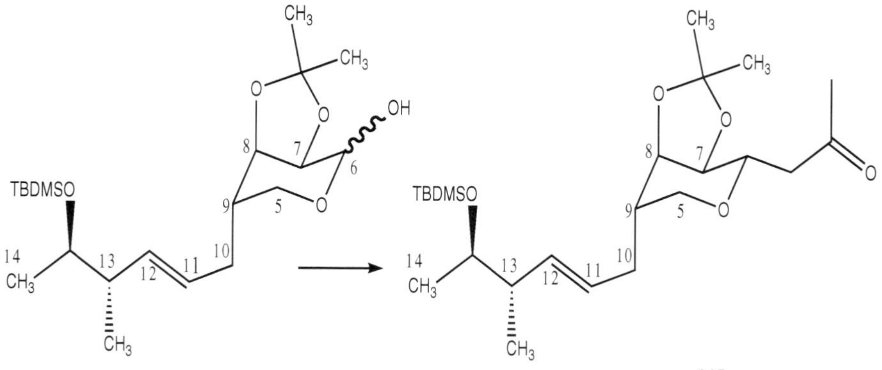

Pseudomonic Acid C

Fig. 13.32

Horner–Wadsworth–Emmons reaction of *217* with (methyldimethylphosphono)acetate *218* [79–83] gave in 75% yield *219* in which the C2, C3 double bond had *E* configuration. Deacetonation of *219* with 85% acetic acid in tetrahydrofuran (THF) gave finally, in 93% yield, the methyl ester of pseudomonic acid *220* (Fig. 13.33).

A much more convergent strategy for the synthesis of (+)-pseudomonic acid C was published by the same authors [84] in 1989. The approach is outlined in Fig. 13.34 and is based on the assumption that a suitably functionalized allyl fragment *221*, with X = SPh, SOPh, or SO$_2$Ph, should couple via an addition–fragmentation mechanism with a carbon-centered radical derived from iodide *222*.

Preparation of iodide *222* started from the known benzyl 2,3-*O*-isopropylidene-L-lyxopyranoside *202* (mixture of anomers) [73]. Mesylation of *202* with methanesulfonyl chloride in pyridine at room temperature gave the corresponding 4-mesylate *225* that on treatment with 1:1 1 N HCl/THF gave diol *226* in 87% overall yield from *202*. Epoxide formation to yield *227* was accomplished in 96% yield by treating *226* with potassium *tert*-butoxide in THF at room temperature for 30 min. Reaction of *227* with 2.0 N HI in acetone at reflux followed by isopropylidenation of the resulting 4-iodo-2,3-diol *228* with dimethoxy 2-propane and *p*-toluenesulfonic acid in acetone gave the desired iodide *222* in 85% overall yield from the epoxide *227* (Fig. 13.35).

Synthesis of sulfone *232* is shown in Fig. 13.36. It starts from the readily available ester *229* which was homologated to allyl alcohol *230* in one-pot operation (65% yield) via reduction and in situ Horner–Wadsworth–Emmons reaction according to the Takacs protocol [85], followed by the addition of 2.1 equivalents of (*i*-Bu)$_2$AlH and workup. The conversion of allyl alcohol *230* to sulfone *231* was initiated by [2, 3] sigmatropic rearrangement of the derived sulfenate (1.0 equivalent of *n*-BuLi, THF, 0°C; PhSCl) via the general procedure of Evans [86] followed by oxidation with oxone [87] to give the desired sulfone *232* as ca. 2:1 mixture of epimers (the configurations of epimeric sulfones were not assigned but they are separable).

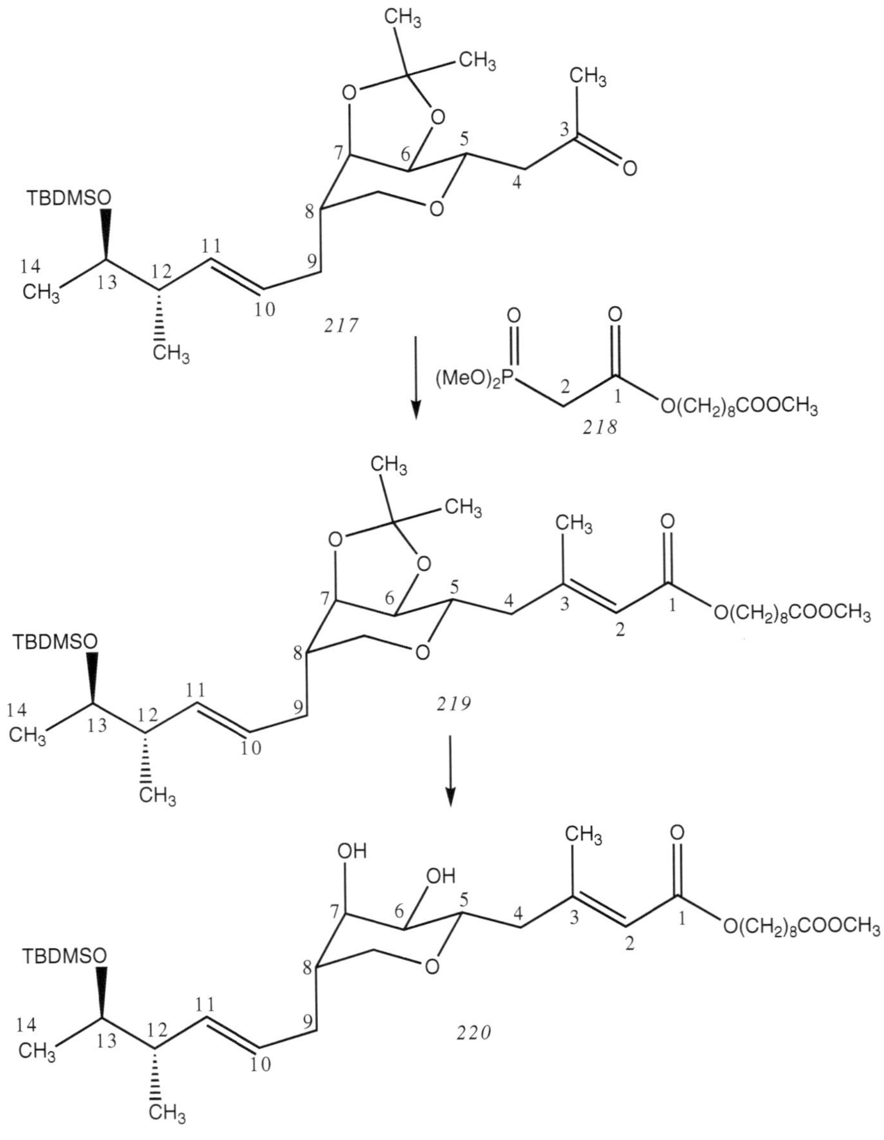

Fig. 13.33

The coupling of *232* and *222* was accomplished in 74% yield by slow addition (syringe pump) of a THF solution of sulfone *232* and hexabutylditin to an irradiated (450-W Hanovia lamp using Pyrex filter) THF solution of iodide *222* and hexabutylditin under argon atmosphere. The addition of the upper right appendage was accomplished as previously described [73].

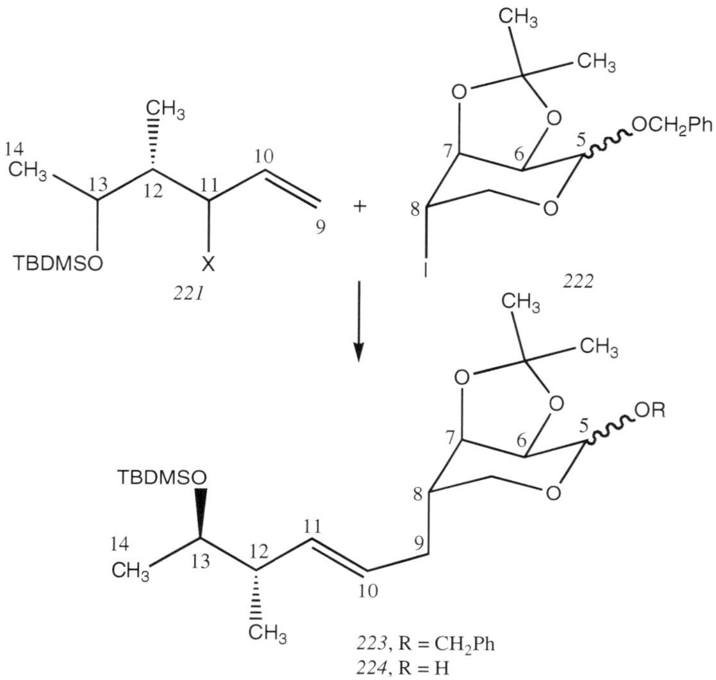

Pseudomonic Acid C

Fig. 13.34

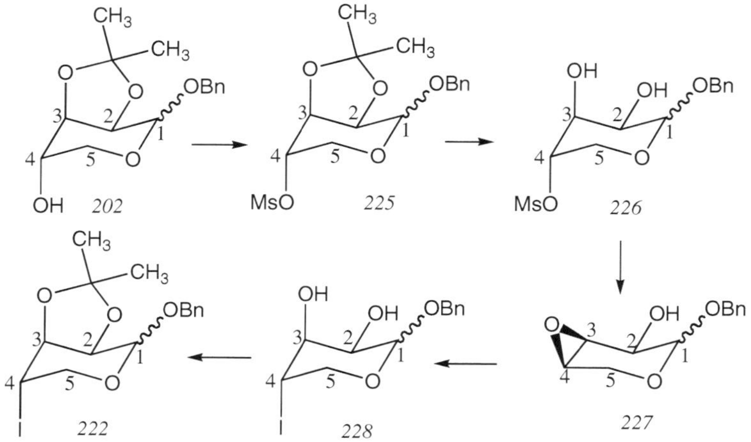

223, R = CH$_2$Ph
224, R = H

Fig. 13.35

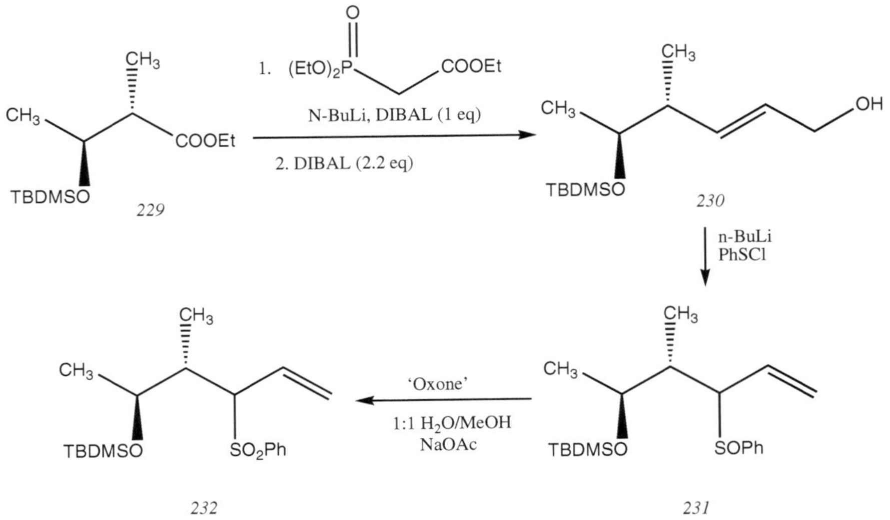

Fig. 13.36

Aplasmomycin

Aplasmomycin 233 (Fig. 13.37) is a boron-containing antibiotic from a marine-derived strain *Streptomyces griseus* that exhibits activity against gram-positive

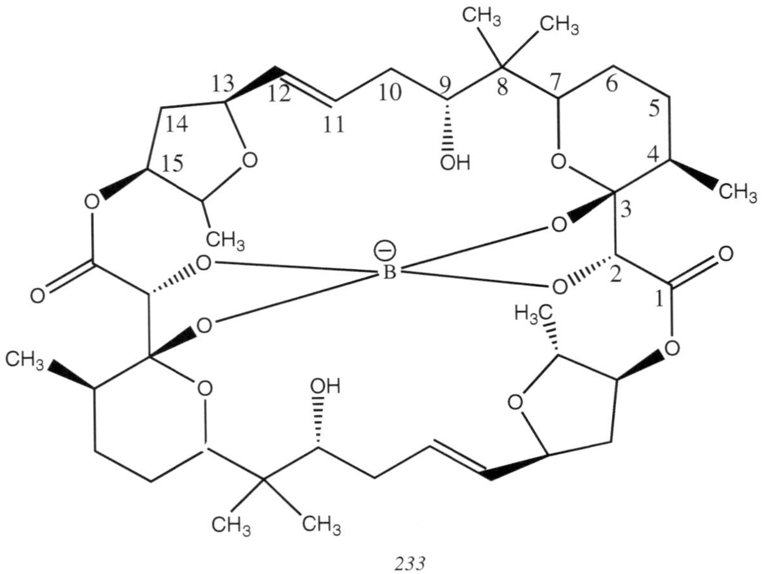

Fig. 13.37

Aplasmomycin

bacteria and Plasmodia [88]. It belongs to the family of borate-bridged antibiotics of which the boromycin was the first known member [89]. Aplasmomycin has C_2 symmetry indicating that it is composed of two identical subunits.

Corey et al. [90, 91] approached the synthesis of aplasmomycin by constructing the precursors corresponding to the C3–C10 fragment *235* starting from (+)-pulegone and the C11–C17 fragment *234* starting from D-mannose shown in Fig. 13.38.

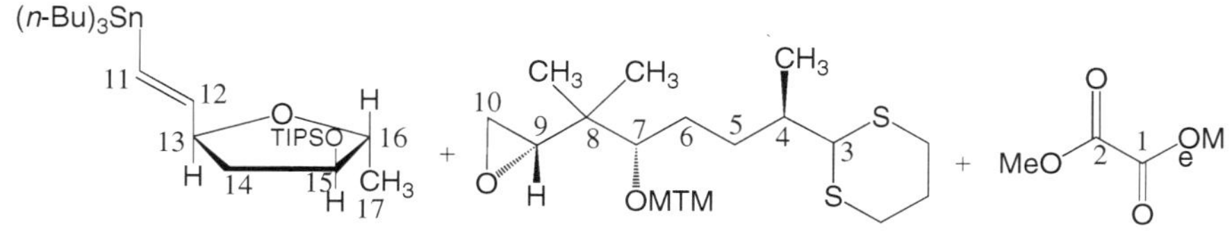

Fig. 13.38

We will describe the synthesis of fragment *234* since this is the part of the synthesis of aplasmomycin where carbohydrate is used as a chiral synthon. We will not describe the synthesis of fragment *235* since it is made from (+)-pulegone and the carbon atoms C1 and C2 are made from dimethyloxalate *236*. We will, however, describe the coupling of fragments *234, 235,* and *236* into C1–C17 fragment of aplasmomycin as well as the coupling of the obtained two identical fragments into macrocyclic lactone aplasmomycin (Fig. 13.39).

The synthesis of the C11–C16 fragment *234* started from 2,3:5,6-di-*O*-isopropylidene-D-mannose *237* which can be easily prepared from D-mannose with 2,2-dimethoxypropane in the presence of *p*-toluenesulfonic acid in acetone. Reaction of *237* with methyllithium in ether at –40°C for 1 h and then at 0°C for 6 h proceeded stereospecifically to give 99% yield of the diol *238*. The *238* was converted to tetrahydrofuran derivative *239* via tosylation in pyridine of the methyl carbinol *238*. The selective hydrolysis of the 5,6-*O*-isopropylidene group was accomplished in 90% yield at 60% conversion with (30:2:1 methanol–water–12 N HCl) at 4°C for 24 h to give the 5,6-diol *240*. Oxidation of 5,6-diol *240* with equimolar amounts of NaIO$_4$ and sodium bicarbonate in aqueous solution at 0°C gave the C5 aldehyde *241* which with bromotrichloromethane and tris (dimethylamino) phosphine at –50°C for 2 h, at –10°C for 1 h, at 5°C for 0.5 h was converted to dichloroolefin *242* in 75% yield. Reaction of *242* with *n*-butyllithium in THF at –78°C for 1.5 h afforded acetylene *243* in 99% yield. The 2,3-*O*-isopropylidene group was cleaved with 10:1 methanol–4 N HCl at 23°C for 24 h giving the 2,3-diol *244* in 92% yield. Selective silylation of the C2 hydroxyl group was accomplished with triisopropylsilyl chloride in the presence of 4-(dimethylamino)pyridine in dichloromethane at 0°C for 18 h. The C3 hydroxyl group of the C2 TIPS ether *245* was then converted to the triflate ester *246* by treating *245* with trifluoromethanesulfonic anhydride in presence of pyridine in dichloromethane solution at –10°C for 5 h (85% yield). Displacement of triflate by iodide using tetra-*n*-butylammonium iodide in benzene at

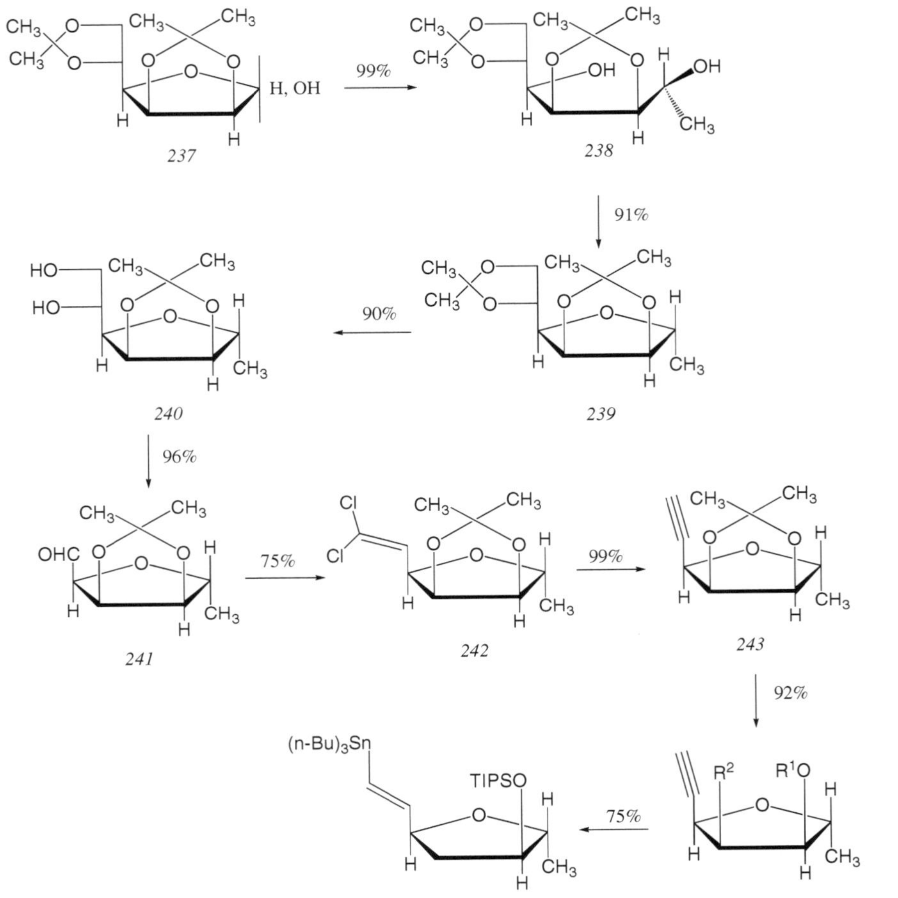

244, R^1 = H; R^2 = OH
245, R^1 = TIPS; R^2 = OH
246, R^1 = TIPS; R^2 = CF_3SO_3
247, R^1 = TIPS; R^2 = I
248, R^1 = TIPS; R^2 = H

Fig. 13.39

reflux for 2 h gave 3-deoxy-3-iodo derivative *247* in 94% yield, which with sodium borohydride and tri-*n*-butyltin chloride in ethanol under sunlamp irradiation gave 3-deoxy derivative *248* in 85% yield. Heating of *248* with tri-*n*-butyltin hydride and azobis(isobutyronitrile) at 90°C for 3 h gave the *trans*-vinylstannane *234* in 75% yield (Fig. 13.40).

The coupling of the vinylstannane fragment *234* with the epoxide *235* (prepared from (+)-pulegone) to form *250* corresponding to the C3–C17 segment of aplasmomycin was carried out as follows (Fig. 13.41). Reaction of *234* with

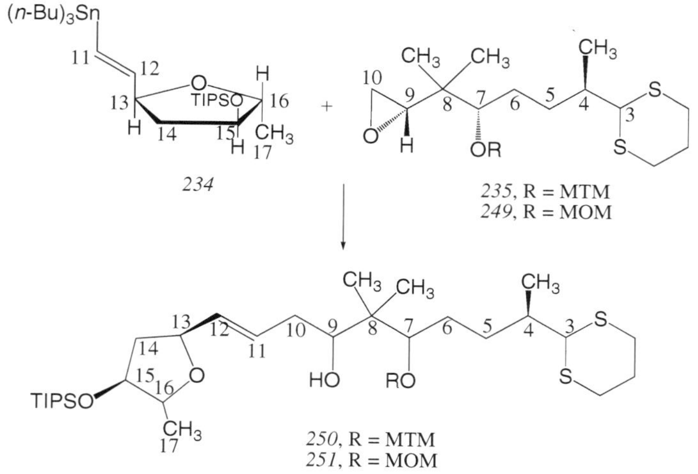

Aplasmomycin

Fig. 13.40

n-butyllithium in THF at −78°C for 1 h and −50°C for 1.5 h produced the lithium reagent corresponding to *234* which was sequentially treated with cuprous cyanide at −78°C for 1 h and the epoxide *235* (at −35°C for 2 h, −25°C for 24 h, and −15°C for 24 h) to give the coupling product *250* in 75% yield. In strictly analogous way the epoxide MOM ether *249* was coupled to *234* to give *251* as the product. The intermediates *250* and *251* that correspond to the C3–C17 segment of the two identical C1–C17 molecular subunits were finally coupled to aplasmomycin using two different synthetic routes.

The MTM ether *250* was converted to C7, C9 bisilylated derivative *252* in 85% overall yield by the following sequence: silylation of the C9 hydroxyl group with *tert*-butyldimethyl-silyl triflate-2,6-lutidine [92], followed by the cleavage of MTM ether using silver nitrate-2,6-lutidine in 4:1 THF–water at room temperature for 2 h [93] and silylation of the C7 hydroxyl group with *tert*-butyldimethyl-silyl triflate-2,6-lutidine, as described above for the C9 hydroxyl group (Fig. 13.40). Metalation of the dithiane *254* was accomplished using *n*-butyllithium

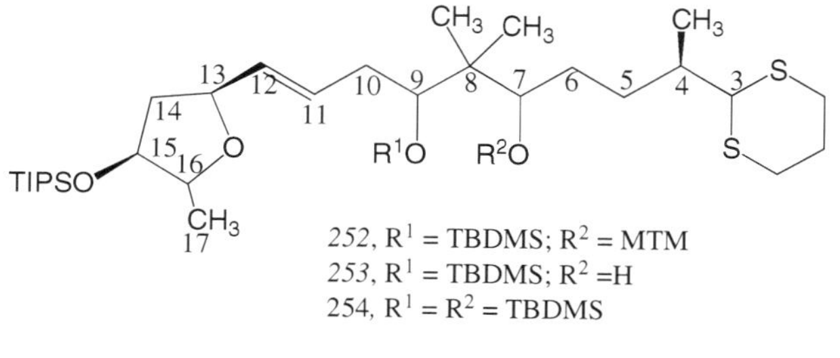

Fig. 13.41

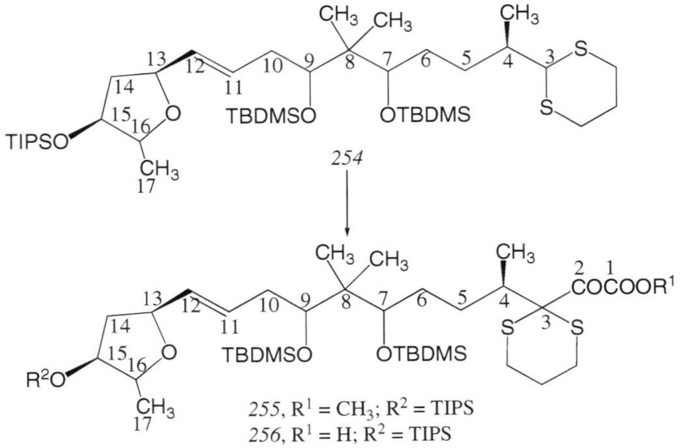

255, R^1 = CH_3; R^2 = TIPS
256, R^1 = H; R^2 = TIPS
257, R^1 = CH_3; R^2 = H

Fig. 13.42

and tetramethylenediamine in THF at –30°C for 2 h to give lithium reagent which was cooled to –78°C, treated with hexamethylphosphorictriamide (HMPA), and then reacted with dimethyl oxalate in THF at –78°C for 30 min, –50°C (30 min), –30°C (30 min), and 0°C (15 min), whereby a ketoester *255* was obtained in 96% yield (Fig. 13.41). Conversion of ketoester *255* to the corresponding keto acid *256* is accomplished in quantitative yield by heating *255* with lithium iodide and 2,6-lutidine in dimethylformamide at 75°C for 18 h. The transformation of *256* to hydroxy ketoester *257* in 97% yield was accomplished by treating *256* with tetra-*n*-butylammonium fluoride in THF at room temperature for 30 min (Fig. 13.42). Reaction of *256* with *257* in the presence of *N,N*-bis[2-oxo3-oxazolinodyl]

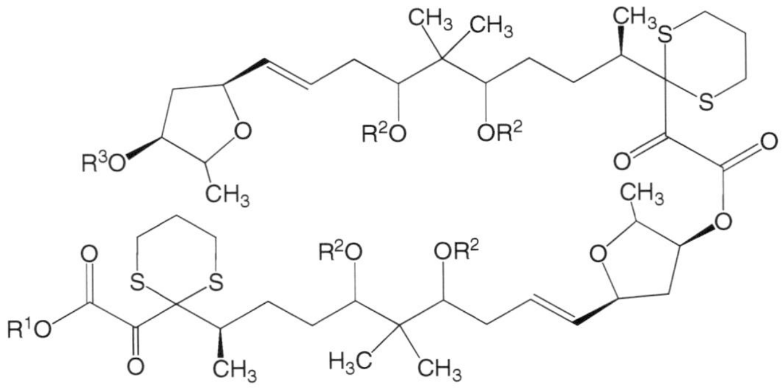

258, R^1 = CH_3; R^2 = TBDMS; R^3 = TIPS
259, R^1 = H; R^2 = TBDMS; R^3 =TIPS
260, R^1 = R^3 = H; R^2 = TBDMS

Fig. 13.43

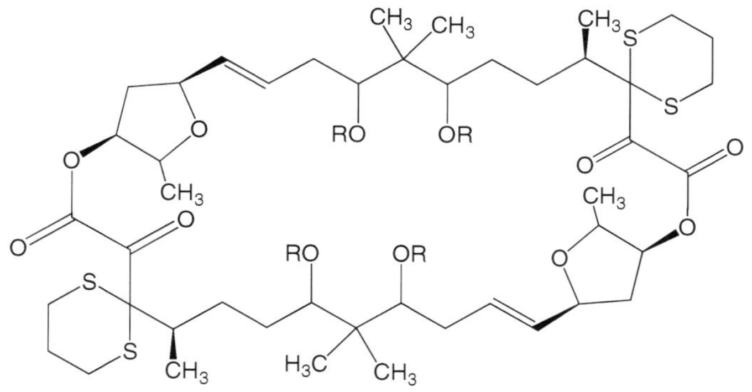

Aplasmomycin

261, R = TBDMS

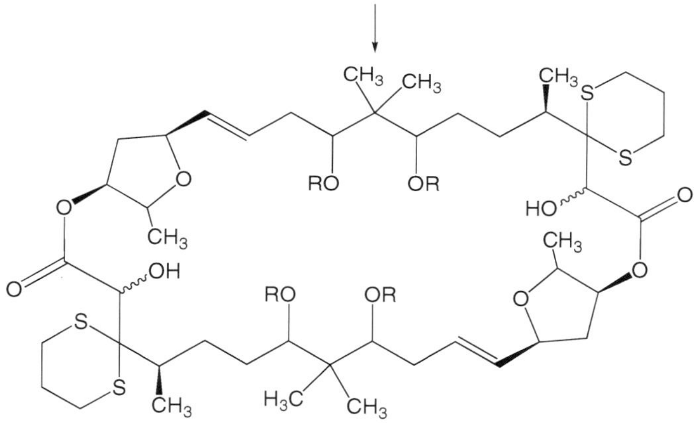

262

262, R = TBDMS
263, R = H

Fig. 13.44

phosphorodiamidic chloride (BOP chloride) [94] and triethylamine in dichloromethane at room temperature for 2 h gave the ester *258* in 98% yield (Fig. 13.43). Cyclization of *258* to the macrocycle *261* was accomplished in the following way. Methyl ester cleavage of *258* was effected in 96% yield with lithium iodide-2,6-lutidine in DMF affording the acid *259*, as previously described; the triisopropylsilyl group of *259* was cleaved in 96% yield with tetra-*n*-butylammonium fluoride in THF at room temperature giving the hydroxy acid *260* and the lactonization of the resulting hydroxy acid *260* in the presence of BOP chloride, as previously described, gave *261* in 71% yield. Reduction of two α-keto groups with sodium borohydride in ethanol at –20°C gave the epimeric mixture of diols *262* in 88% yield. Desilylation of *262* with 7:3 acetonitrile–48% HCl at –10°C for 20 min and at room temperature

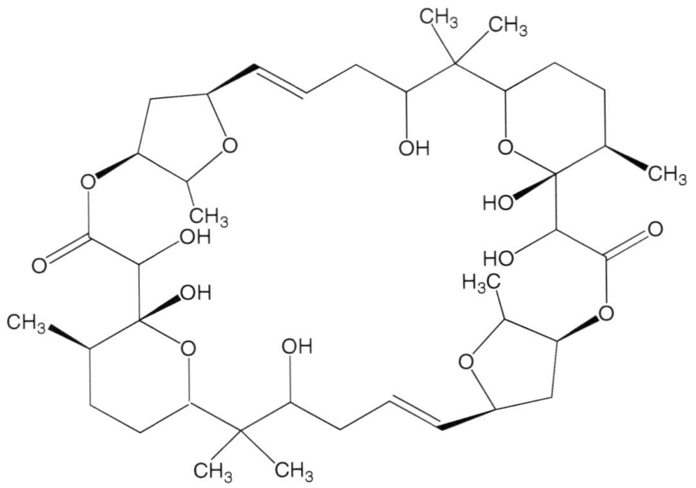

Fig. 13.45

for 2 h gave tetrol *263* in 95% yield. Finally, two dithioacetal groups were removed by treating 263 with mercuric chloride–calcium carbonate in 4:1 acetonitrile–water at room temperature for 9 h giving in 94% yield the "deboro" aplasmomycin *264*, obtained as a mixture of diastereomers differing in configuration at the carbon atom α to the lactone carbonyls (Figs. 13.44 and 13.45).

References

1. Inch, T. D., "*Formation of convenient chiral intermediates from carbohydrates and their use in synthesis*", Tetrahedron (1984) **40**, 3161–3213
2. Nakata, M., "*Formation of Complex Natural Compounds from Monosaccharides*", in *Glycoscience, Chemistry and Chemical Biology*, Fraser-Reid, B.; Tat suta, K.; Thiem, J., Eds, Vol. II, Springer Verlag, New York, 2001, pp. 1175–1213
3. Woodward, R. B., "*Struktur und Biogenese der Makrolide*", Angew. Chem. (1957) **69**, 50–58
4. Miljkovic, M.; Gligorijevic, M.; Satoh, T.; Miljkovic, D., "*Synthesis of macrolide antibiotics. I. Stereospecific addition of methyllithium and methylmagnesium iodide to methyl α-D-xylohexopyranosid-4-ulose derivatives. Determination of the configuration at the branching carbon atom by carbon-13 nuclear magnetic resonance spectroscopy*", J. Org. Chem. (1973) **39**,1379–1384
5. Miljkovic, M.; Gligorijevic, M.; Satoh, T.; Glisin, Dj.; Pitcher, R. G., "*Carbon-13 nuclear magnetic resonance spectra of branched-chain sugars. Configurational assignment of the branching carbon atom of methyl branched-chain sugars*", J. Org. Chem. (1974) **39**, 3847–3850
6. Miljkovic, M.; Glisin, Dj., "*Synthesis of macrolide antibiotics. II. Stereoselective synthesis of methyl 4,6-O-benzylidene-2-deoxy-2-C,3-O-dimethyl-α-D-glucopyranoside. Hydrogenation of the C-2 methylene group of methyl 4,6-O-benzylidene-2-deoxy-2-C-methylene-3-O-methyl-α-and -β-D-arabinohexopyranoside*", J. Org. Chem. (1975) **40**, 3357–3359

7. Miljkovic, M.; Glisin, Dj., "*Synthesis of macrolide antibiotics. III. Stereoselective synthesis of methyl-2,6-dideoxy-2-C,3-O,4-C,6-C-tetramethyl-α-D-glucopyranoside representing the 11-O-methyl derivative of the C-9-C-15-segment of erythronolide A*", Bull Soc. Chim. Beograd (1977) **42**, 659–661
8. Miljkovic, M.; Choong, T. C.; Glisin, Dj., "*Synthesis of macrolide antibiotics. IV. Stereoselective syntheses of the 3-O-methyl and the 11-O-methyl derivatives of the C(1)-C(6) segment of erythronolides A and B and the C(9)-C(15) segment of erythronolide A, respectively*", Croat. Chim. Acta (1985) **58**, 681–698
9. Dalling, D. K.; Grant, D. M., "*Carbon-13 magnetic resonance. IX. Methylcyclohexanes*", J. Am. Chem. Soc. (1967) **89**, 6612–6622
10. Anet, F. A. L.; Bradley, C. H.; Buchanan, G. W., "*Direct detection of the axial con former of methylcyclohexane by 63.1 MHz carbon-13 nuclear magnetic resonance at low temperatures*", J. Am. Chem. Soc. (1971) **93**, 258–259
11. Stothers, J. B., *Carbon-13 NMR Spectroscopy*, Academic Press, New York, NY, 1972, pp. 404, 426
12. Inch, T. D., "*The Use of Carbohydrates in the Synthesis and Configurational Assignments of Optically Active, Non-Carbohydrate Compounds*", Advan. Carbohydr. Chem. Biochem. (1972) **27**, 191–225
13. Burton, J. S.; Overend, W. G.; Williams, N. R., "*Branched-chain sugars. Part III. The introduction of branching into methyl 3,4-O-isopropylidene-β-L-arabinoside and the synthesis of L-hamamelose*", J. Chem. Soc. (1965) 3433–3445
14. Feast, A. A. J.; Overend, W. G.; Williams, N. R., "*Branched-chain sugars. Part VI. The reaction of methyl 3,4-O-isopropylidene-β-L-erythro-pentopyranosidulose with organolithium reagents*", J. Chem. Soc. C (1966) 303–306
15. Flaherty, B.; Overend, W. G.; Williams, N. R., "*Branched-chain sugars. Part VII. The synthesis of D-mycarose and D-cladinose*", J. Chem. Soc. C (1966) 398–403
16. Inch, T. D.; Lewis, G. J.; Williams, N. E., "*A stereochemical comparison of some ad dition reactions to methyl 4,6-O-benzylidene-3-deoxy-3-C-ethyl-α-D-hexopyranosid-2-uloses*", Carbohydr. Res. (1971) **19**, 17–27
17. Inch, T. D., "*Asymmetric synthesis: Part I. A stereoselective synthesis of benzylic centres. Derivatives of 5-C-phenyl-D-gluco-pentose and 5-C-phenyl-L-ido-pentose*" Carbohydr. Res. (1967) **5**, 45–52
18. Guillerm-dron, D.; Capmau, M.-L.; Chodkiewicz, W., "*Assistance du groupe m- thoxyle en α d'un carbonyle dans le cours stérique de l'addition d'organométalliques insaturés*", Tetrahedron Lett. (1972) **13**, 37–40
19. Miljkovic, M.; Gligorijevic, M.; Miljkovic, D., "*Steric and Electrostatic Interactions in Reactions of Carbohydrates. II. Stereochemistry of Addition Reactions to the Carbonyl Group of Glycopyranosiduloses. Synthesis of Methyl 4, 6-O-Benzylidene-3-O- methyl-β-D-mannopyranoside*", J. Org. Chem. (1974) **39**, 2118–2120
20. Hanessian, S.; Rancourt, G., "*Carbohydrates as chiral intermediates in organic synthesis. Two functionalized chemical precursors comprising eight of the ten chiral centers of erythronolide A*", Can. J. Chem. (1976) **55**, 1111–1113
21. Hanessian, S.; Rancourt, G., "*Approaches to the total synthesis of natural products from carbohydrates*", Pure Appl. Chem. (1977) **49**, 1201–1214
22. Hanessian, S.; Rancourt, G.; Guindon, Y., "*Assembly of the carbon skeletal frame work of erythronolide A*", Can. J. Chem. (1978) **56**, 1843–1846
23. Kochetkov, N. K.; Sviridov, A. F.; Ermolenko, M. S.; Zelinsky, N. D., "*Synthesis of macrolide antibiotics. 1. Synthesis of the C1–C6 segment of 14-membered macrolide antibiotics*", Tetrahedron Lett. (1981) **22**, 4315–4318
24. Kochetkov, N. K.; Sviridov, A. F.; Ermolenko, M. S.; Zelinsky, N. D., "*Synthesis of macrolide antibiotics. 2. Synthesis of the C9–C13 segments of erythronolides A, B and oleandonolide*", Tetrahedron Lett. (1981) **22**, 4319–4322

25. Kochetkov, N. K.; Sviridov, A. F.; Ermolenko, M. S.; Yashunskii, D. V., "*Synthesis of macrolide antibiotics. 3. Revised synthesis of C9–C13 segment of erythronolide A*", Tetrahedron Lett. (1984) **25**, 1605–1608
26. Kochetkov, N. K.; Sviridov, A. F.; Ermolenko, M. S.; Yashunskii, D. V.; Borodkin, V. S., "*Stereocontrolled synthesis of erythronolides A and B from 1,6-anhydro-β-D-glucopyranose (levoglucosan). Skeleton assembly in (C9–C13) + (C7–C8) + (C1–C6) sequence*", Tetrahedron (1989) 45(16), 5109–5136
27. Sviridov, A. F.; Yashunskii, D. V.; Ermolenko, M. S.; Kochetkov, N. K., "*New method for the synthesis of 2,4-dideoxy-2,4-di-C-methyl-D-glucopyranose derivatives*", Izv. Akad. Nauk. SSSR, Ser. Khim. (1984) 723–725
28. Kochetkov, N. K.; Sviridov, A. F.; Yashunskii, D. V.; Ermolenko, M. S.; Borodkin, V. S., "*Synthesis of C-methyldeoxysugars: deoxygenation of xanthates of tertiary alcohols and hydrozirconation of exomethylene derivatives of carbohydrates*", Izv. Akad. Nauk SSSR, Ser. Khim. (1986) 441–445
29. Omura, K.; Swern, D., "*Oxidation of alcohols by "activated" dimethyl sulfoxide. A preparative, steric and mechanistic study*", Tetrahedron (1978) **34**, 1651–1660
30. Kelly, A. G.; Roberts, J. S., "*A simple, stereocontrolled synthesis of a thromboxane B2 synthon*", J. Chem. Soc. Chem. Commun. (1980) 228–229
31. Cha, J. K.; Christ, W. J.; Kishi, Y., "*On stereochemistry of osmium tetraoxide oxidation of allylic alcohol systems. Empirical rule*", Tetrahedron (1984) **40**, 2247–2255
32. Ashby, E. C.; Lin, J. J., "*Selective reduction of alkenes and alkynes by the reagent lithium aluminum hydride-transition-metal halide*", J. Org. Chem. (1978) **43**, 2567–2572
33. Gigg, R.; Warren, C. D., "*The allyl ether as a protecting group in carbohydrate chemistry. Part II*", J. Chem. Soc. C (1968) 1903–1911
34. Oikawa, Y.; Yoshioka, T.; Yonemitsu, O., "*Protection of hydroxy groups by intramolecular oxidative formation of methoxybenzylidene acetals with DDQ*", Tetrahedron Lett. (1982) **23**, 889–892
35. Seebach, D.; Ertasogon, M.; Locher, R.; Schweizer, W. B., "*Tritylketone und Tritylenone. Beiträge zur sterisch erzwungenen Michael-Addition und zur diastereoselektiven Aldol-Addition*", Helv. Chim. Acta (1985) **68**, 264–282
36. Nakagawa, I.; Hata, T., "*A convenient method for the synthesis of 5'-S-alkylthio-5'-deoxyribonucleosides*", Tetrahedron Lett. (1975) **16**, 1409–1412
37. Andersen, K. K.; Gaffield, W.; Papanikolaou, N. E.; Foley, J. W.; Perkins, P. I., "*Optically Active Sulfoxides. The Synthesis and Rotatory Dispersion of Some Diaryl Sulfoxides*", J. Am. Chem. Soc. (1964) **86**, 5637–5646
38. Drabowicz, J.; Oae, S., "*Mild Reductions of Sulfoxides with Trifluoroacetic Anhydride/Sodium Iodide System*", Synthesis (1977) 404–404
39. Corey, E. J.; Shibasaki, M.; Knolle, J., "*Simple, stereocontrolled synthesis of thromboxane B2 from D-glucose*", Tetrahedron Lett. (1977) **19**, 1625–1626
40. Hanessian, S.; Lavallee, P., "*A stereospecific, total synthesis of thromboxane B2*", Can. J. Chem. (1977) **55**, 562–565
41. Hanessian, S.; Lavallee, P., "*Total synthesis of (+)-thromboxane B2 from D-glucose. A detailed account*", Can. J. Chem. (1981) **59**, 870–877
42. Wick, A. E.; Felix, D.; Steen, K.; Eschenmoser, A., "*CLAISEN'sche Umlagerungen bei Allyl- und Benzylalkoholen mit Hilfe von Acetalen des N, N-Dimethylacetamids. Vorläufige Mitteilung*", Helv. Chim. Acta (1964) **47**, 2425–2429
43. Felix, D.; Gschwend-Steen, K.; Wick, A. E.; Eschenmoser, A., "CLAISEN'sche Umlagerungen bei Allyl- und Benzylalkoholen mit 1-Dimethylamino-1-methoxyäthen", Hel. Chim. Acta (1969) **52**, 1030–1042
44. Corey, E. J.; Schaaf, T. K.; Huber, W.; Koelliker, U.; Weinschenker, N. M., "*Total Synthesis of Prostaglandins F2α and E2 as the Naturally Occurring Forms*", J. Am. Chem. Soc. (1970) **92**, 397–398

References

45. Colegate, S. M.; Dorling, P. R.; Huxtable, C. R., "*A Spectroscopic Investigation of Swainsonine: An α-Mannosidase Inhibitor Isolated from Swainsona canescens*", Aust. J. Chem. (1979) **32**, 2257–2264
46. Molyneux, R. J.; James, L. F., "*Loco intoxication: indolizidine alkaloids of spotted locoweed (Astragalus lentiginosus)*", Science (1982) **216**, 190–191
47. Schneider, M. J.; Ungemach, F. S.; Broquist, H. P.; Harris, T. M., "*(1S,2R,8R,8aR)- 1,2,8-trihydroxyoctahydroindolizine (swainsonine), an α-mannosidase inhibitor from Rhizoctonia leguminicola*", Tetrahedron (1983) **39**, 29–32
48. Hohenschutz, L. D.; Bell, E. A.; Jewess, P. J.; Leworthy, D. P.; Pryce, R. J.; Arnold, E.; Clardy, J.,"*Castanospermine, A 1,6,7,8-tetrahydroxyoctahydroindolizine alkaloid, from seeds of Castanospermum australe*", Phytochemistry (1981) **20**, 811–814
49. Freer, A. A.; Gardner, D.; Greatbanks, D.; Poyser, J. P.; Sim, G. A., "*Structure of cyclizidine (antibiotic M146791): x-ray crystal structure of an indolizidinediol metabolite bearing a unique cyclopropyl side chain*", J. Chem. Soc. Chem. Commun. (1982) 1160–1162
50. 50, Ali, M. H.; Hough, L.; Richardson, A. C., "*A chiral synthesis of swainsonine from D-glucose*", J. Chem. Soc. Chem. Commun. (1984) 447–448
51. Richardson, A. C., "*Improved preparation of methyl 3-amino-3-deoxy-α-D-mannopyranoside hydrochloride*", J. Chem. Soc. (1962) 373–374
52. Fleet, G. W. J.; Gough, M. J.; Smith, P. W., "*Enantiospecific Synthesis of Swain sonine, (1S, 2R, 8R, 8aR)-1,2,8-trihydroxyoctahydroindolizine, from D-mannose*", Tetrahedron Lett. (1984) **25**, 1853–1856
53. Ohrui, H.; Emoto, S., "*Stereospecific synthesis of (+)-biotin*", Tetrahedron Lett. (1975) **16**, 2765–2766
54. Ogawa, T.; Kawano, T.; Matsui, M., "*A biomimetic synthesis of (+)-biotin from D-glucose*", Carbohydr. Res. (1977) **57**, C31-C35
55. Corey, E. J.; Nicolaou, K. C.; Balanson, R. D.; Machida, Y., "*A Useful Method for the Conversion of Azides to Amines*", Synthesis (1975) 590–591
56. Buchta, E.; Andree, F., "*Eine Partialsynthese des "all"-trans-Methyl-bixins und des "all"-trans-4,4'-Desdimethyl-methyl-bixins*", Chem. Berichte (1959) **92**, 3111–3116
57. Alexander, R. G.; Clayton, J. P.; Luk, K.; Rogers, N. H.; King, T. J., "*The chemistry of pseudomonic acid. Part 1. The absolute configuration of pseudomonic acid A*", J. Chem. Soc. Perkin Trans. I (1978) 561–565
58. Chain, E. B.; Mellows, G., "*Pseudomonic acid. Part 3. Structure of pseudomonic acid B*", J. Chem. Soc. Perkin Trans. I (1977) 318–322
59. Clayton, J. P.; O'Hanlon, P. J.; Rogers, N. H., "*The structure and configuration of pseudomonic acid C*", Tetrahedron Lett. (1980) **21**, 881–884
60. Beau, J.-M.; Abyraki, S.; Pougny, J.-R.; Sinaÿ, P., "*Total synthesis of methyl (+)-pseudomonate C from carbohydrates*", J. Am. Chem. Soc. (1983) **105**, 621–622
61. Helferich, B.; Ost, W., "Synthese einiger -D-Xylopyranoside", Chem. Berichte (1962) **95**, 2612–2615
62. Pougny, J.-R.; Sinaÿ, P. "*(3S,4S)-4-Methylheptan-3-ol, a pheromone component of the smaller European elm bark beetle. Synthesis from D-glucose*", J. Chem. Res. Synop Ses. (1982) 1 1
63. Bernet, B.; Vasella, A., "*Carbocyclische Verbindungen aus Monosacchariden. I. Um setzungen in der Glucosereihe*", Helv. Chim. Acta (1979) **62**, 1990–2016
64. Nakane, M.; Hutchinson, C. R.; Gollman, H., "*A convenient and general synthesis of 5-vinylhexofuranosides from 6-halo-6-deoxypyranosides*", Tetrahedron. Lett. ((1980) **21**, 1213–1216
65. Huynh, C.; Derguini-Boumechal, F.; Linstrumelle, G.," *Copper-catalysed reactions of Grignard reagents with epoxides and oxetane*", Tetrahedron Lett. ((1979) **20**,1503–1506
66. Felkin, H.; Frajerman, C.; Roussi, G., "*Stereochemistry of epoxide ring opening by allylic Grignard reagents*", Bull. Soc. Chim. Fr. (1970) 3704–3710
67. Glaze, W. H.; Duncan, D. P.; Berry, D. J., "*Neopentylallyllithium. 5. Stereochemistry of non-rearrangement reactions with epoxides*", J. Org. Chem. (1977) **42**, 694–697

68. Linstrumelle, G.; Lorne, R.; Dang, H. P., *"Copper-catalysed reactions of allylic Grignard reagents with epoxides"*, Tetrahedron Lett. (1978) **19**, 4069–4072
69. Schlosser, M.; Stähle, M., *"Allylic Compounds of Magnesium, Lithium, and Potassium: σ- or π-Structures?"*, Angew. Chem., Int. Ed. Engl. (1980) **19**, 487–489
70. Bagnell, L.; Jeffery, E. A.; Meisters, A.; Mole, T., *"A new conversion of nitriles into acetyl compounds: Nickel-catalysed methylation by trimethylaluminium"*, Aust. J. Chem. (1974) **27**, 2577–2582
71. Kozikowski, A. P.; Schmiesing, R. J.; Sorgi, K. L., *"Total synthesis of pseudomonic acid C: application of the alkoxyselenation reaction in organic synthesis"*, J. Am. Chem. Soc. (1980) **102**, 6577–6580
72. Clayton, J. P.; Luk, K.; Rogers, N. H., *"The chemistry of pseudomonic acid. Part 2. The conversion of pseudomonic acid A into monic acid A and its esters"*, J. Chem. Soc. Perkin Trans. I (1979) 308–313
73. Keck, G. E.; Kachensky, D. F.; Enholm, E. J., *"Pseudomonic acid C from L-lyxose"*, J. Org. Chem. (1985) **50**, 4317–4325
74. Keck, G. E.; Yates, J. B., *"Carbon-carbon bond formation via the reaction of trialkylallylstannanes with organic halides"*, J. Am. Chem. Soc. (1982) **104**, 5829–5831
75. Schaffer, R., *"2,3-O-Isopropylidene-α-D-lyxofuranose, the monoacetone-D-lyxose of Levene and Tipson"*, J. Res. Natl. Bur. Std. (1961) **65A**, 507–512
76. Fráter, G., *"Über die Stereospezifität der alpha-Alkylierung von -Hydroxycarbon säureestern. Vorläufige Mitteilung"*, Helv. Chim. Acta (1979) **62**, 2825–2828
77. Fráter, G., *"Stereospezifische Synthese von (+)-(3R, 4R)-4-Methyl-3-heptanol, das Enantiomere eines Pheromons des kleinen Ulmensplintkäfers (Scolytus multistria tus)"*, Helv. Chim. Acta (1979) **62**, 2829–2832
78. Ohrui, H.; Jones, G. H.; Moffatt, J. G.; Maddox, M. L.; Christensen, A. T.; Byram, S. K., *"C-Glycosyl nucleosides. V. Unexpected observations on the relative stabilities of compounds containing fused five-membered rings with epimerizable substituents"*, J. Am. Chem. Soc. (1974) **97**, 4602–4613
79. Horner, L.; Hoffman, H.; Wippel, H. G., *"Phosphororganische Verbindungen, XII. Phosphinoxyde als Olefinierungsreagenzien"*, Chem. Berichte (1958) **91**, 61–63
80. Horner, L.; Hoffman, H.; Wippel, H. G.; Klahre, G., *"Phosphororganische Verbindungen, XX. Phosphinoxyde als Olefinierungsreagenzien"*, Chem. Berichte (1959) **92**, 2499–2505
81. Wadsworth, W. S., Jr.; Emmons, W. D., *"The Utility of Phosphonate Carbanions in Olefin Synthesis"*, J. Am. Chem. Soc. (1961) **83**, 1733–1738
82. Wadsworth, W. S., Jr.; Emmons, W. D., Organic Synthesis, Coll. Vol. 5 (1973) 547; ibid. (1965) Vol. 45, 44
83. Wadsworth, W. S., Jr., Organic reactions (1977) **25**, 73–253
84. Keck, G. E.; Tafesh, A. M., *"Free-radical addition-fragmentation reactions in synthesis: a "second generation" synthesis of (+)-pseudomonic acid C"*, J. Org. Chem. (1989) **54**, 5845–5846
85. Takacs, J. M.; Helle, M. A.; Seely, F. L., *"An improved procedure for the two carbon homologation of esters to α, β-unsaturated esters"*, Tetrahedron Lett. (1986) **27**, 1257–1260
86. Evans, D. A.; Andrews, G. L., *"Allylic sulfoxides. Useful intermediates in organic synthesis"*, Acc. Chem. Res. (1974) **7**, 147–155
87. Trost, B. M.; Curran, D. P., *"Chemoselective oxidation of sulfides to sulfones with potassium hydrogen persulfate"*, Tetrahedron Lett. (1981) **22**, 1287–1290
88. Okazaki, T.; Kitahara, T.; Okami, Y., *"Studies on marine microorganisms. IV. A new antibiotic SS-228 Y produced by Chainia isolated from shallow sea mud"*, J. Antibiot. (1975) **28**, 176–184
89. Dunitz, J. D.; Hawley, D. M.; Mikloš, D.; White, D. N. J.; Berlin, Y.; Marušič, R.; Prelog, V., *"Structure of boromycin"*, Helv. Chim. Acta (1971) **54**, 1709–1713
90. Corey, E. J.; Pan, B. C.; Hua, D. H.; Deardof, D. R., *"Total synthesis of aplas momycin. Stereocontrolled construction of the C(3)-C(17) fragment"*, J. Am. Chem. Soc. (1982) **104**, 6816–6818

91. Corey, E. J.; Hua, D. H.; Pan, B. C.; Seitz, S. P., "*Total synthesis of aplasmomycin*", J. Am. Chem. Soc. (1982) **104**, 6818–6820
92. Corey, E. J.; Cho, H.; Rücker, C.; Hua, D. H., "*Studies with trialkylsilyltriflates: new syntheses and applications*", Tetrahedron Lett. (1981) **22**, 3455–3458
93. Corey, E. J.; Bock, M. G., "*Protection of primary hydroxyl groups as methylthio-methyl ether*", Tetrahedron Lett. (1975) **16**, 3269–3270
94. Diago-Mesequer, J.; Palomo-Coll, A. L.; Fernández-Lizarbe, J. R.; Zugaza-Bilbao, A., "*A New Reagent for Activating Carboxyl Groups; Preparation and Reactions of N,N-Bis[2-oxo-3-ox-azolidinyl]phosphorodiamidic Chloride*", Synthesis (1980) 547–550

Chapter 14
Carbohydrate-Based Antibiotics

Antibiotics are most often defined as bacterial or fungal products that inhibit the growth of other microorganisms. There is a broader definition proposed [1] that defines antibiotics as chemical compounds derived from or produced by living organisms which are capable, in small concentration, to inhibit the life processes of microorganisms. This definition, however, does not include a vast number of chemically modified (semisynthetic) and synthetic antibiotics. Antibiotics are isolated from bacteria, yeast, molds, algae, and lichens, as well as from higher plants.

There are many antibiotics isolated from microorganisms and even many more chemically modified natural antibiotics (so-called semisynthetic antibiotics) that have been made in order to increase their activity, improve their selectivity, and decrease their side effects. Chemically, antibiotics belong to many classes of organic compounds, but here we are only interested in antibiotics that are carbohydrate related, i.e., in antibiotics that contain carbohydrates as a part of their structure.

There are generally three types of antibiotics that are carbohydrate related. First group consists of antibiotics in which the carbohydrates are glycosidically linked to cyclitols or aminocyclitols (aminoglycoside antibiotics), such as streptomycin, kanamycin, amikacin. Second group of antibiotics consists of oligosaccharides in which the individual monosaccharides are linked both glycosidically and also with one or more orthoester linkages (orthosomycins). Finally, carbohydrates may be glycosidically linked to noncarbohydrate part of an antibiotic, as is the case in macrolide antibiotics erythromycins, nystatin, etc., and many other classes of carbohydrate-related antibiotics.

Aminoglycoside Antibiotics

Aminoglycoside antibiotics represent a group of carbohydrate-based antibiotics that consist of an aminocyclitol such as streptamine (*1*), 2-deoxy-streptamine (*2*), epistreptamine (the C2 epimer of streptamine) (*3*), streptidine (*4*), or 2-deoxy-streptidine (*5*) to which various amino sugars are glycosidically linked (Fig. 14.1). For example, streptomycin, kanamycin, neomycin, amikacin, gentamicin, tobramycin, netilmicin are some of the members of this family of antibiotics. There are two good reviews on aminoglycoside antibiotics [2, 3].

Streptomycin and dihydrostreptomycin are effective against tuberculosis; neomycins are used locally for skin infections and orally for bacterial enterocolitis; paromomycins are used orally against amoebic dysentery and bacterial enterocolitis; kanamycins are highly effective against *Proteus* infections; gentamicin, sisomicin, netilmicin, tobramycin, and dibekacin are indispensable for treatment of severe *Pseudomonas* infections; amikacin is valuable for severe infections caused by gram-negative microorganisms resistant to tobramycin, dibekacin, sisomicin, netilmicin, gentamicin, and kanamycin, as well as spectinomycin which is active against penicillin-resistant gonorrhea.

1, R^1 = H; R^2 = OH, Streptamine
2, R^1 = R^2 = H, 2-Deoxystreptamine
3, R^1 = OH; R^2 = H, Epistreptamine

4, R = OH, Streptidine
5, R = H, 2-Deoxystreptidine

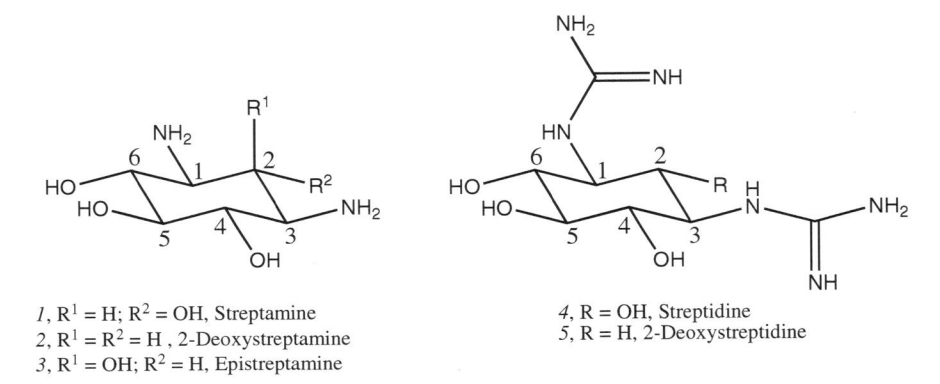

Fig. 14.1

It is interesting to note that aminoglycoside antibiotics are predominantly bactericidal (not bacteriostatic).

Kanamycin

Kanamycin was isolated from fermentation broth of *Streptomyces kanamyceticus* [4] by absorption on Amberlite IRC-50 and elution with 1 N HCl. It has been shown to be the mixture of three components: kanamycin A, kanamycin B, and kanamycin C.

From a study of products of acid hydrolysis of kanamycin A, Cron et al. [5] proposed the structure of kanamycin A to be a trisaccharide-like molecule consisting of two amino sugar moieties glycosidically linked to 2-deoxystreptamine. One hexosamine unit was determined to be 6-amino-6-deoxy-D-glucopyranose [6], and the other hexosamine, named kanosamine, was determined to be 3-amino-3-deoxy-D-glucopyranose [7] and proposal was made that the two hexosamine residues were attached to 2-deoxystreptamine at the C4 and C6 hydroxyl groups. Independent studies on the structure of kanamycin A were conducted by Ogawa et al. [8], and

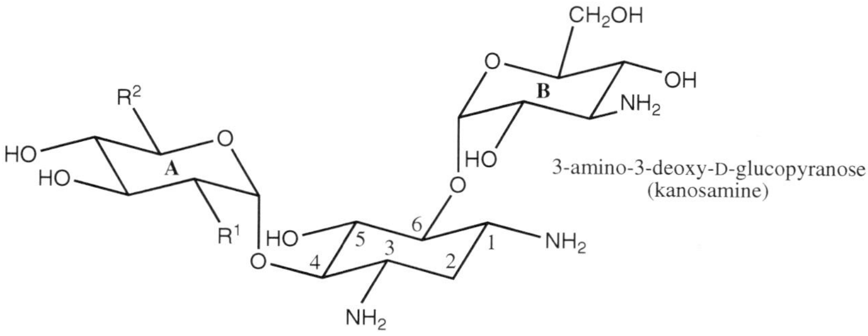

Fig. 14.2

3-amino-3-deoxy-D-glucopyranose (kanosamine)

2-Deoxystreptamine
6, R^1 = OH; R^2 = CH$_2$NH$_2$, Kanamyciin A
7, R^1 = NH$_2$; R^2 = CH$_2$NH$_2$, Kanamycin B
8, R^1 = NH$_2$; R^2 = CH$_2$OH, Kanamycin C

the absolute structure of kanamycin A was determined by Umezawa et al. [9]. The crystal structure of kanamycin A was determined by Koyama et al. [10] and is shown in the Fig. 14.2 (structure 6). The absolute sequence of the C4–C6 substitution was first determined by Hitchens and Rinehart [11] from their studies on neomycin.

The structure of kanamycin B (7) was determined by Ito et al. [12] and the structure of kanamycin C (8) by Murase [13].

The structural difference of kanamycins A, B, and C is in the structure of the amino sugar **A** that is α-glycosidically linked to the C4 hydroxyl group of streptamine. In kanamycin A (6), this sugar is 6-amino-6-deoxy-D-glucopyranose (R^1 = OH, R^2 = CH$_2$NH$_2$); in kanamycin B (7), it is 2, 6-diamino-2,6-dideoxy-D-glucopyranose (R^1 = NH$_2$; R^2 = CH$_2$NH$_2$); and in kanamycin C (8), it is 2-amino-2-deoxy-D-glucopyranose (R^1 = NH$_2$; R^2 = CH$_2$OH) (Fig. 12.2).

Umezawa et al. have reported the first total synthesis of both kanamycin B [14, 15] and kanamycin C [16, 17].

Kanamycin shows inhibitory activity against a wide range of gram-positive and gram-negative bacteria. It has been found to be a particularly valuable chemotherapeutic agent for the treatment of serious gram-negative infections and streptomycin-resistant tuberculosis. The ototoxicity and nephrotoxicity, which are typical side effects of aminoglycoside antibiotics, are much less associated with kanamycin. The mechanism of action of kanamycin, as well as of all other aminoglycoside antibiotics, is the inhibition of protein synthesis at the prokaryotic ribosomal level.

After the first observation of Umezawa et al. [18] that the inactivation of kanamycin by kanamycin-resistant organism is due to O-phosphorylation of kanamycin, it was established that the major inactivation mechanisms of aminoglycoside-resistant organisms are N-acetylation [19–24], O-phosphorylation [25–30], and O-adenylylation [31–33].

One approach to prevent inactivation of aminoglycoside antibiotics was to either remove the functional group that is target for inactivating enzymes or to modify the aminoglycoside antibiotic in such a way to inhibit the inactivating enzyme either at the binding site or at the active site toward the aminoglycoside molecule. Using the first approach, Umezawa et al. [34–36] have synthesized the 3′-deoxykanamycin A and 3′,4′-diedoxykanamycin B, since it has been shown that the 3′-hydroxyl group of aminoglycoside antibiotics is an important target site of phosphorylating enzymes. The modified aminoglycoside antibiotics have shown to be active against kanamycin resistant *Escherichia coli* and *Pseudomonas* species. The product of the second approach is semisynthetic antibiotic amikacin.

Amikacin

Amikacin is a semisynthetic derivative of kanamycin A. It has higher activity against clinical isolates of *Enterobacter* species, *Pseudomonas aeruginosa*, and *Staphylococcus aureus* than gentamicin sulfate, kanamycin sulfate, or tobramycin, and no major toxicity has been observed with normal doses.

Amikacin is kanamycin A that has the C3 amino group of 2-deoxystreptamine (ring B) acylated by L-haba (L-4-amino-2-hydroxybutyric acid) as shown in Fig. 14.3.

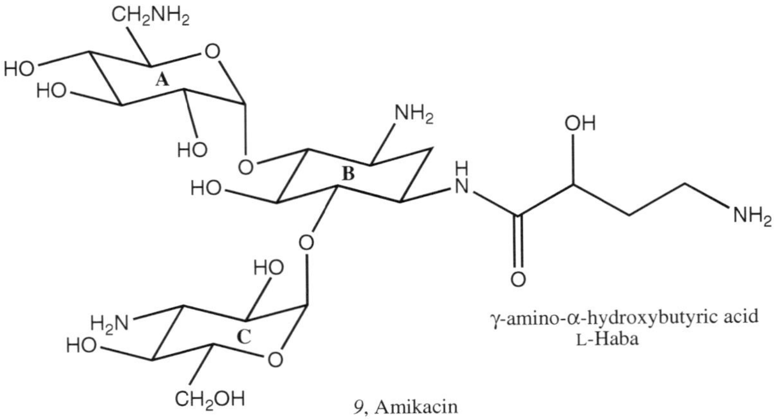

Fig. 14.3

The synthesis of amikacin was accomplished by Kawaguchi et al. [37].

Aminoglycoside Antibiotics

Gentamicins

Weinstein et al. [38] have isolated gentamicins, an antibiotic complex produced by *Micromonospora purpurea* and *M. echinospora* or variants thereof. Gentamicins A, A_1, A_2, A_3, A_4; gentamicins B and B_1; and gentamicins C, C_1, C_2, C_{2b} have been identified in this complex mixture. The structure of gentamicin A (Fig. 14.4) has been elucidated by Maehr and Schaffner [39, 40].

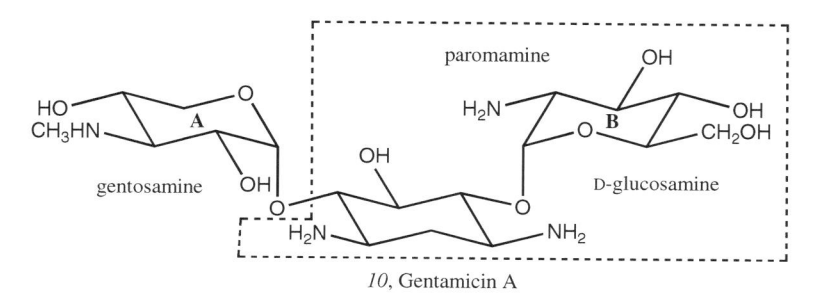

10, Gentamicin A

Fig. 14.4

The structures of gentamicins A_1, A_3, A_4 [41] and of gentamicin A_2 [42] are shown in Figs. 14.4 and 14.5.

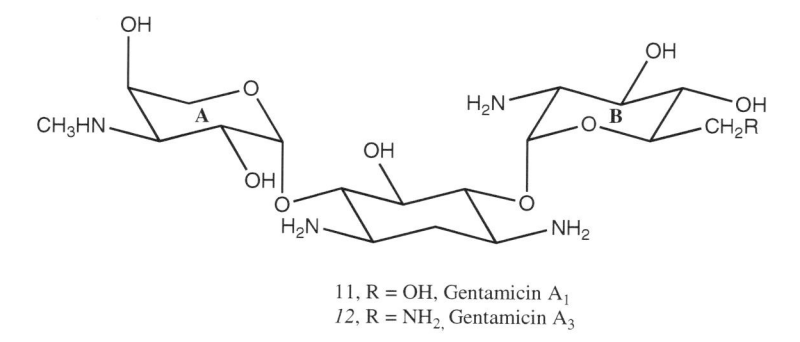

11, R = OH, Gentamicin A_1
12, R = NH_2, Gentamicin A_3

Fig. 14.5

As can be seen from Figs. 14.4–14.6, the structural variations are on both sugar residues.

Gentamicin B (*15* in Fig. 14.7) together with small amounts of gentamicin B_1 (*16*) is produced by a variety *Micromonospora*, and it shows a broad spectrum of activity [43].

Gentamicin C is an aminoglycoside antibiotic that represents a mixture of several closely related and structurally similar compounds of which the most important are Gentamicins C_1 ($R^1 = R^2 = CH_3$), C_{1a} ($R^1 = R^2 = H$), C_2 ($R_1 = CH_3$, $R^2 = H$),

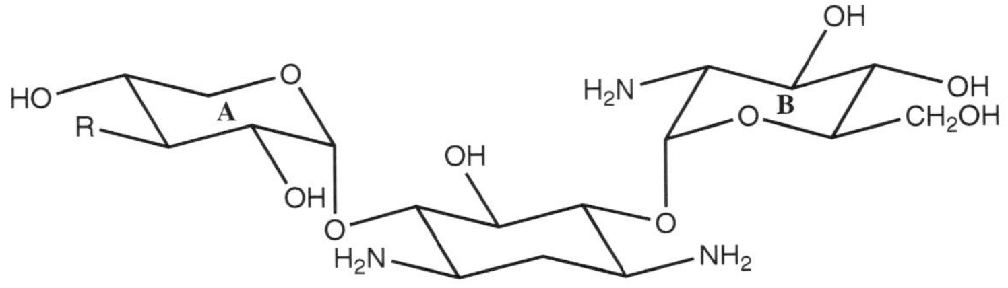

13, R = OH, Gentamicin A$_2$

14, R = N(CH$_3$)—CHO , Gentamicin A$_4$

Fig. 14.6

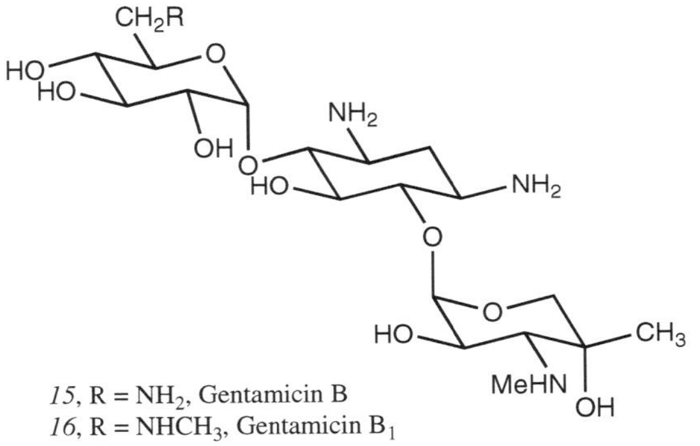

15, R = NH$_2$, Gentamicin B
16, R = NHCH$_3$, Gentamicin B$_1$

Fig. 14.7

and C$_{2b}$ (R^1 = H, R^2 = CH$_3$). Gentamicin C is the broad-spectrum antibiotic and is the most extensively used aminoglycoside antibiotic. The structures of gentamicins C antibiotic have been elucidated by Cooper et al. [44] and are represented in Fig. 14.8.

The synthesis of racemic purpurosamine B (Fig. 14.8) was reported by Chmielewski et al. [45].

Tobramycin (Nebramycin)

Tobramycin (*21* in Fig. 14.9) is a broad-spectrum antibiotic produced by a number of *Streptomyces* species. It is more active than gentamicin against *Pseudomonas aeruginosa* but less active against other gram-negative bacteria. Both the nephrotoxicity and ototoxicity in guinea pigs are less than those of gentamicin but generally greater than for kanamycin. An important feature is the high activity against strains

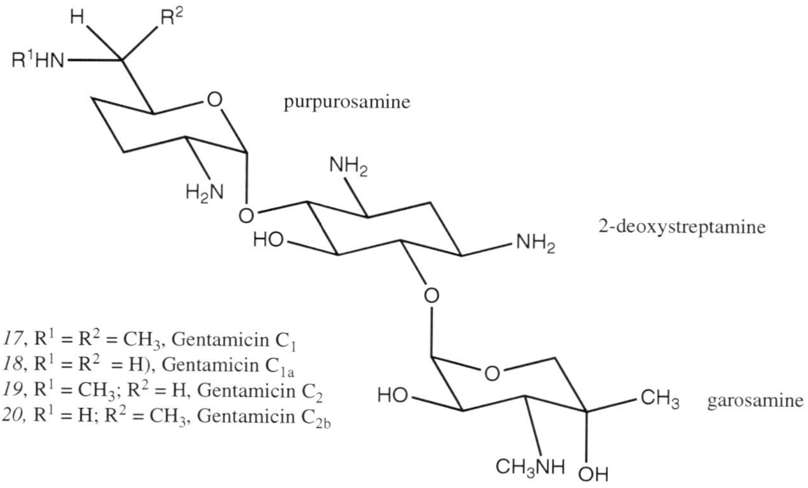

Aminoglycoside Antibiotics

Fig. 14.8

17, $R^1 = R^2 = CH_3$, Gentamicin C_1
18, $R^1 = R^2 = H$), Gentamicin C_{1a}
19, $R^1 = CH_3$; $R^2 = H$, Gentamicin C_2
20, $R^1 = H$; $R^2 = CH_3$, Gentamicin C_{2b}

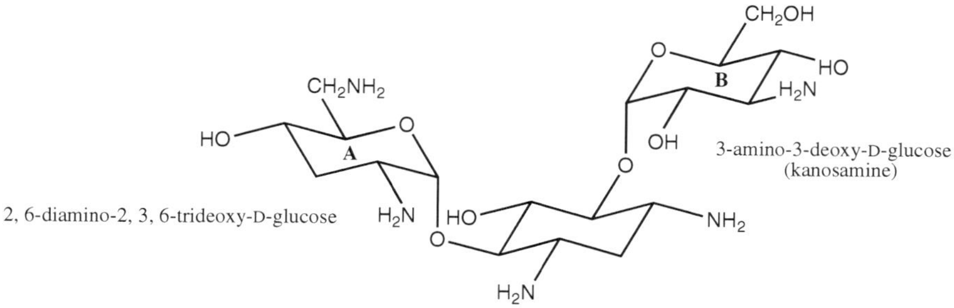

purpurosamine
2-deoxystreptamine
garosamine

Fig. 14.9

21, Tobramycin

2,6-diamino-2,3,6-trideoxy-D-glucose
3-amino-3-deoxy-D-glucose (kanosamine)

of *Pseudomonas* that are resistant to gentamicin. It has been also shown to be active in vitro against *Enterobacter*, *E.coli*, *Klebsiella*, and *S. aureus*.

The structure of tobramycin was determined by Koch and Rhoades [46] and is shown in the Fig. 14.9. As can be seen, it is very similar to the structure of kanamycin B except that the amino sugar A has no C3 hydroxyl group. Synthesis of tobramycin was accomplished by two groups: Takagi et al. [47] and Tanabe et al. [48].

Neomycin B (Actilin, Enterfram, Framecetin, Soframycin)

Neomycin B is obtained from cultures of *Streptomyces fradiae* and subsequently from other *Streptomyces* species such as *Streptomyces coeruleoprunus*. Neomycin is active against gram-positive and gram-negative bacteria, mycobacteria, and actinomycetes. It is active against streptomycin-resistant bacteria, including tuberculosis organisms. It was discovered by Waksman and Lechevalier [49], and the structure of neomycins B and C (Fig. 14.10) was determined by Rinehart et al. [50–53]. The absolute configuration of neomycin C was determined by Hitchens and Rinehart [11]. Total synthesis of neomycin C was reported by Umezawa et al. [54, 55] and of neomycin B by Usui and Umezawa [56, 57].

$R^1 = CH_2NH_2, R^2 = H$, Neosaminee B (Paromose)
$R^1 = H; R^2 = CH_2NH_2$, Neosamine C

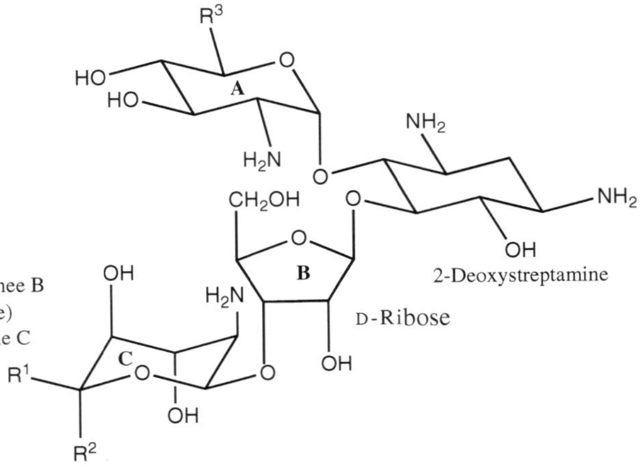

2-Deoxystreptamine
D-Ribose

22, $R^1 = CH_2NH_2, R^2 = H, R^3 = CH_2NH_2$, Neomycin B
23, $R^1 = H; R^2 = CH_2NH_2, R^3 = CH_2NH_2$, Neomycin C
24, $R^1 = CH_2NH_2; R^2 = H; R^3 = CH_2OH$, Paromomyciin I
25, $R^1 = H; R^2 = CH_2NH_2; R^3 = CH_2OH$, Paromomycin II

Fig. 14.10

Paromomycin

Paromomycins I and II (Fig. 14.10) are closely related to neomycins B and C. The only difference between them is in the structure of amino sugar A. In neomycins B and C, the R^3 substituent of amino sugar A is aminomethyl group (CH_2NH_2), whereas in paromomycins I and II it is the hydroxymethyl group (CH_2OH).

Paromomycin (catenulin, hydroxymycin, moenomycin A) has been isolated from cultures of *Streptomyces rimosus* [58], *Streptomyces catenulae* [59], and *Streptomyces chrestomyceticus* [60, 61]. Structure of paromomycin was determined by Haskell et al. [11, 62–65].

Butirosins A and B

Butirosins A and B were produced by *Bacillus circulans*. The structures of both butirosins differ in the structure of pentose (in butirosin A the pentose is D-arabinose, and in butirosin B, it is D-ribose). The structures of both butirosins have been determined by Woo et al. [66–68] from mass spectra and are shown in Fig. 14.11.

Butirosin B was synthesized by Ikeda et al. [69] and by Akita et al. [70].

It is active against a number of gram-positive and gram-negative organisms, particularly *Pseudomonas aeruginosa*.

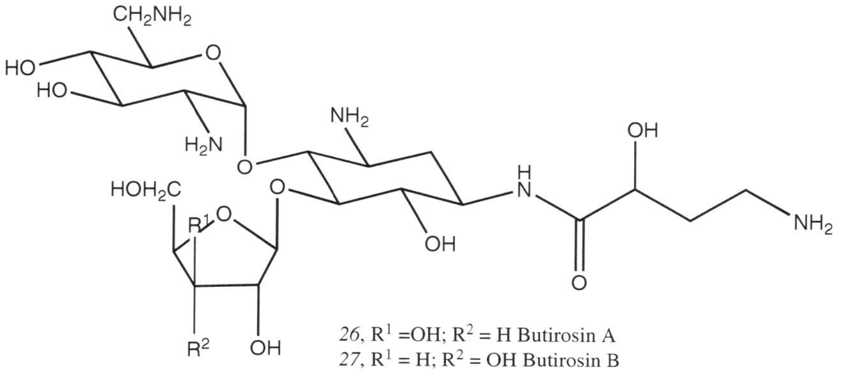

26, R^1 = OH; R^2 = H Butirosin A
27, R^1 = H; R^2 = OH Butirosin B

Fig. 14.11

Streptomycin A

It is one of the two aminoglycosidic antibiotics isolated from cultures of *Streptomyces griseus* [71]. The antibiotic is normally used as the trihydrochloride and is freely soluble in water. It is active against a range of gram-positive and gram-negative bacteria and mycobacteria, particularly against *Mycobacterium tuberculosis*. Structure was determined by Kuehl et al. [72]. It was found that two sugar components are not individually glycosidically linked to streptidine, as is the case in other aminoglycoside antibiotics, but that they are linked as disaccharide streptobiosamine to the C4 carbon atom of streptidine (Figs. 14.12 and 14.13)

The 2-deoxy-2-methylamino-L-glucose was synthesized by Wolfrom and Thompson [73], and streptose was synthesized by Dyer et al. [74]. The total synthesis of streptomycin was accomplished by Umezawa et al. [75].

Orthosomycins

Orthosomycins (also known as oligosaccharide antibiotics) are a family of carbohydrate-based antibiotics which are characterized by the presence of one or

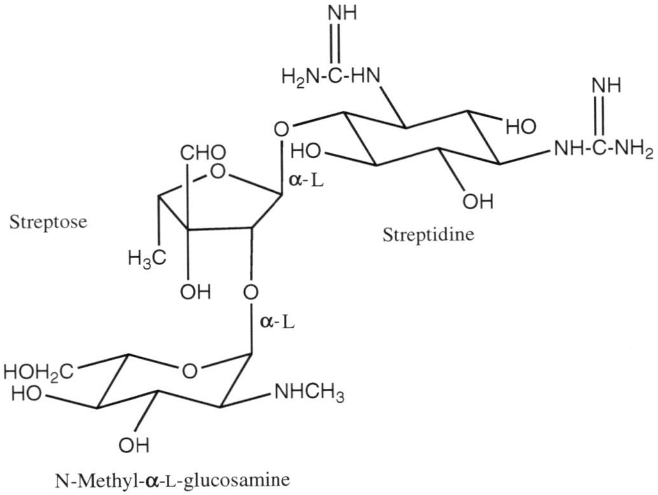

Fig. 14.12

28, Streptomycin

Fig. 14.13

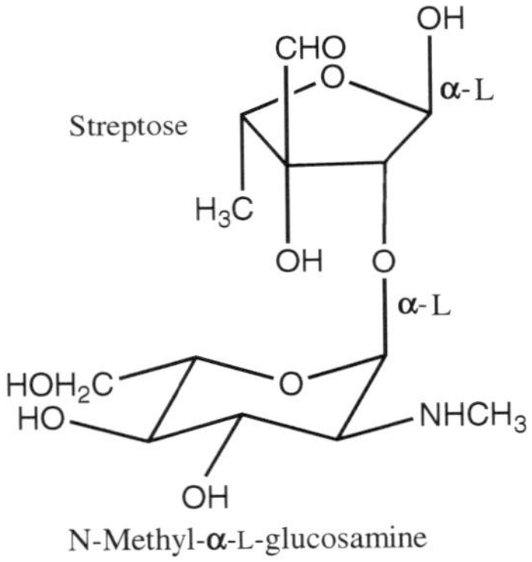

29, Streptobiosamine

more orthoester linkages in their oligosaccharide structure. These antibiotics include Destomycin-**A** [76–78] (*32*), Destomycin-**B** [78, 79] (*33*) and Destomycin-**C** [80, 81] (*34*), Flambamycin [76–80] (*36*), *Everninomycins*: **B** [81] (*37*), **C** [82] (*38*), **D** [83] (*39*), and *Hygromycin* **B** [87] (*35*).

This group of antibiotics can be conveniently subdivided into two distinct subgroups based on additional structural features, namely (a) those which contain an aminocyclitol residue (*30*) (Fig. 14.14) (for example destomycins and hygromycin B) (Fig. 14.15) and (b) those which are esters of dichlorisoeverninic

Aminoglycoside Antibiotics 479

R^1 = H, or Me; R^2 = H, or Me
30

31

Fig. 14.14

acid (*31*) (Fig. 14.14) [for example flambamycin (Fig. 14.16) and everninomicins (Fig. 14.17)].

Destomycin A

Destomycin family of antibiotics consists of destomycin A, B, and C (Fig. 14.15). Destomycin A is the major component isolated from the culture broth of *Streptomyces rimofaciens*. Destomycins are aminoglycoside antibiotics with two sugar units and one 2-deoxy-*N*,*N*-dimethyl-streptidine molecule. In destomycins A and C and in hygromycin B (Fig. 14.15), the amino sugar is a heptose, 6-amino-6-deoxy-L-glycero-D-galacto-heptopyranose. In destomycin B, the amino sugar is 6-amino-6-deoxy-L-glycero-D-glucopyranose. The other sugar component in destomycins A

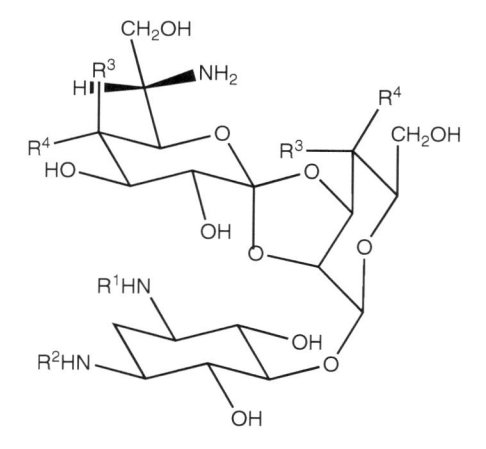

32, Destomycin A, $R^1 = R^4 = H$; $R^2 = CH_3$; $R^3 = OH$
33, Destomycin B, $R^1 = R^2 = CH_3$; $R^3 = H$; $R^4 = OH$
34, Destomycin C, $R^1 = R^2 = CH_3$; $R^3 = OH$; $R^4 = H$
35, Hygromycin B, $R^1 = CH_3$; $R^2 = R^4 = H$; $R^3 = OH$

Fig. 14.15

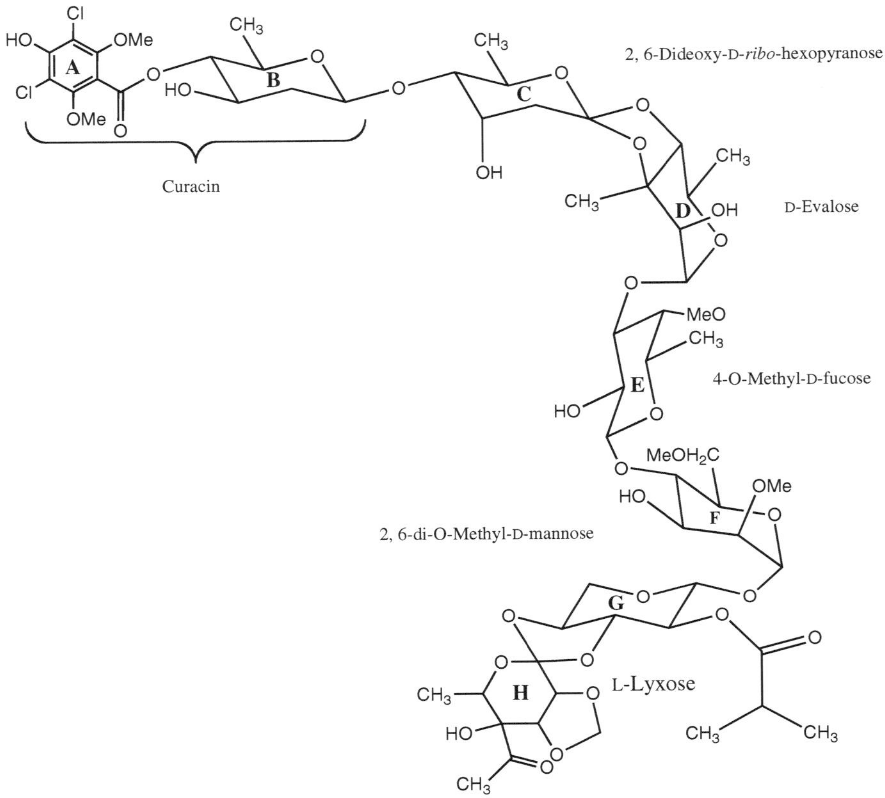

Fig. 14.16

and C is D-mannose, and in destomycin B and hygromycin B, it is D-talose. In all four antibiotics, the heptose and hexose are linked via an orthoester linkage between the C1 of heptose and the C2 and the C3 hydroxyl groups of hexose.

Destomycin family of antibiotics is produced by *S. rimofaciens* and was isolated by Kondo et al. [76]. Structures of destomycins A and B were elucidated by Kondo et al. [77–79] and of destomycin C by Shimura et al. [80]. Synthesis of destomycin C was reported by Tamura et al. [81]. It shows anthelmintic activity in poultry.

Hygromycin B is produced by *Streptomyces hygroscopicus*. It was isolated by Mann and Bromer [82], and the structure was determined by Neuss et al. [83]. Hygromycin A does not belong to the family of orthosomycins, although it is produced by the same microorganism as is the hygromycin B and bears the same name and therefore we will not discuss it.

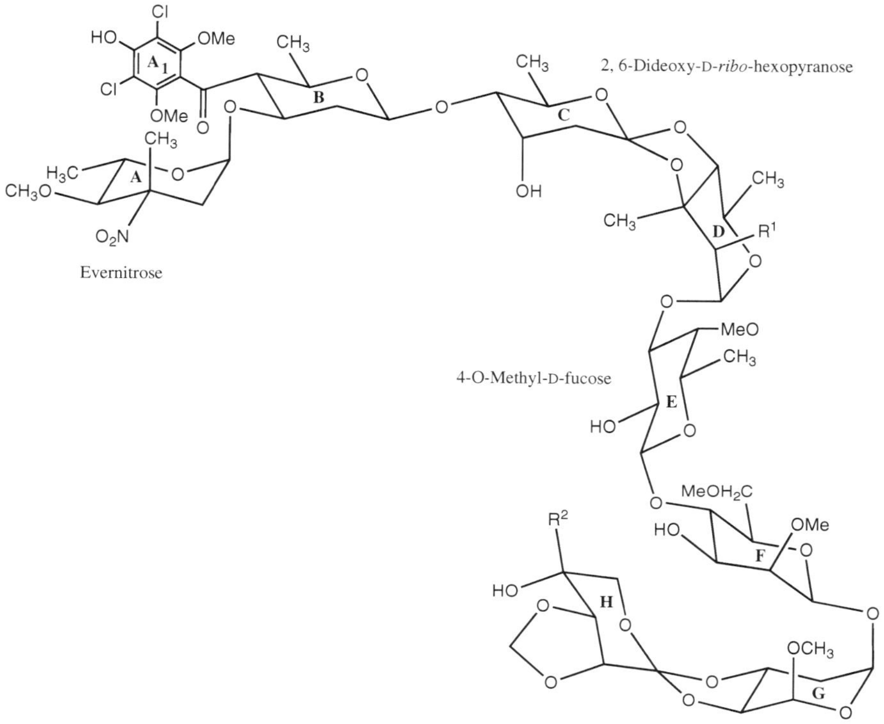

Aminoglycoside Antibiotics

Fig. 14.17

37, Everninomicin-B, R^1 = OH; R^2 = CH(OCH$_3$)CH$_3$
38, Everninomicin-C, R^1 = R^2 = H
39, Everninomicin-D, R^1 = H; R^2 = CH(OCH$_3$)CH$_3$

Flambamycin

The antibiotic flambamycin is produced also by *S. hygroscopicus*, and it was isolated in 1974 by Ninet et al. [84]. The structure of flambamycin was elucidated by Ollis et al. [85–92]. The structure was elucidated on the basis of degradation such as acid hydrolysis, methanolysis, and determination of the structure of degradation products. At first Ollis et al. proposed the incorrect structure for flambamycin [89], but 3 years later they proposed the correct structure [90]. In 1979 Ollis et al. reported the results of their studies on the ^{13}C NMR spectra of flambamycin and its derivatives [91] and on mass spectral studies of flambamycin and its degradation products [92].

Zagar and Scharf reported the synthesis of the terminal A–B–C disaccharide fragment of flambamycin, curamycin, and avilamycin [93].

Everninomicin

The structures of everninomicins B [94, 95], C [96], and D [97] were elucidated by classical degradation studies which included acidic hydrolysis, acidic methanolysis, and permethylation studies. Structural and stereochemical assignments proposed for flambamycin, everninomicins, and all degradation products were based on extensive studies by IR, UV, ^{1}H and ^{13}C NMR spectroscopy, as well as by low- and high-resolution mass spectrometry. In some cases, circular dichroism and X-ray crystallography were used for identification of fragments.

Everninomicin D (*39*, Fig. 14.17) is the major component present in the mixture of everninomicins produced by cultures of *Micromonospora carbonacea*. It is highly active in vitro against a variety of gram-positive bacteria, including penicillin-resistant strains, but inactive against gram-negative organisms, e.g., *E. coli, Klebsiella pneumoniae, Pseudomonas aeruginosa*, and *Salmonella schotmulleri*. Stereocontrolled synthesis of the everninomycin $A_1B(A)C$ ring framework was reported by Nicolaou et al. [98] and the total synthesis of everninomycin was reported by Helen J. Mitchell [99].

References

1. Benedict, R. G.; Langlykke, A. F., "*Antibiotics*", Ann. Rev. Microbiol. (1947) **1**, 193–236
2. Umezawa, S., "*Structures and Synthesis of Aminoglycoside Antibiotics*", Adv. Carbohydr. Chem. Biochem. (1974) **30**, 111–182
3. Umezawa, H.; Hooper, I. R. (Eds.), "*Aminoglycoside Antibiotics*", Springer Verlag, New York, Heidelberg, 1982
4. Umezawa, H.; Ueda, M.; Maeda, K.; Yagishita, K.; Kondo, S.; Okami, Y.; Utahara, R.; Osato, Y.; Nitta, K.; Takeuchi, T., "*Production and isolation of a new antibiotic, kanamycin*", J. Antibiot. (Japan) (1957) **10A**, 181–188
5. Cron, M. J.; Johnson, D. L.; Palermiti, F. M.; Perron, Y.; Taylor, H. D.; Whitehead, D. F.; Hooper, I. R., "*Kanamycin. I. Characterization and Acid Hydrolysis Studies*", J. Am. Chem. Soc. (1958) **80**, 752–753
6. Cron, M. J.; Fardig, O. B.; Johnson, D. L.; Schmitz, H.; Whitehead, D. F.; Hooper, I. R.; Lemieux, R. U., "*Kanamycin. II. The Hexosamine Units*", J. Am. Chem. Soc. (1958) **80**, 2342–2342
7. Cron, M. J.; Evans, D. L.; Palermiti, F. M.; Whitehead, D. F.; Hooper, I. R.; Chu, P.; Lemieux, R. U., "*Kanamycin. V. The Structure of Kanosamine*", J. Am. Chem. Soc. (1958) **80**, 4741–4742
8. Ogawa, H.; Ito, T.; Kondo, S., "Chemistry of Kanamycin. V. The Structure of Kanamycin", J. Antibiotics (Japan) (1958) **11A**, 169–170
9. Umezawa, S.; Tatsuta, K.; Tsuchiya, T., "*Studies of Aminosugars. XII. The Absolute Structure of Kanamycin as Determined by a Copper Complex Method*", Bull Chem. Soc. Japan (1966) **39**, 1244–1248
10. Koyama, G.; Litaka, Y.; Maeda, K.; Umezawa, H., "*The crystal structure of kanamycin*", Tetrahedron Lett. (1968) **9**, 1875–1879
11. Hitchens, M.; Rinehart, K. L., Jr.,, "*Chemistry of the Neomycins. XII. The Absolute Configuration of Deoxystreptamine in the Neomycins, Paromomycins and Kanamycins*", J. Am. Chem. Soc. (1963) **85**, 1547–1548

12. Ito, T.; Nishio, M.; Ogawa, H., "*The Structure of Kanamycin B*", J. Antibiot. (Japan) (1964) **17A**, 189
13. Murase, M., "*Structural studies on kanamycin C*". J. Antibiot. (Japan) (1961) **14A**, 367
14. Umezawa, S.; Koto, S.; Tatsuta, K.; Hineno, H.; Nishimura, Y.; Tsumura, T., "*The total Synthesis of Kanamycin B*", J. Antibiot. (Japan)(1968) **21**, 424–425
15. Umezawa, S.; Koto, S.; Tatsuta, K.; Hineno, H.; Nishimura, Y.; Tsumura, T., "*Studies of Aminosugars. XXIII. The Total Synthesis of Kanamycin B*", Bull. Chem. Soc. (Japan) (1969) **42**, 537–541
16. Umezawa, S.; Koto, S.; Tatsuta, K.; Tsumura, T., "*The total synthesis of kanamycin C*", Bull Chem. Soc. (Japan) (1968) **41**, 533
17. Umezawa, S.; Koto, S.; Tatsuta, K.; Tsumura, T., "*The total synthesis of kanamycin C*", J. Antibiot. (Japan) (1968) **21A**, 162–163
18. Umezawa, H.; Okanishi, M.; Kondo, S.; Hamana, K.; Utahara, R.; Maeda, K.; Mitsuhashi, S., "*Phosphorylative inactivation of aminoglycosidic antibiotics by Escherichia coli carrying R factor*", Science (1967) **157**, 1559–1561
19. Umezawa, H.; Okanishi, M.; Utahara, R.; Maeda, K.; Kondo, S., "*Isolation and structure of kanamycin inactivated by a cell free system of kanamycin resistant E. Coli*", J. Antibiot. (1967) **20**, 136–141
20. Benveniste, R.; Davies, J., "*Enzymatic acetylation of aminoglycoside antibiotics by Escherichia coli carrying R factor*", Biochemistry (1971) **10**, 1787–1796
21. Brzezinska, M.; Benveniste, R.; Davies, J.; Daniels, P.; Weinstein, J. L., "*Gentamicin resistance in strains of Pseudomonas aeruginosa mediated by enzymatic N-acetylation of the deoxystreptamine moiety*", Biochemistry (1972) **11**, 761–765
22. Kawabe, H.; Mitsuhashi, S., "*Acetylation of dideoxykanamycin B by Pseudomonas aeruginosa*", Jpn. J. Microbiol. (1972) **16**, 436–437
23. Yamamoto, H.; Yasigawa, M.; Naganawa, H.; Kondo, S.; Takeuchi, T.; Umezawa, H., "*Kanamycin 6′-acetate and ribostamycin 6′-acetate, enzymatically inactivated products by Pseudomonas aeruginosa*", J. Antibiot. (1972) **25**, 746–747
24. Chevereau, M.; Daniels, P. J. L.; Davies, J.; LeGoffic, F., "*Aminoglycoside resistance in bacteria mediated by gentamicin acetyltransferase II, an enzyme modifying the 2′- amino group of aminoglycoside antibiotics*", Biochemistry (1974) **13**, 598–603
25. Okanishi, M.; Kondo, S.; Utahara, R.; Umezawa, H., "*Phosphorylation and inactivation of aminoglycoside antibiotics by E. coli carrying R factors*", J. Antibiot. (1968) **21**, 13–21
26. Umezawa, H.; Doi, O.; Ogura, M.; Kondo, S.; Tanaka, N., "*Phosphorylation and in activation of kanamycin by Pseudomonas aeruginosa*", J. Antibiotics (1968) **21**, 154–155
27. Ozanne, B.; Benveniste, R.; Tipper, D.; Davies, J., "*Aminoglycoside antibiotics: inactivation by phosphorylation in Escherichia coli carrying R factors*", J. Bacteriol. (1969) **100**, 1144–1146
28. Davies, J.; Brzezinska, M.; Benveniste, R., "*R Factors: biochemical mechanisms of resistance to aminoglycoside antibiotics*", Ann. N. Y. Acad. Sci. (1971) **182**, 226–233
29. Yagisawa, M.; Yamamoto, H.; Naganawa, H.; Kondo, S.; Takeuchi, R.; Umezawa, H., "*A new enzyme in Escherichia coli carrying R factor phosphorylating 3′-hydroxyl of butirosin A, kanamycin, neamine, and ribostamycin*", J. Antibiot. (1972) **25**, 748–750
30. Brzezinska, M.; Davies, J., "*Two enzymes which phosphorylate neomycin and kanamycin in Escherichia coli strains carrying R factors*", Antimicrob. Agents Chemother. (1973) **3**, 266–269
31. Umezawa, H.; Takasawa, S.; Okanishi, M.; Utahara, R., "*Adenylstreptomycin, a product of streptomycin inactivated by E. coli carrying R factor*", J. Antibiot. (1968) **21**, 81–82
32. Benveniste, R; Daviess, J., "*R-factor mediated gentamicin resistance: a new enzyme which modifies aminoglycoside antibiotics*", F.E.B.S, Lett. (1971) **14**, 293–296
33. Yagisawa, M.; Naganawa, H.; Kondo, S.; Hamada, M.; Takeuchi, T.; Umezawa, H., "*Adenyllyldeoxykanamycin B, a product of the inactivation of dideoxy kanamycin B by Escherichia coli carrying R factor*", J. Antibiot. (1971) **24**, 911–912

34. Umezawa, S.; Tsuchiya, T.; Muto, R.; Nishimura, Y.; Umezawa, H., *"Synthesis of 3′-deoxykanamycin effective against kanamycin-resistant Escherichia coli and Pseudomonas aeruginosa"*, J. Antibiot. (1971) **24**, 274–275
35. Umezawa, H.; Umezawa, S.; Tsuchiya, T.; Okazaki, Y., *"3′, 4′-Dideoxykanamycin B active against kanamycin-resistant Escherichia coli and Pseudomonas aeruginosa"*, J. Antibiot. (1971) **24**, 485–487
36. Umezawa, S.; Umezawa, H.; Okazaki, Y.; Tsuchiya, T., *"Studies on aminosugars. XXII. Synthesis of 3′, 4′-dideoxykanamycin B"*. Bull. Soc. Chem. Jpn. (1972) **45**, 3624–3628
37. Kawaguchi, H.; Naito, T.; Nakagawa, S.; Fujisawa, K., *"BB-K8, a new semisynthetic aminoglycoside antibiotic"*, J. Antibiot. ((1972) **25**, 695–708
38. Weinstein, M. J.; Luedemann, G. M.; Oden, E. M.; Wagman, G. H., *"Gentamycin, a New Broad-Spectrum Antibiotic Complex"*, Antimicrob. Agents Chemother. (1963) **161**, 1–7
39. Maehr, H.; Schaffner, C. P., *"Chemistry of the gentamycins. I. Characterization and gross structure of gentamycin A"*, J. Am. Chem. Soc. (1967) **89**, 6787–6788
40. Maehr, H.; Schaffner, C. P., *"Chemistry of the gentamicins. II. Stereochemistry and synthesis of gentosamine. Total structure of gentamicin A"*, J. Am. Chem. Soc. (1970) **92**, 1697–1700
41. Nagabhushan, T. L.; Turner, W. N.; Daniels, P. J. L.; Morton, J. B., *"Gentamicin antibiotics. 7. Structures of the gentamicin antibiotics A1, A3, and A4"*, J. Org. Chem. (1975) **40**, 2830–2834
42. Nagabhushan, T. L.; Daniels, P. J. L.; Jaret, R. S.; Morton, J. B., *"Gentamicin antibiotics. 8. Structure of gentamicin A2"*, J. Org. Chem. (1975) **40**, 2835–2936
43. Waitz, J. A.; Moss, E. L., Jr; Oden, E. M.; Wagman, G. H.; Weinstein, M. J., *"Biological activity of Sch 14342, an aminoglycoside antibiotic coproduced in the gentamicin fermentation"*, Antimicrob. Agents. Chemother. (1972) **2**, 464–469
44. Cooper, D. J.; Daniels, P. J. L.; Yudis, M. D.; Marigliano, H. M.; Guthrie, R. D.; Bukhari, S. T. K., *"The gentamicin antibiotics. Part III. The gross structures of the gentamicin C components"*, J. Chem. Soc. (C) (1971) 3126–3129
45. Chmielewski, M.; Konowal, A.; Zamojski, A., *"The synthesis of racemic purpurosamine B"*, Carbohydr. Res. (1979) **70**, 275–282
46. Koch, K. F.; Rhoades, J. A. *"Structure of Nebramycin Factor 6, a New Aminoglycosidic antibiotic"*, Antimicrob. Agents. Chemother. (1970) 309–313
47. Takagi, Y.; Miyake, T.; Tsuchiya, T.; Umezawa, S., *"Synthesis of 3′-Deoxykanamycin B (Tobramycin)"*, Bull. Chem. Soc. Japan (1976) **49**, 3649–3651
48. Tanabe, M.; Yasuda, D. M.; Detre, G., *"Aminoglycoside antibiotics: synthesis of nebramine, tobramycin and 4′-epi-tobramycin"*, Tetrahedron Lett. (1977) **18**, 3607–3610
49. Waksman, S. A.; Lechevalier, H. A., *"Neomycin, a New Antibiotic Active against Streptomycin-Resistant Bacteria, including Tuberculosis Organisms"*, Science (1949) **109**, 305–307
50. Rinehart, K. L.; Woo, P. W. K.; Argoudelis, A. D.; Giesbrecht, A. M., *"Chemistry of the Neomycins. I. Partial Structure for Neobiosamines B and C"*, J. Am. Chem. Soc. (1957) **79**, 4567–4568
51. Rinehart, K. L.; Chilton, W. S.; Hichens, M., *"Chemistry of the Neomycins. XI.1 N.M.R. Assignment of the Glycosidic Linkages"*, J. Am. Chem. Soc. (1962) **84**, 3216–3218
52. Reinehart, Jr., K. L.; Hitchens, M.; Argoudelis, A. D.; Chilton, W. S.; Carter, H. E.; Georgiadis, M. P.; Schaffner, C. P.; Schillings, R. T., *"Chemistry of the Neomycins. X. Neomycins B and C"*, J. Am. Chem. Soc. (1962) **84**, 3218–3220
53. Rinehart, Jr., K. L.; Chilton, W. S.; Hitchens, M.; von Philipsborn, M., *"Chemistry of the Neomycins. XI.1 N.M.R. Assignment of the Glycosidic Linkages"*, J. Am. Chem. Soc. (1962) **84**, 3216–3218
54. Umezawa, S.; Nishimura, Y., *"Total synthesis of neomycin C"*, J. Antibiot. (1977) **30**, 189–191
55. Umezawa, S.; Harayama, A.; Nishimura, Y., *"The total synthesis of neomycin C"*, Bull. Soc. Chim. Japan (1980) **53**, 3259–3262
56. Usui, T.; Umezawa, S., *"Total synthesis of neomycin B"*, J. Antibiot. (1987) **40**, 1464–1467
57. Usui, T.; Umezawa, S., *"Total synthesis of neomycin B"*, Carbohydr. Res. (1988) **174**, 133–143

58. Frohardt, R. P.; Haskell, T. H.; Ehrlich, J.; Knudsen, M. P., "*Paromomycin*", U. S. Patent 2,916,485 (1959) to Parke Davis
59. Davisson, J. W.; Finlay, A. C., "*Catenulin*", U.S. Patent 2,895,876 to Pfizer
60. Canevazzi, G.; Scotti, T., "*Description of a new species of streptomycetes (Streptomyces chrestomyceticus) producing a new antibiotic, amminosidin*", Giorn. Microbiol. (1959) **7**, 242–250
61. Arcamone, F.; Bertazzoli, C.; Ghione, M.; Scotti, T., "*Amminosidin, a New Oligosaccharide Antibiotic*", Giorn. Microbiol. (1959) **7**, 251–253
62. Haskell, T. H.; French, J. C.; Bartz, Q. R., "*Paromomycin. I. Paromamine, a Glycoside of D-glucosamine*", J. Am. Chem. Soc. (1959) **81**, 3480–3481
63. Haskell, T. H.; French, J. C.; Bartz, Q. R., "*Paromomycin. II. Paromobiosamine, a Diaminohexosyl-D-ribose*", J. Am. Chem. Soc. (1959) **81**, 3481
64. Haskell, T. H.; French, J. C.; Bartz, Q. R., "Paromomycin. III. The Structure of Paromobiosamine", J. Am. Chem. Soc. (1959) **81**, 3481–3482
65. Haskell, T. H.; French, J. C.; Bartz, Q. R., "*Paromomycin. IV. Structural Studies*", J. Am. Chem. Soc. (1959) **81**, 3482–3483
66. Woo, P. W. K.; Dion, H. W.; Bartz, Q. R., "*Butirosins A and B, aminoglycoside antibiotics. I. Structural units*", Tetrahedron Lett. (1971) **12**, 2617–2620
67. 67. Woo, P. W. K., "*Butirosins A and B, aminoglycoside antibiotics. II. Mass spectrometric study*", Tetrahedron Lett. (1971) **12**, 2621–2624
68. Woo, P. W. K.; Dion, H. W.; Bartz, Q. R., "*Butirosins A and B, aminoglycoside antibiotics. III. Structures*", Tetrahedron Lett. (1971) **12**, 2625–2628
69. Ikeda, D.; Tsuchiya, T.; Umezawa, S; Umezawa H., "*Synthesis of butirosin B*", J. Antibiot. (1972) **25**, 741–742
70. 70. Akita, E.; Horiuchi, Y.; Yasuda, S. "*Synthesis of butirosin B and related compounds by an acyl migration method*", J. Antibiot. (1973) **26**, 365–367
71. Schatz, A.; Bugie, E.; Waksman, S. A., "*Streptomycin, a substance exhibiting antibiotic activity against gram-positive and gram-negative bacteria*", Proc. Exptl. Biol. Med. (1944) **55**, 66–69
72. Kuehl, F. A., Jr.; Peck, R. L.; Hoffhine, C. E., Jr.; Peel, E. W.; Folkers, K., "*Streptomyces Antibiotics. XIV. The Position of the Linkage of Streptobiosamine to Streptidine in Streptomycin*", J. Am. Chem. Soc. (1947) **69**, 1234–1234
73. Wolfrom, M. L.; Thompson, A., "*Derivatives of N-Methyl-L-glucosaminic Acid; N-Methyl-l-mannosaminic Acid*", J. Am. Chem. Soc. (1947) **69**, 1847–1849
74. Dyer, J. R.; McGonigal, W. E.; Rice, K. C., "*Streptomycin. II. Streptose*", J. Am. Chem. Soc.(1965) **87**, 654–655
75. Umezawa, S.; Takahashi, Y.; Usu, T.; Tsuchiya, T., "*Total synthesis of streptomycin*", J. Antibiot. (1974) **27**, 979–999
76. Kondo, S.; Sezaki, M.; Koike, M.; Shimura, M.; Akita, E.; Satoh, K.; Hara, T., "*Destomycins A and B, Two New Antibiotics Produced by a Streptomyces*", J. Antibiot. (1965) **18A**, 38–42
77. Kondo, S. I.; Akita, E.; Koike, M., "*The structure of destomycin A*" J. Antibiot. (1966) **19** (Ser. A), 139–140
78. Kondo, S.; Iinuma, K.; Naganawa, H.; Shimura, M.; Sekizawa, Y., "*Structural studies on destomycins A and B*", J. Antibiot. (1975) **28**, 79–82
79. Shimura, M.; Sekizawa, Y.; Iinuma, K.; Naganawa, H.; Kondo, S., "*Structure of destomycin B*", Agr. Biol. Chem. (1976) **40**, 611–618
80. Shimura, M.; Sekizawa, Y.; Iinuma, K.; Naganawa, H.; Kondo, S., "*Destomycin C, a new member of destoycin family antibiotics*", J. Antibiot. (1975) **28**, 83–84
81. Tamura, J.-I.; Horito, S.; Hashimoto, H.; Yoshimura, J., "*The synthesis of destomycin C, a typical pseudo-trisaccharide of destomycin-group antibiotics*", Carbohydr. Res. (1988) **174**, 181–199
82. Mann, R. L.; Bromer, W. W., "*The Isolation of a Second Antibiotic from Streptomyces hygroscopicus*", J. Am, Chem. Soc. (1958) **80**, 2714–2716

83. Neuss, N.; Koch, K. F.; Molloy, B. B.; Day, W.; Huckstep, L. L.; Dorman, D. E.; Roberts, J. D., *"Structure of Hygromycin B, an Antibiotic from Streptomyces hygroscopicus; The Use of CMR Spectra in Structure Determination, I."*, Helv. Chim. Acta (1970) **53**, 2314–2319
84. Ninet, L.; Benazet, F.; Charpentie, Y.; Dubost, M.; Florent, J.; Lunel, J.; Mancy, D.; Preud'Homme, J., *"Flambamycin, a new antibiotic from Streptomyces hygroscopicus DS 23 230"*, Experientia (1974) **30**, 1270–1272
85. Ollis, W. D.; Smith, C., *"Acidic Hydrolysis of Flambamycin"*, J. Chem. Soc. Chem. Commun. (1974) 881–882
86. Ollis, W. D.; Smith, C., *"Methanolysis of Flambamycin. Formation and Constitutions of Flambalactone, Methyl Flambate, Flambatriose Isobutyrate and Flambatetrose Isobutyrate"*, J./Chem. Soc. Chem. Commun. (1974) 882–884
87. Ollis, W. D.; Smith, C., *"Methasnolysis of Flambamycin. The Constitution of Eurekanate"*. J. Chem. Soc. Chem. Commun. (1976) 347–348
88. Ollis, W. D.; Smith, C., *"Hydrolysis of Flambamycin. The Constitutiion of Flambeurekanose"*, J. Chem. Soc. Chem. Commun. (1976) 348–350
89. Ollis, W. D.; Smith, C.; Sutherland, I. O.; Wright, D. E., *"The constitution of the anti biotic flambamycin"*, J. Chem. Soc. Chem. Commun. (1976) 350–351 (incorrect structure)
90. Ollis, W. D.; Smith, C.; Wright, D. E., *"The orthosomycin family of antibiotics—I: The constitution of flambamycin"*, Tetrahedron (1979) **35**, 105–127 (revised structure)
91. Ollis, W. D.; Sutherland, I. O.; Brain, F.; Taylor, B.; Smith, C.; Wright, D. E., *"The orthosomycin family of antibiotics-II: The ^{13}C NMR spectra of flambamycin and its derivatives"*, Tetrahedron (1979) **35**, 993–1001
92. Ollis, W. D.; Jones, S.; Smith, C.; Wright, D. E., *"The orthosomycin family of antibiotics—III: Mass spectral studies of flambamycin and its degradation products"*, Tetrahedron (1979) **35**, 1003–1014
93. Zagar, C.; Scharf, H. D., *"Synthesis of a terminal A-B-C disaccharide fragment of flambamycin, curamycin, and avilamycin"*, Carbohydr. Res. (1993) **248**, 107–118
94. Ganguly, A. K.; Saksena, A. K., *"Hydrolysis products of everninomicin B"*, J. Chem. Soc. Chem. Commun. (1973) 531–532
95. Ganguly, A. K.; Saksena, A. K., *"Structure of everninomicin B"*, J. Antibiot. (1975) **28**, 707–709
96. Ganguly, A. K.; Szmulewicz, S., *"Structure of everninomicin C"*, J. Antibiot. (1975) **28**, 710–712
97. Ganguly, A. K.; Sarre, O.; Greeves, D.; Morton, J., *"Structure of everninomicin D"*, J. Am. Chem. Soc. (1075) **97**, 1982–1985
98. Nicolaou, K. C.; Rodriguez, R. M.; Mitchell, H. J.; Van Delft, F. L., *"Stereocontrolled synthesis of the everninomicin A1B(A)C ring framework"*, Angew. Chem. Internat. Ed. (1998) **37**, 1874–1876
99. Mitchell, H. J., *"The total synthesis of everninomycin 13,384–1"*, Dissertation, (2000) University of California, San Diego, USA. From: Diss. Abstr. Int., B 2000, 61(3), 1417

Chapter 15
Higher-Carbon Monosaccharides

Introduction

Higher-carbon sugars are defined as sugars having more than six carbon atoms in their carbon chain. Thus monosaccharides containing seven or more consecutive carbon atoms belong to this class of monosaccharides, e.g., heptoses, octoses, nonoses, decoses. There are many reviews written on this topic [1–10].

Higher-carbon sugars have been found in Nature, and they often have very important biological functions, as is the case with sialic acids (e.g., *N*-acetyl-neuraminic acid, NANA, *1*) [11–13] (Fig. 15.1).

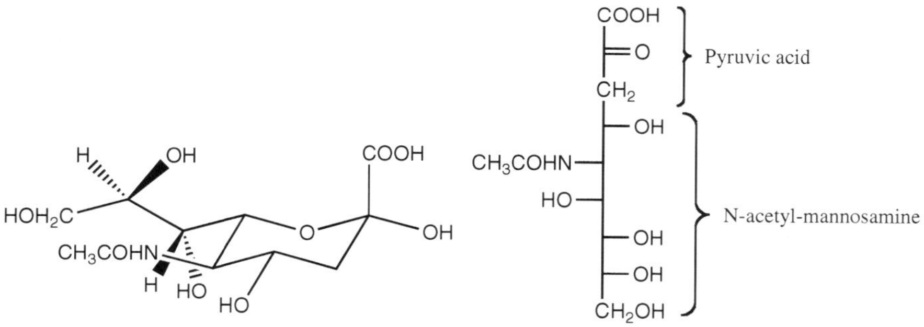

1 *2*

Fig. 15.1

The *N*-acetyl-neuraminic acid (*N*-acetyl-5-amino-3,5-dideoxy-D-glycero-D-galacto-*non*-2-*ulosonic* acid) in Fig. 15.1, shown in the α-D-pyranose form, plays a variety of very important roles in living organisms. One of the most obvious roles is providing the negative charge to glycoproteins on cell membranes, thus influencing the behavior of cells (for example, it has been calculated that $>10^7$ NANA residues are bound to the surface of a single human eryhtrocyte). The importance of this electronegative shield is severalfold. For example, in some cell types, membrane sialic acids prevent aggregation due to electrostatic repulsion in blood platelets, erythrocytes, and carcinoma cells, whereas in others, for example in chick, embryonic

M. Miljković, *Carbohydrates*, DOI 10.1007/978-0-387-92265-2_15,
© Springer Science+Business Media, LLC 2010

muscle cells, aggregation is facilitated, most probably by Ca^{2+} bridges. The repulsive, electrostatic forces of sialic acids contribute also to the rigidity of the cell surface. Sialic acid residues are important Ca^{2+}-binding sites in the muscle cells. Antirecognition effect of sialic acids, i.e., the protection of survival of various serum glycoproteins in blood stream that have their terminal galactosyl residues sialylated, is one of the most fascinating functions of sialic acids. Finally, but not lastly, sialic acids linked to various gangliosides serve often as receptors for various toxins, such as diphtheria, tetanus, cholera, botulism.

Lincosamine [14] (6-amino-6,8-dideoxy-D-galacto-*octose*) (*3*) (Fig. 15.2) is a component of a therapeutically important antibiotic lyncomycin. Another octose, KDO (*4*) (3-deoxy-D-*manno*-2-octulosonic acid) (Fig. 15.2) is a sugar component of lipopolysaccharides and capsular polysaccharides, which occur in the cell surface of gram-negative bacteria, and is an important bridging link in their membrane structure. Octosyl acids [15, 16] are a class of nucleoside antibiotics consisting of a C$_8$ monosaccharide attached to a pyrimidine base (Fig. 15.2). They are unusual eight-carbon bicyclic monosaccharides which are N-glycosydically

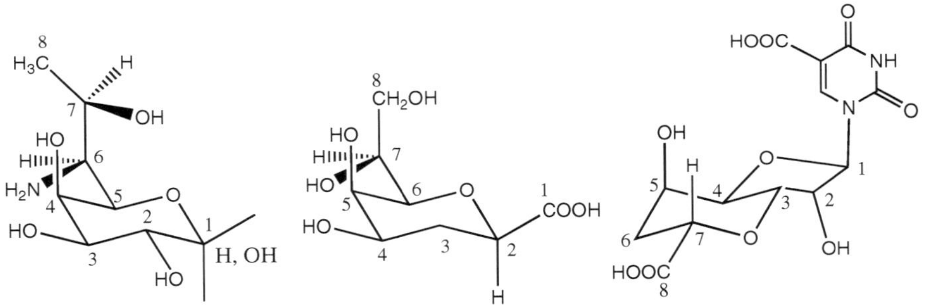

Fig. 15.2

linked to novel pyrimidine bases [15]. Some octosyl acids are powerful phosphodiesterase inhibitors [17, 18]. Hikosamine (4-amino-4-deoxy-D-glycero-D-galacto-D-gluco-*undecose*) *6* (Fig. 15.3) is a higher-sugar component of antibiotic hikizimycin which is active against *Helminthosporium* and numerous other species of plant-pathogenic fungi.

A number of heptoses, heptitols, and heptulose have been found in Nature, for example, bacterial cell wall polysaccharides contain D- and L-glycero-D-*manno*-heptopyranose *7* and *8*, respectively (Fig. 15.4) and 6-deoxy-D-manno- and D-altro-heptopyranose *9* and *10*, respectively (Fig. 15.5).

Unusual C$_9$ branched chain higher-carbon sugar-like structures have been found to be a part of complex lipids of thermoacidophilic bacteria, for example calditol *11* [19–21] (Fig. 15.6).

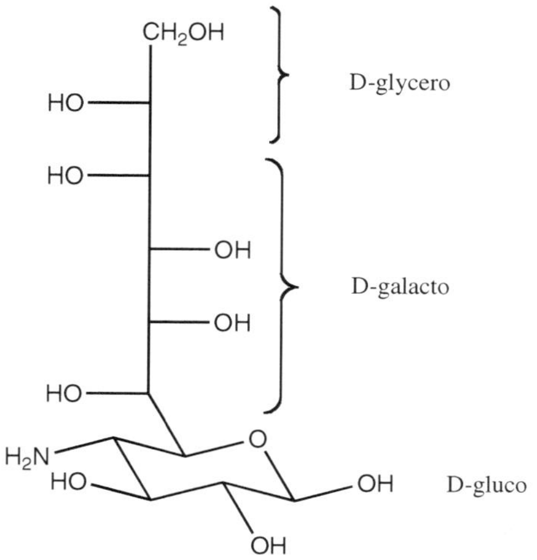

Introduction

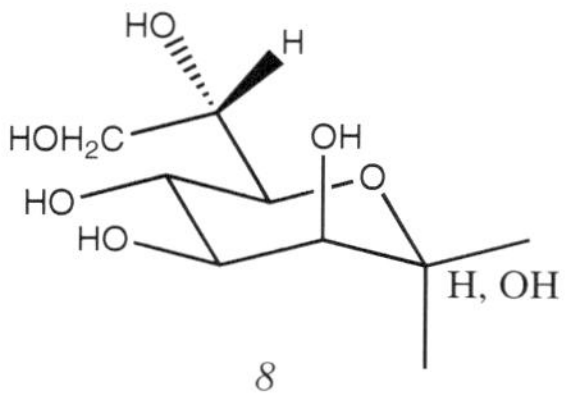

Fig. 15.3

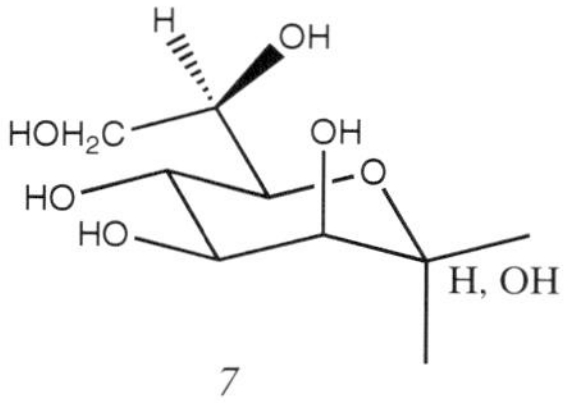

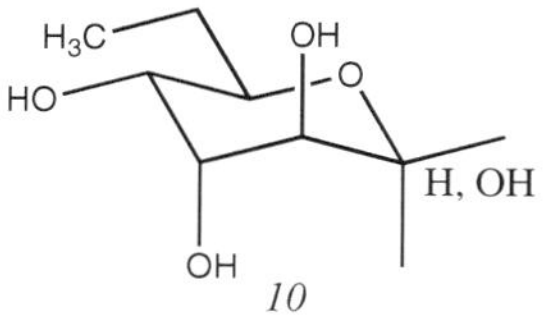

Fig. 15.4

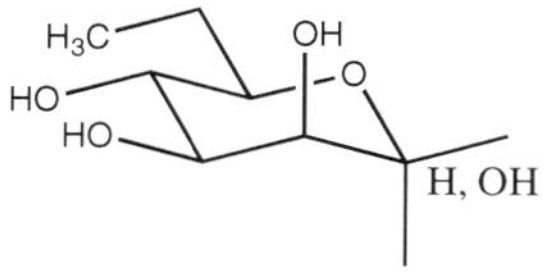

Fig. 15.5

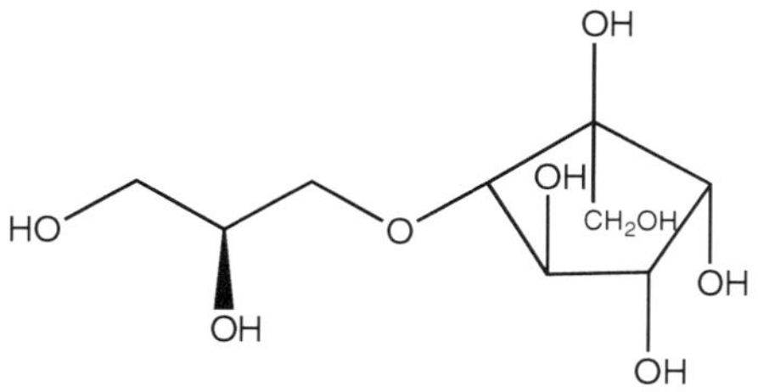

Fig. 15.6

Synthesis of Higher-Carbon Sugars

Since the late 1970s, synthesis of higher-carbon sugars attracted a great interest of a number of organic synthetic chemists. A reason for this rise of synthetic activities in this field was assessing dependence of biological function and activity of the higher-carbon sugars upon their structural modifications. This is the same reason why there has been the synthesis of all 'natural products with biological activity generating such a wide interest in synthetic chemist community for so many years.

There are many approaches for the synthesis of higher-sugars, of which we are going to discuss only a few:

1. Extension of carbohydrate skeleton via Wittig olefination reaction and stereoselective hydroxylation of the generated olefinic bond
2. Aldol condensation and related approaches
3. De novo synthesis.

Wittig Olefination

In the Wittig reaction, an aldehyde or a ketone is treated with a phosphorous ylide (also called phosphorane) to give an olefin [22–24] (Fig. 15.7).

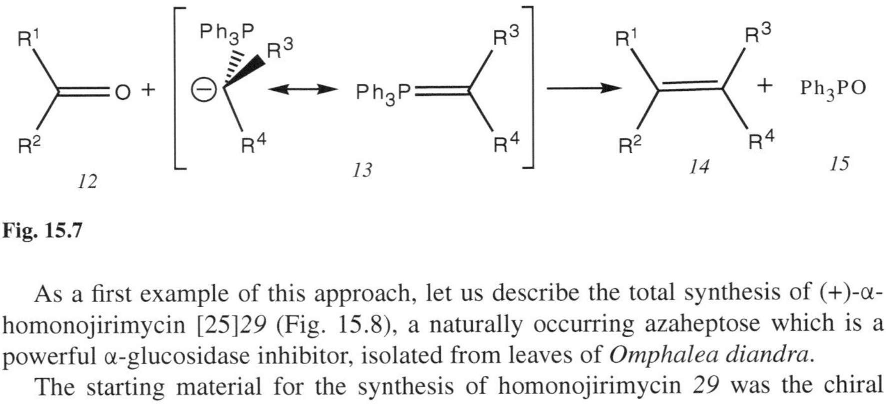

Fig. 15.7

As a first example of this approach, let us describe the total synthesis of (+)-α-homonojirimycin [25]*29* (Fig. 15.8), a naturally occurring azaheptose which is a powerful α-glucosidase inhibitor, isolated from leaves of *Omphalea diandra*.

The starting material for the synthesis of homonojirimycin *29* was the chiral allyl alcohol *16*, which was obtained from ethyl D-tartrate as described by Iida et al. [26] and Aoyagi et al. [27]. The allylic alcohol *16* was converted to *syn*-epoxide by the Sharpless asymmetric epoxidation [28]. Regio- and stereoselective epoxide ring opening was effected by using diethylaluminum amide [29] at 0°C, giving the aminoalcohol *18* as a single stereoisomer with 70% yield (Fig. 15.8). The amino group of *18* was selectively protected with benzyl chloroformate (aq. Na_2CO_3, CH_2Cl_2) with 98% yield. The two hydroxyl groups of the obtained carbamate *19* were methoxymethylated and the *tert*-butyldimethylsilyl group of the obtained *20* was removed with an overall yield of 86%. Swern oxidation of the primary hydroxyl group of *21* gave aldehyde *22* with 98% yield, which was transformed into the

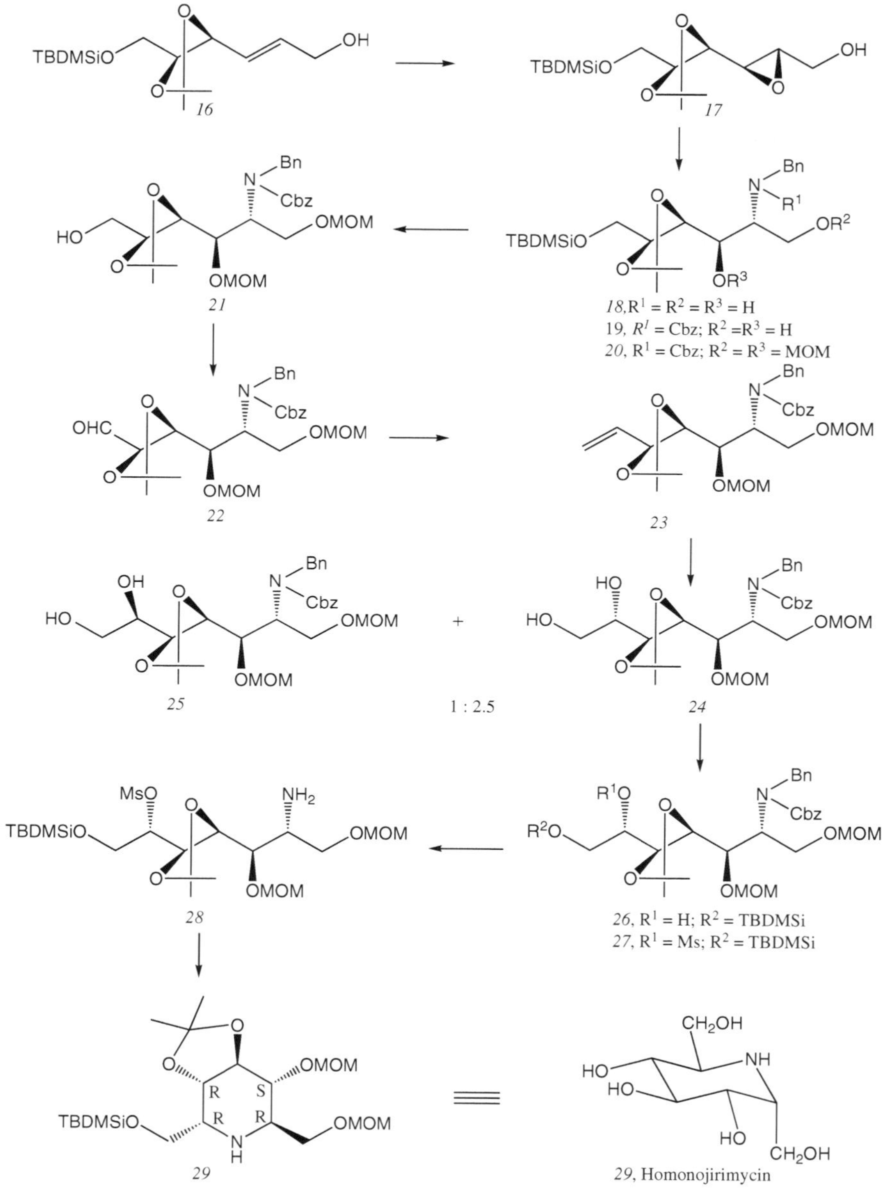

Fig. 15.8

alkene *23* by the Wittig reaction with 84% yield. Hydroxylation of *23* with catalytic amount of osmium tetroxide in the presence of 2 equivalent of *N*-methylmorpholine oxide in aqueous acetone gave the mixture of diols *24* and *25* in which the desired *anti*-stereoisomer *24* predominated in the ratio 2.5:1(total yield 90%). The diol *24* was converted via regioselective silylation with *tert*-butyldimethylsilyl chloride to the TBDMSi ether *26* which was then mesylated into the corresponding mesylate *27*. The simultaneous removal of the carbobenzoxy and benzyl groups was effected by catalytic hydrogenation over palladium hydroxide in methanol. Finally the intramolecular displacement of the mesyl group with the amino group was effected by the heating of the methanolic solution of *28* to which some triethyl amine was added, whereby the protected (+)-homonojirimycin *29* was obtained at 81%.

α-Amino acids bearing a sugar moiety attached by a carbon–carbon bond represent unique substructures in many biologically active molecules [4]. Pyranosidic arrangements are rare and are found, for example, in the antibiotics amipurimycin [30] and miharamycin [31].

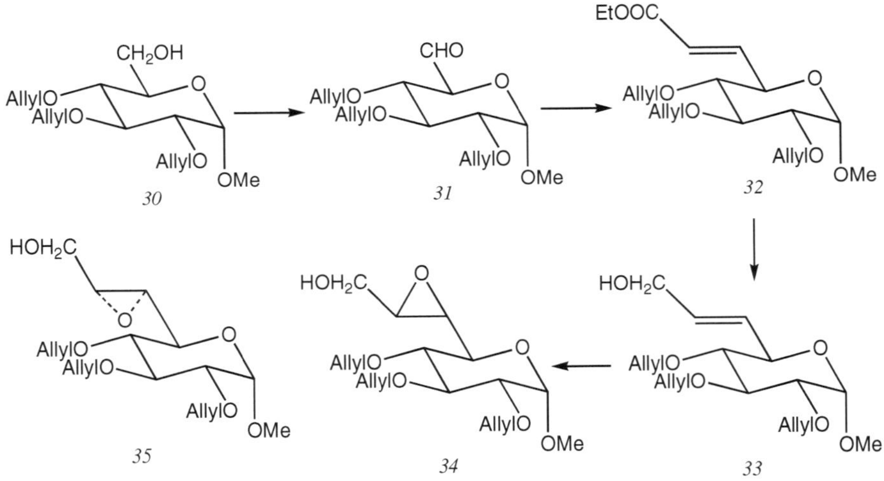

Fig. 15.9

Bessodes et al. [32] have reported chiral synthesis of biologically important terminal α-amino-acyl glycosides. Methyl 2,3,4-tri-*O*-allyl-α-D-glucopyranoside *30* which was used as a starting material for this synthesis was obtained in three steps from methyl α-D-glucopyranoside by selective tritylation of the C6 hydroxyl group, allylation of the remaining free hydroxyl groups, and detritylation by formic acid. Oxidation of *30* with DMSO-oxalyl chloride to 6-aldehydo derivative *31* followed by reaction with ethyl triphenylphosphoranylidene acetate gave the *E*-olefin *32*, with 88% yield. Reduction of the ester group of *32* with DIBAL gave *E*-2,3,4,tri-*O*-allyl-6,7-dideoxy-α-D-*gluco*-oct-6-enopyranoside *33* with 95% yield.

The titanium-catalyzed epoxidation of *33* in the presence of diisopropyl D-tartrate (DIPT) [28] gave epoxide *34*, with 93% yield, as the only product. If the same reaction was performed in the presence of diisopropyl L-tartrate, an inseparable mixture of epoxides *34* and *35* was obtained with 93% yield, in which the ratio

Synthesis of Higher-Carbon Sugars 493

of the desired epoxide *34* to the undesired epoxide *35* was 1:4. The regioselective 6,7-epoxide ring opening was effected by treating the mixture of epoxides *34* and *35* with titanium diisopropoxide diazide [33], and the obtained regio- and diastereoisomers *36*, *37*, *38*, and *39* (Fig. 15.10) were separated by column chromatography. The regioselectivity in the epoxide ring opening of *34* and *35* was estimated to be ca. 80%. The reduction of azide *38* with LiAlH$_4$ followed by acetylation of *36* with

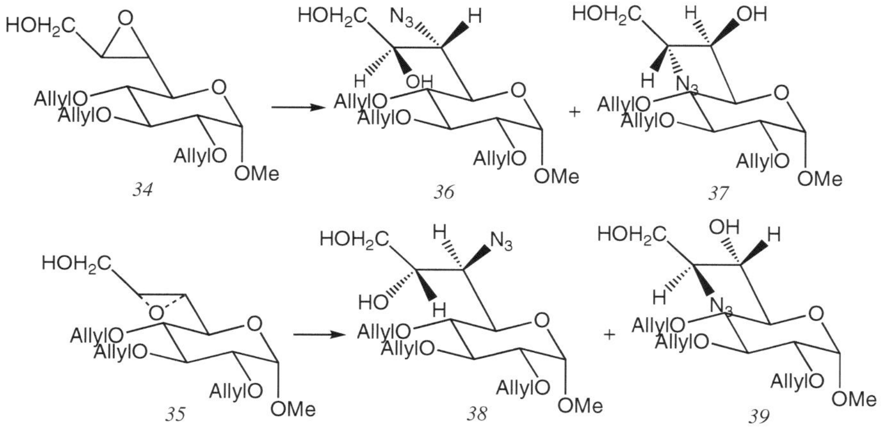

Fig. 15.10

acetic anhydride in methanol gave the 6-acetamido derivative *40* which was then oxidized with periodic acid and potassium dichromate to aminoacyl sugar *42* (Figs. 15.9 and 15.11).

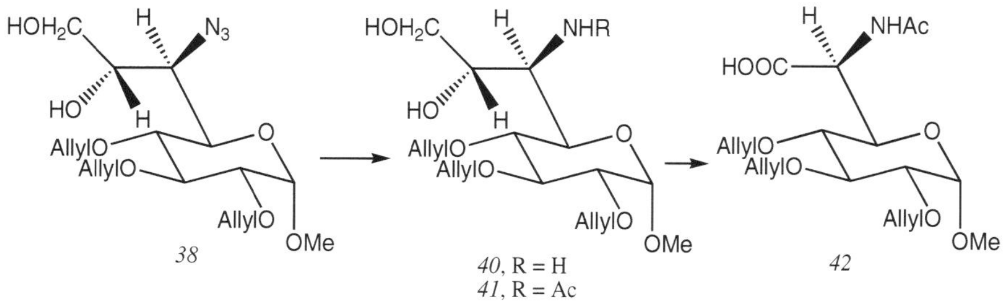

Fig. 15.11

In a series of papers, Brimacombe et al. [34–41] have approached the synthesis of higher-carbon sugars using Wittig olefination (Fig. 15.12) of 1,2: 3,4-di-*O*-isopropylidene-6-*aldehydo*-α-D-galactopyranose *43* followed by catalytic osmylation [42] of obtained unsaturated sugars *44* or *46* either according to Kishi empirical rule [43] or via Sharpless epoxidation with diisopropyl L-(+)-tartrate [28, 44, 45].

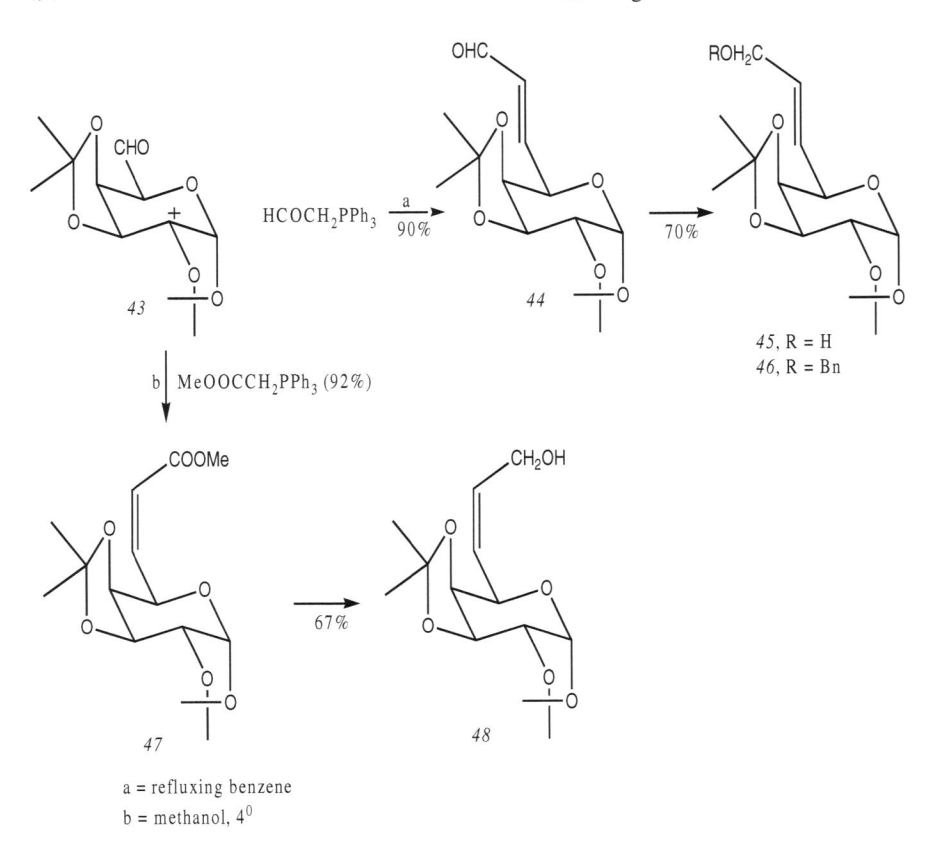

Fig. 15.12

a = refluxing benzene
b = methanol, 4^0

Catalytic osmylation [42] of E-octenopyranose *45* gave a mixture of 1,2:3,4-di-O-isopropylidene-β-L-*threo*-D-*galacto*-octopyranose *49* and α-D-*threo*-D-*galacto* isomer *50* in the ratio 7:1. Catalytic osmylation of the benzylated derivative *46* was less stereoselective, giving a mixture of *49* and *50* (after removal of benzyl group) in the ratio 3:1 (Fig. 15.13). It should be noted that, for the reason of clarity, only the side chain comprising of carbon atoms C6, C7, and C8 was shown in Fig. 15.13.

Catalytic osmylation of Z-octenopyranose *47* produced a mixture containing 1,2:3,4-di-O-isopropylidene-β-L-*erythro*-D-*galacto*-octopyranose *51* and 1,2:3,4-di-O-isopropylidene-α-D-*erythro*-D-*galacto*-octopyranose *49* in the ratio 1:7 (Fig. 15.14).

Titanium-catalyzed asymmetric epoxidation [28, 44, 45] of the E-octenopyranose *46* with di isopropyl L-(+)-tartrate ((+)-DIPT) at −23°C readily gave a single 6,7-oxirane *57* (6,7-anhydro-1,2:3,4-di-O-isopropylidene-β-L-threo-D-galacto-octopyranose) with 66% yield. By contrast, titanium-catalyzed epoxidation of the

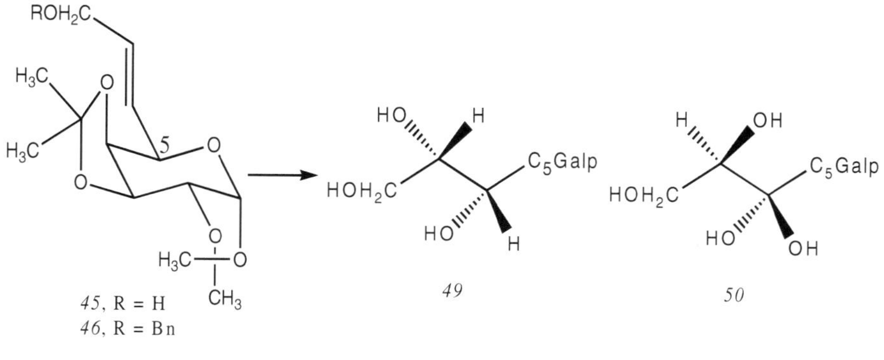

Synthesis of Higher-Carbon Sugars

Fig. 15.13

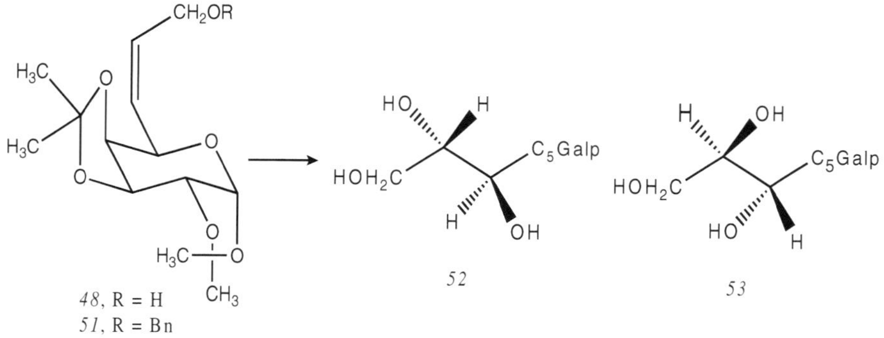

Fig. 15.14

E-octenopyranose *45* with diisopropyl D-(−)-tartrate ((−)-DIPT) at −23°C was incomplete after 8 days and gave a mixture of epoxides *54* and *56* in which the epoxide *56* slightly predominated (Fig. 15.15). Treatment of a solution of epoxide *54* in 1,4-dioxane–water with sodium hydroxide gave finally 1,2:3,4-di-*O*-isopropylidene-α-D-*threo*-D-*galacto*-octopyranose *53* with 64% yield.

It is likely that the 7,8-epoxide *58* is the intermediate in this reaction because of the possibility for the neighboring group participation of the C8 hydroxyl group as shown in Fig. 15.16.

Brimacombe et al. [38, 41] have used the same approach to make decitols and decoses. Acetonation of the C6 and C7 hydroxyl groups, oxidation of the C8 hydroxyl group to aldehyde, another Wittig reaction, and Sharpless epoxidation of the obtained olefin followed by treatment with sodium hydroxide produced the corresponding decitols and/or a decose again as the mixture of isomers.

Secrist and Barnes [46] synthesized methyl peracetyl α-hikosaminide, the undecose portion of the nucleoside antibiotic hikizimycin, by allowing the unstabilized five-carbon carbohydrate phosphorane to react with a six-carbon

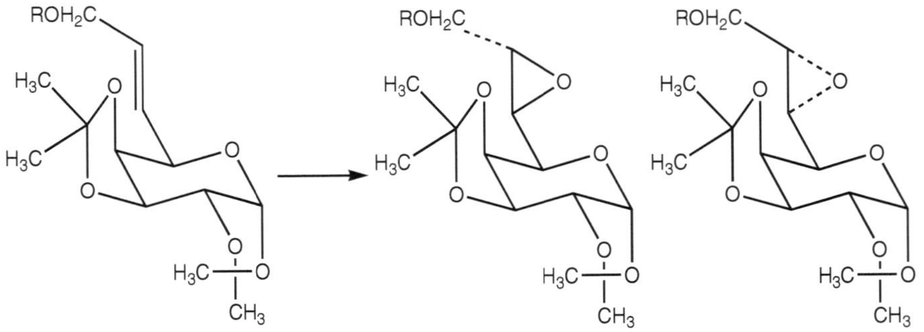

496   15 Higher-Carbon Monosaccharides

Fig. 15.15

45, R = H
46, R = Bn

54, R = H
55, R = Bn

56, R = H
57, R = Bn

Fig. 15.16

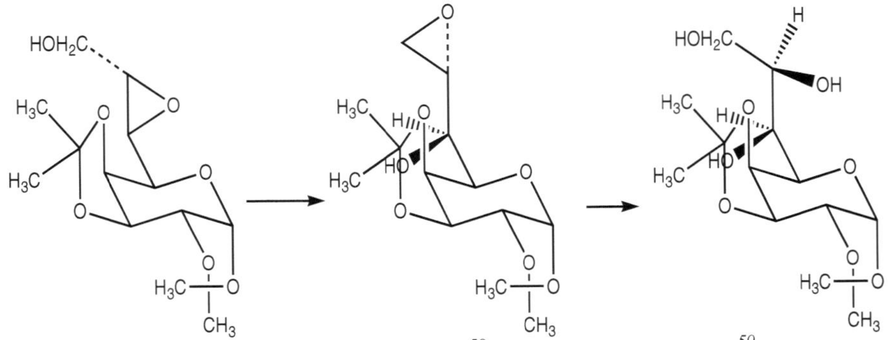

54   58   59

carbohydrate 6-aldehyde. Thus, the reaction of 2,3:4,5-di-*O*-cyclohexyliden-1-deoxy-1-triphenylphosphonio-D-arabinitol iodide *60* (obtained with 67% yield from 2,3:4,5-di-*O*-cyclohexylidene-D-arabinitol) with methyl 4-azido-2,3-di-*O*-benzyl-4-deoxy-α-D-*dialdo*-glucopyranoside *61* gave exclusively the corresponding Z-olefin *62* with 50% yield (no *E*-olefin was present). The C4 azide was then reduced with LiAlH$_4$ to the amino group, and the obtained amino sugar *63* was irradiated in the presence of diphenyl disulfide, whereby a 3:2 mixture of the *E*- and *Z*-olefins was obtained, which was separated by chromatography. This was needed because the *E*-olefin was the right substrate for catalytic hydroxylation. After the amino group was acetylated with acetic anhydride, the olefin *64* was treated with osmium tetroxide and *N*-methylmorpholine in 5:1 THF water to produce one isomer of diol *65* (Fig. 15.17).

Miljkovic and Habash-Marino [47] also approached the synthesis of higher-carbon sugars as precursors for polyhydroxylated macrocyclic lactones by using

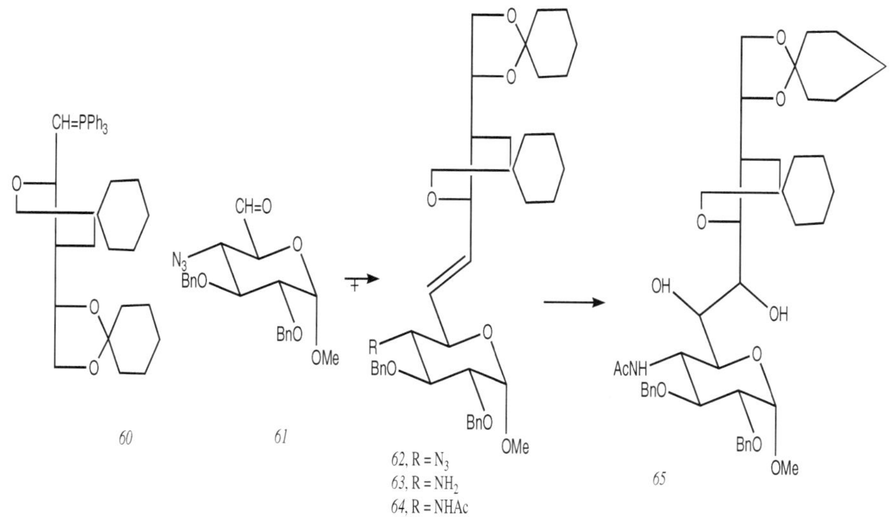

Synthesis of Higher-Carbon Sugars

Fig. 15.17

the Wittig reaction. However, glycopyranosyl triphenylphosphorane was used for coupling with an aldehydo sugar. Thus, 6-deoxy-6-iodo-2,3,4-tri-*O*-methyl-α-D-glucopyranoside was reacted with triphenyl phosphine, whereby the corresponding triphenylphosphonium salt 66 was obtained with a yield of 99% (Fig. 15.18). Reaction of 66 in the presence of *n*-butyl lithium at –60°C with 2,3,4-tri-*O*-methyl-

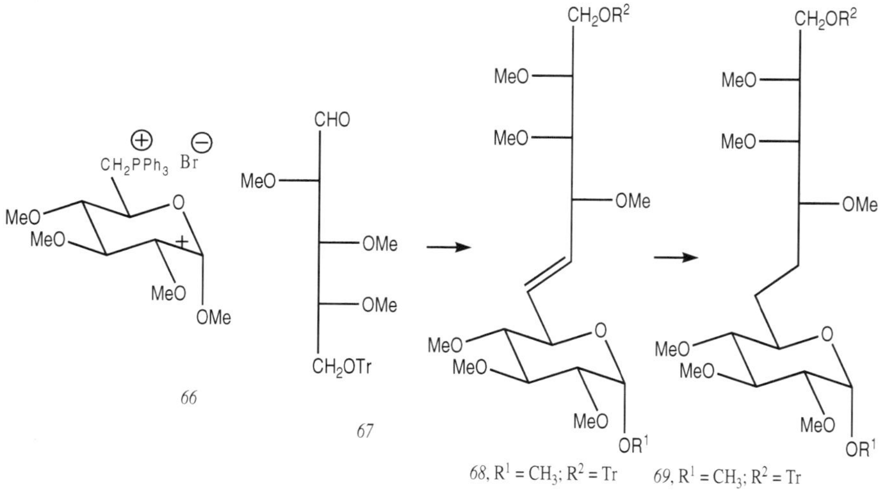

Fig. 15.18

6-O-trityl-*aldehydo*-D-arabinose *67* gave, with 57% yield, methyl *E*-6, 7-diedoxy-2,3,4,8,9,10-hexa-O-methyl-11-O-trityl-D-*arabino*-α-D-*gluco*-undec-6-enopyranoside *68* (the corresponding Z-isomer was not detected). Catalytic hydrogenation of *68* gave methyl 6,7-dideoxy-2,3,4,8,9, 0-hexa-O-methyl-D-*arabino*-D-*gluco*-undecanoside *69* which was converted in several steps to 6,7-dideoxy-2,3,4,5,8,9,10-hepta-O-methyl-D-*arabino*-D-*gluco*-undecanoic acid *71* which was then lactonized [48] to macrocyclic lactone *72* by "double activation" method [49–53] (Fig. 15.19).

The 6,7-olefinic bond could also be hydroxylated via catalytic osmylation or Sharpless epoxidation, if undecanoses were desired, but Miljkovic et al. wanted to explore the possibility of converting higher-carbon sugars into macrocyclic polyhydroxylated lactones.

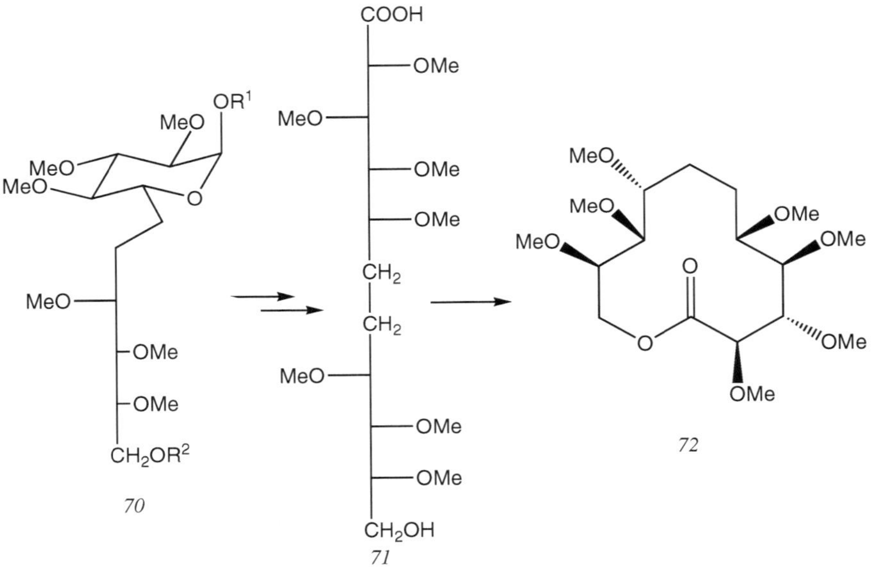

Fig. 15.19

Aldol Condensation

Vasella et al. [54] developed a method for chain elongation of uloses by the base-catalyzed addition of 1-deoxy-1-nitro-aldoses to aldehydes followed by subsequent solvolytic displacement of the nitro group by a hydroxy group. Thus, for example, the addition of 1-deoxy-2,3:5,6-di-O-isopropylidene-1-nitro-α-D-mannofuranose *73* to 6-aldo-1,2:3,4-di-O-isopropylidene-α-D-galactopyranose *74* in the presence of tetrabutylammonium fluoride gave, with 78% yield, the corresponding 7-nitro derivative *75* as single product. Acetylation of the C6 hydroxyl group of *75*

Synthesis of Higher-Carbon Sugars   499

followed by treatment of the C6 acetate 76 with $NaHCO_3$ in formamide at 100°C gave the corresponding hemiacetal 77 with 69% yield as the mixture of two anomers in 3:1 ratio (Fig. 15.20).

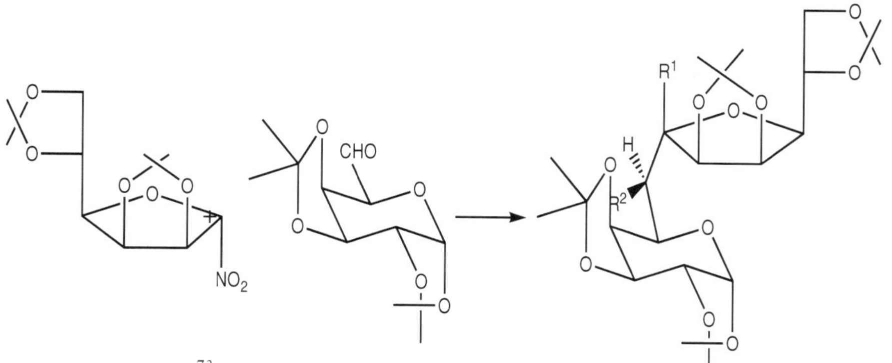

75, $R^1 = NO_2$; $R^2 = OH$
76, $R^1 = NO_2$; $R^2 = OAc$
77, $R^1 = OH$, $R^2 = Ac$

Fig. 15.20

Chapleur reported [55] that lithium enolate 79 of methyl 4,6-O-benzylidene-2-deoxy-α-D-*erythro*-hexopyranosid-3-ulose 78 obtained by treating 78 with *n*-butyl lithium at –30°C reacts with electrophiles to give the C2 alkylated derivatives 80–83 with 40–50% yield, as shown in Fig. 15.21.

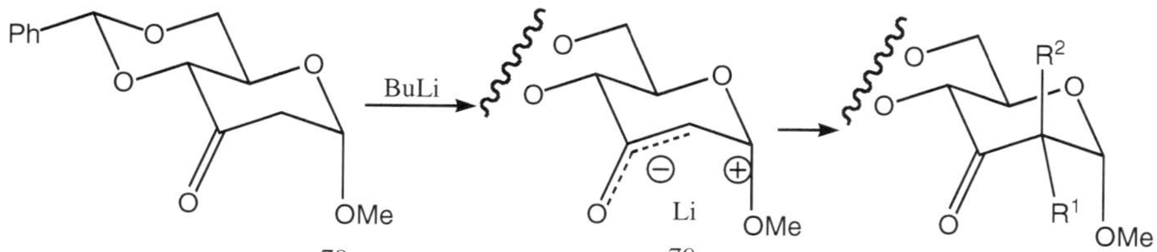

80, R_1 = Me; R_2 = H
81, R_1 = CH_2Ph; R_2 = H
82, R_1 = $CH_2CH=CH_2$; R_2 = H
83, R_1 = CH_2COOEt; R_2 = H
84, R_1 = Me; R_2 = CH_2Ph
85, R_1 = Me; R_2 = CH_2COOEt
86, R_1 = CH_2Ph; R_2 = Me

Fig. 15.21

Alkylation of C2 alkyl derivatives 80 with benzyl bromide or ethyl bromoacetate and 81 with methyl iodide gave the C2 dialkyl derivatives 84, 85, and 86 in which the second alkyl group is axially oriented. This implies that the preferred side for the electrophilic attack is the β-face and that the monoalkyl derivatives *80* and *81* with the C2 alkyl groups equatorially oriented are the results of epimerization of the initially formed axial derivatives. The "axial" alkylation of enolate is due to stereoelectronic control. Namely, the orbital overlap producing a pyranose ring in the chair conformation can only take place via an axial attack on enolate (steric control is due to the axially oriented α-methoxy group). Another important observation made in these studies was that the C1 methoxy group is remarkably resistant to β-elimination.

These findings prompted Fraser-Reid et al. [56–59] to use the principle of double stereodifferentiation [60, 61] to synthesize the higher-carbon sugars using aldol condensation of chiral aldehyde and chiral ketone.

Thus they reacted the enolate of 2-deoxy-3-oxo-pyranoside *87* with a number of sugar aldehydes and found that the aldol addition takes place exclusively from β-face due to the α-anomeric configuration of *87* and that the stereochemistry of the newly created chiral carbon is entirely controlled by the aldehydo sugar (α- or α, β-chelation) (Figs. 15.22 and 15.23).

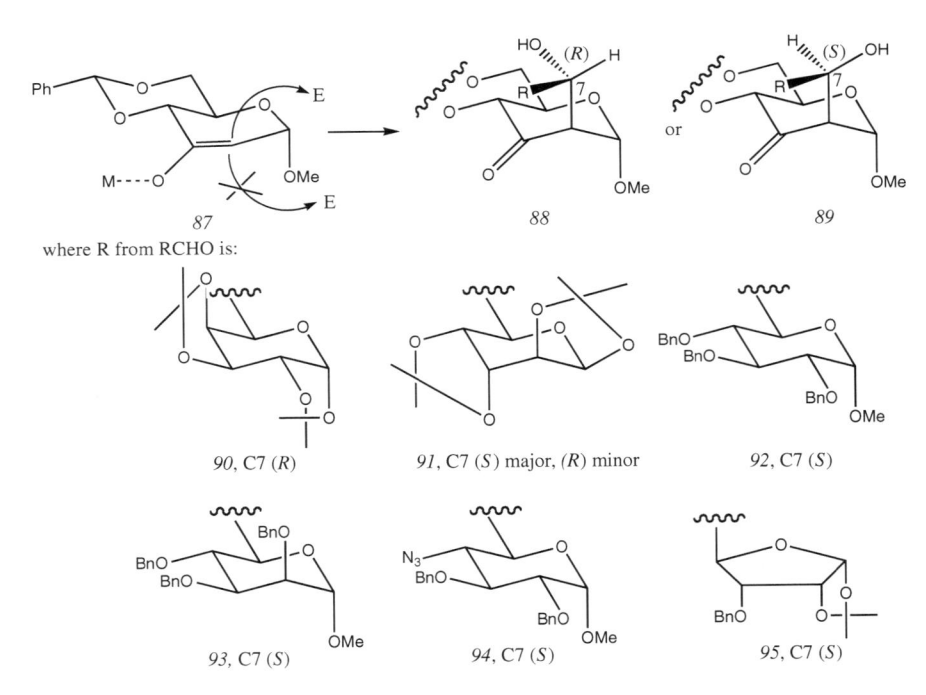

Fig. 15.22

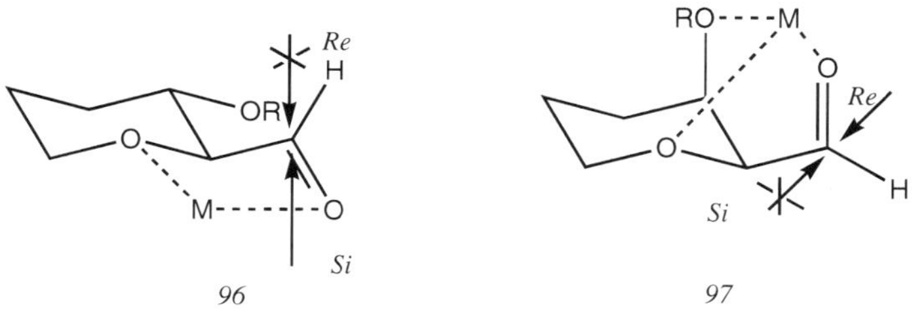

Synthesis of Higher-Carbon Sugars

Fig. 15.23

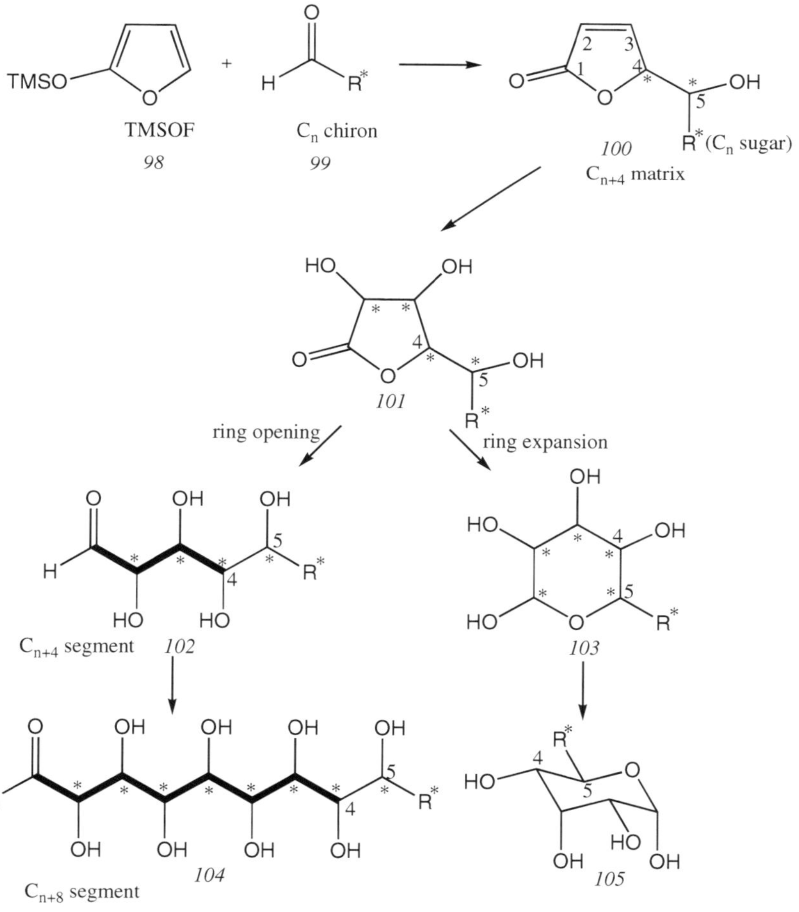

Fig. 15.24

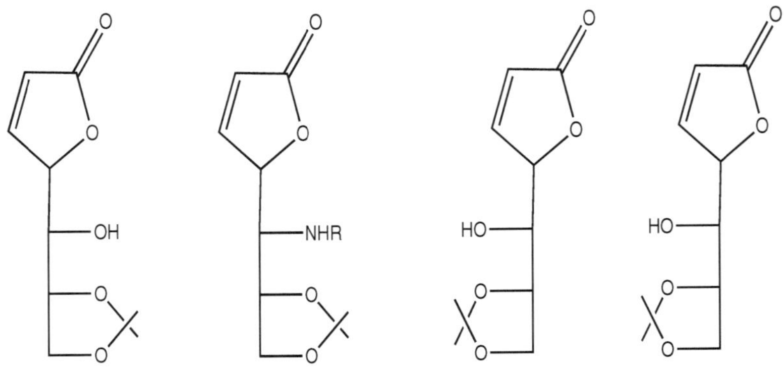

106 (85%) *107* (80%) *108*(81%) *109* (76%)

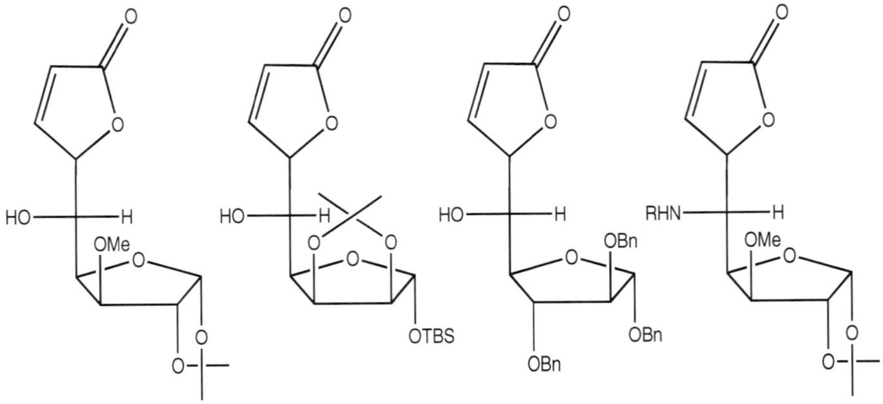

110 (90%) *111* (86%) *112*(74%) *113* (90%)

Fig. 15.25

The selectivities observed in products imply that each of the aldehydes controls the stereochemistry independently from the ketone. This facial stereoselectivity seems to be dependent upon the alkoxy substitution at the α- and/or β-carbons of the aldehyde, if Cram cyclic model is assumed with α-chelation, as in *96,* or α, β-chelation, as in *97* [62]. The results with aldehydes *89* and *90* are consistent with the α-, β-chelation shown in *96.* The literature suggests [63] that aldol reaction of *78* with *91–94*, to give *anti*-Cram products only, can be rationalized by the α-chelation pattern depicted in *95* (Fig. 15.21).

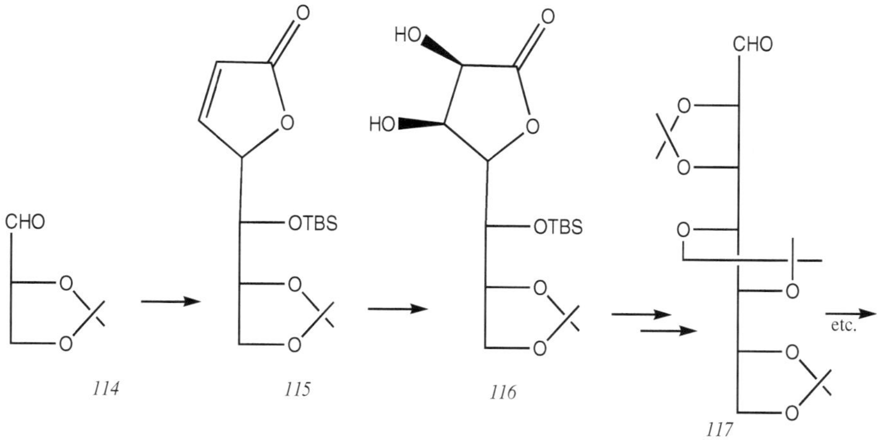

Synthesis of Higher-Carbon Sugars 503

Fig. 15.26

The Butenolide Approach

Jefford pointed out [64–67] that commercially available 2-(trimethylsiloxy)-furan (TMSOF) **98** can be an attractive synthon for the assembly of butenolides. Aldol-type condensation of TMSOF with certain carbonyl compounds, depending on the reaction conditions and catalysts, results in the formation of *erythro* and/or *threo* butenolides in high yield, as is shown in Fig. 15.24.

The stereospecific 4,5-*threo*-5,6-*erythro* (*syn*, *anti*) butenolide matrices such as the C_{n+4} matrix *100* were made by treating an aldehydo sugar precursor in dichloromethane with TMSOF at –80°C in the presence of BF_3 etherate. In this way many butenolide matrices were prepared [68–71](Fig. 15.25).

Figure 15.26 depicts the carbohydrate chain elongation by four carbon atoms, using the butenolide approach.

Total Synthesis of Higher-Carbon Monosaccharides

This approach developed by Danishefsky et al. [2–8] consists of a Diels–Alder reaction and reiterative cyclocondensation.

In classical Diels–Alder reaction, a dienophile consists of two carbon atoms connected by either a double or a triple bond. Danishefsky has developed the synthesis of a new class of highly activated and functionalized siloxydienes [72] (Fig. 15.27) that have been found to be very valuable reagents. They are prepared from the corresponding α, β-unsaturated ketones. The electron donating effects of the 1- and 3-oxygens are synergistic, thus serving to endow such dienes with a high degree of reactivity and apparently total regioselectivity as opposed to electron-withdrawing dienophiles. Moreover, the functionality endowed by such dienes, their corresponding cycloadducts, can be very nicely exploited in the synthesis of polyfunctional target systems. Figure 15.27 shows all-carbon Diels–Alder reaction with siloxydienes, where A (activating group) is CO, CN, or NO_2.

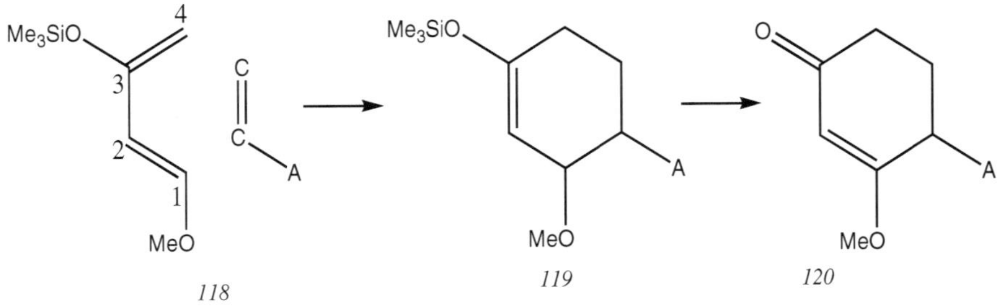

Fig. 15.27

Danishefsky et al. have shown [73] that the cyclocondensation of dienes with aldehydes can be accomplished in the presence of Lewis acids. They have also shown [74, 75] that under appropriate catalysis, a large number of syloxydienes react with virtually any aldehyde. Figure 15.28 shows Lewis acid-catalyzed cyclocondensation of diene *118* with an aldehyde giving the corresponding dihydropyranone *122*.

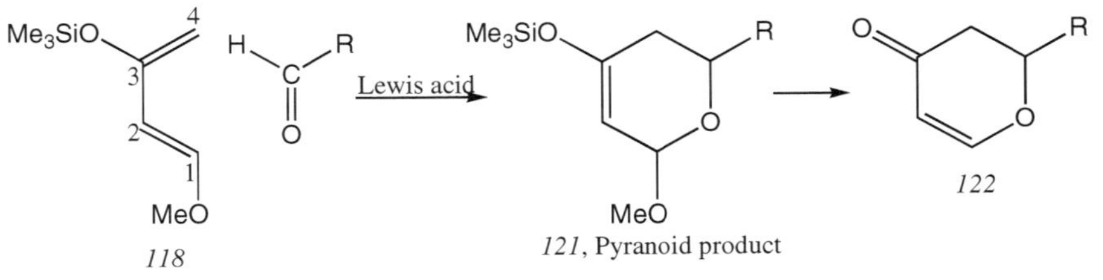

Fig. 15.28

Using heavily oxygenated dienes (e.g., *123* in Fig. 15.29) in cyclocondensations with aldehydes was instrumental in obtaining various types of sugars, such as galactosyl types of sugars [74](Fig. 15.29). A methodology was developed to introduce the oxygen function subsequent to cyclocondensation by oxidation with manganese (III) acetate [76]. In this way, glucose-like stereochemistry is generated at C4.

The reduction of the C3 keto group in *123* or *124* was accomplished with sodium borohydride–Ce (III) chloride (Luche reagent) [77] which is known to give equatorial alcohols. In this way, a very rapid and highly stereoselective routes to the glucal and galactal family of monosaccharide was developed.

The CC double bond of the galactal and glucal precursors can be functionalized either equatorially or axially at C2 with high stereoselectivity. Thus, for example, if the meta-chloroperbenzoic acid (MCPBA) reacts with the glucal-type system that

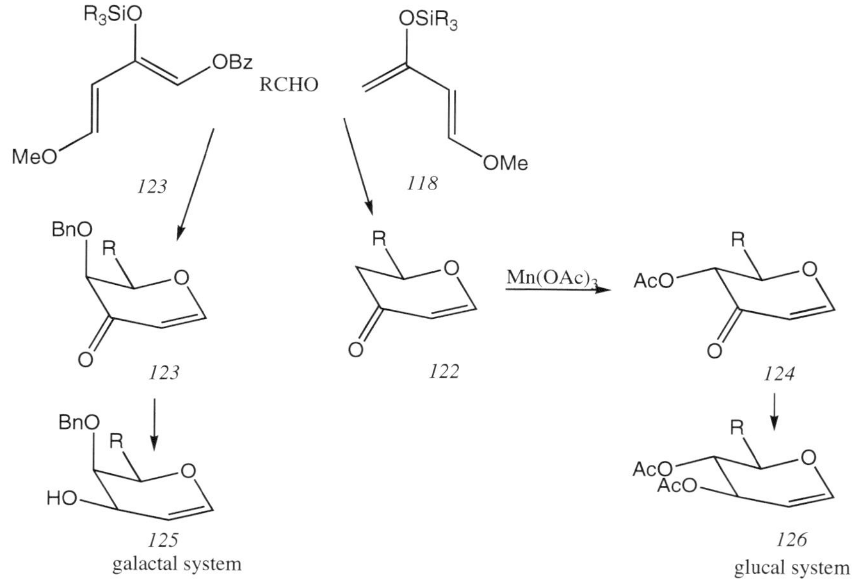

Synthesis of Higher-Carbon Sugars

Fig. 15.29

has unprotected C3 hydroxyl group, β-hydroxyl group at C2 will be introduced, giving mannose-like configuration. However, if MCPBA reacts with the glucal-type system that has protected C3 hydroxyl group, α-hydroxyl group at C2 will be introduced giving glucose-like configuration (Fig. 15.30). Similarly, the galactal-type system will give under the same reaction conditions either the talose-like structure or the galactose-like configuration (Fig. 15.30).

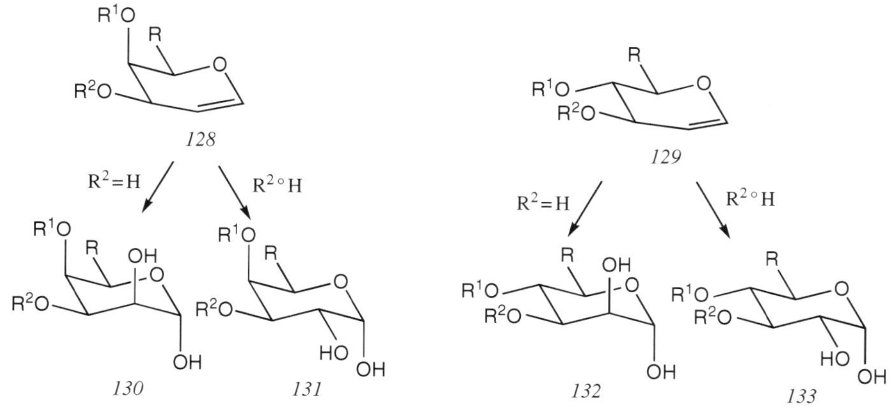

Fig. 15.30

Synthesis of higher-carbon sugars was approached by reiterative cyclocondensation principle, depicted in Fig. 15.31.

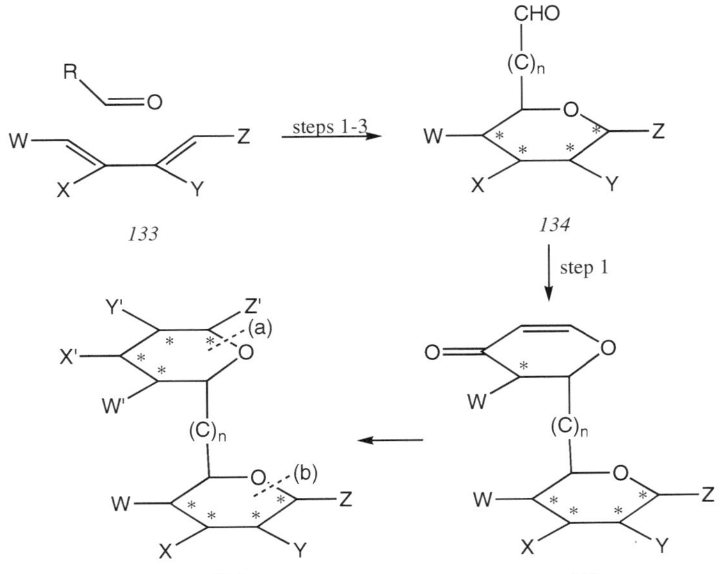

Fig. 15.31

Cyclocondensation of an aldehyde with diene *133* under Lewis acid catalysis leads to pyranoid structure *134* which is then manipulated as described above to achieve the appropriate stereochemistry of chiral carbon atoms. In the following stage of synthesis, a new aldehyde will be made on the side chain of the obtained pyranoid structure (compound *134*). Another cyclocondensation of this aldehyde with diene *133* gives compound *135* which after functionalizing of the ring carbon atoms with appropriate stereochemistry, as described above, can be converted to a higher-carbon sugar *136*.

This reiterative cyclocondensation strategy was applied to the total synthesis of octosyl acid A [78], peracetyltunicamynyluracil [79], and peracetyl-β-methylhikozamide [80].

The total synthesis of octosyl acid A is shown in Fig. 15.32. The "ribose aldehyde" *137* was used as starting material. Although also available by total synthesis [80] for this synthesis, it was prepared from D-ribose.

Cyclocondesation of aldehyde *137* with the diene *118* (Fig. 15.27) in the presence of a Lewis acid catalyst gave, with very high stereoselectivity, the bis-saccharide *138* [81], with 85% yield. Thus, the ribose ring fully controls the chirality at the C5′ carbon in the newly formed pyranone. The stereogenic center at the C7′ was formed by Luche reduction ($NaBH_4$–$CeCl_3$ in methanol, vide supra) of *138*. The obtained alcohol *139* was converted into *p*-methoxybenzyl (PMB) ether *140*. The

Synthesis of Higher-Carbon Sugars

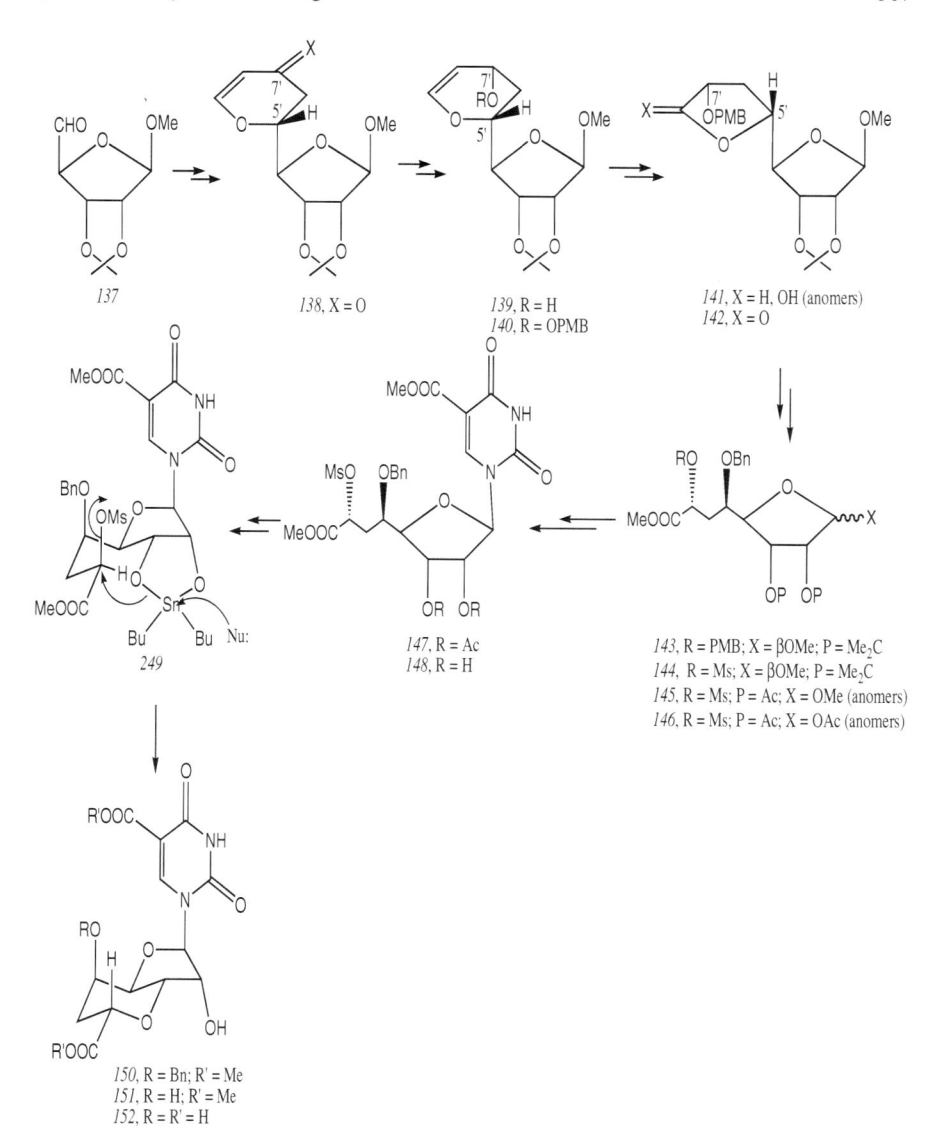

Fig. 15.32

compound *140* was then subjected to degradation with osmium tetroxide (catalytic) and sodium metaperiodate. After cleavage of the resulting formate (K_2CO_3–ethanol, room temperature), lactol anomers *141* were obtained with 93% yield. Oxidation of *141* with Ag_2CO_3–celite–xylene (reflux) afforded lactone *142*. After suitable functional group adjustments (several steps), the pyrimidine base was installed and mesyl-substituted cyclic stannane *249* was made. Fortunately, the otherwise

508 15 Higher-Carbon Monosaccharides

difficult ring closure of the 3-hydroxyl group of the ribose with the C7 carbon of the side chain proved to be possible via the tin derivative [78].

N-Acetylneuraminic acid was subject of great attention of many chemists and biochemists over the years due to its extraordinary biological importance. The first synthesis of N-acetylneuraminic acid was reported by Cornforth et al. [82, 83] by condensation of N-acetylglucosamine with oxaloacetic acid at pH 11 with the yield of 2%. Since the N-acetylneuraminic acid is composed of N-acetylmannosamine and pyruvic acid residues, the fact that it is obtained from N-acetylglucosamine suggests that epimerization of N-acetylglucosamine↔N-acetylmannosamine takes place in strongly alkaline solutions. In 1962 Carroll and Cornforth [84] repeated the synthesis of N-acetylneuraminic acid, but this time from N-acetylmannosamine and sodium oxaloacetate at pH 10. This time they obtained the N-acetylneuraminic acid in the 9 and 10% yield.

Kuhn and Baschang [85] have considerably increased the yield of N-acetylneuraminic acid when 2-N-acetyl-2-deoxy-4, 6-O-benzylidene-D-mannosamine is condensed with the potassium salt of di-tert-butyl-oxalacetate instead of sodium oxaloacetate. The initially obtained lactone is converted to N-acetylneuraminic acid with 34% yield by heating the lactone on water bath (Fig. 15.33).

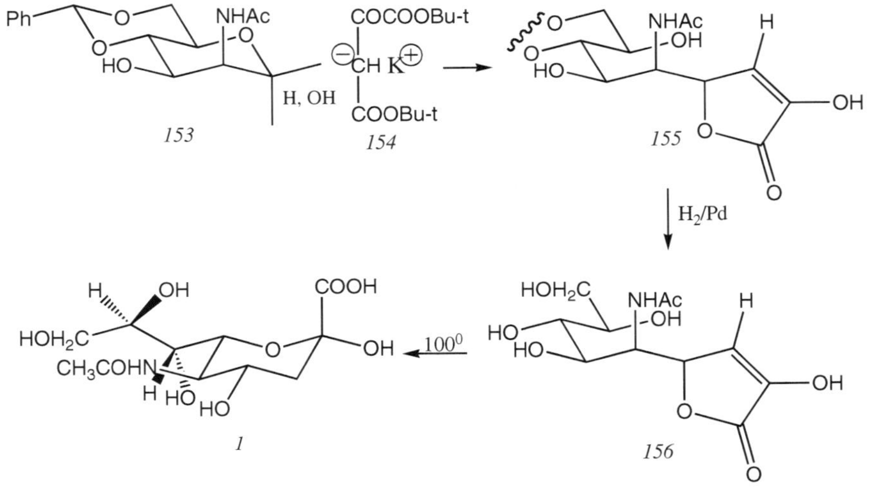

Fig. 15.33

Similar yield of N-acetylneuraminic acid was obtained by using N-acetylmannosamine instead of the corresponding 4,6-O-benzylidene derivative.

A very elegant total synthesis of N-acetylneuraminic acid was described by Danishefsky et al. [86]. The same group [87] described earlier a similar version of the total synthesis of N-acetylneuraminic acid.

Synthesis of Higher-Carbon Sugars 509

The (S)-selenoaldehyde *159* used as dienophil in cyclocondensation with an appropriate diene was prepared from (R)-methyl lactate *157* in three steps: the mesylation of (R)-methyl lactate, displacement of mesyl group by PhSe⁻ to (S)-seleno ester *158*, and the reduction with DIBAL of the ester group to the (S)-seleno aldehyde *159* (Fig. 15.34).

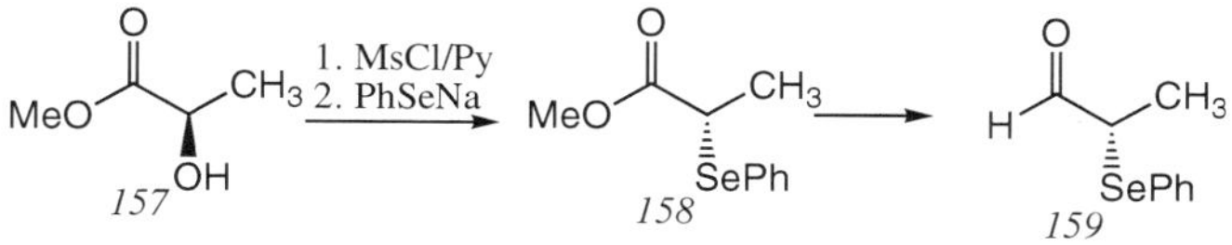

Fig. 15.34

The furyl diene needed for cyclocondensation with (S)-seleno aldehyde *159* was prepared from the mixture of geometrical isomers of furyl enone *160* by methylation with diazomethane (Fig. 15.35). The mixture of two geometrical isomers obtained by methylation was separated by chromatography on silica gel. The enone *161* led to pure E-diene *163*, while enone *162* led to the pure Z-diene *164* (Fig. 15.35). Of

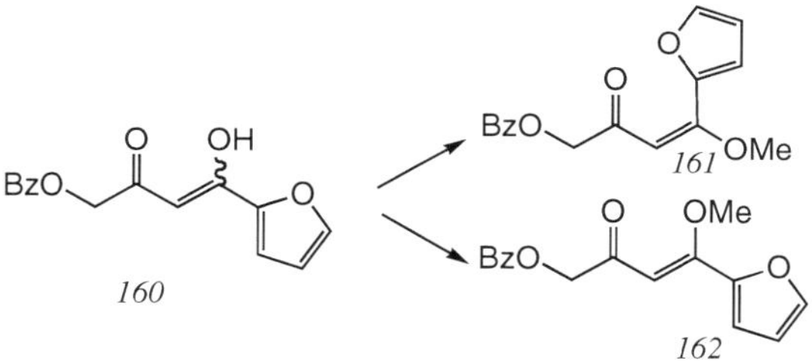

Fig. 15.35

the two dienes *163* and *164*, only diene *164* undergoes cyclocondensation with the (S)-selenoaldehyde *159* (Fig. 15.36) in methylene chloride at −78°C and in the presence of a Lewis catalyst (BF$_3$.OEt$_2$) giving the 5:1 mixture of *cis 165* and *trans*

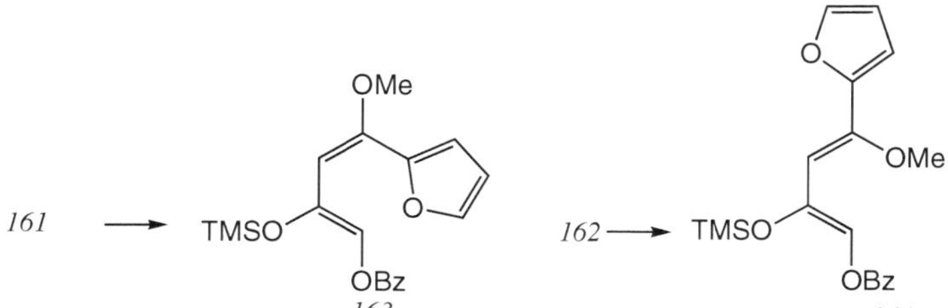

Fig. 15.36

166 dihydropyrones (Fig. 15.37). Optical purities in the range of 95% of *165* were realized when an aqueous workup for isolation of *158(S)* was avoided. Apparently aqueous treatment led to partial racemization of this labile selenoaldehyde.

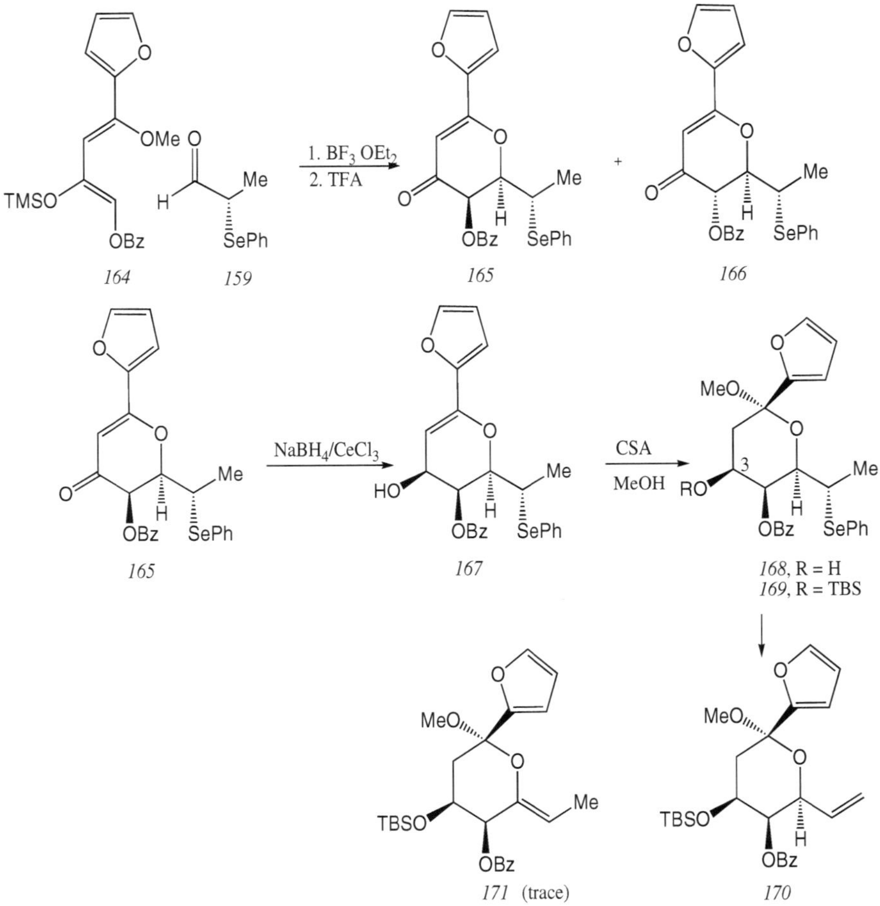

Fig. 15.37

Reduction of keto group of *165* with sodium borohydride in the presence of cerium (III) chloride [88] afforded the equatorial alcohol *167*. Addition of methanol to the double bond was accomplished in the presence of camphorsulfonic acid (CSA) giving axial glycoside *168*. After blocking the C3 hydroxyl group with TBS (*tert*-butyldimethylsilyl group), the phenylseleno group was removed from *169* by oxidative elimination using hydrogen peroxide giving almost exclusively the olefin *170* in 81% yield (olefin *171* was also present but only in traces). Osmium tetroxide

Synthesis of Higher-Carbon Sugars 511

hydroxylation of *170* (Fig. 15.37) gave diol *172* which was then cleaved with lead tetraacetate giving the aldehyde *173*. Condensation of *173* with Still phosphonate [89] gave, with 80% yield, the Z-enoate *174* (with less than 5% of *E*-isomer). The hydroxylation of *174* with OsO$_4$ proceeded with a high stereoselectivity (ca. 20:1) to give the desired product *175* with 90% yield, which was then perbenzoylated to *176*.

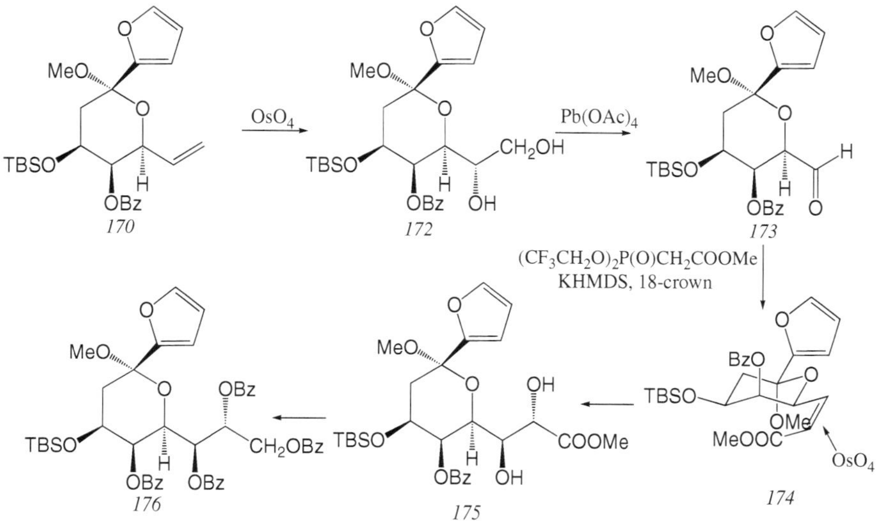

Fig. 15.38

At this point, the furan ring was oxidized with ruthenium tetroxide in the presence of excess of sodium bicarbonate as a buffer [90] (Fig. 15.38). The reaction was complete after 1 min giving the corresponding carboxylic acid that was esterified with diazo-methane to *177* (Fig. 15.39). The TBS protecting group was now removed with HF in methanol. The major product (60%) was the expected 4-hydroxy compound *178*. The other 30% was compound *179* that was obtained from *178* by benzoyl group migration. Reaction of *178* with potassium carbonate induced again benzoyl migration to produce additional amounts of *179*.

The conversion of *179* to the corresponding triflate *180* and displacement of triflate with azide using tetra-*n*-butylammonium azide gave the corresponding azide *181* with 86% yield. Reductive acetylation was accomplished in two steps: the azide group was first reduced with hydrogen in the presence of Lindlar catalyst to the amino group and the obtained amine *182* was acetylated with acetic anhydride to give *183*. Debenzoylation and hydrolysis of the methyl ester finally gave *N*-acetylneuraminic acid *1* (Fig. 15.39).

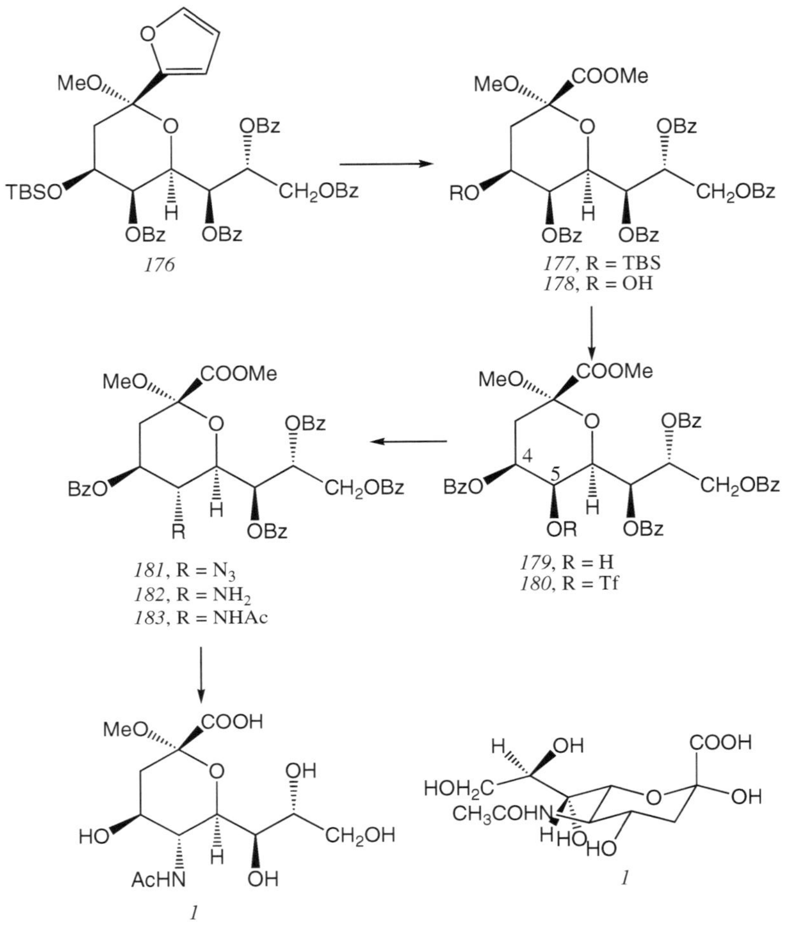

Fig. 15.39

References

1. Secrist, J. A., III; Barnes, K. D.; Wu, S.-R., in *Trends in Synthetic Carbohydrate Chemistry*, ACS Symposium Series 386, Horton, D.; Hawkins, L. D.; McGarvey, G. J. (Eds.), American Chemical Society, Washington, D. C. (1989), p. 93
2. Danishefsky, S. J.; DeNinno, M. P., *"Totally Synthetic Routes to the Higher Monosaccharides"*, Angew. Chem. Int. Ed. Engl. (1987) **26**, 15–23
3. Danishefsky, S. J., *"Cycloaddition and cyclocondensation reactions of highly functionalized dienes: applications to organic synthesis"*, Chemtracts: Organic Chem. (1989) **2**, 273–97
4. Achmatowich, O. in *Organic Synthesis Today and Tomorrow,* Trost, B. M.; Hutchinson, C. R., (Eds.), Pergamon Press, Oxford (1981) p. 307
5. Brimacombe, J. S., in *Studies in Natural Products Chemistry*, Vol 4, Part C, Atta-ur-Rahman (Ed.), Elsevier, Amsterdam (1989) p. 157

References

6. Witchak, Z. J., in *Studies in Natural Products Chemistry*, Vol. 3, Part B, Atta-ur-Rahman (Ed.), Elsevier, Amsterdam (1989) p. 209
7. Garner, P. P. in *Studies in Natural Products Chemistry*, Vol. 1, Part A, Atta-ur-Rahman (Ed.), Elsevier, Amsterdam (1988) p. 397
8. Danishefsky, S. J.; DeNinno, M. P.; Audia, J. E.; Schulte, G., in *Trends in Synthetic Carbohydrate Chemistry*, ACS Symposium Series 386, Horton, D.; Hawkins, L. D.; McGarvey, G. J. (Eds.), American Chemical Society, Washington, D. C. (1989) p. 160
9. Vogel, P.; Auberson, Y.; Bimwala, M.; de Guchteneere, E.; Vieira, E.; Wagner, G., in *Trends in Synthetic Carbohydrate Chemistry*, ACS Symposium Series 386, Horton, D.; Hawkins, L. D.; McGarvey, G. J. (Eds.), American Chemical Society, Washington, D. C. (1989) p. 197
10. Casiraghi, G.; Rassu, G., "*Aspects of Modern Higher Carbon Sugar Synthesis*", in Studies in Natural Products Chemistry, Atta-ur-Rahman (Ed.), Vol 11 (1992), Elsevier Publishers, Amsterdam, pp. 429–480
11. Gottschalk, A., "*N-substituted isoglucosamine released from mucoproteins by the influenza virus enzyme*", Nature (London) (1951) **167**, 845–847
12. Schauer, R., "*Chemistry, Metabolism, and Bilogical Functions of Sialic Acid*", Adv. Carbohydr. Chem. Biochem. (1982) **40**, 131
13. Schauer, R., *Sialic Acids*, Springer Verlag, Vienna/New York, 1982
14. Magerlein, B. J. in Perlman, D. (Ed.), "*Structure Activity Relationships Among the Semisynthetic Antibiotics*", Academic Press, New York, 1977, pp. 601–650
15. Isono, K.; Crain, P. F.; McCloskey, J. A., "*Isolation and structure of octosyl acids. Anhydrooctose uronic acid nucleosides*", J. Am. Chem. Soc. (1975) **97**, 943–945
16. More, J. D.; Finney, N. S., "*Synthesis of the Bicyclic Core of the Nucleoside Antibiotic Octosyl Acid A*", J. Org. Chem. (2006) **71**, 2236–2241
17. Azuma, T.; Isono, K., "*Transnucleosidation: an improved method for transglycosylation from pyrimidines to purines*", Chem. Pharm. Bull. (1977) **25**, 3347–3353
18. Bloch, A., "*Uridine 3′,5′-monophosphate (cyclic UMP). I. Isolation from rat liver ex tracts*", Biochem. Biophys. Res. Commun. (1975) **64**, 210–218
19. De Rosa, M.; De Rosa, S.; Gambacorta, A.; Bu'lock, J. O., "*Structure of calditol, a new branched-chain nonitol, and of the derived tetraether lipids in thermoacidophile archaebacteria of the Caldariella group*", Phytochemistry (1980) **19**, 249–254
20. De Rosa, M.; Esposito, E.; Gambacorta, A.; Nicolaus, B.; Bu'Lock, J. D.,"*Effects of temperature on ether lipid composition of Caldariella acidophila*", Phytochemistry (1980) **19**, 827–831
21. Blériot, Y.; Untersteller, E.; Fritz, B.; Sinaÿ, P., "*Total Synthesis of Calditol: Structural Clarification of this Typical Component of Archaea Order Sulfolobales*", Chem. Eur. J. (2002) **8**, 240–246
22. Cadogan, J. I. (Ed.), *Organophosphorus Reagents in Organic Synthesis*, Academic Press, New York, 1979
23. Bestmann, H. J., "*Synthesis of polyenes via phosphonium ylids*", Pure Appl. Chem. (1979) **51**, 515–533
24. Wadsworth, W. S., Org. Reactions (1978) **25**, 73
25. Aoyagi, S.; Fujimaki, S.; Kibayashi, C., "*Total Synthesis of (+)-α-Homonojirimycin*", J. Chem. Soc. Chem. Commun. (1990) 1457–1459
26. Iida, H.; Yamazaki, N.; Kibayashi, C., "*Total Synthesis if (+)-Nojirimycin and (+)-1-Deoxynojirimycin*", J. Org. Chem. (1987) **52**, 3337–3342
27. Aoyagi, S.; Fujimaki, S.; Yamazaki, N.; Kibayashi, C., "Synthesis of (+)-galactostatin", Heterocycles (1990) **30**, 783–787
28. Katsuki, T.; Sharpless, K. B., "*The first practical method for asymmetric epoxidation*", J. Am. Chem. Soc. (1980) **102**, 5974–5976
29. Overman, L. E.; Flippin, A., "*Facile aminolysis of epoxides with diethylaluminum amides*", Tetrahedron Lett. (1981) **22**, 195–198

30. Goto, T.; Toya, Y.; Ohgi, T.; Kondo, T., "*Structure of amipurimycin, a nucleoside antibiotic having a novel branched sugar moiety*", Tetrahedron Lett. (1982) **23**, 1271–1274
31. Seto, H.; Koyama, M.; Ogino, H.; Tsuruoka, T.; Shigeharu Inouye, S.; Otake, N., "*The structures of novel nucleoside antibiotics, miharamycin A and miharamycin B*", Tetrahedron Lett. (1983) **24**, 1805–1808
32. Bessodes, M.; Komiotis, D.; Antonakis, K., "*Stereoselective Synthesis of Aminoacyl Hepto Glycosides; Synthetic Tools for Biochemical Interactions Studies*", J. Chem. Soc. Perkin Trans. I (1989) 41–45
33. Caron, M.; Sharpless, K. B., "*Titanium isopropoxide-mediated nucleophilic openings of 2,3-epoxy alcohols. A mild procedure for regioselective ring-opening*", J. Org. Chem. (1985) **50**, 1557–1560
34. Brimacombe, J. S.; Kabir, A. K. M. S., "*The synthesis of some seven-carbon sugars via the osmylation of olefinic sugars*", Carbohydr. Res. (1986) **150**, 35–51
35. Brimacombe, J. S.; Kabir, A. K. M. S., "*Convenient Synthesis of L-glycero-D-manno-heptose*", Carbohydr. Res. (1986) **152**, 329–334
36. Brimacombe, J. S.; Kabir, A. K. M. S., "*A synthesis of L-galacto-D-galacto-decose*", Carbohydr. Res. (1986) **152**, 335–338
37. Brimacombe, J. S.; Hanna, R.; Kabir, A. K. M. S.; Bennet, F.; Taylor, I. D., "*Highercarbon Sugars. Part 1. The Synthesis of Some Octose Sugars via the Osmylation of Unsaturated Precursors*", J. Chem. Soc. Perkin Trans.I (1986) 815–821
38. Brimacombe, J. S.; Hanna, R.; Kabir, A. K. M. S., "*Higher-carbon Sugars. Part 2.The Synthesis of Some Decitols via the Osmylation of Unsaturated Precursors*", J. Chem. Soc. Perkin Trans. I (1986) 823–828
39. Brimacombe, J. S.; Kabir, A. K. M. S.; Bennet, F., "*Higher-carbon Sugars. Part 6. The Synthesis of Some Octose Sugars via the Epoxidation and Unsaturated Precursors*", J. Chem. Soc. Perkin Trans. (1986) 1677–1680
40. Brimacombe, J. S.; Kabir, A. K. M. S., "*Higher-carbon Sugars. Part 7. A Synthesis of L-lyxo-L-altro-nonitol, A New Nonitol*", Carbohydr. Res. (1986) **158**, 81–89
41. Brimacombe, J. S.; Hanna, R.; Kabir, A. K. M. S., "*Higher-carbon Sugars. Part 8. The Synthesis of Some Decitols via the Epoxidation of Unsaturated Precursors*", J. Chem. Soc. Perkin Trans. 1 (1987) 2421–2426
42. VanRheenan, V.; Kelly, R. C.; Cha, D. Y., "*An improved catalytic OsO_4 oxidation of olefins to cis-1,2-glycols using tertiary amine oxides as the oxidant*", Tetrahedron Lett. (1976) **17**, 1973–1976
43. Cha, J. K.; Christ, W. J.; Kishi, Y., "*On stereochemistry of osmium tetraoxide oxidation of allylic alcohol systems. Empirical rule*", Tetrahedron (1984) **40**, 2247–2255
44. Sharpless, K. B.; Behrens, C. H.; Katsuki, T.; Lee, A. M. W.; Martin, V. S.; Takatani, M.; Viti, S. M.; Walker, F. J.; Woodard, S. S., "*Stereo and regioselective openings of chiral 2,3-epoxy alcohols. Versatile routes to optically pure natural products and drugs. Unusual kinetic resolutions*", Pure Appl. Chem. (1983) **55**, 589–604
45. Pfenninger, A., "*Asymmetric Epoxidation of Allylic Alcohols: The Sharpless Epoxidation*", Synthesis (1986) 89–116
46. Secrist, J. A., III; Barnes, K. D., "*Synthesis of Methyl Peracetyl α-Hikosaminide, the Undecose Portion of the Nucleoside Antibiotic Hikizymycin*", J. Org, Chem. (1980) **45**, 4526–4528
47. Miljkovic, M.; Habash-Marino, M., "*Synthesis of higher sugars as precursors for the synthesis of chiral polyhydroxylated macrocyclic lactones*", J. Serb. Chem. Soc. (2000) **65**, 497–505
48. Habash-Marino, M., Ph.D. Dissertation, "*Synthesis of Chiral Macrocyclic Lactones from Monosaccharides*", The Pennsylvania State University, 1983
49. Corey, E. J.; Nicolaou, K. C., "*Efficient and mild lactonization method for the synthesis of macrolides*", J. Am. Chem. Soc. (1974) **96**, 5614–5616
50. Corey, E. J.; Nicolaou, K. C.; Melvin, L. S., "*Synthesis of novel macrocyclic lactones in the prostaglandin and polyether antibiotic series*", J. Am. Chem. Soc. (1975) **97**, 653–654

51. Corey, E. J.; Nicolaou, K. C.; Melvin, L. S., "*Synthesis of brefeldin A, carpaine, vertaline, and erythronolide B from nonmacrocyclic precursors*", J. Am. Chem. Soc. (1975) **97**, 654–655
52. Corey, E. J.; Nicolaou, K. C.; Toru, T., "*Total synthesis of (+-)-vermiculine*", J. Am. Chem. Soc. (1975) **97**, 2287–2288
53. Corey, E. J.; Ulrich, P.; Fitzpatrick, J. M., "*A stereoselective synthesis of (+-)-11- hydroxy-trans-8-dodecenoic acid lactone, a naturally occurring macrolide from Cephalosporium recifei*", J. Am. Chem. Soc. (1976) **98**, 222–224
54. Aebischer, B.; Bieri, J. H.; Prewo, R.; Vasella, A., "*Synthese von Ketosen durch Kettenvelängerungen von 1-Desoxy-1-nitro-aldosen. Nucleophile Additionen und Solvolyse von Nitroaethern*", Helv. Chim. Acta (1982) **65**, 2251–2272
55. Chapleur, Y., "*A short route to 2-C-alkyl-2-deoxy-sugars from D-mannose*", J. Chem. Soc. Chem. Commun. (1983) 141–142
56. Jarosz, S.; Fraser-Reid, B., "*Synthesis of a higher carbon sugar via directed aldol condensation*", Tetrahedron Lett. (1989) **30**, 2359–2362
57. Yu, K.-L.; Handa, S.; Tsang, R.; Fraser-Reid, B., "*Carbohydrate-derived partners display remarkably high stereoselectivity in aldol coupling reactions*", Tetrahedron (1991) **47**, 189–204
58. Handa, S.; Tsang, R.; McPhail, A. T.; Fraser-Reid, B., "*The pyranoside ring as a nucleophile in aldol condensations*", J. Org. Chem. (1987) **52**, 3489–3491
59. Yu, K.-L.; Fraser-Reid, B., "*Facial-selective carbohydrate-based aldol additions*", J. Chem. Soc. Chem. Commun. (1989) 1442–1445
60. Heathcock, C. H.; White, C. T.; Morrison, J. J.; Van Derveer, D., "*Acyclic Stereose lection.11. Double Stereodifferentitation as a Method for Achieving Superior Cram's rule selectivity in aldol condensation with chiral Aldehydes*", J. Org. Chem. (1981) **46**, 1296–1309
61. Masamune, S.; Choy, W.; Petersen, J. S.; Sita, L. R., "*Double stereodifferentiation and a new strategy for stereocontrol in organic syntheses*" Angew. Chem. (1985) **97**, 1–31
62. Jurczak, J.; Pikul, S.; Bauer, T., "*Tetrahedron report number 195. (R)- and (S)-2,3-0- isopropylideneglyceraldehyde in stereoselective organic synthesis*", Tetrahedron (1986) **42**, 447–488
63. Frye, S. V.; Eliel, E. L.; Cloux, R., "*Rapid-injection nuclear magnetic resonance investigation of the reactivity of alpha- and beta-alkoxy ketones with dimethyl magnesium: kinetic evidence for chelation*", J. Am. Chem. Soc. (1987) **109**, 1862–1863
64. Jefford, C. W.; Jaggi, D.; Sledeski, A. W.; Boukouvalas, J. "*New methodology for the synthesis of biologically active lactones*", in Studies in Natural Products Chemistry, Vol 3, Part B, Atta-ur-Rahman (Ed.), Elsevier, Amsterdam, 1989, p. 157–171
65. Jefford, C. W.; Sledeski, A. W.; Boukouvalas, J., "*A direct synthesis of (±)- eldanolide via the highly regioselective prenylation of 2-trimethylsiloxyfuran*", Tetrahedron Lett. (1987) **28**, 949–950
66. Jefford, C. W.; Jaggi, D.; Boukouvalas, J., "*Diastereoselectivity in the directed aldol condensation of 2-trimethylsiloxyfuran with aldehydes. A stereodivergent route to threo and erythro δ-hydroxy-γ-lactones*", Tetrahedron Lett. (1987) **28**, 4037–4040
67. Jefford, C. W.; Jaggi, D.; Bernardinelli, G.; Boukouvalas, J., "*The synthesis of (±)- cavernosine*", Tetrahedron Lett. (1987) **28**, 4041–4044
68. Casiraghi, G.; Colombo, L.; Rassu, G.; Spanu, P., "*Synthesis of enantiomerically pure 2,3-dideoxy-hept-2-enono-1,4-lactone derivatives via-diastereoselective addition of 2-(trimethylsiloxy)furan to D-glyceraldehyde and D-serinal-based three-carbon synthons*", Tetrahedron Lett. (1989) **30**, 5325–5328
69. Casiraghi, G.; Colombo, L.; Rassu, G.; Spanu, P.; Gasparri, F.; Ferrari Belicchi, M., "*The four-carbon elongation of three-carbon chiral synthons using 2-(trimethylsiloxy)furan: highly stereocontrolled entry to enantiomerically pure seven-carbon α,β- unsaturated 2,3-dideoxy-aldonolactones*", Tetrahedron (1990) **46**, 5807–5824
70. Casiraghi, G.; Colombo, L.; Rassu, G.; Spanu, P., "*The four-carbon elongation of aldehydo sugars using 2-(trimethylsiloxy)furan: a butenolide route to higher monosaccharides*", J. Org. Chem. (1990) **55**, 2565–2567

71. Gasparri, F. G.; Ferrari Belicchi, M.; Belletti, D.; Casiraghi, G.; Rassu, G., "*Crystal and molecular structure of (-)-1,2-O-isopropylidene-3-O-methyl-7,8-dideoxy-β-L- glycero-D-gluco-non-7-enofuranurono-9,6-lactone, $C_{13}H_{18}O_7$*", J. Crystallog. Spect. Res., (1991) **21**, 261–264
72. Danishefsky, S. J., "*Siloxy dienes in total synthesis*", Acct. Chem. Res. (1981) **14**, 400–406
73. Danishefsky, S. J.; Kerwin, J. F.; Kobayashi, S., "*Lewis acid catalyzed cyclocondensations of functionalized dienes with aldehydes*", J. Am. Chem. Soc. (1982) **104**, 358–360
74. Danishefsky, S. J.; Maring, C., "*A new approach to the synthesis of hexoses: an entry to (.+-.)-fucose and (.+-.)-daunosamine*", J. Am. Chem. Soc. (1985) **107**, 1269–1274
75. Bednarsky, M.; Danishefsky, S. J., "*Mild Lewis acid catalysis: tris(6,6,7,7,8,8,8- heptafluoro-2,2-dimethyl-3,5-octanedionato)europium-mediated hetero-Diels-Alder reaction*", J. Am. Chem. Soc. (1983) **105**, 3716–3717
76. Danishefsky, S. J.; Bednarsky, M., "*On the acetoxylation of 2,3-dihydro-4-pyrones: a concise, fully synthetic route to the glucal stereochemical series*", Tetrahedron Lett. (1985) **26**, 3411–3412
77. Luche, J. L.; Gemal, A. L., "*Lanthanoids in organic synthesis. 5. Selective reductions of ketones in the presence of aldehydes*", J. Am. Chem. Soc. (1979) **101**, 5848–5849
78. Danishefsky, S. J.; Hungate, R., "*The total synthesis of octosyl acid A: a new depar ture in organostannylene chemistry*", J. Am. Chem. Soc. (1986) **108**, 2486–2487
79. Danishefsky, S. J.; Barbachyn, M., "*A fully synthetic route to tunicaminyluracil*", J. Am. Chem. Soc. (1985) **107**, 7761–7762
80. Danishefsky, S. J.; Maring, C., "*A fully synthetic route to hikosamine*", J. Am. Chem. Soc. (1985) **107**, 7762–7764
81. Danishefsky, S. J.; Maring, C. J.; Barbachyn, M. R.; Segmuller, B. E., "*An approach to the Synthesis of Carbon-Carbon Linked Disaccharides*", J. Org. Chem. (1984) **49**, 4564–4565
82. Cornforth, J. W.; Daines, M. E.; Gottschalk, A., "*Synthesis of N-acetylneuraminic acid (lactaminic acid, O-sialic acid)*", Proc. chem. Soc. (London) (1957) 25–26
83. Cornforth, J. W.; Firth, M. E.; Gottschalk, A., "*The synthesis of N-acetylneuraminic acid*", Biochem. J. (1958) **68**, 57–61
84. Carroll, P. M.; Cornforth, J. W., "*Preparation of N-acetylneuraminic acid from N-acetyl-D-mannosamine*", Biochim. Biophys. Acta (1960) **39**, 161–162
85. Kuhn, R.; Baschang, G., "*Aminozucker-Synthesen, XXV. Synthese der Lactaminsäure*", Liebigs Ann. Chem. (1962) **659**, 156–163
86. Danishefsky, S. J.; DeNinno, M. P.; Chen, S., "*Stereoselective Total Synthesis of the Naturally Occurring Enantiomers of N-Acetylneuraminic Acid and 3-Deoxy-D- manno-2-octulosonic Acid. A New and Stereospecific Approach to Sialo and 2- Deoxy-D-manno-octulosonic Acid Conjugates*", J. Am. Chem. Soc. (1988) **110**, 3929–3940
87. Danishefsky, S. J.; DeNinno, M. P., "*The Total Synthesis of (±)-N-Acetylneuraminic Acid (NANA): A Remarkable Hydroxylation of a (Z)-Enoate*", J. Org. Chem. (1986) **51**, 2615–2617
88. Gemal, A. L.; Luche, J. L., "*Lanthanoids in organic synthesis. 6. Reduction of.alpha.-enones by sodium borohydride in the presence of lanthanoid chlorides: synthetic and mechanistic aspects*", J. Am. Chem. Soc. (1981) **103**, 5454–5459
89. Still, W. C.; Gennari, C., "*Direct synthesis of Z-unsaturated esters. A useful modification of the Horner-Emmons olefination*", Tetrahedron Lett. (1983) **24**, 4405–4408
90. Carlsen, P. H. J.; Katsuki, T.; Martin, V. S.; Sharpless, K. B., "*A greatly improved procedure for ruthenium tetroxide catalyzed oxidations of organic compounds*", J. Org. Chem. (1981) **46**, 3936–3938

Author Index

Note: The italicized locators in parenthesis refer to the Reference numbers.

A

Abbas, S. A., 141 (*32*)
Aberg, P.-M., 217 (*82*)
Abraham, R. J., 54 (*33*)
Abyraki, S., 465 (*60*)
Accountius, O. E., 286 (*74*)
Achmatowich, C. R., 512 (*4*)
Acton, E. M., 189 (*29*)
Adkins, H., 25 (*14*), 163 (*9*)
Aebischer, B., 515 (*54*)
Ajisaka, K., 319 (*34*)
Akishin, P. A., 90 (*35*)
Akita, E., 485 (*70, 76, 77*)
Akiya, S., 214 (*26*)
Al Janabi, S. A. S., 219 (*122*)
Albrecht, H. P., 319 (*24*)
Albright, J. D., 285 (*50*), 286 (*55*)
Alder, R. W., 92 (*88*)
Alexander, B. H., 284 (*9*)
Alexander, J., 25 (*8*)
Alexander, R. G., 465 (*57*)
Alexander-Jackson, E., 25 (*8*)
Ali, M. H., 465 (*50*)
Ali, Y., 188 (*19*)
Al-Jeboury, F. S., 164 (*19*)
Allinger, N. L., 55 (*46, 52*), 56 (*79*), 90 (*31, 42, 44*), 91 (*56, 72*), 188 (*12*), 320 (*38, 39*)
Almenningen, A., 54 (*28*)
Alonso, R., 215 (*29*)
Alt, G. H., 217 (*89*)
Altona, C., 90 (*41, 42, 43, 44, 45, 46, 47, 48*), 93 (*105, 106*), 420 (*309*)
Amadori, M., 242 (*65*)
Amatore, C., 415 (*186*)
Ames, G. R., 242 (*58*)
Amyes, T. L., 419 (*282*)
Anbar, M., 288 (*106*)
Andersen, C. B., 89 (*23*)
Andersen, K. K., 464 (*37*)

Andersson, F. O., 416 (*210, 215*)
Andree, F., 465 (*56*)
Andrejević, V., 189 (*36*)
Andrews, G. L., 466 (*86*)
Anet, E. F. L. J., 111 (*45, 46*)
Anet, F. A. L., 54 (*42*), 319 (*27*), 463 (*10*)
Anet, R., 54 (*42*)
Angyal, S. J., 54 (*35*), 55 (*46, 47, 49, 50, 51, 52, 60, 65, 66*), 89 (*7, 8, 25, 26*), 90 (*31*), 91 (*56*), 141 (*37, 38, 39*), 188 (*12*), 218 (*94, 115*), 288 (*110, 111*), 290 (*161*), 320 (*39*)
Anno, K., 414 (*163*)
Anslyn, E. V., 419 (*278*)
Antonakis, K., 287 (*96*), 288 (*97*), 514 (*32*)
Aoyagi, S., 513 (*25, 27*)
Aped, P., 421 (*314*)
Apparu, M., 416 (*209*)
Appel, H., 164 (*32*)
Araki, K., 189 (*26*)
Arcamone, F., 485 (*61*)
Argoudelis, A. D., 484 (*50, 52*)
Arison, B., 285 (*29*)
Aritomi, M., 142 (*68*)
Arlt, v. F., 408 (*46*)
Armour, C., 417 (*238*)
Armstrong, E. F., 408 (*41*)
Armstrong, K. B., 91 (*73*)
Armstrong, R. K., 188 (*14*), 241 (*28*)
Arnold, E., 465 (*48*)
Arrick, R. E., 287 (*80*)
Arth, G. E., 286 (*75*)
Arzoumanian, H., 189 (*29*)
Ashby, E. C., 166 (*78*), 464 (*32*)
Asp, L., 420 (*304*)
Aspinal, G. O., 140 (*14*)
Aston, J. G., 54 (*24, 25*)
Ataie, M., 219 (*120*)
Auberson, Y., 513 (*9*)
Augestad, I., 418 (*252, 253, 254*)

Augustyns, K., 244 (*112*)
Ault, R. G., 165 (*43*)
Austin, P. W., 218 (*106*), 219 (*123*)
Au-Yeung, B. W., 412 (*132, 133, 134*)
Auzanneau, F.-I., 216 (*50*)
Avela, E., 142 (*57, 58, 59, 60, 61, 62*)
Azarnia, N., 53 (*14*)
Azuma, T., 513 (*17*)

B

Baas, J. M. A., 54 (*44, 45*)
Bacquet, C., 412 (*135*)
Bader, R. F. W., 53 (*11*)
Baer, E., 165 (*44*)
Baggett, N., 164 (*17, 18, 19*), 141 (*18, 19*)
Bagnell, L., 466 (*70*)
Bailey, W. F., 420 (*299*)
Bair, R. A., 53 (*4*)
Baker, B. R., 241 (*36*), 285 (*32*)
Baker, D. C., 287 (*80*)
Baker, J. R., 188 (*4*)
Balanson, R. D., 465 (*55*)
Balaram, P., 412 (*132, 133, 134*)
Ball, D. H., 140 (*3, 6, 7*), 189 (*21*)
Ballou, C. E., 214 (*10*)
Banhoeffer, K. F., 110 (*13*)
Banks, B. E. C., 418 (*250*)
Banoub, J., 408 (*33, 34*)
Barbachyn, M., 516 (*79*)
Bardolph, M. P., 214 (*7*)
Barker, I. R. L., 284 (*17, 18*)
Barker, R., 165 (*36*), 190 (*44*), 407 (*9*)
Barker, S. A., 163 (*1*), 284 (*10*)
Barnes, K. D., 512 (*1*), 512 (*46*)
Barr, S. J., 54 (*23*)
Barros Papoula, M. T., 290 (*166*)
Barrows, S. E., 56 (*72*)
Barton, D. H. R., 217 (*89*), 290 (*166*)
Bartz, Q. R., 485 (*62, 63, 64, 65, 66, 68*)
Baschang, G., 516 (*85*)
Bastiansen, O., 54 (*28*)
Batchelor, J. G., 91 (*68*)
Bates, F. J., 89 (*6*)
Bauer, M., 142 (*67*)
Bauer, T., 515 (*62*)
Baum, G., 214 (*17, 19*)
Beau, J.-M., 465 (*60*)
Becker, J., 141 (*47*)
Bednarsky, M., 516 (*75, 76*)
Behrendt, M., 412 (*129*)
Behrens, C. H., 514 (*44*)
Bell, E. A., 465 (*48*)
Bell, R. P., 110 (*15*)

Belletti, D., 516 (*71*)
BeMiller, J. N., 167 (*88*), 244 (*98*), 417 (*226*), 419 (*272*)
Benazet, F., 486 (*84*)
Benedict, R. G., 482 (*1*)
Benn, M. H., 242 (*44*)
Bennet, A. J., 244 (*96*), 419 (*284*)
Bennet, F., 514 (*37, 39*)
Bennis, K., 216 (*50*)
Benveniste, R., 483 (*20, 21, 27, 28, 32*)
Berend, G., 165 (*41*)
Bergmann, M., 216 (*63*), 411 (*112*)
Berlin, P., 411 (*104*)
Berlin, Y., 466 (*89*)
Berman, H. M., 53 (*15, 16*)
Berner, E., 418 (*252, 253, 254*)
Bernet, B., 465 (*63*)
Berry, D. J., 465 (*67*)
Bertazzoli, C., 485 (*61*)
Berti, P. J., 244 (*97, 101, 102*), 419 (*285*)
Bessodes, M., 514 (*32*)
Bestmann, H. J., 513 (*23*)
Betaneli, V. I., 411 (*106, 107*)
Beyler, R. E., 286 (*75*)
Beynon, P. J., 284 (*22, 23*)
Bieder, A., 288 (*119, 120*), 290 (*163*)
Bieri, J. H., 515 (*54*)
Bigeleisen, J., 110 (*18*)
Bigelow, N. M., 141 (*46*)
Bigelow, S. S., 414 (*169*)
Bilodeau, M. T., 414 (*162*)
Bimwala, M., 513 (*9*)
Binkley, R. W., 167 (*85*)
Bird, J. W., 287 (*91*)
Birkofer, L., 141 (*28*)
Bishop, C. T., 54 (*37, 38*), 90 (*37*), 110 (*27*), 406 (*6, 7*), 407 (*11*)
Blanc-Muesser, M., 216 (*49*), 416 (*209*)
Blanksma, J. J., 111 (*39*)
Bleha, T., 421 (*313*)
Blériot, Y., 513 (*21*)
Bloch, A., 513 (*18*)
Blom, J., 418 (*255*)
Blumbergs, P., 188 (*7*), 241 (*22, 23*)
Bochkov, A. F., 407 (*17*), 410 (*93, 94, 95, 96, 97, 98*), 411 (*99, 105, 106, 107, 108, 116*)
Bock, M. G., 467 (*93*)
Boeden, H.-F., 411 (*104*)
Bognar, R., 410 (*93*), 411 (*110*)
Bogusiak, J., 416 (*216*)
Böhm, G., 409 (*76*), 410 (*77, 78*)
Bohme, H., 189 (*30*)
Bohn, E., 407 (*25*)

Bolker, H. I., 167 (*81, 82*)
Bolliger, H. R., 140 (*9*)
Bolton, C. H., 240 (*2, 3*)
Bolz, F., 216 (*57*)
Bonner, T. G., 165 (*50*)
Bonner, W. A., 89 (*9*), 290 (*150*)
Boons, G.-J., 413 (*142*)
Booth, G. E., 90 (*33*)
Booth, H., 91 (*54, 69*), 92 (*94, 97*), 93 (*99, 100*), 420 (*309*)
Boren, H. B., 142 (*54*)
Borodkiin, V. S., 464 (*26, 28*)
Boukouvalas, J., 515 (*64, 65, 66, 67*)
Boullanger, P., 215 (*30*)
Bourne, E. J., 141 (*40*), 163 (*1*), 284 (*10*)
Bouveng, H., 141 (*43*)
Braccini, I., 415 (*188*)
Bradley, C. H., 319 (*27*), 463 (*10*)
Bradley, P. R., 417 (*236*)
Brady, J. W., 55 (*67, 68*)
Brady, R. F., 163 (*4*)
Brain, F., 486 (*91*)
Brandänge, S., 321 (*64*)
Braun, E., 241 (*31, 33*)
Brauns, F., 241 (*33*)
Brenemann, C., 53 (*11*)
Breslow, R., 166 (*63*)
Brewster, J. H., 165 (*55*)
Brice, C., 189 (*35*), 407 (*22*), 416 (*202*)
Brickman, J., 56 (*76*)
Brigl, P., 164 (*29*), 217 (*83*)
Brimacombe, J. S., 165 (*58*), 188 (*17*), 216 (*53*), 217 (*76*), 285 (*52*), 320 (*55, 56*), 321 (*58*), 512 (*5*), 514 (*34, 35, 36, 37, 38, 39, 40, 41*)
Broderick, A. E., 25 (*14*), 163 (*9*)
Brodskii, A. I., 288 (*107*)
Bromer, W. W., 485 (*82*)
Broquist, H. P., 465 (*47*)
Brown, D. M., 244 (*110*)
Brown, H. C., 165 (*55*)
Brown, J. F., 110 (*10*)
Brown, R. K., 90 (*40*), 167 (*80*)
Browne, L. J., 412 (*132, 133, 134*)
Brutcher, F. V., 54 (*23*)
Bruylants, A., 244 (*95*)
Bruzzi, A., 141 (*26*)
Bryan, J. G. H., 285 (*52*)
Brzezinska, M., 483 (*21, 28, 30*)
Buben, I., 218 (*117*)
Buchanan, G. W., 319 (*27*), 463 (*10*)

Buchanan, J. G., 140 (*15*), 141 (*41*), 165 (*46*), 218 (*106*), 219 (*118, 120, 122, 123, 124*), 241 (*18*)
Büchner, E., 288 (*112*)
Buchta, E., 465 (*56*)
Buckles, R. E., 189 (*33*)
Bugie, E., 485 (*71*)
Bujacz, G., 92 (*82*)
Bukhari, S. T. K., 484 (*44*)
Bulgrin, V. C., 288 (*113, 114*)
Bull, H. G., 164 (*15*), 419 (*271*)
Bullock, C., 217 (*77*)
Bu'lock, J. O., 513 (*19, 20*)
Buncell, E., 417 (*236*)
Bunton, C. A., 288 (*115*), 417 (*234, 238*)
Bunzel, H. H., 284 (*13*)
Burgdorf, M., 241 (*27*)
Burkhart, O., 241 (*31*)
Burnouf, C., 320 (*54*)
Burshtein, K. Ya., 411 (*108*)
Burton, J. S., 284 (*25, 26*), 285 (*27*), 287 (*84*), 319 (*16*), 463 (*13*)
Busch, M., 284 (*5*)
Bush, J. D., 286 (*74*)
Buss, D. H., 188 (*2*), 190 (*45*), 285 (*32*)
Butterworth, R. F., 287 (*77*)
Buys, H. R., 90 (*41, 42, 44*)
Byram, S. K., 466 (*78*)

C

Cadas, O., 215 (*30*)
Cadet, J., 243 (*90*)
Cadogan, J. I., 513 (*22*)
Cahn, R. S., 318 (*13*)
Canevazzi, G., 485 (*60*)
Cantley, M., 289 (*138*)
Cao, S., 416 (*215*)
Čapek, K., 416 (*215*)
Caperelli, C. A., 287 (*95*)
Capmau, M.-L., 319 (*32*), 463 (*18*)
Capon, B., 190 (*42*), 243 (*76, 82*), 417 (*225*), 418 (*261*), 420 (*302*)
Card, P. J., 410 (*89*), 412 (*132, 133, 134*), 414 (*172*)
Carey, F. A., 141 (*31*), 320 (*44*)
Carlsen, P. H. J., 516 (*90*)
Carmichael, I., 92 (*87*), 92 (*91*)
Carniero, T. M. G., 92 (*88*)
Caron, M., 514 (*33*)
Carroll, P. M., 516 (*84*)
Carson, J. F., 243 (*74, 75*)
Carter, H. E., 484 (*52*)
Caserio, M. C., 142 (*65*)

Casiraghi, G., 513 (*10*), 515 (*68, 69, 70*), 516 (*71*)
Castillon, S., 217 (*78*)
Caulfield, T. J., 409 (*70*)
Ceder, O., 418 (*262*)
Černy, M., 164 (*27, 28*), 214 (*3*), 215 (*32, 35, 45, 46*), 216 (*56*), 218 (*116, 117*), 289 (*131*), 416 (*213*)
Cha, D. Y., 514 (*42*)
Cha, J. K., 464 (*31*), 514 (*43*)
Chain, E. B., 465 (*58*)
Chalk, R. C., 140 (*6*)
Challis, B. C., 110 (*17*)
Chan, S. S. C., 91 (*78*), 92 (*79*)
Chang, H., 141 (*33*)
Chang, P.-I., 215 (*26*)
Chapleur, Y., 515 (*55*)
Charalambous, G., 241 (*21*)
Chargaff, E., 244 (*108, 109*)
Charpentie, Y., 486 (*84*)
Chastain, B. H., 218 (*109, 110*)
Chattopadhyaya, J., 92 (*85*), 93 (*102, 103, 104, 110, 111*)
Cheeseman, J. R., 53 (*11*)
Chen, C. H., 412 (*132, 133, 134*)
Chen, K.-H., 56 (*79*)
Chen, S., 516 (*86*)
Chênevert, R. B., 167 (*87*), 412 (*134*)
Cheng, Y., 416 (*219*)
Chevereau, M., 483 (*24*)
Chilton, W. S., 484 (*51, 52, 53*)
Chittenden, G. J. F., 141 (*41*), 142 (*69*), 286 (*60, 61*)
Chizov, O. S., 321 (*73*)
Chmielewski, M., 484 (*45*)
Cho, H., 467 (*92*)
Chodkiewicz, W., 319 (*32*), 463 (*18*)
Choong, T. C., 463 (*8*)
Christ, W. J., 464 (*31*), 514 (*43*)
Christensen, A. T., 466 (*78*)
Christensen, J. E., 166 (*66*), 190 (*47*)
Chü, N. J., 55 (*54*), 89 (*16, 17*), 66 (*9*)
Chu, P., 241 (*32*), 482 (*7*)
Chucholowski, A., 409 (*61*)
Cilento, G., 285 (*33*)
Cipera, J. D. T., 411 (*111*)
Claessens, M., 419 (*280*)
Clardy, J., 465 (*48*)
Clark, D. F., 243 (*88*)
Claus, C., 284 (*19*)
Clayton, J. P., 465 (*57, 59*), 466 (*72*)
Cleaver, A. J., 289 (*139*)
Clode, D. M., 163 (*8*), 164 (*20*), 219 (*118*)

Cloran, F., 92 (*91*)
Cloux, R., 515 (*63*)
Clutterbuck, P. W., 288 (*104*)
Colegate, S. M., 465 (*45*)
Coleman, G. H., 214 (*6, 7*)
Colley, A., 408 (*40*)
Collins, D. V., 164 (*34*)
Collins, J. C., 287 (*78*)
Collins, P. M., 164 (*14*), 284 (*22, 23*), 285 (*31*), 287 (*82*)
Colombo, L., 515 (*68, 69, 70*)
Commeyras, A., 419 (*290*), 420 (*291*)
Compton, J., 140 (*4*)
Connett, B. E., 243 (*76*)
Contour, M.-O., 415 (*195*)
Cooper, F. P., 54 (*37*), 90 (*37*), 110 (*27*), 406 (*6, 7*)
Copper, D. J., 484 (*44*)
Cordes, E. H., 164 (*15*), 243 (*81*), 419 (*271*)
Corey, E. J., 89 (*19*), 320 (*35, 36, 49*), 464 (*3, 449*), 465 (*55*), 466 (*90*), 467 (*91, 92, 93*), 514 (*49, 501*), 515 (*51, 52, 53*)
Cornforth, J. W., 516 (*82, 83, 84*)
Cosio, F. P., 288 (*100*)
Courtois, J. E., 288 (*119, 120*), 290 (*163*)
Cowdrey, W. A., 189 (*31*)
Coxon, B., 111 (*42*)
Crain, P. F., 513 (*15*)
Cram, D. J., 319 (*33*)
Cramer, C. J., 56 (*72*), 91 (*76*), 92 (*89*)
Creasey, S. E., 140 (*12*), 141 (*29*)
Cree, G. M., 286 (*72*)
Crich, D., 410 (*92*)
Criegee, R., 288 (*108, 112*), 290 (*155, 156, 157, 158, 159, 160*)
Cron, M. J., 482 (*5, 6, 7*)
Csuk, R., 321 (*65, 67*)
Csürös, Z., 216 (*66, 70*), 408 (*38*)
Cuevas, G., 89 (*5*), 92 (*81, 84*)
Curran, D. P., 466 (*87*)
Czernecki, S., 288 (*101*)

D

Dahlgard, M., 218 (*108, 109, 110*)
Dahlgren, G., 288 (*114*)
Dahlman, O., 321 (*64*)
Daines, M. E., 516 (*82*)
Dakkouri, M., 285 (*35*)
Dallinga, G., 90 (*36*)
Dalling, D. K., 319 (*26*), 463 (*9*)
Damm, W., 56 (*78*)
Dang, H. P., 466 (*68*)
Daniels, P., 483 (*21*)
Daniels, P. J. L., 483 (*24*), 484 (*41, 42, 44*)

Danishefsky, S. J., 413 (*158, 159*), 414 (*161, 162*), 512 (*2, 3*), 513 (*8*), 516 (*72, 73, 74, 75, 76, 78, 79, 80, 81, 86, 87*)
Dansi, A., 242 (*67*)
Danzig, M., 244 (*94*)
Dasgupta, F., 414 (*179*), 415 (*196*), 420 (*292, 293*)
David, S., 287 (*94*), 413 (*145*)
Davidson, E. A., 189 (*36, 37, 38*)
Davies, J., 483 (*20, 21, 24, 27, 28, 30, 32*)
Davison, B. E., 321 (*72*)
Davisson, J. W., 485 (*59*)
Day, J. N. E., 417 (*233*)
Day, W., 486 (*83*)
de Belder, A. N., 163 (*2, 3*)
De Bruyne, C. K., 416 (*207*)
de Guchteneere, E., 513 (*9*)
de Hoog, A. J., 90 (*41*)
de los Angeles Laborde, M., 320 (*54*)
De Nijs, M. P., 215 (*27*)
De Pascual, J., 217 (*88*)
De Rosa, M., 513 (*19, 20*)
De Rosa, S., 513 (*19*)
De Smet, M. D., 416 (*222*)
de Wolf, N., 90 (*48*)
De Wolfe, R. H., 164 (*13*)
Deàk, G., 214 (*24*), 216 (*66*)
Deardorf, D. R., 466 (*90*)
DeBlauwe, F., 216 (*59*)
Dee, K. K., 417 (*244*)
Defaye, J., 216 (*49, 50*), 415 (*195*), 416 (*209*)
Dekker, C. A., 287 (*90*)
Del Re, G., 95 (*21*)
Delbrück, K., 416 (*204*)
Della, E. W., 91 (*71*)
DeNinno, M. P., 512 (*2*), 513 (*8*), 516 (*86, 87*)
Depezay, J.-C., 321 (*57, 59*)
Derevitskaya, V. A., 411 (*102, 103*)
DeRoose, F., 416 (*221*)
Derouet, C., 415 (*188*)
Descotes, G., 215 (*30*)
Deslongchamps, P., 89 (*3*), 91 (*64*), 167 (*86, 87*), 418 (*267, 268*), 419 (*273, 276*), 421 (*312*)
Desmet, T., 419 (*280*)
Dessinges, A., 217 (*78*)
Detre, G., 484 (*48*)
Dextraze, P., 321 (*70, 71*)
Diago-Mesequer, J., 467 (*94*)
Dickinson, M. J., 285 (*28*)
Diehl, H. W., 164 (*23*), 289 (*136*)
Dienes, M. T., 290 (*162*)
Dimant, E., 286 (*53*)

Dimler, R. J., 215 (*33*)
Dinya, Z., 411 (*110*)
Dion, H. W., 485 (*66, 68*)
Djerassi, C., 284 (*21*), 320 (*37, 38*)
Doering, W. von, E., 286 (*70, 71*)
Doerr, I. L., 244 (*111*)
Doerschuk, A. P., 141 (*36*)
Doganges, P. T., 284 (*23*), 285 (*31*)
Doi, O., 483 (*26*)
Dolle, R. E., 409 (*61*)
Donath, W. E., 53 (*22*), 165 (*56*)
Doner, L. W., 241 (*35*)
Dorling, P. R., 465 (*45*)
Dorman, D. E., 486 (*83*)
Dorofenko, G. N., 110 (*20*)
Dory, Y. L., 418 (*267*), 419 (*276*), 421 (*312*)
Doty, P. M., 54 (*24*)
Doyle, E. R., 419 (*272*)
Drabowicz, J., 464 (*38*)
Drew, K., 92 (*87*)
Driguez, H., 216 (*49*), 416 (*209*)
Drisko, R. W., 290 (*150*)
Dubost, M., 486 (*84*)
Dubrunfaut, A. P., 109 (*1*)
Duke, F. R., 288 (*116*)
Dulles, F. J., 56 (*72*)
Duncan, D. P., 465 (*67*)
Dunitz, J. D., 466 (*89*)
Dürr, W., 167 (*90*), 216 (*57*)
Dutcher, J. D., 240 (*4, 5, 6, 7*)
Dutta, S., 243 (*85*)
Duxbury, J. M., 164 (*17*)
Dyer, J. R., 485 (*74*)
Dyfverman, A., 214 (*8*)

E

Ebata, T., 215 (*34, 37*)
Eberstein, K., 219 (*121*)
Ebihara, M., 415 (*185*)
Eby, R., 142 (*63*), 215 (*28*), 408 (*31*)
Edgar, A. R., 219 (*120, 122*)
Edge, C. J., 56 (*73*)
Edward, J. T., 55 (*53*), 89 (*18*)
Egan, L. P., 408 (*48*)
Egron, M.-J., 288 (*97*)
Ehrlich, J., 485 (*58*)
Eichstedt, R., 243 (*71*)
Eisenbraun, E. J., 320 (*37*)
Ekborg, G., 318 (*7*)
Eliel, E. L., 25 (*1*), 55 (*46, 48, 52, 56*), 89 (*22*), 90 (*31, 38, 42, 44, 50*), 91 (*56, 71*), 140 (*19*), 188 (*12*), 320 (*39*), 420 (*299*), 515 (*63*)

Ellencweig, A., 421 (*314*)
Emmons, W. D., 466 (*81, 82*)
Emoto, S., 286 (*67*), 465 (*53*)
Endo, Y., 53 (*7*)
Engle, R. R., 284 (*21*)
Enholm, E. J., 466 (*73*)
Epstein, W. W., 286 (*59*)
Erdmann, E. D., 110 (*3*)
Erickson, J. G., 242 (*59*)
Ermolenko, M. S., 463 (*23, 24*), 464 (*25, 26, 27, 28*)
Ernst, B., 217 (*82*)
Ertasogon, M., 464 (*35*)
Erwig, E., 25 (*3, 4*)
Eschenmoser, A., 287 (*88*), 464 (*42, 43*)
Esnault, J., 415 (*188*)
Esposito, E., 513 (*20*)
Esswein, A., 412 (*126*)
Evans, D. A., 466 (*86*)
Evans, D. L., 482 (*7*)
Evans, M. E., 165 (*47, 48*)
Evans, W. L., 407 (*24*)
Ewig, C. S., 55 (*69*)

F

Fabian, M. A., 92 (*80*)
Fabian, W., 166 (*63*)
Fanton, E., 216 (*50*)
Fantoni, R., 53 (*8*)
Farazdel, A., 56 (*74*)
Fardig, O. B., 482 (*6*)
Farkas, I., 411 (*110*)
Farral, M. J., 416 (*222*)
Fasman, G. D., 244 (*110*)
Faulkner, I. J., 110 (*8, 19*)
Favolle, M., 25 (*6*)
Feast, A. A. J., 463 (*14*)
Feather, M. S., 407 (*19*)
Felix, D., 464 (*42, 43*)
Felkin, H., 465 (*66*)
Felsenstein, A., 286 (*53*)
Fenselau, A. H., 285 (*51*)
Fernández-Lizarbe, J. R., 467 (*94*)
Ferrari Belicchi, M., 515 (*69*), 516 (*71*)
Ferrier, R. J., 110 (*25*), 318 (*14*), 407 (*10*), 414 (*166*), 415 (*199, 200*)
Fields, D. L., 408 (*43*)
Fife, T. H., 164 (*12*), 418 (*264, 265*)
Finan, P. A., 141 (*44*)
Finch, P., 91 (*66*)
Fink, H. L., 54 (*24, 25*)
Finlay, A. C., 485 (*59*)
Finney, N. S., 513 (*16*)

Firth, M. E., 516 (*83*)
Fischer, E., 25 (*2, 5*), 141 (*35*), 241 (*25*), 406 (*4, 5*), 408 (*41, 42*), 411 (*112*), 416 (*204*)
Fischer, H. O. L., 165 (*44*)
Fitzpatrick, J. M., 515 (*53*)
Flaherty, B., 287 (*83*), 320 (*52*), 463 (*15*)
Fleet, G. W. J., 465 (*52*)
Fleming, B. I., 167 (*81, 82*)
Flemming, I., 412 (*121*)
Flesch, H., 217 (*71*)
Fletcher, H. G., 164 (*23, 24*), 166 (*65*), 286 (*62, 66, 68*), 289 (*136*), 290 (*162, 165*), 407 (*12, 15, 16*), 408 (*44, 49, 50*)
Fleury, P. F., 288 (*118, 119, 120*), 290 (*163*)
Fliri, A., 412 (*134*)
Florent, J., 486 (*84*)
Foley, J. W., 464 (*37*)
Folkers, K., 485 (*72*)
Ford, R. A., 91 (*72*)
Foster, A. B., 164 (*17, 18, 19*), 165 (*58*), 218 (*105*), 240 (*2, 3*), 289 (*139*), 417 (*241*)
Fox, J. J., 241 (*37*), 244 (*111*), 287 (*93*)
Frajerman, C., 465 (*66*)
Franchimont, A. P. N., 242 (*47*)
Frank, F. J., 287 (*78*)
Frank, R. W., 419 (*279*)
Fraser-Reid, B., 56 (*73*), 91 (*67*), 215 (*29, 40*), 217 (*81*), 412 (*136, 137, 138, 139*), 413 (*140, 141, 142, 143, 144, 157*), 419 (*274*), 420 (*300*), 462 (*2*), 515 (*56, 57, 58, 59*)
Fráter, G., 466 (*76, 77*)
Fréchet, J. M. J., 416 (*222*)
Fredenhagen, H., 110 (*13*)
Freer, A. A., 465 (*49*)
French, A. D., 56 (*72*)
French, J. C., 485 (*62, 63, 64, 65*)
Freudenberg, K., 165 (*39*), 167 (*90*), 216 (*57, 58, 67*), 241 (*31, 33*)
Friebolin, H., 319 (*15*)
Friedmann, M., 91 (*63*), 420 (*298*)
Friesen, R. W., 413 (*158, 159*)
Fritz, B., 513 (*21*)
Frobel, K., 412 (*134*)
Frohardt, R. P., 485 (*58*)
Frontera, A., 56 (*78*)
Frush, H. L., 91 (*74*), 242 (*49, 50, 51*), 417 (*229*)
Frye, S. V., 515 (*57*)
Fuchs, B., 92 (*90*), 421 (*314*)
Fügedi, P., 166 (*71, 72, 73, 75, 77, 79*), 414 (*177*), 415 (*187*)
Fuji, T., 243 (*83*)
Fujimaki, S., 513 (*25, 27*)

Author Index

Fujisawa, K., 484 (*37*)
Fundiler, I. N., 411 (*108*)
Furneaux, R. H., 215 (*39*), 415 (*200*)
Fürst, A., 190 (*46*), 240 (*11*)
Furuhata, K., 415 (*185*)
Füstner, A., 321 (*65, 67*)

G

Gaffield, W., 464 (*37*)
Gais, H.-J., 412 (*134*)
Galkowski, T. T., 289 (*132*)
Galoyan, A. A., 409 (*64*)
Gambacorta, A., 513 (*19, 20*)
Ganguly, A. K., 486 (*94, 95, 96, 97*)
Ganguly, B., 81 (*77*)
Gao, J., 92 (*90*)
Gardner, D., 465 (*49*)
Garegg, P. J., 141 (*27*), 142 (*54*), 166 (*67*), 167 (*84*), 287 (*86*), 414 (*171, 177, 179*), 415 (*187, 189, 192, 196*)
Garner, P. P., 513 (*7*)
Garratt, D. G., 412 (*134*)
Garrett, E. R., 243 (*92*)
Gasparri, F., 515 (*69, 71*)
Gates, K. S., 243 (*85*)
Geerlings, P., 244 (*112*)
Gelas, J., 216 (*50*)
Geller, L. E., 320 (*37, 38*)
Gemal, A. L., 516 (*77*), 516 (*88*)
Gennari, C., 516 (*89*)
George, R. W., 414 (*164*)
Georges, M., 215 (*40*)
Georgiadis, M. P., 484 (*52*)
Georgoulis, C., 288 (*101*)
Gerecs, A., 214 (*15*), 216 (*68*), 217 (*71, 73*)
Germain, E. A., 419 (*290*), 420 (*291*)
Gero, S. D., 319 (*25*)
Ghini, A. A., 320 (*54*)
Ghione, M., 485 (*61*)
Giesbrecht, A. M., 484 (*50*)
Gigg, R., 464 (*33*)
Gilham, P. T., 218 (*94, 115*), 244 (*106*)
Giza, C. A., 90 (*38*)
Glaudemans, C. P. J., 408 (*44*)
Glaze, W. H., 465 (*67*)
Glen, W. L., 165 (*38*)
Glennon, T. M., 56 (*71*)
Gligorijevi, M., 189 (*24, 39*), 318 (*1*), 319 (*29, 30*), 413 (*156*), 462 (*4, 5*), 463 (*19*)
Glinski, R. P., 188 (*7*), 241 (*22, 23, 39*)
Glišin, Dj., 189 (*24, 39*), 190 (*41*), 318 (*8*), 413 (*156*), 462 (*5, 6*), 463 (*7, 8*)
Goddard, W. A., 53 (*4*)

Goewey, G. S., 167 (*85*)
Goldman, L., 285 (*50*), 286 (*55*)
Gollman, H., 465 (*64*)
Gonzalez, A., 288 (*98, 100*)
Goodman, L., 166 (*66*), 189 (*29*), 190 (*47*), 241 (*36*), 242 (*41*), 286 (*65*)
Goody, R. S., 241 (*37*)
Gordy, W., 53 (*3*)
Goto, T., 514 (*30*)
Gottschalk, A., 240 (*1*), 513 (*11*), 516 (*82, 83*)
Gough, M. J., 465 (*52*)
Goutarel, R., 285 (*43*)
Graczyk, P., 92 (*82*)
Grahe, G., 243 (*88*)
Grand-Maître, C., 416 (*210*)
Grant, D. M., 319 (*26*), 463 (*9*)
Grant, G. A., 165 (*38*)
Gray, G. R., 190 (*44*)
Greatbanks, D., 465 (*49*)
Green, D. F., 92 (*92*)
Green, J. W., 284 (*14*), 408 (*36*)
Green, R. D., 111 (*44*)
Greeves, D., 486 (*97*)
Grein, F., 91 (*64, 65*)
Greshnykh, R. D., 242 (*57*)
Griesser, R., 243 (*87*)
Griffith, C. F., 140 (*8*)
Grisebach, H., 318 (*11*), 319 (*15*)
Groen, S. H., 286 (*58*)
Grüner, H., 164 (*29*)
Grünewald, H., 416 (*206*)
Gschwend-Steen, K., 464 (*43*)
Guillerm-dron, D., 319 (*32*), 463 (*18*)
Guindon, Y., 463 (*22*)
Guler, L. P., 56 (*80*)
Gunner, S. W., 287 (*81*)
Gupta, R. B., 419 (*275*)
Gupta, S. K., 241 (*22, 23*)
Gut, V., 215 (*46*)
Guthrie, J. P., 92 (*93*)
Guthrie, R. D., 140 (*12*), 141 (*29*), 190 (*45*), 215 (*31*), 321 (*72*), 420 (*294*), 484 (*44*)
Gutsche, C. D., 320 (*46*)
Guttman, S., 288 (*106*)
Györgdeák, Z., 91 (*63*)
Gyr, M., 218 (*97*)

H

Ha, S., 55 (*68*)
Haapakka, K., 420 (*305*)
Habash-Marino, M., 420 (*311*), 514 (*47, 48*)
Haga, M., 214 (*21*), 416 (*214*)
Hagel, P., 165 (*53*), 190 (*48*)

Hageman, H. J., 90 (*30*)
Hagemann, G., 242 (*55, 56*)
Hagopian, L., 418 (*265*)
Haines, A. H., 140 (*2*), 141 (*32*), 165 (*58*)
Haines, S. R., 410 (*91*)
Halcomb, R. L., 414 (*161*)
Hall, A. N., 418 (*249*)
Hall, H. K., 216 (*59*)
Hall, L. D., 54 (*33*), 188 (*2*)
Hall, M. W., 25 (*12*)
Hällgren, C., 415 (*189*)
Halsall, T. G., 288 (*117*)
Hamada, M., 483 (*33*)
Hamana, K., 483 (*18*)
Hamil, W. H., 110 (*11*)
Han, R.-J. L., 218 (*109, 110*)
Hanajima, M., 318 (*5*)
Handa, S., 515 (*57, 58*)
Hanessian, S., 167 (*83*), 188 (*8*), 189 (*28*), 240 (*8*), 287 (*77*), 320 (*40*), 321 (*70, 71*), 408 (*33, 34*), 412 (*135*), 463 (*20, 21, 22*), 464 (*40, 41*)
Hann, R. M., 165 (*59*), 217 (*91*)
Hanna, R., 321 (*58*), 514 (*37, 38, 41*)
Haq, S., 216 (*69*)
Hara, T., 485 (*76*)
Haradahira, T., 216 (*55*)
Harangi, J., 166 (*72*)
Haraszthy-Papp, M., 216 (*66*)
Harayama, Y., 484 (*55*)
Hardegger, E., 217 (*88*)
Harris, J. F., 407 (*19*)
Harris, T. M., 465 (*47*)
Hartman, F. C., 190 (*44*)
Hasegawa, A., 165 (*49*), 415 (*190, 191*)
Hashimoto, H., 189 (*26*), 320 (*41*), 485 (*81*)
Hashimoto, S., 409 (*58*), 410 (*84*)
Hashimoto, Y., 409 (*57*)
Haskell, T. H., 240 (*8*), 485 (*58, 62, 63, 64, 65*)
Hassid, W. Z., 165 (*35*)
Hata, T., 464 (*36*)
Hatton, L. R., 110 (*25*), 407 (*10*)
Havinga, E., 90 (*41, 42, 43, 44*)
Hawley, D. M., 466 (*89*)
Haworth, W. N., 25 (*11*), 141 (*42, 45*), 165 (*43*), 217 (*87*), 240 (*16*), 417 (*227, 232*)
Hay, R. W., 414 (*166*)
Hayakawa, K., 412 (*134*)
Hayami, J.-I., 408 (*52*)
Hayami, J.-Y., 89 (*11*)
Hayashi, M., 409 (*58*), 410 (*84*)
Haynes, L. J., 407 (*21*)

Head, F. S. H., 289 (*127, 133*)
Heard, D. D., 407 (*9*)
Hecht, S. M., 215 (*26*)
Hedgley, E. J., 289 (*139*)
Heggie, W., 412 (*134*)
Heineman, R., 284 (*7*)
Heischkeil, R., 142 (*67*)
Heitman, J. A., 411 (*109*)
Helferich, B., 141 (*46, 47*), 142 (*52*), 164 (*32*), 241 (*27*), 242 (*43*), 407 (*25*), 408 (*35, 37*), 411 (*115*), 416 (*206*), 465 (*61*)
Helle, M. A., 466 (*85*)
Hendrickson, J. B., 54 (*41*)
Hendriks, K. B., 91 (*61*)
Henrichson, C., 414 (*171*)
Herscovici, J., 287 (*96*), 288 (*97*)
Hervé de Penhoat, C., 415 (*188*)
Hess, B. A., 166 (*64*)
Hess, K., 217 (*79*), 241 (*38*)
Hess, W. W., 287 (*78*)
Hesson, D. P., 412 (*134*)
Heumann, K. E., 217 (*79*)
Hevesi, L., 244 (*95*)
Heyns, K., 215 (*44*), 243 (*71, 72, 73*), 284 (*7, 8*)
Hibbert, H., 165 (*54*)
Hichens, M., 484 (*51*)
Hickinbottom, W. J., 217 (*86, 87*), 407 (*20*)
Hilbert, G. E., 216 (*65*)
Hill, J., 188 (*5, 6*)
Hineno, H., 483 (*14, 15*)
Hirano, S., 142 (*53*), 286 (*63, 69*)
Hirota, E., 53 (*7*)
Hirst, E. L., 141 (*42, 45*), 165 (*43*), 288 (*117*), 417 (*232*)
Hitchens, M., 482 (*11*), 484 (*52, 53*)
Hixon, R. M., 165 (*39*), 242 (*52*)
Hocket, R. C., 142 (*50, 51*), 164 (*34*), 290 (*162, 165*)
Hodge, J. E., 242 (*48, 53, 54, 66*)
Hodgson, K. O., 141 (*31*), 320 (*44*)
Hoffhine, C. E., 485 (*72*)
Hoffman, H., 466 (*79, 80*)
Hofheinz, W., 319 (*15*)
Höger, E., 290 (*160*)
Hohenschutz, L. D., 465 (*48*)
Holla, E. W., 216 (*52*)
Hollenberg, D. H., 287 (*93*)
Hollingshead, S., 418 (*249*)
Holly, F. W., 285 (*28, 29*)
Holly, S., 214 (*24*)
Holmbom, B., 142 (*57, 58*)

Holness, N. J., 55 (*55*), 89 (*21*)
Holst, K. A., 418 (*257*)
Holum, J. R., 287 (*76*)
Honeyman, J., 165 (*51*), 218 (*95*), 289 (*123, 124*)
Hong, N., 320 (*42*)
Hooper, I. R., 482 (*3, 5, 6, 7*)
Hoppe, D., 412 (*134*)
Hoppe, I., 412 (*134*)
Horito, S., 485 (*81*)
Horiuchi, Y., 485 (*70*)
Horner, L., 286 (*56*), 466 (*79, 80*)
Horton, D., 53 (*13*), 91 (*57*), 188 (*15*), 286 (*54*), 287 (*80*), 319 (*25*), 320 (*51*), 415 (*193*), 416 (*203, 211*), 417 (*241*)
Hosono, S., 410 (*79, 80*)
Hossenlopp, I. A., 54 (*27*)
Hotta, K., 415 (*191*)
Hough, L., 111 (*42*), 163 (*6*), 188 (*2, 5, 6, 18*), 189 (*22*), 190 (*45*), 217 (*77*), 289 (*126, 134, 138*), 290 (*152*), 407 (*8*), 465 (*50*)
House, H. O., 318 (*2*)
How, M. J., 188 (*17*)
Howarth, G. B., 320 (*45*)
Hua, D. H., 466 (*90*), 467 (*91, 92*)
Huang, X., 419 (*284*)
Huber, G., 217 (*84, 85*), 240 (*15*), 290 (*160*), 407 (*22*)
Huber, W., 464 (*44*)
Huckstep, L. L., 486 (*83*)
Hudson, C. S., 109 (*2*), 110 (*5*), 142 (*50, 51*), 165 (*59*), 215 (*38, 41*), 217 (*91, 92*), 284 (*11, 12*), 289 (*122*), 318 (*9*), 407 (*15, 16*), 408 (*49*)
Huggard, A. J., 141 (*40*)
Hughes, E. D., 189 (*31*)
Hughes, G., 289 (*133*)
Hughes, N. A., 188 (*9*), 190 (*43*)
Hultberg, H., 166 (*67*)
Hummel, C. W., 409 (*71*)
Hunedy, F., 216 (*53*)
Hungate, R., 516 (*78*)
Hunsberger, I. M., 285 (*42*)
Hunter, F. D., 53 (*18*)
Hurd, C. D., 25 (*13*)
Hürzeler-Jucker, E., 289 (*121*)
Husain, A., 285 (*52*)
Hutchinson, C. R., 465 (*64*)
Hutson, D. H., 415 (*193*)
Huxtable, C. R., 465 (*45*)
Huynh, C., 465 (*65*)
Hyatt, J. A., 412 (*134*)

I

Idel, K., 141 (*28*)
Iinuma, K., 485 (*78, 79, 80*)
Ikeda, D., 412 (*134*), 485 (*69*)
Imre, J., 166 (*76*)
Inatome, M., 217 (*72*)
Inch, T. D., 319 (*31*), 321 (*68, 69*), 462 (*1*), 463 (*12, 16, 17*)
Ingold, C. K., 164 (*10*), 189 (*31*), 318 (*13*), 417 (*233*)
Ingraham, L. L., 189 (*34*)
Inokawa, S., 321 (*74*)
Isaacs, N. S., 218 (*99*), 240 (*14*)
Isbell, H. S., 89 (*24*), 91 (*74*), 111 (*28, 29*), 242 (*49, 50, 51*), 284 (*11, 12, 15*), 417 (*229*)
Ishida, H., 415 (*191*)
Ishido, Y., 189 (*25*)
Ishij, T., 419 (*277*)
Ishizu, A., 419 (*277*)
Isono, K., 513 (*15, 17*)
Ito, T., 482 (*8*), 483 (*12*)
Ito, Y., 414 (*181*)
Ivanetich, K. M., 164 (*13*)
Iwabuchi, Y., 409 (*71*)

J

Jacobi, P. A., 412 (*134*)
Jaggi, D., 515 (*64, 66, 67*)
James, L. F., 465 (*46*)
James, M. N. G., 91 (*59*)
James, S. P., 217 (*90*)
Jánossy, L., 166 (*72, 76*)
Janson, J., 214 (*12*)
Jao, L. K., 164 (*12*), 418 (*264*)
Jardetzky, C. D., 54 (*30, 31, 32*)
Jaret, R. S., 484 (*42*)
Jarosz, S., 515 (*56*)
Jarreau, F. X., 285 (*43*)
Jarý, J., 188 (*1*), 219 (*119*), 241 (*20, 24*), 241 (*40*)
Jasinski, T., 243 (*77, 78*)
Jeanloz, D. A., 140 (*10*), 188 (*10*), 241 (*19*)
Jeans, A., 216 (*65*)
Jeffery, E. A., 419 (*289*), 466 (*70*)
Jefford, C. W., 515 (*64, 65, 66, 67*)
Jeffrey, G. A., 53 (*14, 16, 17*)
Jencks, W. P., 243 (*81*), 244 (*99*), 419 (*281*)
Jennings, H. J., 414 (*178*), 415 (*197*)
Jensen, W. E., 242 (*46*)
Jewess, P. J., 465 (*48*)
Jeyaraman, R., 413 (*160*)
Jin, H., 414 (*172*)
Jodál, I., 166 (*69, 70, 72*)

Johnson, A. P., 285 (46)
Johnson, C. R., 285 (38, 39, 40, 41)
Johnson, D. L., 482 (5, 6)
Johnston, B. D., 92 (92), 420 (296)
Johnston, J. C., 167 (85)
Jokić, A., 189 (36, 38)
Jonás, J., 91 (70)
Jones, A. S., 242 (44, 45, 46), 287 (85)
Jones, D. N., 285 (44)
Jones, G. H., 466 (78)
Jones, J. K. N., 140 (22), 285 (30), 287 (91), 288 (117), 320 (45)
Jones, P. G., 92 (86)
Jones, S., 486 (92)
Jorgensen, W. L., 56 (78)
Josephson, K., 141 (48), 216 (64)
Jozefowicz, M. L., 91 (69)
Juaristi, E., 89 (5), 92 (81, 84)
Jungius, C. L., 110 (26)
Jurczak, J., 515 (62)

K

Kabayama, M. A., 140 (24)
Kabir, A. K. M. S., 514 (34, 35, 36, 37, 38, 39, 40, 41)
Kachensky, D. F., 466 (73)
Kahne, D., 413 (147), 416 (219, 220)
Kai, Y., 216 (55)
Kaiser, P., 286 (56)
Kalvoda, L., 215 (35)
Kampf, A., 286 (53)
Kampf, G., 243 (87)
Kanemitsu, K., 318 (5)
Kanie, O., 214 (25)
Kankaanperä, A., 164 (11), 418 (263), 420 (305)
Kapinos, L. E., 243 (87)
Kaplan, L., 244 (111)
Karl, H., 413 (150, 151)
Karlstrom, O., 243 (86)
Karplus, M., 53 (10), 55 (68)
Karrer, P., 214 (13), 289 (125)
Kashimura, N., 286 (63, 69)
Katano, K., 215 (26)
Kataoka, H., 409 (70)
Katsuki, M., 409 (68)
Katsuki, T., 512 (28), 514 (44), 516 (90)
Kawabe, H., 483 (22)
Kawaguchi, H., 484 (37)
Kawakami, H., 215 (34, 37)
Kawamura, S., 286 (57)
Kawano, T., 465 (54)
Kawasaki, T., 142 (68)

Kawashiro, I., 289 (140, 141, 142, 143, 144, 145, 146), 290 (147)
Keck, G. E., 466 (73, 74), 466 (84)
Keller-Schierlein, W., 319 (17, 18, 20)
Kelly, A. G., 464 (30)
Kelly, R. C., 464 (42)
Kennedy, J., 92 (87)
Kent, L. H., 240 (10)
Kent, P. W., 188 (3), 218 (113)
Kenttämaa, H. I., 56 (80)
Kenyon, W. O., 419 (288)
Kern, C. W., 53 (10)
Kerwin, J. F., 516 (73)
Khorana, H. G., 244 (105, 106)
Khorlin, A. Ya., 410 (94, 97, 98), 411 (99, 105, 116)
Kibayashi, C., 513 (25)
Kilpatrick, J. E., 53 (21)
Kilpatrick, M., 418 (258)
Kim, H. S., 53 (14, 17)
Kim, K. S., 412 (134)
Kim, S., 141 (33)
Kim, W. J., 141 (33)
Kim, W.-S., 410 (79, 80, 81)
King, T. A., 242 (58)
King, T. J., 465 (57)
Kinsman, R. G., 219 (120)
Kirby, A. J., 88 (2), 93 (98), 418 (269), 419 (270, 273)
Kirsanov, A. V., 410 (86)
Kishi, Y., 464 (31), 514 (43)
Kiso, M., 165 (49), 415 (190, 191)
Kitagawa, S., 142 (53)
Kitahara, K., 188 (13)
Kitahara, T., 466 (88)
Kitao, T., 286 (57)
Kitaoka, Y., 286 (57)
Kitos, T. E., 244 (96)
Kivelevich, D., 166 (63)
Kiyomoto, A., 319 (21)
Klaffke, W., 216 (52), 413 (154)
Klavins, J. E., 55 (51)
Klein, R. S., 287 (93)
Klemer, A., 111 (47), 214 (11, 16, 18, 19)
Klimov, E. M., 411 (102, 103), 414 (180)
Kloosterman, M., 215 (27)
Klyagina, A. P., 218 (107)
Klyne, W., 53 (12)
Knell, M., 288 (109)
Knippers, W., 53 (8)
Knobler, C., 90 (90)
Knoeber, M. C., 55 (48)
Knolle, J., 464 (39)

Knorr, E., 407 (*14*)
Knudsen, M. P., 485 (*58*)
Kobayashi, S., 409 (*73*), 516 (*73*)
Kobuke, Y., 412 (*134*)
Koch, K. F., 484 (*46*), 486 (*83*)
Koch, von, F. K., 216 (*63*)
Koch, W., 243 (*73*)
Kochetkov, N. K., 166 (*60*), 218 (*107*), 321 (*73*), 410 (*93, 94, 95, 96, 97, 98*), 411 (*99, 102, 103, 105, 106, 107, 116*), 414 (*180*), 463 (*23, 24*), 464 (*25, 26, 27, 28*)
Kocourek, J., 215 (*32*)
Kodama, H., 320 (*42*)
Koelliker, U., 464 (*44*)
Koenigs, W., 25 (*3, 4*)
Koide, K., 409 (*73*)
Koike, M., 485 (*76, 77*)
Kojima, K., 412 (*134*)
Kojima, M., 216 (*55*)
Kolář, Č., 215 (*47*)
Kolarikol, A., 285 (*31*)
Kolp, G., 410 (*85*)
Komarov, I. V., 92 (*86*)
Komiotis, D., 514 (*32*)
Konda, Y., 319 (*34*)
Kondo, S., 482 (*4, 8*), 483 (*18, 19, 23, 25, 26, 29, 33*), 485 (*76, 77, 78, 79, 80*)
Kondo, T., 514 (*30*)
Kondo, Y., 140 (*13, 23*), 142 (*53, 56*)
Königs, W., 407 (*14*)
Konoval, A., 484 (*45*)
Konradsson, P., 412 (*137, 138*), 413 (*144*)
Kopecky, K. R., 319 (*33*)
Kops, J., 217 (*75*)
Korytnyk, W., 410 (*88*), 411 (*113*)
Kosche, W., 242 (*43*)
Koseki, K., 215 (*34, 37*)
Koto, S., 90 (*49*), 91 (*60*), 216 (*60, 61*), 483 (*14, 15, 16, 17*)
Kovář, J., 241 (*20, 24*)
Koyama, G., 482 (*10*)
Koyama, M., 514 (*31*)
Kozikowski, A. P., 466 (*71*)
Kraft, L., 290 (*159*)
Kraus, A., 242 (*70*)
Kremann, R., 408 (*47*)
Krieble, V. K., 418 (*256, 257*)
Kronzer, F. J., 407 (*30*)
Krowicki, K., 412 (*134*)
Kruck, P., 290 (*160*)
Krüger, W., 407 (*27*), 411 (*101*)
Kubo, K., 320 (*41, 42*)
Kudryashov, L. I., 166 (*60*), 218 (*107*)

Kuehl, F. A., 485 (*72*)
Kuhn, R., 242 (*67, 68*), 516 (*85*)
Kuhn, W., 216 (*57*)
Kulkarni, V. R., 415 (*182*)
Kulonpaa, A., 420 (*306*)
Kunz, H., 409 (*62, 63*)
Kurzynski, M., 419 (*283*)
Kuzuhara, H., 216 (*54*), 286 (*62, 66, 67*)

L

La Mer, V. K., 110 (*11*)
Lacombe, J. M., 416 (*218*)
Lafont, D., 215 (*30*)
Laidig, K. E., 53 (*11*)
Lake, W. H. G., 240 (*16*)
Lameignére, E., 320 (*54*)
Lampen, J. O., 289 (*137*)
Langenhoff, F., 416 (*206*)
Langlykke, A. F., 482 (*1*)
Lartey, P. A., 410 (*90*)
Lavallee, P., 464 (*40, 41*)
Lawton, B. T., 285 (*30*)
Le Grand, S. M., 55 (*70*)
Leake, C. D., 285 (*36*)
Lechevalier, H. A., 484 (*49*)
Lee, A. M. W., 514 (*44*), 487 (*40*)
Lee, C.-H., 241 (*22*)
Lee, E. E., 141 (*26*)
Lee, J. B., 289 (*130*)
Lee, J. R., 420 (*300*)
Lee, V. J., 412 (*134*)
Lee, W. W., 286 (*65*)
Leggetter, B. E., 167 (*80*)
LeGoffic, F., 483 (*24*)
Lehong, N., 412 (*135*)
Leiniger, P. M., 418 (*258*)
Lemal, D. M., 319 (*19*)
Lemieux, R. U., 54 (*29, 34, 36, 40*), 55 (*54, 63*), 89 (*10, 11, 17, 20*), 90 (*49*), 91 (*54, 58, 60, 61*), 92 (*94*), 110 (*22*), 140 (*25*), 163 (*7*), 189 (*35*), 214 (*9*), 217 (*84, 85*), 241 (*32*), 407 (*22, 23*), 408 (*52*), 411 (*111*), 413 (*148, 149, 157*), 416 (*201, 202*), 420 (*301, 307, 309, 310*), 482 (*6, 7*)
Leroi, G. E., 53 (*6*)
Lester, D. J., 290 (*166*)
Leung, Y. N., 91 (*75*)
Leutert, T., 412 (*134*)
Levene, P. A., 164 (*31*)
Levine, S., 413 (*148*)
Levy, L. W., 290 (*153*)
Lewis, G. J., 321 (*68, 69*), 463 (*16*)
Lewis, T. A., 417 (*234*)

Lewis, W. L., 111 (*43, 44*)
Leworthy, D. P., 465 (*48*)
Li, S., 418 (*267*), 419 (*273, 276*)
Lieser, Th., 140 (*17, 18*)
Lii, J.-H., 56 (*79*)
Lim, J. J., 413 (*147*)
Lim, L. B. L., 410 (*92*)
Lin, J. J., 464 (*32*)
Lindahl, T., 243 (*84, 86*)
Lindberg, B., 89 (*13, 14, 15*), 141 (*43*), 214 (*8, 12*), 318 (*7*), 420 (*303, 304*)
Lindegren, C. R., 189 (*34*)
Link, K. P., 165 (*42*)
Linstrumelle, G., 465 (*65*), 466 (*68*)
Lipkin, D., 164 (*18*)
Lippert, B., 243 (*87*)
Lipták, A., 166 (*61, 68, 69, 70, 71, 72, 73, 74, 75, 76, 77, 79*)
Liras, J. L., 419 (*278*)
Litaka, Y., 482 (*10*)
Little, M., 415 (*195*)
Liu, D., 287 (*95*)
Liu, H. M., 318 (*6*)
Llewellyn, D. R., 417 (*234*)
Lobry de Bruyn, C. A., 111 (*30, 31, 32, 33, 34, 35, 36, 37, 38*), 242 (*47*)
Locher, R., 464 (*35*)
Logusch, E., 412 (*132, 134*)
Long, F. A., 110 (*17, 18*), 417 (*235*)
Long, L., 140 (*6, 7*), 165 (*48*)
Lönn, H., 414 (*173, 174, 175, 176*)
Lonnberg, H., 420 (*305, 306*)
Lönngren, J., 318 (*7*)
Lopez, C., 288 (*100*)
López, J. C., 320 (*54*)
Loring, H. S., 290 (*153*)
Lorne, R., 466 (*68*)
Lorthiois, E., 419 (*287*)
Los, J. M., 25 (*15*)
Loverix, S., 244 (*112*)
Lowary, T. L., 217 (*81*)
Lowe, J. P., 53 (*1, 2*)
Lowry, T. M., 110 (*6, 7, 8, 9, 19*)
Lubineau, A., 413 (*145*)
Lucas, T. J., 408 (*32*)
Luche, J. L., 516 (*77, 88*)
Luedemann, G. M., 484 (*38*)
Luederitz, O., 140 (*11*)
Luk, K., 465 (*57*), 466 (*72*)
Lukach, C. A., 140 (*19*)
Lukacs, C. A., 55 (*56*)
Lukacs, G., 217 (*78*)
Lukowski, H., 111 (*47*)

Lundt, I., 413 (*153*)
Lunel, J., 486 (*84*)
Lutskii, A. E., 142 (*70*)
Lyssikatou, M., 219 (*120*)

M

McCann, 244 (*102*)
McCarthy, J. F., 420 (*294*)
McCarty, K. S., 25 (*8*)
McCloskey, C. M., 214 (*6*)
McCloskey, J. A., 513 (*14*)
McCormick, J. E., 407 (*13*)
McCullough, J. P., 54 (*26, 27*)
McDevitt, R. E., 412 (*137*)
Macdonald, C. G., 141 (*39*)
MacDonald, D. L., 165 (*36*)
McDonald, E. J., 286 (*64*)
McGinn, C. J., 289 (*128*)
McGonigal, W. E., 485 (*74*)
McHugh, D. J., 55 (*49, 50*)
McInnes, A. G., 140 (*25*)
McIntyre, D., 417 (*235*)
McLauchlan, K. A., 54 (*33*)
McLeod, J. M., 215 (*48*)
McNaughton, M., 244 (*112*)
McPhail, A. T., 515 (*58*)
McPhail, D. R., 420 (*300*)
Machida, Y., 465 (*55*)
Mackie, D. W., 55 (*62*), 286 (*72*)
Maclay, W. D., 165 (*59*)
Madaj, J., 214 (*22*)
Maddox, M. L., 466 (*78*)
Madsen, R., 412 (*139*), 413 (*140, 141*)
Maeda, K., 482 (*4, 10*), 483 (*18, 19*)
Maeda, M., 216 (*55*)
Maehr, H., 484 (*39, 40*)
Maeta, H., 409 (*65, 66, 67, 69, 72*)
Magerlein, B. J., 513 (*14*)
Magrath, D. I., 244 (*110*)
Mahon, M. F., 219 (*120*)
Malaprade, L., 288 (*102, 103*)
Malchenko, S., 412 (*134*)
Malleron, A., 287 (*94*)
Mallet, J.-M., 415 (*186, 188*)
Malysheva, N. N., 414 (*180*)
Manatt, S. L., 166 (*62*)
Mancy, D., 486 (*84*)
Mann, R. L., 485 (*82*)
Manson, L. A., 289 (*137*)
Marigliano, H. M., 484 (*44*)
Maring, C. J., 516 (*74, 80, 81*)
Markovski, L. N., 410 (*86*)
Marktscheffel, F., 290 (*160*)

Marra, A., 415 (*186*)
Marshall, H., 189 (*34*)
Marshall, R. D., 417 (*237*)
Martens, J., 412 (*134*)
Martin, F. S., 284 (*20*)
Martin, V. S., 514 (*44*), 516 (*90*)
Maruşič, R., 466 (*89*)
Masse, R., 321 (*71*)
Masuda, F., 286 (*69*)
Mather, A. M., 216 (*53*), 320 (*55, 56*), 321 (*58*)
Mathews, A. P., 284 (*13*)
Mathews, R. S., 412 (*134*)
Matsuhira, N., 318 (*5*)
Matsui, M., 142 (*64*), 415 (*198*), 465 (*54*)
Matsumoto, K., 215 (*34, 37*)
Matsumoto, T., 409 (*65, 66, 67, 68, 69, 72*)
Matsushita, H., 215 (*34, 37*)
Mattlock, G. L., 407 (*18*)
Mazurek, M., 411 (*114*)
Mehltretter, C. L., 284 (*9*)
Meinecke, K.-H., 243 (*72*)
Meinwald, Y., 418 (*250*)
Meisters, A., 466 (*70*)
Melander, B., 148 (*58, 61*)
Mellows, G., 465 (*58*)
Melrose, G. J. H., 141 (*37, 38*)
Melvin, L. S., 514 (*50*), 515 (*51*)
Mentch, F., 244 (*100*)
Mereyalla, H. B., 415 (*182*)
Merrer, Y. L., 321 (*57*)
Merritt, J. R., 413 (*140, 143*)
Merz, K. M., 55 (*70*), 56 (*71*)
Meyer zu Reckendorf, W., 241 (*29, 30*)
Meyer, B., 56 (*77*)
Meyyappan, M., 419 (*287*)
Michael, A., 406 (*1, 2, 3*)
Micheel, F., 214 (*11, 14, 16, 17, 18, 19*), 242 (*55, 56*)
Michel, J., 412 (*122, 123, 125, 128*)
Michelson, A. M., 244 (*103, 104, 107*)
Michon, V., 415 (*188*)
Middleton, W. J., 410 (*87*)
Miethchen, R., 410 (*85*)
Mikloš, D., 466 (*89*)
Mikolajczyk, M., 92 (*82, 83*)
Miljković, D., 189 (*36*), 318 (*1*), 319 (*30*), 462 (*4*), 463 (*19*)
Miljković, M., 165 (*53*), 189 (*24, 36, 37, 38, 39*), 190 (*41, 48*), 318 (*1, 8*), 319 (*29, 30*), 413 (*156*), 420 (*311*), 421 (*312*), 462 (*4, 5, 6*), 463 (*7, 8, 19*), 514 (*47*)

Millar, A., 215 (*26*)
Miller, K. J., 241 (*18*)
Miller, R., 413 (*147*)
Mills, J. A., 218 (*101*), 411 (*113*)
Minkin, V. I., 110 (*20*), 142 (*70*)
Minshall, J., 217 (*76*)
Minster, D. K., 215 (*26*)
Mitchell, H. J., 486 (*98, 99*)
Mitchell, M. J., 166 (*63*)
Mitra, A. K., 140 (*7*)
Mitsuhashi, S., 483 (*18, 22*)
Mitts, E., 242 (*52*)
Miyake, T., 484 (*47*)
Miyazaki, N., 286 (*69*)
Moelwyn-Hughes, E. A., 417 (*239*), 418 (*246, 247, 259*)
Moffatt, J. G., 285 (*47, 48, 49, 51*), 319 (*24*), 466 (*78*)
Moldovány, L., 287 (*88*)
Mole, T., 466 (*70*)
Molloy, B. B., 486 (*83*)
Molyneux, R. J., 465 (*46*)
Monagle, J. J., 285 (*45*)
Montgomery, E. M., 215 (*41*)
Mootoo, D. R., 412 (*137*), 413 (*144*), 419 (*274*)
Morazain, J. G., 165 (*54*)
Mörch, L., 321 (*64*)
More, J. D., 513 (*16*)
Moreau, C., 167 (*86, 87*)
Morgan, A. R., 91 (*58*), 407 (*23*), 413 (*149*)
Morgan, J. W. W., 218 (*95*)
Morrison, G. A., 55 (*46, 52*), 90 (*31*), 91 (*56*), 188 (*12*), 320 (*39*)
Morrison, J. J., 515 (*60*)
Morton, J. B., 484 (*41, 42*)
Morton, J., 486 (*97*)
Moss, E. L., 484 (*43*)
Moss, L. K., 290 (*153*)
Motherwell, W. B., 290 (*166*)
Mowlam, R. W., 92 (*88*)
Moy, B. F., 242 (*48*)
Mueller, E., 142 (*67*)
Mukaiyama, T., 408 (*56*), 409 (*57*), 410 (*82*), 414 (*167*)
Müller, A., 218 (*111, 112*)
Müller, E., 142 (*66*)
Munavu, R. M., 141 (*34*)
Murai, Y., 408 (*56*)
Murase, M., 483 (*13*)
Murphy, D., 190 (*45*)
Murray, R. W., 413 (*160*)
Murray, T. F., 419 (*288*)

Muto, R., 484 (*34*)
Myers, G. S., 165 (*38*)

N

Nace, H. R., 285 (*45*)
Nagabhushan, T. L., 484 (*41, 42*)
Nagahama, T., 415 (*190*)
Nagai, W., 216 (*58*)
Naganawa, H., 483 (*23, 29, 33*), 485 (*78, 79, 80*)
Nagarajan, M., 214 (*23*)
Nagarajan, R., 54 (*40*)
Nagpurkar, A. G., 91 (*66*)
Nagy, J. B., 244 (*95*)
Nagy, O. B., 244 (*95*)
Naisang, E., 413 (*143*)
Naito, T., 484 (*37*)
Nakadate, M., 286 (*54*)
Nakagawa, I., 464 (*36*)
Nakagawa, S., 484 (*37*)
Nakajima, M., 188 (*13*)
Nakane, M., 465 (*64*)
Nakano, J., 417 (*230*), 419 (*277*)
Nakasaka, T., 243 (*83*)
Nakata, H., 284 (*24*)
Nakata, M., 462 (*2*)
Nakatsuka, T., 414 (*167*)
Nambiar, K. P., 412 (*132, 134*)
Nánási, P., 166 (*61, 68, 69, 70, 71, 72, 73, 75, 76, 77, 79*)
Nath, R. L., 418 (*248*)
Nayak, U. G., 241 (*34*)
Neeman, M., 142 (*65*)
Nerinckx, W., 419 (*280*)
Ness, R. K., 289 (*136*), 407 (*16*), 408 (*49, 50*)
Neszmélyi, A., 166 (*61, 77*)
Nettleton, D. E., 164 (*21*)
Neuss, N., 486 (*83*)
Newth, F. H., 218 (*100, 102, 114*), 240 (*13*), 407 (*21*)
Nicolaou, K. C., 409 (*61, 70, 71*), 414 (*170*), 416 (*221*), 465 (*55*), 486 (*98*), 514 (*49, 50*), 515 (*51, 52*)
Nicolaus, B., 513 (*20*)
Nicolet, B. H., 288 (*105*)
Nicolle, J., 110 (*12*)
Ninet, L., 486 (*84*)
Nishimura, Y., 483 (*14, 15*), 484 (*34*), 484 (*54, 55*)
Nishio, M., 483 (*12*)
Nitta, K., 482 (*4*)
Nokami, T., 217 (*80*)
Nooner, T., 243 (*85*)

Norberg, T., 414 (*171*), 415 (*194*)
Norita, T., 142 (*59*)
Noumi, K., 142 (*53*)
Noyori, R., 409 (*58*), 410 (*84*)
Nutt, R. F., 285 (*28, 29*)
Nyberg, B., 243 (*84*)

O

O'Brien, E., 141 (*26*)
O'Colla, P. S., 141 (*26*)
O'Hanlon, P. J., 465 (*59*)
O'Meara, D., 165 (*52*)
O'Neil, A. N., 289 (*132*)
Oae, S., 286 (*57*), 464 (*38*)
Oakes, E. M., 218 (*106*), 219 (*123*)
Oden, E. M., 484 (*38, 43*)
Oehler, E., 321 (*66*)
Oelfke, W. C., 53 (*3*)
Ogawa, H., 482 (*8*), 483 (*12*)
Ogawa, T., 142 (*64*), 409 (*59, 60*), 414 (*181*), 415 (*198*), 465 (*54*)
Ogihara, Y., 214 (*25*)
Ogino, H., 514 (*31*)
Oguchi, N., 286 (*67*)
Ogura, H., 415 (*185*)
Ogura, M., 483 (*26*)
Ohgi, T., 215 (*26*), 514 (*30*)
Ohgo, Y., 319 (*34*)
Ohle, H., 165 (*41*), 167 (*91*), 216 (*62*), 218 (*98*)
Ohno, M., 409 (*73*)
Ohrui, H., 286 (*67*), 465 (*53*), 466 (*78*)
Oikawa, Y., 464 (*34*)
Okami, Y., 466 (*88*), 482 (*4*)
Okanishi, M., 483 (*18, 19, 25, 31*)
Okazaki, T., 466 (*88*)
Okazaki, Y., 484 (*35, 36*)
Okhi, H., 415 (*190*)
Okuda, T., 319 (*21*)
Okui, S., 290 (*151*)
Okuno, Y., 318 (*5*)
Olesker, A., 217 (*78*), 320 (*54*)
Ollis, W. D., 486 (*85, 86, 87, 88, 89, 90, 91, 92*)
Omura, K., 464 (*29*)
Ong, B. S., 412 (*134*)
Ong, K.-S., 319 (*22, 23*)
Onodera, K., 286 (*63, 69*)
Orpen, A. G., 92 (*88*)
Osborn, J., 92 (*91*)
Oscarson, S., 415 (*187*)
Ossawa, T., 241 (*26*)
Ossowski, P., 413 (*152*)
Ost, W., 465 (*61*)

Ostroumov, Yu. A., 110 (*20*)
Otake, N., 514 (*31*)
Ott, K.-H., 56 (*77*)
Otterbach, H., 140 (*5*)
Ottosson, H., 412 (*135*), 413 (*140*)
Ouellette, R. J., 90 (*23*)
Overend, W. G., 110 (*25*), 218 (*105*), 240 (*12*), 284 (*17, 18, 22, 23, 25, 26*), 285 (*27, 31*), 287 (*81, 82, 83, 84*), 289 (*139*), 318 (*14*), 319 (*16*), 320 (*52*), 417 (*225, 231*), 463 (*13, 14, 15*)
Ozanne, B., 483 (*27*)

P

Pacák, J., 164 (*27, 28*), 215 (*32, 35, 46*), 216 (*56*), 218 (*116, 117*), 416 (*213*)
Pacht, P. D., 319 (*19*)
Pacsu, E., 89 (*12*), 141 (*49*), 408 (*45*), 411 (*100*), 416 (*212*)
Paguaga, E., 413 (*147*)
Pakulski, Z., 416 (*217*)
Palermiti, F. M., 482 (*5, 7*)
Palmacci, E. R., 417 (*223*)
Palomo, C., 288 (*98, 100*)
Palomo-Coll, A. L., 467 (*94*)
Pan, B. C., 466 (*90*), 467 (*91*)
Papahatjis, D. P., 414 (*170*)
Papanikolaou, N. E., 464 (*37*)
Parham, W. E., 286 (*58*)
Parikh, J. R., 286 (*70, 71*)
Parish, C., 56 (*75*)
Park, Y. J., 53 (*14*)
Parker, R. E., 218 (*99*), 240 (*14*)
Parkin, D. W., 244 (*100*)
Parrish, F. W., 140 (*3*), 165 (*48*), 189 (*21*)
Pashinnik, V. E., 410 (*86*)
Pastor, J., 416 (*221*)
Patai, S., 417 (*238*)
Patterson, D., 140 (*24*)
Paulsen, H., 91 (*63*), 215 (*44, 47*), 219 (*121*), 243 (*71*), 284 (*1, 2, 3, 4*), 320 (*53*), 412 (*130*), 420 (*298*)
Pavia, A. A., 416 (*218*)
Pearson, N., 54 (*23*)
Peat, S., 214 (*1*), 240 (*16*)
Peck, R. L., 485 (*72*)
Pedretti, V., 416 (*208*)
Peel, E. W., 485 (*72*)
Pekár, F., 166 (*72*)
Pelter, A., 285 (*46*)
Pennington, R. E., 54 (*27*)
Percival, E., 241 (*21*)
Perelman, M., 164 (*21*)

Perkins, P. I., 464 (*37*)
Perlin, A. S., 55 (*61, 62*), 286 (*72*), 289 (*135*), 290 (*154*), 411 (*114*)
Perrin, C. L., 91 (*73*), 92 (*80*)
Perron, Y., 482 (*5*)
Perry, N. F., 164 (*13*)
Petillo, P. A., 92 (*88*)
Pfaehler, K., 289 (*125*)
Pfenninger, A., 514 (*45*)
Pfitzner, K. E., 285 (*47, 48, 49*)
Philips, B. E., 164 (*18*)
Philips, W. G., 285 (*38, 39*)
Phillips, G. O., 407 (*18*)
Pickles, V. A., 55 (*65*)
Pierozynski, D., 416 (*217*)
Pierson, G. O., 90 (*39*)
Pigman, W. W., 89 (*24*), 111 (*28, 29*), 188 (*16*), 284 (*15, 16*)
Pikul, S., 515 (*62*)
Pinto, B. M., 91 (*75*), 92 (*92*), 420 (*296*)
Piret, E. L., 25 (*12*)
Pitcher, R. G., 319 (*29*), 462 (*5*)
Pittet, A. O., 289 (*138*)
Pitzer, K. S., 53 (*5, 20, 21, 22*), 93 (*101*), 165 (*56*)
Pitzer, R. M., 53 (*9, 10*)
Planje, M. C., 90 (*32, 36*)
Plante, O. J., 417 (*223*)
Plattner, P. A., 190 (*46*), 240 (*11*)
Plavec, J., 92 (*85*), 93 (*102, 103, 104, 110, 111*)
Ploeser, J. Mc. T., 290 (*153*)
Pocker, Y., 110 (*17*)
Polavarapu, P. P., 55 (*69*)
Poos, G. I., 286 (*75*)
Posner, G. H., 410 (*91*)
Pougny, J.-R., 465 (*60, 62*)
Poyser, J. P., 465 (*49*)
Pozsgay, V., 215 (*26*), 415 (*197*)
Prahst, A., 413 (*153*)
Praly, J.-P., 420 (*310*)
Prather, J., 166 (*78*)
Pratt, J. W., 215 (*36*), 289 (*122*)
Pravdic, N., 286 (*68*)
Prelog, V., 53 (*12*), 318 (*13*), 466 (*89*)
Press, J. B., 412 (*134*)
Preud'Homme, J., 486 (*84*)
Prewo, R., 515 (*54*)
Price, C. C., 285 (*34*), 288 (*109*)
Prins, D. A., 140 (*9*)
Prior, A. M., 140 (*12*)
Promé, D., 216 (*50*)

Pryce, R. J., 465 (48)
Purlee, E. L., 110 (16)
Purves, C. B., 417 (228), 418 (260)

R

Rabe, A., 411 (112)
Rafferty, G. A., 318 (14)
Rajan Babu, T. V., 412 (134)
Rakotomanomana, N., 416 (218)
Rammler, D. H., 287 (90)
Ramsden, H. E., 290 (162)
Rånby, B. G., 417 (230)
Rancourt, G., 320 (40), 463 (20, 21, 22)
Randall, J. L., 409 (61)
Randell, K. D., 92 (92)
Ranganathan, R., 189 (27)
Rank, B., 290 (159)
Rao, C. S., 413 (140)
Rao, M. V., 214 (23)
Rapin, A. M. C., 216 (51)
Rappoport, D. A., 165 (35)
Rassu, G., 513 (10), 515 (68, 69, 70), 516 (71)
Ratcliffe, A. J., 91 (67), 419 (274)
Ratcliffe, R., 287 (79)
Rathke, M. W., 321 (63)
Raymond, A. D., 56 (73)
Raymond, A. L., 164 (31)
Reddy, G. V., 415 (182)
Redlich, H., 320 (53)
Rees, C. W., 284 (18), 417 (231)
Reeves, R. E., 55 (58, 59), 89 (27), 90 (28, 29)
Reformatsky, S., 321 (60, 61)
Reichstein, T., 167 (89), 215 (43), 218 (96, 97)
Reid, J., 240 (6, 7)
Reiding, S., 56 (76)
Reininger, K., 321 (66)
Reist, E. J., 188 (4), 241 (36)
Reuss, J., 53 (8)
Reuter, F., 288 (104)
Reynolds, D. D., 407 (24)
Rhind-Tutt, A. J., 418 (250)
Rhoades, J. A., 484 (46)
Rice, K. C., 484 (74)
Richards, E. M., 110 (7, 19)
Richards, G. F., 411 (109)
Richards, G. N., 218 (102, 103, 104), 417 (243)
Richardson, A. C., 140 (16, 20), 163 (6), 188 (5, 6, 11, 18, 19), 189 (20, 22, 23), 190 (45), 217 (77), 407 (8), 465 (50, 51)
Richtmyer, N. K., 164 (25), 215 (36, 38, 41, 42), 217 (92), 289 (122)
Riiber, C. N., 417 (245)
Riley, D. A., 410 (90)

Rinehart, K. L., 482 (11), 484 (50, 51, 52, 53)
Rist, C. E., 242 (53, 54), 284 (9)
Roberts, C., 413 (140)
Roberts, J. D., 142 (65), 166 (62), 486 (83)
Roberts, J. S., 464 (30)
Roberts, T., 54 (23)
Robertson, G. J., 140 (8), 217 (93)
Robertson, R. E., 243 (79)
Roccaronek, J., 287 (88)
Rodehorst, R., 287 (79)
Rodriguez, R. M., 486 (98)
Rogers, N. H., 465 (57, 59), 466 (72)
Röhle, G., 407 (26, 27, 28, 29), 412 (131)
Roldan, F., 288 (98)
Rolle, M., 243 (71)
Romers, C., 90 (42, 43, 44, 46, 47, 48)
Roncari, G., 319 (17, 18, 20)
Roos, M., 412 (125)
Roseman, S., 111 (41)
Rosén, G., 415 (187)
Rosenbrook, W., 410 (90)
Rosenstein, R. D., 53 (15, 16, 17, 18)
Rosenthal, A., 319 (22, 23)
Ross, G. W., 242 (45, 46)
Ross, S. D., 189 (32)
Roth, W., 188 (16)
Rousseau, G., 412 (134)
Roussi, G., 465 (66)
Roy, R., 416 (210, 215)
Rücker, C., 467 (92)
Rudrum, M., 55 (64)
Ruediger, G., 284 (3)
Ruiz, N., 409 (75)
Rundell, W., 142 (66)
Runquist, O. A., 90 (39)
Rupitz, K., 419 (286)
Ryan, K. J., 189 (29)
Rydon, H. N., 418 (248, 249)

S

Sadekov, I. D., 142 (70)
Saeed, M. A., 285 (44)
Sager, W., 409 (62)
Saito, S., 53 (7)
Saito, T., 243 (83)
Saito, Y., 189 (26)
Sakairi, N., 189 (25)
Sakan, K., 412 (132, 134)
Sakari, N., 216 (54)
Sakata, M., 416 (214)
Saksena, A. K., 486 (94, 95)
Salomaa, P., 164 (11), 418 (263)
Saluja, S. S., 91 (62)

Salzner, U., 91 (*77*)
Samek, P. N., 241 (*40*)
Samuelson, B., 287 (*86*), 288 (*99*)
Sandstrom, J., 54 (*43*)
Sapp, J. B., 285 (*41*)
Sarre, O., 486 (*97*)
Sarret, L. H., 286 (*75*)
Sasai, H., 410 (*79, 80, 81*)
Sasaki, H., 321 (*74*)
Satchell, D. P. N., 419 (*289*)
Sato, K.-I., 320 (*41, 42, 43, 47, 48*)
Sato, N., 415 (*185*)
Sato, Y., 318 (*6*)
Satoh, C., 319 (*21*)
Satoh, K., 485 (*76*)
Satoh, T., 319 (*29, 30*), 462 (*4, 5*)
Saunders, R. M., 165 (*46*)
Saunders, W. H., 25 (*13*)
Sauter, H. M., 412 (*134*)
Scattergood, A., 164 (*34*)
Schaad, L. J., 166 (*64*)
Schaaf, T. K., 464 (*44*)
Schaffer, R., 466 (*75*)
Schaffner, C. P., 241 (*22*), 484 (*39, 40, 52*)
Scharf, H. D., 486 (*93*)
Schatz, A., 485 (*71*)
Schauer, R., 513 (*11, 12*)
Schechter, H., 165 (*55*)
Schellenberger, H., 290 (*160*)
Schier, O., 240 (*15*)
Schillings, R. T., 484 (*52*)
Schlenkrich, M., 56 (*76*)
Schleyer, P. v. R., 91 (*77*)
Schlosser, M., 466 (*69*)
Schlubach, H. H., 408 (*51*)
Schmid, R., 318 (*11*)
Schmidt, O. Th., 165 (*37*), 167 (*88*)
Schmidt, R. R., 411 (*117, 118, 120*), 412 (*122, 123, 124, 125, 126, 128, 129*)
Schmidt, U., 361 (*66*), 407 (*28*)
Schmiesing, R. J., 466 (*71*)
Schmitz, H., 482 (*6*)
Schneider, M. J., 465 (*47*)
Schneider, W., 416 (*205*)
Schoorl, M. N., 242 (*42*)
Schramm, V. L., 244 (*100, 101*)
Schreiber, J., 287 (*88*)
Schroeder, L. R., 215 (*48*), 408 (*36*), 411 (*109*)
Schuerch, C., 214 (*2*), 215 (*28*), 217 (*75*), 407 (*30*), 408 (*31, 32*)
Schuerch, K., 142 (*63*)
Schuhmacher, M., 412 (*127*)
Schulz, M., 411 (*104*)

Schumann, S. C., 54 (*24, 25*)
Schwarz, J. C. P., 140 (*15*)
Schweizer, R., 140 (*17, 18*)
Schweizer, W. B., 464 (*35*)
Schwentner, J., 413 (*150*)
Scotti, T., 485 (*60, 61*)
Secrist, J. A., 512 (*1*), 514 (*46*)
Seebach, D., 320 (*49, 50*), 464 (*35*)
Seeberger, P. H., 217 (*80*), 417 (*223*)
Seely, F. L., 466 (*85*)
Seib, P. A., 215 (*48*)
Seitz, S. P., 414 (*170*), 467 (*91*)
Sekizawa, Y., 485 (*78, 79, 80*)
Sell, H. M., 165 (*42*)
Sell, K., 189 (*30*)
Selman, L. H., 417 (*238*)
Senderowitz, H., 56 (*75*)
Sepp, D. T., 89 (*23*), 90 (*34*)
Sepp, J., 416 (*205*)
Sepulchre, A. M., 319 (*25*)
Seqeueira, J. S., 417 (*231*)
Sergeyev, N. M., 55 (*57*)
Serianni, A. S., 92 (*87, 91*)
Sessler, P., 164 (*30*)
Seto, H., 514 (*31*)
Seydel, J. K., 243 (*92*)
Sezaki, M., 485 (*76*)
Shaban, M. A. E., 318 (*3, 4*)
Shafizadeh, F., 188 (*14*), 215 (*39*), 241 (*28*), 318 (*10*), 417 (*224*)
Shapiro, H. S., 244 (*108, 109*)
Shapiro, R., 243 (*93*), 244 (*94*)
Sharma, M., 410 (*88*)
Sharpen, A. J., 243 (*92*)
Sharpless, K. B., 513 (*28*), 514 (*33, 44*), 516 (*90*)
Sharpless, T. W., 243 (*88*)
Shashkov, A. S., 321 (*73*)
Shaw, C. J. G., 289 (*123, 124*)
Shaw, D. F., 55 (*64*)
Sheft, I., 418 (*250*)
Shen Han, T. M., 188 (*14*), 241 (*28*)
Shepherd, D. M., 165 (*52*)
Shibasaki, M., 410 (*79, 80, 81*), 464 (*39*)
Shibata, H., 188 (*13*)
Shigeharu Inouye, S., 514 (*31*)
Shimizu, J., 415 (*185*)
Shimizu, M., 408 (*55*)
Shimura, M., 485 (*76, 78, 79, 80*)
Shinn, L. A., 288 (*105*)
Shmyrina, A. Ya., 321 (*73*)
Shoda, S., 408 (*56*), 409 (*57*), 414 (*167*)
Shriner, R. L., 321 (*62*)

Shutzberg, B. A., 56 (*71*)
Sicher, J., 91 (*70*)
Sigel, H., 243 (*87*)
Silanpaa, R., 142 (*59*)
Silwanis, B. A., 415 (*187*)
Sim, G. A., 465 (*49*)
Simon, H., 242 (*70*)
Simpson, L. B., 25 (*15*)
Sinaÿ, P., 415 (*186, 188*), 416 (*208*), 465 (*60, 62*), 513 (*21*)
Singh, P. P., 420 (*292, 293*)
Sinnot, M. L., 419 (*281*)
Sinnwell, V., 216 (*52*)
Sirokman, F., 188 (*7*), 241 (*39*)
Sisler, H. H., 286 (*74*)
Sivakumaran, T., 140 (*22*)
Skancke, P. N., 54 (*28*)
Skorupowa, E., 214 (*22*)
Skraup, Zd. H., 408 (*47*)
Sledeski, A. W., 515 (*64, 65*)
Smiataczowa, K., 243 (*77, 78*)
Smirnoff, A. P., 214 (*13*)
Smirnyagin, V., 54 (*38*), 407 (*11*)
Smith, C., 486 (*85, 86, 87, 89, 90, 91, 92*)
Smith, F., 217 (*90*)
Smith, G. F., 110 (*9*)
Smith, J. C., 54 (*27*)
Smith, P. W., 465 (*52*)
Smith, S. G., 285 (*37*)
Smith, V. H., 56 (*74*)
Snyatkova, V. J., 410 (*95*)
Soff, K., 216 (*67*)
Sokolova, N. P., 90 (*35*)
Sokolovskaya, T. A., 410 (*95*)
Sokolowski, J., 214 (*22*), 243 (*78*)
Sondheimner, S. J., 215 (*28*)
Sørensen, N. A., 417 (*245*)
Sorgi, K. L., 466 (*71*)
Sorkin, E., 215 (*43*), 218 (*96*)
Sorokin, B., 242 (*60, 61, 62, 63*)
Sovers, O. J., 53 (*10*)
Spanu, P., 515 (*68, 69, 70*)
Speakman, P. R. H., 188 (*9*), 190 (*43*)
Speck, J. C., 111 (*40*)
Spencer, C. C., 289 (*128*)
Spencer, M., 54 (*39*)
Spencer, R. R., 188 (*4*), 241 (*36*)
Spencker, K., 216 (*62*)
Spitzer, R., 53 (*21*), 93 (*101*)
Spivak, C. T., 111 (*41*)
Springmann, H., 140 (*5*)
Squires, T. G., 408 (*48*)

Srivastava, H. C., 142 (*55*), 420 (*292, 293*)
Srivastava, V. K., 142 (*55*)
Staab, H. A., 53 (*19*), 141 (*30*)
Stacey, M., 217 (*90*), 218 (*105*), 240 (*2, 3*), 284 (*10*), 285 (*52*), 417 (*241*)
Stähle, M., 466 (*69*)
Stanek, J., 289 (*131*)
Steen, K., 464 (*42, 43*)
Steinbrunn, G., 216 (*57*)
Stening, T. C., 165 (*51*)
Stensio, K. E., 297 (*92*)
Stenzel, H., 241 (*38*)
Stenzel, W., 215 (*47*), 421 (*38*)
Stepanenko, B. N., 242 (*57*)
Sterk, H., 321 (*67*)
Stevens, C. L., 188 (*7*), 241 (*22, 23, 39*), 288 (*101*)
Stevens, J. D., 55 (*63*)
Stewart, L. C., 251 (*42*)
Steyaert, J., 222 (*112*)
Stick, R. V., 91 (*61*)
Stiehler, O., 416 (*205*)
Still, W. C., 56 (*75*), 516 (*89*)
Stirm, S., 140 (*11*)
Stoddart, J. F., 163 (*5*)
Stoeckel, O., 284 (*8*)
Stothers, J. B., 319 (*28*), 463 (*11*)
Strange, R. E., 240 (*9, 10*)
Streitwieser, A., 190 (*40*)
Stumpp, M., 412 (*124*)
Stylianides, N. A., 409 (*70*)
Subbotin, O. A., 55 (*57*)
Sugihara, J. M., 139 (*1*), 287 (*89*)
Sutherland, I. O., 486 (*89, 91*)
Sutra, R., 164 (*33*)
Suzuki, A., 415 (*185*)
Suzuki, K., 409 (*65, 66, 67, 68, 69, 72*)
Suzuki, M., 412 (*134*)
Sviridov, A. F., 321 (*73*), 463 (*23, 24*), 464 (*25, 26, 27, 28*)
Swain, C. G., 110 (*10*)
Sweat, F. W., 286 (*59*)
Sweet, F., 90 (*40*)
Swern, D., 464 (*29*)
Swiderski, J., 110 (*23*)
Szabo, I. F., 411 (*110*)
Szarek, W. A., 56 (*74*), 88 (*1*), 90 (*49*), 91 (*78*), 92 (*79*), 285 (*30*), 320 (*45*)
Szeja, W., 140 (*21*), 416 (*216*)
Szejtli, J., 166 (*72, 79*)
Szmant, H. H., 141 (*34*)
Szmulewicz, S., 486 (*96*)

T

Tafesh, A. M., 466 (84)
Taha, M. I., 290 (152)
Takacs, J. M., 466 (85)
Takagi, Y., 484 (47)
Takahashi, S., 216 (54)
Takahashi, Y., 142 (64), 409 (60), 485 (75)
Takasawa, S., 483 (31)
Takashi, S., 188 (13)
Takatani, M., 514 (44)
Takeda, K., 415 (185)
Takeda, T., 214 (25)
Takeuchi, K., 410 (82)
Takeuchi, T., 482 (4), 483 (23, 33)
Tamura, J.-I., 485 (81)
Tanabe, H., 289 (140)
Tanabe, M., 484 (48)
Tanaka, K. S. E., 244 (97), 419 (284, 285)
Tanaka, N., 483 (26)
Tanghe, L. J., 414 (164)
Tanret, C., 25 (9), 214 (4, 5)
Tarasiejska-Glazer, Z., 241 (19)
Tartakovsky, E., 421 (314)
Tatlow, J. C., 141 (40)
Tatsuta, K., 217 (81), 412 (134), 482 (9), 483 (14, 15, 16, 17)
Taylor, B., 486 (91)
Taylor, H. D., 482 (5)
Taylor, I. D., 514 (37)
Taylor, K. G., 188 (7), 241 (22, 23, 39)
Taylor, T. J., 289 (126)
Tchoubar, B., 285 (43)
Teece, E. G., 141 (42, 45)
Tejima, S., 214 (21), 416 (214)
Temeriusz, A., 110 (23, 24)
Teoule, R., 243 (90)
Thacker, D., 190 (42), 418 (261)
Thang, T. T., 217 (78)
Thatcher, G. R. J., 89 (4), 91 (65, 75, 78), 92 (79)
Theander, O., 141 (43), 167 (93)
Thibaudeau, C., 92 (85), 93 (111)
Thiel, H., 218 (98)
Thielebeule, W., 164 (26)
Thiem, J., 217 (81), 413 (142, 150, 151, 152, 153, 154, 155), 462 (2)
Thomas, G. H. S., 289 (126)
Thompson, A., 188 (15), 217 (72), 289 (132), 485 (73)
Tichý, M., 91 (70)
Tidor, B., 55 (68)
Tien, J. M., 285 (42)
Timell, T. E., 417 (242, 244), 418 (251)
Tipper, D., 483 (27)
Tipson, R. S., 165 (40), 466 (75)
Tobie, W. C., 25 (7)
Todd, A. R., 244 (104, 110)
Toepfer, A., 412 (129)
Togo, H., 408 (55)
Tolbert, L. M., 412 (134)
Tollens, B., 25 (10)
Toneman, L. H., 90 (36)
Tong, G. L., 286 (65)
Torii, K., 415 (185)
Toru, T., 515 (52)
Toshima, K., 408 (53)
Toya, Y., 514 (30)
Tronchet, J. M. J., 286 (54)
Tropper, F. D., 416 (210, 215)
Trost, B. M., 412 (121), 466 (87), 512 (4)
Truesdale, E. A., 412 (134)
Truhlar, D. G., 56 (72)
Tsai, T. Y. R., 414 (172)
Tsang, R., 515 (57, 58)
Tsuboyama, K., 415 (185)
Tsuchihashi, G., 337 (65, 66, 67)
Tsuchiya, T., 408 (54), 482 (9), 484 (47, 34, 35, 36), 485 (69, 75)
Tsuda, Y., 318 (5, 6)
Tsumura, T., 483 (14, 15, 16, 17)
Tsuruoka, T., 514 (31)
Tucker, L. C. N., 216 (53), 217 (76)
Turner, A. F., 244 (105)
Turner, R. B., 164 (21)
Turner, W. N., 91 (57), 484 (41)
Tvaroška, I., 421 (313)
Typke, V., 285 (35)

U

Uchida, I., 412 (134)
Uchida, T., 216 (60, 61)
Udodong, U. E., 412 (136, 138), 413 (140, 144)
Ueda, M., 482 (4)
Ueda, Y., 412 (134)
Ueno, Y., 216 (54)
Ulrich, P., 515 (53)
Umezawa, H., 482 (3, 4, 10), 483 (18, 19, 23, 25, 26, 29, 31, 33), 484 (69, 34, 35, 36)
Umezawa, S., 482 (2, 9), 483 (14, 15, 16, 17), 484 (34, 35, 36, 47, 54, 55, 56, 57), 485 (69, 75)
Ungemach, F. S., 465 (47)
Untersteller, E., 513 (21)
Urech, F., 110 (4)
Usov, A. I., 166 (60)

Usui, T., 484 (*56, 57*)
Utahara, R., 482 (*4*), 483 (*18, 19, 25, 31*)
Uyehara, T., 412 (*134*)

V

Vachon, D. J., 92 (*88*)
Vaino, A. R., 92 (*79*)
Valatin, T., 214 (*15*), 217 (*73*)
Van Boom, J. H., 215 (*27*), 415 (*183, 184*)
van Cleve, J. W., 414 (*168*)
Van de Graaf, B., 54 (*44, 45*)
Van Delft, F. L., 486 (*98*)
van Eikeren, P., 418 (*266*)
van Ekenstein, W. A., 111 (*31, 32, 33, 34, 35, 36, 37, 38, 39*)
Van Engen, D. J., 416 (*219*)
van Helroort, K., 53 (*8*)
van Leeuwen, S. H., 415 (*183*)
van Roon, J. D., 165 (*57*)
Vandemeulebroucke, A., 244 (*112*)
VanRheenan, V., 514 (*42*)
Vargha, L., 290 (*164*)
Vasela, A., 419 (*287*)
Vatèle, J.-M., 189 (*28*), 413 (*145*)
Vaughan, G., 240 (*12*)
Veenemann, G. H., 415 (*183, 184*)
Venner, H., 231 (*89, 91*)
Vercellotti, J. R., 408 (*48*)
Vernon, C. A., 417 (*234, 238*), 418 (*250*)
Verses, W., 244 (*112*)
Vethaviyasar, N., 414 (*166*)
Veyriéres, A., 416 (*208*)
Vieira, E., 513 (*9*)
Vilkov, L. V., 90 (*35*)
Villiers, A., 25 (*6*)
Viscontini, M., 289 (*121*)
Vite, G. D., 215 (*29*)
Viti, S. M., 514 (*44*)
Vizsolyi, J. P., 244 (*105*)
Vladuchik, W. C., 412 (*134*)
Vocadlo, D. J., 419 (*286*)
Vogel, P., 513 (*9*)
Voisin, D., 90 (*49*)
von Philipsborn, M., 484 (*53*)
von Saltza, M. H., 240 (*6, 7*)
Vozny, V. Ya., 409 (*64*)
Vysotskaya, N. A., 288 (*107*)

W

Wachtmeister, C. A., 287 (*92*)
Wade, P. A., 412 (*134*)
Wadsworth, W. S., 466 (*81, 82, 83*), 513 (*24*)
Wagman, G. H., 484 (*38*), 484 (*43*)

Wagner, G., 513 (*9*)
Waisbrot, S. W., 414 (*164*)
Waitz, J. A., 484 (*43*)
Waksman, S. A., 484 (*49*), 485 (*71*)
Waldmann, H., 409 (*63, 76*), 410 (*77, 78*)
Walker, F. J., 514 (*44*)
Walker, S., 416 (*219*)
Wall, H. M., 318 (*14*)
Wallin, N. H., 142 (*54*)
Walters, D. R., 240 (*4, 5*)
Walther, W., 288 (*112*)
Walton, E. J., 285 (*28, 29*)
Wander, J. D., 53 (*13*), 319 (*25*), 320 (*51*)
Wang, F., 216 (*54*)
Ward, D. E., 412 (*132, 133, 134*)
Ward, P. F. V., 218 (*113*)
Warren, C. D., 141 (*44*), 464 (*33*)
Warsi, S. A., 289 (*129*)
Watanabe, K. A., 92 (*85*), 241 (*37*)
Weakley, T. J., 321 (*58*)
Webber, J. M., 164 (*17, 18, 19*), 240 (*2, 3*)
Webster, C., 419 (*274*)
Webster, K. T., 142 (*63*)
Wedemeyer, K.-F., 408 (*35*)
Weidman, H., 321 (*65, 67*)
Weigner, E., 418 (*252*)
Weinstein, J. L., 483 (*21*)
Weinstein, M. J., 484 (*38*), 484 (*43*)
Weisbuch, F., 110 (*12*)
Weisman, G. R., 92 (*88*)
Weiss, K., 411 (*115*)
Weiss, S., 53 (*6*)
Welsh, P. M., 219 (*120*)
Wempen, I., 244 (*111*)
Wendel, K., 166 (*63*)
Wepster, B. M., 54 (*44*)
Werz, D. B., 217 (*80*)
Wessel, H. P., 409 (*75, 74*)
Westheimer, F. H., 287 (*87, 88*)
Westphal, O., 140 (*11*)
Weyer, J., 215 (*44*), 284 (*3*)
Weygand, F., 242 (*64, 68, 69*), 414 (*165*)
Whelan, W. J., 216 (*69*), 289 (*129*)
Whistler, R. L., 165 (*37*), 167 (*88*), 241 (*34, 35*), 407 (*12*)
White, D. N. J., 446 (*89*)
White, J. M., 92 (*88*)
Whitehead, D. F., 482 (*5, 6, 7*)
Whitehead, W., 217 (*93*)
Wiberg, K. B., 53 (*11*)
Wick, A. E., 464 (*42, 43*)

Wicki, J., 419 (286)
Wieczorek, M. W., 92 (82)
Wiesner, K., 25 (15), 414 (172)
Wiggins, L. F., 165 (45), 217 (90), 218 (102, 103, 104), 241 (17)
Wilham, C. A., 216 (65)
Willi, A. V., 243 (79, 80)
Williams, J. M., 140 (16, 20), 188 (11)
Williams, N. E., 463 (16)
Williams, N. R., 284 (25, 26), 287 (81, 83, 84), 318 (14), 319 (16), 463 (13, 14, 15), 320 (52)
Williams, R. M., 412 (134)
Williams, T. H., 91 (71)
Williamson, A. R., 287 (85)
Winkler, S., 407 (25)
Winkley, M., 287 (85)
Winssinger, N., 416 (221)
Winstein, S., 55 (55), 89 (21), 189 (33, 34), 285 (37)
Winterfeldt, E., 412 (121)
Wintersteiner, O., 240 (4, 5, 6, 7)
Wippel, H. G., 466 (79, 80)
Wise, W. S., 218 (104)
Wiśniewski, A., 214 (22)
Witchak, Z. J., 513 (6)
Withers, S. G., 419 (286)
Wolfrom, M. L., 111 (43), 165 (37), 167 (88), 188 (14, 15), 217 (72), 241 (28), 289 (132), 407 (12), 408 (43), 414 (163, 164), 485 (73)
Wolfson-Davidson, E., 244 (95)
Wong, E., 415 (195)
Wong, N.-C., 412 (134)
Woo, P. W. K., 484 (50), 485 (66, 67, 68)
Wood, H. B., 164 (23)
Woodard, S. S., 514 (44)
Woods, B. M., 289 (126, 134)
Woods, R. J., 56 (73, 74)
Woodward, R. B., 319 (19), 412 (132, 133, 134), 462 (3)
Wothers, P. D., 92 (86), 93 (98)
Wright, D. E., 486 (89, 90, 91, 92)
Wu, G. Y., 287 (89)
Wu, J., 92 (87)
Wu, S.-R., 512 (1)
Wu, Z., 412 (136), 413 (140)
Wulf, G., 411 (101)
Wuts, P. G. M., 414 (169)

Y

Yagisawa, M., 483 (29, 33)
Yagishita, K., 482 (4)
Yajima, T., 286 (69)
Yamamoto, H., 321 (74), 483 (23, 29)
Yamamoto, K., 214 (21)
Yan, L., 416 (220)
Yang, D., 413 (147)
Yashunskii, D. V., 464 (25, 26, 27, 28)
Yasigawa, M., 483 (23)
Yasuda, D. M., 484 (48)
Yasuda, S., 485 (70)
Yates, J. B., 466 (74)
Yde, M., 416 (207)
Yeagley, D., 421 (312)
Yokoyama, M., 408 (55), 410 (83)
Yonemitsu, O., 464 (34)
Yoshimura, J., 318 (12), 319 (34), 320 (41, 42, 43, 47, 48), 485 (81)
Yoshimura, Y., 189 (26)
Yoshioka, T., 464 (34)
Young, R. J., 288 (110, 111), 290 (161)
Yu, K.-L., 515 (57, 59)
Yu, Y.-Q., 56 (80)
Yudis, M. D., 484 (44)

Z

Zach, K., 241 (25)
Zachystalova, D., 416 (213)
Zagar, C., 486 (93)
Zaikov, G. E., 407 (17)
Zamojski, A., 416 (217), 484 (45)
Zàrà-Kacziàn, E., 214 (24)
Zelinsky, N. D., 463 (23, 24)
Zemplén, G., 214 (15), 216 (68, 70), 217 (71, 73), 408 (38, 39)
Zen, S., 216 (60, 61)
Zerhusen, F., 111 (47)
Zervas, L., 164 (22, 30)
Zhdanov, Yu. A., 110 (20)
Zheng, Y.-J., 55 (70), 56 (71)
Zhu, Y., 92 (91)
Ziemann, H., 414 (165)
Zinner, H., 164 (26)
Zirner, J., 408 (37)
Zoltewicz, J. A., 243 (88)
Zottola, M. A., 215 (29)
Zugaza-Bilbao, A., 467 (94)
Zweifel, G., 140 (14)

Subject Index

A

Acetalation of monosaccharides, 158
Acetolysis of glycosides, 396
Acetonation of monosaccharides, 150
Acyclic forms of monosaccharides, naming, 17
Acyl migrations, 133
Addition of carbon nucleophiles to glycopyranosiduloses (synthesis of branched-chain sugars), 297
Addition of diazomethane to glycopyranosiduloses, 307
Addition of hydride ion (reduction) glycopyranosiduloses, 291
Addition of nucleophiles to glycopyranosiduloses, 291
Addition of organometals to glycopyranosid-4-ulose, 298
Addition of organometals to glycopyranosiduloses, 298
Aldaric acids, 22
Aldgarose, 298
Aldonic acids, 21
Alternate chair conformation, 37
Alternative names for substituted derivatives, 11
Amadori rearrangement, 232
Amikacin, 472
Aminoglycoside antibiotics, 469
Amino-monosaccharides, 15
Amino sugars, 221
Ammonolysis of oxiranes, 221
Anchimeric assistance, 185
Anhydrosugars, 191
1,2-Anhydrosugars, 200
1,4-Anhydrosugars, 198
1,6-Anhydrosugars (glycosans), 193
Anhydrosugars not involving the anomeric carbon, 203
Anomeric carbon atom, 5

Anomeric effect (AE), 59
Anomeric effect, quantum-mechanical explanation of, 63
Anomeric effect in systems O–C–N, 84
Anomeric hydroxyl group, reactivity, 113
Anomerization, 105
Anomers and anomeric configurational symbols ("α" or "β"), 18
anti-clinal conformation, 30
anti-conformation, 31
anti-periplanar conformation, 31
Apiose, 297
Aplasmomycin, 456
Asymmetric epoxidation, titanium catalyzed, 494
Automated oligosaccharide synthesis, 374
Axial (*a*) ligands, 36
1,3-*syn*–axial interaction, 32, 38

B

Baeyer ring strain, 33
Benzoylimidazole as acylating agent, 130
1-(Benzoyloxy) benzotriazole as acylating agent, 131
Benzylidenation of monosaccharides, 148
Bimolecular nucleophilic substitution in sugars, 169
Biotin, 443
Blastmycinone, 297
Boat (B) conformation of pyranoses, 36
Boat conformation of cyclohexane, 36
Bowsprit (*b*) orientation, 36
Branched-chain sugars with functionalized branched chains, 308
Bromine oxidation of monosaccharides, 247
Butirosins A and B, 477

C

C_2 conformation, 34
Calculation of conformational energies, 41
Carbohydrate based antibiotics, 469
Catalytic osmylation of olefinic bond, 498
Catalytic oxidation of monosaccharides, 245
Chair conformation (C) of pyranoses, 39
Chair conformation of cyclohexane, 36
Chemistry of glycosidic bond, 323
Chiral carbon, 2
Chromium trioxide–acetic acid oxidation of monosaccharides, 270
Chromium trioxide oxidation of monosaccharides, 266
Chromium trioxide–pyridine oxidation of monosaccharides, 266
Cladinose, 297
Cleavage of glycosidic bond, 374
Configurational determination of the tertiary carbon atom, 299
Configurational symbols and prefixes, 11
Conformational analysis of acyclic forma of monosaccharides, 31
Conformational analysis of acyclic hydrocarbons, 28
Conformational analysis of alditol, 31
Conformational analysis of cyclic forms of monosaccharides, 33
Conformational analysis of furanoses, 33
Conformational analysis of monosaccharides, 27
Conformational analysis of pyranoses, 36
Conformation, definition, 27
$^1C_4$ Conformation of pyranoses, 41
$^4C_1$ Conformation of pyranoses, 40
Conventions, 10
Conversion of Fischer projection to Haworth perspective formulae, 8
C_s conformation, 35
Cyclic acetals and ketals, 24
Cyclic acetals of monosaccharides, 144
Cyclic ketals of monosaccharides, 144

D

Deoxy-monosaccharides, 14
Desosamine, 222
Destomycin A, 479
Dibutylstannylene derivatives of pyranoses, selective benzoylation, 131
Dimethyl sulfoxide (DMSO) oxidation of monosaccharides, 254
1-Dimethoxy-2-lithio-2-propene as carbon nucleophile, 312

Displacement of the C2 sulfonyloxy group in pyranoses, 174
DMSO–acetic anhydride oxidation of monosaccharides, 258
DMSO–DCC (Pfitzner–Moffatt) oxidation of monosaccharides, 255
DMSO–phosphorus pentoxide oxidation of monosaccharides, 263
DMSO–sulfurtrioxide–pyridine oxidation of monosaccharides, 264

E

Eclipsed conformation, 29
$\Delta 2$ effect, 49
Endo-anomeric effect, 57
Endocyclic secondary hydroxyl group, reactivity, 113
Envelope conformation of furanoses, 34
Epoxide migrations, 211
Epoxides, 209
Equatorial (*e*) ligands, 36
Erythronolides A and B, 423
Ethylidenation of monosaccharides, 149
Everninomicin, 482
Exo-anomeric effect, 67
Exocyclic secondary hydroxyl group, reactivity, 113

F

Fischer glycosidation, 325
Fischer projection formulae, 2
Fischer projection formulae of cyclic forms of monosaccharides, 7
Fischer projection formulae, rules, 4
Flagpole (*f*) orientation, 36
Flambamycin, 481
Furanoses, 6

G

Garosamine, 297
Gauche conformation, 29
Generalized anomeric affect, 69
Gentamicin, 473
Glycals as glycosyl donors, 357
Glycofuranosides, acid-catalyzed hydrolysis, 385
Glycopyranosides, acid-catalyzed hydrolysis, 377
Glycosans, 193
Glycoside hydrolysis, 324
Glycoside hydrolysis, recent developments, 389
Glycosides, 19
Glycosides, acid-catalyzed hydrolysis, 374

Subject Index

Glycoside synthesis *via* remote activation, 353
Glycosylamines, 20
Glycosylamines, 231
Glycosyl fluorides as glycosyl donors, 336
Glycosyl phosphatidylinositol(GPI) linked proteins, 229
Glycosyl radicals, 20
Glycosyl sulfoxides as glycosyl donors, 368

H
Half-chair (H) conformation, 38
Half-chair (H) conformation of pyranoses, 39
Hamamelose, 297
Haworth perspective formulae of cyclic forms of monosaccharides, 7
Heyns rearrangement, 233
Higher-carbon monosaccharides, 487
Higher-carbon sugars, synthesis, 490
Higher-carbon sugars, synthesis *via* aldol condensation, 498
Higher-carbon sugars, synthesis *via* butenolides, 503
Higher-carbon sugars, synthesis *via* Wittig olefination, 490
Higher-carbon sugars, total synthesis, 503
Hikosamine, 488
Hydrazinolysis, 228
Hydrogen peroxide, conformation, 27
Hygromycin, 480

I
Interconversion of pyranose conformations, 41
Intramolecular anhydrides, 24
Isomerization of cyclic acetals and ketals, 154
Isomerizations of epoxides, 212
Isomerization of sugars, 95
Isopropylidenation of monosaccharides, 150

K
Kanamycin, 470
KDO, 488
Ketalation of monosaccharides, 150
Ketoses, 13
Klyne–Prelog proposal for description of stereochemistry across a single bond, 30
Königs–Knorr synthesis of glycosides, 330

L
Lead tetraacetate oxidation of carbohydrates, 277
Lincosamine, 488
2-Lithio-1,3-dithiane as carbon nucleophile, 309

Lobry de Bruyn–Alberda van Eckenstein transformation, 108

M
Macrolide antibiotics, 423
Methoxyvinyl lithium as carbon nucleophile, 312
Methynolide, 424
Migration of acetal and ketal groups, 155
Monosaccharides, acyclic forms, 2
Monosaccharides, classification, 1
Monosaccharides, configuration of, 2
Monosaccharides, cyclic forms, 4
Monosaccharides, definition, 1
Monosaccharides, general formula, 1
Monosaccharides, numbering of the carbon atoms in its chain, 1
Monosaccharides, representation of, 2
Monosaccharides, stereochemistry of, 2
Muramic acid, 222
Mutarotation, 95
Mutarotation, bifunctional catalyst for, 98
Mutarotation, complex, 97
Mutarotation, concerted mechanism of, 100
Mutarotation constant, 95
Mutarotation, simple, 96
Mycosamine, 222

N
N-3,4,5-Trimethoxy-benzoyl imidazole as acylating agent, 129
N-Acetyl-D-galactosamine, 222
N-Acetyl-D-glucosamine, 222
N-Acetyl-neuraminc acid, total synthesis, 508
N-Acetyl-neuraminic acid (NANA), 487
N-Acetyl-neuraminic acid (NANA), synthesis from mannosamine, 508
Narbonolide, 423
Neighboring group participation, 180
Neomycin B (Actilin, Eterfram, Framecetin, Soframycin), 476
Neosamine B, 221
Neosamine C, 221
N-Glycosides, 231
Nicotine dichromate oxidation of monosaccharides, 272
N-Linked glycoproteins, 231
Nomenclature of monosaccharides, 9
Nonbonded interactions, 35
Nonselective oxidation of monosaccharides, 248
North (N) conformation of nucleosides, 87
Noviose, 297
n-Pentenyl glycosides as glycosyl donors, 354

Nucleophiles, 170
Nucleophilic displacement of primary tosylates with azide, 227
Nucleophilic displacement of secondary tosylates with azide, 227
Nucleophilic displacement in sugars, 169
Nucleophilic displacement of sulfonates (or halides) with nitrogen nucleophiles, 227
Nucleosides, acid-catalyzed hydrolysis, 237
Nucleosides (or nucleotides), 236

O

γ-octose, 297
Octosyl acid A, 488
Octosyl acid A, total synthesis, 506
O-Linked glycoproteins, 229
Opening of oxiranes with carbon nucleophiles, 316
Ortho esters as glycosyl donors, 340
Orthosomycins, 477
O-Substitution, 15
Oxidation of carbohydrates with the cleavage of carbohydrate chain, 273
Oxidation of monosaccharides, 245
Oxiranes, 203

P

Parent monosaccharides, choice of, 10
Paromomycin, 476
Pentavalent organobismuth oxidation of carbohydrates, 283
Periodate oxidation of carbohydrates, 273
Picronolide, 423
Pillarose, 297
Pitzer strain, 29
Planar conformation of furanoses, 33
Primary hydroxyl group, reactivity, 113
Pseudomonic acid C, 445
Pseudorotation in furanoses, 34
Puckered conformation, 34
Purine bases, 236
Purine nucleosides, 237
Pyranoses, 6
Pyridinium chlorochromate oxidation of monosaccharides, 270
Pyridinium dichromate–acetic anhydride oxidation of monosaccharides, 272
Pyrimidine bases, 236
Pyrimidine nucleosides, 236

R

Rearrangements of anhydrosugars, 211
Relative reactivity of hydroxyl groups in monosaccharides, 113
Removal of benzylidene group, 157
Removal of isopropylidene group, 163
Reverse anomeric effect (RAE), 74
Ruthenium tetroxide oxidation of monosaccharides, 252

S

Selective acetylation of monosaccharides, 127
Selective acylation (esterification) of monosaccharides, 114
Selective alkylation of metal complexes of pyranoses, 138
Selective alkylation of pyranoses, 136
Selective arylation of pyranoses, 136
Selective benzoylation of monosaccharides, 120
Selective benzylation of pyranoses, 139
Selective oxidation of monosaccharides, 245
Selective tosylation and mesylation of monosaccharides, 118
Semiempirical method for calculation of conformational energies, 41
Sialic acid, 222, 488
Sickle (bent) conformation, 32
Skew-boat conformation of cyclohexane, 36
Skewed conformation, 30
Solid phase synthesis of oligosaccharides, 369
South (S) conformation of nucleosides, 87
Staggered conformation, 30
Stem and systematic names, 9
Streptomycin A, 477
Streptose, 297
Swainsonine, 441
syn-clinal conformation, 30
syn-periplanar conformation, 31
Synthesis of acylated glycosyl chloride and bromides, 335
Synthesis of glycosyl fluorides, 338
Synthesis of polychiral natural products from carbohydrates, 423
Synthesis of thioglycosides, 368

T

Thioglycosides as glycosyl donors, 367
Thromboxane B_2, 438
Tobramycin (Nebramycin), 474
Torsional strain, 29
Transacetalation of monosaccharides, 152
Transketalation of monosaccharides, 152
Trichloroacetimidates as glycosyl donors, 350
Tridentate complexes of inositols, 43

Subject Index

2,4,6-Trimethylbenzenesulfonyl chloride as acylating agent, 130
Tritylation of monosaccharides, 136
Trivial names, 9
Twist-boat (S) conformation of pyranoses, 39
Twist conformation of cyclohexane, 36
Twist conformation of furanoses, 34

U
Uronic acids, 22

V
Vinyl carbanion as carbon nucleophile, 312

Z
Zigzag conformation, 31